Mechanosensitivity of the Heart

Mechanosensitivity in Cells and Tissues

Volume 3

Series Editors

A. Kamkin
Department of Fundamental and Applied Physiology, Russian State Medical University, Ostrovitjanova Str. 1, 117997 Moscow, Russia

I. Kiseleva
Department of Fundamental and Applied Physiology, Russian State Medical University, Ostrovitjanova Str.1, 117997 Moscow, Russia

For further volumes:
http://www.springer.com/series/7878

Andre Kamkin · Irina Kiseleva
Editors

Mechanosensitivity of the Heart

Foreword by Vadim V. Fedorov

Springer

Editors
Prof. Andre Kamkin
Russian State Medical University
Dept. Fundamental &
Applied Physiology
Ostrovitjanova Str. 1
Moskva
Russia 117997
kamkin.a@g23.relcom.ru

Prof. Irina Kiseleva
Russian State Medical University
Dept. Fundamental &
Applied Physiology
Ostrovitjanova Str. 1
Moskva
Russia 117997

Editorial Assistant
Dr. Ilya Lozinsky
Department of Fundamental &
Applied Physiology
Russian State Medical University
Ostrovitjanova Str.1
Moskva
Russia 117997

ISBN 978-94-007-3089-2 e-ISBN 978-90-481-2850-1
DOI 10.1007/978-90-481-2850-1
Springer Dordrecht Heidelberg London New York

Softcover reprint of the hardcover 1st edition 2009

Printed on acid-free paper

Springer is part of Springer Science+Business Media (www.springer.com)

Foreword

Investigation of the mechanisms of cellular response to different mechanical stimuli, as well as mechano-electrical feedback (MEF) in the intact heart is one of the main topics in fundamental and clinical cardiology. The present volume of "Mechanosensitivity in Cells and Tissues: Mechanosensitivity of the Heart" combines excellent reviews written by worldwide leaders in this field. The 3rd volume is a great addition to this excellent series of books edited by Andre Kamkin and Irina Kiseleva. This volume successfully combines reviews, aimed at academic, physiology and clinical cardiology communities, devoted to mechanosensitivity of the normal and diseased heart at the ion channel, cell, tissue and organ levels.

Kamkin and Kiseleva have made significant contributions to the investigation of mechanosentive ion channels in cardiomyocytes and fibroblasts. Their background, in addition to extensive collaborations helped them to find and consolidate valuable research findings from prominent specialists in the field of cardiac mechanosensitivity.

In the last decade, interest in the role of MEF in the heart has increased significantly. MEF within cardiac tissue is a complex phenomenon in which electrophysiological changes are triggered by myocardial stretch. This phenomenon has been studied in the clinical community for over a century and may have both pro-rhythmic and arrhythmogenic consequences. While significant advances have been made in understanding of the effects of mechanical forces on cardiac cells, many questions remain regarding the mechanisms whereby mechanical forces are transduced into changes which alter the behavior of various cardiac cells.

The discussion of mechanosensitivity of the heart in this volume is divided into three main parts based on molecular, cell and tissue and organ investigational levels. The first part of the volume provides an excellent overview of the fundamental mechanisms, underlying stretch-activated mechano-signaling cascades in cardiac myocytes, fibroblast and stem-cell derived cells, presenting most recent advances in this increasingly important field. The second part discusses the importance of stretch-activated ion channels as mechano-transducers and the role of stretch-activated ion channels in modification of the electrical activity of the heart. Hence, mechanosensitive ion channels are one of the important targets for pharmacological therapy of different arrhythmias. Moreover this part introduces the latest findings, employing intracellular recordings of the bioelectrical activity of cardiomyocytes during mechanical stretch of healthy and diseased tissues from animals and humans. Finally, the reader can find very interesting reviews, describing contribution of MEF to electrical heterogeneity and arrhythmogenesis in the whole heart based on computer simulation studies. The uniqueness of this volume is that it collects and integrates a broad range of information, which will be of use to molecular biologists, physiologists, cardiologists and computer modelers seeking an understanding of the role of the mechanosensitivity in the cardiac physiology.

In conclusion, the molecular, biochemical, electrophysiological, functional and computer simulation approaches, employed by distinguished scientists worldwide, and collected in this volume will provide the reader with a global picture of mechanically-induced changes in the heart. It should stimulate the curiosity of cardiologists in gaining insight into the mechanisms of MEF in the heart.

This brilliant collection of in depth reviews is a valuable contribution to the spheres of scientific research and education and serves as a great guide to the mechanosensitivity of the heart.

St. Loius, Missouri, March 19, 2009 Vadim V. Fedorov

Contents

Editorial

Mechanically Gated Channels and Mechanotransduction: The Contribution to Electrical Heterogeneity and Cardiac Abnormality

Andre Kamkin and Irina Kiseleva

Mechanical deformation of the cell triggers electrophysiological and cellular signaling responses in cells. One of the mechanisms, through which the cell responds to mechanical stress, employs ion channels, which react to membrane deformation. Such channels were originally called mechanosensitive channels (MSCs), and recently redefined as mechanically gated channels (MGCs). MGCs convert mechanical force exerted on the cell membrane into electrical signals. Cellular signaling in response to mechanical stress starts rapid induction of immediate-early genes, which act as transcription factors, and trigger long-term changes in gene expression. However, the plasma membrane remains the primary target for mechanical stimulation. It responds to variable physical stress, with changes of the open probability of MGCs. MGCs can produce considerable currents in cells and therefore play an important role in forming their electric response.

Currently, we are witnessing remarkable progress in the investigation of ion channels which respond to stress. It refers very much to MGCs, their molecular organization and problems related to leak channels. Discussion of these issues was not planned to be presented in the volume devoted to mechanosensitivity of the heart cells. Still we'll give a very short coverage of those problems in this Editorial to let the reader get an idea about them, especially that the findings from a number of studies allow us to critically reconsider some controversial issues from this field of research. For example, it is well known that single channel recordings of MGCs showed that they selectively respond only to membrane tension, which can be changed by application of negative or positive pressure to the patch pipette (Gustin et al., 1988; Sokabe and Sachs, 1990; Sokabe et al., 1991). On-the-other-hand, there is a number of publications describing ion channels which react to changes of the membrane curvature. Below, we address this controversy in the light of recent findings (Honoré et al., 2006; Suchyna and Sachs, 2007).

Now it seems important to discuss a number of topics, linked to the role of K_{2P} channels, which determine the basal level of leak K^+ ions and therefore influence resting potential, since a number of such channels is sensitive to mechanical stress.

After that, it is worthy to discuss questions of molecular organization of MGCs since the majority of earlier works on this topic were performed employing X-ray crystallography, while it was recently reported that this method has significant limitations, which can be overcome only by means of electron paramagnetic resonance studies. We draw your attention to the fact that the combination of X-ray crystallography and electron paramagnetic resonance is the most efficient means for such investigations and that several studies have recently yielded insights into the structures of both the closed and open states of MGC (Ursell et al., 2008), which were previously impossible to obtain.

Although MGCs convert mechanical force exerted on the cell membrane into electrical signals, until recently molecular mechanisms of mechanotransduction in cells remained unknown. This was partially due to the fact that it is exceptionally difficult to adequately describe the membrane in terms of its mechanical organization. The complications begin with the fact that deformation of the membrane is distributed among various components, which are associated with the cortex of the cell (Akinlaja and Sachs, 1998). The lipid bilayer is far from being homogeneous in content (Lillemeier et al., 2006), let alone the actual distribution of stress. The cytoskeleton applies forces parallel and normal to the bilayer (Suchyna and Sachs, 2007). The forces in the extracellular matrix are unknown. Thus, when for convenience the term "membrane" or "deformation of the cell" is used in the following discussion, the reader must recognize that it does not apply exclusively to the lipid bilayer, unless otherwise stated, and actually relates in general to a complex of various components of the cell, which participate in mechanotransduction. This is why we later discuss the contribution of the bilayer and cytoskeleton to the transduction of mechanical energy to the channels. The first part of the volume deals with these very issues, i.e. how mechanical energy is transferred to MGCs or to MSCs. Under consideration is the membrane with intracellular components and extracellular matrix tethers, i.e. the basics of the so-called "Tethered model" (see below).

Up to date, few experimental studies of MGCs and mechanically induced whole-cell currents, allowed to link them with a particular single cellular function, never mind that on the level of the tissue. On the other hand there are several papers, which shed light on the role of mechanically induced whole-cell currents in tissue functioning under normal and pathological conditions. Second part of this volume is devoted to those issues. The third part of this volume considers a number of problems of mechano-electrical feedback in the whole heart and a computer simulation study. Finally the fourth part is a review dealing with the problem of arteries as a source of myogenic contractile activity. Much attention is given to ionic mechanisms.

MSCs and MGCs

MSC is a channel, which opens in response to membrane deformation. At the same time, back in 1998 Sachs and Morris noted that among MSCs under investigation there are channels, that recognize mechanical deformation as a proper physiological signal, and those, that react to mechanical stimulation with slight changes in kinetics. For the latter channels the authors introduced the term – channels with weak mechanosensitivity (Sachs and Morris, 1998). Recent findings call for new definitions and new specific terminology. Ability of MSCs to change their spatial organization from closed to open state during the transition period presently is considered to be their major distinctive characteristic, while permeability modulation of voltage gated channels and ligand gated channels during mechanical stress can no longer be used for defining a channel as one properly responding to mechanical stress (Sachs and Morris, 1998). Therefore all the channels that recognize mechanical deformation as a proper physiological signal have been called – mechanically gated channels (MGC) – by a number of authors (Hamill and Martinac, 2001; Kamkin and Kiseleva, 2008; White, 2006; Zhang and Hamill, 2000). Many authors divide mechanically gated channels (MGCs) into stretch activated channels (SACs), that underlie mechanically gated whole-cell current during cell stretching and are registered under pipette (although their identity as SACs has not been proven), and volume activated channels (VAC) – for the second corresponding category of chan-

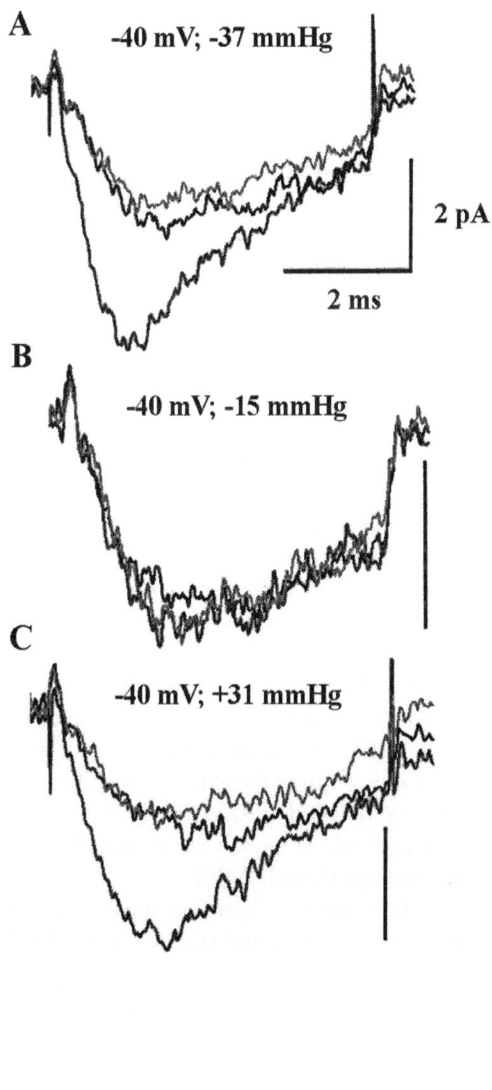

Fig. 1 Human heart Nav1.5 pore α-subunit was expressed in oocytes. Figure demonstrates reversible stretch effects with negative (**a** and **b**) or positive (**c**) pipette pressure. Three sets (before/during/after stretch) from a patch during steps to –40 mV. Reproduced from Moris and Juranka (2007) with copyright permission of the Biophys Soc and Biophys J

nels (Baumgarten, 2005). Recent works by Honoré et al. (2006) and Sunchya and Sachs (2007) show that SICs, isolated earlier into a separate group, are most probably typical SAC channels in pre-stretch state. As for PACs, their existence remains questionable considering implications of Laplace's law.

Mechanosensitive channels (MSCs), as a term, is recently usually used to describe channels that only modulate their permeability in response to mechanical stress (Morris et al., 2006; Moris and Juranka, 2007; Morris and Laitko, 2007). For example mechanosensitivity of Na_V channels (Moris and Juranka, 2007), Ca_V channels (Calabrese et al., 2002), and K_V channels (Piao et al., 2006) has been convincingly shown (See Fig. 1 as an example of Na_V channel). Definition of this term in such way allows further division of MSCs into two groups – mechanosensitive voltage-gated channels (MSC_{VG}) and mechanosensitive ligand-gated channels (MSC_{LG}) (Kamkin and Kiseleva, 2008).

Are All MGCs Stretch-Activated Channels?

Guharay and Sachs (1984), who first characterized MGCs as SACs in chick skeletal muscle, have comprehensively reviewed work from their lab. Most studies of MGCs were focused on SACs, which are the most frequently encountered type of MSCs in recordings from membrane patches.

Later studies showed that MGC activity, that is present at rest tension and is inhibited by the application of a pipette suction (Franco-Obregón and Lansman, 1994), should be isolated into a separate group and should be attributed to stretch-inactivated-channels (SICs) (Franco-Obregón and Lansman, 2002). Those authors showed differences in the suction dependence of stretch-activated and stretch-inactivated gating. According to the authors, the experiments show that SACs have a very low open probability at rest tension. In response to suction, they rapidly open and then close quickly, when the pressure stimulus is terminated. By contrast, SICs have a very high open probability at rest tension, close rapidly, when suction is applied, and reopen after releasing the suction. But recently Honoré et al. (2006) and later Sunchya and Sachs (2007) based on their own experiments gave a different interpretation of this data. They believe, that SICs do not exist, and inactivation of channel activity in response to suction can be explained by the activity of pre-stressed SACs (Honoré et al., 2006). The origin of that phenomenon was analyzed in detail, and lies in the fact, that the studied channels are sometimes open, when no mechanical stimulus is applied (Suchyna and Sachs, 2007). In this case authors observe SACs, which are opening in pre-stretch condition, but not SICs. In this respect it is worth noting that gramicidin A, which exhibits mechanosensitivity in lipid bilayers (Goulian et al., 1998; Hamill and Martinac, 2001), could behave as a stretch-activated or stretch-inactivated channel, depending on the bilayer thickness (Martinac and Hamill, 2002).

Similar concerns regarding pressure-activated cation channel (PACs), which were previously reported (Köhler et al., 1998, 2001a, b), arose after publication of reports by Honoré et al. (2006) and Sunchya and Sachs (2007). According to those studies this type of channels is activated by application of positive pressure, that changes the membrane curvature in the direction opposite to the base of the pipette, depending on the volume of applied positive pressure (Köhler et al., 1998). This point of view is not shared by all researchers. First of all, the authors, postulating the PACs presence, believe that under pipette pressure the patch-clamped membrane becomes spherical (Köhler et al., 1998), which might be responsible for the activation of PACs. Meanwhile, a recent research in this field shows a variety of different structures, many of which are dynamic, and only some of which persist as spherical caps (Hamill, 2006; Honoré et al., 2006; Sunchya and Sachs, 2007). Secondly, based on channel activity recordings during application of positive pressure, the authors postulate that SACs and hypothetical SICs did not show any distinct response to positive pipette pressure. Therefore they conclude that the activation mechanism of PACs seems to be different from other MGCs (Köhler et al., 1998). Unfortunately there are no other reports of asymmetric responses to positive and negative pressure, while several groups on the contrary report symmetric responses (Akinlaja and Sachs, 1998; Hamill, 2006; Suchyna and Sachs, 2007; Suchyna et al., 2004). For example, (Hamill, 2006) shows that a suction step in a membrane patch before, during, and after it, and a similar pressure step, in both cases, activated a 50-pA inward current (Zhang and Hamill, 2000; Zhang et al., 2000a). It was shown that both flexions of the patch outward (suction) or inward (pressure) resulted in inward current responses to suction and pressure pulses (Hamill, 2006). In the oocyte, suctions or pressures of approximately 20 mmHg produced saturating responses, so it was assumed that any channel opening, caused by an increase in suction or pressure of at least 20 mmHg would involve reopening of channels that had just closed. According to the Laplace's equation, positive or negative pressures should make equal contributions to the stress (Akinlaja and Sachs, 1998). Let us remind that membrane curvature, per se, cannot account for channel activation, and we favor the traditional interpretation that the primary stimulus for activation is tension rather then curvature (Honoré et al., 2006).

Therefore taking into consideration the above described discussion it is necessary to note that this resolves one of the key questions of the MGCs research. It looks quite likely that known MGCs are actually SACs. From our point of view this qualifies to be the most significant finding of the last two years.

Leak Channels – MGCs

About 10 years ago, the identification of K_{2P} channels experimentally proved the existence of "leak channels", which were predicted to underlie the basal leakage of K^+ by Hodgkin and Huxley in 1952.

Recently a new member of this family is identified. It consists of two 2TM/1P region-containing subunits, linked in tandem, and its functional channel is a dimer of the 4TM/2P subunits (Fig. 2.)

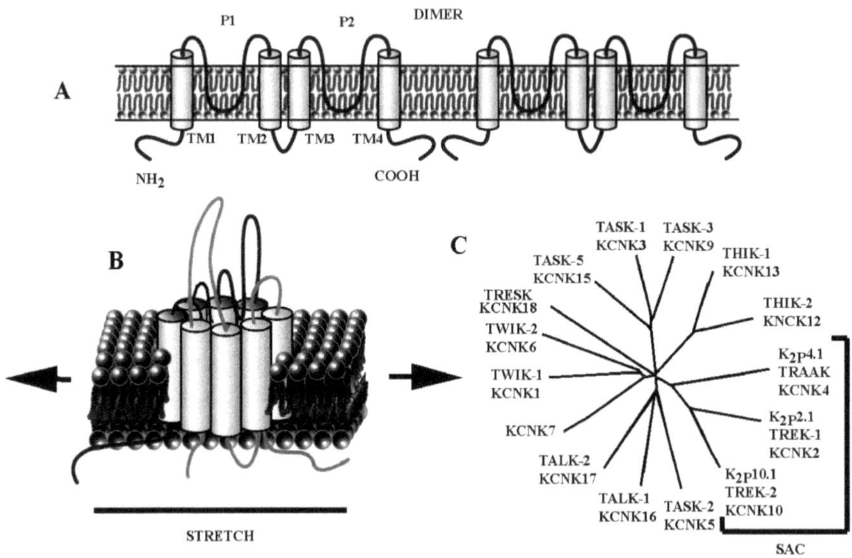

Fig. 2 Mammalian 2P domains K^+ channels (K_{2P}). (**a**) – Topology of a 2P domains K^+ channel (dimmer). (**b**) – Model of activation of K_{2P} channels by membrane stretch. (**c**) – Phylogenetic tree of 2P domain K^+ channels

Several K_{2P} channels have been recently cloned. Three of them: $K_{2P}2.1$ (TREK-1), $K_{2P}10.1$ (TREK-2), $K_{2P}4.1$ (TRAAK) can be activated by mechanical stress. Previously it has been shown in the inside–out patch configuration that positive pressure is significantly less effective, compared with negative pressure, in opening of channels, suggesting that a specific membrane deformation (convex curving) preferentially opens these channels (Maingret et al., 1999; Patel et al., 1998). However it is quite likely that this line of argument has the same limitations, which we already discussed earlier. This was shown for the K_{2P} MGCs (Honoré et al., 2006).

At the whole-cell level, $K_{2P}2.1$ and $K_{2P}4.1$ are modulated by cellular volume. For example, hyperosmolarity closes the channels (Maingret et al., 2000a, b; Patel and Honoré, 2001; Patel et al., 1998). Both the number of active channels and the sensitivity to mechanical stretch are strongly enhanced after treating the cell-attached patches with the cytoskeleton disrupting agents, colchicines and cytochalasin D (Maingret et al., 1999). This suggests that mechanical force might be transmitted directly to the channel via the lipid bilayer and does not require the integrity of the cytoskeleton (Maingret et al., 1999; Patel et al., 1998; 2001). Both $K_{2P}2.1$ and $K_{2P}4.1$ are blocked by amiloride and Gd^{3+} (Maingret et al., 1999, 2000a, b).

In general the observation that channels, which are responsible for generation K^+-leak current, can be activated by mechanical stress, suggests their direct contribution to regulation of cellular functions, one of which would be the maintenance of the resting membrane potential.

Molecular Organization of Channels

Investigation of molecular organization of MGCs is the most rapidly progressing part of the field, although it is extremely important to keep in mind particular limitations of each method employed.

MscL was the first of the bacterial MGCs to be cloned and sequenced. To date it remains the best characterized mechanosensitive channel. Researchers have the advantage of knowing a crystal structure and having detailed models for how the channel senses and responds to mechanical forces (Blount et al., 2008). A major advance in the study of MscL came when a crystal structure was derived for a homologue from *M. tuberculosis* (Chang et al., 1998). It was shown that it has a homopentameric structure. The pentameric composition was later confirmed by independent approaches (Sukharev et al., 1999). As anticipated (Blount et al., 1996), the crystal structure suggested that each subunit within the complex has two α-helical transmembrane domains. Because there is only a small opening at the pore (~4 Å), the channel appears to be in a closed, or nearly closed, conformation (Blount et al., 2008). However, at the time the crystal structure was published, no functional information was available (Blount et al., 2008). Further, the lack of crystal structures in the open and closed states of many channel proteins means we cannot be sure where each residue moves, which are exposed to the surrounding lipids and which are facing the hydrated internal pore (Ursell et al., 2008). Several experiments using relatively independent approaches performed with the *E. coli* MscL channel now suggest that the crystal structure does not represent the fully closed state found in membranes (Blount et al., 2008).

Although the above mentioned findings look very interesting, the actual conformation of the channel in the membrane will need to be clarified, because in the crystal the membrane is replaced by a blanket of detergent molecules (Sigworth, 2003). For example, the pioneering work of the MacKinnon group presented the first crystal structure of K_V, a voltage gated channel (Jiang et al., 2003). The crystal was obtained in detergent and in the presence of Fab fragments, and now there is a wide consensus, that it is distorted and that it does not represent a conformation, found in the native bilayer. Even with the crystal structure of the transmembrane (TM) segments (i.e. without the pore region), which was presented by the same group (Jiang et al., 2003), the crystals available today do not address the structural issues of how channels are assembled and, to a lesser extent, how they operate. A recent paper by Cuello et al. (2004) of the Perozo group approached the study of the structure of the channels in its open-inactivated conformation in a bilayer environment using electron paramagnetic resonance (Bezanilla, 2005).

A recent electron paramagnetic resonance study of K_V, a prokaryotic voltage-gated channel, in its lipid native environment has revealed the location of the TM segments, the connecting loops and the relative position of the voltage-sensing charges. The results confirm that the previously reported crystal structure does not represent a native conformation and gives us structural constraints that will help in determining the molecular structure of the voltage sensor (Bezanilla, 2005). For example, the arrangement of the TM segments of voltage-gated channels, which was not determined by the crystal structures of K_V, are now better defined by electron paramagnetic resonance studies on the same channel, at least for the opening activated state of the channel (Bezanilla, 2005).

For MGCs structural changes in MscL, induced by bilayer-deformation forces, were studied by combination of cysteine-scanning mutagenesis with site-directed spin labeling and electron paramagnetic resonance spectroscopy. Those studies were complemented by analysis of channel function by means of the patch-clamp technique (Perozo et al., 2002b). The open state of MscL has a water-filled pore of > 25 Å in diameter that is lined by the TM1 helices from the five subunits (Perozo et al., 2002a), and several studies show that the channel undergoes a large conformational change when opening and closing (Gullingsrud et al., 2001, 2003; Sukharev et al., 2001). The study by Perozo and coworkers (2002a) demonstrates that hydrophobic mismatch is not the driving force that triggers MscL opening, although specific mismatch levels could stabilize intermediate states along the kinetic path towards the open state. A combination of X-ray crystallography and electron paramagnetic resonance studies yielded insights into the structures of both the closed and open states of MscL (Perozo et al., 2001, 2002a).One of the outcomes of this structural analysis is the idea that the structure can be roughly approximated as a cylinder, making it amenable to mechanical modeling (Ursell et al., 2008).

How Mechanical Energy is Transferred to MGC or to MSC

So far we focused on discussion of how mechanical energy is transferred to MGC or MSC: through the lipid bilayer of the membrane or through the cytoskeleton and which of the mechanisms prevails. Several groups reported that MGCs are activated by the stretch of the lipid bilayer, while other papers focused on the role of the cytoskeleton in MGCs activation. Some authors brought up the role of extracellular matrix, although the forces distribution in the extracellular matrix remains unknown.

Two current models describe the MGC gating: the bilayer model and the more speculative, tethered model (Hamill and McBride Jr, 1997). In the bilayer model, initially proposed for gating of MGC in *E. coli* giant spheroplasts (Martinac et al., 1990), lipid bilayer tension alone is sufficient to gate the MGS channels directly, because purified MscL, MscS and other prokaryotic MGC are still mechanosensitive when reconstituted into liposomes (Häse et al., 1995; Kloda and Martinac, 2001a, b, c; Martinac, 2001; Perozo and Rees, 2003; Sukharev et al., 1994).

Bilayer Model

The lipid bilayer is far from being homogeneous in content (Lillemeier et al., 2006), let alone stress. But even studies, employing experimenting with a pure bilayer, face certain problems with data interpretation. Planar bilayer has serious limitations for reconstitution and testing of the MGC/MSC activity. The measurements of membrane tension suggest that in-plane tension in the bilayer is too weak to have a significant effect on MGC gating (Dai and Sheetz, 1995). When several authors inserted MSC into planar bilayer and showed their activation with stress (A cloned rat epithelial Na^+ channel in planar lipid bilayers (Ismailov et al, 1996a, b, c), they reported presence of an artifact. MSC does not sense tension of the planar lipid bilayer. Moreover planar lipid bilayer is under resting tension of about 1–5 dyne/cm (Elliott et al., 1983; Gruen and Wolfe, 1982; Ring, 1992; Ring and Sandblom, 1988) (for review see Sachs and Morris (1998)). In this case MSC in the planar lipid bilayer maintains resting activity. Therefore the results by Awayda et al. (1995), who studied ENaC in planar lipid bilayer, are difficult to interpret. Finally, planar bilayers are tension clamped and therefore cannot be used.

Similar problems arise from using vesicles. Only vesicles with a fixed amount of lipid can be studied. But, despite this limitation, research studies targeted on the role of cytoskeleton in gating of MGCs can be carried out. It was demonstrated that lipid bilayer tension alone is sufficient to gate the MGCs directly, because purified MscL, MscS and other prokaryotic MGCs are still mechanically gated when reconstituted into liposomes (Kloda and Martinac, 2001a, b, c; Martinac, 2001, 2004; Martinac et al., 1990; Perozo and Rees, 2003).

The membrane proteins forming specific MGCs were identified. Based on the finding that bacterial MGCs remain functional upon reconstitution into liposomes (Delcour et al., 1989) a novel strategy was developed by Sukharev et al. (2001) towards molecular identification of MscL that involved detergent-solubilizing and fractionating membrane proteins, reconstituting the protein fractions in liposomes, and then assaying the fractions for stretch sensitivity, using the patch clamp recording. This technique was used to identify a variety of MSGs proteins in bacteria and archaea (Kloda and Martinac, 2001a; Martinac and Kloda, 2003; Sukharev, 2002) and, most recently, the TRPC1 channel protein was identified as the MscCa channel in *Xenopus* oocytes (Maroto et al., 2005). But now Gottlieb et al. (2008) report that the amplitude of the mechano-sensitive current is not significantly altered by overexpression of TRPC1 or TRPC6 subunits.

Using liposomes allows to address another old question. It is clear that stretching the bilayer will tend to decrease its lipid packing density and thickness. If the channel proteins experience a change, occurring in the membrane, they will respond with changes in the distribution between closed and open channel conformations (Hamill and Martinac, 2001; Kung, 2005; Markin and Sachs 2004a, b). Introducing into liposome membranes lysophospholipids and amphipathic molecules may cause local changes in tension and curvature at the lipid–protein interface and thereby can shift the channel distribution (for review see Hamill, 2006). Thus we have to consider the MGCs mediated responses of the bilayer in terms of their activation by amphipaths or lysophospholipids.

In many reviews (Hamill, 2006; Morris and Homann, 2001; Martinac, 2004) it is demonstrated that MGCs of various ionic selectivities were found in spheroplasts, prepared from bacteria and fungi, in protoplasts prepared from plants, and in a multitude of animal cells of vertebrate and invertebrate origin. Macroscopic currents can be activated in spheroplasts formed from *E. coli*, yeast, and other microbial cells (Cui et al., 1995; Hamill and Martinac, 2001).

Recent reports that $K_{2P}2.1$ and $K_{2P}4.1$ retain stretch sensitivity in cytoskeleton-free membrane blebs indicate that they are also bilayer-gated channels (Honoré et al., 2006).

Tethered Model

This is another interesting experimental model, which allows to study how MGCs sense mechanical forces. The basic assumption, underlying it, is that intracellular (e.g. cytoskeletal components) and extracellular (e.g. extracellular matrix) tethers interact directly with portions of the channel protein, and it is that these tethers draw the channel components in different directions upon mechanical perturbation, thus supplying the energy for channel gating (Blount et al., 2008). Early publications suggested that the most likely source for the stimulus, required to gate a mechanosensitive channel, would be originating from such cytoskeletal elements (Guharay and Sachs, 1984; Sachs, 1988).

The responses of whole cells to the mechanical stress are easier to understand under the assumption that the energy of local compression is transferred by cytoskeletal elements to the cardiomyocyte channel protein (Kamkin et al., 2003a). Treatment of the isolated cardiomyocytes (Kamkin et al., 2003a) and cardiac fibroblasts (Kamkin et al., 2003b, c) with cytochalasin D, which is thought to disrupt F-actin, reduces the amplitude of stretch-activated whole-cell currents during continuous stretch. When the cells are pre-treated before application of stretch, the mechanosensitivity is reduced or abolished, i.e. the mechanical stimuli (stretch or pressure) become ineffective (Kamkin et al., 2003a, b, c). Nearly identical results were obtained in case of cell dialysis with 5 μM colchicin. In this case depolymerisation of tubulin reduced or abolished the stretch-activated whole-cell currents (Isenberg et al., 2003). From these observations, we can conclude that an intact cytoskeleton is necessary for the mechanosensitive gating of MGCs. We might see the cytoskeleton as part of a pathway that transforms the exogenous mechanical energy into activation energy of membrane channels. This hypothesis is in line with the increase of stretch-activated whole-cell currents in hypertrophied cells (Kamkin et al., 2000a, b; Kiseleva et al., 2000) with enhanced stiffness, resulting from a pathological high expression of tubulin in the cytoskeletal cortex (Watson et al., 1996). Mechanical stretch of isolated cardiac fibroblasts directly stimulates expression of extracellular matrix components and proliferation, both hallmarks of fibrosis. The linkage between the extracellular matrix, integrin receptors and the cytoskeleton undoubtedly plays a critical role in this process (Carver and Fuseler, 2009). Authors have recently shown that mechanical stretch induces rapid changes in cardiac fibroblast morphology and the organization of the actin cytoskeleton. The Rho family of small GTPases has received considerable attention in their role in organizing the actin cytoskeleton. These data are presented in detail in the first part of this volume (Carver and Fuseler, 2009).

The mechanisms by which myocardial cells convert mechanical stimuli into cellular signals remain to be completely understood. First of all titin and titin-associated proteins in myocardial stress-sensing are considered in the first part of the volume (Linke, 2009). Titin is the most abundant protein of the intrasarcomeric cytoskeleton. Titin and titin-based protein complexes are now recognized as integral parts of the mechanosensitive protein network and as critical components in cardiomyocyte stress/stretch signalling (Linke, 2009). The author concludes that titin together with some of its direct and indirect ligands in the Z-disk and M-band regions, and the N2B, N2A, and PEVK domains in the I-band region, could act as a "tensiometer" that when stretched, triggers downstream signalling events leading to changes in muscle-gene expression. Role of titin as a cytoskeletal network of proteins, other structural proteins and related signaling pathways in mechanotransduction and heart failure (Jacot et al., 2009) are discussed in detail in the second part of this volume as well. Also discussed are the regulation of structural and junctional proteins by stretch, the role of the cytoskeleton in mechanotransduction and heart failure, signaling pathways involved in mechanotransduction and load-induced hypertrophic responses, and the role of substrate stiffness in stem cell differentiation and maturation of excitation-contraction coupling (Jacot et al., 2009). In the continued first part of this volume considered are the molecular signaling mechanisms of myocardial stretch (Lal et al., 2009). Integrins, caveolae and focal adhesions have been shown to have important mechanosensing roles in cardiac myocytes. Downstream effectors activated by mechanosensors include guanine-nucleotide binding proteins (G-proteins), mitogen-activated protein (MAP) kinases, Janus-associated kinase/signal transducers and activators of transcription (JAK/Stat), protein kinase C (PKC) and protein kinase B/Akt pathways. Multiple levels of crosstalk exist between these pathways (Lal et al., 2009). Recent studies suggest that acute mechanical stretch activates protective pathways including, c-jun N-terminal kinase (JNK) and akt as a tolerance response, rather than injury-related signaling cascades such as p38 MAP kinase. This review provides an overview of the fundamental mechanisms, underlying stretch-activated mechano-signaling cascades in myocardial cells, presenting most recent advances in our understanding of this increasingly important field (Lal et al., 2009). The first part of this volume finishes with the discussion that mechanical stress induces cardiomyocyte hypertrophy through agonist-independent activation of Angiotensin II type 1 receptor (Akazawa and Komuro, 2009). The authors show that in response to stretch stimulation, the AT_1 receptor undergoes a specific switch in the receptor conformation without the involvement of AngII. It is conceptually novel that the AT_1 receptor, a member of G protein-coupled receptor, is a mechanical force-transducing molecule and mediates mechanical stress-induced cellular responses. The authors discuss molecular and structural bases for mechanosensation by the AT_1 receptor and inverse agonism at the AT_1 receptor (Akazawa and Komuro, 2009).

Mechanically Induced Currents and Potentials in Isolated Cardiomyocytes and Cell Tissues

Although direct mechanical stretching of a cell is the simplest form of mechanical stimulation, integration of this approach into experimental design of a cardiomyocyte study proved to be methodologically challenging. Therefore the second part of this volume starts with a review discussing the methods to apply mechanical stimuli to isolated cells and tissues, methods for patterned growth of cells, effects of stretch and shear stress on cellular function and tissue electrophysiology (Jacot et al., 2009). In the following review of the second part of this volume the authors discuss the importance of stretch-activated ion channels as mechano-transducers in the heart, with emphasis on their contribution to the regulation of contractile performance. As well, the role of stretch-activated channels in modifying the electrical activity of the heart is also discussed (Ward and Allen, 2009). The authors consider cardiac response to stretch, discussing three different cellular mechanisms – Increased overlap between the thick and thin filaments, Increased Ca^{2+} sensitivity of the contractile machinery, Increased Ca^{2+} transients (the systolic rise in Ca^{2+} which activates the contractile proteins) which gradually become larger over some minutes after the stretch. Further on the authors dwell on the possible mechanisms of the slow force response (SFR), considering the role of the sarcoplasmic reticulum in the SFR, stretch-activation of Na^+/H^+ exchanger, stretch-activated channels and the SFR (Ward and Allen, 2009). The most part of the review is devoted to stretch-sensitive channels in the heart and stretch-induced arrhythmias. In this connection the authors (Ward and Allen, 2009) often refer to works by the group of A. Kamkin, I. Kiseleva and G. Isenberg, that were continued by Prof. A. Kamkin's colleagues, first Dr. D. Kondratev and Dr. V. Kazansky, and then Dr. V. Dyachenko in the laboratory of Prof. G. Isenberg. The detailed analysis of that material is presented in this part of this volume (see: Lozinsky and Kamkin, 2009). In the next review of the second part of this volume Banderali et al. (2009) discuss the effects of applied stretch on native and recombinant cardiac Na^+ currents (alpha subunit, $Na_V1.5$). The authors demonstrate that in both native mammalian myocytes and in the heterologous expression system, applied stretch causes the Na^+ current to activate at more negative membrane potentials. Stretch also significantly increases the Na^+ current density. In the mammalian heart the effects of stretch on conventional time- and voltage-dependent intrinsic Na^+ currents need to be taken into account when attempting to understand either the basis for or the consequences of mechanoelectrical feedback (Banderali et al., 2009).

In recent years the problem of direct axial stretch of isolated cardiomyocytes from adult animals and humans was solved and studies have been performed characterizing whole-cell currents under artificial axial stretch or squeeze. In general, it was believed that direct stretching of isolated cardiomyocytes would cause huge problems (Sachs and Morris, 1998). A breakthrough report appeared in 2000, when the successful stretching of atrial (Zhang et al., 2000b) and ventricular cardiomyocytes (Kamkin et al., 2000a, 2003a; Zeng et al., 2000) was reported. For example,

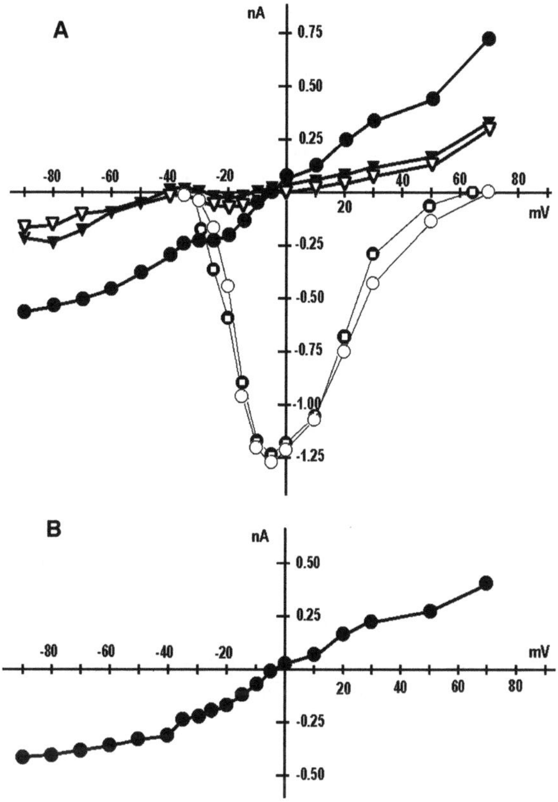

Fig. 3 I_{SAC} and I_{Ca} after cell dialysis with 5 mM BAPTA. (**a**) – Late currents at control (*open triangles*), during 10 μm stretch (*dots*) and during stretch 5 min after addition of 5 μM Gd^{3+} (*filled triangles*). Note: in the presence of BAPTA, I_{Ca} at control (*circles*) and during stretch (*thick circles*) does not differ. (**b**) – Stretch-activated difference currents (Stretch minus Control), E_{rev}=0 mV. Reproduced from Kamkin et al. (2003a) with copyright permission of the Springer and Pflügers Arch – Eur J Physiol

stretch by 10 μm shifted the late currents to more negative values if the clamp step potentials were negative (filled circles in Fig. 3a). Stretch shifted E_0 to –5 mV. At positive clamp steps, it shifted the currents into the outward direction. The stretch activated difference current I_{SAC} had an almost linear voltage dependence and reversed about 0 mV (Fig. 3b) (Kamkin et al., 2000a, 2003a).

The following review of the second part of this volume discusses in detail the effects of stretch and compression at the tissue and cell levels (Lozinsky and Kamkin, 2009). It is shown that stretch of the single cells includes changes in mechano-gated channels, mechanosensitive whole-cell currents which lead to membrane depolarization which is equal to decrease in the resting membrane potential and elicited stretch-induced depolarizations, that appear in the different time of repolarization phase of cardiomyocytes. Membrane depolarization and stretch-induced depolarization in action potentials provoke extra-action potentials when the stretch-induced depolarizations reach a threshold potential. Shown are the origin mechanisms of mechanosensitive whole-cell currents during stretch and compression of isolated cardiomyocytes which stick to the bottom of the perfusion chamber in two different positions: edgewise, staying on the narrow side, or broad-wise. In recent years, a good correlation between the results from experiments, carried out as whole-cell recordings from isolated cardiomyocytes, and the results, obtained from microelectrode recordings in stretched fragments of atria and ventricles from healthy animals and humans, and those with hypertrophied myocardium was recognized. Before, it was believed that the latter task was practically not feasible. Similarly to cardiomyocyte stretching by a glass stylus, the technical problem of keeping a floating microelectrode in the tissue for a considerably long time, was solved. The most important finding of these experiments was the discovery of a highly increased sensitivity to stretch in cells from hypertrophied myocardium. This increased sensitivity to stretch could be explained by altered signaling pathways or increased channel expression in hypertrophied myocardium. These data particularly allow to explain the origin mechanisms of mechanically-induced arrhythmias. Besides, it was shown that reactions to stretch or compression are different and the response to compression was different in the same cell, subject to its spatial position (Lozinsky and Kamkin, 2009).

The next review of the second part of this volume is devoted to cardiac fibroblasts (Kamkin et al., 2009). Fundamentally new aspects in studying mechanosensitivity of heart cells are revealed by investigation of mechanosensitivity of cardiac fibroblasts, their intercellular interaction with each other and with cardiomyocytes. At present, it has been proved that cardiac fibroblasts act as mechano-electrical transducers in the heart and they can participate in regulation of electrical activities of both healthy and hypertrophied hearts. Nevertheless to the best of our knowledge there are no reports of investigation of the influence of stretch on gap junction (Kamkin et al., 2009).

Final in the second part of this volume is the article by Max Lab (2009) devoted to scanning ion conductance microscopy for imaging and mechanosensitive activation of selected areas of live cells. The review considers non-contact nanoscale method for applying force to selected areas on the surface of living cells. The method applies hydrostatic pressure through a nanopipette, the operative probe of a scanning ion conductance microscope. The pipette is kept above the cell surface using distance feedback. This prevents surface contact, and promotes non-invasive mechanical probing. First the microscope scans and images a living cell surface at high resolution – no applied pressure. Subsequently the authors apply pressure to areas selected from the scanned image for mechanosensitive studies, as well as studies of their nanomechanical properties (Lab, 2009).

Mechano-Electrical Feedback in the Whole Heart

The third part of this volume deals with that problem. The part opens with a review devoted to contribution of mechano-electric feedback (MEF) to electrical heterogeneity and arrhythmogenesis (Saint et al., 2009). Stretch of the myocardium can alter action potential morphology, propagation velocity and intracellular calcium handling, all of which can contribute to arrhythmogenesis. In particular, it is now becoming clear that MEF is not homogeneous in the heart. It is also clear that MEF is altered in some diseases such as hypertrophy, where it may explain the propensity to arrhythmias in these diseases. The authors discuss the evidence that MEF is heterogeneous in the heart, in the same way that other electrophysiological properties are heterogeneous (Saint et al., 2009).

The following review deals with atrial flutter which is a supraventricular arrhythmia, based on a reentrant mechanism, which presents small fluctuations in cycle length. The authors report on studies in humans and animals which disclosed the nature of these variations and supported their mechanical origin. The sources of the spontaneous variability of atrial flutter cycle length were identified in ventricular contraction and respiration, which cause phasic variations in atrial interval. The phase-response curves were shown to be closely related to atrial volume changes during ventricular and respiratory activities and oscillations in cycle length were reported to be independent of autonomic tone. All this evidence led to the formulation of the MEF paradigm, which suggests that changes in atrial volume directly affect atrial flutter cycle length variability via direct alteration of the reentrant circuit size and mechano-electrical modulation of conduction velocity (Ravelli and Masè, 2009).

The MEF subject is continued in the chapter by Cingolani et al. (2009). In this chapter the enhanced activity of the cardiac Na^+/H^+ exchanger (NHE-1) after myocardial stretch is considered a key step of the intracellular signaling pathway leading to the slow force response to stretch as well as an early signal for the development of cardiac hypertrophy. The authors propose that the chain of events triggered by stretch begins with the release of small amounts of angiotensin II which in turn induce the release/formation of endothelin. The actions of these hormones trigger the production of mitochondrial reactive oxygen species that enhances NHE-1 activity, causing an increment in the intracellular Na^+ concentration which promotes the increase in intracellular Ca^{2+} concentration ($[Ca^{2+}]_i$) through the Na^+/Ca^{2+} exchanger. This $[Ca^{2+}]_i$ increase would trigger cardiac hypertrophy by activation of widely recognized Ca^{2+}-dependent intracellular signaling pathways (Cingolani et al., 2009).

The following review is devoted to stretch-induced inotropy in atrial and ventricular myocardium. It is shown that in human heart both atrial and ventricular myocardium exhibit a stretch-dependent slow force response that is likely to serve as adjustment mechanism regulating cardiac output in case of increased preload. The authors believe that the stretch-induced slow force response is a universal phenomenon in both human atrium and ventricle. Although of comparable amplitude, underlying signal transduction mechanisms differ significantly (von Lewinski et al., 2009).

Under further discussion are the effects of wall stress on the dynamics of ventricular fibrillation. The review considers computer simulation of mechanoelectric feedback (Hirabayashi et al., 2009). The model formulated the biophysics of specific ionic currents, excitation-contraction coupling, anisotropic non-linear deformation of the myocardium, and mechanoelectric feedback through stretch-activated channels. The authors' model suggested that sustained stretches shorten action potential duration (APD) and flatten the electrical restitution curve, whereas stretches applied at the wavefront prolong APD. The wavefront around the core was highly stretched, even at lower pressures, resulting in a prolongation of APD and extension of the refractory area in the wavetail. This simulation study indicated that mechanical loading promotes meandering and wave breaks of spiral re-entry through mechanoelectric feedback. Mechanical loading in pathological conditions may contribute to the maintenance of VF through these mechanisms.

Problems of electromechanical modelling of cardiac tissue (Cherubini et al., 2009) are discussed in the final review of the third part. The authors present an electromechanical model of myocardium tissue, coupling finite elasticity, endowed with the capability of describing muscle contractions, with a FitzHugh – Nagumo type system, describing the electrical activity proper to excitable media. They exploit a novel point of view which introduces the notion of active deformation as opposed to that of active stress. The high degree of deformability of the medium makes mandatory to set the diffusion process in a moving domain, thereby producing a direct influence of the deformation on the electrical activity.

Arteries as a Source of Myogenic Contractile Activity

Part IV of the Volume is devoted to arteries as a source of myogenic contractile activity. In this connection ionic mechanisms are considered in great detail. These issues are discussed by Nakayama et al. (2009) in the review devoted to specific mechanotransduction signaling involved in myogenic responses of the cerebral arteries. Stretching and intraluminal pressurization induce many different responses, including contraction, activation of various kinases and ionic channels, production of various vasoactive substances, gene expression, and phenotype changes. The authors discuss specific mechanotransduction signaling pathways involved in the myogenic responses of cerebral arteries. All in all the authors consider three basic problems – structural and functional characteristics of the cerebral arteries in the source of stretch/pressure-induced contraction; ionic mechanisms for myogenic contractile response to mechanical stretch; multiple phosphorylation of 20-kDa myosin light chain of cerebral artery smooth muscle cells as a self-inhibitory mechanism in stretch-induced contraction.

Conclusions and Perspectives

Taking into consideration the above mentioned we think that among recent reports, devoted to the topic of investigation of the response of ion channels towards mechanical stimulation, at least several of them provide new insights, which allow to discriminate between true mechanically gated channels and other mechanosensitive channels, which can be voltage gated or/and ligand gated. In the experiments, employing application of positive or negative pressure through pipettes tip, it seems reasonable to assume that MGCs respond to mechanical forces translated to them via changes in membrane tension. All those channels most likely belong to stretch-activated channels (SACs). Recently researchers conducted throughput investigation of several types of lipid bilayers, which made possible current outbreak of intensive investigation of basic principles and mechanisms, responsible for transduction of mechanical energy directly to the channels. In this lane more and more attention is paid to the tethered model treating integrins, caveolae and focal adhesions as means playing important mechanosensing roles, in particular in cardiac myocytes. It has been shown that the linkage between the extracellular matrix, integrin receptors and the cytoskeleton undoubtedly plays a critical role in this process. It has been discovered how important are titin and titin-based protein complexes that are presently recognized as integral parts of the mechanosensitive protein network and as critical components in cardiomyocyte stress/stretch signalling.

It has been convincingly shown that some K_{2P} channels react to mechanical stress. Up to date several groups identified genes, which encode MGC. Furthermore the molecular organization of several channels has been revealed. And finally a new method of cardiomyocyte stretching was developed in addition to widely used methods of stretching of different cells, which allowed characterization of mechanically gated whole cell currents. One of the major findings of the last years was the identification of fibroblasts (and in particular cardiac fibroblasts) as effective mechanoelectrical transducers.

It is necessary to indicate that up to date several stretch-activated signaling pathways have been identified. Current research in this field goes beyond investigation of the stretch-activated signaling cascades into advocating the possibility that mechanoelectro-chemical transduction forms a part of a network of mechanically linked crosstalk (Mechanically Mediated Crosstalk: MMC) (Lammerding et al., 2004; Lab, 2005). MMC can shape downstream signals leading to alterations of intracellular Ca^{2+} signaling. Some authors examine the downstream cellular response to mechanically-activated Ca^{2+} signaling and its importance (Pingguan-Murphy and Knight, 2008). MMC can also span other regulatory systems and processes, such as the autonomic nervous system, and in addition, can operate through the whole heart as an integrative system (Lab, 2005). MMC can be perceived in several ways. Firstly, in terms of discussion of mechanotransduction, which is defined as the biochemical response of cells to mechanical stimulation. There is also evidence pointing toward participation of signaling cascades in modulation of mechanosensitive channels function on the one hand, and possibility of regulation of intracellular signaling transduction pathways by mechanosensitive ion channels (Boriek and Kumar, 2008) on the other hand. It looks quite likely that local stretching of freshly isolated and cultured cells will allow a detailed description of those mechanisms.

In addition the prospects of investigation of cellular interaction during stretch look extremely interesting. The potential of investigation of such topic in the whole tissue is limited due to the fact that if the resistance of the electrode will exceed the resistance of the cell this will result in the drop of the resistance in the electrode, and not on the cell. However, for such purposes, double patch clamp whole cell recording of the interacting cell pair during stretching of one of them seems to be the method of choice.

References

Akazawa H and Komuro I (2009) Mechanical stress induces cardiomyocyte hypertrophy through agonist-independent activation of angiotensin II type 1 receptor. In: Mechanosensitivity in Cells and Tissues. Mechanosensitivity of the Heart. A Kamkin and I Kiseleva (eds.) Springer, pp. 83–95.

Akinlaja J and Sachs F (1998) The breakdown of cell membranes by electrical and mechanical stress. Biophys J 75:247–254.

Awayda MS, Ismailov II, Berdiev BK, Benos DJ (1995) A cloned renal epithelial Na^+ channel protein displays stretch activation in planar lipid bilayers. Am J Physiol 268(6 Pt 1):C1450–C1459.

Banderali U, Clark RB, Morris CE, Fink M, Giles WR (2009) Effects of applied stretch on native and recombinant cardiac Na^+ currents. In: Mechanosensitivity in Cells and Tissues. Mechanosensitivity of the Heart. A Kamkin and I Kiseleva (eds.) Springer, pp. 169–184.

Baumgarten CM (2005) Cell volume-sensitive ion channels and transporters in cardiac myocytes. In: Cardiac Mechano-Electrical Feedback and Arrhythmias: From Pipette to Patient. P Kohl, MR Franz, F Sachs (eds.) Saunders, Philadelphia, pp. 21–32.

Bezanilla F (2005) The voltage-sensor structure in a voltage-gated channel. TRENDS Biochem Sci 30(4):166–168.

Blount P, Li Y, Moe PC, Iscla I (2008) Mechanosensitive channels gated by membrane tension: bacteria and beyond. In: Mechanosensitivity in Cells and Tissues. Mechanosensitive Ion Channels. A Kamkin and I Kiseleva (eds.) Springer, pp. 71–101.

Blount, P, Sukharev, SI, Moe, PC, Schroeder, MJ, Guy, HR, and Kung, C (1996) Membrane topology and multimeric structure of a mechanosensitive channel protein of Escherichia coli. EMBO J 15:4798–4805.

Boriek AM and Kumar A (2008) Regulation of intracellular signal transduction pathways by mechanosensitive ion channels. In: Mechanosensitivity in Cells and Tissues. Mechanosensitive Ion Channels. A Kamkin and I Kiseleva (eds.) Springer, pp. 303–327.

Calabrese B, Tabarean IV, Juranka P, Morris CE (2002) Mechanosensitivity of N-type calcium channel currents. Biophys J 83(5):2560–2574.

Carver W and Fuseler JW (2009) Mechanical stretch-induced reorganization of the cytoskeleton and the small gtpase rac-1 in cardiac fibroblasts. In: Mechanosensitivity in Cells and Tissues. Mechanosensitivity of the Heart. A Kamkin and I Kiseleva (eds.) Springer, pp. 35–54.

Chang G, Spencer RH, Lee AT, Barclay MT, Rees DC (1998) Structure of the MscL homolog from mycobacterium tuberculosis: a gated mechanosensitive ion channel. Science 282(5397): 2220–2226.

Cherubini C, Filippi S, Nardinocchi P, Teresi L (2009) Electromechanical modelling of cardiac tissue. In: Mechanosensitivity in Cells and Tissues. Mechanosensitivity of the Heart. A Kamkin and I Kiseleva (eds.) Springer, pp. 421–449.

Cingolani HE, Pérez NG, Caldiz CI, Garciarena CD, De Giusti VC, Correa MV, Villa-Abrille MC, Yeves AM, Ennis IL, de Cingolani GC, Aiello EA (2009) Early hypertrophic signals after myocardial stretch: role of reactive oxygen species and the sodium/hydrogen exchanger. In: Mechanosensitivity in Cells and Tissues. Mechanosensitivity of the Heart. A Kamkin and I Kiseleva (eds.) Springer, pp. 327–371.

Cuello LG, Cortes DM, Perozo E (2004) Molecular architecture of the KvAP voltage dependent K^+ channel in a lipid bilayer. Science 306:491–495.

Cui C, Smith DO, Adler J (1995) Characterization of mechanosensitive channels in Escherichia coli cytoplasmic membrane by whole-cell patch-clamp recording. J Membr Biol 144:31–42.

Dai J and Sheetz MP (1995) Regulation of endocytosis, exocytosis, and shape by membrane tension. Cold Spring Harb Symp Quant Biol 60:567–571. Review.

Delcour AH, Martinac B, Adler J, Kung C (1989) Modified reconstitution method used in patch-clamp studies of Escherichia coli ion channels. Biophys J 56(3):631–636.

Elliott JR, Needham D, Dilger JP, Haydon DA (1983) The effects of bilayer thickness and tension on gramicidin single-channel lifetime. Biochim Biophys Acta 735:95–103.

Franco-Obregón A and Lansman JB (1994). Mechanosensitive ion channels in skeletal muscle from normal and dystrophic mice. J Physiol (Lond) 481:299–309.
Franco-Obregón A and Lansman JB (2002) Changes in mechanosensitive channel gating following mechanical stimulation in skeletal muscle myotubes from the mdx mouse. J Physiol (Lond) 539(2):391–407.
Gottlieb P, Folgering J, Maroto R, Raso A, Wood TG, Kurosky A, Bowman C, Bichet D, Patel A, Sachs F, Martinac B, Hamill OP, Honoré E (2008) Revising TRPC1 and TRPC6 mechanosensitivity. Eur J Physiol – Pflugers Arch 455(6):1097–1103.
Goulian M, Mesquita ON, Fygenson DK, Nielsen C, Andersen OS, Libchaber A (1998) Gramicidin channel kinetics under tension. Biophys J 74(1):328–337.
Gruen DW and Wolfe J (1982) Lateral tensions and pressures in membranes and lipid monolayers. Biochim Biophys Acta 688:572–580.
Guharay F and Sachs F (1984) Stretch-activated single ion channel currents in tissue cultured embryonic chick skeletal muscle. J Physiol (Lond) 352:685–701.
Gullingsrud J, Kosztin D, Schulten K (2001) Structural determinants of MscL gating studied by molecular dynamics simulations. Biophys J 80(5):2074–2081
Gullingsrud J, Schulten K (2003) Gating of MscL studied by steered molecular dynamics. Biophys J 85(4):2087–2099.
Gustin MC, Zhou XL, Martinac B, Kung C (1988) A mechanosensitive ion channel in the yeast plasma membrane. Science 242:762–765.
Hamill OP (2006) Twenty odd years of stretch-sensitive channels. Pflügers Arch – Eur J Physiol 453:333–351.
Hamill OP and Martinac B (2001) Molecular basis of mechanotransduction in living cells. Physiol Revs 81:685–740.
Hamill OP and McBride DW Jr (1997) Induced membrane hypo/hyper-mechanosensitivity: a limitation of patch-clamp recording. Annu Rev Physiol 59:621–631.
Häse CC, Le Dain AC and Martinac B (1995) Purification and functional reconstitution of the recombinant large mechanosensitive ion channel (MscL) of Escherichia coli. J Biol Chem 270:18329–18334.
Hirabayashi S, Inagaki M, Hisada T, Sugimachi M (2009) Effects of wall stress on the dynamics of ventricular fibrillation: a computer simulation study of mechanoelectric feedback. In: Mechanosensitivity in Cells and Tissues. Mechanosensitivity of the Heart. A Kamkin and I Kiseleva (eds.) Springer, pp. 387–419.
Honoré E, Patel AJ, Chemin J, Suchyna T, Sachs F (2006) Desensitization of mechano-gated K2P channels. Proc Natl Acad Sci USA 103(18):6859–6864.
Isenberg G, Kazanski V, Kondratev D, Gallitelli MF, Kiseleva I, Kamkin A (2003) Differential effects of stretch and compression on membrane currents and $[Na^+]c$ in ventricular myocytes. Prog Biophys Mol Biol 82(1–3):43–56.
Ismailov II, Awayda MS, Berdiev BK, Bubien JK, Lucas JE, Fuller CM, Benos DJ (1996a) Triple-barrel organization of ENaC, a cloned epithelial Na^+ channel. J Biol Chem 271(2):807–816.
Ismailov II, Awayda MS, Jovov B, Berdiev BK, Fuller CM, Dedman JR, Kaetzel M, Benos DJ (1996b) Regulation of epithelial sodium channels by the cystic fibrosis transmembrane conductance regulator. J Biol Chem 271(9):4725–4732.
Ismailov II, Berdiev BK, Bradford AL, Awayda MS, Fuller CM, Benos DJ (1996c) Associated proteins and renal epithelial Na^+ channel function. J Membr Biol 149(2):123–132.
Jacot JG, Raskin AJ, Omens JH, McCulloch AD, Tung L (2009) Mechanostransduction in cardiac and stem-cell derived cardiac cells. In: Mechanosensitivity in Cells and Tissues. Mechanosensitivity of the Heart. A Kamkin and I Kiseleva (eds.) Springer, pp. 99–139.
Jiang Y, Lee A, Chen J, Ruta V, Cadene M, Chait BT, MacKinnon R (2003) X-ray structure of a voltage-dependent K^+ channel. Nature 423:33–41.
Kamkin A, Kiseleva I, Isenberg G (2000a) Stretch-activated currents in ventricular myocytes: amplitude and arrhythmogenic effects increase with hypertrophy. Cardiovasc Res 48: 409–420.

Kamkin A, Kiseleva I, Isenberg G (2003a) Ion selectivity of stretch-activated cation currents in mouse ventricular myocytes. Pflügers Arch – Eur J Physiol 446(2):220–231.

Kamkin A, Kiseleva I, Isenberg G (2003b) Activation and inactivation of a non-selective cation conductance by local mechanical deformation of acutely isolated cardiac fibroblasts. Cardiovasc Res 57:793–803.

Kamkin A, Kiseleva I, Isenberg G, Wagner KD, Günther J, Theres H, Scholz H (2003c) Cardiac fibroblasts and the mechanoelectric feedback mechanism in healthy and diseased hearts. Prog Biophys Mol Biol 82:111–120.

Kamkin A, Kiseleva I, Lozinsky I (2009) The role of mechanosensitive fibroblasts in the heart: evidence from acutely isolated single cells, cultured cells and from intracellular microelectrode recordings on multicellular preparations from healthy and diseased cardiac tissue. In: Mechanosensitivity in Cells and Tissues. Mechanosensitivity of the Heart. A Kamkin and I Kiseleva (eds.) Springer, pp. 239–266.

Kamkin A, Kiseleva I, Wagner KD, Leiterer KP, Theres H, Scholz H, Günther J, Lab MJ (2000b) Mechanoelectric feedback in right atrium after left ventricular infarction in rats. J Mol Cell Cardiol 32:465–477.

Kamkin A and Kiseleva I (2008) Mechanically gated channels and mechanosensitive channels. In: Mechanosensitivity in Cells and Tissues. Mechanosensitive Ion Channels. A Kamkin and I Kiseleva (eds.) Springer, pp. xiii–xviii.

Kiseleva I, Kamkin A, Wagner KD, Theres H, Ladhoff A, Scholz H, Günther J, Lab MJ (2000) Mechano-electric feedback after left ventricular infarction in rats. Cardiovasc Res 45:370–378.

Kloda A and Martinac B (2001a) Mechanosensitive channel in thermoplasma a cell wall-less archaea: cloning and molecular characterization. Cell Biochem Biophys 34:321–347.

Kloda A and Martinac B (2001b) Molecular identification of a mechanosensitive channel in archaea. Biophys J 80:229–240.

Kloda A and Martinac B (2001c). Structural and functional similarities and differences between MscMJLR and MscMJ, two homologous MS channels of M. jannashii. EMBO J 20: 1888–1896.

Köhler R, Distler A, Hoyer J (1998) Pressure-activated cation channel in intact rat endocardial endothelium. Cardiovasc Res 38:433–440.

Köhler R, Grundig A, Brakemeier S, Rothermund L, Distler A, Kreutz R, Hoyer J (2001a) Regulation of pressure-activated channel in intact vascular endothelium of stroke-prone spontaneously hypertensive rats. Am J Hypertension 14:716–721.

Köhler R, Kreutz R, Grundig A, Rothermund L, Yagli C, Yagli Y, Pries AR, Hoyer J (2001b) Impaired function of endothelial pressure-activated cation channel in salt-sensitive genetic hypertension. J Am Soc Nephrol 12:1624–1629.

Kung C (2005) A possible unifying principle for mechanosensation. Nature 436:647–654.

Lab MJ (2005) Mechanically mediated crosstalk in heart. In: Mechanosensitivity in Cells and Tissues. A Kamkin and I Kiseleva (eds.) Academia Publishing House Ltd, Dordrecht, pp. 58–78.

Lab MJ (2009) Scanning ion conductance microscopy for imaging and mechanosensitive activation of selected areas of live cells. In: Mechanosensitivity in Cells and Tissues. Mechanosensitivity of the Heart. A Kamkin and I Kiseleva (eds.) Springer, pp. 267–272.

Lal H, Verma SK, Golden HB, Foster DM, Holt AM, Dostal DE (2009) Molecular signaling mechanisms of myocardial stretch: implications for heart disease. In: Mechanosensitivity in Cells and Tissues. Mechanosensitivity of the Heart. A Kamkin and I Kiseleva (eds.) Springer, pp. 55–81.

Lammerding J, Kamm PD, Lee RT (2004) Mechanotransduction in cardiac myocytes. Ann NY Acad Sci 1015:53–70.

Lillemeier BF, Pfeiffer JR, Surviladze Z, Wilson BS, Davis MM (2006) Plasma membrane-associated proteins are clustered into islands attached to the cytoskeleton. Proc Natl Acad Sci USA 103(50):18992–18997.

Linke WA (2009) Titin and titin-associated proteins in myocardial stress-sensing and mechanical dysfunction. In: Mechanosensitivity in Cells and Tissues. Mechanosensitivity of the Heart. A Kamkin and I Kiseleva (eds.) Springer, pp. 3–34.

Lozinsky I and Kamkin A (2009) Mechanosensitive alterations of action potentials and membrane currents in healthy and diseased cardiomyocytes: cardiac tissue and isolated cell. In: Mechanosensitivity in Cells and Tissues. Mechanosensitivity of the Heart. A Kamkin and I Kiseleva (eds.) Springer, pp. 185–238.

Maingret F, Fosset M, Lesage F, Lazdunski M, Honore E (1999) TRAAK is a mammalian neuronal mechano-gated K^+ channel. J Biol Chem 274:1381–1387.

Maingret F, Lauritzen I, Patel AJ, Heurteaux C, Reyes R, Lesage F, Lazdunski M, Honore E (2000a) TREK-1-1 is a heat-activated background K^+ channel. EMBO J 19:2483–2491.

Maingret F, Patel AJ, Lesage F, Lazdunski M, Honoré E (2000b) Lysophospholipids open the two P domain mechano-gated K^+ channels TREK-1 and TRAAK. J Biol Chem 275:10128–10133.

Markin VS and Sachs F (2004a) Thermodynamics of mechanosensitivity: lipid shape, membrane deformation and anesthesia. Biophysical J 86:370A.

Markin VS and Sachs F (2004b) Thermodynamics of mechanosensitivity. Phys Biol 1:110–124.

Maroto R, Raso A, Wood TG, Kurosky A, Martinac B, Hamill OP (2005) TRPCI forms the stretch-activated cation channel in vertebrate cells. Nat Cell Boil 7(2):179–185.

Martinac B (2001) Mechanosensitive channels in prokaryotes. Cell Physiol Biochem 11:61–76.

Martinac B (2004) Mechanosensitive ion channels: molecules of mechanotransduction. J Cell Sci 117:2449–2460.

Martinac B and Hamill OP (2002) Gramicidin A channels switch between stretch activation and stretch inactivation depending on bilayer thickness. Proc Natl Acad Sci USA 99:4308–4312.

Martinac B and Kloda A (2003) Evolutionary origins of mechanosensitive ion channels. Prog Biophys Mol Biol 82:11–24.

Martinac B, Adler J and Kung C (1990) Mechanosensitive ion channels of E. coli activated by amphipaths. Nature 348:261–263.

Moris CE and Juranka PF (2007) Na_V channel mechanosensitivity: activation and inactivation accelerate reversibly with stretch. Biophys J 93(3):822–833.

Morris CE and Homann U (2001) Cell surface area regulation and membrane tension. J Membr Biol 179(2):79–102.

Morris CE and Laitko U (2007) The mechanosensitivity of voltage-gated channels may contribute to cardiac mechano-electric feedback. In: Cardiac Mechano-Electrical Feedback and Arrhythmias: From Pipette to Patient. P Kohl, MR Franz, and F Sachs (eds.) Saunders, Philadelphia, pp. 33–41.

Morris CE, Juranka PF, Lin W, Morris TJ, Laitko U (2006) Studying the mechanosensitivity of voltage-gated channels using oocyte patches. Methods Mol Biol 322:315–329.

Nakayama K, Obara K, Ishikawa T, Nishizawa S (2009) Specific mechanotransduction signaling involved in myogenic responses of the cerebral arteries. In: Mechanosensitivity in Cells and Tissues. Mechanosensitivity of the Heart. A Kamkin and I Kiseleva (eds.) Springer, pp. 453–481.

Patel AJ and Honoré E (2001) Properties and modulation of mammalian 2P domain K^+ channels. TRENDS in Neurosciences 24(6):339–345.

Patel AJ, Honoré E, Maingret F, Lesage F, Fink M, Duprat F, Lazdunski M (1998) A mammalian two pore domain mechano-gated s-like K^+ channel. EMBO J 17:4283–4290.

Patel AJ, Lazdunski M, Honore E (2001) Lipid and mechano-gated 2P domain K+ channels. Curr Opin Cell Biol 13:422–428.

Perozo E and Rees DC (2003) Structure and mechanism in prokaryotic mecahnosensitive channels. Curr Opin Struct Biol 13:432–442.

Perozo E, Cortes DM, Sompornpisut P, Kloda A, Martinac B (2002a) Open channel structure of MscL and the gating mechanism of mechanosensitive channels. Nature 418(6901):942–948.

Perozo E, Kloda A, Cortes DM, Martinac B (2001) Site-directed spin-labeling analysis of reconstituted MscL in the closed state. J Gen Physiol 118(2):193–206

Perozo E, Kloda A, Cortes DM, Martinac B (2002b) Physical principles underlying the transduction of bilayer deformation forces during mechanosensitive channel gating. Nat Struct Biol 9(9):696–703.

Piao L, Li HY, Park CK, Cho IH, Piao ZG, Jung SJ, Choi SY, Lee SJ, Park K, Kim JS, Oh SB (2006) Mechanosensitivity of voltage-gated K^+ currents in rat trigeminal ganglion neurons. J Neurosci Res 83(7):1373–1380.

Pingguan-Murphy B and Knight MM (2008) Mechanosensitive purinergic calcium signalling in articular chondrocytes. In: Mechanosensitivity in Cells and Tissues. Mechanosensitive Ion Channels. A Kamkin and I Kiseleva (eds.) Springer, pp. 235–251.

Ravelli F and Masè M (2009) Mechanical modulation of a reentrant arrhythmia: the atrial flutter case. In: Mechanosensitivity in Cells and Tissues. Mechanosensitivity of the Heart. A Kamkin and I Kiseleva (eds.) Springer, pp. 301–325.

Ring A (1992) Monitoring the surface tension of lipid membranes by a bubble method. Pflügers Arch – Eur J Physiol 420:264–268.

Ring A and Sandblom J (1988) Evaluation of surface tension and ion occupancy effects on gramicidin A channel lifetime. Biophys J 53:541–548.

Sachs F (1988) Mechanical transduction in biological systems. Crit Rev Biomed Eng 16(2): 141–169.

Sachs F and Morris CE (1998) Mechanosensitive ion channels in nonspecialized cells. Rev Physiol Biochem Pharmacol 132:1–77.

Saint DA, Kelly D and Mackenzie L (2009) The contribution of MEF to electrical heterogeneity and arrhythmogenesis. In: Mechanosensitivity in Cells and Tissues. Mechanosensitivity of the Heart. A Kamkin and I Kiseleva (eds.) Springer, pp. 275–300.

Sigworth FJ (2003) Voltage-gated ion channels control electrical activity in nerve, muscle and many other cell types. The crystal structure of a bacterial voltage-gated channel reveals the astonishingly simple design of its voltage sensor. Nature 423:21–22.

Sokabe M and Sachs F (1990). The structure and dynamics of patch clamped membrane, a study using differential interference contrast microscopy. J Cell Biol 111:599–606.

Sokabe M, Sachs F, Jing Z (1991) Quantitative video microscopy of patch clamped membranes, stress, strain, capacitance and stretch channel activation. Biophys J 59:722–728.

Suchyna TM and Sachs F (2007) Mechanosensitive channel properties and membrane mechanics in mouse dystrophic myotubes. J Physiol (Lond) 581(1):369–387.

Suchyna TM, Tape SE, Koeppe RE, Andersen OS, Sachs F, Gottlieb PA (2004) Bilayer-dependent inhibition of mechanosensitive channels by neuroactive peptide enantiomers. Nature 430: 235–240.

Sukharev S (2002) Purification of the small mechanosensitive channel in Escherichia coli (MscS): the subunit structure, conduction and gating characteristics. Biophys J 83:290–298.

Sukharev S, Betanzos M, Chiang CS, Guy HR (2001) The gating mechanism of the large mechanosensitive channel MscL. Nature 409(6821):720–724.

Sukharev SI, Blount P, Martinac B, Blattner FR and Kung C (1994) A large mechanosensitive channel in E. coli encoded by mscL alone. Nature 368:265–268.

Sukharev SI, Schroeder MJ and McCaslin DR (1999) Stoichiometry of the large conductance bacterial mechanosensitive channel of E. coli. A biochemical study. J Membr Biol 171:183–193.

Ursell T, Phillips R, Kondev J, Reeves D, Wiggins PA (2008) The role of lipid bilayer mechanics in mechanosensation. In: Mechanosensitivity in Cells and Tissues. Mechanosensitive Ion Channels. A Kamkin and I Kiseleva (eds.) Springer, pp. 37–70.

Von Lewinski D, Kockskämper J, Khafaga M, Gasser R, Pieske B (2009) Stretch-induced inotropy in atrial and ventricular myocardium. In: Mechanosensitivity in Cells and Tissues. Mechanosensitivity of the Heart. A Kamkin and I Kiseleva (eds.) Springer, pp. 373–385.

Ward M-L and Allen DG (2009) Stretch-activated channels in the heart: contribution to cardiac performance. In: Mechanosensitivity in Cells and Tissues. Mechanosensitivity of the Heart. A Kamkin and I Kiseleva (eds.) Springer, pp. 141–167.

Watson PA, Hannan R, Carl LL, Giger KE (1996) Contractile activity and passive stretch regulate tubulin mRNA and protein content in cardiac myocytes. Am J Physiol – Cell Physiol 271:C684–C689.

White E (2006) Mechanosensitive channels: therapeutic targets in the myocardium? Curr Pharm Des 12(28):3645–3663.

Zeng T, Bett GCL, Sachs F (2000) Stretch-activated whole cell currents in adult rat cardiac myocytes. Am J Physiol 278:H548–H557.

Zhang Y and Hamill OP (2000) On the discrepancy between membrane patch and whole cell mechanosensitivity in Xenopus oocytes. J Physiol (Lond) 523(1):101–115.

Zhang Y, Gao F, Popov V, Wan J, Hamill OP (2000a) Mechanically-gated channel activity in cytoskeleton deficient blebs and vesicles from Xenopus oocytes. J Physiol (Lond) 523(1): 117–129.

Zhang YH, Youm JB, Sung HK, Lee SH, Ryu SY, Ho WK, Earm YE (2000b) Stretch-activated and background non-selective cation channels in rat atrial myocytes. J Physiol (Lond) 523(Pt 3):607–619.

Contributors

Ernesto A. Aiello Facultad de Ciencias Médicas, Centro de Investigaciones Cardiovasculares, Universidad Nacional de La Plata, Calle 60 y 120, 1900 La Plata, Argentina, aaiello@atlas.med.unlp.edu.ar

Hiroshi Akazawa Department of Cardiovascular Science and Medicine, Chiba University Graduate School of Medicine, 1-8-1 Inohana, Chuo-ku, Chiba 260-8670, Japan, akazawah-tky@umin.ac.jp

David G. Allen School of Medical Sciences, Institute for Biomedical Sciences, University of Sydney F13, NSW 2006, Australia, davida@physiol.usyd.edu.au

Umberto Banderali Faculty of Kinesiology, University of Calgary, 2500 University Drive NW Calgary, AB, T2N 1N4, Canada, ubanderali@kin.ucalgary.ca

Claudia I. Caldiz Facultad de Ciencias Médicas, Centro de Investigaciones Cardiovasculares, Universidad Nacional de La Plata, Calle 60 y 120, 1900 La Plata, Argentina

Wayne Carver Department of Cell Biology and Anatomy, University of South Carolina, School of Medicine, Columbia, SC 29209, USA, Wayne.Carver@uscmed.sc.edu

Christian Cherubini Laboratory of Nonlinear Physics and Mathematical Modeling, Università Campus Bio-Medico, Rome, Italy, C.Cherubini@unicampus.it

Gladys Chiappe de Cingolani Facultad de Ciencias Médicas, Centro de Investigaciones Cardiovasculares, Universidad Nacional de La Plata, Calle 60 y 120, 1900 La Plata, Argentina

Horacio E. Cingolani Facultad de Ciencias Médicas, Centro de Investigaciones Cardiovasculares, Universidad Nacional de La Plata, Calle 60 y 120, 1900 La Plata, Argentina, cicmes@infovia.com.ar

Robert B. Clark Faculty of Kinesiology, University of Calgary, 2500 University Drive NW Calgary, AB, T2N 1N4, Canada, rclar@ucalgary.ca

María V. Correa Facultad de Ciencias Médicas, Centro de Investigaciones Cardiovasculares, Universidad Nacional de La Plata, Calle 60 y 120, 1900 La Plata, Argentina

Verónica C. De Giusti Facultad de Ciencias Médicas, Centro de Investigaciones Cardiovasculares, Universidad Nacional de La Plata, Calle 60 y 120, 1900 La Plata, Argentina

David E. Dostal Division of Molecular Cardiology, College of Medicine, Scott & White, Cardiovascular Research Institute, The Texas A&M University System Health Science Center, Central Texas Veterans Health Care System, Temple, TX, USA, ddostal@medicine.tamhsc.edu

Irene L. Ennis Facultad de Ciencias Médicas, Centro de Investigaciones Cardiovasculares, Universidad Nacional de La Plata, Calle 60 y 120, 1900 La Plata, Argentina

Vadim V. Fedorov Department of Biomedical Engineering, Washington University, Campus Box 1097 One Brookings Drive, St. Louis, MO 63130-4899 USA, vadimfed@gmail.com; vfedorov@biomed.wustl.edu

Simonetta Filippi Laboratory of Nonlinear Physics and Mathematical Modeling, Università Campus Bio-Medico, Rome, Italy

Martin Fink Department of Physiology, Anatomy and Genetics, Oxford University, Oxford, UK, martin.fink@dpag.ox.ac.uk

Donald M. Foster Central Texas Veterans Health Care System, Temple, TX 76504, USA

John W. Fuseler Department of Cell Biology and Anatomy, School of Medicine, University of South Carolina, Columbia, SC 29209, USA, fuseler@gw.med.sc.edu

Carolina D. Garciarena Facultad de Ciencias Médicas, Centro de Investigaciones Cardiovasculares, Universidad Nacional de La Plata, Calle 60 y 120, 1900 La Plata, Argentina

Robert Gasser Abteilung Kardiologie, Medizinische Universität Graz, Auenbruggerplatz 15, 8036 Graz, Austria

Wayne R. Giles Faculty of Kinesiology, University of Calgary, 2500 University Drive NW Calgary, AB, T2N 1N4, Canada, wgiles@ucalgary.ca

Honey B. Golden Division of Molecular Cardiology, College of Medicine, Scott & White, Cardiovascular Research Institute, The Texas A&M University System Health Science Center, Central Texas Veterans Health Care System, Temple, TX, USA

Satoko Hirabayashi Biomechanics Laboratory, Department of Mechanical Science & Engineering, Graduate School of Engineering, Nagoya University, Furo-cho, Chikusa, Nagoya, 464-8603, Japan, hirabayashi@mech.nagoya-u.ac.jp

Toshiaki Hisada Computational Biomechanics Laboratory, Department of Human and Engineered Environmental Studies, Graduate School of Frontier Sciences, The University of Tokyo, Hongo 7-3-1, Bunkyo-ku, Tokyo 113-8656, Japan, hisada@mech.t.u-tokyo.ac.jp

April M. Holt Division of Molecular Cardiology, College of Medicine, Scott & White, Cardiovascular Research Institute, The Texas A&M University System Health Science Center, Temple, TX, USA

Masashi Inagaki National Cardiovascular Center Research Institute, Osaka, Japan, masashii@ri.ncvc.go.jp

Tomohisa Ishikawa Department of Pharmacology, School of Pharmaceutical Sciences, University of Shizuoka, Shizuoka City, Shizuoka 422-8526, Japan

Jeffrey G. Jacot Department of Bioengineering, Rice University; Congenital Heart Surgery, Texas Children's Hospital. Houston, TX, USA, jeff.jacot@rice.edu

Andre Kamkin Department of Fundamental and Applied Physiology, Russian State Medical University, Ostrovitjanova Str.1, 117997 Moscow, Russia, Kamkin.A@g23.relcom.ru; andre.kamkin@gmail.com

Douglas Kelly The School of Molecular and Biomedical Science, University of Adelaide, Adelaide, SA 5005, Australia

Mounir Khafaga Abteilung Kardiologie, Medizinische Universität Graz, Auenbruggerplatz 15, 8036 Graz, Austria

Irina Kiseleva Department of Fundamental and Applied Physiology, Russian State Medical University, Ostrovitjanova Str.1, 117997 Moscow, Russia

Jens Kockskämper Abteilung Kardiologie, Medizinische Universität Graz, Auenbruggerplatz 15, 8036 Graz, Austria

Issei Komuro Department of Cardiovascular Science and Medicine, Chiba University Graduate School of Medicine, 1-8-1 Inohana, Chuo-ku, Chiba 260-8670, Japan, komuro-tky@umin.ac.jp

Max J. Lab National Heart and Lung Institute, Imperial College London, London SW3 6LY, UK, m.lab@imperial.ac.uk

Hind Lal Division of Molecular Cardiology, College of Medicine, Scott & White, Cardiovascular Research Institute, The Texas A&M University System Health Science Center, Temple, TX, USA

Wolfgang A. Linke Physiology and Biophysics Unit, University of Muenster, Schlossplatz 5, D-48149 Muenster, Germany, wlinke@uni-muenster.de; Department of Cardiovascular Physiology, Faculty of Medicine, Ruhr University Bochum, MA 2/156, D-44780 Bochum, Germany, wolfgang.linke@rub.de

Ilya Lozinsky Department of Fundamental and Applied Physiology, Russian State Medical University, Ostrovitjanova Str.1, 117997 Moscow, Russia, Ilya.Lozinsky@gmx.net

Lorraine Mackenzie The School of Molecular and Biomedical Science, University of Adelaide, Adelaide, SA 5005, Australia

Michela Masè Laboratory of Biophysics and Biosignals, Department of Physics, Faculty of Science, University of Trento,Via Sommarive 14, 38050 Povo – Trento, Italy

Andrew D. McCulloch Department of Bioengineering, University of California San Diego, La Jolla, CA, USA

Catherine E. Morris Neuroscience, Ottawa Hospital Research Institute, 451 Smyth Rd., Ottawa, Ontario, K1H 8M5, Canada, cmorris@uottawa.ca

Koichi Nakayama Department of Molecular and Cellular Pharmacology, Faculty of Pharmaceutical Sciences, Iwate Medical University, Yahaba, Iwate 028-3694, Japan, nakyamk@iwate-med.ac.jp

Paola Nardinocchi Department of Structural Engineering and Geotechnics, Università degli Studi di Roma "La Sapienza", Rome, Italy

Shigeru Nishizawa Department of Neurosurgery, University of Occupational and Environmental Health, Kitakyushu, Fukuoka 807-8555, Japan

Kazuo Obara Department of Pharmacology, School of Pharmaceutical Sciences, University of Shizuoka, Shizuoka City, Shizuoka 422-8526, Japan

Jeffrey H. Omens Department of Medicine and Bioengineering, University of California, San Diego. San Diego, CA, USA.

Néstor G. Pérez Facultad de Ciencias Médicas, Centro de Investigaciones Cardiovasculares, Universidad Nacional de La Plata, Calle 60 y 120, 1900 La Plata, Argentina

Burkert Pieske Abteilung Kardiologie, Medizinische Universität Graz, Auenbruggerplatz 15, 8036 Graz, Austria

Anna J. Raskin Saint Jude Medical, Los Angeles, CA, USA

Flavia Ravelli Laboratory of Biophysics and Biosignals, Department of Physics, Faculty of Science, University of Trento,Via Sommarive 14, 38050 Povo – Trento, Italy, ravelli@science.unitn.it

David A. Saint The School of Molecular and Biomedical Science, University of Adelaide, Adelaide, SA 5005, Australia, david.saint@adelaide.edu.au

Masaru Sugimachi National Cardiovascular Center Research Institute, Osaka, Japan, su91mach@ri.ncvc.go.jp

Luciano Teresi SMFM-Mathematical Structures of Materials Physics, Università Roma Tre, Rome, Italy

Leslie Tung Department of Biomedical Engineering, The Johns Hopkins University, Traylor Building, 720 Rutland Ave., Baltimore, MD 21205, USA, ltung@jhu.edu

Suresh K. Verma Division of Molecular Cardiology, College of Medicine, Scott & White, Cardiovascular Research Institute, The Texas A&M University System Health Science Center, Temple, TX, USA

María C. Villa-Abrille Facultad de Ciencias Médicas, Centro de Investigaciones Cardiovasculares, Universidad Nacional de La Plata, Calle 60 y 120, 1900 La Plata, Argentina

Dirk Von Lewinski Abteilung Kardiologie, Medizinische Universität Graz, Auenbruggerplatz 15, 8036 Graz, Austria, dirk.von-lewinski@meduni-graz.at

Marie-Louise Ward Department of Physiology, Faculty of Medical and Health Sciences, University of Auckland, Private Bag 92019, Auckland, New Zealand, m.ward@auckland.ac.nz

Alejandra M. Yeves Facultad de Ciencias Médicas, Centro de Investigaciones Cardiovasculares, Universidad Nacional de La Plata, Calle 60 y 120, 1900 La Plata, Argentina

Part I
Molecular Mechanisms of Mechanotransduction in Cardiac Cells

Chapter 1
Titin and Titin-Associated Proteins in Myocardial Stress-Sensing and Mechanical Dysfunction

Wolfgang A. Linke

Abstract Mechanical stress signals transmitted through the heart walls during hemodynamic loading are sensed by the myocytes. These signals play an important role in physiological heart development and hypertrophy, but disruption of the well-balanced stress-sensing machinery causes mechanical dysregulation, cardiac remodelling, and heart failure. In cardiomyocytes, nodal points of force transmission and mechanosensing reside in the Z-disk, M-band, and I-band regions of the sarcomeres. Longitudinal linkage of these regions is provided by the titin filament and several "hot spots" along this giant protein may be, along with some of its > 20 ligands, pivotal to the myofibrillar stretch response. This review outlines the known interaction partners of titin and highlights the putative stress/stretch sensor complexes at titin's NH_2 and COOH termini and their role in myopathies. Another focus is the elastic I-band titin section, which interacts with a diverse number of proteins and whose main function is as a determinant of diastolic distensibility and passive stiffness. The discussion summarizes recent insights into the plasticity, mechanical role, and regulation of the elastic titin springs in cardiac development and human heart disease. Titin and titin-based protein complexes are now recognized as integral parts of the mechanosensitive protein network and as critical components in cardiomyocyte stress/stretch signalling.

Keywords Elasticity · Stress-sensing · Passive tension · Cardiomyopathy · Diastolic function · Connectin

1.1 Introduction

Mechanical stresses play a central role in the regulation of physiological processes, and the heart is no exception. Physical forces promote cardiac development and hypertrophy, but dysregulation of mechanical signalling can lead to chronic heart

W.A. Linke (✉)
Physiology and Biophysics Unit, University of Muenster, Muenster, Germany
e-mail: wlinke@uni-muenster.de

A. Kamkin, I. Kiseleva (eds.), *Mechanosensitivity of the Heart*,
Mechanosensitivity in Cells and Tissues 3, DOI 10.1007/978-90-481-2850-1_1,

diseases such as hypertrophic (HCM) or dilated cardiomyopathy (DCM). Research on mechanotransduction in normal and diseased heart is aimed at elucidating the molecular mechanisms by which myocardial structures sense physical loads and transduce them into biochemical signals to alter gene expression and modify cellular structure and function (Frey and Olson, 2003; Heineke and Molkentin, 2006; Olson, 2006). The propagation and sensing of mechanical forces in myocardium involves many different components, including, but not restricted to, the extracellular matrix (ECM) (MacKenna et al., 2000; Schellings et al., 2004), the costameric protein network at focal adhesions (Ervasti, 2003; Samarel, 2005; Romer et al., 2006; Brancaccio et al., 2006), the adherens junction (Perriard et al., 2003) and transitional junction (Bennett et al., 2006) at intercalated disks, as well as protein complexes associated with the contractile units of striated muscle, the sarcomeres (Knoll et al., 2003; Miller et al., 2004; Lange et al., 2006; Hoshijima, 2006; Linke, 2008). A notable feature of the mechanotransduction network in the heart is that external force signals, such as those imposed during hemodynamic load, are transmitted from the ECM to the cardiomyocyte cytoskeleton, while at the same time the sarcomeres themselves generate active and passive forces, which propagate in the opposite direction. This bidirectional force transduction is mediated by highly specialized nodal points of mechanosignalling: the Z-disks (Hoshijima, 2006; Pyle and Solaro, 2004; Frank et al., 2006; Linke, 2008) and M-bands (Agarkova and Perriard, 2005; Linke, 2008). Although we are far from understanding how these sarcomeric protein complexes accomplish the feat of stress-sensing, recent studies have shed light on some of the hitherto elusive mechanisms.

Like a myofibrillar backbone, > 1-μm-long filaments of titin (also named connectin) run from the Z-disk to the M-band or center of the sarcomere (Fig. 1.1). Judged by their layout in the sarcomere it is conceivable that titin strands are part of the stress-responsive machinery that senses the propagating mechanical signals (Linke, 2008). During diastolic distension of the ventricular walls, when actin and myosin are largely detached, titin filaments are stretched in their elastic I-band region (Fig. 1.1a) and behave as passive-force generators in parallel with the contractile apparatus. These titin-spring forces contribute substantially to diastolic wall stiffness in mammalian (Linke et al., 1994) as well as human heart (Neagoe et al., 2002). Moreover, titin is attached to the thin filaments at the Z-disk and runs along the thick filaments, bound to myosin in the A-band and M-band regions (Fig. 1.1a). These sections of the titin molecule, in concert with some of their interaction partners (Fig. 1.1b), could sense the forces generated in the sarcomere during both diastolic stretch and systolic contraction. In addition, the multiple linkages involving those titin regions may provide a means for sensing stresses from various directions. Hence, present concepts about the Z-disk- and M-band-associated protein-complexes acting as mechanosensors in the cardiomyocyte must include the NH_2 and COOH terminal portions of titin (Miller et al., 2004; Lange et al., 2006; Hoshijima, 2006; Linke, 2008). This review deals with the functions of titin, the known titin ligands, and putative "hot spots" of stress-sensing along the titin molecule. Disease-causing mutations in these titin regions are outlined and the

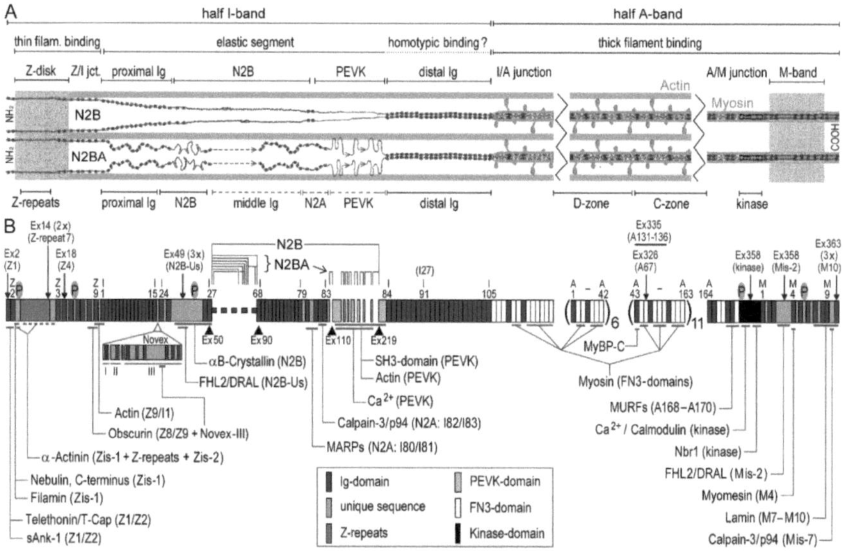

Fig. 1.1 Titin architecture and binding partners. (**a**) Layout of cardiac titin-isoforms, N2B and N2BA, co-expressed in the same half-sarcomere. *Dashed lines* indicate differentially spliced segments in titin's elastic spring region. (**b**) Domain structure of human cardiac titin and binding sites of known titin ligands. Binding sites are listed in *parentheses* behind the ligand name (for full names of titin ligands, see legend to Fig. 1.2). Illustrated are splice pathways for human N2B and N2BA titin-isoforms between exons (Ex) 50 and 219. Novex I, II, and III, are exons 45, 46, and 48, respectively, in the titin genomic sequence. *Arrows*/text indicate mutations found in human titin. Domain names (nomenclature from Bang et al., 2001a) are shown for selected Ig/FN3 modules above the titin sequence. Ig-domain I91 is sometimes called I27 (old nomenclature from Labeit and Kolmerer, 1995). P, titin-phosphorylation site. (Figure taken from Linke, 2008. With permission from Cardiovascular Research)

plasticity, mechanical role, and regulation of the elastic titin springs in heart development and disease are discussed.

1.2 The Titin Gene and Tissue Specificity of Titin Expression

Membersof the titin family are regarded as the largest known proteins. The human titin gene on chromosome 2q31 has a size of 294 kilobases and encompasses 363 exons predicted to code for a total number of 38,138 amino-acid residues or a polypeptide with a maximum molecular mass of 4,200 kDa (Bang et al., 2001a). There is only a single gene for titin coding for both the cardiac and the skeletal-muscle isoforms in human, mouse (also on the long arm of chromosome 2), and probably other mammals as well. Interestingly, zebrafish contains two orthologous titin genes, ttna and ttnb, which are located in tandem-array on chromosome 9 and have distinct functions (Seeley et al., 2007). In the zebrafish heart, ttna (but not ttnb) is required for sarcomere assembly and the establishment of cardiac contractility and ttna was found to be the earliest sarcomeric mRNA expressed, suggesting it is an early molecular marker for cardiomyocyte differentiation. A mutation (*pickwick*) in the elastic segment of zebrafish titin causes DCM (Xu et al., 2002). The largest human titin sequenced to date is from the adult soleus skeletal muscle (3,700 kDa) (Labeit and Kolmerer, 1995), but some titins in fetal mammalian hearts reach similar (Opitz et al., 2004; Warren et al., 2004; Kruger et al., 2006) or even bigger (Greaser et al., 2005) sizes.

Titins are also found in human smooth-muscle tissues: aorta, bladder, carotid, and stomach each express 80–100 of the 363 titin exons encoding parts of the Z-disk, I-band, and A-band titin regions (Labeit et al., 2006). The titin proteins are of relatively small size with molecular masses of no more than approximately 1,000 kDa. The functional role(s) of the smooth-muscle titins still await exploration and the relationship between these titins and a previously described smooth-muscle titin-like protein, smitin (Kim and Keller, 2002), also remains to be shown. Furthermore, human non-muscle cells, such as fibroblasts and platelets, express multiple isoforms of cellular titin (c-titin), products of the human titin gene that are associated with stress-fibers and apparently contain many of the titin-domains found in striated muscles (Cavnar et al., 2007). It will also be interesting to follow up on scattered reports hinting at the presence of titin as a nuclear protein in non-muscle cells potentially providing elasticity and structural flexibility to chromosomes (Machado et al., 1998; Machado and Andrew, 2000; Zastrow et al., 2006). However, the extent and nature of titin's nuclear roles remain unknown or even controversial (Wernyj et al., 2001).

1.3 Titin Isoform Diversity and Functions in the Heart

Approximately 90% of the mass of the titin molecule is made up of globular domains of the immunoglobulin (Ig) or fibronectin-type-III (FN3) like folds; the remainder is unique sequence insertions (Labeit and Kolmerer, 1995). All domain types are involved in numerous protein-protein interactions (Fig. 1.1b). Nearly all Z-disk and A-band/M-band titin domains, and some of the I-band domains, are constitutively expressed in the human cardiac (and skeletal-muscle) titin-isoforms. Differential splicing occurs mainly in the I-band titin segment (~800 to ~1,500 kDa) (Freiburg et al., 2000) and gives rise to the presence of two distinct isoforms in mammalian heart, N2B (~3 MDa) and N2BA (3.2–3.7 MDa) (Fig. 1.1). Both isoforms have a "proximal" (I1-I15) and a "distal" (I84-I105) region composed of tandemly arranged Ig-domains, a so-called "N2B"-domain (titin-exon 49; this domain is not expressed in skeletal muscles) encompassing a long unique sequence (N2B-Us) flanked by Ig-domains, and a "PEVK"-domain, so called for its high content in proline (P), glutamic acid (E), valine (V), and lysine (K) residues (Fig. 1.1). In human N2B-titin, exon 50 coding for Ig-domain I27 is spliced directly to exon 219 in the PEVK-segment (Fig. 1.1), skipping many exons coding for Ig-domains and unique sequences. Note that Ig-domain I27 is not to be confused with Ig-domain I91 (Fig. 1.1b), which was initially named I27 (Labeit and Kolmerer, 1995) but was later renamed to I91 (Bang et al., 2001a). N2BA-isoforms contain, in addition, a "middle-Ig-domain" region (I28-I79), a so-called 'N2A'-domain (exons 102–109) containing four Ig-domains plus a few unique sequences, and a PEVK-segment that varies greatly in length among these isoforms. Extensive alternative splicing occurs in the middle-Ig region between exons 50 and 90 and in the PEVK-domain between exons 110 and 219 (Fig. 1.1b), resulting in innumerable potential N2BA-isoforms, only some of which have been sequenced so far (Freiburg et al., 2000). The N2BA-isoforms are co-expressed with N2B at the level of the half-sarcomere (Linke et al., 1996; Trombitas et al., 2001) and in adult human left ventricle, the normal N2BA:N2B isoform ratio is ~30:70 to ~40:60 (Neagoe et al., 2002).

Titins are well known for their mechanical functions. The titin springs endow the sarcomeres with long-range elasticity and are main determinants of myocardial passive tension (PT) and stiffness, together with the ECM-based collagen fibers (Linke et al., 1994). The elasticity of titin may support elastic recoil in early diastole (Helmes et al., 1996, 2003) and early systolic shortening (Opitz et al., 2003). Titin is thought to help center the A-band in the middle of the sarcomere during activation (Horowits, 1999). Moreover, titin could be a factor in determining the length-dependence of Ca^{2+}-activated force development, the molecular basis for the Frank-Starling mechanism (Fukuda and Granzier, 2005). Apart from these mechanical roles, titin has been suggested to serve as a blueprint for sarcomere assembly (Miller et al., 2004; Trinick and Tskhovrebova, 1999; Clark et al., 2002). Via direct protein–protein interactions (Fig. 1.1b) titin aligns structural, regulatory, and contractile proteins within the sarcomere (Linke, 2008) and may coordinate the precise assembly of myofibrils during muscle development and hypertrophy (van der Ven et al., 2000). Direct and indirect links to various signalling molecules, along with the presence of phosphorylation sites in the Z-disk, I-band, and M-band segments (Labeit et al., 1997; Tskhovrebova and Trinick, 2004) (Fig. 1.1b), have implicated titin as player in myocardial signalling (Miller et al., 2004; Lange et al., 2006). Further, there is a kinase domain near the COOH-terminus of titin (Mayans et al., 1998), the function of which is becoming increasingly clear (Gotthardt et al., 2003; Lange et al., 2005b; Grater et al., 2005; Weinert et al., 2006; Musa et al., 2006; Peng et al., 2007).

1.4 Interactions and Properties of Z-Disk Titin

Z-disk titin is encoded by exons 1–28 and has nine Ig-domains (Z1-Z9) separated by unique sequences (Fig. 1.1b). Up to Ig-domain Z4, this titin segment is an integral part of the Z-disk, whereas the remainder, up to Ig-domain Z9, reaches beyond the Z-disk edge but is still tightly associated with the thin filament (Fig. 1.1a) (Linke et al., 1997; Trombitas and Granzier, 1997). Actin-binding was confirmed for the domains Z9-I1 (Linke et al., 1997).

1.4.1 Ig-Domains Z1/Z2 and the Z-Disk-Based Mechanosensor

Titin's most NH_2-terminal domains, Z1/Z2, interact with the sarcoplasmic reticulum (SR) membrane protein, small-ankyrin-1 (sAnk-1) (Fig. 1.1b) (Kontrogianni-Konstantopoulos and Bloch, 2003), which has links to spectrin, desmin, and obscurin (Fig. 1.2) (Bagnato et al., 2003; Kontrogianni-Konstantopoulos et al., 2003; Kontrogianni-Konstantopoulos et al., 2006). The ~800-kDa-protein obscurin contains ~50 Ig-modules and several signalling domains (SH3, DH, PH, kinase) and is expressed in different cardiac isoforms (Fukuzawa et al., 2005), the composition of which can be altered in human DCM (Makarenko et al., 2004). Obscurin, which binds more efficiently to sAnk-1 than titin-Z1/Z2 does (Armani et al., 2006), also associates with the Ig-domains Z8/Z9 more distal in Z-disk titin (Young et al., 2001), suggesting connectivity between the SR, the Z-disk, and other cytoskeletal structures.

Additionally, the Z-disk-titin Ig-domains, Z1/Z2, bind to telethonin (also called T-Cap) (Mues et al., 1998; Gregorio et al., 1998). The recently determined atomic model of the titin Z1/Z2-telethonin complex showed that telethonin assembles two titin filaments entering the Z-disk from the same half-sarcomere into a tightly packed anti-parallel sandwich structure (Zou et al., 2006). This Z1/Z2-telethonin complex is highly resistant to stretching forces, a property conferred by multiple hydrogen bonds that cross-link beta-strands of the two proteins (Lee et al., 2006). Telethonin may have a role in titin assembly and telethonin-based Z-disk anchorage of titin filaments may be a pre-requisite for the proper functioning of the putative Z-disk mechanosensor. It remains to be seen whether the robust Z1/Z2-telethonin complex is the critical anchoring site for titin′s NH_2-terminus or whether other attachment sites of titin in the Z-disk (see below) are similarly important. Other structural and signalling proteins are targeted via telethonin to the Z-disk (Fig. 1.2). These proteins, covered in reviews elsewhere (Lange et al., 2006;Hoshijima, 2006; Frank et al., 2006; Granzier and Labeit, 2004), include the muscle-growth-factor myostatin, the potassium-channel-subunit minK (I_{ks}), protein-kinase-D (PKD), muscle-specific RING-finger proteins 1 and 2 (MURF1 and MURF2), ankyrin-repeat-domain protein-2 (Ankrd2), calsarcin (calsarcin-2 is also called FATZ or myozenin), and muscle-LIM protein (MLP), also known as cysteine-rich protein-3 (CRP3). Mutations in telethonin that modulate the interaction with titin, MLP, or calsarcin have been reported to cause either HCM or DCM in patients (Hayashi et al., 2004; Bos et al., 2006). Together with titin's NH_2-terminus and MLP, telethonin is believed to be central to the Z-disk-based mechanosensor (Knoll et al., 2002).

MLP is highly expressed in myocardium where it interacts with numerous other proteins (reviewed in Hoshijima, 2006; Frank et al., 2006), such as the cytoskeletal proteins, β-spectrin, α-actinin, zyxin, and nebulin-related anchoring-protein (N-RAP), and the Ca^{2+}-calmodulin-dependent phosphatase, calcineurin, which activates the transcription factor, nuclear-factor-of-activated-T-cells (NFAT) (Fig. 1.2). Calcineurin and NFAT, which are downstream effectors of MLP (Heineke et al., 2005), are part of a main hypertrophic response pathway in heart (Heineke and Molkentin, 2006). MLP itself can translocate from the Z-disk, cytosol, or

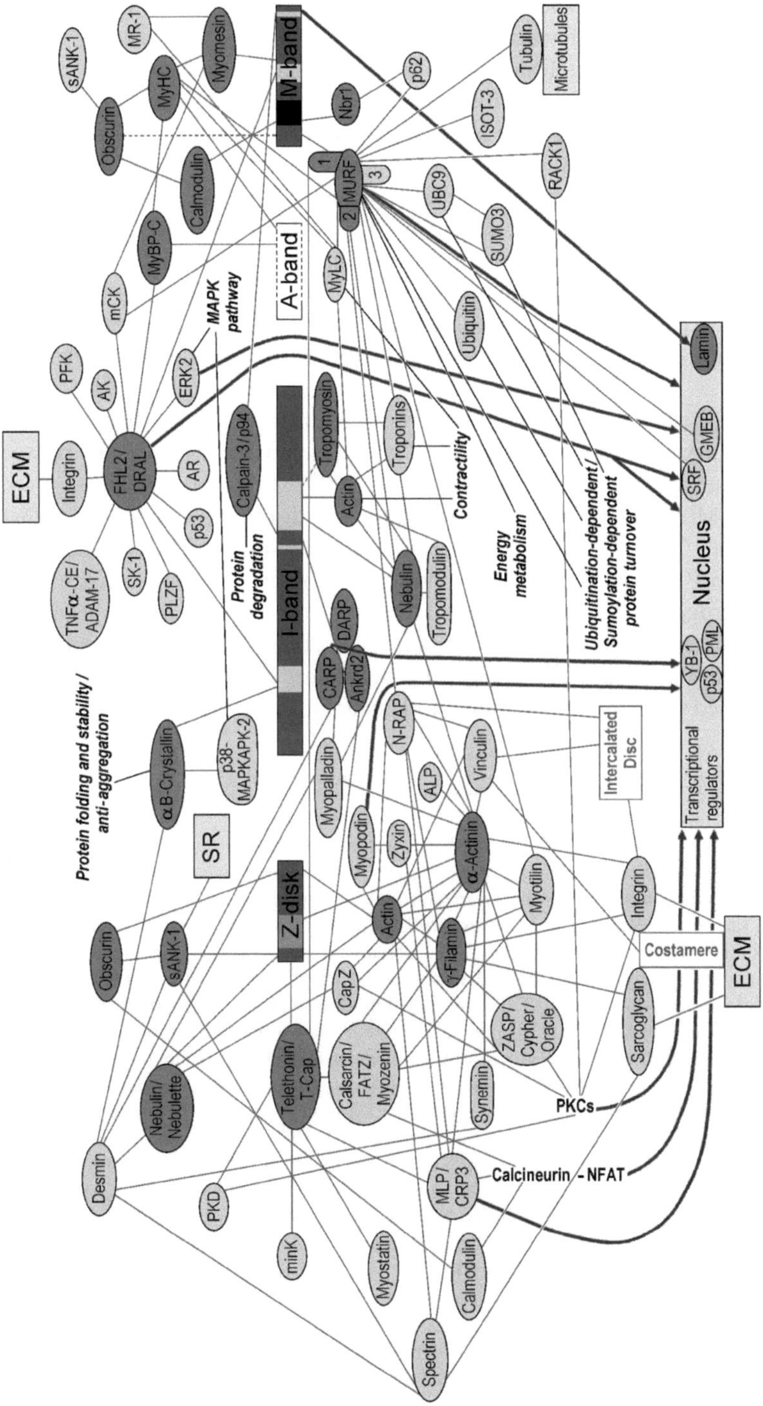

Fig. 1.2 (continued)

intercalated disk to the nucleus, where it associates with the muscle transcriptional regulators, myogenic-differentiation-antigen (MyoD), muscle-regulatory-factor-4 (MRF4), and myogenin (Hoshijima, 2006). Point mutations have been found in human MLP at sites involved in protein-protein interactions (reviewed in Hoshijima, 2006; Frank et al., 2006) and the affected patients develop either dilated or hypertrophic cardiomyopathy. Biochemical evidence suggested that the MLP mutations reduce the binding affinity for the respective ligand, possibly rendering the putative Z-disk stress-sensor less effective. Another hint at the importance of MLP came from the finding that this protein is substantially down-regulated in chronically ischemic rat hearts (Wilding et al., 2006), as well as in failing hearts of both dilated and ischemic cardiomyopathy patients (Zolk et al., 2000; Boateng et al., 2007), an effect apparently mediated through nitric-oxide-based signalling (Heineke et al., 2003).

An MLP-deficient mouse model has provided insights into the function of the putative Z-disk mechanosensor. MLP-null mice develop a cardiomyopathy between two and four weeks after birth (Arber et al., 1997). The cardiomyocytes of 6–9-month-old MLP-null mice showed widening and disorganization of the Z-disks (Knoll et al., 2002). Functional changes in juvenile MLP-null hearts included a reduction in PT of isolated papillary muscles (Knoll et al., 2002), decreased passive myocardial stiffness and prolonged relaxation time, but no alterations in most systolic characteristics (Lorenzen-Schmidt et al., 2005). These findings suggested altered elastic properties of titin in MLP-null mice; however, titin has not been directly tested yet in this mouse model with regard to changes in isoform expression

Fig. 1.2 Current state of the titin-interactome. Summary of known connections between titin (Z/I/A/M-band region) and its direct (*darker ovals*) and indirect (*lighter ovals*) ligands. *Grey lines* show interaction between proteins, *arrows* indicate capacity for cytoplasmic-nuclear shuttling. SR, sarcoplasmic reticulum; ECM, extracellular matrix. AK, adenylate-kinase; ALP, actinin-associated LIM-protein; Ankrd2, ankyrin-repeat-domain protein-2; AR, androgen-receptor; CARP, cardiac ankyrin-repeat protein; DARP, diabetes-related ankyrin-repeat protein; ERK2, extracellular signal-regulated kinase-2; FATZ, filamin-/actinin-/telethonin-binding protein of the Z-disk; FHL2 (DRAL), four-and-a-half-LIM-domain protein (down-regulated-in-rhabdomyosarcoma LIM-domain protein); GMEB, glucocorticoid-modulatory-element binding-protein; ISOT-3, isopeptidase-T-3; MAPK, mitogen-activated protein-kinase; mCK, muscle creatine-kinase; minK, potassium-channel subunit; MLP (CRP3), muscle-LIM-protein (cysteine-rich protein-3); MR-1, myofibrillogenesis regulator-1; MURF, muscle-specific RING-finger protein; MyHC, myosin-heavy-chain; MyLC, myosin-light-chain; Nbr1, neighbor-of-BRCA1 gene-1; NFAT, nuclear-factor-of-activated-T-cells; N-RAP, nebulin-related anchoring-protein; p38-MAPKAPK-2, p38-activated MAPK-activated protein-kinase-2; PFK, phosphofructokinase; PKCs, isoforms of protein-kinase-C; PKD, protein-kinase-D; PLZF, promyelocytic leukemia-zinc-finger protein; PML, promyelocytic leukaemia-protein; RACK1, receptor-for-activated-C-kinase-1; sANK-1, small-ankyrin-1 isoform; SK-1, sphingosine-kinase-1; SRF, serum-response factor; SUMO3, small ubiquitin-related modifier-3; TNFα-CE (ADAM-17), tumor-necrosis-factor-alpha converting-enzyme (A-disintegrin-and-metalloprotease-17 protein); UBC9; ubiquitin-conjugating enzyme-9; YB-1, Y-box-binding protein-1; ZASP, Z-band alternatively-spliced PDZ-motif protein. (Figure taken from Linke, 2008. With permission from Cardiovascular Research)

or elasticity. Notably, when neonatal cardiomyocytes were cultured on a flexible membrane and exposed to 10% equibiaxial stretch for 24 h, the "stress marker" brain-natriuretic-peptide (BNP) was upregulated in wildtype but not in MLP-null cells (Knoll et al., 2002). However, BNP could still be induced by pharmacological stimulation using endothelin-1, consistent with a role for MLP specific to mechanosignalling.

How could a stress-sensing mechanism via MLP and the titin-Z1/Z2-telethonin complex work? Extracellular or sarcomere-generated forces transmitted via the Z-disk may affect the fraction of Z-disk-bound versus nuclear MLP, and increased MLP in the nucleus may activate transcriptional (co)-factors. This could evoke a graded response in terms of variable changes in muscle-gene expression depending on the force level. Alternatively, mechanical forces may alter the interaction between the MLP-telethonin-titin ternary complex and a ligand, thus triggering a hypertrophy pathway, e.g., via activation of calcineurin-NFAT or PKC; again, muscle-gene expression would be modified depending on the applied force level. Possibly, the force level is somehow fed back to the Z-disk sensor to alter its stress-responsive behavior. Experimental evidence for such a force-feedback mechanism would now be needed to substantiate the concept of the Z-disk-based stress-sensor.

1.4.2 Connectivity Provided by Unique Sequence Insertions at Titin's NH_2 -Terminus

Several interactions involve a sequence insertion adjacent to the titin Z2-domain, Zis-1 (Fig. 1 1b). Titin-Zis-1 associates with an SH3-domain at the C-terminus of nebulin (Witt et al., 2006), a giant polypeptide (600–900 kDa) expressed not only in skeletal muscle but at low levels also in cardiomyocytes (Kazmierski et al., 2003; McElhinny et al., 2005; Bang et al., 2006) – although earlier work and a recent study (Witt et al., 2006) have not confirmed its presence in heart. Unlike in skeletal muscle, nebulin cannot be a molecular ruler of the thin filaments in cardiac cells (Witt et al., 2006; Bang et al., 2006), but it might regulate actin-filament dynamics (Fowler et al., 2006; Horowits, 2006), stabilize cytoskeletal linkages to the Z-disk by interacting with actin, desmin, CapZ, and myopalladin (Fig. 1.2) (Pyle and Solaro, 2004; Bang et al., 2006), and possibly specify Z-disk width (Witt et al., 2006). The SH3-domain in nebulin that binds to titin-Zis-1 is present also in human nebulette, a 109-kDa cardiac-specific protein that shares extensive similarity with the C-terminal portion of human nebulin (Moncman and Wang, 1995; Millevoi et al., 1998). Interaction is therefore likely also between nebulette and titin-Zis-1.

Titin-Zis-1 also binds γ-filamin (Labeit et al., 2006), a striated muscle-specific filamin with multiple links in the myocardial stress-sensing pathway. The NH_2-terminus of titin is coupled via filamin to structural and signalling proteins (Fig. 1.2), such as integrin and sarcoglycan at the costameres (Samarel, 2005; Brancaccio et al., 2006), α-actinin, actin, myotilin, ZASP (cypher/oracle) and calsarcin at the Z-disks (Ervasti, 2003), and N-RAP at the intercalated disks (Lu et al., 2003). The filamin-mediated connection between titin and the focal-adhesion complex is particularly interesting, as external forces transmitted from the ECM via the costameres to the cytoskeleton are known to initiate a mechano-chemical signalling cascade involving proteins such as vinculin, melusin, talin, focal-adhesion kinase (FAK), integrin-linked kinase (ILK), Src-tyrosine kinase, zyxin, paxillin, protein-kinase-C_{ε}, and members of the Rho-family GTPases (Samarel, 2005; Brancaccio et al., 2006; Mitra et al., 2005). Further downstream, the mitogen-activated protein-kinase (MAPK) and AKT/PKB signalling pathways are activated, thus promoting gene expression and cardiomyocyte growth (Heineke and Molkentin, 2006).

A third ligand of the Zis-1-titin region is α-actinin (Labeit et al., 2006). This interaction exists in addition to the known α-actinin binding sites within the seven 45-amino-acid repeats known as "titin Z-repeats" (exons 8–14) (Gautel et al., 1996; Ohtsuka et al., 1997; Sorimachi et al., 1997; Young et al., 1998; (Joseph et al., 2001) and within the sequence insertion Zis-2 (Young et al., 1998) (Fig. 1.1b). The Z-repeats are alternatively spliced between exons 9 and 12, generating a variable number of titin-α-actinin links depending on muscle type. The Z-repeats were suggested to be a factor in specifying the variable number of actin – α-actinin cross-links found in different muscle types, but evidence on the contrary has been presented (Luther and Squire, 2002). In any case, via α-actinin the titin NH_2-terminus is cross-linked not only to actin but to a vast network of Z-disk-associated proteins (Fig. 1.2) providing additional mechanical stability. To conclude, by interacting with multiple ligands, the unique sequence between titin-Ig-domains Z2 and Z3 supports Z-disk assembly and structure, force transmission, and perhaps mechanical signalling.

1.5 I-Band Titin: Interactions and Multifaceted Roles in Normal and Diseased Heart

The mechanically active element of titin in the I-band (encoded by exons 28–251) begins approximately 100 nm away from the center of the Z-disk (Linke et al., 1997; Trombitas and Granzier, 1997). Mounting evidence suggests that I-band titin not only generates passive force but also associates with multiple ligands and might serve as a "tensiometer" in the sarcomere.

1.5.1 Extensible Elements in I-Band Titin

Distinct subsegments in I-band titin have been identified that contribute differentially to the extensibility and passive-force generation of the cardiac myocyte (Fig. 1.1a). When the sarcomere is stretched from slack length, initially the proximal and distal Ig-domain regions (and the middle-Ig region in N2BA-titin) extend by straightening out their inter-domain linkers, whereas unfolding of individual Ig-domains is a rare event (Linke et al., 1999; Trombitas et al., 1999; Li et al., 2002). The distal Ig-domains may interact with those of other parallel titin molecules emanating from the thick-filament tip (Liversage et al., 2001) and this homotypic interaction, along with the low unfolding probability of the constituent Ig-domains (Li et al., 2002), may cause a rather limited extensibility of the distal Ig-region in situ (Bennett et al., 1997; Linke and Fernandez, 2002). Once the extensibility of the (folded) Ig-domain regions is largely exhausted, first the PEVK-domain and then also the 572-amino-acid N2B-Us begin to extend, while passive force now rises much more steeply than during Ig-segment extension (Linke et al., 1999; Trombitas et al., 1999; Li et al., 2002). Thus, a step-wise titin-extension model has emerged (also see Fig. 1.3c), in which straightening of I-band Ig-segments at low stretch forces is followed by extension of the long unique sequence insertions at higher forces (also see Fig. 1.3c).

1.5.2 Novex Domains

Exons 45, 46, and 48 in the titin genomic sequence, just COOH-terminal to the proximal Ig-domain region (Fig. 1.1b), are known as Novex I, II, and II, respectively (Bang et al., 2001a), because they were discovered years after the first report on the I-band-titin sequence. These exons are not expressed in the main cardiac isoforms, N2B and N2BA. Novex-I and Novex-II code for sequence in the ~3-MDa Novex-1/N2B and Novex-2/N2B titin-isoforms, respectively, which are both expressed at very low levels in heart. Novex-III codes for a large domain that binds obscurin and can function as an alternative COOH-terminus in cardiac and skeletal muscle, thus generating a (low-abundance) Novex-3 titin-isoform of ~650 kDa that integrates into the Z-disk lattice but is too short to reach the A-band (Bang et al., 2001a). The functional role of the Novex-domains and Novex-isoforms is still obscure.

1.5.3 Cardiac-Specific N2B-Domain: Molecular Spring and Ligand-Binding Site

The N2B-domain (encoded by titin-exon 49) encompasses the Ig-domains I24-I26 and intervening N2B-Us. The latter associates with four-and-a-half-LIM-domain protein (FHL2), also called down-regulated-in-rhabdomyosarcoma LIM-domain protein (DRAL), which in turn binds creatine-kinase, adenylate-kinase, and phosphofructokinase (Fig. 1.2), thus targeting these metabolic enzymes to the sarcomere (Lange et al., 2002). Another interaction site for FHL2 exists in the M-band titin region. FHL2, which is abundantly expressed in heart, has > 50 binding partners belonging to different functional classes, including receptors, structural proteins, signal transducers, transcription factors and cofactors, splicing factors, and DNA replication and repair enzymes (Johannessen et al., 2006; Canault et al., 2006; Sun et al., 2006; McGrath et al., 2006). Only some of them, with confirmed cardiac localization, are illustrated in Fig. 1.2. FHL2 itself and several ligands, e.g., serum-response factor (SRF) and extracellular signal-regulated kinase-2 (ERK2),

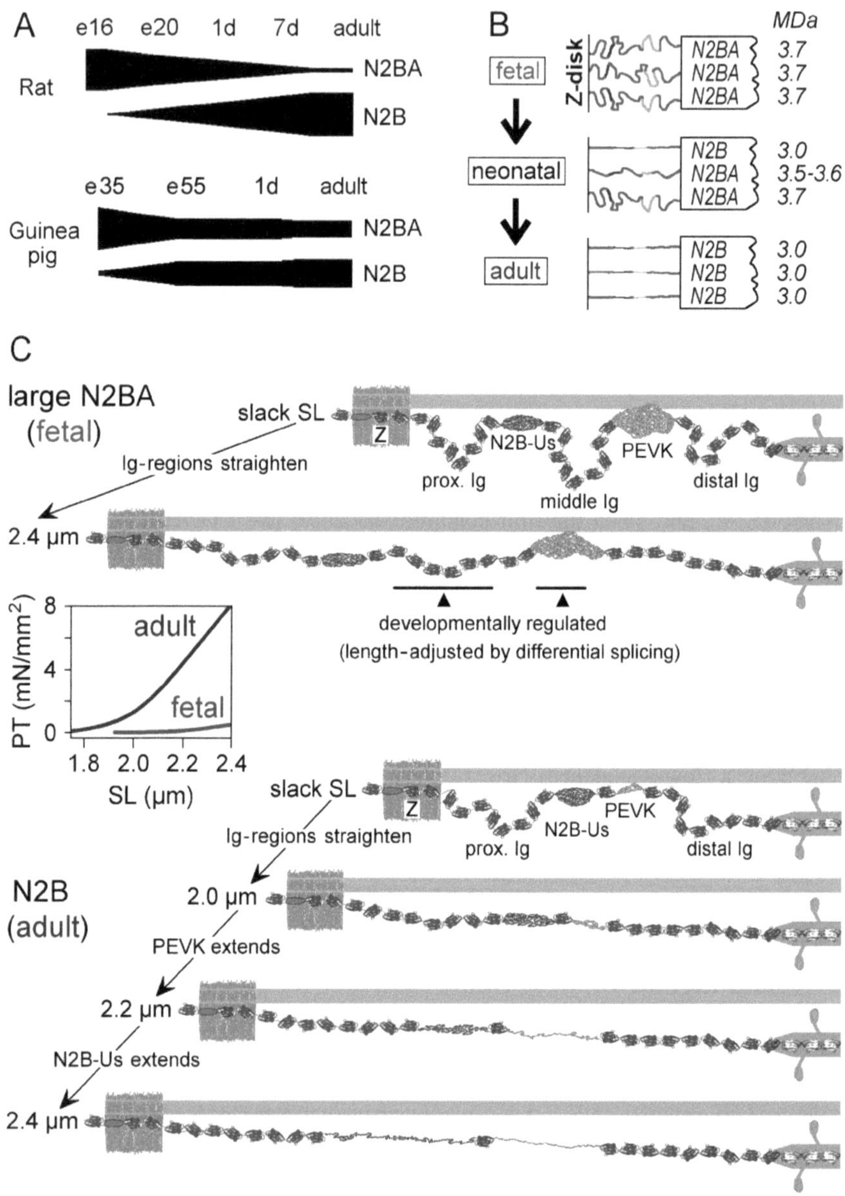

Fig. 1.3 Developmental isoform transitions in cardiac titin and consequences for passive tension. (**a**) Shifts observed in N2BA versus N2B titin-isoform composition during fetal/perinatal development of rat and guinea-pig hearts. e, embryonic day; d, postnatal day. (**b**) Scheme illustrating the changes in titin-isoform composition of a half-sarcomere during rat-heart development. (**c**) Model explaining the extension of the elastic I-band segments in long N2BA-titin (predominant in fetal heart) and short N2B-titin (predominant in adult heart), in the physiological sarcomere-length (SL) range. For clarity, the schemes do not show the real number of titin-Ig-domains (see Fig. 1.1). Inset: Passive tension (PT)-SL relationships of isolated cardiac myofibrils from fetal (e16) and adult rat heart. (Figure taken from Linke, 2008. With permission from Cardiovascular Research)

can translocate to the nucleus where they act as modifiers of gene expression. FHL2 links I-band titin to MAPK stress-signalling via ERK-binding (Purcell et al., 2004). FHL2 also ties the titin N2B-region to integrin and integrin-related mechanotransduction pathways (Samson et al., 2004).

Titin's N2B-Us, as well as the Ig-domains I26/I27, also associate with αB-crystallin (Fig. 1.1b), a member of the small heat-shock-protein family (Golenhofen et al., 2002; Bullard et al., 2004). In the heart, αB-crystallin moves to the myofibrils under conditions of stress, such as ischemia (Barbato et al., 1996; van de Klundert et al., 1998), and may act as a protector of cytoskeletal proteins (Launay et al., 2006), including titin and desmin (Wang et al., 2003). N2B-titin domains might be protected from stretch-induced unfolding when αB-crystallin binds to intermediate folding states (Bullard et al., 2004). The stress-protective effect of αB-crystallin is phosphorylation-activated and mediated by p38-dependent MAPK-activating protein-kinase-2 (Launay et al., 2006; Hoover et al., 2000), again linking a titin ligand to the MAPK signalling pathway (Fig. 1.2).

Knock-down of titin-exon 49 in mouse hearts leads to cardiac atrophy and diastolic dysfunction due to increased diastolic wall stress, and cardiomyocytes deficient in N2B-domain generate higher-than-normal PT and have reduced slack sarcomere length (SL) (Radke et al., 2007). In these knockout mice, FHL2, but not αB-crystallin, is downregulated suggesting the FHL2-N2B-Us connection is important for the cardiac hypertrophic response. One can speculate that, if the affinity between FHL2 and the springy N2B-Us were dependent on the stretch state, this interaction could represent a *bona fide* stretch-sensor.

1.5.4 N2A-Domain: A Stress-Sensing Element?

The N2A-region encompasses the stretch of Ig-domains I80-I83 interspersed with a few unique sequences (Fig. 1.1) (Labeit and Kolmerer, 1995). The Ig-domains I80/I81 (with an intervening sequence) interact with the three homologous muscle-ankyrin-repeat proteins (MARPs) (Fig. 1.1b), cardiac-ankyrin-repeat protein (CARP), diabetes-related ankyrin-repeat protein (DARP), and ankyrin-repeat-domain protein-2 (Ankrd2; also known as Arpp) (Miller et al., 2003; Witt et al., 2004). MARPs were previously identified by their induction after cardiac injury and muscle denervation (CARP) (Kuo et al., 1999; Tsukamoto et al., 2002), during recovery following metabolic challenge (DARP) (Ikeda et al., 2003), and after skeletal muscle stretch or eccentric contraction (Ankrd2) (Kemp et al., 2000; Hentzen et al., 2006), suggesting they could be part of muscle stress-response pathways. Ankrd2 also associates with telethonin and additionally, with the three transcription factors, Y-box-binding protein-1 (YB-1), promyelocytic-leukaemia protein (PML), and p53 (Kojic et al., 2004) (Fig. 1.2), which hints at the potential of MARPs to act as nuclear regulators of transcription. CARP binds to myopalladin (Bang et al., 2001b) and desmin (Witt et al., 2005b), likely via a potential coiled-coil dimerization motif that also mediates homo-dimer formation of other MARPs (Witt et al., 2005b). End-stage failing human DCM hearts showed increased expression levels of MARPs (Nagueh et al., 2004) and in cultured cardiomyocytes, CARP and DARP could be induced by cyclic stretch both in the nucleus and at the sarcomeric I-bands (Miller et al., 2003). Thus, MARPs may indeed provide a link between myofibrillar stress-response and muscle gene expression, implicating the N2A-domain in a mechano-chemical signalling pathway.

Another ligand in the N2A-region is the Ca^{2+}-dependent muscle protease, calpain-3/p94, which interacts with the Ig-domains I82/I83, but also has a second binding site in titin's M-band region (Sorimachi et al., 1995; Ojima et al., 2005) (Fig. 1.1b). Binding of calpain-3 to the N2A-region inhibits autolytic activation and disassembly of the protease (Taveau et al., 2003; Ono et al., 2006). Calpain-3 is not expressed in adult heart but is important in skeletal muscle, as loss-of-function mutations in the calpain-3 gene cause limb-girdle muscular dystrophy type-2A (LGMD2A) in humans (Duguez et al., 2006). A mouse model with a deletion mutation in the calpain-3-binding site of the N2A-region develops muscular dystrophy with myositis (MDM), presumably from the loss of the titin-calpain-3 interaction (Garvey et al., 2002), and appears to be a useful model to study the molecular basis of LGMD2A (Huebsch et al., 2005). As expected, the MDM mouse shows no cardiac phenotype (Witt et al., 2004). In the MDM mouse, ligands of the N2A-domain, such as CARP and Ankrd2, are strongly upregulated (Witt et al., 2004), suggesting feedback between titin-N2A, calpain-3, and MARPs in a signalling complex associated with the central I-band region. Finally, another protease, calpain-1, which is expressed also in heart, was found to associate with myofibrillar I-bands and with fragments of I-band titin (Raynaud et al., 2005), indicating that skeletal and cardiac titins may be targets of proteolytic cleavage by this calpain, possibly in the N2A-region.

1.5.5 PEVK-Domain: Signalling and Mechanical Functions

Almost one third of the exons in the human titin gene (exons 110–225) code for the PEVK-domain, a segment made up of conserved alternating motifs of 26–28 amino-acid repeats (PPAKs) separated by regions rich in glutamic-acid residues (polyE-motifs) (Greaser, 2001; Ma et al., 2001). The PEVK-segment was confirmed to interact with nebulin SH3-domains and suggested to bind to SH3-domains of other proteins as well (Figs. 1.1b and 1.2), possibly implying a still unappreciated role for this region in signalling processes during sarcomere assembly or even in mechanosensing (Ma et al., 2006). Within the PEVK-domain three conformational states have been identified, polyproline-II helix, beta-turn and unordered coil (Ma and Wang, 2003), which may be important for the mechanical properties of this domain. PEVK-titin elasticity is thought to be largely based upon an entropic-spring mechanism (Linke et al., 2002) and in-situ extension of this segment is associated with a significant rise in sarcomeric PT (Linke et al., 1996).

The passive mechanical properties of the sarcomere are modified by interaction of the PEVK-domain with actin filaments (Fig. 1.1). This interaction was suggested in one study to be regulated by Ca^{2+}/S100A1 (Yamasaki et al., 2001), but found to be independent of Ca^{2+}/S100 and Ca^{2+}/calmodulin in another study (Kulke et al., 2001). Actin-binding propensity is characteristic of both the constitutively expressed PEVK-titin in the cardiac N2B as well as N2BA-isoforms (encoded by exons 219–225) and the differentially spliced PEVK-titin in the N2BA (and skeletal N2A) isoforms (Linke et al., 2002; Gutierrez-Cruz et al., 2001; Nagy et al., 2004). Biochemically, polyE-motifs showed a stronger apparent actin-binding than PPAKs (Nagy et al., 2004). The interaction between actin filaments and the PEVK-domain is rather weak and might be further alleviated by physiological levels of Ca^{2+} (Linke et al., 2002) and the presence of tropomyosin, which itself binds to I-band titin (Raynaud et al., 2004). Still, the titin – thin-filament interaction imposes a viscoelastic load on the passively stretched (Kulke et al., 2001) or actively contracting (Opitz et al., 2003) myocardium. Further, the stiffness of the PEVK-segment is increased by Ca^{2+}-binding (Labeit et al., 2003), an effect mediated by the differentially spliced, but not the constitutively expressed, PEVK-titin (Fujita et al., 2004). Somewhat unexpectedly then, addition of Ca^{2+} did not alter various biophysical and biochemical properties of a PEVK-motif encoded by titin exon 115 (Duan et al., 2006). In summary, the PEVK-segment is an intriguing titin region with a major function as a molecular spring that is tunable by ligand binding.

1.5.6 Plasticity of Titin in Cardiac Development

Differential splicing of titin's I-band segment not only generates great diversity of titin-isoforms in adult skeletal muscles (Prado et al., 2005) and in different compartments of the adult heart (Cazorla et al., 2000; Neagoe et al., 2003), but also leads to dramatic length changes of the titin springs during fetal and perinatal heart development (Fig. 1.3). At mid-gestational stages, the hearts of rodents and pigs express a unique fetal N2BA-isoform of ~3.7 MDa but no N2B-titin (Opitz et al., 2004; Warren et al., 2004; Opitz and Linke, 2005). The large N2BA-isoform is gradually replaced later during development by smaller N2BA-titins co-expressed with the N2B-isoform (Fig. 1.3b) (Opitz et al., 2004; Warren et al., 2004; Lahmers et al., 2004). The length differences between these developmentally regulated N2BA-isoforms result from differential splicing of the middle Ig-region and PEVK-domain (Opitz et al., 2004; Lahmers et al., 2004). N2B is the predominant titin-isoform in the adult left ventricles of small mammalian species (e.g., rodents) and also humans, whereas the N2BA-titins prevail over N2B in the adult hearts of large mammals (e.g., cow, goat) (Cazorla et al., 2000; Neagoe et al., 2003). The developmental titin-isoform switching is particularly fast in mice or rats, where it occurs perinatally within 1–2 weeks (Fig. 1.3a, top). In other species, e.g., guinea pig which has a comparatively long gestation period, the switching takes longer but is nearly completed before birth (Fig. 1.3a, bottom) (Kruger et al., 2006). These fetal/perinatal transitions from high to low N2BA:N2B ratios (Fig. 1.3a) cause the myofibrillar PT to be much higher in adult than in fetal myocardium (Fig. 1.3c, inset).

The molecular basis behind this phenomenon is illustrated in Fig. 1.3c. Stretching a long N2BA-isoform in the physiological sarcomere-length (SL) range of

~1.8–2.4 μm straightens out the Ig-domain regions, but does not significantly extend the unique sequences, PEVK and N2B-Us (Fig. 1.3c, top). Titin-based passive force therefore remains low. In contrast, when the short N2B-isoform is stretched to the same SLs, the strain on the I-band segment is much higher and also the PEVK-domain and the N2B-Us elongate (Fig. 1.3c, bottom), causing passive force to rise steeply (Trombitas et al., 2001; Linke et al., 1999; Li et al., 2002). The overall passive stiffness then depends on the proportion of compliant N2BA *versus* stiff N2B springs expressed in the sarcomere and stretched in parallel (Trombitas et al., 2001; Linke and Fernandez, 2002). The dramatic alterations in N2BA:N2B ratio during heart development serve (in concert with collagen modifications) to adjust the passive stiffness of myocardium to the changing hemodynamic situation during cardiac growth and the increased power requirements of newborn hearts.

The triggers for the developmental titin-isoform switch are beginning to be uncovered. They include growth hormones, particularly thyroid hormone (triiodo-L-thyronine, T3), but also angiotensin-II (Ang-II) and mechanical factors (Kruger et al., 2008). Both T3 and Ang-II alter titin-isoform composition by activating the phosphatidylinositol-3-kinase/Akt (Protein kinase B) pathway (Kruger et al., 2008). Downstream effectors of Akt are known to promote cardiac hypertrophy and suppress muscle wasting, but in the case of titin (and other muscle proteins) they may also affect alternative splicing.

1.5.7 Mechanical Function of Titin in Human Heart Disease

Since titin is a major contributor to diastolic wall stiffness, along with collagen in the ECM (Linke et al., 1994; Wu et al. 2000), what role do the titin springs play in the passive stiffening of the cardiac walls in chronic human-heart disease? Earlier electron microscopical and immunohistological studies of human end-stage failing hearts with DCM showed altered distribution and loss of titin (Hein et al., 2000). When the N2BA:N2B titin-isoform ratio was analyzed in chronically ischemic left ventricles of coronary-artery-disease (CAD) patients with end-stage systolic heart failure (HF), the mean N2BA-isoform percentage was found to be elevated to nearly 50%, up from ~30% in the left ventricles of control donor patients (Fig. 1.4a) (Neagoe et al., 2002). These titin-isoform changes were associated with decreased myofibrillar PT (Fig. 1.4b). The failing hearts showed increased fibrosis and collagen accumulation and it was proposed that the shift towards more compliant N2BA-isoforms occurs in response to elevated ECM-based stiffness, thus counteracting the global passive-stiffness increase to some degree (Neagoe et al., 2002)

Subsequently, evidence was presented that titin may elevate passive stiffness in myocardium of a DCM patient via lowered N2BA:N2B expression ratio (Wu et al., 2002). However, analyses of larger cohorts of explanted non-ischemic DCM hearts (mean ejection fraction, EF, ~20%) again demonstrated increased proportions of N2BA-isoforms (Fig. 1.4a) (Makarenko et al., 2004; Nagueh et al., 2004), in particular upregulation of long N2BA-isoforms larger than 3.3 MDa (Makarenko et al., 2004). Like in the human CAD hearts, the titin-isoform switching lowered

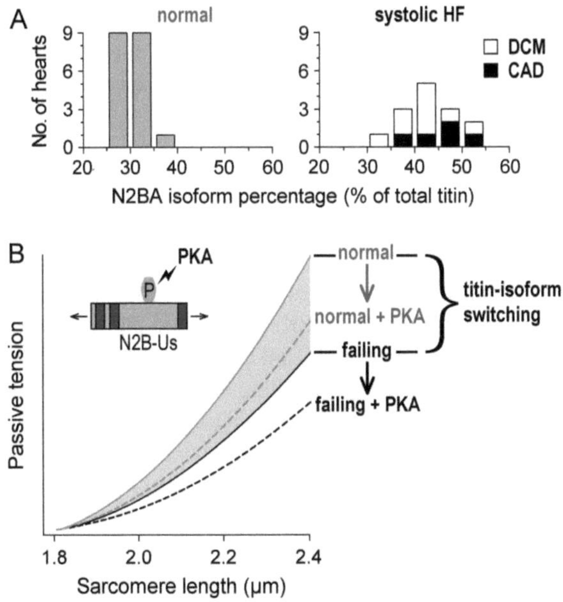

Fig. 1.4 Regulation of titin spring force in normal and failing human hearts by titin-isoform switching and protein-kinase-A (PKA) mediated titin phosphorylation. (**a**) Proportion of N2BA-titin isoform expressed in hearts from normal donors and from patients with dilated cardiomyopathy (DCM) or coronary artery disease (CAD) diagnosed with systolic heart failure (HF) (compiled from Neagoe et al., 2002 and Makarenko et al., 2004). (**b**) Graph demonstrating the role of titin-isoform composition and PKA-mediated titin phosphorylation for myofibrillar passive tension-SL relationships of normal and end-stage failing human hearts. *Inset* highlights the PKA-phosphorylation site on titin's N2B-unique sequence. (Figure taken from Linke, 2008. With permission from Cardiovascular Research)

passive myocyte stiffness in comparison to normal donor hearts (Fig. 1.4b) and these changes affected diastolic filling. The findings suggest that the hearts of systolic HF patients express increased proportions of compliant N2BA-titin isoforms, which helps reduce wall stiffness, thus benefiting diastolic function. A drawback, however, may be that a reduced titin spring force in heart failure could compromise systolic function through impairment of the Frank-Starling mechanism and myocyte stretch-sensing. Interestingly, a recent analysis of left-ventricular biopsies from patients with diastolic HF (mean EF, 62%) reported N2BA:N2B ratios that were much lower than those found in systolic HF patients, while myocyte passive stiffness was high (van Heerebeek et al., 2006). However, the number of hearts in which titin expression could be measured was too low to infer that a reduced N2BA:N2B ratio causing high titin-based stiffness is a general feature of human HF with preserved EF. In summary, cardiac titin-isoform composition and titin-based stiffness are altered in chronic human-heart disease. The increased N2BA:N2B ratios observed so far in human systolic HF may occur in response to global fibrosis, altered loading conditions, or perhaps altered humoral status (Wu et al., 2007). Future work will need to better define the triggers for the titin-isoform switch and establish possible differences in the direction of the switch in systolic *versus* diastolic HF. It will also be important to establish whether the titin switch is a cause or a consequence of heart failure.

1.5.8 Regulation of Titin Stiffness by Phosphorylation

The stiffness of the cardiac titin springs can also be dynamically regulated on a faster time-scale by protein-kinase-A (PKA) mediated phosphorylation (Fig. 1.4b) (Yamasaki et al., 2002; Kruger and Linke, 2006). PKA targets many different proteins in the cardiomyocyte, including the myofibrillar proteins troponin-I and myosin-binding protein-C (MyBP-C). Titin phosphorylation induced by β-adrenergic stimulation was only recently recognized to lower titin-based stiffness in rat and cow heart (Yamasaki et al., 2002; Fukuda et al., 2005). PKA decreased the PT also in skinned cardiomyocytes from failing human hearts and interestingly, this effect was much more pronounced in cells from diastolic HF patients than in those from systolic HF patients or control donors (van Heerebeek et al., 2006; Borbely et al., 2005). Thus, an abnormal titin-phosphorylation state may contribute to altered diastolic stiffness in diastolic HF.

In skinned muscle strips from normal donor hearts, PKA reduced the PT by ~20–40% (Fig. 1.4b) and the mechanical changes were associated with phosphorylation of both the N2B and the N2BA titin-isoforms (Kruger and Linke, 2006). The PKA-dependent PT-drop in human-heart preparations was substantially larger when titin was first de-phosphorylated, suggesting that inherent phosphorylation of titin is important for the basal myocardial PT level (Kruger and Linke, 2006). PKA specifically targets Ser/Thr phosphorylation sites in the N2B-Us of the cardiac-specific N2B-domain (Fig. 1.4b, inset) and the PKA-induced PT-decrease is therefore seen in myocardial preparations but not in skeletal muscle (Yamasaki et al., 2002; Fukuda et al., 2005; Kruger and Linke, 2006). Although the molecular basis of the PKA-effect on titin-spring force is still unresolved (Leake et al., 2006), the effect is interesting from a therapeutic point of view, because raising myocardial PKA activity by beta-adrenoceptor stimulation could improve left-ventricular diastolic function in patients with diastolic HF (van Heerebeek et al., 2006). To conclude, modifications in titin-based passive stiffness triggered by titin phosphorylation represent a novel mechanosensitive signalling event in heart muscle, which is worth exploring in follow-up studies.

1.6 The Scaffolding Role of A-Band Titin

The segment of titin at the I-band/A-band junction and in the A-band is composed of Ig and FN3 modules that are mainly arranged in a super-repeat pattern with either six (6×) or eleven (11×) domains (Fig. 1.1) (Labeit et al., 1990). The ~2-MDa

A-band-titin is functionally inextensible, since it is tightly associated with the thick filaments (Houmeida et al., 1995), presumably at a stoichiometry of six titin molecules per half-thick filament (Liversage et al., 2001). Titin-binding to the shaft region of myosin is mediated by FN3-domains (Fig. 1.1b) (Houmeida et al., 1995). Moreover, within the 11-domain super-repeats, the first Ig-domain interacts with MyBP-C (Freiburg and Gautel, 1996). Because A-band titin provides regularly spaced binding sites for other thick filament proteins, it is viewed as a molecular blueprint which controls the precise assembly and exact length of the myosin filaments (Miller et al., 2004; Trinick and Tskhovrebova, 1999).

1.7 Structural and Signalling Complexes of M-Band Titin

The ~200-kDa COOH-terminal end of titin (encoded by exons 355–363) is at the A-band/M-band junction and the M-band (Fig. 1.1). This titin segment is involved in numerous protein-protein interactions (Miller et al., 2004; Lange et al., 2006; Agarkova and Perrriard, 2005) and also contains a stretch-activated Ser/Thr kinase-domain (Labeit and Kolmerer, 1995; Lange et al., 2005b; Grater et al., 2005), which implicates M-band-titin in myofibrillar signal-transduction pathways and stress-sensing. Targeted homozygous deletion of the entire M-band-titin region in cardiomyocytes prevents sarcomere formation (Musa et al., 2006), demonstrating the importance of this titin segment for thick-filament assembly, M-band formation and even maturation of other parts of the sarcomere.

1.7.1 The Titin-Kinase Region: A Putative Stretch-Sensor Complex

The domains A168-170 just NH_2-terminal to the titin-kinase domain provide a binding site for MURF-1 (Fig. 1.1b) (Centner et al., 2001; McElhinny et al., 2002; Mrosek et al., 2007), which in turn associates with MURF-2. The MURFs can homo- and hetero-oligomerize, allegedly via their coiled-coil domains, and have both been detected ultra-structurally at the M-line region (McElhinny et al., 2002; McElhinny et al., 2004). MURF-1 acts as an E3 ubiquitin ligase (Willis and Patterson, 2006) and binds to various other muscle proteins, including troponins, myosin-light-chain, myotilin, telethonin, N-RAP, and nebulin (Witt et al., 2005a), probably controlling their proteasomal degradation (Fig. 1.2) (Adams et al., 2007). MURF-2 also associates with these muscle proteins and titin-domain A168 (Witt et al., 2005a). In contrast, a third member of the MURF family, MURF-3, does not interact with those muscle proteins but is microtubule-associated (as is MURF-2) (Witt et al., 2005a; Gregorio et al., 2005). MURF-1 is important for nuclear signalling and may co-regulate gene expression, as it can translocate to the nucleus and bind to other proteins with nuclear functions (Fig. 1.2) (McElhinny et al., 2002; Gregorio et al., 2005). MURF-1 may also have a role in energy metabolism by interacting with many enzymes involved in ATP production (Witt et al., 2005a). Recent work on MURF-1-deficient mice suggested that MURF-1 is dispensable for normal cardiac development, but has an inhibitory role in cardiac hypertrophy, likely by its direct association with the transcriptional co-factor SRF (Willis et al., 2007) and perhaps also via inhibition of PKCε-activity through interaction with RACK1 (Fig. 1.2) (Arya et al., 2004). Thus, the titin-MURF-1 linkage could be at the heart of a stress-dependent signalling pathway acting via the sarcomeric M-band.

The titin-kinase domain (encoded by titin-exon 358) was initially shown to be activated by phosphorylation of a tyrosine and subsequent binding of Ca^{2+}/calmodulin to the regulatory tail (Mayans et al., 1998). The titin-kinase phosphorylates telethonin, but whether this is important for myofibrillogenesis, as suggested earlier (Mayans et al., 1998), remains controversial (Weinert et al., 2006). Deletion of the titin-exons 358 and 359 in a conditional knockout mouse caused sarcomeric disassembly in both skeletal and cardiac muscle and early death (Gotthardt et al., 2003). In a conventional knockout-mouse model containing the same deletion, the initial assembly of sarcomeres was unaffected, but impaired cardiac hypertrophy resulted in late-embryonic lethality (Weinert et al., 2006), thus underscoring the prominent role of the titin-kinase for normal myocardial function. A tamoxifen-inducible deletion of the titin-kinase region in adult mouse hearts produced severe cardiac hypertrophy and congestive heart failure, associated with an attenuated response to adrenergic stimulation and extracellular Ca^{2+} (Peng et al., 2007). Surprisingly, despite the deletion of the MURF-1-binding site in this mouse model, MURF-1 was upregulated and PKCε and troponin-I were unchanged, which raises new questions about the role of the titin-MURF-1 signalling pathway in cardiac hypertrophy.

A model has been proposed in which the sarcomeric M-band region acts as a bona fide stress-sensor through the titin-kinase domain (Lange et al., 2005b). In this model, stretch-induced conformational changes in the titin-kinase (Grater et al., 2005) lead to its activation, thus allowing interaction with Nbr1 (neighbor-of-BRCA1 gene-1), a protein that associates with p62, which in turn binds to MURF-2 (Fig. 1.2). MURF-2 then interacts with SRF to inhibit its nuclear localization and transcriptional activity, hence suppressing hypertrophic responses elicited by mechanical forces. Although this model is attractive, a recent study on MURF-2-deficient mice (Willis et al., 2007) showed no changes in the hypertrophic response of the hearts to experimentally induced pressure-overload, compared to wildtype hearts. These results suggest that the titin-MURF-2 signalling axis may be dispensable for normal cardiac response to mechanical stress. In conclusion, it is likely that the titin-kinase domain is centrally involved in myocyte stress-sensing, but the molecular mechanisms remains to be demonstrated unambiguously.

1.7.2 Interactions and Function of COOH-Terminal Titin Domains

Titin exons 358 and 359 code also for the Ig-domains M1-M7 and several intervening unique sequences, the largest of which is Mis-2 (Fig. 1.1). Mis-2 contains a second binding site for FHL2/DRAL (Lange et al., 2002), thus linking both the

M-band and the I-band titin segments to a multitude of proteins belonging to different functional classes (see Section 5.3). Just COOH-terminal of Mis-2, the Ig-domain M4 interacts with myomesin (Obermann et al., 1997), a myosin-binding protein that forms dimers which cross-link the thick filaments and titin's COOH-terminus in an elastic manner (Lange et al., 2005a). Elasticity of myomesin may be important to rectify force imbalances between parallel thick filaments during active muscle contraction (Schoenauer et al., 2005). Structurally similar to myomesin is M-protein, which contains the same titin-binding domains as myomesin and is thus likely to bind to titin-M4 as well (Agarkova and Perriard, 2005). Myomesin associates with muscle creatine-kinase and myofibrillogenesis-regulatory-factor-1 (MR-1) and is regulated in its affinity to titin by phosphorylation (Obermann et al., 1997). The myomesin-titin-myosin complex is most likely the critical structure that maintains the stability of the M-band (Agarkova and Perriard, 2005).

At the extreme COOH-terminus of titin, domains M7-M10 bind to A- and B-type lamins (Fig. 1.1b), proteins that form structural filaments in the nucleus (Zastrow et al., 2006). The nuclear location of lamins excludes an interaction with myofibrillar titin and suggests that nuclear forms of titin may be the binding partner. Nuclear titin, which has been found in non-muscle cells (Machado and Andrew, 2000), could contribute via its lamin-binding properties to nuclear organization during interphase. Finally, the last unique sequence (Mis-7) encoded by exon 363 (Mex5), offer a second binding site for the protease calpain-3/p94 (Sorimachi et al., 1995). Targeting of this protease to the center of the sarcomere may have a significant role in degradation and turnover of M-band-associated proteins in skeletal muscle. In summary, association of titin's COOH-terminus with structural and signalling molecules implicates this titin region in thick-filament assembly, confers mechanical stability but also flexibility to the M-band, and is a pre-requisite for a potential stress-sensor complex.

1.8 Human Titin as a Candidate Gene for Hereditary Myopathies

The titin gene locus on chromosome 2q31 has long been recognized as a strong candidate for familial DCM (Siu et al., 1999). Currently, all segments in human titin (Z-disk, I-band, A-band, M-band) are known to be affected by mutations causing various forms of hereditary myopathies. Among them are DCM as well as HCM, but also skeletal muscle diseases, such as tibial muscular dystrophy (TMD), also called limb-girdle muscular dystrophy type-2 J (LGMD2J), and hereditary myopathy with early respiratory failure (HMERF). The respective locations of the currently known mutations in human titin are highlighted in Fig. 1.1b and the mutations are further explained in Table 1.1. Although relatively few mutations in titin (8 DCM; 2 HCM; 4 muscular dystrophies) have been reported so far, the huge size of this molecule and the prevalence of the mutations already found suggest that titin mutations may be a more common cause of human DCM and muscular dystrophy (Hein and Schaper,

Table 1.1 Disease-associated mutations in human titin (also see Fig. 1.1b)

Phenotype	Location on titin	Mutation	Remarks	References
DCM	Z1 (exon 2)	Val54Met point mutation	Decreased binding to telethonin	Itoh-Satoh et al. (2002)
DCM	Z-repeat 7 (exon 14)	Ala743Val point mutation	Decreased binding to α-actinin	Itoh-Satoh et al. (2002)
DCM	Z4 (exon 18)	Trp930Arg missense mutation	Predicted to disrupt IgZ4-fold	Gerull et al. (2002)
DCM	N2B-Us (exon 49)	Gln4053ter nonsense mutation	Predicted to generate truncated titin	Itoh-Satoh et al. (2002)
DCM	N2B-Us (exon 49)	Ser4465Asn missense mutation	Mutation in FHL2-binding site	Itoh-Satoh et al. (2002)
DCM	A67 (exon 326)	2-basepair insertion, frameshift mutation	Predicted to generate truncated A-band titin	Gerull et al. (2002)
DCM	A131-A136 (exon 335)	62890delG1 1-basepair deletion, frameshift mutation	Predicted to generate truncated A-band titin	Gerull et al. (2006)
DCM	Mis-2 (exon 358)	Arg25618Gln point mutation	Mutation in FHL2-binding site	Matsumoto et al. (2005)
HCM	Z-repeat 7 (exon 14)	Ala740Leu point mutation	Increased binding to α-actinin	Satoh et al. (1999)
HCM	N2B-Us (exon 49)	Ser3799Tyr point mutation	Increased binding to FHL2	Itoh-Satoh et al. (2002); Matsumoto et al. (2005)
HMERF	Titin kinase (exon 358)	Arg279Trp in exon 358; point mutation	Mutation in Nbr1-binding site	Lange et al. (2005b)
TMD/ LGMD2J	M10 (exon 363)	Complex 11-bp deletion-insertion	Mutation near calpain-3-binding site; found in Finnish population	Udd et al. (2005); Hackman et al. (2002)
TMD/ LGMD2J	M10 (exon 363)	Iso293329Asp point mutation	Found in Belgian family	van den Bergh et al. (2003)
TMD/ LGMD2J	M10 (exon 363)	Leu293357Pro point mutation	Found in French family	Hackman et al. (2002)

2002; Udd et al., 2005). Many titin mutations are predicted to alter the interaction with a ligand (Table 1.1), suggesting they could affect the putative stress-sensing function of those titin regions.

1.9 Conclusions and Perspectives

Increasing evidence suggests that the myocardial stress-response machinery extends to the sarcomeres where distinct regions, including "hot spots" along the giant titin molecules, participate in a mechano-chemical coupling. Titin's functional roles were once thought to be restricted to molecular scaffolding and providing myofibrillar elasticity, but this protein may have additional important duties as a stress-sensor. Titin together with some of its direct and indirect ligands in the Z-disk and M-band regions, and the N2B, N2A, and PEVK domains in the I-band region, could act as a 'tensiometer' that when stretched, triggers downstream signalling events (e.g., activation of transcriptional (co)-factors) leading to changes in muscle-gene expression and cardiac hypertrophy. Conversely, a compromised stress-response function of the titin-signalosome, for instance caused by mutations in protein-protein interaction sites, can result in mechanical dysregulation and congestive heart failure.

Future work may aim at detecting novel titin-ligands that participate in the mechano-chemical coupling and uncovering their stress-dependent interaction to the atomic detail. It will be useful to explore the involvement of titin-based "hot spots" in the stress-sensing network of the cardiomyocyte by gene knock-down and functional tests. Additional studies should also identify the triggers that cause the large changes in titin-isoform composition during heart development and disease, which greatly affect myocardial passive stiffness and possibly, stress-dependent signalling. An intriguing property that warrants further research is the dynamic regulation of titin-based passive stiffness and titin's putative tensiometer function by phosphorylation of the N2B-domain. Some of the mysteries of the sensitive molecular giant titin have now been revealed, but we are likely to see many more of its secrets uncovered in the years to come.

Acknowledgements I would like to thank the Deutsche Forschungsgemeinschaft for financial support.

References

Adams V, Linke A, Wisloff U, Doring C, Erbs S, Krankel N, et al. (2007) Myocardial expression of Murf-1 and MAFbx after induction of chronic heart failure: effect on myocardial contractility. Cardiovasc Res 73:120–129.

Agarkova I, Perriard JC (2005) The M-band: an elastic web that crosslinks thick filaments in the center of the sarcomere. Trends Cell Biol 15:477–485.

Arber S, Hunter JJ, Ross J Jr, Hongo M, Sansig G, Borg J, et al. (1997) MLP-deficient mice exhibit a disruption of cardiac cytoarchitectural organization, dilated cardiomyopathy, and heart failure. Cell 88:393–403.

Armani A, Galli S, Giacomello E, Bagnato P, Barone V, Rossi D, et al. (2006) Molecular interactions with obscurin are involved in the localization of muscle-specific small ankyrin1 isoforms to subcompartments of the sarcoplasmic reticulum. Exp Cell Res 312:3546–3558.

Arya R, Kedar V, Hwang JR, McDonough H, Li HH, Taylor J, et al. (2004) Muscle ring finger protein-1 inhibits PKC{epsilon} activation and prevents cardiomyocyte hypertrophy. J Cell Biol 167:1147–1159.

Bagnato P, Barone V, Giacomello E, Rossi D, Sorrentino V (2003) Binding of an ankyrin-1 isoform to obscurin suggests a molecular link between the sarcoplasmic reticulum and myofibrils in striated muscles. J Cell Biol 160:245–253.

Bang ML, Centner T, Fornoff F, Geach AJ, Gotthardt M, McNabb M, et al. (2001a) The complete gene sequence of titin, expression of an unusual approximately 700-kDa titin isoform, and its interaction with obscurin identify a novel Z-line to I-band linking system. Circ Res 89:1065–1072.

Bang ML, Li X, Littlefield R, Bremner S, Thor A, Knowlton KU, et al. (2006) Nebulin-deficient mice exhibit shorter thin filament lengths and reduced contractile function in skeletal muscle. J Cell Biol 173:905–916.

Bang ML, Mudry RE, McElhinny AS, Trombitas K, Geach AJ, Yamasaki R, et al. (2001b) Myopalladin, a novel 145-kilodalton sarcomeric protein with multiple roles in Z-disc and I-band protein assemblies. J Cell Biol 153:413–427.

Barbato R, Menabo R, Dainese P, Carafoli E, Schiaffino S, Di Lisa F (1996) Binding of cytosolic proteins to myofibrils in ischemic rat hearts. Circ Res 78:821–828.

Bennett PM, Hodkin TE, Hawkins C (1997) Evidence that the tandem Ig domains near the end of the muscle thick filament form an inelastic part of the I-band titin. J Struct Biol 120:93–104.

Bennett PM, Maggs AM, Baines AJ, Pinder JC (2006) The transitional junction: a new functional subcellular domain at the intercalated disc. Mol Biol Cell 17:2091–2100.

Boateng SY, Belin RJ, Geenen DL, Margulies KB, Martin JL, Hoshijima M, et al. (2007) Cardiac dysfunction and heart failure are associated with abnormalities in the subcellular distribution and amounts of oligomeric muscle LIM protein. Am J Physiol Heart Circ Physiol 292: H259–H269.

Borbely A, van der Velden J, Papp Z, Bronzwaer JG, Edes I, Stienen GJ, Paulus WJ (2005) Cardiomyocyte stiffness in diastolic heart failure. Circulation 111:774–781.

Bos JM, Poley RN, Ny M, Tester DJ, Xu X, Vatta M, et al. (2006) Genotype-phenotype relationships involving hypertrophic cardiomyopathy-associated mutations in titin, muscle LIM protein, and telethonin. Mol Genet Metab 88:78–85.

Brancaccio M, Hirsch E, Notte A, Selvetella G, Lembo G, Tarone G (2006) Integrin signalling: the tug-of-war in heart hypertrophy. Cardiovasc Res 70:422–433.

Bullard B, Ferguson C, Minajeva A, Leake MC, Gautel M, Labeit D, et al. (2004) Association of the chaperone alphaB-crystallin with titin in heart muscle. J Biol Chem 279:7917–7924.

Canault M, Tellier E, Bonardo B, Mas E, Aumailley M, Juhan-Vague I, et al. (2006) FHL2 interacts with both ADAM-17 and the cytoskeleton and regulates ADAM-17 localization and activity. J Cell Physiol 208:363–372.

Cavnar PJ, Olenych SG, Keller TC 3rd (2007) Molecular identification and localization of cellular titin, a novel titin isoform in the fibroblast stress fiber. Cell Motil Cytoskeleton 64:418–433.

Cazorla O, Freiburg A, Helmes M, Centner T, McNabb M, Wu Y, et al. (2000) Differential expression of cardiac titin isoforms and modulation of cellular stiffness. Circ Res 86:59–67.

Centner T, Yano J, Kimura E, McElhinny AS, Pelin K, Witt CC, et al. (2001) Identification of muscle specific ring finger proteins as potential regulators of the titin kinase domain. J Mol Biol 306:717–726.

Clark KA, McElhinny AS, Beckerle MC, Gregorio CC (2002) Striated muscle cytoarchitecture: an intricate web of form and function. Annu Rev Cell Dev Biol 18:637–706.

Duan Y, DeKeyser JG, Damodaran S, Greaser ML (2006) Studies on titin PEVK peptides and their interaction. Arch Biochem Biophys 454:16–25.

Duguez S, Bartoli M, Richard I (2006) Calpain 3: a key regulator of the sarcomere? FEBS J 273:3427–3436.

Ervasti JM (2003) Costameres: the Achilles' heel of Herculean muscle. J Biol Chem 278:13591–13594.

Fowler VM, McKeown CR, Fischer RS (2006) Nebulin: does it measure up as a ruler? Curr Biol 16:R18–R20.

Frank D, Kuhn C, Katus HA, Frey N (2006) The sarcomeric Z-disc: a nodal point in signalling and disease. J Mol Med 84:446–468.

Freiburg A, Gautel M (1996) A molecular map of the interactions between titin and myosin-binding protein C. Implications for sarcomeric assembly in familial hypertrophic cardiomyopathy. Eur J Biochem 235:317–323.

Freiburg A, Trombitas K, Hell W, Cazorla O, Fougerousse F, Centner T, et al. (2000) Series of exon-skipping events in the elastic spring region of titin as the structural basis for myofibrillar elastic diversity. Circ Res 86:1114–1121.

Frey N, Olson EN (2003) Cardiac hypertrophy: the good, the bad, and the ugly. Annu Rev Physiol 65:45–79.

Fujita H, Labeit D, Gerull B, Labeit S, Granzier HL (2004) Titin isoform-dependent effect of calcium on passive myocardial tension. Am J Physiol Heart Circ Physiol 287:H2528–H2534.

Fukuda N, Granzier HL (2005) Titin/connectin-based modulation of the Frank-Starling mechanism of the heart. J Muscle Res Cell Motil 26:319–323.

Fukuda N, Wu Y, Nair P, Granzier HL (2005) Phosphorylation of titin modulates passive stiffness of cardiac muscle in a titin isoform-dependent manner. J Gen Physiol 125:257–271.

Fukuzawa A, Idowu S, Gautel M (2005) Complete human gene structure of obscurin: implications for isoform generation by differential splicing. J Muscle Res Cell Motil 26:427–434.

Garvey SM, Rajan C, Lerner AP, Frankel WN, Cox GA (2002) The muscular dystrophy with myositis (mdm) mouse mutation disrupts a skeletal muscle-specific domain of titin. Genomics 79:146–149.

Gautel M, Goulding D, Bullard B, Weber K, Furst DO (1996) The central Z-disk region of titin is assembled from a novel repeat in variable copy numbers. J Cell Sci 109:2747–2754.

Gerull B, Atherton J, Geupel A, Sasse-Klaassen S, Heuser A, Frenneaux M, et al. (2006) Identification of a novel frameshift mutation in the giant muscle filament titin in a large Australian family with dilated cardiomyopathy. J Mol Med 84:478–483.

Gerull B, Gramlich M, Atherton J, McNabb M, Trombitas K, Sasse-Klaassen S, et al. (2002) Mutations of TTN, encoding the giant muscle filament titin, cause familial dilated cardiomyopathy. Nat Genet 30:201–204.

Golenhofen N, Arbeiter A, Koob R, Drenckhahn D (2002) Ischemia-induced association of the stress protein alpha B-crystallin with I-band portion of cardiac titin. J Mol Cell Cardiol 34:309–319.

Gotthardt M, Hammer RE, Hubner N, Monti J, Witt CC, McNabb M, et al. (2003) Conditional expression of mutant M-line titins results in cardiomyopathy with altered sarcomere structure. J Biol Chem 278:6059–6065.

Granzier HL, Labeit S (2004) The giant protein titin: a major player in myocardial mechanics, signaling, and disease. Circ Res 94:284–295.

Grater F, Shen J, Jiang H, Gautel M, Grubmuller H (2005) Mechanically induced titin kinase activation studied by force-probe molecular dynamics simulations. Biophys J 88:790–804.

Greaser M (2001) Identification of new repeating motifs in titin. Proteins 43:145–149.

Greaser ML, Krzesinski PR, Warren CM, Kirkpatrick B, Campbell KS, Moss RL (2005) Developmental changes in rat cardiac titin/connectin: transitions in normal animals and in mutants with a delayed pattern of isoform transition. J Muscle Res Cell Motil 26:325–332.

Gregorio CC, Perry CN, McElhinny AS (2005) Functional properties of the titin/connectin-associated proteins, the muscle-specific RING finger proteins (MURFs), in striated muscle. J Muscle Res Cell Motil 26:389–400.

Gregorio CC, Trombitas K, Centner T, Kolmerer B, Stier G, Kunke K, et al. (1998) The NH2 terminus of titin spans the Z-disc: its interaction with a novel 19-kD ligand (T-cap) is required for sarcomeric integrity. J Cell Biol 143:1013–1027.

Gutierrez-Cruz G, Van Heerden AH, Wang K (2001) Modular motif, structural folds and affinity profiles of the PEVK segment of human fetal skeletal muscle titin. J Biol Chem 276: 7442–7449.
Hackman P, Vihola A, Haravuori H, Marchand S, Sarparanta J, De Seze J, et al. (2002) Tibial muscular dystrophy is a titinopathy caused by mutations in TTN, the gene encoding the giant skeletal-muscle protein titin. Am J Hum Genet 71:492–500.
Hayashi T, Arimura T, Itoh-Satoh M, Ueda K, Hohda S, Inagaki N, et al. (2004) Tcap gene mutations in hypertrophic cardiomyopathy and dilated cardiomyopathy. J Am Coll Cardiol 44:2192–2201.
Hein S, Kostin S, Heling A, Maeno Y, Schaper J (2000) The role of the cytoskeleton in heart failure. Cardiovasc Res 45:273–278.
Hein S, Schaper J (2002) Weakness of a giant: mutations of the sarcomeric protein titin. Trends Mol Med 8:311–313.
Heineke J, Kempf T, Kraft T, Hilfiker A, Morawietz H, Scheubel RJ, et al. (2003) Downregulation of cytoskeletal muscle LIM protein by nitric oxide: impact on cardiac myocyte hypertrophy. Circulation 107:1424–1432.
Heineke J, Molkentin JD (2006) Regulation of cardiac hypertrophy by intracellular signalling pathways. Nat Rev Mol Cell Biol 7:589–600.
Heineke J, Ruetten H, Willenbockel C, Gross SC, Naguib M, Schaefer A, et al. (2005) Attenuation of cardiac remodeling after myocardial infarction by muscle LIM protein-calcineurin signaling at the sarcomeric Z-disc. Proc Natl Acad Sci USA 102:1655–1660.
Helmes M, Lim CC, Liao R, Bharti A, Cui L, Sawyer DB (2003) Titin determines the Frank-Starling relation in early diastole. J Gen Physiol 121:97–110.
Helmes M, Trombitas K, Granzier H (1996) Titin develops restoring force in rat cardiac myocytes. Circ Res 79:619–626.
Hentzen ER, Lahey M, Peters D, Mathew L, Barash IA, Friden J, et al. (2006) Stress-dependent and -independent expression of the myogenic regulatory factors and the MARP genes after eccentric contractions in rats. J Physiol 570:157–167.
Hoover HE, Thuerauf DJ, Martindale JJ, Glembotski CC (2000) alphaB-crystallin gene induction and phosphorylation by MKK6-activated p38. A potential role for alphaB-crystallin as a target of the p38 branch of the cardiac stress response. J Biol Chem 275:23825–23833.
Horowits R (1999) The physiological role of titin in striated muscle. Rev Physiol Biochem Pharmacol 138:57–96.
Horowits R (2006) Nebulin regulation of actin filament lengths: new angles. Trends Cell Biol 16:121–124.
Hoshijima M (2006) Mechanical stress-strain sensors embedded in cardiac cytoskeleton: Z disk, titin, and associated structures. Am J Physiol Heart Circ Physiol 290:H1313–H1325.
Houmeida A, Holt J, Tskhovrebova L, Trinick J (1995) Studies of the interaction between titin and myosin. J Cell Biol 131:1471–1481.
Huebsch KA, Kudryashova E, Wooley CM, Sher RB, Seburn KL, Spencer MJ, et al. (2005) Mdm muscular dystrophy: interactions with calpain 3 and a novel functional role for titin's N2A domain. Hum Mol Genet 14:2801–2811.
Ikeda K, Emoto N, Matsuo M, Yokoyama M (2003) Molecular identification and characterization of a novel nuclear protein whose expression is up-regulated in insulin-resistant animals. J Biol Chem 278:3514–3520.
Itoh-Satoh M, Hayashi T, Nishi H, Koga Y, Arimura T, Koyanagi T, et al. (2002) Titin mutations as the molecular basis for dilated cardiomyopathy. Biochem Biophys Res Commun 291: 385–393.
Johannessen M, Moller S, Hansen T, Moens U, Van Ghelue M (2006) The multifunctional roles of the four-and-a-half-LIM only protein FHL2. Cell Mol Life Sci 63:268–284.
Joseph C, Stier G, O'Brien R, Politou AS, Atkinson RA, Bianco A, et al. (2001) A structural characterization of the interactions between titin Z-repeats and the alpha-actinin C-terminal domain. Biochemistry 40:4957–4965.

Kazmierski ST, Antin PB, Witt CC, Huebner N, McElhinny AS, Labeit S, et al. (2003) The complete mouse nebulin gene sequence and the identification of cardiac nebulin. J Mol Biol 328:835–846.
Kemp TJ, Sadusky TJ, Saltisi F, Carey N, Moss J, Yang SY, et al. (2000) Identification of Ankrd2, a novel skeletal muscle gene coding for a stretch-responsive ankyrin-repeat protein. Genomics 66:229–241.
Kim K, Keller TC 3rd (2002) Smitin, a novel smooth muscle titin-like protein, interacts with myosin filaments in vivo and in vitro. J Cell Biol 156:101–111.
Knoll R, Hoshijima M, Chien K (2003) Cardiac mechanotransduction and implications for heart disease. J Mol Med 81:750–756.
Knoll R, Hoshijima M, Hoffman HM, Person V, Lorenzen-Schmidt I, Bang ML, et al. (2002) The cardiac mechanical stretch sensor machinery involves a Z-disc complex that is defective in a subset of human dilated cardiomyopathy. Cell 111:943–955.
Kojic S, Medeot E, Guccione E, Krmac H, Zara I, Martinelli V, et al. (2004) The Ankrd2 protein, a link between the sarcomere and the nucleus in skeletal muscle. J Mol Biol 339:313–325.
Kontrogianni-Konstantopoulos A, Bloch RJ (2003) The hydrophilic domain of small ankyrin-1 interacts with the two N-terminal immunoglobulin domains of titin. J Biol Chem 278: 3985–3991.
Kontrogianni-Konstantopoulos A, Catino DH, Strong JC, Sutter S, Borisov AB, Pumplin DW, et al. (2006) Obscurin modulates the assembly and organization of sarcomeres and the sarcoplasmic reticulum. FASEB J 20:2102–2111.
Kontrogianni-Konstantopoulos A, Jones EM, Van Rossum DB, Bloch RJ (2003) Obscurin is a ligand for small ankyrin 1 in skeletal muscle. Mol Biol Cell 14:1138–1148.
Kruger M, Kohl T, Linke WA (2006) Developmental changes in passive stiffness and myofilament Ca2+ sensitivity due to titin and troponin-I isoform switching are not critically triggered by birth. Am J Physiol Heart Circ Physiol 291:H496–H506.
Kruger M, Linke WA (2006) Protein kinase-A phosphorylates titin in human heart muscle and reduces myofibrillar passive tension. J Muscle Res Cell Motil 27:435–444.
Kruger M, Sachse C, Zimmermann WH, Eschenhagen T, Klede S, Linke WA (2008) Thyroid hormone regulates developmental titin isoform transitions via the phosphatidylinositol-3-kinase/AKT pathway. Circ Res 102:439–447.
Kulke M, Fujita-Becker S, Rostkova E, Neagoe C, Labeit D, Manstein DJ, et al. (2001) Interaction between PEVK-titin and actin filaments: origin of a viscous force component in cardiac myofibrils. Circ Res 89:874–881.
Kuo H, Chen J, Ruiz-Lozano P, Zou Y, Nemer M, Chien KR (1999) Control of segmental expression of the cardiac-restricted ankyrin repeat protein gene by distinct regulatory pathways in murine cardiogenesis. Development 126:4223–4234.
Labeit D, Watanabe K, Witt C, Fujita H, Wu Y, Lahmers S, et al. (2003) Calcium-dependent molecular spring elements in the giant protein titin. Proc Natl Acad Sci USA 100:13716–13721.
Labeit S, Barlow DP, Gautel M, Gibson T, Holt J, Hsieh CL, et al. (1990) A regular pattern of two types of 100-residue motif in the sequence of titin. Nature 345:273–276.
Labeit S, Kolmerer B (1995) Titins: giant proteins in charge of muscle ultrastructure and elasticity. Science 270:293–296.
Labeit S, Kolmerer B, Linke WA (1997) The giant protein titin. Emerging roles in physiology and pathophysiology. Circ Res 80:290–294.
Labeit S, Lahmers S, Burkart C, Fong C, McNabb M, Witt S, et al. (2006) Expression of distinct classes of titin isoforms in striated and smooth muscles by alternative splicing, and their conserved interaction with filamins. J Mol Biol 362:664–681.
Lahmers S, Wu Y, Call DR, Labeit S, Granzier H (2004) Developmental control of titin isoform expression and passive stiffness in fetal and neonatal myocardium. Circ Res 94:505–513.
Lange S, Auerbach D, McLoughlin P, Perriard E, Schafer BW, Perriard JC, et al. (2002) Subcellular targeting of metabolic enzymes to titin in heart muscle may be mediated by DRAL/FHL-2. J Cell Sci 115:4925–4936.

Lange S, Ehler E, Gautel M (2006) From A to Z and back? Multicompartment proteins in the sarcomere. Trends Cell Biol 16:11–18.
Lange S, Himmel M, Auerbach D, Agarkova I, Hayess K, Furst DO, et al. (2005a) Dimerisation of myomesin: implications for the structure of the sarcomeric M-band. J Mol Biol 345:289–298.
Lange S, Xiang F, Yakovenko A, Vihola A, Hackman P, Rostkova E, et al. (2005b) The kinase domain of titin controls muscle gene expression and protein turnover. Science 308:1599–1603.
Launay N, Goudeau B, Kato K, Vicart P, Lilienbaum A (2006) Cell signaling pathways to alphaB-crystallin following stresses of the cytoskeleton. Exp Cell Res 312:3570–3584.
Leake MC, Grutzner A, Kruger M, Linke WA (2006) Mechanical properties of cardiac titin's N2B-region by single-molecule atomic force spectroscopy. J Struct Biol 155:263–272.
Lee EH, Gao M, Pinotsis N, Wilmanns M, Schulten K (2006) Mechanical strength of the titin Z1Z2-telethonin complex. Structure 14:497–509.
Li H, Linke WA, Oberhauser AF, Carrion-Vazquez M, Kerkvliet JG, Lu H, et al. (2002) Reverse engineering of the giant muscle protein titin. Nature 418:998–1002.
Linke WA (2008) Sense and stretchability: the role of titin and titin-associated proteins in myocardial stress-sensing and mechanical dysfunction. Cardiovasc Res 77:637–648.
Linke WA, Fernandez JM (2002) Cardiac titin: molecular basis of elasticity and cellular contribution to elastic and viscous stiffness components in myocardium. J Muscle Res Cell Motil 23:483–497.
Linke WA, Ivemeyer M, Labeit S, Hinssen H, Ruegg JC, Gautel M (1997) Actin-titin interaction in cardiac myofibrils: probing a physiological role. Biophys J 73:905–919.
Linke WA, Ivemeyer M, Olivieri N, Kolmerer B, Ruegg JC, Labeit S (1996) Towards a molecular understanding of the elasticity of titin. J Mol Biol 261:62–71.
Linke WA, Kulke M, Li H, Fujita-Becker S, Neagoe C, Manstein DJ, et al. (2002) PEVK domain of titin: an entropic spring with actin-binding properties. J Struct Biol 137:194–205.
Linke WA, Popov VI, Pollack GH (1994) Passive and active tension in single cardiac myofibrils. Biophys J 67:782–792.
Linke WA, Rudy DE, Centner T, Gautel M, Witt C, Labeit S, et al. (1999) I-band titin in cardiac muscle is a three-element molecular spring and is critical for maintaining thin filament structure. J Cell Biol 146:631–644.
Liversage AD, Holmes D, Knight PJ, Tskhovrebova L, Trinick J (2001) Titin and the sarcomere symmetry paradox. J Mol Biol 305:401–409.
Lorenzen-Schmidt I, Stuyvers BD, ter Keurs HE, Date MO, Hoshijima M, Chien KR, et al. (2005) Young MLP deficient mice show diastolic dysfunction before the onset of dilated cardiomyopathy. J Mol Cell Cardiol 39:241–250.
Lu S, Carroll SL, Herrera AH, Ozanne B, Horowits R (2003) New N-RAP-binding partners alpha-actinin, filamin and Krp1 detected by yeast two-hybrid screening: implications for myofibril assembly. J Cell Sci 116:2169–2178.
Luther PK, Squire JM (2002) Muscle Z-band ultrastructure: titin Z-repeats and Z-band periodicities do not match. J Mol Biol 319:1157–1164.
Ma K, Forbes JG, Gutierrez-Cruz G, Wang K (2006) Titin as a giant scaffold for integrating stress and Src homology domain 3-mediated signaling pathways: the clustering of novel overlap ligand motifs in the elastic PEVK segment. J Biol Chem 281:27539–27556.
Ma K, Kan L, Wang K (2001) Polyproline II helix is a key structural motif of the elastic PEVK segment of titin. Biochemistry 40:3427–3438.
Ma K, Wang K (2003) Malleable conformation of the elastic PEVK segment of titin: non-co-operative interconversion of polyproline II helix, beta-turn and unordered structures. Biochem J 374:687–695.
Machado C, Andrew DJ (2000) Titin as a chromosomal protein. Adv Exp Med Biol 481:221–232.
Machado C, Sunkel CE, Andrew DJ (1998) Human autoantibodies reveal titin as a chromosomal protein. J Cell Biol 141:321–333.
MacKenna D, Summerour SR, Villarreal FJ (2000) Role of mechanical factors in modulating cardiac fibroblast function and extracellular matrix synthesis. Cardiovasc Res 46:257–263.

Makarenko I, Opitz CA, Leake MC, Neagoe C, Kulke M, Gwathmey JK, et al. (2004) Passive stiffness changes caused by upregulation of compliant titin isoforms in human dilated cardiomyopathy hearts. Circ Res 95:708–716.

Matsumoto Y, Hayashi T, Inagaki N, Takahashi M, Hiroi S, Nakamura T, et al. (2005) Functional analysis of titin/connectin N2-B mutations found in cardiomyopathy. J Muscle Res Cell Motil 26:367–374.

Mayans O, van der Ven PF, Wilm M, Mues A, Young P, Furst DO, et al. (1998) Structural basis for activation of the titin kinase domain during myofibrillogenesis. Nature 395:863–869.

McElhinny AS, Kakinuma K, Sorimachi H, Labeit S, Gregorio CC (2002) Muscle-specific RING finger-1 interacts with titin to regulate sarcomeric M-line and thick filament structure and may have nuclear functions via its interaction with glucocorticoid modulatory element binding protein-1. J Cell Biol 157:125–136.

McElhinny AS, Perry CN, Witt CC, Labeit S, Gregorio CC (2004) Muscle-specific RING finger-2 (MURF-2) is important for microtubule, intermediate filament and sarcomeric M-line maintenance in striated muscle development. J Cell Sci 117:3175–3188.

McElhinny AS, Schwach C, Valichnac M, Mount-Patrick S, Gregorio CC (2005) Nebulin regulates the assembly and lengths of the thin filaments in striated muscle. J Cell Biol 170: 947–957.

McGrath MJ, Cottle DL, Nguyen MA, Dyson JM, Coghill ID, Robinson PA, et al. (2006) Four and a half LIM protein 1 binds myosin-binding protein C and regulates myosin filament formation and sarcomere assembly. J Biol Chem 281:7666–7683.

Miller MK, Bang ML, Witt CC, Labeit D, Trombitas C, Watanabe K, et al. (2003) The muscle ankyrin repeat proteins: CARP, ankrd2/Arpp and DARP as a family of titin filament-based stress response molecules. J Mol Biol 333:951–964.

Miller MK, Granzier H, Ehler E, Gregorio CC (2004) The sensitive giant: the role of titin-based stretch sensing complexes in the heart. Trends Cell Biol 14:119–126.

Millevoi S, Trombitas K, Kolmerer B, Kostin S, Schaper J, Pelin K, et al. (1998) Characterization of nebulette and nebulin and emerging concepts of their roles for vertebrate Z-discs. J Mol Biol 282:111–123.

Mitra SK, Hanson DA, Schlaepfer DD (2005) Focal adhesion kinase: in command and control of cell motility. Nat Rev Mol Cell Biol 6:56–68.

Moncman CL, Wang K (1995) Nebulette: a 107 kD nebulin-like protein in cardiac muscle. Cell Motil Cytoskeleton 32:205–225.

Mrosek M, Labeit D, Witt S, Heerklotz H, von Castelmur E, Labeit S, Mayans O (2007) Molecular determinants for the recruitment of the ubiquitin-ligase MuRF-1 onto M-line titin. FASEB J 21:1383–1392.

Mues A, van der Ven PF, Young P, Furst DO, Gautel M (1998) Two immunoglobulin-like domains of the Z-disc portion of titin interact in a conformation-dependent way with telethonin. FEBS Lett 428:111–114.

Musa H, Meek S, Gautel M, Peddie D, Smith AJ, Peckham M (2006) Targeted homozygous deletion of M-band titin in cardiomyocytes prevents sarcomere formation. J Cell Sci 119: 4322–4331.

Nagueh SF, Shah G, Wu Y, Torre-Amione G, King NM, Lahmers S, et al. (2004) Altered titin expression, myocardial stiffness, and left ventricular function in patients with dilated cardiomyopathy. Circulation 110:155–162.

Nagy A, Cacciafesta P, Grama L, Kengyel A, Malnasi-Csizmadia A, Kellermayer MS (2004) Differential actin binding along the PEVK domain of skeletal muscle titin. J Cell Sci 117: 5781–5789.

Neagoe C, Kulke M, del Monte F, Gwathmey JK, de Tombe PP, Hajjar RJ, et al. (2002) Titin isoform switch in ischemic human heart disease. Circulation 106:1333–1341.

Neagoe C, Opitz CA, Makarenko I, Linke WA (2003) Gigantic variety: expression patterns of titin isoforms in striated muscles and consequences for myofibrillar passive stiffness. J Muscle Res Cell Motil 24:175–189.

Obermann WM, Gautel M, Weber K, Furst DO (1997) Molecular structure of the sarcomeric M band: mapping of titin and myosin binding domains in myomesin and the identification of a potential regulatory phosphorylation site in myomesin. EMBO J 16:211–220.
Ohtsuka H, Yajima H, Maruyama K, Kimura S (1997) The N-terminal Z repeat 5 of connectin/titin binds to the C-terminal region of alpha-actinin. Biochem Biophys Res Commun 235:1–3.
Ojima K, Ono Y, Hata S, Koyama S, Doi N, Sorimachi H (2005) Possible functions of p94 in connectin-mediated signaling pathways in skeletal muscle cells. J Muscle Res Cell Motil 26:409–417.
Olson EN (2006) Gene regulatory networks in the evolution and development of the heart. Science 313:1922–1927.
Ono Y, Torii F, Ojima K, Doi N, Yoshioka K, Kawabata Y, et al. (2006) Suppressed disassembly of autolyzing p94/CAPN3 by N2A connectin/titin in a genetic reporter system. J Biol Chem 281:18519–18531.
Opitz CA, Kulke M, Leake MC, Neagoe C, Hinssen H, Hajjar RJ, et al. (2003) Damped elastic recoil of the titin spring in myofibrils of human myocardium. Proc Natl Acad Sci USA 100:12688–12693.
Opitz CA, Leake MC, Makarenko I, Benes V, Linke WA (2004) Developmentally regulated switching of titin size alters myofibrillar stiffness in the perinatal heart. Circ Res 94:967–975.
Opitz CA, Linke WA (2005) Plasticity of cardiac titin/connectin in heart development. J Muscle Res Cell Motil 26:333–342.
Peng J, Raddatz K, Molkentin JD, Wu Y, Labeit S, Granzier H, et al. (2007) Cardiac hypertrophy and reduced contractility in hearts deficient in the titin kinase region. Circulation 115: 743–751.
Perriard JC, Hirschy A, Ehler E (2003) Dilated cardiomyopathy: a disease of the intercalated disc? Trends Cardiovasc Med 13:30–38.
Prado LG, Makarenko I, Andresen C, Kruger M, Opitz CA, Linke WA (2005) Isoform diversity of giant proteins in relation to passive and active contractile properties of rabbit skeletal muscles. J Gen Physiol 126:461–480.
Purcell NH, Darwis D, Bueno OF, Muller JM, Schule R, Molkentin JD (2004) Extracellular signal-regulated kinase 2 interacts with and is negatively regulated by the LIM-only protein FHL2 in cardiomyocytes. Mol Cell Biol 24:1081–1095.
Pyle WG, Solaro RJ (2004) At the crossroads of myocardial signaling: the role of Z-discs in intracellular signaling and cardiac function. Circ Res 94:296–305.
Radke MH, Peng J, Wu Y, McNabb M, Nelson OL, Granzier H, et al. (2007) Targeted deletion of titin N2B-region leads to diastolic dysfunction and cardiac atrophy. Proc Natl Acad Sci USA 104:3444–3449.
Raynaud F, Astier C, Benyamin Y (2004) Evidence for a direct but sequential binding of titin to tropomyosin and actin filaments. Biochim Biophys Acta 1700:171–178.
Raynaud F, Fernandez E, Coulis G, Aubry L, Vignon X, Bleimling N, et al. (2005) Calpain 1-titin interactions concentrate calpain 1 in the Z-band edges and in the N2-line region within the skeletal myofibril. FEBS J 272:2578–2590.
Romer LH, Birukov KG, Garcia JG (2006) Focal adhesions: paradigm for a signalling nexus. Circ Res 98:606–616.
Samarel AM (2005) Costameres, focal adhesions, and cardiomyocyte mechanotransduction. Am J Physiol Heart Circ Physiol 289:H2291–H2301.
Samson T, Smyth N, Janetzky S, Wendler O, Muller JM, Schule R, et al. (2004) The LIM-only proteins FHL2 and FHL3 interact with alpha- and beta-subunits of the muscle alpha7beta1 integrin receptor. J Biol Chem 279:28641–28652.
Satoh M, Takahashi M, Sakamoto T, Hiroe M, Marumo F, Kimura A (1999) Structural analysis of the titin gene in hypertrophic cardiomyopathy: identification of a novel disease gene. Biochem Biophys Res Commun 262:411–417.
Schellings MW, Pinto YM, Heymans S (2004) Matricellular proteins in the heart: possible role during stress and remodeling. Cardiovasc Res 64:24–31.

Schoenauer R, Bertoncini P, Machaidze G, Aebi U, Perriard JC, Hegner M, et al. (2005) Myomesin is a molecular spring with adaptable elasticity. J Mol Biol 349:367–379.

Seeley M, Huang W, Chen Z, Wolff WO, Lin X, Xu X (2007) Depletion of zebrafish titin reduces cardiac contractility by disrupting the assembly of Z-discs and A-bands. Circ Res 100: 238–245.

Siu BL, Niimura H, Osborne JA, Fatkin D, MacRae C, Solomon S, et al. (1999) Familial dilated cardiomyopathy locus maps to chromosome 2q31. Circulation 99:1022–1026.

Sorimachi H, Freiburg A, Kolmerer B, Ishiura S, Stier G, Gregorio CC, et al. (1997) Tissue-specific expression and alpha-actinin binding properties of the Z-disc titin: implications for the nature of vertebrate Z-discs. J Mol Biol 270:688–695.

Sorimachi H, Kinbara K, Kimura S, Takahashi M, Ishiura S, Sasagawa N, et al. (1995) Muscle-specific calpain, p94, responsible for limb girdle muscular dystrophy type 2A, associates with connectin through IS2, a p94-specific sequence. J Biol Chem 270:31158–31162.

Sun J, Yan G, Ren A, You B, Liao JK (2006) FHL2/SLIM3 decreases cardiomyocyte survival by inhibitory interaction with sphingosine kinase-1. Circ Res 99:468–476.

Taveau M, Bourg N, Sillon G, Roudaut C, Bartoli M, Richard I (2003) Calpain 3 is activated through autolysis within the active site and lyses sarcomeric and sarcolemmal components. Mol Cell Biol 23:9127–9135.

Trinick J, Tskhovrebova L (1999) Titin: a molecular control freak. Trends Cell Biol 9:377–380.

Trombitas K, Freiburg A, Centner T, Labeit S, Granzier H (1999) Molecular dissection of N2B cardiac titin's extensibility. Biophys J 77:3189–3196.

Trombitas K, Granzier H (1997) Actin removal from cardiac myocytes shows that near Z line titin attaches to actin while under tension. Am J Physiol 273:C662–C670.

Trombitas K, Wu Y, Labeit D, Labeit S, Granzier H (2001) Cardiac titin isoforms are coexpressed in the half-sarcomere and extend independently. Am J Physiol Heart Circ Physiol 281:H1793–H1799.

Tskhovrebova L, Trinick J (2004) Properties of titin immunoglobulin and fibronectin-3 domains. J Biol Chem 279:46351–46354.

Tsukamoto Y, Senda T, Nakano T, Nakada C, Hida T, Ishiguro N, et al. (2002) Arpp, a new homolog of carp, is preferentially expressed in type 1 skeletal muscle fibers and is markedly induced by denervation. Lab Invest 82:645–655.

Udd B, Vihola A, Sarparanta J, Richard I, Hackman P (2005) Titinopathies and extension of the M-line mutation phenotype beyond distal myopathy and LGMD2J. Neurology 64:636–642.

van de Klundert FA, Gijsen ML, van den IJssel PR, Snoeckx LH, de Jong WW (1998) Alpha B-crystallin and hsp25 in neonatal cardiac cells–differences in cellular localization under stress conditions. Eur J Cell Biol 75:38–45.

Van den Bergh PY, Bouquiaux O, Verellen C, Marchand S, Richard I, Hackman P, et al. (2003) Tibial muscular dystrophy in a Belgian family. Ann Neurol 54:248–251.

van der Ven PF, Bartsch JW, Gautel M, Jockusch H, Furst DO (2000) A functional knock-out of titin results in defective myofibril assembly. J Cell Sci 113:1405–1414.

van Heerebeek L, Borbely A, Niessen HW, Bronzwaer JG, van der Velden J, Stienen GJ, et al. (2006) Myocardial structure and function differ in systolic and diastolic heart failure. Circulation 113:1966–1973.

Wang X, Klevitsky R, Huang W, Glasford J, Li F, Robbins J (2003) AlphaB-crystallin modulates protein aggregation of abnormal desmin. Circ Res 93:998–1005.

Warren CM, Krzesinski PR, Campbell KS, Moss RL, Greaser ML (2004) Titin isoform changes in rat myocardium during development. Mech Dev 121:1301–1312.

Weinert S, Bergmann N, Luo X, Erdmann B, Gotthardt M (2006) M-line-deficient titin causes cardiac lethality through impaired maturation of the sarcomere. J Cell Biol 173:559–570.

Wernyj RP, Ewing CM, Isaacs WB (2001) Multiple antibodies to titin immunoreact with AHNAK and localize to the mitotic spindle machinery. Cell Motil Cytoskeleton 50:101–113.

Wilding JR, Lygate CA, Davies KE, Neubauer S, Clarke K (2006) MLP accumulation and remodelling in the infarcted rat heart. Eur J Heart Fail 8:343–346.

Willis MS, Ike C, Li L, Wang DZ, Glass DJ, Patterson C (2007) Muscle ring finger 1, but not muscle ring finger 2, regulates cardiac hypertrophy in vivo. Circ Res 100:456–459.
Willis MS, Patterson C (2006) Into the heart: the emerging role of the ubiquitin-proteasome system. J Mol Cell Cardiol 41:567–579.
Witt CC, Burkart C, Labeit D, McNabb M, Wu Y, Granzier H, et al. (2006) Nebulin regulates thin filament length, contractility, and Z-disk structure in vivo. EMBO J 25:3843–3855.
Witt CC, Ono Y, Puschmann E, McNabb M, Wu Y, Gotthardt M, et al. (2004) Induction and myofibrillar targeting of CARP, and suppression of the Nkx2.5 pathway in the MDM mouse with impaired titin-based signaling. J Mol Biol 336:145–154.
Witt SH, Granzier H, Witt CC, Labeit S (2005a) MURF-1 and MURF-2 target a specific subset of myofibrillar proteins redundantly: towards understanding MURF-dependent muscle ubiquitination. J Mol Biol 350:713–722.
Witt SH, Labeit D, Granzier H, Labeit S, Witt CC (2005b) Dimerization of the cardiac ankyrin protein CARP: implications for MARP titin-based signaling. J Muscle Res Cell Motil 26: 401–408.
Wu Y, Cazorla O, Labeit D, Labeit S, Granzier H (2000) Changes in titin and collagen underlie diastolic stiffness diversity of cardiac muscle. J Mol Cell Cardiol 32:2151–2162.
Wu Y, Labeit S, LeWinter MM, Granzier H (2002) Titin: an endosarcomeric protein that modulates myocardial stiffness in DCM. J Card Fail 8(6 Suppl):S276–S286.
Wu Y, Peng J, Campbell KB, Labeit S, Granzier H (2007) Hypothyroidism leads to increased collagen-based stiffness and re-expression of large cardiac titin isoforms with high compliance. J Mol Cell Cardiol 42:186–195.
Xu X, Meiler SE, Zhong TP, Mohideen M, Crossley DA, Burggren WW, et al. (2002) Cardiomyopathy in zebrafish due to mutation in an alternatively spliced exon of titin. Nat Genet 30: 205–209.
Yamasaki R, Berri M, Wu Y, Trombitas K, McNabb M, Kellermayer MS, et al. (2001) Titin-actin interaction in mouse myocardium: passive tension modulation and its regulation by calcium/S100A1. Biophys J 81:2297–2313.
Yamasaki R, Wu Y, McNabb M, Greaser M, Labeit S, Granzier H (2002) Protein kinase-A phosphorylates titin's cardiac-specific N2B domain and reduces passive tension in rat cardiac myocytes. Circ Res 90:1181–1188.
Young P, Ehler E, Gautel M (2001) Obscurin, a giant sarcomeric Rho guanine nucleotide exchange factor protein involved in sarcomere assembly. J Cell Biol 154:123–136.
Young P, Ferguson C, Banuelos S, Gautel M (1998) Molecular structure of the sarcomeric Z-disk: two types of titin interactions lead to an asymmetrical sorting of alpha-actinin. EMBO J 17:1614–1624.
Zastrow MS, Flaherty DB, Benian GM, Wilson KL (2006) Nuclear titin interacts with A- and B-type lamins in vitro and in vivo. J Cell Sci 119:239–249.
Zolk O, Caroni P, Bohm M (2000) Decreased expression of the cardiac LIM domain protein MLP in chronic human heart failure. Circulation 101:2674–2677.
Zou P, Pinotsis N, Lange S, Song YH, Popov A, Mavridis I, et al. (2006) Palindromic assembly of the giant muscle protein titin in the sarcomeric Z-disk. Nature 439:229–233.

Chapter 2
Mechanical Stretch-Induced Reorganization of the Cytoskeleton and the Small GTPase Rac-1 in Cardiac Fibroblasts

Wayne Carver and John W. Fuseler

Abstract Mechanical forces play important roles in development and disease of most tissues. In vivo studies have illustrated that increased mechanical load as seen during neonatal development or in the hypertensive adult promote a fibrotic response in the heart. In vitro studies have established that mechanical stretch of isolated cardiac fibroblasts directly stimulates expression of extracellular matrix components and proliferation, both hallmarks of fibrosis. While significant advances have been made in understanding the effects of mechanical forces on cardiac fibroblasts, many questions remain regarding the mechanisms whereby mechanical forces are transduced into changes in cellular phenotype. The linkage between the extracellular matrix, integrin receptors and the cytoskeleton undoubtedly plays a critical role in this process. We have recently shown that mechanical stretch induces rapid changes in cardiac fibroblast morphology and the organization of the actin cytoskeleton. The Rho family of small GTPases has received considerable attention in their role in organizing the actin cytoskeleton. Data is presented herein providing quantitative analysis of alterations in the activation and subcellular organization of the small GTPase Rac-1 following equibiaxial stretch of isolated cardiac fibroblasts.

Keywords Mechanical stretch · Cytoskeleton · GTPase · Fibroblast · Cardiac

2.1 Introduction

Mechanical forces play important regulatory roles in tissue and organ development, homeostasis and disease. The critical nature of these forces in morphogenetic patterning is evident from the earliest stages of embryogenesis as tensional forces generated by cells promote compaction of the embryonic morula (Reithmacher et al.,

J.W. Fuseler (✉)
Department of Cell Biology and Anatomy, School of Medicine Columbia, University of South Carolina, Columbia, SC, USA
e-mail: fuseler@gw.med.sc.sdu

A. Kamkin, I. Kiseleva (eds.), *Mechanosensitivity of the Heart*,
Mechanosensitivity in Cells and Tissues 3, DOI 10.1007/978-90-481-2850-1_2,

1995). Later in development, shear stress generated by fluid flow plays a fundamental role in blood vessel formation and remodeling (Le Noble et al., 2004; Lucitti et al., 2007). In the mature organ, most cells are constantly responding to changes in the mechanical environment. In the cardiovascular system, increased workload as seen during hypertension or aortic stenosis, initiates an adaptive process that includes myocardial hypertrophy and fibrosis.

The bulk of the myocardium is made up of myocytes, whose responses to mechanical forces have been relatively extensively studied (for recent reviews see Catalucci et al., 2008; Lammerding et al., 2004). The myocytes are surrounded by an elaborate extracellular matrix (ECM) that supports and interconnects cells of the myocardium (Borg and Caulfield, 1981). The myocardium also includes fibroblasts, endothelial cells, smooth muscle cells and transient inflammatory cells. Fibroblasts comprise the most numerous cells in the heart and are arranged so that each cardiac myocyte is closely enveloped by a network of fibroblasts and their associated endomysium (Camelliti et al., 2005). The fibroblasts synthesize the majority of the ECM of the heart and interact with this ECM to modulate cellular and organ activity (Carver et al., 1993; Goldsmith et al., 2004; Grinnell, 2000). As mentioned above, increased cardiovascular load results in substantial remodeling of the heart including enhanced synthesis and deposition of ECM by the fibroblasts.

In vitro studies have illustrated that mechanical stretch elicits a pro-fibrotic response from cardiac fibroblasts including enhanced production of ECM components and increased proliferation (Butt et al., 1995; Carver et al., 1991; Lee et al., 1999; MacKenna et al., 1998). Several studies have begun to elucidate the mechanisms of this response; however, exactly how cardiac fibroblasts recognize and respond to their mechanical environment and how these mechanical forces induce major changes in cellular phenotype remain relatively obscure. These cells undergo rapid changes in shape and cytoskeletal organization in response to mechanical forces (Fuseler et al., 2007), but the mechanisms regulating rearrangements of their actin filamentous network have been under-investigated.

2.2 The Cytoskeleton, Rho GTPases and Mechanotransduction

Mechanical forces are transmitted to cells, at least in part, through the physical interactions of the cell with the surrounding ECM (Geiger et al., 2001; Matthews et al., 2006; Janmey and Weitz, 2004). The ECM is physically linked to the cytoskeleton via cell surface receptors primarily of the integrin family. Thus, alterations in the mechanical environment can be transmitted from the ECM to the cell via integrin receptors (Bershadsky et al., 2006; Sanchez-Esteban et al., 2006). Numerous studies have illustrated that the physical network involving the ECM, integrins and the cytoskeleton is critical to the response of cells to mechanical forces.

In most cells, Rho small GTPases, particularly RhoA, Rac-1 and Cdc42 isoforms, are indispensable regulators of actin cytoskeletal organization (Hall, 1998; Manser, 2005; Schmidt and Hall, 1998). Each of the small GTPases has specialized functions, acting as molecular switches to modulate formation of actin stress fibers

(Rho), filapodia extrusion (Cdc42) and the production of lamellipodia or membrane ruffles (Rac-1). Activity of the GTPases is controlled by effector proteins, including GTPase activators, guanine nucleotide dissociation inhibitors, guanine nucleotide exchangers, and guanine nucleotide dissociation factors. Downstream of activation, Rho small GTPases interact with a variety of target proteins that in turn directly affect actin reorganization. Different Rho type GTPases are activated by specific extracellular stimuli. The particular effector and target proteins involved also vary greatly from cell to cell. Finally, specific intracellular localization (Michaelson et al., 2005) and interactions between the Rho family GTPases are important for their functioning (Yuan et al., 2003), with shuttling between cytoplasmic and membrane-bound forms of the GTPases engendered differentially by various stimuli.

The Rho family GTPases have begun to receive considerable attention in cardiac development and disease, particularly with regards to the cardiomyocyte (for recent review see Brown et al., 2006). Recent studies have suggested that Cdc42 plays an important signaling role in stretch-induced hypertrophy of myocytes (Pan et al., 2005). Several studies have also implicated the RhoA/Rho kinase (ROCK) pathway in cardiac hypertrophy and in the transition from hypertrophy to overt heart failure (Hu and Lee, 2003; Ren and Fang, 2005). Activation of RhoA in cardiac myocytes results in the expression of hypertrophy-related genes. This response is dependent upon interactions between the β1 integrins and the actin cytoskeleton (Kawamura et al., 2003; Wei et al., 2001). Studies have also suggested that Rac-1 is involved in the development of cardiac hypertophy through its mediation of stress-induced activation of p38 mitogen-activated protein kinase (Aikawa et al., 2001). Importantly, Rac-1 has been shown to be required for the activation of NAD(P)H oxidase, which initiates a redox-dependent signal transduction pathway leading to the activation of NF-κB and subsequent expression of genes related to cardiac hypertrophy and cellular transformation (Sulciner et al., 1996; Perona et al., 1997). While studies have begun to focus on the Rho GTPases in heart myocytes, less is known about the function of these proteins in cardiac fibroblasts.

It has been demonstrated that the Rho GTPases are indeed important in fibroblasts and that their activity is dependent upon the stimulus encountered (Grinnell, 2003; Lee et al., 2003). Both Rac and Rho GTPases have been implicated in the response of cardiac fibroblasts to ECM and to external soluble factors. For example, Rac-1 and RhoA, are activated differentially in cardiac fibroblasts by lysophosphatidic acid and platelet-derived growth factor (Abe et al., 2003). The involvement of Rho GTPases in cardiac disease has also been implicated. Human atrial myofibroblast proliferation, pertaining to heart remodeling following infarction, is inhibited by simvastatin via a mechanism involving RhoA (Porter et al., 2004). Recent studies have also demonstrated roles for RhoA and its downstream acceptor ROCK in cardiac dysfunction, where expression of RhoA was elevated in the myocardia of both rats and dogs (Satoh et al., 2003; Suematsu et al., 2001).

It is becoming clear, therefore, that fibroblasts of the heart employ GTPases of the Rho family to regulate their actin cytoskeleton; that this involvement is dependent on specific stimuli; and that alterations in shape are determinants of cardiac fibroblast function. However, most reported studies have implicated GTPases only indirectly; for example, by using inhibitory pharmacological agents and thereby inferring involvement via loss of function (Yuan et al., 2003). Few detailed measurements of precise levels of total versus activated Rho-type GTPases have been reported. Investigations of the intracellular localizations of GTPases and how multiple isoforms interact within the fibroblast cytoplasm in response to differing stimuli have not been reported. To begin to investigate the alterations in activation and in the subcellular distribution of Rho GTPase family members in response to mechanical forces, experiments have been performed to investigate the response of Rac-1 in cardiac fibroblasts subjected to equibiaxial stretch.

2.3 Analysis of the Actin Cytoskeleton and Rac-1 GTPase in Mechanically Stretched Cardiac Fibroblasts

2.3.1 Model Systems to Study the Effects of Mechanical Forces

At the cellular level, mechanical forces have been clearly demonstrated to modulate differentiation, proliferation, gene expression and survival of many cell types. While it is clear that mechanical forces impact cellular, tissue and whole organ function, many questions remain regarding the transduction of mechanical signals by cells and the integration of these signals with biochemical stimuli. An obvious obstacle to fully understanding the effects of the mechanical environment on cells and tissues is the complexity of the in vivo milieu. For this reason, investigations focused on mechanobiology have relied heavily on in vitro model systems. These systems have varied greatly in their complexity and ability to produce a homogeneous and consistent mechanical microenvironment.

Early investigations in the mechanobiology field relied on relatively simple, non-quantitative systems. For instance, Gluckmann (1939) utilized a chick tibia explant culture system to begin to examine the effects of compressive forces on endosteal cells. A number of studies utilized a hanging-drop culture system to examine the effects of tensile forces on connective tissue cells (Bassett and Hermann, 1961). Studies by Rodan and colleagues began to utilize more precisely quantifiable systems to examine the effects of mechanical forces on cells and tissues. Innovative systems to apply tensile strain to rat calvarial cells cultured on ribbons of collagen (Yeh and Rodan, 1984) and compressive forces to chick long bones (Rodan et al., 1975) were developed by these investigators.

Since these early studies, several model systems have been developed to apply uniaxial distension or stretch to cells and tissues. These systems date back several decades to studies conducted on smooth muscle cells that were cultured on deformable elastin matrices (Leung et al., 1976; 1977). These studies were among the first to demonstrate that cyclic mechanical loading promotes the production of extracellular matrix components by vascular smooth muscle cells. Many modifications to this original concept have been made largely in attempts to produce more homogeneous and predictable strains across the deformable substrate. Most of the systems produced utilize a flexible, ECM-coated substrate upon which cells are attached. The substrate and hence the cells are subjected to a controlled static or dynamic deformation regimen. While the cells are being subjected to stretch in one direction in these uniaxial systems, they are also being subjected to compressive strain perpendicular to the axis of loading and to shear forces near the ends of the substrate (Hung and Williams, 1994). More recently developed systems have attempted to minimize shear stress generated in these systems by careful design of the stretch apparatus and membrane geometry (Yost et al., 2000).

Realizing that most cells are subjected to complex mechanical forces, not typically along a single axis, systems have been developed more recently that generate mechanical deformation in multiple axes (Schaffer et al., 1994; Lee et al., 1996). Some of these systems utilize vacuum pressure to indent a clamped circular flexible membrane. In these devices, the generated strains are not homogeneous over the membrane surface and may vary from 0 to 30% depending on the region of the membrane examined. More recent modifications of these systems have incorporated mechanisms to promote homogeneous biaxial (equibiaxial) strain to the flexible substratum and attached cells (Lee et al., 1996). Static and cyclic equibiaxial systems are now being widely used to understand the effects of stretch on isolated cells and the mechanisms of these effects. As fibroblasts in the heart likely respond to mechanical forces from multiple directions, a static equibiaxial system has been utilized in our recent work (Fuseler et al., 2007) and in the present studies.

2.4 Alterations in Cardiac Fibroblast Morphology in Response to Equibiaxial Stretch

Cardiac fibroblasts were isolated from 3 day old neonatal Sprague Dawley rat hearts and cultured in Dulbecco's Modified Eagle's Medium supplemented with 10% fetal bovine serum, 5% neonatal calf serum and antibiotics (Borg et al., 1984; Carver et al., 1991). Cells were plated onto laminin-coated (10 μg/ml) silastic membranes (Specialty Manufacturing; Saginaw, MI) and grown to approximately 80–90% confluency. The fibroblasts were subjected to 5% equibiaxial mechanical stretch (Lee et al., 1996: Atance et al., 2004) for 0, 1.25, 2.5, 5.0, 10.0, 20.0, and 40.0 min. Cells were stained with rhodamine phalloidin for analysis of f-actin and overall cell morphology. Cardiac fibroblasts exhibited a rapid initial response to externally applied mechanical stretch. Under the influence of initial stretch, the cells exhibited significant ruffling at the cell margins, which is consistent with observations in other cell types (Yoshogi et al., 2003) responding to stretch. The altered gross morphology of the cell was characterized by an initial, very rapid, decrease in both cell area and perimeter (Fig. 2.1), accompanied by cellular rounding, elongation and retraction of cytoplasmic processes.

With continuously applied equibiaxial stretch cardiac fibroblasts shorten more rapidly parallel to the major axis than to the minor axis then proceed to attain a more rounded morphology. In such a morphology, the cells present a minimal surface

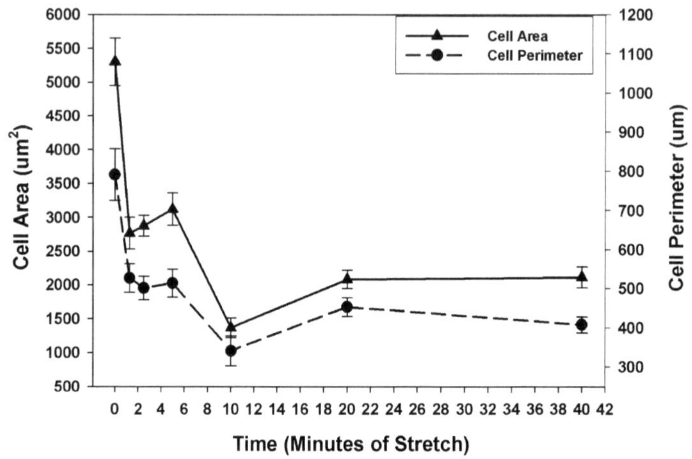

Fig. 2.1 Response of cardiac fibroblasts to 5% equibiaxial stretch. Cardiac fibroblasts undergo an immediate decrease in cell area which reaches a minimum by 10 min of applied stretch, then remains unchanged in the presence of prolonged stretch. The change in cell perimeter in response to stretch parallels the changes observed for cell area. These parallel decreases cell area and perimeter indicated that cardiac fibroblasts appear to react to mechanical stretch by pulling against it, resulting in the cells becoming smaller and less flattened

area to the omni-directional stretch forces and establish an equilibrium state to the magnitude of the applied force. The changes in morphological parameters of cell area and perimeter appear jointly coupled and regulated as the correlation between changes in area and perimeter over the range of applied stretch is linear (Fig. 2.2).

These morphological changes in cardiac fibroblasts induced by time dependent mechanical stretch are consistent with our pervious study (Fuseler et al., 2007). These changes in membrane activity at the cell margin suggest altered microtubule integrity (Vasiliev 1991) and the participation of activated small GTPases, in particular Rac-1.

2.5 Biochemical Analysis of Rac-1 GTPase Activation by Equibiaxial Stretch

The total amount of Rac-1 in cardiac fibroblasts was determined by western blot analysis. The proportion of GTP-bound Rac-1 was determined by a pull-down method as an indicator of Rac-1 activation (Pierce Biotech; Rockford, IL). The overall expression of Rac-1 in neonatal cardiac fibroblasts was not affected by equibiaxial mechanical stretch at the time-points assayed here (Fig. 2.3). This suggests that short term mechanical stretch does not induce or make available additional Rac-1 protein to the cell.

The activation of Rac-1 protein rapidly increased in response to mechanical stretch (Fig. 2.3). Maximum activation of Rac-1 protein occurred by 1 min

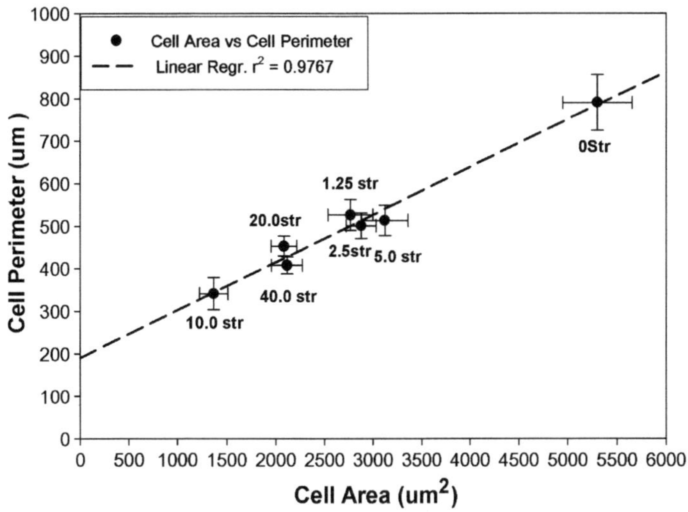

Fig. 2.2 The cell area and perimeter maintain a linear correlation over the range of times in which 5% equibiaxial stretch was applied to cardiac fibroblasts. This suggest that the morphological changes cardiac fibroblasts undergo in the presence of external mechanical stretch at this magnitude of force may be series of uniformly directed events in response to an adverse change in the environment

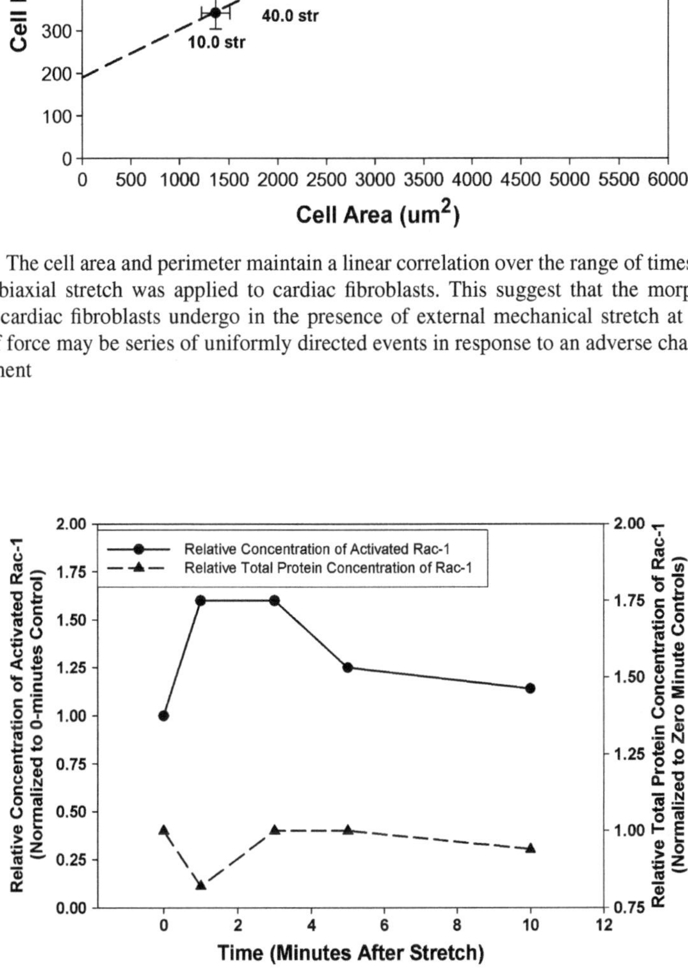

Fig. 2.3 Stretch induced activation of Rac-1 by mechanical stretch in neonatal cardiac fibroblasts. Rac-1 is rapidly activated by mechanical stretch reaching a maximum level within 1 min. With continued applied mechanical stretch the level of activation of Rac-1 remains elevated before declining at 5 min. The concentration of activated Rac-1 slowly declines in the presence of prolonged stretch and slowly approaching control values at 10 min of stretch. Mechanical stretch has no effect on the total amount of Rac-1 protein present in neonatal cardiac fibroblasts

after the initiation of stretch. Active Rac-1 remained elevated before declining by approximately 50% at 5 min. These data suggests that in cardiac fibroblasts, the small GTPases, are highly sensitive to externally applied mechanical forces transmitted through the ECM-cell membrane-cytoskeletal axis. The GTPases may function as early responsive sensors of such externally applied forces. Since regional changes in GTPase activity are not discernable by biochemical analysis of whole cell lysates, immunofluorescent analyses were performed to examine stretch-induced changes in the actin cytoskeleton and Rac-1 localization and morphology.

2.6 Morphological Alterations in Cytoskeletal Actin and Rac-1 in Response to Equibiaxial Stretch: Fractal Analysis

To analyze changes in GTPase morphology and subcellular distribution, fibroblasts were subjected to equibiaxial stretch and immunocytochemically stained for Rac-1. Silastic membranes containing cells were fixed in 3% paraformaldehyde and subsequently permeablized in phosphate-buffered saline containing 0.1% Triton X-100 and 10 mM glycine. The cells were double labeled for Rac-1 GTPase and F-actin. Images were captured using a Zeiss LSM 510 confocal microscope. Changes in distribution and morphology of Rac-1 were analyzed by using the integrated morphometry subroutine of MetaMorph 6.1 (Universal Imaging Corp, Downingtown, PA). Briefly, the confocal images were color separated into two channels, red (rhodamine conjugates for the actin cytoskeleton) and green (FITC labeled Rac-1), and converted into 16 bit monochrome images. Each complete cell in the image field was isolated as a region of interest (ROI). From the ROI the fluorescent GTPase granules were thresholded for analysis. The descriptors of area (A), perimeter (P), the integrated optical density (IOD), and fractal dimension (**D**) were measured by the MetaMorph 6.1 software for thresholded GTPase granules in each isolated cardiac fibroblast (Fuseler et al., 2007).

The IOD of the region of fluorescence delineated by the thresholded boundaries may be considered the "mass" of the region and a measurement of the total amount of labeled material in the region (Walter and Berns, 1986; Fuseler et al., 2006, 2007; Rogers and Fuseler, 2007). The IOD is the weighted sum of the image histogram in which each term in the histogram is multiplied by the gray value it represents. When applied to thresholded boundaries the IOD is expressed as:

$$\mathrm{IOD}(\mathrm{T}_1,\mathrm{T}_2) = \sum_{\mathrm{GV}=T_1}^{\mathrm{T}_2} \mathrm{H}(\mathrm{GV})$$

Where GV is the gray value of each pixel and H(GV) is the gray level histogram, and T_1 and T_2 are the upper and lower thresholds defining the region of interest in the histogram. Values of IOD are calculated directly from the integrated morphometry subroutine of MetaMorph image analysis software. Using the software's optical calipers, the measurements are refined by setting specific boundary conditions for area and IOD for acceptance of the fluorescent signal from the labeled cells and to minimize or eliminate the contributions of any non-specific and background auto-fluorescence.

The changes in the morphological characteristics of cellular organelles undergoing activation in response to adverse mechanical and environmental stresses have only been superficially described. Here we demonstrate that the morphological changes the small GTPase Rac-1 undergoes during activation and subsequent cytoplasmic translocation in response to application of mechanical stretch can be analyzed and quantified by application of fractal geometry. Because of the irregularity and complexity of the Rac-1 cytoplasmic array it cannot be characterized or defined by regular Euclidean geometry or dimensions. The regular Euclidean dimension assigns an integer to each point or set of points in Euclidean space and includes the familiar geometrical descriptors or numbers; 0 to a point, 1 to a straight line, 2 to a plane surface, and 3 to a volume or three dimensional figure. These integer descriptors are exponents of power functions which describe these objects. No Macro-or micro-anatomical structure because of their complexity can be described by or correspond to a regular Euclidean geometric figure. The dimension of complex or irregular structures can be described by non-integer numbers, with values falling between two integer topological dimensions. These non-integer numbers are described as non-Euclidean and define the fractal dimension (**D**) of an object. The concept of fractals currently provides a useful method to quantify the inherent irregularity or complexity of phenomena (Zhang et al., 2005), or changes in a phenomena as it becomes more organized or undergoes randomization or increases in chaos. In general, a fractal may be considered to be any rough and irregular object consisting of parts that are in some way similar to the whole. That is self-similarity is present at the various levels or dimensions of the object. As a descriptor of shape complexity **D** has been applied to a wide spectrum of studies in biological and physical sciences.

Computations of **D** values have been widely used in describing complex biological systems. In neuroanatomical studies, **D** values have been used to describe neuronal growth (Behar 2001; Borodinsky and Fiszman, 2001; Bernard et al., 2001) and arborization patterns (Caserta et al., 1995; Ristanovic et al., 2002). The morphological distribution of bronchial capillaries around large airways (Anderson et al., 2005) and capillary branching during angiogenesis (Grizzi et al., 2005) have been characterized by **D** values. Additionally, **D** values have been useful in describing patterns of growth associated with tumors (Gazit et al., 1997, 1995), angiogenesis associated with tumors (Heymans et al., 1999) and other related pathologies (Cross, 1997; Baish and Jain 2000 for reviews), including alterations in trabecular bone patterns associated with inflammatory arthritis (Caldwell et al., 1998). Fractal analysis has been applied in tracking the directional migration of epithelial cells (Wick et al., 2003) and cellular growth patterns (Sedivy et al., 2002). Cytoskeletal changes in endothelial cells resulting from chemical stress (DeMeester et al., 1998) and in cardiac myocytes due to increased volume load (Thomason et al., 1996) have been quantified and characterized by determination of the fractal dimension of these systems.

We have applied the fractal dimension in combination with image analysis to quantify the changes in the actin cytoskeleton of cardiac fibroblasts responding to equibiaxial mechanical stretch (Fuseler et al., 2007). The actin cytoskeleton in cardiac fibroblasts undergoes rapid and dramatic changes in response to equibiaxial mechanical stretch as indicated by significant changes in the descriptors, of the fractal dimension (**D**) and IOD associated with the f-actin cytoskeleton and cytoplasmic g-actin (Fuseler et al., 2007). During the first 2 min following application of mechanical stretch the actin cytoskeleton undergoes a rapid disorganization or significant increase in overall complexity or chaos as indicated by a significant increase in fractal **D** values. Concurrent with this increase in complexity or **D**, there is a rapid decrease in IOD of the f-actin fluorescence and simultaneous increase in g-actin IOD. Since IOD reflects a measure of the mass of the fluorescent signal and rhodamine-phalloidin binds only to f-actin, the decrease in IOD represent significant depolymerization, fragmentation or erosion of the f-actin filaments without replacement. Similarly, since DNase-1-A488 binds only to g-actin, the concurrent increase in g-actin IOD indicates a significant increase in the pool of cytoplasmic g-actin presumably from the depolymerization of the f-actin filaments in response to mechanical stretch (Fuseler et al., 2007). Ultrastructural observations of cardiac fibroblasts responding to mechanical stretch are consistent with the **D** values and IOD results observed at the light level. These observations further support the contention that cytoplasmic f-actin stress fibers bundles and long, intact actin cytoplasmic filaments undergo rapid fragmentation and depolymerization as immediate responses to mechanical stretch.

The actin cytoskeleton appears to initiate some reorganization In the presence of sustained equibiaxial stretch. Between 20 and 40 min of stretch, the actin cytoskeleton exhibits an increase in IOD of the f-actin fluorescence and decrease in **D** values for the entire structure. The increase in IOD associated with f-actin is indicative of the formation or polymerization of new f-actin filaments. The concurrent decreases in **D** values further indicate that the newly forming f-actin filaments possess a more regular organization and are less chaotically arranged than those seen at the earlier time points.

This response of cardiac fibroblasts to mechanical stress may be considered a dynamic, adaptation to altered environmental conditions induced by mechanical stress to maintain a viable cell (Thomason et al., 1996). The complex changes associated with the actin cytoskeleton described by qualitative observations and quantified by fractal and optical image analysis may reflect adaptive restructuring and not necessarily disruption associated with pathology or elevated chaos.

At present, application of the fractal dimension has not been applied to the analysis of morphological rearrangements of cellular organelles undergoing activation in response to changes in environmental condition. In this study, we apply image analysis and non-Euclidian geometry (fractal dimensional analysis), using the box-counting method of fractal analysis to mathematically describe the changes induced in the Rac-1 cytoplasmic array in cardiac fibroblasts in response to mechanical stress. Values of $\mathbf{D_b}$ for describing the changes in organization of the Rac-1 cytoplasmic array in response to mechanical stretch were determined using HarFA software (Nezadal et al., 2001. [http://www.fch.vutbr.cz/lectures/imagesci]).

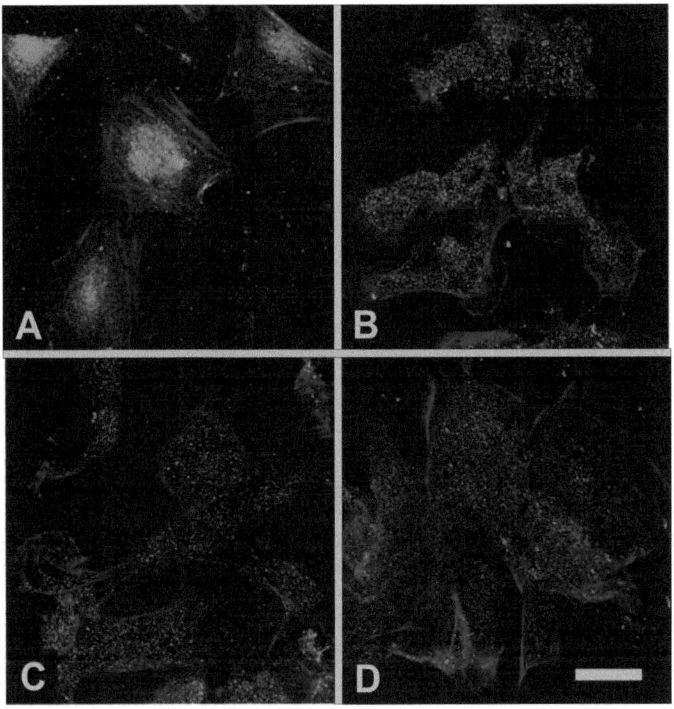

Fig. 2.4 Morphological changes of Rac-1 in cardiac fibroblasts induced by 5% equibiaxial stretch. The bright punctate staining represents Rac-1 (green in the original color images) and the lighter, filamentous staining indicates filamentous actin (red in the original color images). (**a**) Control. No stretch. Rac-1 appears as small granules predominately localized in the perinuclear cytoplasm. (**b**) 1.25 min of stretch. The small granules of Rac-1 have dispersed throughout the cytoplasm. (**c**) 2.5 min of stretch. The granules of Rac-1 remain small, pronounced, and dispersed through the cell including the cellular margins. (**d**) 20 min of stretch. The Rac-1 granules remain small, and dispersed throughout all regions of the cytoplasm. The weak *green fluorescence* of the small Rac1 granules was enhanced by altering the color balance of the images using color balance subroutine of Corel Photo Paint 3X. Scale Bar = 50 μm

In quiescent, control cardiac fibroblasts, the Rac-1 organelle array was characterized by numerous small granules (gA = $1.25 \pm 0.19\ \mu m^2$) which were predominately localized in clusters in the perinuclear cytoplasm, with only a few granules being localized in the peripheral or marginal cytoplasm (Fig. 2.4a).

In response to stretch, Rac-1 granules rapidly translocated from the perinuclear region and assumed an apparent uniform distribution throughout the cytoplasm including the peripheral regions (Fig. 2.4b, 1.25 min of stretch). During this time period, concurrent with translocation, Rac-1 granules decreased in size measured as mean area of the granules (Fig. 2.5) and underwent a rapid and significant decrease in IOD (Fig. 2.6). Interestingly, the $\mathbf{D_b}$ value for Rac-1 during this time period (1–2 min) remained high and unchanged. Once the Rac-1 granules attained minimal size between 2 and 5 min of stretch, $\mathbf{D_b}$ decreased indicating the smaller,

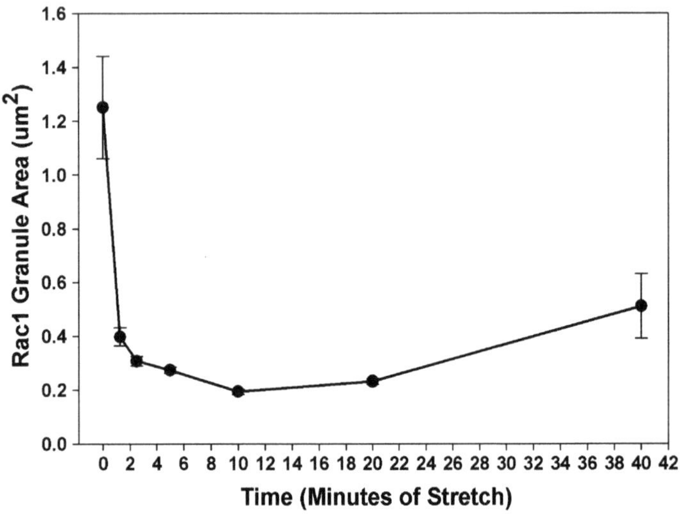

Fig. 2.5 Response of Rac-1 morphology to mechanical stretch. Rac1 granules undergo a significant decrease in size which in 1–2 min following the application of 5% equibiaxial stretch. This decrease in granules size is concurrent with the activation of Rac1 and translocation to associate with the activation of NAD(P)H oxidase. The Rac1 granules remain small and relatively constant in mean size (area) for the duration of the application of stretch (40 min)

activated Rac-1 particles have become more organized (Fig. 2.6). In response to prolonged stretch of up to 40 min, the activated Rac-1 GTPase array showed no further changes in morphology or distribution (Figs. 2.4c–d, and 2.6). However, the organized array of Rac-1 granules seen earlier underwent progressive disorganization and an increase in randomness as revealed by the large increase in $\mathbf{D_b}$ during these later time periods (Fig. 2.6).

As observed upon activation, Rac-1 underwent a rapid and significant decrease in granule size (Fig. 2.5). This response has also been seen with RhoA and Cdc42 in response to stretch of cardiac fibroblasts (unpublished observations). In all three of the small GTPases, the decrease in granule size appears to occur before or concurrent with onset of granule translocation from the perinuclear cytoplasm to the distal target sites on the plasma membrane and membrane – associated cytoskeleton.

2.7 Morphometric Analysis of Rac-1 in Response to Cellular Activation by Tumor Necrosis Factor-Alpha

Additional experiments were performed to compare changes in Rac-1 seen with mechanical stretch to those induced by the cytokine, tumor necrosis factor-α (TNF-α). Similar changes in the morphological distribution, IOD and $\mathbf{D_b}$ for the Rac-1 cytoplasmic array were observed when cardiac fibroblasts were activated by tumor necrosis factor-alpha as those described above for mechanical stretch

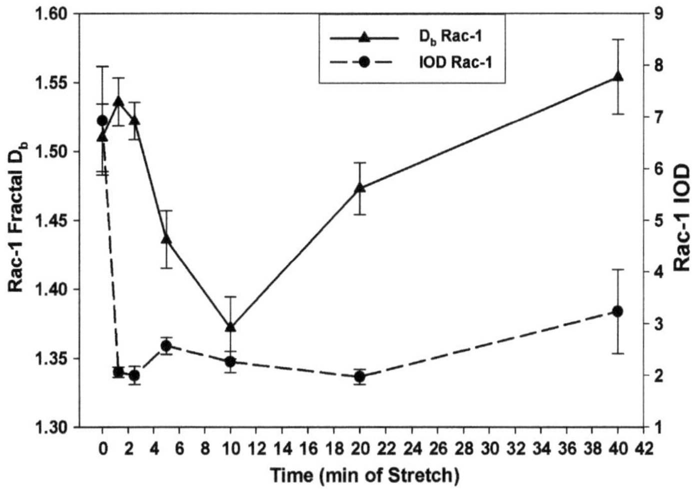

Fig. 2.6 Response of Rac-1 to activation by mechanical stretch: Changes in IOD and fractal dimension (**D$_b$**). Changes in IOD and fractal dimension (**D$_b$**), of Rac-1 granules in response to mechanical stress reveal altered morphology not apparent in cell images. Initially Rac-1 consists of large granules clustered in a disordered array in the perinuclear cytoplasm as indicated by high values of IOD and (**D$_b$**). Application of initial stretch (between 1 and 2 min) results in an immediate and rapid decrease in Rac-1 granule IOD while (**D$_b$**) remains constant, indicating that during activation, Rac-1 is transformed into a population of smaller granules which remain disorganized. In the presence of continued stretch, the IOD remains constant while (**D$_b$**) steadily decreases reaching a minimum at 10 min, indicating the granules remain small but become highly organized. At prolonged stretch beyond 10 min, the IOD remains at a minimum concurrent with an increase in (**D$_b$**), indicating the small Rac-1 granules are characterized by progressive levels of increasing disorder

(TNF-α, 10 ng/ml, Fig. 2.7). Within 1 min of treatment of fibroblasts with TNF-α, the IOD of the Rac-1 cytoplasmic array dropped to a minimum value (between 1 and 2 IOD units). The Rac-1 IOD remained within this range for the duration of the observations (out to 30 min). The Rac-1 granules, in response to TNF-α activation, also underwent a rapid and progressive decrease in **D$_b$** indicating the activated granules are more organized and less space filling than the inactive form which were concentrated in the perinuclear cytoplasm.

Here again, the decrease in size of the Rac-1 granules may facilitate their translocation through the cytoplasm and may also facilitate their association with NAD(P)H oxidase. Similar to the response to mechanical stretch, the TNF-α activated Rac-1 granules underwent an increase in disorganization and increase in chaos between 20 and 30 min as indicated by a significant increase in **D$_b$**. The essential equivalent activation, morphological changes and translocation of Rac-1 both by receptor-mediated cytokine activation of the cell and activation by mechanical deformation of the cell suggest a common pathway in the response to both mechanical and biochemical stress on the cell. The nature this common pathway is uncertain and requires further studies. The common feature of this

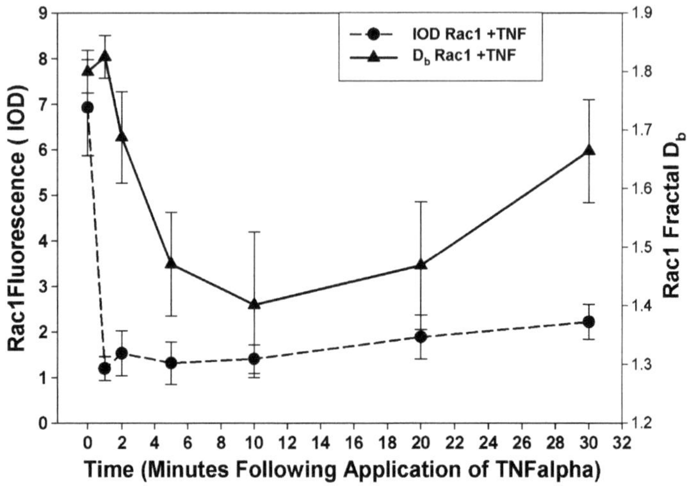

Fig. 2.7 Response of Rac-1 to activation by TNF-α: Changes in IOD and fractal dimension (**D**$_b$). Activation of Rac-1 by TNF-α expresses similar time course of changes in IOD and (**D**$_b$) as those seen when Rac-1 is activated by mechanical stretch. Following treatment with TNF-α there is an immediate (between 1 and 2 min) and rapid decrease in Rac-1 granule IOD which remains relatively constant during prolonged stretch to 30 min. Rac-1 (**D**$_b$) remains constant during the initial 2 min following TNF-α activation then begins to decrease reaching a minimal value by 10 min. This would indicate that during TNF-α activation, Rac-1 is transformed into a population of smaller granules which initially remain disorganized then in the presence of continued stretch, the IOD remains constant while steadily decreases indicating the granules remain small but become highly organized. The activation response of Rac-1 to TNF-α is parallel to that induced by mechanical stretch. In the presence of TNF-α the Rac-1 IOD remains at a minimum concurrent with an increase in (**D**$_b$), indicating that small Rac-1 granules beginning to undergo deactivation are characterized by progressive levels of disorder

proposed pathway is that it must be capable of leading to activation of Rac-1 within 1–2 min following cellular stimulation.

2.8 Activation of NF-κB in Cardiac Fibroblasts by Equibiaxial Stretch and Proinflammatory Cytokines

Production of superoxide radical is generated by the NAD(P)H complex in a Rac-1 dependent manner (Moldovan L et al., 2000, 2006; Ushio-Fukai M et al., 2002). The production of superoxide radical can lead to a change in the redox potential of the cell, activating redox sensitive transcription factors in particular Nuclear Factor-kappa B (NF-κB). In quiescence cardiac fibroblasts (Fig. 2.8a), NF-κB is predominantly present in the cytoplasm with a minimal presence in the nuclei. Following cellular activation by TNF-α, NF-κB undergoes activation and the activated forms of NF-κB (p65 and p50 subunits) measured by fluorescence IOD appears maximally in the nucleus between 25 and 30 min (Fig. 2.8b) (Fuseler et al., 2006).

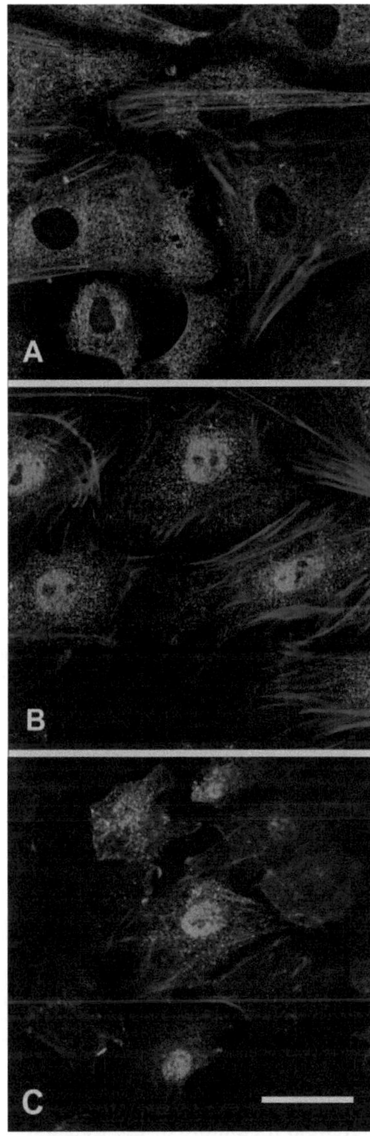

Fig. 2.8 Activation of nuclear factor – kappa B (NF-κB) by equibiaxial stretch and TNF-α in cardiac fibroblasts. The bright, granular staining indicates NF–kB (green in the original images) and the filamentous staining indicates the actin cytoskeleton (red in the original image). (**a**) Quiescent cardiac fibroblasts. NF-κB (*green fluorescence*) is predominantly localized in the cytoplasm with minimal presence in the nuclei. Actin cytoskeleton (*red fluorescence*). (**b**) Cardiac fibroblasts activated with TNF-α (10 ng/ml) for 30 min. NF-κB has been activated and translocated into the nuclei of most of the cells. (**c**) Cardiac fibroblasts activated by 5% equibiaxial stretch for 30 min. NF-κB has been activated and translocated into the nuclei in the majority of the cells. Scale bar = 50 μm

The activation and translocation of NF-κB mediated by cytokines and other biological agents is characterized by a change in the redox potential of the cell. Following mechanical stretch of cardiac fibroblasts, NF-κB is activated and translocates into the nucleus. The time of activation and nuclear translocation of NF-κB in response to mechanical stretch is the same as seen following cellular activation by TNF-α. This suggests that mechanical stretch can activate and induce nuclear translocation presumably by changing the redox potential of the cell following the

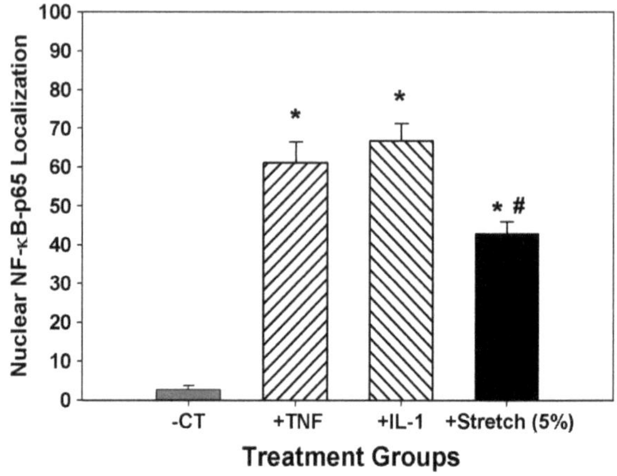

Fig. 2.9 Quantification of the magnitude of the nuclear translocation of NF-κB in cardiac fibroblasts activated by 5% equibiaxial stretch and the proinflammatory cytokines, TNF-α and IL-1β. Nuclear NF-κB localization is expressed as {[nuclear NF-κB-p65 IOD/cytoplasmic NF-κB-p65 IOD] ×100} to normalize the values for the shape and size of the cells and their nuclei. In quiescencecardiac fibroblasts (-CT) the presence of NF-κB in the nucleus is minimal. Following activation by TNF-α (10 ng/ml) or IL-1β (10 ng/ml) NF-κB-p65 reaches a maximal value in the nucleus. In cardiac fibroblasts subjected to 5% equibiaxial stretch for 30 min, NF-κB-p65 reaches a maximal values in the nucleus which is significantly greater than the that in the quiescence cells but ~70% of that seen in the cells maximally activated by cytokines. * = significant]difference ($P < 0.05$) from quiescence cells (negative controls, -CT), # = significant difference ($P < 0.05$) from TNF-α and IL-1β treated groups (maximal activation)

activation of Rac-1 leading to the generation of superoxide free radical by NAD(P)H oxidase. The efficiency of the activation of NF-κB by mechanical stretch in cardiac fibroblasts is less than the maximal activation induced by the proinflammatory cytokines, TNF-α and IL-1β (Fig. 2.9).

2.9 Conclusions and Perspectives

The Rho family of small GTPases has received significant attention recently as potential mediators of the transduction of mechanical stimuli into alterations in cell phenotype and gene expression. The principle site of action of activated Rac-1 is its association with NAD(P)H oxidase complex located on the cell membrane. Rac-1 is essential for the activation of NAD(P)H oxidase and the production of superoxide radicals. The production of superoxide radicals can lead to alterations in the redox status of the cell and also can induce or regulate actin-mediated cellular motility. Altered redox status of the cell can further lead to activation of various nuclear transcription factors, especially NF-κB. These in turn can induce expression and upregulation of their dependent gene products leading to phenotypic transformation of the cardiac fibroblasts which may lead to a pathological state.

The present studies suggest that for Rac-1 to interact with the subunits comprising membrane localized NAD(P)H oxidase, Rac-1 granules must be significantly reduced in size before becoming more widely distributed (disordered) throughout the cell. Further studies will be required to address the functional significance of this decrease in Rac-1 granule size. However, the significant decrease in the size of the granules, which occurs during activation may serve to facilitate their translocation through the cytoplasm and ability to associate with the NAD(P)H oxidase complex. Further studies will also be required to determine the functional role of Rac-1 activation and translocation on alterations in fibroblast morphology and gene expression in response to mechanical stretch.

References

Abe M, Ho C-H, Kamm KE, Grinnell F (2003) Different molecular motors mediate platlet-derived growth factor and lysophosphatidic acid-stimulated floating collagen matrix contraction. J Biol Chem 278: 47707–47712.

Aikawa R, Nagai T, Tanaka M, Zou Y, Ishihara T, Takano H, Hasegawa H, Akazawa H, Mizukami M, Nagai R, Komuro I (2001) Reactive oxygen species in mechanical stress-induced cardiac hypertrophy. Biochem Biophys Res Commun 289: 901–907.

Anderson JC, Babb AL, Hlastala MP (2005) A fractal analysis of the radial distribution of bronchial capillaries around large airways. J Appl Physiol 98: 850–855.

Atance J, Yost MJ, Carver W (2004) Influences of the extracellular matrix on the regulation of cardiac fibroblast behavior by mechanical stretch. J Cell Physiol 200: 377–386.

Baish JW, Jain RK (2000) Fractals and cancer. Cancer Res 60: 3683–3688.

Bassett CAL, Hermann L (1961) Influence of oxygen concentration and mechanical factors on differentiation of connective tissue in vitro. Nature 193: 460–461.

Behar TN (2001) Analysis of fractal dimension of O2A glial cells differentiating in vitro. Methods 24: 331–339.

Bernard F, Bossu JL, Gaillard S (2001) Identification of living oligodendrocyte developmental stages by fractal analysis of cell morphology. J Neurosci Res 65: 439–445.

Bershadsky A, Kozlov M, Geiger B (2006) Adhesion-mediated mechanosensitivity: a time to experiment, and a time to theorize. Curr Opin Cell Biol 18: 472–481.

Borg TK, Caulfield JB (1981) The collagen matrix of the heart. Fed Proc 40: 2037–2041.

Borg TK, Rubin K, Lundgren E, Borg K, Obrink B (1984) Recognition of extracellular matrix components by neonatal and adult cardiac myocytes. Dev Biol 104: 86–96.

Borodinsky LN, Fiszman ML (2001) A single-cell model to study changes in neuronal fractal dimension. Methods 24: 341–345.

Brown JH, Del Re DP, Sussman MA (2006) The Rac and Rho hall of fame: a decade of hypertrophic signaling hits. Circ Res 98: 730–742.

Butt RP, Laurent GJ, Bishop JE (1995) Collagen production and replication by cardiac fibroblasts is enhanced in response to diverse classes of growth factors. Eur J Cell Biol 68: 330–335.

Caldwell CB, Moran EL, Bogoch ER (1998) Fractal dimension as a measure of altered trabecular bone in experimental inflammatory arthritis. J Bone Miner Res 13: 978–985.

Camelliti P, Borg TK, Kohl P (2005) Structural and functional characterization of cardiac fibroblasts. Cardiovasc Res 65: 40–51.

Carver W, Nagpal ML, Nachtigal M, Borg TK, Terracio L (1991) Collagen expression in mechanically stimulated cardiac fibroblasts. Circ Res 69: 116–122.

Carver W, Terracio L. Borg TK (1993) Expression and accumulation of interstitial collagen in the neonatal rat heart. Anat Rec 511–520.

Caserta F, Eldred WD, Fernandez E, Hausman RE, Stanford LR, Bulderev SV, Schwarzer S, Stanley HE (1995) Determination of fractal dimension of physiologically characterized neurons in two and three dimensions. J Neurosci Methods 56: 133–144.
Catalucci D, Latronico MV, Ellingsen O, Condorelli G (2008) Physiological myocardial hypertrophy: how and why? Front Biosci 13: 312–324.
Cross SS (1997) Fractals in pathology. J Pathol 182: 1–8.
DeMeester SL, Cobb JP, Hotchkiss RS, Osborne DF, Karl IE, Tinsley KW, Buchman TG (1998). Stress-induced fractal rearrangement of the endothelial cell cytoskeleton causes apoptosis. Surgery 124: 362–371.
Fuseler JW, Merrill DM, Rogers JA, Grisham MB, Wolf RE (2006) Analysis and quantitation of NF-kappa B nuclear translocation in tumor necrosis factor alpha (TNF-alpha) activated vascular endothelial cells. Microsc Microanal 12: 269–276
Fuseler JW, Millette CF, Davis JM, Carver W (2007) Fractal and image analysis of morphological changes in the actin cytoskeleton of neonatal cardiac fibroblasts in response to mechanical stretch. Microsc Microanal 13: 133–143.
Gazit Y, Berk DA, Leunig M, Baxter LT, Jain, R K (1995) Scale-invariant behavior and vascular network formation in normal and tumor tissue. Phys Rev Lett 75: 2428–2431.
Gazit Y, Baish JW, Safabakhsh N, Leunig M, Baxter LT, Jain RK (1997) Fractal characteristics of tumor vascular architecture during tumor growth and regression. Microcirculation 4: 395–402.
Geiger B, Bershadsky A, Pankov R, Yamada KM (2001) Transmembrane crosstalk between the extracellular matrix- cytoskeleton crosstalk. Nat Rev Mol Cell Biol 2: 793–805.
Glucksmann A (1939) Studies on bone mechanics in vitro: II. The role of tension and pressure in chondrogenesis. Anat Rec 73: 39–56.
Goldsmith EC, Hoffman A, Morales MO, Potts JD, Price RL, McFadden A, Rice M, Borg TK (2004) Organization of fibroblasts in the heart. Develop Dyn 230: 787–794.
Grinnell F (2000) Fibroblast-collagen matrix contraction: growth factor signaling and mechanical loading. TRENDS Cell Biol 10: 362–365.
Grinnell F (2003) Fibroblast biology in three-dimensional collagen matrices. TRENDS Cell Biol 13: 264–269.
Grizzi F, Russo C, Colombo P, Franceschini B, Frezza EE, Cobos E, Chiriva-Internati M (2005) Quantitative evaluation and modeling of two-dimensional neovascular network complexity: the surface fractal dimension. BMC Cancer 5: 14–23.
Hall A. (1998) Rho GTPases and the actin cytoskeleton. Science 279: 509–514.
Heymans O, Blacher S, Brouers F, Pierard GE (1999) Fractal quantification of the microvasculature heterogeneity in cutaneous melanoma. Dermatology 198: 212–217.
Hu E, Lee D (2003) Rho kinase inhibitors as potential therapeutic agents for cardiovascular diseases. Curr Opin Investig Drugs 4: 1065–1075.
Hung CT, Williams JL (1994) A method for inducing equibiaxial and uniform strains in elastomeric membranes used as cell substrates. J Biomechanics 27: 227–232.
Janmey PA, Weitz DA (2004) Dealing with mechanics: mechanisms of force transduction in cells. Trends Biochem Sci 29: 364–370.
Kawamura S, Miyamoto S, Brown JH (2003) Initiation and transduction of stretch-induced RhoA and Rac1 activation through caveolae: cytoskeletal regulation of ERK translocation. J Biol Chem 278: 31111–31117.
Lammerding J, Kamm RD, Lee RT (2004) Mechanotransduction in cardiac myocytes. Ann NY Acad Sci 1015: 53–70.
Le Noble F, Moyon D, Pardanaud L, Yuan L, Djonov V, Matthijsen R, Breant C, Fleury V, Eichman A (2004) Flow regulates arterial-venous differentiation in the chick embryo yolk sac. Development 131: 361–375.
Lee AA, Delhaas T, McCulloch AD, Villarreal FJ (1999) Differential responses of adult cardiac fibroblasts to in vitro biaxial strain patterns. J Mol Cell Cardiol 31: 1833–1843.
Lee AA, Delhaas T, Waldman LK, MacKenna DA, Villarreal FJ, McCulloch AD (1996) An equibiaxial strain system for cultured cells. Am J Physiol 271: C1400–C1408.

Lee DJ, Ho C-H, Grinnell F (2003) LPA-stimulated fibroblast contraction of floating collagen matrices does not require Rho kinase activity or retraction of fibroblast extensions. Expt Cell Res 289: 86–94.

Leung DY, Glagov S and Matthews MB (1976) Cyclic stretching stimulates synthesis of matrix components by arterial smooth muscle cells in vitro. Science 191: 475–477.

Leung DYM, Glagof S, Matthews MB (1977) A new in vitro system for studying cell response to mechanical stimulation. Expt Cell Res 109: 285–298.

Lucitti JL, Jones EA, Huang C, Chen J, Fraser SE, Dickinson ME (2007) Vascular remodeling of the mouse yolk sac requires hemodynamic force. Development 134: 3317–3326.

MacKenna DA, Dolfi F, Vuori K, Ruoslahti E (1998) Extracellular signal-regulated kinase and c-Jun NH2-terminal kinase activation by mechanical stretch is integrin-dependent and matrix-specific in rat cardiac fibroblasts. J Clin Invest 101: 301–310.

Manser E. (2005) Rho GTPases in the organization of the actin cytoskeleton.In: Symons M, (ed.) Rho GTPases, New York: Kluwer Academic/Plenum Publishers, pp. 107–114.

Matthews BD, Overby DR, Mannix R, Ingber DE (2006) Cellular adaptation to mechanical stress: role of integrins, Rho, cytoskeletal tension and mechanosensitive ion channels. J Cell Sci 119: 508–518.

Michaelson D, Rush M, Philips MR (2005) Intracellular targeting of Rho family GTPases: implications of localization on function. In: Symons M, (ed.) Rho GTPases, New York: Kluwer Academic/Plenm Publishers, pp. 17–31.

Moldovan L, Moldovan NI, Sohn RH, Parikh SA, Goldschmidt-Clermont PJ (2000) Redox changes of cultured endothelial cells and actin dynamics. Circ Res 86: 549–557.

Moldovan L, Mythreye K, Goldschmidt-Clermont PJ, Satterwhite LL (2006) Reactive oxygen species in vascular endothelial cell motility. Roles of NAD(P)H oxidase and Rac1. Cardiovasc Res 71: 236–246.

Nezadal M, Zemeskal O, Buchnicek M (2001) The box-counting: critical study, 4th conference on prediction, synergetic and more... The Faculty of Management, Institute of Information Technologies, Faculty of Technology, Tomas Bata University in Zlin, October 25–26. p18. HarFA software (http://www.fch.vutbr.cz/lectures/imagesci).

Pan J, Singh US, Takahashi T, Oka Y, Palm-Leis A, Herbelin BS, Baker KM. (2005) PKC mediates cyclic stretch-induced cardiac hypertrophy through Rho family GTPases and mitogen-activated protein kinases in cardiomyocytes. J Cell Physiol 202: 536–553.

Perona R, Montaner S, Saniger L, Sanchez-Perez I, Bravo R, Lacal JC (1997) Activation of the nuclear factor-kappaB by Rho, CDC42, and Rac-1 proteins. Genes Dev 11: 463–475.

Porter KE, Turner NA, O'regan DJ, Balmforth AJ, Ball SG (2004) Simvastatin reduces human atrial myofibroblast proliferation independently of cholesterol lowering via inhibition of RhoA. Cardiovasc Res 61: 745–755.

Reithmacher D, Brinkmann V, Birchmeier C (1995) A targeted mutation in the mouse e-cadherin gene results in defective preimplantation development. Proc Natl Acad Sci USA 92: 855–859.

Ren J, Fang CX (2005) Small guanine nucleotide-binding protein Rho and myocardial function. Acta Pharmacol Sin 26: 279–285.

Ristanovic D, Nedeljkov V, Stefanovic BD, Milosevic NT, Grgurevic M, Stulic V (2002) Fractal and nonfractal analysis of cell images: comparison and application to neuronal dendritic arborization. Biol Cybern 87: 278–288.

Rodan GA, Mensi T, Harvey A (1975) A quantitative method for application of compressive forces to bone in tissue culture. Calcified Tissue Res 18: 125–131.

Rogers JA, Fuseler JW (2007) Regulation of NF-kappa B activation and nuclear translocation by exogenous nitric oxide (NO) donors in TNF-alpha activated vascular endothelial cells. Nitric Oxide 16: 379–391.

Sanchez-Esteban J, Wang Y, Filardo EJ, Rubin LP, Ingber DE (2006) Integrins beta1, alpha6, and alpha3 contribute to mechanical strain-induced differentiation of fetal lung type II epithelial cells via distinct mechanisms. Am J Physiol Lung Cell Mol Physiol 290: L343–L350.

Satoh S, Ueda Y, Koyanagi M, Kadokami T, Sugano M, Yoshikawa Y, Makino N (2003) Chronic inhibition of Rho kinase blunts the process of left ventricular hypertrophy leading to cardiac contractile dysfunction in hypertension-induced heat failure. J Mol Cell Cardiol 35: 59–70.

Schaffer JL, Rizen M, L'Italien GJ, Benbrahim A, Megerman J, Gerstenfeld LC, Gray ML (1994) Device for the application of biaxially uniform and isotropic strain to a flexible cell culture membrane. J Orthop Res 12: 709–719.

Schmidt A, Hall MN. (1998) Signalling to the actin cytoskeleton. Annu Rev Cell Dev Biol 14: 305–338.

Sedivy R, Thurner S, Budinsky AC, Kostler WJ, Zielinski CC (2002) Short-term rhythmic proliferation of human breast cancer cell lines: surface effects and fractal growth patterns. J Pathol 197: 163–169.

Sulciner DJ, Irani K, Yu ZX, Ferrans VJ, Goldschmidt-Clermont P, Finkel T (1996) Rac1 regulates a cytokine-stimulated, redox-dependent pathway necessary for NF-kappaB activation. Mol Cell Biol 16: 7115–7121.

Suematsu N, Satoh S, Kinugawa S, Tsutsui H, Hayashidani S, Nakamura R, Egashira K, Makino N, Takeshita A (2001) Alpha 1-adrenoceptor-Gq-RhoA signaling is upregulated to increase myofribillar Ca^{+2} sensitivity in failing hearts. Am J Physiol Heart Circ Physiol 281: H637–H646.

Thomason DB, Anderson O 3rd, Menon V (1996) Fractal analysis of cytoskeleton rearrangement in cardiac muscle during head-down tilt. J Appl Physiol 81: 1522–1527.

Ushio-Fukai M, Tang Y, Fukai T, Dikalov SI, Ma Y, Fujimoto M, Quinn MT, Pagano PJ, Johnson C, Alexander RW (2002) Novel role of gp91(phox)-containing NAD(P)H oxidase in vascular endothelial growth factor-induced signaling and angiogenesis. Circ Res 91: 1160–1167.

Vasiliev JM (1991) Polarization of pseudopodial activities: cytoskeletal mechanisms. J Cell Sci 98: 1–4.

Walter Jr., RJ, Berns MW (1986) Digital image processing and analysis. In Inoue S, (ed.) Video Microscopy, New York and London: Plenum Press, pp. 327–392.

Wei L, Roberts W, Wang L, Yamada M, Zhang S, Zhao Z, Rivkees SA, Schwartz RJ, Imanaka-Yoshida K (2001) Rho kinases play an obligatory role in vertebrate embryonic organogenesis. Development 128: 2953–2962.

Wick N, Thurner S, Paiha K, Sedivy R, Vietor I, Huber LA (2003) Quantitative measurement of cell migration using time-lapse videomicroscopy and non-linear system analysis. Histochem Cell Biol 119: 15–20.

Yeh C, Rodan GA (1984) Tensile forces enhance PGE synthesis in osteoblasts grown on collagen ribbon. Calcif Tissue Int 36: 67–71.

Yoshigi M, Clark EB, Yost HJ (2003) Quantification of stretch-induced cytoskeletal remodeling in vascular endothelial cells by image processing. Cytometry A 55: 109–118.

Yost MJ, Simpson D, Wrona K, Ridley S, Ploehn HJ, Borg TK, Terracio L (2000) Design and construction of a uniaxial stretcher. Am J Physiol Heart Circ Physiol 279: H3124–H3130.

Yuan X-B, Jin M, Xu X, Song Y-Q, Wu C-P, Poo M-M, Duan S (2003) Signalling and crosstalk of Rho GTPases in mediating axon guidance. Nature Cell Biol 5: 38–43.

Zhang L, Liu JZ, Dean D, Sahgal V, Yue GH (2005) A three dimensional fractal analysis method for quantifying white matter structure in human brain. J Neurosci Methods 150: 242–253.

Chapter 3
Molecular Signaling Mechanisms of Myocardial Stretch: Implications for Heart Disease

Hind Lal, Suresh K. Verma, Honey B. Golden, Donald M. Foster, April M. Holt, and David E. Dostal

Abstract With every heartbeat, myocardial cells are subjected to substantial mechanical stretch. Stretch is a potent stimulus for growth, differentiation, migration, remodeling and gene expression. Mechanical load is a major cause of cardiac hypertrophy. Since the initial observation of stretch-induced growth, our understanding of this complex field has been steadily growing, but remains incomplete. The mechanisms by which myocardial cells convert mechanical stimuli into biochemical signals that result in physiologic and pathological changes remain to be completely understood. Integrins, caveolae and focal adhesions have been shown to have important mechanosensing roles in cardiac myocytes. Downstream effectors activated by mechanosensors include guanine-nucleotide binding proteins (G-proteins), mitogen-activated protein (MAP) kinases, Janus-associated kinase/signal transducers and activators of transcription (JAK/Stat), protein kinase C (PKC) and protein kinase B/Akt pathways. Multiple levels of crosstalk exist between these pathways. Early studies have implicated most of these pathways in cardiac injury and growth response, however, more recent advancements in the development of kinase-specific inhibitors and genetically-engineered animal models have revealed significant new insights. Recent studies suggest that acute mechanical stretch activates protective pathways including c-jun N-terminal kinase (JNK) and Akt as a tolerance response, rather than injury-related signaling cascades such as p38 MAP kinase. However, chronic stretch/mechanical load creates an imbalance that favors the injury related pathway by an unknown mechanism in the myocardium. The following chapter provides an overview of the fundamental processes of stretch-activated mechano-signaling in myocardial cells, and recent advances in our understanding of this increasingly important field.

D. E. Dostal (✉)
Division of Molecular Cardiology, College of Medicine, Scott & White, Cardiovascular Research Institute, The Texas A&M University System Health Science Center, Central Texas Veterans Health Care System, Temple, TX, USA
e-mail: ddostal@medicine.tamhsc.edu

A. Kamkin, I. Kiseleva (eds.), *Mechanosensitivity of the Heart*,
Mechanosensitivity in Cells and Tissues 3, DOI 10.1007/978-90-481-2850-1_3,

Keywords Mechanotransduction · Extracellular matrix · Integrins · Angiotensin · Focal adhesion kinase · Integrin-linked kinase · Rho family · Protein kinase C · Transduction cascades

3.1 Introduction

We describe herein an organ system, the heart, whose principle product is to deliver a physical force, i.e., the force of contraction. Normal cardiac function begins with a mechanical event which becomes converted into chemical events which in turn are converted back into a mechanical event. As such, the heart is an organ that is highly organized to sense and respond to physical forces in every aspect of its activities under normal conditions. When special conditions arise through mechanical overload or other factors, this highly sensitive organ must respond with adaptive processes which can sometimes become maladaptive, leading to a variety of disease states including hypertrophy. Below we describe important aspects of the structural and mechanosensing systems in the myocardium and how these relate to activation of signal transduction systems and cardiac function.

3.1.1 Extracellular Matrix and Mechanotransduction

The extracellular matrix (ECM) is vital for the transmission and sensing of mechanical forces across the myocardium. The transmission of mechanical force is one of the principal functions of the connective tissue network in the heart. The organization of this network has been well described in normal hearts and in a variety of cardiac disease models (Bishop and Laurent, 1995). Myocytes and the other cardiac cell types are connected by an intricate ECM lattice, which provides structural integrity to the tissue and a means for optimal vectoral transmission of force (Baicu et al., 2003). This intimate association with the ECM enables cardiac cells to detect and respond to changes in mechanical, chemical and electrical signals within the myocardium, thus facilitating the myocardial response to changes in cardiac workload. The ECM lattice is composed of a complex network of structural proteins (collagen and elastic fibres) and adhesive proteins (fibronectin, laminin) (Hein and Schaper, 2001; Jane-Lise et al., 2000), in which the ECM synthesis and turnover is primarily regulated by cardiac fibroblasts (Manabe et al., 2002). Fibrillar collagen types I and III coexist to form the collagenous network of the myocardium comprising 85 and 11%, respectively, of the collagen in healthy mammalian hearts (Linehan et al., 2001). The nonfibrillary collagen types IV, V, VI and VIII are less abundant in the myocardium. These molecules assemble into an open network within basement membranes, rather than as fibrils (Madri et al., 1980). To date, most mechanical signaling studies have been performed using collagen I, thereby neglecting the established role of other forms of collagens in disease progression. Subsequent studies are needed to provide a better understanding of how these less abundant collagens participate in structural support and mechanical signaling in the myocardium.

The adhesive proteins fibronectin, laminin and collagen type IV not only constitute the structural support of the basement membranes, but are involved in signaling and are vital in coupling the myocyte to the collagenous ECM scaffold. These molecules primarily serve as ligands to integrin receptors, which have roles in cellular adhesion and signaling. Pathologic conditions which alter levels of these proteins are likely to significantly affect not only myocyte adhesion to the basement membrane, but cellular signaling and ultimately contractile force.

3.2 Mechanosensors Implicated in Cardiac Pathophysiology

3.2.1 Integrins

Integrins, a class of membrane receptors, are major players in transmitting the mechanical force across the plasma membrane and sensing the mechanical load in cardiac myocytes and fibroblasts (Lal et al., 2007a, 2007b; Ross, 2004). Integrins, together with a number of associated cytoskeletal proteins, connect the contractile apparatus to the extracellular matrix across the plasma membrane and trigger intracellular signaling pathways that regulate cellular function (Goldsmith et al., 2004). These molecules trigger a coordinated downstream signaling cascade involving proteins that execute the biochemical programs leading to cardiac hypertrophy and myocardial remodeling. Because integrins mediate key steps in the pathogenesis of several disease states including heart disease and stroke (Lal et al., 2007a), there has been considerable interest in understanding how integrins couple to signal transduction systems and integrate with other receptor systems. However, a complete understanding of underlying molecular mechanism and downstream signaling cascades remain to be realized.

3.2.1.1 Structure and Function of Integrins

Integrins are the main receptors for extracellular matrix proteins like collagen, fibronectin and laminin. These molecules are heterodimeric receptors comprised of noncovalently associated α and ß subunits. The human integrin family now includes at least 18 known α-subunits and eight known ß-subunits. A given α-subunit may interact with more than one ß-subunit, resulting in 24 different heterodimers identified to date. The specificity of integrin signaling is made possible by α and β subunits that form the heterodimeric pair. The α subunit generally confers ligand specificity

(Hynes, 2002b), whereas the β subunit interacts with the cytoplasmic environment. Some α-subunits contain an inserted I-domain, which is a major ligand-binding site (Takagi and Springer, 2002). The integrin α/β heterodimer is believed to convert between the bent structure of the resting state and the extended arrangement of the activated state. Ligand binding to the extracellular integrin domain induces conformational changes and integrin clustering for activation of signaling cascades and recruitment of multiprotein complexes to focal adhesions (Schoenwaelder and Burridge, 1999). The ability of integrins to dynamically change their molecular conformation in response to intracellular signals (inside-out) enables them to modify their ligand binding affinity level from low to high, and vice-versa (Mould and Humphries, 2004). Integrin–ligand interactions are also dependent on divalent cations such as Mg^{2+} or Mn^{2+}, the latter of which stabilizes a high-affinity conformation. By contrast, Ca^{2+} is inhibitory for these interactions and stabilizes a low-affinity conformation. The existence of multiple affinity states of integrins predicts the existence of multiple ligand-activated signaling states. Some conformational states are likely to be more suited to a transient adhesion (e.g. cell migration), others to stable adhesion (Lu et al., 2004).

3.2.1.2 Integrin Expression in the Myocardium

Integrin function is required for proper cardiac development and myocyte attachment to extracellular matrix, growth and viability (Valencik et al., 2002). Integrin-dependent pathways also mediate hypertrophic responses to mechanical stimuli associated with cardiac myocyte strain (Aikawa et al., 2002; Lal et al., 2007b) and are required for stimulation of hypertrophy by phenylephrine (PE) or endothelin-1 (ET-1) (Heidkamp et al., 2002; Ross, 2002). Cardiac myocytes express integrins α_1, α_3, α_5, α_6, α_7, α_9, α_{10}, β_1, β_3, and β_5 (Ross, 2004), many of which are regulated by developmental and pathological stimuli (Ross and Borg, 2001). In the embryonic heart, cardiac myocytes primarily express $\alpha_5\beta_1$ (fibronectin receptor) and the $a_6\beta_1$ (laminin receptor), (Brancaccio et al., 1998; van der Flier et al., 1997), whereas $\alpha_1\beta_1$ (collagen IV receptor) and $\alpha_7\beta_1$ (laminin receptor) are the major integrins expressed in the neonate and adult mycardium, respectively. The primary β integrin subunit found in myocytes is β_1. Different splice variants are expressed in the embryonic (β_{1A}) and adult myocytes (β_{1D}) (Pham et al., 2000), which differ in specific amino acid sequences at the cytoplasmic domain and their interaction with cytoskeletal and signaling molecules (Belkin et al., 1997). Cardiac fibroblasts express integrins α_1, α_2, α_3, α_5, α_8, α_{10}, β_1, β_3 and β_5 (Burgess et al., 1994; Kawano et al., 2000; Stawowy et al., 2005). Integrin subunits including α_1, α_4, α_5, αv, β_1 and β_3 have been directly implicated in the cardiac pathophysiology (Ren et al., 2007; Shai et al., 2002; Valencik and McDonald, 2001; Valencik et al., 2006). Angiotensin II (Ang II) and other growth factors stimulate cardiac fibroblast contraction and adhesion via β_1 and $\alpha_v\beta_3$ integrins, which involve inside-to-outside signaling mechanisms (Burgess et al., 1994; Kawano et al., 2000; Stawowy et al., 2005).

3.2.1.3 Integrins as Mechanotransducers

Mechanical load applied to integrin ligands (ECM) triggers the assembly and growth of focal contacts (Bershadsky et al., 2003) and activation of several downstream second messenger systems, including Rho GTPases, serine/threonine kinases, phosphatases, MAP kinases, Akt, and PKC (Fig. 3.1). Stretch-induced conformational

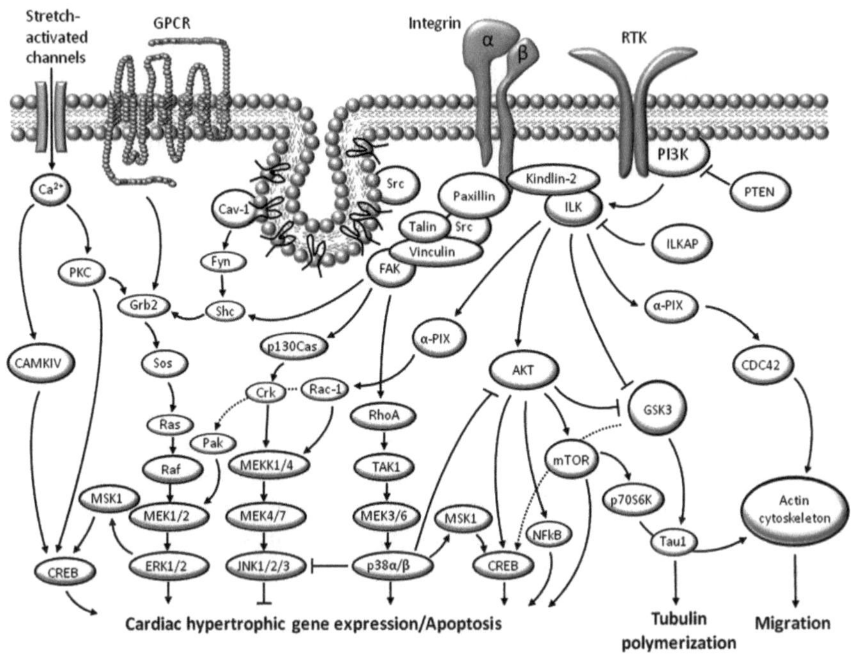

Fig. 3.1 Integrin mediated MAP kinase signaling. Mechanical forces are detected by mechanosensors (integrins, RTKs, stretch-activated channels) which activate signal transduction cascades involving Rho GTPases (RhoA, Rac1, Cdc42), MAP kinases (ERK1/2, JNK, p38) and subsequent transcription factors which regulate the function of cardiac myocytes, fibroblasts. Activation of the Erk cascade is one target for the activated Ras through Raf and MEK1/2. Association of Src-family kinases with FAK potentiates the tyrosine phosphorylation of p130Cas, which leads to activation of the JNK MAP kinase cascade. The p38 pathway can be activated by several integrin dependent upstream signaling molecules or crosstalk such as FAK//Rac1/p38 or FAK/RhoA/p38, which lead to cardiac hypertrophy and apoptosis. Integrin and growth factor receptors can activate Akt and GSK3 in an ILK dependent manner. The activity of ILK is upregulated by PI3K and down regulated by the ILK associated phosphatase (ILKAP). PTEN is a negative regulator of PI3K, thus down-regulating the activities of ILK and PKB/Akt. By stimulating the phosphorylation of Akt at Ser^{473}, ILK stimulates several signaling pathways like mTOR, NFkappa-B and CREB, leading to the cardiac hypertrophic gene expression. PTEN: Posphatase and tensin homologue deleted on chromosome 10; Akt: Akt8 virus oncogene cellular homolog; GSK: Glycogen synthase kinase; mTOR: Mammalian target of rapamycin; NF kappa-B: Nuclear factor kappa-B; CREB: cAMP response element-binding protein; Cav-1: caveolin-1; Sos: Son of sevenless guanine nucleotide exchange factor; Pak: p21-activated kinase; MEK: MAPK/Erk kinase; TAK: TGF beta-activated kinase; ERK: Extracellular signal-regulated kinase; JNK: Jun N-terminal kinase; SAPK: Stress-activated protein kinase; Src: Rous sarcoma oncogene cellular homolog; PI3K: phosphatidylinositol 3-kinase; Shc: SH2-containing collagen-related proteins; Cas: Crk associated substrate; GRB2: Growth factor receptor-bound protein

changes in the ECM may alter integrin structure, resulting in activation of liganded integrin receptors and focal contact-associated secondary messenger pathways in the cell, such as FAK, Src family kinases, Abl and integrin-linked kinase (ILK) (Li et al., 1999; Liu et al., 2000). Interactions with ECM, cytoskeletal and intracellular signaling cascades enables integrins to mediate both "outside-in" and "inside-out" signaling (Hynes, 2002a; Ross, 2004). Binding of integrins to extracellular ligands (ECM) produces intracellular signals (i.e., outside-in signals) such as changes in intracellular signaling events and cytoskeletal reorganization that critically influence cell shape, migration, growth, and survival (Hynes, 2002a). Inside-out signaling occurs when specific intracellular signals impinge on integrin cytoplasmic domains, triggering changes in conformation and ligand-binding affinity in the extracellular domain.

3.2.2 *Angiotensin II Type I (AT_1) As a Mechanotransducer*

Mechanical stress can also induce cardiac signal transduction pathways both in vitro and in vivo through the angiotensin type I receptor (AT_1) without the involvement of Ang II (Zou et al., 2004). This appears to be a unique property of the AT_1, as other myocardial G-protein coupled receptors, such as ET-1 and β-adrenergic receptors are not activated by mechanical stretch (Zou et al., 2004). Candesartan, an inverse agonist which stabilizes the AT_1 receptor in an inactive conformation, suppresses AT_1 activation by both mechanical stress and Ang II. This suggests that mechanical stress may activate the AT_1 by directly changing the conformation of the receptor. The mechanisms responsible for AT_1 activation remain to be investigated. Unlike α-adrenergic stimulation, in which mechanical effects on cardiac myocyte hypertrophy require the β_1-integrin function (Pham et al., 2000), the AT_1 can activate hypertrophic growth pathways independent of β_1 integrin. However, AT_1 can modulate β_1 integrin signaling (Lal et al., 2007b) and β_1 integrin expression (Jia et al., 2003).

3.3 Proximal Effectors of Cardiac Mechanosensing

3.3.1 *Focal Adhesion Kinase (FAK)*

Recent studies indicate that FAK, a 125 kDa non-receptor kinase, is important for transducing mechanical stimuli in isolated cardiac myocytes and in mechanically overloaded myocardium. FAK directly binds to the cytoplasmic tail of β-integrin subunits thereby playing a major role in integrin-mediated signaling (Samarel, 2005). The biological importance of FAK-mediated signal transduction is underscored by its fundamental roles in embryonic development (Ilic et al., 1997), control of cell migration (Owen et al., 1999) and cell cycle progression (Zhao et al., 2003). FAK is an essential kinase, as null mice are embryonically lethal. FAK is highly expressed in cardiac myocytes and undergoes phosphorylation at Tyr^{397}, Tyr^{861} and

Tyr^{925} in response to mechanical loading (Lal et al., 2007b; Samarel, 2005; Torsoni et al., 2003). In neonatal rat ventricular myocytes, FAK is rapidly activated by cyclic stretch and translocated from the perinuclear area to costameres (Torsoni et al., 2003). Consistent with in vitro findings, increased LV pressure causes an increase in tyrosine phosphorylation of FAK, as well as an increased association of FAK with Src and Grb2 (Domingos et al., 2002). Stretch experiments, performed using cardiac myocytes isolated from AT_{1A} receptor knockout mice (Kudoh et al., 1998) and with AT_1 receptor blocker (Torsoni et al., 2003), indicate that mechanical stretch alone is sufficient to activate FAK signaling. Aside from its well-established role in mediating integrin signaling, FAK also participates in signal transduction by G protein – coupled receptors, such as Ang II (Salazar and Rozengurt, 2001), ET-1 (Eble et al., 2000) and phenylephrine (PE) (Taylor et al., 2000). In addition, vascular epidermal growth factor can induce activation and subcellular translocation of FAK from perinuclear sites to focal adhesions in cultured neonatal cardiomyocytes (Takahashi et al., 1999). Thus, mechanical stretch, together with autocrine release of factors activate FAK in cardiac myocytes. Taken together, these data suggest that FAK may function as a converging point in the signaling pathways triggered by integrin, G protein – coupled, and growth factor receptors that are important in the regulation of cardiomyocyte function. The molecular mechanisms by which FAK is activated by mechanical signals require further exploration. FAK activation could result from integrin activation and/or conformational changes due to stretching of the FAK molecule, such as in the case of p130Cas (Sawada et al., 2006) (Fig. 3.1).

Analysis of FAK function using total knockout embryos and in vitro systems, suggests a potential role of FAK in heart development and function. It is established that FAK has important roles in mediating fibroblast migration and differentiation into myofibroblasts. These are key events involved the deleterious remodeling process that occurs along with exaggerated production of connective tissue following cardiac injury (Greenberg et al., 2006; Mimura et al., 2005; Thannickal et al., 2003). In transgenic mice, cardiac specific FAK gene inactivation results in a lethal embryonic phenotype with major defects in the axial mesoderm and cardiovascular system (Furuta et al., 1995; Ilic et al., 1995). The role of FAK in the development of pathologic hypertrophy appears complex. Recently, two contradictory studies using ventricular myocyte-restricted FAK-inactivated transgenic mouse models have been reported (DiMichele et al., 2006; Peng et al., 2006). One study advocates that inactivation of FAK promotes eccentric cardiac hypertrophy (Peng et al., 2006), whereas the other suggests that it attenuates pressure overload-induced hypertrophy (DiMichele et al., 2006). A recent in vivo gene delivery study has demonstrated that targeting FAK with small interfering RNA prevents and reverses load-induced cardiac hypertrophy in mice (Clemente et al., 2007). Although results from human studies are limited, FAK expression has been shown to be elevated in biopsies from patients with mitral regurgitation (Lopes et al., 2007). Thus, it remains unclear as to whether FAK promotes or prevents cardiac hypertrophy. The precise role of FAK in controlling hemodynamic load induced cardiac hypertrophy will be an important issue to resolve due to its potential clinical relevance.

3.3.2 Integrin-Linked Kinase (ILK)

Integrin-linked kinase (ILK) links integrins with the force-generating actin cytoskeleton and is therefore a candidate molecule in the transduction of mechanical signals initiated by altered loading conditions affecting the heart. ILK is a protein serine/threonine (Ser/Thr) kinase that binds to the cytoplasmic domains of β_1, β_2-, and β_3-integrin subunits (Hannigan et al., 1996; Pasquet et al., 2002). Bendig et al., 2006 demonstrated that ILK is a novel component of the cardiac mechanical stretch sensor and activates PKB/AKT to regulate the response to stretch. ILK has a low basal activity, which is markedly increased by growth factors and integrin clustering (Dedhar et al., 1999). Phosphatidylinositol 3 kinase (PI3-kinase), PIP_3 lipid phosphatase (PTEN), and integrin-linked kinase-associated phosphatase 2C (ILKAP) are upstream regulators, whereby PI3K and PTEN regulate ILK activity by affecting PIP_3 binding to the pleckstrin-homology domain of ILK (Delcommenne et al., 1998; Kumar et al., 2004; Leung-Hagesteijn et al., 2001; Wu and Dedhar, 2001). ILK directly couples to Akt and glycogen synthase kinase–3β (GSK-3β) (Delcommenne et al., 1998; Persad et al., 2001; Troussard et al., 2003). Activated Akt and GSK-3β further phosphorylate downstream signaling cascades mTOR, NF-κB and CREB, which have been implicated in cardiac cell growth (Fig. 3.1). The mechanism by which ILK couples to these effectors is complex. Recent studies suggest that ILK is more important as an adaptor than a kinase, by recruiting kinases into a multi-protein complex, which in turn phosphorylates Akt and GSK-3β (Grashoff et al., 2003; Hill et al., 2002; Lynch et al., 1999). However, the importance of catalytic and noncatalytic functions of ILK may be cell-dependent and requires further investigation.

In the few studies performed, results indicate that ILK plays an integral role in cardiovascular signaling mechanisms associated with integrin signaling. Targeted ablation of ILK from murine heart in cardiac myocytes results in dilated cardiomyopathy and spontaneous heart failure (White et al., 2006). ILK appears to primarily activate signaling pathways associated with survival, development and adaptive cardiac hypertrophy (physiologic hypertrophy), rather than maladaptive hypertrophy (pathological hypertrophy) (Hannigan et al., 2007). In a pressure-overload mouse model of cardiac hypertrophy, there is a significant increase in ILK mRNA (Johnatty et al., 2000). Thymosin β_4 protein regulates cardiac cell migration and survival through activation of ILK. In a coronary artery ligated mouse model, thymosin β_4 treatment resulted in upregulation of ILK and Akt activity in the heart, enhanced early myocyte survival and improved cardiac function (Bock-Marquette et al., 2004). Given the central role of ILK in heart physiology, future studies in this area may prove to be important for a healthy heart.

3.3.3 Rho Family of GTPases

The Rho GTPase protein subfamily is comprised of a multitude of monomeric proteins (over 20 members in human) with a relative molecular mass of 20–30 kDa. Rho GTPases regulate a wide variety of cellular processes such as cell division, migration and differentiation. On the basis of their amino acid sequences and functional similarities, Rho GTPases are divided into Rac1, Cdc42 and RhoA subgroups (Wennerberg et al., 2005). Most of the Rho GTPases are post-translationally modified by prenylation at their carboxy terminus which helps to anchor the protein to membranes (Adamson et al., 1992). Rho GTPases act as GDP-GTP- regulated molecular switches and are activated by guanine nucleotide exchange factors (GEFs) in response to diverse extracellular stimuli, however, the complete mechanism for activation of small GTPase still needs more attention (Lezoualc'h et al., 2008). In the GTP active state they interact with a number of effector molecules to activate intracellular signaling cascades. The revolutionary studies in Rho GTPase field were performed by Ridley and Hall (1992). They showed that Rac1 activation leads to the formation of actin rich lamellipodia and membrane ruffles at cell periphery whereas Cdc42 induces the formation of filopodia (Nobes and Hall, 1995). Activation of RhoA leads to the formation of actin stress fibers and focal adhesion complexes in fibroblasts (Ridley and Hall, 1992). In the past decade, the Rho family of GTPases, which link integrins and other cell surface proteins to the actin cytoskeleton and orchestrate fundamental cellular processes (Lu et al., 2006) have become recognized as important regulators in the cardiovascular system.

The cellular signaling events mediated by small GTPases are either actin-dependent or independent. Once activated and translocated to specific subcellular locations, Rho proteins interact with downstream effector molecules to engage specific signaling cascades (Jaffe and Hall, 2005). To date, more than 70 potential effectors have been identified for members of the Rho/Rac family (Bustelo et al., 2007). The effects of RhoA and Rac1 on the actin cytoskeleton and cell morphology are mediated through stimulation of downstream effector kinases by the activated (GTP-liganded) Rho protein. For RhoA, the best known effectors are Rho kinase (ROCK) and mammalian diaphanous (mDia). Two isoforms of ROCK (ROCK1 and ROCK2), have been identified (Fukata et al., 2001). Although both isoforms are ubiquitously expressed, ROCK2 is highly expressed in brain and heart, whereas ROCK1 is preferentially expressed in lung, liver, spleen kidney and testis. ROCK phosphorylates the myosin binding subunit of MLC phosphatase, resulting in increased myosin phosphorylation and contraction (Kimura et al., 1996) (Fig. 3.2).

3.3.3.1 RhoA and Rac1

Of the 20 known Rho family gene products, RhoA and Rac1 have been most extensively studied in the context of cardiovascular signaling. Rho-associated protein kinase increases the sensitivity of vascular smooth muscle to calcium in hypertension (Uehata et al., 1997) and coronary spasm (Katsumata et al., 1997). They are also involved in pressure overload induced cardiac hypertrophy (Satoh et al., 2006; Wang et al., 1997). Recent studies have shown that ROCK1 deficient mice preserved compensatory hypertrophic response, but showed reduced perivascular fibrosis and interstitial fibrosis in response to pressure overload (Zhang et al., 2006). Increased ROCK activation has been observed in a mouse model of myocardial

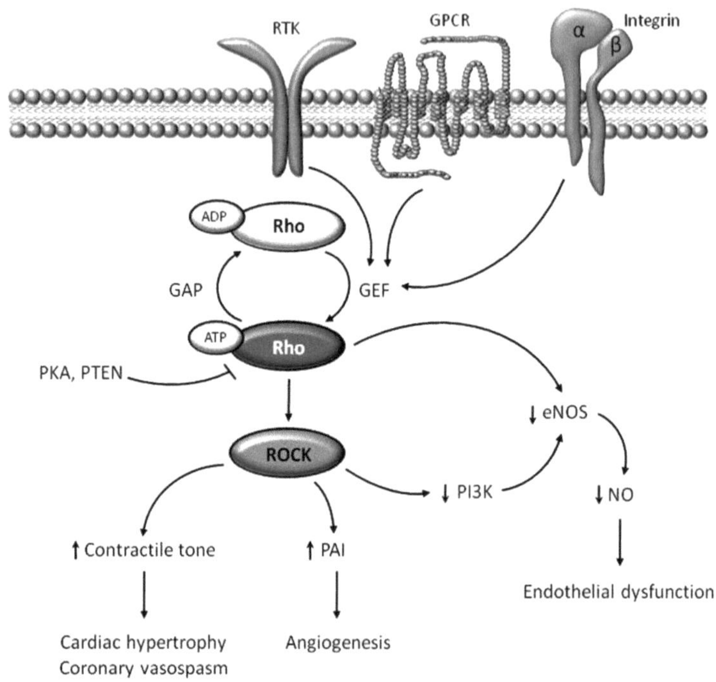

Fig. 3.2 Integrin-mediated Rho GTPase signaling. Various extracellular stimuli such as hormones, growth factors, neuromediators, and integrin interaction with extracellular matrix (ECM) activate guanine exchange factors (GEFs) leading to activation of Rho. Conversely, Rho may be inhibited by PTEN or PKA. GTP-bound Rho subsequently activates ROCK to phosphorylate several substrates leading to various cellular responses that directly and/or indirectly cause cardiovascular diseases. PTEN, phosphatase and tensin homolog; ROCK, Rho kinase; PI3-K, phosphoinositide 3 kinase; eNOS, endothelial nitric oxide synthase; NO, nitric oxide; PAI, Plasmogen activator inhibitor; GEF, guanine nucleotide exchange factor; GAP, GTP-ase activating protein; ECM, extracellular matrix; GPCRs, G-protein coupled receptors

infarction, as indicated by an increase in ezrin/radixin/moesin (downstream targets of ROCK) phosphorylation, fibrosis, hypertrophy and inflammation in the left ventricle following coronary artery ligation (Hattori et al., 2004). Expression of constitutively active Rac1 produces hypertrophic remodeling of cultured cardiac myocytes and dilated cardiomyopathy in vivo (Sussman et al., 2000). Although integrin coupling to Cdc42 has been examined in non-cardiovascular cell types (Price et al., 1998), its role remains to be determined in the heart and vasculature.

3.3.4 Protein Kinase C (PKC)

Several PKC family members have been implicated in mediating cardiac myocyte responses to mechanical forces (Pan et al., 2005). However, only α, δ, ε and ζ isoforms are consistently expressed in cardiac myocytes (Mackay and Mochly-Rosen,

2001). The major PKC isoforms expressed in the heart are PKCα and PKCε (Sabri and Steinberg, 2003). A limited number of studies have addressed how PKC isoenzyme expression and/or phosphorylation change during the induction of cardiac hypertrophy and its progression to heart failure. PKCε has been recently shown to be a major regulator of cardiac contractility with a propensity toward heart failure (Braz et al., 2004). In cardiac myocytes, calcineurin and PKCε have been shown to cooperate during stretch with complex interactions, in which calcineurin is necessary for stretch-induced PKCε translocation. Translocated PKCε is associated with calcineurin in a signaling complex at the level of the perinuclear membrane. The kinase-phosphatase interaction at the perinuclear membrane may serve to more finely modulate the degree of phosphorylation of targets, such as p38, which are regulated in the hypertrophic process. Recent in vitro evidence (Bullard et al., 2007) suggests that changes in PKC expression and phosphorylation is a mechanism by which cardiac myocytes can distinguish between the nature, direction and intensity of mechanical stretch. This implies that each PKC isozyme has specific roles in mechanotransduction and the transition from hypertrophy to heart failure. Cardiac fibroblasts are the cell type primarily responsible for homeostatic maintenance of the extracellular matrix in the myocardium (Brown et al., 2005). Pharmacological inhibition of PKCε attenuates progression of cardiac fibrosis in hypertensive Dahl rats suggest the direct involvement of PKCε in this processes (Inagaki et al., 2008). Further, treatment of hypertensive Dahl salt-sensitive rats with the PKCβ_{II} specific inhibitor, but not with the PKCβ specific inhibitor, also greatly delays the development of heart failure and death, suggesting that PKCβ_{II} also has a negative role in heart failure (Takeishi et al., 1998). Taken together, the above studies indicate that PKCα, PKCε, and PKCβ_{II} have major roles in mediating cardiac hypertrophy and/or cardiac fibrosis. Therefore targeting these sites, using pharmacological inhibitors, may be beneficial for the treatment of maladaptive cardiac hypertrophy.

3.4 Mechanosensitive Signal Transduction Cascades

3.4.1 Mitogen-Activated Protein (MAP) Kinase Cascades

MAP kinase signaling cascades provide important links between integrins and the nucleus via phosphorylation and regulation of multiple transcription factors. On the basis of sequence homology, MAP kinases have been divided into three major subfamilies: extracellular signal-regulated kinase (ERK), p38 and JNK. Although MAP kinases are ~60–70% identical, these molecules differ in sequence, size of their activation loop, as well as in activation responses to different stimuli. Each MAP kinase subfamily consists of several isoforms and members, each with distinct functions. ERK1/2, which is primarily activated by humoral stimuli, has been most widely studied, with respect to activation mechanisms. JNK and p38, originally identified as stress-activated protein kinases (SAPKs), are now known to belong to different signaling pathways, with different upstream activators and downstream targets (Sugden and Clerk, 1998). In addition to signal amplification, upstream kinases serve to integrate extracellular stimuli and orchestrate the correct balance of ERK, JNK, and p38 phosphorylation. Once activated, MAP kinases phosphorylate a variety of transcription factors and other proteins that regulate gene transcription, mRNA stability, and gene translation. An extensive number of studies have documented ERK, JNK and p38 activation in the pressure overloaded myocardium and in cardiac myocytes exposed to mechanical stress and various types of humoral stimuli (Wang, 2007). This initially led to the postulate that all three branches of the MAP kinase pathway are involved in mediating myocyte hypertrophy. However, more recent studies suggest that both ERK (Thorburn et al., 1997) and JNK (Nemoto et al., 1998) activate anti-hypertrophic signaling pathways in the heart, therefore opposing p38 effects on cardiac growth. Interestingly, over-expression of MAP kinase phosphatase-1, which inhibits all the three major branches, blocks agonist-induced pressure-overload induced cardiac hypertrophy (Bueno et al., 2001). This suggests MAP kinase family members have significantly different roles in the cardiac hypertrophic signaling, which remain to be clarified by future studies.

3.4.1.1 Extracellular Regulated Kinase (ERK) Cascade

Mechanical stress; mediated through integrins; can activate ERK1/2 via FAK dependent (Lee and Juliano, 2004) and FAK-independent mechanisms (Lal et al., 2007b). Certain α integrins bind to the membrane protein Cav-1 through their external and trans-membrane domains (Wary et al., 1998). The FAK independent activation of ERK by integrins appears to involve PI3-kinase and PKC activation (Lin et al., 1997). β_1-integrin mediates PE-induced myocyte hypertrophy (Pham et al., 2000), which is blocked by ERK1/2 antisense oligonucleotides (Glennon et al., 1996). Similarly, the MEK1 inhibitor PD98059 reduces sarcomeric organization induced by hypertrophic agonists (Clerk et al., 1998), and a dominant negative MEK1-encoding adenovirus reduced agonist-mediated hypertrophy in cultured cardiac myocytes (Ueyama et al., 2000). With respect to a gain-of-function, expression of activated MEK1 in cultured neonatal cardiac myocytes by adenoviral gene transfer induces a prominent growth response (Bueno and Molkentin, 2002).

Recent molecular studies have examined the roles of ERK signaling in the context of pressure-overload induced cardiac hypertrophy. Interestingly, transgenic *Erk1*$^{-/-}$ and *Erk2*$^{+/-}$ mice display no reduction in pathologic or physiologic stimulus-induced cardiac growth in vivo (Purcell et al., 2007). However, blockade or deletion of cardiac ERK1/2 predisposes the heart to decompensation and failure after long-term pressure overload in conjunction with an increase in myocyte apoptosis. The physiologic roles of ERK1/2 dephosphorylation in vivo was further investigated by using mice with cardiac targeted disruption of the gene encoding *dual-specificity phosphatase 6* (*Dusp6*) (Maillet et al., 2008). However, mice lacking *Dusp6* had larger hearts, which was associated with greater rates of myocyte proliferation during embryonic development and in the early postnatal period, resulting in cardiac hypercellularity. This increase in myocyte content in the heart was protective against decompensation and hypertrophic cardiomyopathy following long-term pressure overload and myocardial infarction injury in adult mice. Also in vivo, MEK transgenic hearts are protected from ischemia/reperfusion injury and myocyte apoptosis (Lips et al., 2004). These findings suggest that ERK1/2 signaling is not required for mediating physiologic or pathologic cardiac hypertrophy in vivo, although it does play a protective role in response to pathologic stimuli (Maillet et al., 2008; Purcell et al., 2007). The molecular mechanisms in ERK-mediated cardiac protection remain to be fully understood. The functional role of ERK1/2 in the heart remains a dilemma to any heart disease therapies targeting this pathway. Uncontrolled activation of Ras-Raf-MEK-ERK signaling can trigger hypertrophic cardiomyopathy, whereas inhibiting the pathway can render hearts more vulnerable to stress-induced myocyte death.

3.4.1.2 p38 Cascade

Mechanical load is a potent inducer of p38 in isolated cardiac myocytes and the pressure overloaded myocardium (Lal et al., 2007b; Wang, 2007). There is overwhelming evidence that prolonged activation of p38 accelerates myocardial injury. Studies with constitutively active mutants of specific upstream kinases p38, MKK3 and MKK6 to achieve specific activation of the p38 pathway in cardiac myocytes suggest that p38 activation is sufficient to induce characteristic changes in cardiac hypertrophy and cell death (Wang et al., 1998; Zechner et al., 1997). Liao and colleagues (2001) have studied the effects of p38 on the intact heart in transgenic mice. They achieved targeted activation of p38 in ventricular myocytes in vivo by using a gene-switch transgenic strategy with activated mutants of upstream kinases MKK3bE and MKK6bE. Transgene expression resulted in significant induction of p38 activity and premature death at 7–9 weeks. Both groups of transgenic hearts exhibited marked interstitial fibrosis and expression of fetal marker genes characteristic of cardiac failure, but no significant hypertrophy at the organ level. Echocardiographic and pressure-volume analyses revealed a similar extent of systolic contractile depression and restrictive diastolic abnormalities related to markedly increased passive chamber stiffness. However, MKK3bE-expressing hearts had increased end-systolic chamber volumes and a thinned ventricular wall, associated with heterogeneous myocyte atrophy, whereas MKK6bE hearts had reduced end-diastolic ventricular cavity size, a modest increase in myocyte size, and no significant myocyte atrophy. These data provided in vivo evidence for a negative inotropic and restrictive diastolic effect from p38 activation in ventricular myocytes and revealed specific roles of the p38 pathway in the development of ventricular end-systolic remodeling. However, subsequent studies using transgenic mice revealed very different effects in vivo. Targeted activation of p38 in the mouse heart did not produce any significant degree of cardiac hypertrophy. On the contrary, pressure overload – induced cardiac hypertrophy appeared to be enhanced further by dominant-negative mutants of p38, revealing an inhibitory function of p38 on cardiac hypertrophy (Braz et al., 2003). Therefore, further studies with inducible expression of wild-type and dominant-negative p38 are needed to delineate the detrimental and protective role of p38 in a stressor-specific manner.

3.4.1.3 JNK Cascade

JNK is named after the immediate-early gene c-jun, the first substrate identified (Kyriakis et al., 1994). The JNKs are encoded by three genes, *jnk1*, *jnk2* and *jnk3*, which are differentially spliced to yield four JNK1 isoforms, four JNK2 isoforms and two JNK3 isoforms (Gupta et al., 1996). Only JNK1 and JNK2 isoforms are expressed in the myocardium (Sugden and Clerk, 1998). The alternative forms of each JNK1, 2 and 3 appear to differ in their ability to bind and phosphorylate different substrate proteins (Johnson and Nakamura, 2007). Whether JNK mediates myocyte injury or protection is controversial. Ventricular myocyte targeted activation of JNK has been shown to induce restrictive cardiomyopathy and conduction defects in transgenic mice, suggesting that JNK can induce heart failure. Likewise, ablation of the MEKK1 gene, which abrogates JNK activation, has been reported to have cardioprotective effects in the left ventricular pressure-overloaded mouse heart (Sadoshima et al., 2002). Recent in vivo studies have challenged the previously proposed role of JNK as pro hypertrophic signaling effector in cardiac myocyte and suggest that it may actually serve as a negative regulator of this response in the adult heart (Ni et al., 2007).

3.4.2 Phosphoinositide 3-kinase/AKT/mTOR/FOXO Cascade

Akt is a serine/threonine-specific protein kinase and a fundamental regulator of myocyte growth (both physiologic and pathologic), as well as survival, metabolism, and gene transcription. The three known Akt isoforms (Akt1/PKBα, Akt2/PKBβ and Akt3/PKBγ) are derived from distinct genes, but maintain approximately 80% homology. Akt1 and Akt2 are the most abundant isoforms expressed in the heart and vasculature (Heineke and Molkentin, 2006). Akt1 is required for physiological hypertrophy in response to exercise training and IGF1 stimulation (DeBosch et al., 2006). Although total Akt1 is acutely upregulated during exercise (Zhang et al., 2007), chronic activation of the PI3-kinase/Akt pathway has also been shown in failing hearts (Haq et al., 2001). In addition, nuclear-targeted Akt in transgenic mice can lead to enhanced contractility (Rota et al., 2005), as well as protection from ischemic injury (Shiraishi et al., 2004).

3.4.2.1 Mechanical Regulation of Akt

Akt activity is modulated by mechanical stretch and humoral factors, as well as feedback regulation through the FOXO transcription factors and phosphatases, PP2A and calcineurin (Fig. 3.3). Akt1 activity is regulated by phosphorylation of Thr^{308}, located in the central kinase domain and Ser^{473}, located in the C-terminal regulatory domain. Although phosphorylation at both sites is synergistic and required for full Akt activation, phosphorylation of Thr^{308} alone can lead to activation of several downstream targets (mTORC1, TSC2, and GSK3β), whereas Ser^{473} phosphorylation specifically affects the activity of the Forkhead transcription factors FOXO4 and FOXO1/3a. The Ser^{473} regulatory site is typically associated with mechanosensitivity, however, stretch can indirectly activate Thr^{308}, due to release of humoral factors, such as ET-1 and Ang II (de Jonge et al., 2007; van Wamel et al., 2001).

Akt activation at Thr^{308} proceeds through phosphorylation by phosphoinositide-dependent protein kinase-1 (PDK1). Activation of Akt at Ser^{473} has proven to be more complex than Thr^{308}, in terms of overall regulation and potential binding partners. Ser^{473} activation by a potential "PDK2" has been controversial, since several kinases can phosphorylate this Akt regulatory site, such as the rapamycin-insensitive complex consisting of mammalian target of rapamycin (mTOR), rictor, protor-1 and Sin1 (mTORC2) (Sarbassov et al., 2005b), as well as ILK (Lynch et al., 1999) and PKC βII (Kawakami et al., 2004). Several lines of evidence point to mTORC2 as the prominent Ser^{473} regulator. Although rictor or Sin1 knockdown significantly inhibits Ser^{473} phosphorylation (Jacinto et al., 2006), Ser^{473} is also phosphorylated by other molecules. The details of Ser^{473} activation by mechanism are unknown; Sin1 may act as scaffold not only for mTOR, but also JNK and p38 (Jacinto et al., 2006), posing as an adaptor for MAP kinase/Akt crosstalk. In addition, PDK1 has been shown to phosphorylate both Thr^{308} and Ser^{473} (Balendran et al., 1999), and autophosphorylation at Ser^{473} has also been reported (Toker et al., 2000). In terms of mechanosensitivity, however, phosphorylation at Ser^{473} by ILK is perhaps the most intriguing. At the focal adhesion complex, FAK and Src activation may lead to activation of ILK. Additionally, ILK may be directly activated by β-integrin through the adaptor molecule kindlin-2 (Dowling et al., 2008). Mechanosignaling at focal adhesion complexes may also directly affect phosphorylation of Thr^{308}, since Src has also been shown to activate PI3-kinase (Laser et al., 1998). Thus, Akt may be fully activated at the plasma membrane by mechanosignaling through focal adhesion complexes, and modulated by autocrine factors released through mechanical stretch.

3.4.2.2 mTOR and Regulation of Cardiac Growth

Signaling through mTOR is necessary for mechanical load-induced growth of cardiac and skeletal muscles (Hornberger et al., 2006; Kemi et al., 2008). Although the exact signaling mechanisms downstream of Akt that distinguish healthy physiologic growth from pathologic growth in the myocardium are not clearly defined, a role for mTOR may exist in both processes. mTOR forms two complexes, mTORC1 and mTORC2, by associating with specific proteins that determine the biological function of mTOR (Wullschleger et al., 2006). mTORC1 is a downstream effector of Akt and consists of mLST8 and raptor, whereas mTORC2 consists of Sin1, mLST8, rictor and protor, and is an activator of Akt by phosphorylating Ser^{473}. Reduced expression of the specific proteins targeting the two mTOR complexes have defined a role for mTORC2 in spatial growth through cytoskeletal regulation (Jacinto et al., 2004) and temporal growth induction for mTORC1 through enhanced protein synthesis (McMullen et al., 2004). Both mTOR complexes have been reported in quiescent cells (Sarbassov et al., 2005a), and can be pharmacologically distinguished by the high sensitivity of mTORC1 to rapamycin, an immunosuppressant drug (Loewith et al., 2002). In the volume-overloaded mouse myocardium, rapamycin treatment

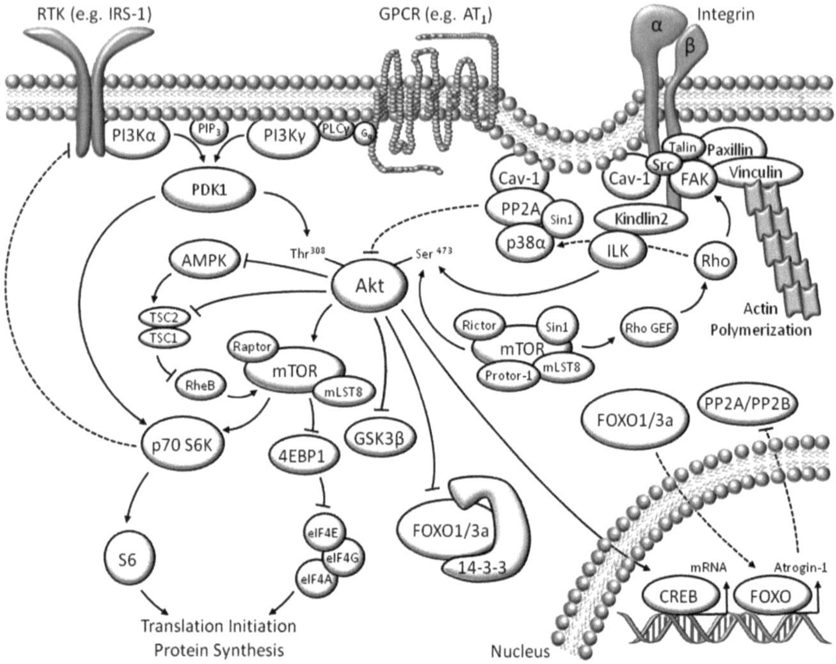

Fig. 3.3 Activation and downstream signaling mechanisms of Akt. Humoral (RTK or GPCR) factors lead to activation of Akt at Thr^{308}, inducing cellular growth through mTORC1 activation. Akt maintains several levels of control over mTORC1, both directly and indirectly. By inhibiting AMPK and TSC1/2, Akt releases the TSC1/2-mediated inhibition of RheB, a ras-family GTP-binding protein. The effects of mTORC1 activation are translation initiation by release of 4EBP1-mediated inhibition of the eIF4E translation initiation factor, and protein synthesis through activation of p70 S6K, which activates the 40S ribosomal protein S6. Hyper-phosphorylation of p70 S6K leads to inactivation of the RTK and inactivation of Akt. Activation at Akt at Ser^{473} is influenced by mechanical signaling through ILK at focal adhesion complexes, and by mTORC2, which may also signal to the actin cytoskeleton through the activation the Rho GEF. Rho signaling to FAK at the focal adhesion site may then lead to actin polymerization or ILK activation through the adaptor kindlin-2, which would potentiate Akt phosphorylation at Ser^{473}. Additionally, Rho activation induces p38α activity, which may interact with PP2A in caveolae to inhibit Akt (Sin1 acts as a molecular scaffold). Although phosphorylation of Akt at Ser^{473} by mTORC2 may preferentially lead to the targeted inhibition of FOXO by 14-3-3 protein, dephosphorylation of Akt by PP2A may result in FOXO activation. Active FOXO1/3a participates in a negative feedback loop, whereby transcription of the ubiquitin ligase atrogin-1 targets PP2A and calcineurin (PP2B) for proteasomal degradation. Thus, Akt integrates information from both mechanical and humoral stimuli to coordinate spatial and temporal cellular growth. Feedback pathways are indicated by dashed lines. RTK, receptor tyrosine kinse; GPCR, G-protein coupled receptor; AMPK, AMP-activated kinase; 4EBP1, 4-eukaryotic binding protein-1; eIF4E, eukaryotic initiation factor 4E; Rho GEF, Rho Guanine nucleotide exchange factor

attenuates cardiac hypertrophy by inhibiting mTORC1 (McMullen et al., 2004). Akt may directly activate mTORC1 through phosphorylation at Ser^{2448}, or indirectly through phosphorylation of the TSC1/TSC2 heterodimer, an upstream integrator of primarily nutritional and growth related signals, that influence mTOR activation (Sarbassov et al., 2005a; Wullschleger et al., 2006) (Fig. 3.3).

3.4.2.3 Regulation of FOXO Transcription Factors

The O subfamily of Forkhead/winged helix transcription factors, FOXO1, FOXO3a, and FOXO4 are inactivated by Akt, leading to nuclear exclusion and inhibition of the pro-apoptotic Forkhead transcriptional program (Matsuzaki et al., 2003). As noted above, phosphorylation of Akt at Ser^{473} by mTORC2 may preferentially target Akt to the FOXO transcription factors. FOXO transcription factors have been specifically implicated in regulating survival, as well as differentiation, proliferation, and metabolism. FOXO activity is regulated not only by Akt, but also through protein processing mechanisms including acetylation, protylic cleavage, proteasomal activation (Tremblay and Giguere, 2008). Akt activation suppresses specific atrogenes, a set of genes regulated during rapid loss in muscle mass, such as cathepsin L, ubiquitination factor E4B, and atrogin-1 (Skurk et al., 2005). Atrogin-1 targets Type 2 protein phosphatases PP2A and calcineurin (PP2B), two phosphatases that dephosphorylate Akt at Thr^{308} and Ser^{473}, for proteasomal degradation (Ni et al., 2007), providing a negative feedback loop for FOXO-mediated increases in Akt activity. Although there are few known mechanisms for PP2A or PP2B phosphatase regulation, other kinases such as p38α have also been shown to inhibit Akt activity through PP2A, through interaction with Cav-1 (Zuluaga et al., 2007). It is possible that the phosphatase regulatory subunits are targeted to specific intracellular compartments under tightly-controlled mechanisms.

3.4.3 Janus-Associated Kinase (JAK)/Signal Transducers and Activators of Transcription (STATs)

Excessive mechanical load initially induces compensatory hypertrophy, which ultimately leads to irreversible decompensation in cardiac function. The major challenge has been determination of the molecular mechanisms responsible for the development of cardiac hypertrophy, as well as its transition into heart failure. The common receptor component of the interleukin-6 (IL-6) family of cytokines, gp130, has been demonstrated to play an important role in cardiac hypertrophy and heart failure (Hirota et al., 1999; Negoro et al., 2001). The gp130 cytokines, such as cardiotrophin-1 (CT-1) and leukemia inhibitory factor (LIF), are potent inducers of cardiomyocyte hypertrophy (Pan et al., 1998) and promote myocyte survival (Negoro et al., 2001). In the heart, CT-1 and LIF are induced by the biomechanical stress of mechanical stretch or aortic banding (Pan et al., 1999; Wang et al., 2001), and clinical studies have demonstrated elevated plasma levels of CT-1 in patients with congestive heart failure. (Talwar et al., 2000). In addition, mice with cardiac restricted knockout of gp130 demonstrate a rapid-onset dilated cardiomyopathy and massive myocyte apoptosis during the biomechanical stress associated with transverse-aortic constriction (Hirota et al., 1999).

The binding of ligands to the gp130 and LIF receptor complex triggers dimerization and results in the activation of janus kinase (JAK). Upon activation, JAKs rapidly phosphorylate tyrosine residues of these receptors, resulting in recruitment of various signaling molecules to the receptor complex. One of these factors is signal transducer and activators of transcription-3 (STAT3), which upon activation by JAK, forms STAT dimers and translocates to the nucleus, where it leads to transcriptional activation of downstream target genes (Dostal et al., 1997). Apart from promoting survival and stimulating hypertrophic responses in cardiac myocytes, the JAK-STAT pathway also stimulates cardiac fibroblast growth and proliferation and seems to be involved in wound healing after myocardial infarction (Fischer and Hilfiker-Kleiner, 2007). Evidence indicates that CT-1 induces cardiac myocyte hypertrophy by upregulation of angiotensinogen mRNA expression in myocytes, via STAT3 binding to the St-domain of the angiotensinogen gene promoter (Fukuzawa et al., 2000). The cardiac effects of gp130 and Stat3, are countered by suppressor cytokine signaling 3 (SOCS3). SOCS3 is a mechanical stress-inducible gene in cardiac muscle cells and that it directly modulates stress-induced gp130 cytokine receptor signaling as the key molecular switch for a negative feedback circuit for both myocyte hypertrophy and survival (Yasukawa et al., 2001). In cardiac myocytes SOCS3 overexpression has been shown to suppress LIF induced activation of MEK1-ERK1/2 and Akt, in addition to STAT3 (Yasukawa et al., 2001). In this context, SOCS3 activation appears to have the deleterious effect of leading the heart into failure, due to suppression of the cardioprotective effects of the Stat3. Although mechanical stretch rapidly activates components of the JAK-Stat pathway, the potential role of integrins and other mechanosensors remains to be investigated.

3.5 Conclusions and Perspectives

In summary, we have focused on major signaling pathways whose induction begins with mechanical forces as experimentally determined by myocardial stretch. These pathways have proximal mechano-sensitive receptors located within and spanning the cell membrane which transduce the mechanical signals into chemical signaling cascades that lead to many types of responses including mechanical forces of contraction, cell growth, differentiation, cell cycle regulation, apoptosis, remodeling, hypertrophy, and other adaptive and maladaptive states. A substantial amount of work is required to clarify the mechanisms by which mechanoreceptors couple to proximal effectors and cross-talk with other mechanosensing and growth factor receptors in the various cardiac cell types under pathophysiological conditions. Novel theoretical and experimental methodologies will be required to unravel the precise details of these mechanisms. As our understanding of integrins as multifunctional adhesion and signaling molecules has grown, so has their recognition as potential therapeutic targets in human diseases. A better understanding of the functional role of the initiator elements is of key importance to developing novel strategies to control cardiac hypertrophy and prevent heart failure induced by hemodynamically overload.

Acknowledgements This work was supported by grants from National Institutes of Health (HL-68838) and Scott and White Memorial Hospital.

References

Adamson P, Marshall CJ, Hall A, Tilbrook PA (1992) Post-translational modifications of p21rho proteins. J Biol Chem 267: 20033–20038.

Aikawa R, Nagai T, Kudoh S, Zou Y, Tanaka M, Tamura M, Akazawa H, Takano H, Nagai R, Komuro I (2002) Integrins play a critical role in mechanical stress-induced p38 MAPK activation. Hypertension 39: 233–238.

Baicu CF, Stroud JD, Livesay VA, Hapke E, Holder J, Spinale FG, Zile MR (2003) Changes in extracellular collagen matrix alter myocardial systolic performance. Am J Physiol Heart Circ Physiol 284: H122–H132.

Balendran A, Casamayor A, Deak M, Paterson A, Gaffney P, Currie R, Downes CP, Alessi DR (1999) PDK1 acquires PDK2 activity in the presence of a synthetic peptide derived from the carboxyl terminus of PRK2. Curr Biol 9: 393–404.

Belkin AM, Retta SF, Pletjushkina OY, Balzac F, Silengo L, Fassler R, Koteliansky VE, Burridge K, Tarone G (1997) Muscle beta1D integrin reinforces the cytoskeleton-matrix link: modulation of integrin adhesive function by alternative splicing. J Cell Biol 139: 1583–1595.

Bendig G, Grimmler M, Huttner IG, Wessels G, Dahme T, Just S, Trano N, Katus HA, Fishman MC, Rottbauer W (2006) Integrin-linked kinase, a novel component of the cardiac mechanical stretch sensor, controls contractility in the zebrafish heart. Genes Dev 20: 2361–2372.

Bershadsky AD, Balaban NQ, Geiger B (2003) Adhesion-dependent cell mechanosensitivity. Annu Rev Cell Dev Biol 19: 677–695.

Bishop JE, Laurent GJ (1995) Collagen turnover and its regulation in the normal and hypertrophying heart. Eur Heart J 16 Suppl C: 38–44.

Bock-Marquette I, Saxena A, White MD, Dimaio JM, Srivastava D (2004) Thymosin beta4 activates integrin-linked kinase and promotes cardiac cell migration, survival and cardiac repair. Nature 432: 466–472.

Brancaccio M, Cabodi S, Belkin AM, Collo G, Koteliansky VE, Tomatis D, Altruda F, Silengo L, Tarone G (1998) Differential onset of expression of alpha 7 and beta 1D integrins during mouse heart and skeletal muscle development. Cell Adhes Commun 5: 193–205.

Braz JC, Bueno OF, Liang Q, Wilkins BJ, Dai YS, Parsons S, Braunwart J, Glascock BJ, Klevitsky R, Kimball TF, Hewett TE, Molkentin JD (2003) Targeted inhibition of p38 MAPK promotes hypertrophic cardiomyopathy through upregulation of calcineurin-NFAT signaling. J Clin Invest 111: 1475–1486.

Braz JC, Gregory K, Pathak A, Zhao W, Sahin B, Klevitsky R, Kimball TF, Lorenz JN, Nairn AC, Liggett SB, Bodi I, Wang S, Schwartz A, Lakatta EG, DePaoli-Roach AA, Robbins J, Hewett TE, Bibb JA, Westfall MV, Kranias EG, Molkentin JD (2004) PKC-alpha regulates cardiac contractility and propensity toward heart failure. Nat Med 10: 248–254.

Brown RD, Ambler SK, Mitchell MD, Long CS (2005) The cardiac fibroblast: therapeutic target in myocardial remodeling and failure. Annu Rev Pharmacol Toxicol 45: 657–687.

Bueno OF, De Windt LJ, Lim HW, Tymitz KM, Witt SA, Kimball TR, Molkentin JD (2001) The dual-specificity phosphatase MKP-1 limits the cardiac hypertrophic response in vitro and in vivo. Circ Res 88: 88–96.

Bueno OF, Molkentin JD (2002) Involvement of extracellular signal-regulated kinases 1/2 in cardiac hypertrophy and cell death. Circ Res 91: 776–781.

Bullard TA, Hastings JL, Davis JM, Borg TK, Price RL (2007) Altered PKC expression and phosphorylation in response to the nature, direction, and magnitude of mechanical stretch. Can J Physiol Pharmacol 85: 243–250.
Burgess ML, Carver WE, Terracio L, Wilson SP, Wilson MA, Borg TK (1994) Integrin-mediated collagen gel contraction by cardiac fibroblasts. Effects of angiotensin II. Circ Res 74: 291–298.
Bustelo XR, Sauzeau V, Berenjeno IM (2007) GTP-binding proteins of the Rho/Rac family: regulation, effectors and functions in vivo. Bioessays 29: 356–370.
Clemente CF, Tornatore TF, Theizen TH, Deckmann AC, Pereira TC, Lopes-Cendes I, Souza JR, Franchini KG (2007) Targeting focal adhesion kinase with small interfering RNA prevents and reverses load-induced cardiac hypertrophy in mice. Circ Res 101: 1339–1348.
Clerk A, Michael A, Sugden PH (1998) Stimulation of the p38 mitogen-activated protein kinase pathway in neonatal rat ventricular myocytes by the G protein-coupled receptor agonists, endothelin-1 and phenylephrine: a role in cardiac myocyte hypertrophy? J Cell Biol 142: 523–535.
de Jonge HW, Dekkers DH, Houtsmuller AB, Sharma HS, Lamers JM (2007) Differential signaling and hypertrophic responses in cyclically stretched vs endothelin-1 stimulated neonatal rat cardiomyocytes. Cell Biochem Biophys 47: 21–32.
DeBosch B, Treskov I, Lupu TS, Weinheimer C, Kovacs A, Courtois M, Muslin AJ (2006) Akt1 is required for physiological cardiac growth. Circulation 113: 2097–2104.
Dedhar S, Williams B, Hannigan G (1999) Integrin-linked kinase (ILK): a regulator of integrin and growth-factor signalling. Trends Cell Biol 9: 319–323.
Delcommenne M, Tan C, Gray V, Rue L, Woodgett J, Dedhar S (1998) Phosphoinositide-3-OH kinase-dependent regulation of glycogen synthase kinase 3 and protein kinase B/AKT by the integrin-linked kinase. Proc Natl Acad Sci USA 95: 11211–11216.
DiMichele LA, Doherty JT, Rojas M, Beggs HE, Reichardt LF, Mack CP, Taylor JM (2006) Myocyte-restricted focal adhesion kinase deletion attenuates pressure overload-induced hypertrophy. Circ Res 99: 636–645.
Domingos PP, Fonseca PM, Nadruz W, Jr., Franchini KG (2002) Load-induced focal adhesion kinase activation in the myocardium: role of stretch and contractile activity. Am J Physiol Heart Circ Physiol 282: H556–H564.
Dostal DE, Hunt RA, Kule CE, Bhat GJ, Karoor V, McWhinney CD, Baker KM (1997) Molecular mechanisms of angiotensin II in modulating cardiac function: intracardiac effects and signal transduction pathways. J Mol Cell Cardiol 29: 2893–2902.
Dowling JJ, Vreede AP, Kim S, Golden J, Feldman EL (2008) Kindlin-2 is required for myocyte elongation and is essential for myogenesis. BMC Cell Biol 9: 36.
Eble DM, Strait JB, Govindarajan G, Lou J, Byron KL, Samarel AM (2000) Endothelin-induced cardiac myocyte hypertrophy: role for focal adhesion kinase. Am J Physiol Heart Circ Physiol 278: H1695–H1707.
Fischer P, Hilfiker-Kleiner D (2007) Survival pathways in hypertrophy and heart failure: the gp130-STAT3 axis. Basic Res Cardiol 102: 279–297.
Fukata Y, Amano M, Kaibuchi K (2001) Rho-Rho-kinase pathway in smooth muscle contraction and cytoskeletal reorganization of non-muscle cells. Trends Pharmacol Sci 22: 32–39.
Fukuzawa J, Booz GW, Hunt RA, Shimizu N, Karoor V, Baker KM, Dostal DE (2000) Cardiotrophin-1 increases angiotensinogen mRNA in rat cardiac myocytes through STAT3: an autocrine loop for hypertrophy. Hypertension 35: 1191–1196.
Furuta Y, Ilic D, Kanazawa S, Takeda N, Yamamoto T, Aizawa S (1995) Mesodermal defect in late phase of gastrulation by a targeted mutation of focal adhesion kinase, FAK. Oncogene 11: 1989–1995.
Glennon PE, Kaddoura S, Sale EM, Sale GJ, Fuller SJ, Sugden PH (1996) Depletion of mitogen-activated protein kinase using an antisense oligodeoxynucleotide approach downregulates the phenylephrine-induced hypertrophic response in rat cardiac myocytes. Circ Res 78: 954–961.
Goldsmith EC, Hoffman A, Morales MO, Potts JD, Price RL, McFadden A, Rice M, Borg TK (2004) Organization of fibroblasts in the heart. Dev Dyn 230: 787–794.

Grashoff C, Aszodi A, Sakai T, Hunziker EB, Fassler R (2003) Integrin-linked kinase regulates chondrocyte shape and proliferation. EMBO Rep 4: 432–438.

Greenberg RS, Bernstein AM, Benezra M, Gelman IH, Taliana L, Masur SK (2006) FAK-dependent regulation of myofibroblast differentiation. Faseb J 20: 1006–1008.

Gupta S, Barrett T, Whitmarsh AJ, Cavanagh J, Sluss HK, Derijard B, Davis RJ (1996) Selective interaction of JNK protein kinase isoforms with transcription factors. Embo J 15: 2760–2770.

Hannigan GE, Coles JG, Dedhar S (2007) Integrin-linked kinase at the heart of cardiac contractility, repair, and disease. Circ Res 100: 1408–1414.

Hannigan GE, Leung-Hagesteijn C, Fitz-Gibbon L, Coppolino MG, Radeva G, Filmus J, Bell JC, Dedhar S (1996) Regulation of cell adhesion and anchorage-dependent growth by a new beta 1-integrin-linked protein kinase. Nature 379: 91–96.

Haq S, Choukroun G, Lim H, Tymitz KM, del Monte F, Gwathmey J, Grazette L, Michael A, Hajjar R, Force T, Molkentin JD (2001) Differential activation of signal transduction pathways in human hearts with hypertrophy versus advanced heart failure. Circulation 103: 670–677.

Hattori T, Shimokawa H, Higashi M, Hiroki J, Mukai Y, Tsutsui H, Kaibuchi K, Takeshita A (2004) Long-term inhibition of Rho-kinase suppresses left ventricular remodeling after myocardial infarction in mice. Circulation 109: 2234–2239.

Heidkamp MC, Bayer AL, Kalina JA, Eble DM, Samarel AM (2002) GFP-FRNK disrupts focal adhesions and induces anoikis in neonatal rat ventricular myocytes. Circ Res 90: 1282–1289.

Hein S, Schaper J (2001) The extracellular matrix in normal and diseased myocardium. J Nucl Cardiol 8: 188–196.

Heineke J, Molkentin JD (2006) Regulation of cardiac hypertrophy by intracellular signalling pathways. Nat Rev Mol Cell Biol 7: 589–600.

Hill MM, Feng J, Hemmings BA (2002) Identification of a plasma membrane Raft-associated PKB Ser473 kinase activity that is distinct from ILK and PDK1. Curr Biol 12: 1251–1255.

Hirota H, Chen J, Betz UA, Rajewsky K, Gu Y, Ross J, Jr., Muller W, Chien KR (1999) Loss of a gp130 cardiac muscle cell survival pathway is a critical event in the onset of heart failure during biomechanical stress. Cell 97: 189–198.

Hornberger TA, Chien S (2006) Mechanical stimuli and nutrients regulate rapamycin-sensitive signaling through distinct mechanisms in skeletal muscle. J Cell Biochem 97: 1207–1216.

Hynes RO (2002a) Integrins: bidirectional, allosteric signaling machines. Cell 110: 673–687.

Hynes RO (2002b) A reevaluation of integrins as regulators of angiogenesis. Nat Med 8: 918–921.

Ilic D, Damsky CH, Yamamoto T (1997) Focal adhesion kinase: at the crossroads of signal transduction. J Cell Sci 110 (Pt 4): 401–407.

Ilic D, Furuta Y, Kanazawa S, Takeda N, Sobue K, Nakatsuji N, Nomura S, Fujimoto J, Okada M, Yamamoto T (1995) Reduced cell motility and enhanced focal adhesion contact formation in cells from FAK-deficient mice. Nature 377: 539–544.

Inagaki K, Koyanagi T, Berry NC, Sun L, Mochly-Rosen D (2008) Pharmacological inhibition of epsilon-protein kinase C attenuates cardiac fibrosis and dysfunction in hypertension-induced heart failure. Hypertension 51: 1565–1569.

Jacinto E, Facchinetti V, Liu D, Soto N, Wei S, Jung SY, Huang Q, Qin J, Su B (2006) SIN1/MIP1 maintains rictor-mTOR complex integrity and regulates Akt phosphorylation and substrate specificity. Cell 127: 125–137.

Jacinto E, Loewith R, Schmidt A, Lin S, Ruegg MA, Hall A, Hall MN (2004) Mammalian TOR complex 2 controls the actin cytoskeleton and is rapamycin insensitive. Nat Cell Biol 6: 1122–1128.

Jaffe AB, Hall A (2005) Rho GTPases: biochemistry and biology. Annu Rev Cell Dev Biol 21: 247–269.

Jane-Lise S, Corda S, Chassagne C, Rappaport L (2000) The extracellular matrix and the cytoskeleton in heart hypertrophy and failure. Heart Fail Rev 5: 239–250.

Jia N, Okamoto H, Shimizu T, Chiba S, Matsui Y, Sugawara T, Akino M, Kitabatake A (2003) A newly developed angiotensin II type 1 receptor antagonist, CS866, promotes regression of cardiac hypertrophy by reducing integrin beta1 expression. Hypertens Res 26: 737–742.

Johnatty SE, Dyck JR, Michael LH, Olson EN, Abdellatif M (2000) Identification of genes regulated during mechanical load-induced cardiac hypertrophy. J Mol Cell Cardiol 32: 805–815.
Johnson GL, Nakamura K (2007) The c-jun kinase/stress-activated pathway: regulation, function and role in human disease. Biochim Biophys Acta 1773: 1341–1348.
Katsumata N, Shimokawa H, Seto M, Kozai T, Yamawaki T, Kuwata K, Egashira K, Ikegaki I, Asano T, Sasaki Y, Takeshita A (1997) Enhanced myosin light chain phosphorylations as a central mechanism for coronary artery spasm in a swine model with interleukin-1beta. Circulation 96: 4357–4363.
Kawakami Y, Nishimoto H, Kitaura J, Maeda-Yamamoto M, Kato RM, Littman DR, Leitges M, Rawlings DJ, Kawakami T (2004) Protein kinase C betaII regulates Akt phosphorylation on Ser-473 in a cell type- and stimulus-specific fashion. J Biol Chem 279: 47720–47725.
Kawano H, Cody RJ, Graf K, Goetze S, Kawano Y, Schnee J, Law RE, Hsueh WA (2000) Angiotensin II enhances integrin and alpha-actinin expression in adult rat cardiac fibroblasts. Hypertension 35: 273–279.
Kemi OJ, Ceci M, Wisloff U, Grimaldi S, Gallo P, Smith GL, Condorelli G, Ellingsen O (2008) Activation or inactivation of cardiac Akt/mTOR signaling diverges physiological from pathological hypertrophy. J Cell Physiol 214: 316–321.
Kimura K, Ito M, Amano M, Chihara K, Fukata Y, Nakafuku M, Yamamori B, Feng J, Nakano T, Okawa K, Iwamatsu A, Kaibuchi K (1996) Regulation of myosin phosphatase by Rho and Rho-associated kinase (Rho-kinase). Science 273: 245–248.
Kudoh S, Komuro I, Hiroi Y, Zou Y, Harada K, Sugaya T, Takekoshi N, Murakami K, Kadowaki T, Yazaki Y (1998) Mechanical stretch induces hypertrophic responses in cardiac myocytes of angiotensin II type 1a receptor knockout mice. J Biol Chem 273: 24037–24043.
Kumar AS, Naruszewicz I, Wang P, Leung-Hagesteijn C, Hannigan GE (2004) ILKAP regulates ILK signaling and inhibits anchorage-independent growth. Oncogene 23: 3454–3461.
Kyriakis JM, Banerjee P, Nikolakaki E, Dai T, Rubie EA, Ahmad MF, Avruch J, Woodgett JR (1994) The stress-activated protein kinase subfamily of c-Jun kinases. Nature 369: 156–160.
Lal H, Guleria RS, Foster DM, Lu G, Watson LE, Sanghi S, Smith M, Dostal DE (2007a) Integrins: novel therapeutic targets for cardiovascular diseases. Cardiovasc Hematol Agents Med Chem 5: 109–132.
Lal H, Verma SK, Smith M, Guleria RS, Lu G, Foster DM, Dostal DE (2007b) Stretch-induced MAP kinase activation in cardiac myocytes: differential regulation through beta1-integrin and focal adhesion kinase. J Mol Cell Cardiol 43: 137–147.
Laser M, Kasi VS, Hamawaki M, Cooper Gt, Kerr CM, Kuppuswamy D (1998) Differential activation of p70 and p85 S6 kinase isoforms during cardiac hypertrophy in the adult mammal. J Biol Chem 273: 24610–24619.
Lee JW, Juliano R (2004) Mitogenic signal transduction by integrin- and growth factor receptor-mediated pathways. Mol Cells 17: 188–202.
Leung-Hagesteijn C, Mahendra A, Naruszewicz I, Hannigan GE (2001) Modulation of integrin signal transduction by ILKAP, a protein phosphatase 2C associating with the integrin-linked kinase, ILK1. Embo J 20: 2160–2170.
Lezoualc'h F, Metrich M, Hmitou I, Duquesnes N, Morel E (2008) Small GTP-binding proteins and their regulators in cardiac hypertrophy. J Mol Cell Cardiol 44: 623–632.
Li F, Zhang Y, Wu C (1999) Integrin-linked kinase is localized to cell-matrix focal adhesions but not cell-cell adhesion sites and the focal adhesion localization of integrin-linked kinase is regulated by the PINCH-binding ANK repeats. J Cell Sci 112 (Pt 24): 4589–4599.
Liao P, Georgakopoulos D, Kovacs A, Zheng M, Lerner D, Pu H, Saffitz J, Chien K, Xiao RP, Kass DA, Wang Y (2001) The in vivo role of p38 MAP kinases in cardiac remodeling and restrictive cardiomyopathy. Proc Natl Acad Sci USA 98: 12283–12288.
Lin TH, Aplin AE, Shen Y, Chen Q, Schaller M, Romer L, Aukhil I, Juliano RL (1997) Integrin-mediated activation of MAP kinase is independent of FAK: evidence for dual integrin signaling pathways in fibroblasts. J Cell Biol 136: 1385–1395.
Linehan KA, Seymour AM, Williams PE (2001) Semiquantitative analysis of collagen types in the hypertrophied left ventricle. J Anat 198: 83–92.

Lips DJ, Bueno OF, Wilkins BJ, Purcell NH, Kaiser RA, Lorenz JN, Voisin L, Saba-El-Leil MK, Meloche S, Pouyssegur J, Pages G, De Windt LJ, Doevendans PA, Molkentin JD (2004) MEK1-ERK2 signaling pathway protects myocardium from ischemic injury in vivo. Circulation 109: 1938–1941.

Liu S, Calderwood DA, Ginsberg MH (2000) Integrin cytoplasmic domain-binding proteins. J Cell Sci 113 (Pt 20): 3563–3571.

Loewith R, Jacinto E, Wullschleger S, Lorberg A, Crespo JL, Bonenfant D, Oppliger W, Jenoe P, Hall MN (2002) Two TOR complexes, only one of which is rapamycin sensitive, have distinct roles in cell growth control. Mol Cell 10: 457–468.

Lopes MM, Ribeiro GC, Tornatore TF, Clemente CF, Teixeira VP, Franchini KG (2007) Increased expression and phosphorylation of focal adhesion kinase correlates with dysfunction in the volume-overloaded human heart. Clin Sci (Lond) 113: 195–204.

Lu C, Shimaoka M, Salas A, Springer TA (2004) The binding sites for competitive antagonistic, allosteric antagonistic, and agonistic antibodies to the I domain of integrin LFA-1. J Immunol 173: 3972–3978.

Lu H, Fedak PW, Dai X, Du C, Zhou YQ, Henkelman M, Mongroo PS, Lau A, Yamabi H, Hinek A, Husain M, Hannigan G, Coles JG (2006) Integrin-linked kinase expression is elevated in human cardiac hypertrophy and induces hypertrophy in transgenic mice. Circulation 114: 2271–2279.

Lynch DK, Ellis CA, Edwards PA, Hiles ID (1999) Integrin-linked kinase regulates phosphorylation of serine 473 of protein kinase B by an indirect mechanism. Oncogene 18: 8024–8032.

Mackay K, Mochly-Rosen D (2001) Localization, anchoring, and functions of protein kinase C isozymes in the heart. J Mol Cell Cardiol 33: 1301–1307.

Madri JA, Dreyer B, Pitlick FA, Furthmayr H (1980) The collagenous components of the subendothelium. Correlation of structure and function. Lab Invest 43: 303–315.

Maillet M, Purcell NH, Sargent MA, York A, Bueno OF, Molkentin JD (2008) Dusp6 (MKP3) null mice show enhanced ERK1/2 phosphorylation at baseline and increased myocyte proliferation in the heart affecting disease susceptibility. J Biol Chem 283: 31246–31255.

Manabe I, Shindo T, Nagai R (2002) Gene expression in fibroblasts and fibrosis: involvement in cardiac hypertrophy. Circ Res 91: 1103–1113.

Matsuzaki H, Daitoku H, Hatta M, Tanaka K, Fukamizu A (2003) Insulin-induced phosphorylation of FKHR (Foxo1) targets to proteasomal degradation. Proc Natl Acad Sci USA 100: 11285–11290.

McMullen JR, Sherwood MC, Tarnavski O, Zhang L, Dorfman AL, Shioi T, Izumo S (2004) Inhibition of mTOR signaling with rapamycin regresses established cardiac hypertrophy induced by pressure overload. Circulation 109: 3050–3055.

Mimura Y, Ihn H, Jinnin M, Asano Y, Yamane K, Tamaki K (2005) Constitutive phosphorylation of focal adhesion kinase is involved in the myofibroblast differentiation of scleroderma fibroblasts. J Invest Dermatol 124: 886–892.

Mould AP, Humphries MJ (2004) Regulation of integrin function through conformational complexity: not simply a knee-jerk reaction? Curr Opin Cell Biol 16: 544–551.

Negoro S, Kunisada K, Fujio Y, Funamoto M, Darville MI, Eizirik DL, Osugi T, Izumi M, Oshima Y, Nakaoka Y, Hirota H, Kishimoto T, Yamauchi-Takihara K (2001) Activation of signal transducer and activator of transcription 3 protects cardiomyocytes from hypoxia/reoxygenation-induced oxidative stress through the upregulation of manganese superoxide dismutase. Circulation 104: 979–981.

Nemoto S, Sheng Z, Lin A (1998) Opposing effects of Jun kinase and p38 mitogen-activated protein kinases on cardiomyocyte hypertrophy. Mol Cell Biol 18: 3518–3526.

Ni YG, Wang N, Cao DJ, Sachan N, Morris DJ, Gerard RD, Kuro OM, Rothermel BA, Hill JA (2007) FoxO transcription factors activate Akt and attenuate insulin signaling in heart by inhibiting protein phosphatases. Proc Natl Acad Sci USA 104: 20517–20522.

Nobes CD, Hall A (1995) Rho, rac, and cdc42 GTPases regulate the assembly of multimolecular focal complexes associated with actin stress fibers, lamellipodia, and filopodia. Cell 81: 53–62.

Owen JD, Ruest PJ, Fry DW, Hanks SK (1999) Induced focal adhesion kinase (FAK) expression in FAK-null cells enhances cell spreading and migration requiring both auto- and activation loop phosphorylation sites and inhibits adhesion-dependent tyrosine phosphorylation of Pyk2. Mol Cell Biol 19: 4806–4818.
Pan J, Fukuda K, Kodama H, Sano M, Takahashi T, Makino S, Kato T, Manabe T, Hori S, Ogawa S (1998) Involvement of gp130-mediated signaling in pressure overload-induced activation of the JAK/STAT pathway in rodent heart. Heart Vessels 13: 199–208.
Pan J, Fukuda K, Saito M, Matsuzaki J, Kodama H, Sano M, Takahashi T, Kato T, Ogawa S (1999) Mechanical stretch activates the JAK/STAT pathway in rat cardiomyocytes. Circ Res 84: 1127–1136.
Pan J, Singh US, Takahashi T, Oka Y, Palm-Leis A, Herbelin BS, Baker KM (2005) PKC mediates cyclic stretch-induced cardiac hypertrophy through Rho family GTPases and mitogen-activated protein kinases in cardiomyocytes. J Cell Physiol 202: 536–553.
Pasquet JM, Noury M, Nurden AT (2002) Evidence that the platelet integrin alphaIIb beta3 is regulated by the integrin-linked kinase, ILK, in a PI3-kinase dependent pathway. Thromb Haemost 88: 115–122.
Peng X, Kraus MS, Wei H, Shen TL, Pariaut R, Alcaraz A, Ji G, Cheng L, Yang Q, Kotlikoff MI, Chen J, Chien K, Gu H, Guan JL (2006) Inactivation of focal adhesion kinase in cardiomyocytes promotes eccentric cardiac hypertrophy and fibrosis in mice. J Clin Invest 116: 217–227.
Persad S, Troussard AA, McPhee TR, Mulholland DJ, Dedhar S (2001) Tumor suppressor PTEN inhibits nuclear accumulation of beta-catenin and T cell/lymphoid enhancer factor 1-mediated transcriptional activation. J Cell Biol 153: 1161–1174.
Pham CG, Harpf AE, Keller RS, Vu HT, Shai SY, Loftus JC, Ross RS (2000) Striated muscle-specific beta(1D)-integrin and FAK are involved in cardiac myocyte hypertrophic response pathway. Am J Physiol Heart Circ Physiol 279: H2916–H2926.
Price LS, Leng J, Schwartz MA, Bokoch GM (1998) Activation of Rac and Cdc42 by integrins mediates cell spreading. Mol Biol Cell 9: 1863–1871.
Purcell NH, Wilkins BJ, York A, Saba-El-Leil MK, Meloche S, Robbins J, Molkentin JD (2007) Genetic inhibition of cardiac ERK1/2 promotes stress-induced apoptosis and heart failure but has no effect on hypertrophy in vivo. Proc Natl Acad Sci U S A 104: 14074–14079.
Ren J, Avery J, Zhao H, Schneider JG, Ross FP, Muslin AJ (2007) Beta3 integrin deficiency promotes cardiac hypertrophy and inflammation. J Mol Cell Cardiol 42: 367–377.
Ridley AJ, Hall A (1992) The small GTP-binding protein rho regulates the assembly of focal adhesions and actin stress fibers in response to growth factors. Cell 70: 389–399.
Ross RS (2002) The extracellular connections: the role of integrins in myocardial remodeling. J Card Fail 8: S326–331.
Ross RS (2004) Molecular and mechanical synergy: cross-talk between integrins and growth factor receptors. Cardiovasc Res 63: 381–390.
Ross RS, Borg TK (2001) Integrins and the myocardium. Circ Res 88: 1112–1119.
Rota M, Boni A, Urbanek K, Padin-Iruegas ME, Kajstura TJ, Fiore G, Kubo H, Sonnenblick EH, Musso E, Houser SR, Leri A, Sussman MA, Anversa P (2005) Nuclear targeting of Akt enhances ventricular function and myocyte contractility. Circ Res 97: 1332–1341.
Sabri A, Steinberg SF (2003) Protein kinase C isoform-selective signals that lead to cardiac hypertrophy and the progression of heart failure. Mol Cell Biochem 251: 97–101.
Sadoshima J, Montagne O, Wang Q, Yang G, Warden J, Liu J, Takagi G, Karoor V, Hong C, Johnson GL, Vatner DE, Vatner SF (2002) The MEKK1-JNK pathway plays a protective role in pressure overload but does not mediate cardiac hypertrophy. J Clin Invest 110: 271–279.
Salazar EP, Rozengurt E (2001) Src family kinases are required for integrin-mediated but not for G protein-coupled receptor stimulation of focal adhesion kinase autophosphorylation at Tyr-397. J Biol Chem 276: 17788–17795.
Samarel AM (2005) Costameres, focal adhesions, and cardiomyocyte mechanotransduction. Am J Physiol Heart Circ Physiol 289: H2291–H2301.
Sarbassov DD, Ali SM, Sabatini DM (2005a) Growing roles for the mTOR pathway. Curr Opin Cell Biol 17: 596–603.

Sarbassov DD, Guertin DA, Ali SM, Sabatini DM (2005b) Phosphorylation and regulation of Akt/PKB by the rictor-mTOR complex. Science 307: 1098–1101.
Satoh M, Ogita H, Takeshita K, Mukai Y, Kwiatkowski DJ, Liao JK (2006) Requirement of Rac1 in the development of cardiac hypertrophy. Proc Natl Acad Sci USA 103: 7432–7437.
Sawada Y, Tamada M, Dubin-Thaler BJ, Cherniavskaya O, Sakai R, Tanaka S, Sheetz MP (2006) Force sensing by mechanical extension of the Src family kinase substrate p130Cas. Cell 127: 1015–1026.
Schoenwaelder SM, Burridge K (1999) Bidirectional signaling between the cytoskeleton and integrins. Curr Opin Cell Biol 11: 274–286.
Shai SY, Harpf AE, Babbitt CJ, Jordan MC, Fishbein MC, Chen J, Omura M, Leil TA, Becker KD, Jiang M, Smith DJ, Cherry SR, Loftus JC, Ross RS (2002) Cardiac myocyte-specific excision of the beta1 integrin gene results in myocardial fibrosis and cardiac failure. Circ Res 90: 458–464.
Shiraishi I, Melendez J, Ahn Y, Skavdahl M, Murphy E, Welch S, Schaefer E, Walsh K, Rosenzweig A, Torella D, Nurzynska D, Kajstura J, Leri A, Anversa P, Sussman MA (2004) Nuclear targeting of Akt enhances kinase activity and survival of cardiomyocytes. Circ Res 94: 884–891.
Skurk C, Izumiya Y, Maatz H, Razeghi P, Shiojima I, Sandri M, Sato K, Zeng L, Schiekofer S, Pimentel D, Lecker S, Taegtmeyer H, Goldberg AL, Walsh K (2005) The FOXO3a transcription factor regulates cardiac myocyte size downstream of AKT signaling. J Biol Chem 280: 20814–20823.
Stawowy P, Margeta C, Blaschke F, Lindschau C, Spencer-Hansch C, Leitges M, Biagini G, Fleck E, Graf K (2005) Protein kinase C epsilon mediates angiotensin II-induced activation of beta1-integrins in cardiac fibroblasts. Cardiovasc Res 67: 50–59.
Sugden PH, Clerk A (1998) 'Stress-responsive' mitogen-activated protein kinases (c-Jun N-terminal kinases and p38 mitogen-activated protein kinases) in the myocardium. Circ Res 83: 345–352.
Sussman MA, Welch S, Walker A, Klevitsky R, Hewett TE, Price RL, Schaefer E, Yager K (2000) Altered focal adhesion regulation correlates with cardiomyopathy in mice expressing constitutively active rac1. J Clin Invest 105: 875–886.
Takagi J, Springer TA (2002) Integrin activation and structural rearrangement. Immunol Rev 186: 141–163.
Takahashi N, Seko Y, Noiri E, Tobe K, Kadowaki T, Sabe H, Yazaki Y (1999) Vascular endothelial growth factor induces activation and subcellular translocation of focal adhesion kinase (p125FAK) in cultured rat cardiac myocytes. Circ Res 84: 1194–1202.
Takeishi Y, Chu G, Kirkpatrick DM, Li Z, Wakasaki H, Kranias EG, King GL, Walsh RA (1998) In vivo phosphorylation of cardiac troponin I by protein kinase Cbeta2 decreases cardiomyocyte calcium responsiveness and contractility in transgenic mouse hearts. J Clin Invest 102: 72–78.
Talwar S, Squire IB, Downie PF, O'Brien RJ, Davies JE, Ng LL (2000) Elevated circulating cardiotrophin-1 in heart failure: relationship with parameters of left ventricular systolic dysfunction. Clin Sci (Lond) 99: 83–88.
Taylor JM, Rovin JD, Parsons JT (2000) A role for focal adhesion kinase in phenylephrine-induced hypertrophy of rat ventricular cardiomyocytes. J Biol Chem 275: 19250–19257.
Thannickal VJ, Lee DY, White ES, Cui Z, Larios JM, Chacon R, Horowitz JC, Day RM, Thomas PE (2003) Myofibroblast differentiation by transforming growth factor-beta1 is dependent on cell adhesion and integrin signaling via focal adhesion kinase. J Biol Chem 278: 12384–12389.
Thorburn J, Xu S, Thorburn A (1997) MAP kinase- and Rho-dependent signals interact to regulate gene expression but not actin morphology in cardiac muscle cells. Embo J 16: 1888–1900.
Toker A, Newton AC (2000) Akt/protein kinase B is regulated by autophosphorylation at the hypothetical PDK-2 site. J Biol Chem 275: 8271–8274.
Torsoni AS, Constancio SS, Nadruz W, Jr., Hanks SK, Franchini KG (2003) Focal adhesion kinase is activated and mediates the early hypertrophic response to stretch in cardiac myocytes. Circ Res 93: 140–147.

Tremblay ML, Giguere V (2008) Phosphatases at the heart of FoxO metabolic control. Cell Metab 7: 101–103.
Troussard AA, Mawji NM, Ong C, Mui A, St -Arnaud R, Dedhar S (2003) Conditional knock-out of integrin-linked kinase demonstrates an essential role in protein kinase B/Akt activation. J Biol Chem 278: 22374–22378.
Uehata M, Ishizaki T, Satoh H, Ono T, Kawahara T, Morishita T, Tamakawa H, Yamagami K, Inui J, Maekawa M, Narumiya S (1997) Calcium sensitization of smooth muscle mediated by a Rho-associated protein kinase in hypertension. Nature 389: 990–994.
Ueyama T, Kawashima S, Sakoda T, Rikitake Y, Ishida T, Kawai M, Yamashita T, Ishido S, Hotta H, Yokoyama M (2000) Requirement of activation of the extracellular signal-regulated kinase cascade in myocardial cell hypertrophy. J Mol Cell Cardiol 32: 947–960.
Valencik ML, Keller RS, Loftus JC, McDonald JA (2002) A lethal perinatal cardiac phenotype resulting from altered integrin function in cardiomyocytes. J Card Fail 8: 262–272.
Valencik ML, McDonald JA (2001) Cardiac expression of a gain-of-function alpha(5)-integrin results in perinatal lethality. Am J Physiol Heart Circ Physiol 280: H361–H367.
Valencik ML, Zhang D, Punske B, Hu P, McDonald JA, Litwin SE (2006) Integrin activation in the heart: a link between electrical and contractile dysfunction? Circ Res 99: 1403–1410.
van der Flier A, Gaspar AC, Thorsteinsdottir S, Baudoin C, Groeneveld E, Mummery CL, Sonnenberg A (1997) Spatial and temporal expression of the beta1D integrin during mouse development. Dev Dyn 210: 472–486.
van Wamel AJ, Ruwhof C, van der Valk-Kokshoom LE, Schrier PI, van der Laarse A (2001) The role of angiotensin II, endothelin-1 and transforming growth factor-beta as autocrine/paracrine mediators of stretch-induced cardiomyocyte hypertrophy. Mol Cell Biochem 218: 113–124.
Wang F, Seta Y, Baumgarten G, Engel DJ, Sivasubramanian N, Mann DL (2001) Functional significance of hemodynamic overload-induced expression of leukemia-inhibitory factor in the adult mammalian heart. Circulation 103: 1296–1302.
Wang SM, Tsai YJ, Jiang MJ, Tseng YZ (1997) Studies on the function of rho A protein in cardiac myofibrillogenesis. J Cell Biochem 66: 43–53.
Wang Y (2007) Mitogen-activated protein kinases in heart development and diseases. Circulation 116: 1413–1423.
Wang Y, Huang S, Sah VP, Ross J, Jr., Brown JH, Han J, Chien KR (1998) Cardiac muscle cell hypertrophy and apoptosis induced by distinct members of the p38 mitogen-activated protein kinase family. J Biol Chem 273: 2161–2168.
Wary KK, Mariotti A, Zurzolo C, Giancotti FG (1998) A requirement for caveolin-1 and associated kinase Fyn in integrin signaling and anchorage-dependent cell growth. Cell 94: 625–634.
Wennerberg K, Rossman KL, Der CJ (2005) The Ras superfamily at a glance. J Cell Sci 118: 843–846.
White DE, Coutu P, Shi YF, Tardif JC, Nattel S, St Arnaud R, Dedhar S, Muller WJ (2006) Targeted ablation of ILK from the murine heart results in dilated cardiomyopathy and spontaneous heart failure. Genes Dev 20: 2355–2360.
Wu C, Dedhar S (2001) Integrin-linked kinase (ILK) and its interactors: a new paradigm for the coupling of extracellular matrix to actin cytoskeleton and signaling complexes. J Cell Biol 155: 505–510.
Wullschleger S, Loewith R, Hall MN (2006) TOR signaling in growth and metabolism. Cell 124: 471–484.
Yasukawa H, Hoshijima M, Gu Y, Nakamura T, Pradervand S, Hanada T, Hanakawa Y, Yoshimura A, Ross J, Jr., Chien KR (2001) Suppressor of cytokine signaling-3 is a biomechanical stress-inducible gene that suppresses gp130-mediated cardiac myocyte hypertrophy and survival pathways. J Clin Invest 108: 1459–1467.
Zechner D, Thuerauf DJ, Hanford DS, McDonough PM, Glembotski CC (1997) A role for the p38 mitogen-activated protein kinase pathway in myocardial cell growth, sarcomeric organization, and cardiac-specific gene expression. J Cell Biol 139: 115–127.

Zhang KR, Liu HT, Zhang HF, Zhang QJ, Li QX, Yu QJ, Guo WY, Wang HC, Gao F (2007) Long-term aerobic exercise protects the heart against ischemia/reperfusion injury via PI3 kinase-dependent and Akt-mediated mechanism. Apoptosis 12: 1579–1588.
Zhang YM, Bo J, Taffet GE, Chang J, Shi J, Reddy AK, Michael LH, Schneider MD, Entman ML, Schwartz RJ, Wei L (2006) Targeted deletion of ROCK1 protects the heart against pressure overload by inhibiting reactive fibrosis. Faseb J 20: 916–925.
Zhao J, Bian ZC, Yee K, Chen BP, Chien S, Guan JL (2003) Identification of transcription factor KLF8 as a downstream target of focal adhesion kinase in its regulation of cyclin D1 and cell cycle progression. Mol Cell 11: 1503–1515.
Zou Y, Akazawa H, Qin Y, Sano M, Takano H, Minamino T, Makita N, Iwanaga K, Zhu W, Kudoh S, Toko H, Tamura K, Kihara M, Nagai T, Fukamizu A, Umemura S, Iiri T, Fujita T, Komuro I (2004) Mechanical stress activates angiotensin II type 1 receptor without the involvement of angiotensin II. Nat Cell Biol 6: 499–506.
Zuluaga S, Alvarez-Barrientos A, Gutierrez-Uzquiza A, Benito M, Nebreda AR, Porras A (2007) Negative regulation of Akt activity by p38alpha MAP kinase in cardiomyocytes involves membrane localization of PP2A through interaction with caveolin-1. Cell Signal 19: 62–74.

Chapter 4
Mechanical Stress Induces Cardiomyocyte Hypertrophy Through Agonist-Independent Activation of Angiotensin II Type 1 Receptor

Hiroshi Akazawa and Issei Komuro

Abstract The angiotensin II (AngII) type 1 (AT_1) receptor is a seven-transmembrane G protein-coupled receptor that plays a crucial role in the development of load-induced cardiac hypertrophy. Besides systemically and locally generated AngII, mechanical stress can activate the AT_1 receptor, and induce cardiac hypertrophy in vivo. In response to stretch stimulation, the AT_1 receptor undergoes a specific switch in the receptor conformation without the involvement of AngII. The agonist-independent activation of the AT_1 receptor can be inhibited by inverse agonists, but not by neutral antagonists, through the specific drug-receptor interactions. It is conceptually novel that the AT_1 receptor, a member of G protein-coupled receptor, is a mechanical force-transducing molecule and mediates mechanical stress-induced cellular responses. In addition, inverse agonist activity emerges as an important pharmacological parameter for the AT_1 receptor blockers that determines the efficacy to prevent organ damage in cardiovascular diseases. In this section, molecular and structural bases for mechanosensation by the AT_1 receptor and inverse agonism at the AT_1 receptor will be discussed.

Keywords Cardiac hypertrophy · G protein-coupled receptor · Inverse agonist · Mechanical stress · Receptor conformation

4.1 Introduction

Mechanical stress is a stimulus of profound significance to cardiomyocytes, because it modulates multiple cellular responses such as protein synthesis, gene expression, ion channel function, sarcomere assembly, cell size and shape, and ultimately cell

I. Komuro (✉)
Department of Cardiovascular Science and Medicine, Chiba University Graduate School of Medicine, Chiba, Japan
e-mail: komuro-tky@umin.ac.jp

A. Kamkin, I. Kiseleva (eds.), *Mechanosensitivity of the Heart*,
Mechanosensitivity in Cells and Tissues 3, DOI 10.1007/978-90-481-2850-1_4,

survival (Komuro and Yazaki, 1993; Orr et al., 2006). In a variety of pathological conditions (e.g. hypertension, valvular heart disease, myocardial infarction, and cardiomyopathy) that impose hemodynamic overload on the heart, cardiomyocytes undergo hypertrophic growth. According to the law of Laplace, cardiac hypertrophy is an adaptive process in a sense that it reduces wall stress. However, prolongation of this process leads to deleterious outcomes such as congestive heart failure, arrhythmia, and sudden death (Lorell and Carabello, 2000). Therefore, the elucidation of molecular mechanisms underlying the development of cardiac hypertrophy is an important subject of cardiovascular research.

Ingeniously designed experiments have demonstrated that mechanical stretching of cardiomyocytes is sufficient for induction of hypertrophic responses, but concomitant increase in the actions of aurocrine and/or paracrine neurohumoral factors may also participate in this process. Therefore, it is challenging to solve how mechanical stress is perceived by cardiomyocytes as a stimulus (mechanosensation) and how mechanical force is converted by cardiomyocytes to biochemical intracellular signals to induce biological responses (mechanotransduction). Several mechanosensors have been reported to date such as muscle LIM protein within the Z-disc (Knoll et al., 2002), integrin-linked kinase (Bendig et al., 2006; White et al., 2006) and melusin (Brancaccio et al., 2003) within the costameres (band-like structures linking sarcolemmal membrane to Z-discs), and stretch-sensitive ion channels (Kung, 2005; Orr et al., 2006). However, the precise mechanisms of mechanosensation and mechanotransduction involving these sensors remain undetermined, especially during the process of load-induced hypertrophic growth.

The angiotensin II (AngII) type 1 (AT_1) receptor is a typical member of the G protein-coupled receptor (GPCR) family, the structure of which is characterized by seven transmembrane-spanning α-helices with an extracellular N-terminus and a cytoplasmic C-terminus (Gether, 2000; Gether and Kobilka, 1998; Miura et al., 2003a). Canonically, AT_1 receptor is activated upon binding to AngII, the specific and endogenous agonist. It is widely recognized that activation of AT_1 receptor contributes to load-induced cardiac hypertrophy, and that pharmacological interference of AT_1 receptor activation by using AT_1 receptor blockers (ARBs) or angiotensin converting enzyme (ACE) inhibitors can reduce hypertrophic growth in patients with hemodynamic overload. But, recent studies revised the paradigm by providing compelling evidence that AT_1 receptor can also be activated by mechanical stress independently of AngII, and that mechanical stress induces cardiac hypertrophy both in vitro and in vivo without the involvement of AngII (Yasuda et al., 2008; Zou et al., 2004). These observations have in turn led to identification of the ligands that are able to inhibit agonist-independent receptor activation, i.e. inverse agonists (Bond and Ijzerman, 2006; Milligan, 2003; Strange, 2002), and now prompt us to re-evaluate pharmacological actions of ARBs.

In this section, we will review the current understanding of mechanical stress-induced AT_1 receptor activation, especially focusing on the structure-function relationship and the pathophysiological or therapeutical relevance in the development of cardiac hypertrophy.

4.2 Activation of AT_1 Receptor in the Development of Cardiac Hypertrophy

A growing body of evidence has shown that various humoral factors such as vasoactive peptides, sympathetic nervous system, cytokines and growth factors contribute to the development of cardiac hypertrophy. Especially, activation of renin-angiotensin system (RAS) plays a central role, and it is well established that pharmacological inhibition of RAS can prevent progression of cardiac hypertrophy and reduce the morbidity and mortality in patients with heart failure (Jessup and Brozena, 2003). In addition to the systemic effects including elevation of blood pressure, sodium and water retention, and activation of sympathetic nervous system, the RAS has unfavorable direct effects on the hearts, especially through a system of local activation in tissues (Paul et al., 2006; Re, 2004). AngII is the pivotal bioactive molecule of RAS, and activates AT_1 and AT_2 receptors, which show 30% of sequence similarity (de Gasparo et al., 2000). Most of the pathophysiological actions of AngII in the cardiovascular system are mainly mediated through AT_1 receptor (de Gasparo et al., 2000). AngII infusion in rats induced cardiac hypertrophy via AT_1 receptor, independently of blood pressure elevation (Dostal and Baker, 1992), and cardiac-specific overexpression of AT_1 receptor also induced cardiac hypertrophy in mice (Hein et al., 1997; Paradis et al., 2000). These results suggest that activation of AT_1 receptor is sufficient for inducing cardiac hypertrophy. Currently, several kinds of non-peptide compounds that selectively inhibit AT_1 receptor activation are available for clinical use as ARBs (Zaman et al., 2002). According to a meta-analysis that evaluated the effects of antihypertensive therapy on cardiac hypertrophy, ARB is the most effective drug class for reducing left ventricular mass in patients with essential hypertension (Klingbeil et al., 2003). In addition, a randomized controlled trial of the Losartan Intervention for European Reduction in Hypertension (LIFE) study provided evidence that ARB conferred benefits to reduce left ventricular mass beyond blood pressure lowering (Kjeldsen et al., 2002). These clinical data strongly support the idea that AT_1 receptor plays a crucial role in the development of cardiac hypertrophy.

A large number of in vitro experiments have demonstrated that activation of AT_1 receptor via $G_{q/11}$ protein coupling stimulates diverse intracellular signaling pathways and enhances production of reactive oxygen species, which consequently evokes hypertrophic responses in cardiomyocytes (Hunyady and Catt, 2006; Kim and Iwao, 2000). Especially, AT_1 receptor signals the mitogen-activated protein (MAP) kinase family such as extracellular signal-regulated protein kinases (ERKs) (Yamazaki et al., 1995; Zou et al., 1996), c-Jun NH(2)-terminal kinase (JNK) (Kudoh et al., 1997) and p38 mitogen-activated protein (MAP) kinase (Nishida et al., 2005). Although the signaling pathways linking AT_1 receptor to ERKs vary according to cell-types, protein kinase C (PKC) and Raf-1 kinase are critically important as the upstream elements of ERKs cascade in cardiomyocytes (Zou et al., 1996). Activated ERKs promote protein synthesis by enhancing p70 S6 kinase activity and ribosomal RNA transcription (Bueno and Molkentin, 2002). In addition,

ERKs phosphorylate and activate several transcription factors such as GATA-4 and STATs or transcriptional coactivators such as p300 and CBP, and thereby enhance gene expression associated with hypertrophic response (Bueno and Molkentin, 2002). Activation of AT_1 receptor also stimulates G-protein-independent signaling pathways such as activation of Jak/STAT pathway and β-arrestin-mediated activation of ERKs (Hunyady and Turu, 2004).

4.3 Mechanical Stress in the Development of Cardiac Hypertrophy

Mechanical stress, in accompany with neurohumoral factors, is the primary stimulus for cardiac hypertrophy. Previously, it was reported that hemodynamic overload induced cardiac hypertrophy through the actions by catecholamines (Siri and McNamara, 1987). However, in pressure-overloaded feline right ventricle, hypertrophic growth was not observed in a single papillary muscle, which was unloaded by transection of chordae tendinae within an otherwise normally loaded ventricle (Cooper et al., 1985). Furthermore, in isolated hearts perfused as Langendorff preparations, the increase in protein synthesis was most closely related to stretching of ventricular wall as a consequence of increased afterload (Kira et al., 1984). An increase in protein synthesis was also observed, when cardiomyocytes cultured on deformable silicone rubber dishes underwent passive stretch even in serum-free media (Mann et al., 1989). These results clearly indicate that mechanical stress itself induces hypertrophic responses primarily by stretching cardiomyocytes without the involvement of neurohumoral factors. Utilizing this kind of device for stretching cultured cells, we and others have demonstrated that mechanical stretching of cultured cardiomyocytes induced hypertrophic responses such as activation of many protein kinases including ERKs and induction of immediate early response genes or fetal-type genes (Komuro and Yazaki, 1993; Sadoshima and Izumo, 1997). Intriguingly, the intracellular signals elicited by mechanical stretch are similar to those by AT_1 receptor activation in cardiomyocytes. For example, the protein kinase cascade, PKC/Raf-1/ERKs, mediates reprogramming of gene expression induced by mechanical stretch as well as by AngII stimulation (Komuro and Yazaki, 1993; Sadoshima and Izumo, 1997).

4.4 Mechanical Stress-Induced Activation of AT_1 Receptor

We recently demonstrated that mechanical stress activates AT_1 receptor independently of AngII (Yasuda et al., 2008; Zou et al., 2004). As mentioned above, activation of AT_1 receptor is profoundly involved in the development of load-induced cardiac hypertrophy. Importantly, pretreatment with ARBs significantly attenuated hypertrophic responses in cardiomyocytes induced by stretching (Sadoshima et al., 1993; Yamazaki et al., 1995). These results indicate that mechanical stress induces cardiac hypertrophy through the activation of AT_1 receptor. However, it has been a challenging problem to solve how AT_1 receptor is activated by mechanical stress. One possibility is that AngII is stored in cardiomyocytes, and that mechanical stretch induces the secretion of stored AngII into the culture medium, resulting in the induction of cardiomyocyte hypertrophy by the autocrine mechanism (Sadoshima et al., 1993). However, direct measurement of AngII concentration in the medium conditioned by stretching cardiomyocytes did not reveal a significant increase in AngII concentration (Zou et al., 2004). Furthermore, a neutralizing antibody to AngII did not suppress the stretch-induced ERKs activation in cardiomyocytes, although the antibody abolished AngII-induced ERKs activation (Zou et al., 2004). These results suggest that that AngII, even if secreted from cardiomyocytes, plays a marginal role in stretch-induced ERKs activation, and raise quite a different possibility that mechanical stress can activate the AT_1 receptor without the involvement of AngII (Fig. 4.1).

In human embryonic kidney (HEK) 293 cells showing no detectable expression of AT_1 receptor and angiotensinogen, neither AngII nor mechanical stretch activated ERKs, but forced expression of AT_1 receptor conferred the ability to activate ERKs in response to both AngII and mechanical stretch (Fig. 4.2). Interestingly, candesartan, as an inverse agonist, inhibited the ERKs activation induced not only by AngII but also by mechanical stretch in HEK293 cells expressing AT_1 receptor. Stretch stimuliation also activated ERKs in HEK293 cells expressing AT1 mutant which did not bind AngII (Yamano et al., 1992) and in cardiomyocytes prepared from *angiotensinogen*-deficient mice (Tanimoto et al., 1994) (Fig. 4.1), and these activations were inhibited by candesartan (Zou et al., 2004). Furthermore, mechanical stress can induce cardiac hypertrophy in vivo through the AT_1 receptor

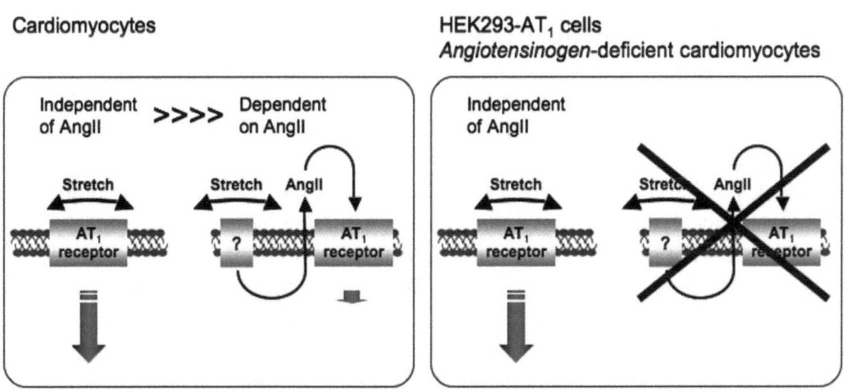

Fig. 4.1 Mechanical stress-induced AT_1 receptor activation in cardiomyocytes, HEK293- AT_1 cells and *angiotensinogen*-deficient cardiomyocytes

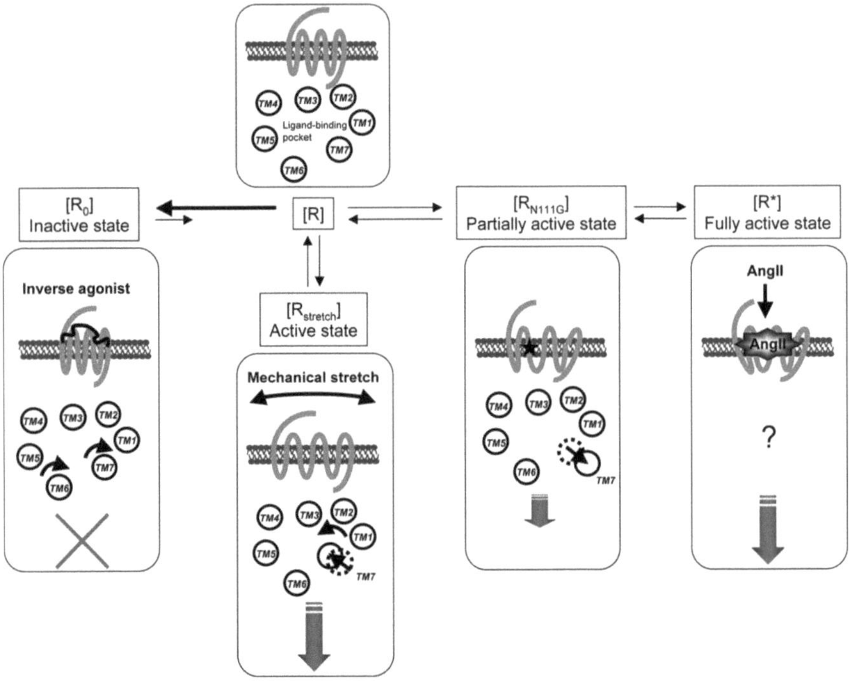

Fig. 4.2 Distinct conformations of the AT_1 receptor. R is an unaligned inactive state, and $[R_0]$ is an inactive state stabilized by an inverse agonist. [R*] is an active state stabilized by the agonist AngII. $[R_{N111G}]$ is an partially active state of AT_1-N111G mutant receptor. Mechanical stretch stabilizes AT_1 receptor to another active state $[R_{stretch}]$, independently of AngII. An inverse agonist forcibly induces a distinct transition from [R] to $[R_0]$, and prevent a shift of equilibrium to [R*] or $[R_{stretch}]$

in the absence of AngII, because pressure overload induced cardiac hypertrophy in *angiotensinogen*-deficient mice as well as in wild-type mice, which was significantly inhibited by candesartan. These experimental data provided compelling evidence that AT_1 receptor is activated in the absence of AngII both in vitro and in vivo, and that this AngII-independent activation of AT_1 receptor is inhibited by candesartan.

Besides AT_1 receptor, several GPCRs, such as the receptors of endothelin 1 (ET-1) and catecholamines, also contribute to induction of cardiomyocyte hypertrophy (Yamazaki et al., 1996; Zou et al., 1999). However, mechanical stretch did not induce significant activation of ERKs in COS7 cells expressing either ET-1 type A receptor or β_2-adrenoceptor in a ligand-independent manner. These results suggest that the activation of GPCRs by mechanical stretch without the involvement of their agonists is not a general phenomenon but specific to some GPCRs including the AT_1 receptor.

4.5 Conformational Change of AT_1 Receptor During Mechanical Stress-Induced Activation

Since AT_1 receptor is activated by mechanical stress, AT_1 receptor is supposed to undergo a conformational switch that couples mechanical stress-induced activation. We recently demonstrated by substituted cysteine accessibility mapping (SCAM) that mechanical stretch increased the accessibility of Cys^{289} in transmembrane (TM) 7 to the ligand-binding pocket in a time-dependent manner (Yasuda et al., 2008). The SCAM study is used to probe relative conformational changes of GPCRs by validating the presence of Cys residues within the ligand-binding pocket (Boucard et al., 2003; Chen et al., 2002; Jongejan et al., 2005; Lemaire et al., 2004; Martin et al., 2007; Miura and Karnik, 2002; Miura et al., 2003b). According to the results of a series of SCAM experiments using mutant receptors with substitution of the TM7 residue ranging from Thr^{287} to Asn^{295} to Cys one at a time, TM7 undergoes a counterclockwise rotation and a shift into the ligand-binding pocket in response to mechanical stretch (Yasuda et al., 2008) (Fig. 4.2). It is probable that the stabilizing interactions involving TM7 in AT_1 receptor are disrupted by mechanical stress independently of AngII and that counterclockwise rotation of TM7 may cause activation of intracellular signaling pathways. A shift of TM7 to inside the ligand-binding pocket during mechanical stress-induced activation contrasts well with the helical movement observed in a constitutively active AT_1-N111G receptor, which contains an Asn^{111} to Gly mutation, because TM7 shifts apart from the ligand-binding pocket in this mutant receptor (Boucard et al., 2003). Since AT_1-N111G receptor mimics the state of WT receptor partially activated by AngII (Le et al., 2003; Miura and Karnik, 2002), an active conformation of AT_1 receptor induced by mechanical stress may be substantially different from that by AngII-dependent receptor activation (Fig. 4.2).

Next obvious question is how the AT_1 receptor senses mechanical stress and undergoes a conformational switch. First, membrane tension may directly induce the conformational change of AT_1 receptor. Reconstitution of mechanosensitive channel of large conductance from *Escherichia coli* in synthetic phosphatidylcholines with different chain lengths revealed that thin bilayer favored the open state of channels while thick bilayer stabilized the closed state (Perozo et al., 2002). Likewise, membrane tension may induce thinning of the lipid bilayer, which triggers tilting of TM7 of AT_1 receptor to avoid hydrophobic mismatch and to rectify lateral pressure profile (Orr et al., 2006). If so, it follows that AT_1 receptor, a GPCR, functions as a receptor for mechanosensation. It will be intriguing, because GPCRs are involved in mediating senses of vision, olfaction and much of gustation, of Aristotle's five senses (Kung, 2005). Second, mechanical stretch may activate unspecified mechanosensors, which secondarily activate AT_1 receptor. It will be a great challenge to elucidate the precise mechanism of mechanosensing by AT_1 receptor.

4.6 Inverse Agonism on Stretch-Induced Activation of AT_1 Receptor

Before the early 1990s, GPCR ligands were simply classified as agonists or antagonists (Bond and Ijzerman, 2006; Milligan, 2003; Strange, 2002). Both agonists and antagonists bind to the cognate GPCR with high affinity, but only agonists can activate the receptor. Therefore, agonists possess both high affinity and positive efficacy, whereas antagonists posses high affinity without intrinsic efficacy. However, some compounds, originally described as antagonists, have been demonstrated to produce effects opposite to those by agonists. Such ligands are classified as "inverse agonists" that have negative efficacy. An inverse agonist stabilizes inactive conformation of the receptor and reduces constitutive activity of the receptor or the agonist-independent receptor activity (Fig. 4.3).

ARBs share a common action, namely blocking AngII-mediated responses, but show a unique pattern of pharmacological properties (Oparil, 2000). The inverse agonist activity of ARBs could be of clinical advantage to inhibition of both AngII-dependent and -independent receptor activation, and thus be a novel and important pharmacological parameter defining the beneficial effects on organ protection. Candesartan reduces the basal activation of *c-fos* gene promoter by AT_1 receptor or a constitutively active AT_1-N111G mutant receptor (Boucard et al., 2003), suggesting that candesartan is an ARB with potent inverse agonist activity (Yasuda et al., 2008). As mentioned above, candesartan suppressed mechanical stretch-induced helical movement of AT_1 receptor (Yasuda et al., 2008), and thereby inhibited receptor activation (Zou et al., 2004). Inverse agonism of candesartan is especially relevant to its ability to attenuate load-induced cardiac hypertrophy, because pressure overload by constricting the transverse aorta induced cardiac hypertrophy even in *angiotensinogen*-deficient mice as well as in wild-type mice, which was significantly inhibited by candesartan (Zou et al., 2004).

Although the inverse agonist activity of individual ARBs has not been fully evaluated, we should consider that the distinctive activity of inverse agonism is primarily determined by chemical structure of the drug. Most of ARBs have a biphenyl-tetrazole ring structure in common, which interacts with Lys^{199} and His^{256} in the AT_1 receptor (Noda et al., 1995). It was reported that the carboxyl group at the

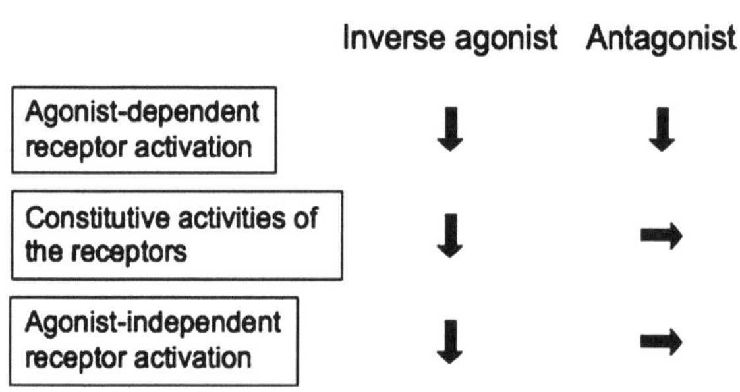

Fig. 4.3 Classification of GPCR ligands as inverse agonists and antagonists

benzimidazole ring of candesartan is an important structure for insurmountable inhibition of AngII-induced receptor activation (Noda et al., 1993; Takezako et al., 2004). We recently found that the bindings of the carboxyl group of candesartan to Gln^{257} in TM6 and Thr^{287} in TM7 are responsible for the potent inverse agonism in inhibiting mechanical stretch-induced activation of AT_1 receptor (Yasuda et al., 2008). It is reasonable that the tight binding to AT_1 receptor is prerequisite for an inverse agonist to stabilize the receptor in an inactive conformation, as well as to exert insurmountable inhibition of AngII-induced receptor activation.

According to a sequential binding and conformational model for the molecular mechanism of ligand action on GPCRs (Gether, 2000; Perez and Karnik, 2005), the unaligned receptor in a state [R] can undergo transition to at least two other stabilized states [R_0] and [R*]. [R_0] is an inactive state stabilized by an inverse agonist, and [R*] is an active state stabilized by an agonist. It is consistent with the result of a recent study using a fluorescence resonance energy transfer approach, demonstrating that agonists and inverse agonists for α_{2A}-adrenergic receptor induced distinct conformational changes of the receptor (Vilardaga et al., 2005). With regard to AT_1 receptor, mechanical stretch stabilizes the receptor to another active state [$R_{stretch}$] (Fig. 4.2). Molecular modeling using the crystal structure of bovine rhodopsin (Palczewski et al., 2000) as a template indicates that, in the inactive state [R_0] in the presence of candesartan, TM6 and 7 move with clockwise rotation, as a consequence of the bindings of the carboxyl group of candesartan to Gln^{257} in TM6 and Thr^{287} in TM7 (Yasuda et al., 2008) (Fig. 4.2). Therefore, candesartan, as an inverse agonist, forcibly induces a distinct transition from [R] to an inactive conformation [R_0], and prevents a shift of equilibrium to an active conformation [$R_{stretch}$] or [R*] (Fig. 4.2).

4.7 Concluding Remarks

The structure-function analyses of the AT_1 receptor have advanced our understanding of the molecular mechanism underlying mechanical stress-induced receptor activation. Although the structural flexibility of AT_1 receptor, like other GPCRs, may underlie the AngII-independent activation, mechanical stress-induced activation seems to be a phenomenon specific to AT_1 receptor. Future investigations with biophysical, biochemical, and pharmacological approaches will elucidate the precise mechanism of mechanosensing and mechanotransduction by AT_1 receptor. Although inverse agonism is now a well-recognized phenomenon in the field of receptor pharmacology, clinical importance of inverse agonist activity is still speculative. At least, in an experimental animal model, inverse agonist activity of ARBs is relevant to its ability to attenuate load-induced cardiac hypertrophy (Zou et al., 2004). Given that inverse agonist activity is a potential determinant of clinical benefits, molecular dissection of the structure-activity relationship will contribute to the development of a novel and desirable ARB.

Recently, crystal structures of β_1- and β_2-adrenergic receptors (Cherezov et al., 2007; Rasmussen et al., 2007; Rosenbaum et al., 2007; Warne, 2008, p. 390) have been obtained, and they reveal several key differences with that of bovine rhodopsin. Indeed the generation of GPCR crystals is exceedingly difficult because of structural instability, but hopefully, crystal structural information of AT_1 receptor in the conformation of $[R_{stretch}]$ will improve our understanding of receptor activation at a molecular level.

Acknowledgements The authors are supported in part by grants from the Japanese Ministry of Education, Science, Sports, and Culture, from Health and Labor Sciences Research Grants, Japan Health Sciences Foundation (to IK and HA); Takeda Medical Research Foundation, Takeda Science Foundation, Uehara Memorial Foundation, Kato Memorial Trust for Nambyo Research, Japan Medical Association (to IK); from Mochida Memorial Foundation, Japanese Heart Foundation/Novartis Research Award on Molecular and Cellular Cardiology, Japan Intractable Diseases Research Foundation, Kowa Life Science Foundation (to HA).

References

Bendig G, Grimmler M, Huttner IG, Wessels G, Dahme T, Just S, Trano N, Katus HA, Fishman MC, Rottbauer W (2006) Integrin-linked kinase, a novel component of the cardiac mechanical stretch sensor, controls contractility in the zebrafish heart. Genes Dev 20: 2361–2372.

Bond RA, Ijzerman AP (2006) Recent developments in constitutive receptor activity and inverse agonism, and their potential for GPCR drug discovery. Trends Pharmacol Sci 27: 92–96.

Boucard AA, Roy M, Beaulieu ME, Lavigne P, Escher E, Guillemette G, Leduc R (2003) Constitutive activation of the angiotensin II type 1 receptor alters the spatial proximity of transmembrane 7 to the ligand-binding pocket. J Biol Chem 278: 36628–36636.

Brancaccio M, Fratta L, Notte A, Hirsch E, Poulet R, Guazzone S, De Acetis M, Vecchione C, Marino G, Altruda F, Silengo L, Tarone G, Lembo G (2003) Melusin, a muscle-specific integrin beta1-interacting protein, is required to prevent cardiac failure in response to chronic pressure overload. Nat Med 9: 68–75.

Bueno OF, Molkentin JD (2002) Involvement of extracellular signal-regulated kinases 1/2 in cardiac hypertrophy and cell death. Circ Res 91: 776–781.

Chen S, Lin F, Xu M, Graham RM (2002) Phe(303) in TMVI of the alpha(1B)-adrenergic receptor is a key residue coupling TM helical movements to G-protein activation. Biochemistry 41: 588–596.

Cherezov V, Rosenbaum DM, Hanson MA, Rasmussen SG, Thian FS, Kobilka TS, Choi HJ, Kuhn P, Weis WI, Kobilka BK, Stevens RC (2007) High-resolution crystal structure of an engineered human beta2-adrenergic G protein-coupled receptor. Science 318: 1258–1265.

Cooper Gt, Kent RL, Uboh CE, Thompson EW, Marino TA (1985) Hemodynamic versus adrenergic control of cat right ventricular hypertrophy. J Clin Invest 75: 1403–1414.

de Gasparo M, Catt KJ, Inagami T, Wright JW, Unger T (2000) International union of pharmacology. XXIII. The angiotensin II receptors. Pharmacol Rev 52: 415–472.

Dostal DE, Baker KM (1992) Angiotensin II stimulation of left ventricular hypertrophy in adult rat heart. Mediation by the AT1 receptor. Am J Hypertens 5: 276–280.

Gether U (2000) Uncovering molecular mechanisms involved in activation of G protein-coupled receptors. Endocr Rev 21: 90–113.

Gether U, Kobilka BK (1998) G protein-coupled receptors. II. Mechanism of agonist activation. J Biol Chem 273: 17979–17982.

Hein L, Stevens ME, Barsh GS, Pratt RE, Kobilka BK, Dzau VJ (1997) Overexpression of angiotensin AT1 receptor transgene in the mouse myocardium produces a lethal phenotype associated with myocyte hyperplasia and heart block. Proc Natl Acad Sci USA 94: 6391–6396.

Hunyady L, Catt KJ (2006) Pleiotropic AT1 receptor signaling pathways mediating physiological and pathogenic actions of angiotensin II. Mol Endocrinol 20: 953–970.

Hunyady L, Turu G (2004) The role of the AT1 angiotensin receptor in cardiac hypertrophy: angiotensin II receptor or stretch sensor? Trends Endocrinol Metab 15: 405–408.

Jessup M, Brozena S (2003) Heart failure. N Engl J Med 348: 2007–2018.

Jongejan A, Bruysters M, Ballesteros JA, Haaksma E, Bakker RA, Pardo L, Leurs R (2005) Linking agonist binding to histamine H1 receptor activation. Nat Chem Biol 1: 98–103.

Kim S, Iwao H (2000) Molecular and cellular mechanisms of angiotensin II-mediated cardiovascular and renal diseases. Pharmacol Rev 52: 11–34.

Kira Y, Kochel PJ, Gordon EE, Morgan HE (1984) Aortic perfusion pressure as a determinant of cardiac protein synthesis. Am J Physiol 246: C247–C258.

Kjeldsen SE, Dahlof B, Devereux RB, Julius S, Aurup P, Edelman J, Beevers G, de Faire U, Fyhrquist F, Ibsen H, Kristianson K, Lederballe-Pedersen O, Lindholm LH, Nieminen MS, Omvik P, Oparil S, Snapinn S, Wedel H (2002) Effects of losartan on cardiovascular morbidity and mortality in patients with isolated systolic hypertension and left ventricular hypertrophy: a Losartan Intervention for Endpoint Reduction (LIFE) substudy. JAMA 288: 1491–1498.

Klingbeil AU, Schneider M, Martus P, Messerli FH, Schmieder RE (2003) A meta-analysis of the effects of treatment on left ventricular mass in essential hypertension. Am J Med 115: 41–46.

Knoll R, Hoshijima M, Hoffman HM, Person V, Lorenzen-Schmidt I, Bang ML, Hayashi T, Shiga N, Yasukawa H, Schaper W, McKenna W, Yokoyama M, Schork NJ, Omens JH, McCulloch AD, Kimura A, Gregorio CC, Poller W, Schaper J, Schultheiss HP, Chien KR (2002) The cardiac mechanical stretch sensor machinery involves a Z disc complex that is defective in a subset of human dilated cardiomyopathy. Cell 111: 943–955.

Komuro I, Yazaki Y (1993) Control of cardiac gene expression by mechanical stress. Annu Rev Physiol 55: 55–75.

Kudoh S, Komuro I, Mizuno T, Yamazaki T, Zou Y, Shiojima I, Takekoshi N, Yazaki Y (1997) Angiotensin II stimulates c-Jun NH2-terminal kinase in cultured cardiac myocytes of neonatal rats. Circ Res 80: 139–146.

Kung C (2005) A possible unifying principle for mechanosensation. Nature 436: 647–654.

Le MT, Vanderheyden PM, Szaszak M, Hunyady L, Kersemans V, Vauquelin G (2003) Peptide and nonpeptide antagonist interaction with constitutively active human AT1 receptors. Biochem Pharmacol 65: 1329–1338.

Lemaire K, Van de Velde S, Van Dijck P, Thevelein JM (2004) Glucose and sucrose act as agonist and mannose as antagonist ligands of the G protein-coupled receptor Gpr1 in the yeast Saccharomyces cerevisiae. Mol Cell 16: 293–299.

Lorell BH, Carabello BA (2000) Left ventricular hypertrophy: pathogenesis, detection, and prognosis. Circulation 102: 470–479.

Mann DL, Kent RL, Cooper Gt (1989) Load regulation of the properties of adult feline cardiocytes: growth induction by cellular deformation. Circ Res 64: 1079–1090.

Martin SS, Holleran BJ, Escher E, Guillemette G, Leduc R (2007) Activation of the angiotensin II type 1 receptor leads to movement of the sixth transmembrane domain: analysis by the substituted cysteine accessibility method. Mol Pharmacol 72: 182–190.

Milligan G (2003) Constitutive activity and inverse agonists of G protein-coupled receptors: a current perspective. Mol Pharmacol 64: 1271–1276.

Miura S, Karnik SS (2002) Constitutive activation of angiotensin II type 1 receptor alters the orientation of transmembrane Helix-2. J Biol Chem 277: 24299–24305.

Miura S, Saku K, Karnik SS (2003a) Molecular analysis of the structure and function of the angiotensin II type 1 receptor. Hypertens Res 26: 937–943.

Miura S, Zhang J, Boros J, Karnik SS (2003b) TM2-TM7 interaction in coupling movement of transmembrane helices to activation of the angiotensin II type-1 receptor. J Biol Chem 278: 3720–3725.

Nishida M, Tanabe S, Maruyama Y, Mangmool S, Urayama K, Nagamatsu Y, Takagahara S, Turner JH, Kozasa T, Kobayashi H, Sato Y, Kawanishi T, Inoue R, Nagao T, Kurose H (2005) G alpha 12/13- and reactive oxygen species-dependent activation of c-Jun NH2-terminal kinase and p38 mitogen-activated protein kinase by angiotensin receptor stimulation in rat neonatal cardiomyocytes. J Biol Chem 280: 18434–18441.

Noda K, Saad Y, Kinoshita A, Boyle TP, Graham RM, Husain A, Karnik SS (1995) Tetrazole and carboxylate groups of angiotensin receptor antagonists bind to the same subsite by different mechanisms. J Biol Chem 270: 2284–2289.
Noda M, Shibouta Y, Inada Y, Ojima M, Wada T, Sanada T, Kubo K, Kohara Y, Naka T, Nishikawa K (1993) Inhibition of rabbit aortic angiotensin II (AII) receptor by CV-11974, a new nonpeptide AII antagonist. Biochem Pharmacol 46: 311–318.
Oparil S (2000) Newly emerging pharmacologic differences in angiotensin II receptor blockers. Am J Hypertens 13: 18S–24S.
Orr AW, Helmke BP, Blackman BR, Schwartz MA (2006) Mechanisms of mechanotransduction. Dev Cell 10: 11–20.
Palczewski K, Kumasaka T, Hori T, Behnke CA, Motoshima H, Fox BA, Le Trong I, Teller DC, Okada T, Stenkamp RE, Yamamoto M, Miyano M (2000) Crystal structure of rhodopsin: A G protein-coupled receptor. Science 289: 739–745.
Paradis P, Dali-Youcef N, Paradis FW, Thibault G, Nemer M (2000) Overexpression of angiotensin II type I receptor in cardiomyocytes induces cardiac hypertrophy and remodeling. Proc Natl Acad Sci USA 97: 931–936.
Paul M, Poyan Mehr A, Kreutz R (2006) Physiology of local renin-angiotensin systems. Physiol Rev 86: 747–803.
Perez DM, Karnik SS (2005) Multiple signaling states of G-protein-coupled receptors. Pharmacol Rev 57: 147–161.
Perozo E, Cortes DM, Sompornpisut P, Kloda A, Martinac B (2002) Open channel structure of MscL and the gating mechanism of mechanosensitive channels. Nature 418: 942–948.
Rasmussen SG, Choi HJ, Rosenbaum DM, Kobilka TS, Thian FS, Edwards PC, Burghammer M, Ratnala VR, Sanishvili R, Fischetti RF, Schertler GF, Weis WI, Kobilka BK (2007) Crystal structure of the human beta2 adrenergic G-protein-coupled receptor. Nature 450: 383–387.
Re RN (2004) Mechanisms of disease: local renin-angiotensin-aldosterone systems and the pathogenesis and treatment of cardiovascular disease. Nat Clin Pract Cardiovasc Med 1: 42–47.
Rosenbaum DM, Cherezov V, Hanson MA, Rasmussen SG, Thian FS, Kobilka TS, Choi HJ, Yao XJ, Weis WI, Stevens RC, Kobilka BK (2007) GPCR engineering yields high-resolution structural insights into beta2-adrenergic receptor function. Science 318: 1266–1273.
Sadoshima J, Izumo S (1997) The cellular and molecular response of cardiac myocytes to mechanical stress. Annu Rev Physiol 59: 551–571.
Sadoshima J, Xu Y, Slayter HS, Izumo S (1993) Autocrine release of angiotensin II mediates stretch-induced hypertrophy of cardiac myocytes in vitro. Cell 75: 977–984.
Siri FM, McNamara JJ (1987) Effects of sympathectomy on heart size and function in aortic-constricted rats. Am J Physiol 252: H442–H447.
Strange PG (2002) Mechanisms of inverse agonism at G-protein-coupled receptors. Trends Pharmacol Sci 23: 89–95.
Takezako T, Gogonea C, Saad Y, Noda K, Karnik SS (2004) "Network leaning" as a mechanism of insurmountable antagonism of the angiotensin II type 1 receptor by non-peptide antagonists. J Biol Chem 279: 15248–15257.
Tanimoto K, Sugiyama F, Goto Y, Ishida J, Takimoto E, Yagami K, Fukamizu A, Murakami K (1994) Angiotensinogen-deficient mice with hypotension. J Biol Chem 269: 31334–31337.
Vilardaga JP, Steinmeyer R, Harms GS, Lohse MJ (2005) Molecular basis of inverse agonism in a G protein-coupled receptor. Nat Chem Biol 1: 25–28.
Warne T, Serrano-Vega MJ, Baker JG, Moukhametzianov R, Edwards PC, Henderson R, Leslie AG, Tate CG, Schertler GF (2008) Structure of a β_1-adrenergic G-protein-coupled receptor. Nature 454: 486–491.
White DE, Coutu P, Shi YF, Tardif JC, Nattel S, St Arnaud R, Dedhar S, Muller WJ (2006) Targeted ablation of ILK from the murine heart results in dilated cardiomyopathy and spontaneous heart failure. Genes Dev 20: 2355–2360.
Yamano Y, Ohyama K, Chaki S, Guo DF, Inagami T (1992) Identification of amino acid residues of rat angiotensin II receptor for ligand binding by site directed mutagenesis. Biochem Biophys Res Commun 187: 1426–1431.

Yamazaki T, Komuro I, Kudoh S, Zou Y, Shiojima I, Hiroi Y, Mizuno T, Maemura K, Kurihara H, Aikawa R, Takano H, Yazaki Y (1996) Endothelin-1 is involved in mechanical stress-induced cardiomyocyte hypertrophy. J Biol Chem 271: 3221–3228.

Yamazaki T, Komuro I, Kudoh S, Zou Y, Shiojima I, Mizuno T, Takano H, Hiroi Y, Ueki K, Tobe K (1995) Angiotensin II partly mediates mechanical stress-induced cardiac hypertrophy. Circ Res 77: 258–265.

Yasuda N, Miura S, Akazawa H, Tanaka T, Qin Y, Kiya Y, Imaizumi S, Fujino M, Ito K, Zou Y, Fukuhara S, Kunimoto S, Fukuzaki K, Sato T, Ge J, Mochizuki N, Nakaya H, Saku K, Komuro I (2008) Conformational switch of angiotensin II type 1 receptor underlying mechanical stress-induced activation. EMBO Rep 9: 179–186.

Zaman MA, Oparil S, Calhoun DA (2002) Drugs targeting the renin-angiotensin-aldosterone system. Nat Rev Drug Discov 1: 621–636.

Zou Y, Akazawa H, Qin Y, Sano M, Takano H, Minamino T, Makita N, Iwanaga K, Zhu W, Kudoh S, Toko H, Tamura K, Kihara M, Nagai T, Fukamizu A, Umemura S, Iiri T, Fujita T, Komuro I (2004) Mechanical stress activates angiotensin II type 1 receptor without the involvement of angiotensin II. Nat Cell Biol 6: 499–506.

Zou Y, Komuro I, Yamazaki T, Aikawa R, Kudoh S, Shiojima I, Hiroi Y, Mizuno T, Yazaki Y (1996) Protein kinase C, but not tyrosine kinases or Ras, plays a critical role in angiotensin II-induced activation of Raf-1 kinase and extracellular signal-regulated protein kinases in cardiac myocytes. J Biol Chem 271: 33592–33597.

Zou Y, Komuro I, Yamazaki T, Kudoh S, Uozumi H, Kadowaki T, Yazaki Y (1999) Both Gs and Gi proteins are critically involved in isoproterenol-induced cardiomyocyte hypertrophy. J Biol Chem 274: 9760–9770.

Part II
Mechanically Induced Potentials and Currents of the Cardiac Cells in Healthy and Diseased Myocardium

Chapter 5
Mechanostransduction in Cardiac and Stem-Cell Derived Cardiac Cells

Jeffrey G. Jacot, Anna J. Raskin, Jeffrey H. Omens, Andrew D. McCulloch, and Leslie Tung

Abstract This review focuses on mechanoelectric feedback and mechanotransduction in cardiac cells and stem-cell derived cardiomyocyte progenitor cells. Topics include: methods to apply mechanical stimuli to isolated cells and tissues; methods for patterned growth of cells; effects of stretch and shear stress on cellular function and tissue electrophysiology; regulation of structural and junctional proteins by stretch; the role of the cytoskeleton in mechanotransduction and heart failure; signaling pathways involved in mechanotransduction and load-induced hypertrophic responses; and the role of substrate stiffness in stem cell differentiation and maturation of excitation-contraction coupling.

Keywords Cardiomyocytes · Mechanical stimulation · Mechanotransduction · Signaling pathways

5.1 Introduction

More than maybe any other cell type, cardiomyocytes exist in an environment with extreme dynamic changes in stress and strain. In the intact heart, researchers have observed relationships between chamber distension and contractile force generation since the early nineteenth century. This relationship was popularly accepted, and subsequently referred to as the "Frank-Starling Law of the Heart" after a lecture by Ernest Starling in 1918, specifically noting that that cardiac tissue responds in a beat-to-beat fashion to load by increasing the ejection of blood in response to tissue distention (diastolic filling), maintaining blood volume homeostasis in the heart (Katz, 2002). Later studies realized that the Frank-Starling phenomenon resulted from the control of contractile force based on precontractile distension at the level of

L.Tung (✉)
Department of Biomedical Engineering, The Johns Hopkins University, Baltimore, MD, USA
e-mail: ltung@jhu.edu

A. Kamkin, I. Kiseleva (eds.), *Mechanosensitivity of the Heart*,
Mechanosensitivity in Cells and Tissues 3, DOI 10.1007/978-90-481-2850-1_5,

each individual cell – this effect has been measured down to the level of single heart fibers (Allen and Kentish, 1985). Today, magnetic resonance imaging has revealed the complex three-dimensional deformation of both normal and diseased hearts with exquisite detail. Researchers have documented many more responses of individual cardiomyocytes to a variety of loading conditions including shear and tensile stress and strain. For example, in the well-documented hypertrophic response, cardiomyocytes will increase their mass through a mechanotransductive response to increasing loads. In addition, mechanotransductive responses affect the development of cardiac structure, the electrical coupling of the cardiomyocytes, the beating frequency, the duration of the contraction, the electrophysiology, and the fine tuning of the cellular response to other endocrine and nervous signals.

In this chapter, we focus on in vitro studies of mechanotransduction in cardiomyocytes and the mechanisms involved in this mechanotransduction. We first describe the techniques used to apply various types of mechanical stress and strain to isolated cells and cells in culture and to control the alignment and patterning of those cultures, followed by an overview of the effects of those applied mechanical signals on the electrophysiology, growth (or hypertrophy) and function, and differentiation and development of cardiomyocytes. Finally, we discuss cytoskeletal, membrane and nuclear signaling for mechanotransduction, the most likely intracellular locations where these mechanical signals are processed.

5.2 Mechanical Signals Applied In Vitro to Cardiac Cells

Many approaches have been developed to apply mechanical stimuli to a host of cell types in vitro (for review, see (Brown, 2000)). With isolated cardiomyocytes, the methods used include axial stretch, cell indentation, substrate deformation or fluid shear. The first two approaches are practical only for studies of small numbers of cells, and have been used primarily to study electrophysiological changes at the single cell level. The latter two approaches are adaptable to large numbers of cells and the study of tissue-level electrophysiology and changes in gene and protein expression. Other forms of mechanical stimulation such as osmotic swelling (Baumgarten and Clemo, 2003) or intracellular pressure (Hagiwara et al., 1992) have been applied to single cardiac cells (Hu and Sachs, 1997), but may not be the same as those evoked by mechanical strain (Sasaki et al., 1992) and therefore are not included in this review (Fig. 5.1).

5.2.1 Direct Stretch

Single, isolated cardiac cells can be axially stretched by any of a number of methods (Garnier, 1994). The most versatile method (also used for studies of muscle mechanics of mammalian cardiomyocytes) consists of a pair of carbon fibers, which are pressed gently on widely separated regions of the cell to which they subsequently attach (Le Guennec et al., 1990). Separation of the carbon fibers produces axial

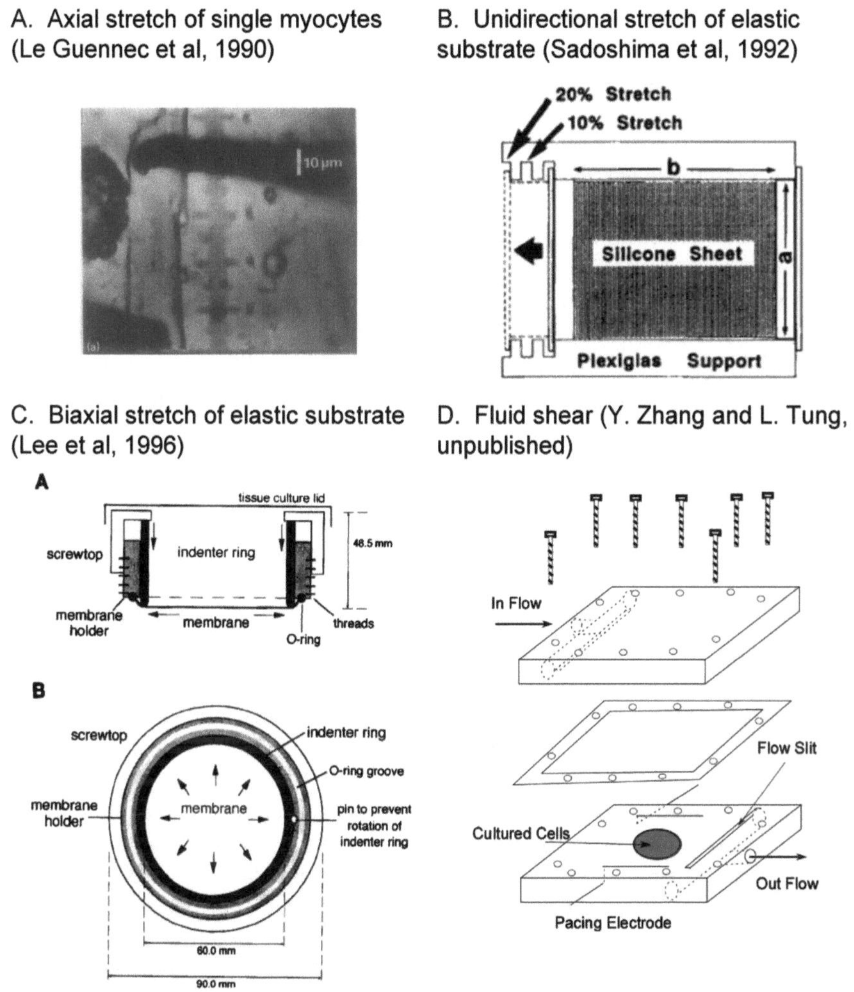

Fig. 5.1 Mechanical inputs applied to single cells and cell monolayers. (**a**) Single cell attached to two carbon fibers is stretched as the fiber separation is widened. Reprinted from Le Guennec et al., Copyright 1990, with permission from Elsevier. (**b**) Elastic substrate on which cells are attached is stretched along a single axis. (**c**) Biaxial stretch is achieved by pressing an indenter ring against an elastic membrane on which cells are attached. (**d**) Cells undergo shear stress by fluid flow when placed inside and immobilized in a parallel plate flow chamber

strain, and if the compliance of one of the fibers is known, the axial stress applied to the cell can be calculated. Portions of the cell can be locally stretched by attaching a glass stylus to the free surface of a cardiomyocyte adhering to a rigid substrate and displacing the stylus along the long axis of the cell (Kamkin et al., 2000). Alternatively, single cardiomyocytes can be seeded on a compliant material to which they can adhere (Tatsukawa et al., 1997), and the material stretched as described below.

For multicellular preparations, when a suspension of cultured cardiomyocytes, typically neonatal rat cardiomyocytes (NRCMs), is plated on a material coated with an appropriate extracellular matrix protein (ECMP) such as collagen, fibronectin or laminin, they adhere and grow on the surface. If the material is elastic, the material can be stretched along 1 or 2-dimensions (Camelliti et al., 2006). Uniaxial strain can be achieved by clamping opposite ends of a thin rectangular elastomeric membrane on which cells are cultured to a frame that is then extended in a static or cyclic manner (Terracio et al., 1988; Komuro et al., 1990; Sadoshima et al., 1992; Vandenburgh et al., 1995; Zhuang et al., 2000). Biaxial strain can also be applied to circular membranes. If negative or positive pressure is applied to one side of the membrane, the membrane bulges axisymetrically, and a graded biaxial strain results (Banes et al., 1985; Terracio et al., 1990). The addition of a central platen over which the membrane is deformed results in a homogeneous, in-plane biaxial strain at the center area of the platen. However, because the membranes are typically clamped in a circular boundary, the edges experience only radial strain, with an area of non-homogeneous strain that must be considered in some studies. Depending on the shape of the platen, equibiaxial (for a circular shape) or uniaxial (for an arctangle shape) can be obtained, and this is the method used by the Flexercell system (Flexcell International Corp.). Cyclic rates up to ~1 Hz and radial strains up to ~15% can be obtained. Alternatively, the membrane can be stretched by displacing the platen in a direction normal to the membrane. Again, by varying the shape of the platen, equibiaxial (for a circular shape) or anisotropic biaxial (for an elliptical shape) strain patterns have been achieved under static stretch conditions (Vandenburgh et al., 1995; Lee et al., 1996; Gopalan et al., 2003; Camelliti et al., 2006; Rana et al., 2008).

Another method has been used for three-dimensional cultures of cardiomyocytes. Cardiomyocytes from embryonic chicken or neonatal rat are embedded in a collagen gel that is cast as either planar constructs or a circular ring. The ends of the planar construct are attached by Velcro to two parallel rods (Eschenhagen et al., 1997), whereas the ring can simply be slipped over the two rods (Zimmermann et al., 2002). The spacing between the rods is then incremented statically or varied cyclically either in a culture dish (Zimmermann et al., 2002; Zhao et al., 2005) or in a standard 6-well tissue culture plate (Shimko and Claycomb, 2008). Cylindrical strands have been made from chicken embryonic or fetal ventricular cells (Tobita et al., 2006) and circular rings from mouse embryonic stem cells (Guo et al., 2006).

5.2.2 Mechanical Indentation

If cardiomyocytes are attached to the surface of a stiff material, a glass pipette or stylus can be pressed against the side of the cell to produce cell indentation. The indentation can be either in a direction parallel to the surface and transverse to the longitudinal cell axis (Sigurdson et al., 1992), or in a direction normal to the surface, producing transverse cell compression (Isenberg et al., 2003).

5.2.3 Fluid Shear Stress

Parallel plate flow chambers are commonly used to study endothelial cell responses, and apply laminar flow and uniform fluid shear stress over the surface of the cultured cell monolayers (Chiu et al., 2003). Similar chambers have been designed to apply fluid shear to cardiomyocyte cultures. A simple flow chamber can be fabricated by placing a thin gasket between two glass slides (Lorenzen-Schmidt et al., 2006). Fluid can be introduced and withdrawn through inlet and outlet ports, and produce a uniform laminar flow field. Larger sized flow chambers can also be fabricated (Zhang et al., 2008). Alternatively, by utilizing fluid flow emanating from the tip of a glass micropipette, a fluid stream can be directed normal to the surface of the cell monolayer, resulting in fluid shear that peaks and then decays with distance from the micropipette (Kong et al., 2005). It should be noted that with this approach, transverse compression (normal stress) of the cells also occurs and decays with distance from the micropipette. Finally, uniform fluid shear stress can also be applied to cultured cell monolayers by using a cone plate apparatus (Dewey et al., 1981).

5.2.4 Substrate Elasticity

The mechanical load on a contracting cell can also be varied by altering the elastic modulus of the cell substrate. On a softer substrate, a cardiomyocyte shortens further and, due to the Frank-Starling relationship between force and length, produces less force and feels less force on its bonds to the extracellular matrix. Substrate stiffness can be modified through changes in monomer to crosslinker ratio in crosslinked polymers, including polyacrylamide hydrogels (Pelham and Wang, 1998), polydimethylsiloxane (PDMS) gels (Brown et al., 2005), alginate gel (Genes et al., 2004), polyethlyeneglycol (PEG) (Peyton et al., 2006), and many other crosslinked polymers (Wong et al., 2004). Other methods of controlling substrate stiffness include varying agarose concentrations in agarose gels (Balgude et al., 2001) and varying the porosity of porous gels such as Poly(1,8-octanediol-co-citric acid) (POC) (Hidalgo-Bastida et al., 2007). In addition to just controlling the overall gradient of a cell substrate, studies have demonstrated methods of generating patterns or gradients in substrate stiffness using a microfluidic channel device (Zaari et al., 2004) or patterned photopolymerization under a mask or partial shield (Wong et al., 2003; Lin et al., 2007a). Variations in substrate stiffness have been shown to affect the behavior of many anchorage-dependent cell types, including neurites, fibroblasts, myocytes, endothelial cells and mesenchymal stem cells, as reviewed previously (Discher et al., 2005; Peyton et al., 2007).

5.3 Patterned Growth and Cell Alignment

Cardiomyocytes are structurally organized along a common axis to provide efficiency in force generation and tissue contraction. Therefore, the study of tissue-level physiological and pathophysiological function requires that the cells be grown in aligned, anisotropic arrays. To achieve this condition, various methods have been utilized, including contact guidance, microcontact printing, and dielectrophoresis. When cells are aligned and grown to confluence, they adopt a tissue-like morphology, with rod-like shapes and coherent sarcomere patterns.

5.3.1 Contact Guidance

The orientation of cells can be guided either from surface topology or from the extracellular matrix. In the former case, the culture substrate can be simply abraded with fine lapping paper, leaving microscopic grooves on the surface (Bursac et al., 2002). Groove patterns can be formed more precisely by using photolithography (Deutsch et al., 2000) or acoustic micromachining (Entcheva and Bien, 2005). After the substrate is coated with ECM, cardiomyocytes grow on top in an aligned fashion. The depth and size of the grooves influence the degree of alignment, and it is possible to alter the orientation of cells within the cell monolayer by changing the direction of the grooves. With the ECM approach, collagen I gel is poured on the surface of the culture substrate. Tilting of the culture dish during polymerization, or spreading with a cell scraper, causes collagen fibril alignment (Simpson et al., 1994). Cardiomyocytes grown on this surface become aligned.

5.3.2 Microcontact Printing

Using photolithography, patterns made of photoresist can be formed on glass cover slips, which are then coated with adhesive ECMP, commonly collagen or fibronectin. Following lift-off of the photoresist, a complementary pattern of ECMP is left behind, to which cardiomyocytes will attach and grow in patterns (Rohr et al., 2003). This method has been successfully used to grow cardiomyocytes as linear strands or strands with expansions, bends or branches (Rohr et al., 1999). In a more recent development, soft lithography can be used to manufacture stamps made of polydimethylsiloxane (PDMS) with photolithographically defined patterns that can then be used to transfer ECMP onto cell culture substrates (Tan et al., 2004). This method has been used to produce parallel arrays of linear strands (Bursac et al., 2002; McDevitt et al., 2002) or zigzag patterns (Bian and Tung, 2006).

5.3.3 Microfluidics

With this approach, a silicon master (or alternatively, photoplastic on a silicon wafer) is microfabricated to serve as a mold on which PDMS is poured and cured. Using photolithography and an etching process, a complementary relief pattern can be etched into the silicon, which is then transferred as channels into the PDMS. The flow of ECMP through the channels deposits the ECMP in the desired patterns via protein adsorption, and subsequent treatment of the membrane surface assures non-adhesion of the cells outside the channels (Folch and Toner, 1998). This method has been used successfully to grow neonatal rat cardiomyocytes as linear strands or hatched patterns on silicone membranes (Gopalan et al., 2003; Camelliti et al., 2006).

5.3.4 Dielectrophoresis

Dielectrophoresis is a method which takes advantage of differences between the dielectric constants of cells and their surrounding medium to use strong gradients in electric field to position cells (Lin et al., 2006). By using electrode arrays that are microfabricated on glass slides, cardiomyocytes can be concentrated at desired locations (Yang et al., 2007).

5.4 Stretch and Shear Stress Effects on Cardiomyocyte Electrophysiology in Culture

Mechanical stimuli result in biological responses that occur on different time scales, from minutes to hours to days (Kamm and Kaazempur-Mofrad, 2004). The earliest changes in cellular function are revealed electrophysiologically as a result of perturbations in membrane currents, cellular ion concentrations, and junctional protein expression. This section focuses on electrophysiological changes that accompany stretch and fluid shear of single cardiomyocytes and two-dimensional cultures of cardiomyocytes, which include changes in spontaneous activity, alterations in cell–cell coupling (connexins), conduction velocity, and action potential duration.

5.4.1 Direct Stretch

Studies of direct stretch effects on single cardiac cells are extensive and have been comprehensively summarized in previous reviews of mechanoelectric feedback (Hu and Sachs, 1996; Kohl and Sachs, 2001). Stretch generally reduces the resting potential (and as a secondary effect, enhances inactivation of the sodium channel, reduces excitability and slows conduction), shortens the early phase of repolarization, and prolongs late repolarization (Bett and Sachs, 1997; Kamkin et al., 2000), although late repolarization can also shorten (White et al., 1993). Strong levels of stretch induce extrasystoles and spontaneous activity (White et al., 1993; Kamkin et al., 2003; Riemer and Tung, 2003). Several mechanisms have been proposed by which stretch may affect ionic membrane currents. These include direct effects on ion channel conductance and kinetics or indirect effects secondary to changes in intracellular Ca^{2+} (Kohl and Sachs, 2001). Two classes of stretch-activated channels having linear current-voltage relations and reversal potentials between –70 and +10 mV have been identified in the heart – nonselective cationic and K^+-selective (Hu and Sachs, 1997). Time-independent, cation-selective currents with linear current-voltage relations were obtained by 20% axial stretch of single guinea pig ventricular myocytes (Sasaki et al., 1992), 20% stretch of rat atrial cells (Zhang et al., 2000) 20% stretch of rat ventricular cells (Zeng et al., 2000) and 8 μm local stretch of guinea pig ventricular cells and human heart failure ventricular cells (Kamkin et al., 2000) with reversal potentials of –15, –6, –6 and 0 mV, respectively. In some cases, the current gradually diminished with time during maintained stretch. Certain voltage-gated or ligand-gated ion channels have also been shown to be sensitive to stretch of the cell membrane. These include the pacemaker current I_f (Lin et al., 2007b), which governs spontaneous activity; the ATP-sensitive K current I_{KATP} (Van Wagoner, 1993), which shortens action potential duration during metabolic stress, and the muscarinic current K_{Ach} (Pleumsamran and Kim, 1995), which serves to modulate heart rate under autonomic control. At the level of the whole cell, the L-type Ca^{2+} current and inward rectifier K^+ current were not found to be stretch-sensitive in single guinea pig (Sasaki et al., 1992) and rat (Hongo et al., 1996) ventricular myocytes. On the other hand, the L-type Ca^{2+} current was found to decrease with 8 μm local stretch in single human atrial myocytes (Kamkin et al., 2003) and single human heart failure ventricular myocytes (Kamkin et al., 2000), but because the decrease was eliminated when intracellular calcium was buffered by BAPTA, this effect may be secondary to a stretch-induced change in intracellular Ca^{2+} levels. Stretch-induced changes in intracellular Ca^{2+} are well documented with axial stretch, and can arise via the flux of Ca^{2+} through stretch-activated channels and by changes in Ca^{2+} binding to intracellular buffers such as troponin C (also linked to the stretch sensitivity of contraction) (Calaghan et al., 2003).

In neonatal rat atrial cells, 13% biaxial strain for 24 h decreased Kv4.2 and KChIP2 gene expression, increased Kv1.5 gene expression, and increased Kir2.1 and Kir2.3 gene expression, with parallel changes in their corresponding ionic currents – I_{to} (transient outward K^+ current), I_{Kur} (ultra rapid delayed rectifier K^+ current) and I_{K1} (inward rectifier K^+ current), respectively (Saygili et al., 2007; Rana et al., 2008). Action potential duration measured at 50 and 90% repolarization was also reduced. The changes in gene expression and ionic currents were suppressed by losartan, a specific angiotensin II (Ang II) receptor-antagonist, suggesting that Ang II may act in a paracrine and/or autocrine manner to regulate these currents. Transmission of the mechanical signals to the ion channel proteins can be passed through the cytoskeleton (Cazorla et al., 1999; Calaghan et al., 2004) and other structural proteins (see Section 5.7).

In multicellular cultures of ventricular NRCMs (NRVMs) grown on silicone membranes, cyclic as well as static stretch upregulates the expression of connexin-43, the major gap junctional protein in the ventricle that permits cell–cell communication between adjacent myocardiac cells (Wang et al., 2000; Zhuang et al., 2000; Shyu et al., 2001; Pimentel et al., 2002; Shanker et al., 2005; Yamada et al., 2005). Furthermore, the expression of junctional proteins like connexin-43 has been coupled to the development of pathological hypertrophy (Peters et al., 1993; Kaprielian et al., 1998; Kostin et al., 2003, 2004). Using a Flexercell system, Wang et al. found that 20% stretch of NRVMs caused a 3-fold increase in Cx43 mRNA levels in 2 h and Cx43 protein levels in 4 h which lasted another 16 h, changes that were accompanied by activation of the Na-H exchanger. Using a similar system, Shyu et al. found that 20% elongation of NRVMs at 1 Hz caused a rise in Cx43 mRNA and in Cx43 protein level as early as 2 h, reaching a 6-fold increase over control at 24 h and remaining at that level for another 24 h. Both increases were accompanied by an increased release of Ang II and both were blocked by the AT1 antagonist, losartan (Shyu et al., 2001). These results suggest that the increase in Cx43 expression with cyclic stretch is mediated by AT1 receptors.

Uniaxial cyclic stretch with a custom system (20% elongation, 0.5 Hz, 24 h) was found to align NRVMs when started 3 h after plating and to localize Cx43 at the longitudinal cell termini in a manner that is regulated by the Rac1 signaling pathway acting downstream of N-cadherin (Matsuda et al., 2006). When uniaxial cyclic stretch (10% elongation at 3 Hz) was applied to patterned strands of cardiomyocytes using a custom designed chamber (Zhuang et al., 2000), Cx43 protein expression increased after 1 h of stretch and even further after 6 h, and was maintained for at least 24 h. Similar, although smaller, changes were observed with static stretch. Using optical mapping, conduction velocity (CV) was measured by Zhuang et al. After 1 h of uniaxial, cyclic stretch (10% elongation, 3 Hz), CV increased from 27 to 35 cm/s and after 6 h to 37 cm/s, consistent with the measured increases in Cx43 protein expression. In a follow up study, Saffitz's lab found that the stretch-induced upregulation of Cx43 was mediated by vascular endothelial growth factor (VEGF), which acts downstream of transforming growth factor-beta (TGF-β), is secreted by the myocytes during stretch, and acts in an autocrine manner (Pimentel et al., 2002). Later, the same lab found that the VEGF secretion depends on focal adhesion kinase (FAK) dependent signaling (Yamada et al., 2005). In a different follow-up study, the Saffitz lab found that the stretch-induced upregulation of Cx43 depended on the composition of the extracellular matrix, being present with type I collagen but not with fibronectin or denatured collagen (Shanker et al., 2005).

Interestingly, when static, biaxial anisotropic (10%:5% strain) stretch was applied for 24 h to patterned strands of NRVMs, Cx43 protein expression did not significantly increase when the principle strain was aligned along the axis of the cells but did increase when it was aligned transverse to the cells (Gopalan et al., 2003), presumably acting through different signaling pathways, or acting differently through the same pathways in ways that are as yet unknown. It has also been shown that stretch applied transverse to the cell axis suppresses contractile protein turnover and increases the accumulation of the contractile proteins, but has little effect when applied parallel to the cell axis in aligned monolayers of NRCMs (Simpson et al., 1999).

On the other hand, anisotropic static stretch of circular, isotropic NRVM monolayers (10%:5% strain) showed prolongation (mean of 6.9%) in action potential duration (APD) and slowing (mean 7.5%) of CV within 10 min, although longer term changes in these parameters were not monitored (Zhang et al., 2008). These observations suggest that acute load-dependent conduction slowing, as observed in the intact ventricle (Sung et al., 2003), is an intrinsic multicellular response that does not occur secondary to other phenomena such as alterations in tissue perfusion, interstitial flow or extracellular resistance to action currents. Furthermore, the duration of loading is likely to have been too brief for changes in Cx43 protein expression and localization to have occurred, which may explain the difference between these results and those of Zhuang et al.

5.4.2 Mechanical Indentation

Transverse indentation of single chick embryonic cells of the order of 10% parallel to the substrate surface initiates a wave of Ca^{2+} and contraction that spreads throughout the cell from the site of stimulation (Sigurdson et al., 1992). The calcium response was prevented either by eliminating extracellular Ca^{2+} or by adding Gd^{3+}, a stretch-activated channel blocker. Indentation of these cells in a direction perpendicular to the substrate surface activates a whole cell current with linear current-voltage relation and reversal potential of –17 mV, consistent with the presence of a combination of cation-selective and potassium-selective stretch-activated channels (Hu and Sachs, 1996).

Indentation of unspecified mammalian ventricular cells by about 15% in a direction perpendicular to the substrate surface suppresses I_{K1} (inward rectifying K^+ current) and I_{Ko} (outward rectifier K^+ current) when the cell is laying on its narrow edge, whereas indentation of the cell by about 50% suppresses these currents as well as I_{ns} (a non-specific current with linear current-voltage relation) when the cell is laying on its broad edge (Isenberg et al., 2003). Mechanosensitivity of the currents was diminished or abolished with the addition of cytochalasin D or colchicine, disruptors of cytoskeletal actin and tubulin filaments, respectively.

5.4.3 Fluid Shear

Deformations of cells under shear stress conditions may be subcellular in length scale (Barbee, 2002) and may be fundamentally different in nature than those involving axial stretch. It has been hypothesized that shear stress influences cell function through different mechanosensitive structures in the membrane, such as adhesion receptors and associated signaling complexes (Janmey and McCulloch, 2007). The phenomenon of fluid shear stress having an effect on myocardial contractile function was first evidenced in embryonic mouse hearts, which increased and decreased their heart rate in response to an increase or decrease in perfusate flow rate (Tanaka et al., 1997). Although the response of many cell types to fluid shear has been studied in the literature, a response of cardiac myocytes to shear has only recently been suggested. NRVM monolayers subjected to continuous low fluid shear rates (5–50/s) in a flow chamber showed an immediate, graded and reversible increase in their spontaneous beating rate (up to 500%) (Lorenzen-Schmidt et al., 2006). The chronotropic response was not affected by streptomycin, a blocker of stretch-activated channels, but was abolished in serum-free media. It also was substantially attenuated either in the presence of isoproterenol or following incubation with integrin-blocking RGD peptides, suggesting that the β-adrenergic signaling pathway and integrin activation are involved in the shear stress response.

In other flow chamber experiments, fluid shear stress (1.1 dyn/cm^2) slowed CV and increased action potential duration (APD) in a reversible fashion in anisotropic monolayers of NRVMs created by microabrasion (Zhang et al., 2008). Furthermore, the time course of changes was not immediate and followed a monotonically increasing time course over a period of 4–6 mins. The magnitudes of CV and APD changes were modest, and may been blunted because serum-containing solution was not used for the experiments. The time delay in shear stress effects might be related to the mechanical properties of cytoskeletal networks which can exhibit viscous characteristics (Cooper, 2006). Zhang et al. also found that among the several hundred recording sites distributed over the monolayer (1.8 cm in diameter), several sites acted as outliers, with particularly prolonged or shortened APD. These outliers may be indicative of mechanically hypersensitive cells, as suggested previously by axial stretch experiments on single cardiac cells (Riemer and Tung, 2003).

Fluid shear arising from fluid flow from the tip of a micropipette caused an increase in spontaneous beating rate, similar to that measured in flow chambers (Lorenzen-Schmidt et al., 2006). Kong et al. used optical mapping to show that fluid pulses from a micropipette can excite propagating electrical wave fronts, much like electrical excitation from a point electrode, although in a stochastic fashion (Kong et al., 2005). The incidence of mechanical excitation increased with higher fluid jet velocity and time between pulses, and was attenuated by the addition of streptomycin or gadolinium, blockers of stretch-activated channels, but not by nifedipine, a blocker of L-type Ca channels. Recently, pulsed fluid jets applied to single rat atrial myocytes were found to trigger spontaneous calcium sparks (the unitary release of Ca^{2+} from the sarcoplasmic reticulum, SR) in the periphery of the cell where the putative mechanosensor may lie, and at high flow rates to trigger longitudinally spreading Ca^{2+} waves (Woo et al., 2007). In another study, application of seconds-long fluid pulses reversibly suppressed the L-type Ca current in single rat ventricular cells and accelerated the inactivation of the current (Lee et al., 2008). The authors attributed the effect to fluid pressure, although fluid shear was also present. The fluid pulse effect was reduced when Ca^{2+} release from the SR was prevented by ryanodine and thapsigargin, suggesting that the primary effect of the fluid pulse is on Ca-induced Ca^{2+} release from the SR, which in turn modulates the L-type Ca^{2+} current in these cells.

5.4.4 Substrate Stiffness

Substrate stiffness can regulate myocyte contraction: one study found that the percentage of beating cardiomyocytes increased with decreasing elastic modulus (Jacot et al., 2008). This result was confirmed by another study on much softer materials that used PEGylated fibrinogen gels with varying concentrations of reactants and a diacrylate crosslinker in order to create varying elastic moduli with a shear modulus range from 8 to 340 Pa (tensile modulus around 20–1,000 Pa, depending on the material Poisson ratio). After 4 weeks of culture on those gels, neonatal rat ventricular myocytes on the softest moduli had the highest percentage of beating cells and the highest correlation of beating times and frequencies across the constructs (Shapira-Schweitzer and Seliktar, 2007).

5.5 Myocardial Stress and/or Strain Regulates Cardiac Muscle Growth and Function

The fetal heart is different from the adult heart in that immature myocardium can increase its mass by two main processes: hypertrophy, an increase in myocyte size, and/or hyperplasia, an increase in myocyte number (Rakusan, 1984). Unlike myocytes from fetal hearts, myocytes that compose the adult myocardium generally no longer retain the capacity to divide. Consequently, the growth of mature myocytes is restricted to one main process: hypertrophy. In general, cardiac hypertrophy is recognized as a useful physiological adaptation of the heart in response to an increase in stress (Heineke and Molkentin, 2006). The hypertrophied myocardium of patients with pressure overload, for example, may be a compensatory mechanism which normalizes myocardial systolic wall stress and enhances left ventricular function (i.e. contractility and peak systolic pressure) to counterbalance the increase in afterload (Grossman et al., 1975). Cardiac hypertrophy is also a major independent risk factor for sudden death, underling various cardiovascular maladies including heart failure, dilated cardiomyopathy and ischemic heart disease (Levy et al., 1990; Koren et al., 1991; Bolognese et al., 1994; Vakili et al., 2001; Frey and Olson, 2003). Myocardial hypertrophy is characterized by an increase in cardiomyocyte size, increased protein synthesis, and activation of genes for natriuretic peptides (atrial natriuretic peptide and B-type natriuretic peptide) (Ruwhof and van der Laarse, 2000; Frey and Olson, 2003). Currently, the proposed mechanisms in the regulation of stress-induced cardiac hypertrophy include, but are not limited to, the involvement of growth and neurohumoral factors (endothelin-1, Ang II, insulin like growth factor, vascular endothelial growth factor), apoptosis, ischemia, and biochemical signaling events induced by mechanical stimuli (Gupta et al., 2007). Mechanical load is believed to be one of the primary stimuli for the development of hypertrophy (Ruwhof and van der Laarse, 2000).

The most common cause of hypertrophy in the human population is an increase in blood pressure, which occurs in patients with hypertension or stenosis of the aortic valve (Asher and Klein, 2002; Gupta et al., 2007). Although a variety of stimuli play a role in the development of hypertrophy in these patients, the increase in myocardial systolic stress, which is commonly associated with an increase in myocardial diastolic stress, is believed to be a major cause for thickening of the ventricle. In vivo animal studies further indicate that chronically overloading the left ventricle by coarctation of the ascending aorta results in an increase in left ventricular mass (hypertrophy), while chronically unloading right ventricular papillary muscles by transecting the cordae tendineae results in a decrease in tissue mass (atrophy) (Tomanek and Cooper, 1981; Rockman et al., 1991; Barbosa et al., 2005). In combination these observations suggest that mechanical load plays a critical role in the regulation of myocardial mass and the development of hypertrophy.

5.5.1 Fiber and Cross-Fiber Loading

The actual stress and strain imposed on a myocyte during, for example, pressure overload, are unknown and the mechanical loads on the myocytes in the intact tissue are undoubtedly complex. In order to dissect the mechanisms of load-sensing in myocytes, various experimental preparations have been employed to apply more simplified external loading conditions to myocytes and study their responses. Consistent with the role mechanical load plays in regulating cardiac hypertrophy in vivo, stretch models have been used to demonstrate that mechanical load applied in the fiber and/or cross fiber direction of cardiac myocytes regulate the development of hypertrophy in vitro. Many labs have utilized a cell culture model and have demonstrated in neonatal ventricular myocytes that static stretch of 10–20% can induce the in vivo hypertrophic phenotype, characterized by an increase in protein synthesis, induction of immediate early genes, and re-expression of genes for contractile proteins and natriuretic peptides (Sadoshima et al., 1992; Yamazaki et al., 1995; Komuro et al., 1996; van Wamel et al., 2000; Gopalan et al., 2003). In general, cross-fiber loading tends to increase the mechanotransduction response more than fiber loading (Gopalan et al., 2003). In most cases, the isolated tissue response to increased loading is similar: an increase in axial tissue stress and/or strain of excised right ventricular papillary muscles results in a rapid increase in protein synthesis and alterations in gene expression (Peterson and Lesch, 1972; Cooper et al., 1989; Jarygin et al., 1994). In chronically cultured trabeculae muscles the application of high mechanical loads resulted in changes in protein expression patterns and upregulation of structural, myofilament proteins, events which were correlated with an increase in myocyte and tissue size (Bupha-Intr et al., 2007). Like in vivo experiments where transverse aortic constriction was utilized as a mechanism to pressure overload the heart, the application of chronic axial mechanical loads to trabeculae muscles also resulted in an increase in developed systolic stress (after 6 h of culture), diastolic stress (after 30 h of culture), and myocardial relaxation (after 30 h of culture) of the tissue. Together these data suggest that an increase in axial stress and/or strain alone is capable of initiating hypertrophic biochemical events, as well as hypertrophy at the cellular and tissue levels.

Based on this evidence, cardiomyocytes are dynamic systems that respond to mechanical load through the mechanisms of mechanotransduction, where physical forces acting on transmembrane and/or intracellular mechanosensors are converted into biochemical information that regulates gene expression, protein synthesis, and the development of hypertrophy (Ruwhof and van der Laarse, 2000; Knöll et al., 2002). In most cases the transduction of physical forces into biochemical signaling is directly mediated through changes in cell and/or protein conformation, and not the applied force (or stress) itself (Cooper et al., 1989; Omens, 1998; Kamm and Kaazempur-Mofrad, 2004; Vogel, 2006). Posttranslational phosphorylation of proteins, for example, alters protein structure thereby modulating enzymatic activity and protein–protein interactions involved in biochemical signaling. Identical to this process, force induced effects on the deformation of proteins or cell structures, which are commonly associated with conformational changes, may represent a general mechanism of mechanotransduction (Kamm and Kaazempur-Mofrad, 2004; Groban et al., 2006; Orr et al., 2006). As of now, a stress sensor has not been confirmed to exist, further supporting the idea that mechanotransduction may be directly mediated by a deformation or strain signal, rather than a force or a stress (Arts et al., 1995). There are three primary and distinct signal transduction pathways that are induced by the application of axial mechanical loads to cardiac myocytes in vitro: the mitogen-activated protein kinase (MAPK) pathway, the Janus associated kinase mediated gp130 pathway (JAK/STAT), and the pathway that involves activation of the Ca^{2+}/calmodulin-dependent phosphatase calcineurin (Ruwhof and van der Laarse, 2000; Lammerding et al., 2004). The MAPK pathway can be further subdivided. The three best described MAPK cascades are: the extracellular regulatory kinases (ERK), the c-Jun N-terminal kinases (JNKs), and the p38 MAPK cascade. Even though there are several interactions between these three major stress inducible hypertrophic pathways, each has been evidenced to be activated by a different mechanosensor or group of mechanosensors. For example, one way the MAPK cascade is triggered is through the activation of G-protein coupled receptors found in the sarcolemma, which in turn activate G-proteins and indirectly activate proteins Raf1 and MEK1/2 (all parts of the ERK signaling cascade). Integrins that 'sense' an increased mechanical load activate FAK, also involved in the MAPK signaling cascade. The JAK/STAT pathway is directly activated by way of the gp130 receptor, whose activity is also dependant on the presence of the cardiotropin-1 cytokine. Calcineurin has also been associated with the development of hypertrophy, and the calcineurin-dependent pathway may link increased concentration of intracellular calcium with the induction of hypertrophy.

5.5.2 *Shear Stress and Strain*

In addition to normal (fiber/cross-fiber) stress and strain, myocytes are subjected to membrane shear stresses which can be sensed by the cells and may play a role in hypertrophy and heart disease. Myocytes may partially detach from the extracellular matrix, for example, and may be more prone to slippage, resulting in elevated shear conditions, which could play a role in regulating cardiac muscle growth and/or function (Paul, 2003). In addition to fluid shearing effects on electrical behavior and beating rate in cardiomyocytes (see Section 5.4.3), fluid shearing can also regulate hypertrophic markers signaling in these isolated cells (Fig. 5.2). The proposed

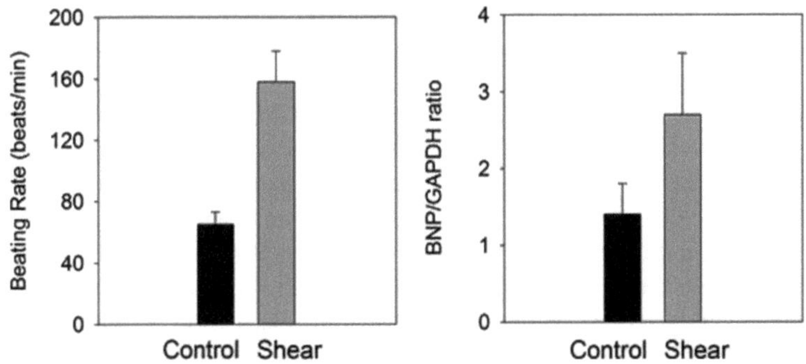

Fig. 5.2 Effect of fluid shear on cultured myocytes. (*Left*) Fluid shear on myocytes results in an immediate increase in beating rate, presumably through membrane sensors for shear stress. (*Right*) Long term fluid shearing of myocytes (1 h) results in upregulation of BNP gene expression, indicating the start of a hypertrophic response in the myocytes. Unpublished data from JHO lab

mechanism in the regulation of myocyte beating frequency by shear rate may include integrin mediated biochemical signaling, since integrin masking RGD peptides almost completely abolished the shear response of neonatal myocytes. Integrins have been evidenced to play an essential role in hypertrophy of the adult heart in vivo and mechanotransduction in vitro (see Section 5.7.2). Consequently, it is possible that integrin mediated shear responses may also exist in mature cardiac myocytes of the adult heart. It is also tempting to speculate that an alteration of the fluid shear environment of mature myocytes may stimulate ectopic action potentials resulting in premature ventricular contractions, which frequently arise in patients with left ventricular hypertrophy of the heart.

5.5.3 Regulation of Junctional Proteins by Mechanical Load

Mutations in proteins that compose intercalated disks are commonly associated with the development of various cardiomyopathies and arrhythmias (Kaplan et al., 2004a, b). Specifically, mutations in desmoplakin and plakoglobin are implicated in the pathogenesis of human cardiomyopathies associated with sudden cardiac death. An increase in the number and changes in the structure of intercalated disks have been found in the hypertrophied myocardium of dogs, humans, and mice (Laks et al., 1970; Maron and Ferrans, 1973; Ding et al., 2000). In vitro studies have demonstrated a positive effect of mechanical load on the expression of intercalated disk proteins (see Section 5.4.1). In particular, the expression of connexin-43, desmoplakin, and plakoglobin has been evidenced to increase in neonatal cardiac myocytes subject to pulsatile stretch. The formation of more mechanical junctions between individual neonatal cardiac myocytes is associated with an increase in electromechanical coupling and appears to be mediated by an integrin/ FAK dependant inside out mechanism.

5.6 Role of Mechanics in Stem Cell Differentiation and Maturation of Stem Cell-Derived Cardiomyocytes

Spontaneously beating, cardiac-like cells have been differentiated from a variety of self-replicating cell sources including embryonic stem cells (Boheler et al., 2002), bone marrow-derived mesenchymal stem cells (Fukuda, 2003; Xu et al., 2004; Yamada et al., 2007; Antonitsis et al., 2008), bone-marrow stromal cells (Makino et al., 1999), umbilical cord blood derived stem cells (Pereira et al., 2008) and adult cardiac stem cells (Bearzi et al., 2007; Smith et al., 2007). In addition to these native multipotent cell types, recent studies found that the expression of four transcription factors, Oct3/4, Sox2, c-myc and Klf4, can induce both human and mouse fibroblasts to become pluripotent cells, which are indistinguishable from embryonic stem cells in many assays of pluripotency and commonly termed induced pluripotent cells (IPCs) (Takahashi and Yamanaka, 2006; Takahashi et al., 2007). Groups have generated spontaneously-beating and cardiac marker-expressing cardiomyocytes from mouse IPCs using growth factor induction and co-culturing with other cell types (Mauritz et al., 2008; Narazaki et al., 2008). Furthermore, cardiac fibroblasts and striated muscle cells have been shown to transdifferentiate into cardiac-like cells (Iijima et al., 2003). It is interesting that many cardiac differentiation studies, and the majority of studies of non-embryonic stem cell differentiation into cardiomyocytes, achieved differentiation through co-culture with primary cardiac cells. Because studies have found that cardiac cells can fuse with stem cells and fibroblasts and that the fused cells are often difficult to distinguish from a differentiated cell (Nygren et al., 2004; Rodic et al., 2004), some claims of cardiac differentiation must be viewed with skepticism. In this section, we will discuss studies that used mechanical means in the absence of co-culture to achieve cardiac cell differentiation or to drive further maturation and development in cells populations selected from stem cells for specific cardiac markers or properties.

5.6.1 Mechanical Influences on Differentiation of Cardiac Myocytes from Embryonic Stem Cells

Embryonic stem cells can spontaneously differentiate into cardiomyocytes in serum-containing media and can be driven toward differentiation into the major components of heart muscle tissue or the conduction system. In general, cardiogenesis in embryonic stem cell cultures is indicated by spontaneous beating, the shape of action potentials and calcium transients, the presence of specific ion currents, and by the expression of specific cardiac cell markers. The differentiation into cardiac tissue is denoted by the termination of certain pluripotency markers (such as Oct-3/4, fibroblast growth factor-5 (FGF-5) and Nodal), the expression of early cardiac markers (such as the transcription factors Nkx2.5 and GATA-4, and sarcoplasmic/endoplasmic reticular calcium ATPase 2a (SERCA2a)) and the expression of some late-stage cardiac markers (such as α- and β-MHC, the ryanodine receptor, cardiac troponin-T, and calsequestrin). An overview of differentiation times and markers has been previously reviewed for mouse embryonic stem cells (Boheler et al., 2002).

One recent study showed that mouse embryonic stem cell embryoid bodies increased the percentage of beating cells and the percentage of cells expressing sarcomeric α-actinin when statically stretched for 2 h and that this effect was graded over 5, 10, 15 and 20% radial strain. These cells also increased expression of cardiac markers MEF2c and GATA-4 when stretched by 10%. The demonstrated cardiogenesis was inhibited with free radical scavengers vitamin E and N-(2-mercaptopropionyl)-glycine, though these treatments further enhanced the upregulation of GATA-4. Interestingly, angiogenesis, indicated by the formation of capillary-like structures and the expression of PECAM-1, increased with increasing strain up to 10%, then decreased with further strain back to basal levels at 20% strain (Schmelter et al., 2006).

Mechanical stretch has been shown to inhibit differentiation as well. At low frequencies of stretch (10 cycles/min), 10% stretch tended to decrease the differentiation of human embryonic stem cells and keep them in a pluripotent state (Saha et al., 2006).

Furthermore, the application of shear stress has been shown to induce the early cardiac and smooth muscle cell markers vascular endothelial growth factor receptor 2 (VEGFR-2), smooth muscle actin, smooth muscle protein 22-α, MEF2c, α-sarcomeric actin, and PECAM, all downstream of a remodeling of chromatin structure (Illi et al., 2005).

Several studies have also found effects of mechanical activation on the maturation of cardiomyocyte-like cells that had already differentiated from embryonic stem cells. One study used mouse embryonic stem cells that were hand-selected for beating colonies, which were then verified for expression of cardiac α-MHC, cardiac α-actin, GATA-4 and Nkx2.5 mRNA. These cells were then seeded onto poly(lactide-co-caprolactone) (PLCL) elastic scaffolds. Cells on scaffolds that had been cyclically stretched for 2 weeks at 10% strain and 1.0 Hz had increased expression of cardiac α-MHC, cardiac α-actin, GATA-4 and Nkx2.5 mRNA compared to control unstretched cultures. These stretched cultures also integrated electrically into the myocardium of infarcted rat hearts, beating in synchrony with the heart, while unstretched cultures did not have synchronous beating (Gwak et al., 2008). Another study found that contractile markers in murine embryonic stem cell-derived cardiomyocytes, selected by transfection of an α-myosin heavy chain (MHC)-promoter-driven gene conferring resistance to Genetecin (G418) and embedded in a collagen-fibronectin scaffold are highly sensitive to the frequency of 10% mechanical stretch. While the expression of α-cardiac actin increased with frequency of stretch of 1, 2, or 3 Hz, the expression of α-skeletal actin, α-MHC, and β-MHC decreased after 3 days of 1 Hz stretch but increased after 3 days of 3 Hz stretch. The transcription factor GATA-4 decreased with 1 Hz stretch but was not significantly different after higher stretch frequencies (Shimko and Claycomb, 2008). One study used stretch in order to both condition and align stem cell-derived cardiomyocytes, though these were not compared to unstretched samples so the added benefit of stretch is difficult to determine (Guo et al., 2006).

Studies of myocytes cultured from embryos have shown that stretch can aid in both proliferation of cells and maturation of functional properties of these myocytes. Embryonic (day 7) white Leghorn chicken cardiomyocytes attached to collagen-

Table 5.1 Effects of mechanical stimulation on the differentiation of embryonic stem cell-derived cardiomyocytes

Study	Cell type	Mechanical stimulation	Measured indicator of cardiogenesis	Result
Schmelter et al. (2006)	Mouse embryonic stem cells	5–20% static strain for 2 h	Sarcomeric α-actin, MEF2c and GATA4 expression	Expression increased in a graded manner with increasing strain
Illi et al. (2005)	Mouse embryonic stem cells	Laminar shear stress of 10 dynes/cm^{2}s^{-1} for 60 min	MEF2c and Sarcomeric α-actin expression	Increased expression with shear
Shimko and Claycomb (2008)	Mouse embryonic stem cell-derived cardiomyocytes	10% strain at 1–3 Hz for 3 days	α-skeletal actin, α-MHC, β-MHC, GATA4 expression	All markers decreased with 1 Hz stretch, all but GATA4 increased with 3 Hz stretch
Gwak et al. (2008)	Mouse embryonic stem cell-derived cardiomyocytes	10% strain at 1 Hz for 14 days	Cardiac α-MHC, cardiac α-actin, GATA-4, Nkx2.5 mRNA expression	Increased expression with strain

coated rubber and radially stretched by 20% at 2 Hz doubled their proliferation, measured by cell number and BrdU uptake (Miller et al., 2000). Embryonic (day 7) or fetal (day 14) White Leghorn chicken ventricular cells embedded in Type I collagen gel and uniaxially stretched at 0.5 Hz by 8% (embryonic) or 4% (fetal) had increased active stress compared to unstretched cells. The constructs also had decreased cross-sectional areas and increased passive stress in fetal constructs and proliferation in embryonic constructs. Stretch did not increase the calcium sensitivity, response to isoproterenol or upregulation of the cardiac markers α-actinin or β-actin in these cells (Tobita et al., 2006).

Table 5.1 presents a summary of the studies specifically examining the effect of mechanical stimulation on cardiac marker expression in pluripotent cells.

5.6.2 Differentiation of Cardiac Myocytes and Mechanical Induction of Differentiation of Other Myocytes from Mesenchymal Stem Cells

Few studies show mesenchymal stem cell differentiation into cardiomyocytes, and we know of no studies that investigated mechanical load, shear or substrate stiffness effects on differentiation of cardiomyocytes. However, studies do show that mesenchymal stem cells differentiate into myoblasts on substrates with an elastic modulus between 2 and 20 kPa, in serum-containing media and in the absence of other directors of myocytes differentiation (Engler et al., 2006). Additionally, culture of bone marrow derived mesenchymal stem cells in a bioreactor that provides mechanical stretch greatly enhances cell proliferation, collagen deposition and tissue strength in a tissue engineered heart valve system (Engelmayr et al., 2006).

The application of static strain of only 3–5% has been shown to increase the expression of osteogenic markers in human mesenchymal stem cells, as well as increasing matrix mineralization of the collagen matrix surrounding these cells (Ward et al., 2007). Cyclic strain of 10% at 1 Hz has been shown to have similar effects on mouse embryonic mesenchymal stem cells as in other mouse embryonic stem cells, increasing the expression of the vascular smooth muscle cell markers smooth muscle α-actin and smooth muscle myosin heavy chain (Riha et al., 2007).

5.6.3 Mechanical Influences on the Transdifferentiation of Skeletal Myocytes into Cardiac-Like Myocytes

One study found that coculture of mouse skeletal myocytes with beating neonatal rat cardiomyocytes induces transdifferentiation of the skeletal myocytes into cardiac-like myocytes, as measured by the expression of Nkx2.5, GATA-4, cardiac troponin T and atrial natriuretic peptide. When the cardiomyocytes were inhibited from beating with the addition of nifedipine, these cells did not express troponin T. However, expression of troponin T was recovered when cells were stretched at 12% strain at 1 Hz for 48 h, suggesting that the cyclic strain is necessary for transdifferentiation of skeletal myocytes into cardiac myocytes (Iijima et al., 2003).

5.6.4 Substrate Stiffness

No study has linked substrate stiffness directly to cardiomyocyte differentiation from stem cells or other precursors. However, one study showed that substrate stiffness alone can affect the differentiation of mesenchymal stem cells into myogenic cells, as shown through cell morphology, presence of striations and the expression of several myogenic markers including Myogenesis Differentiation Protein I (MyoD1), which has a peak in expression in cells on gels with an elastic modulus of 10 kPa and is nearly undetectable in cell on gels with elastic moduli above 20 kPa or below 2 kPa (Engler et al., 2006). Additionally, one study found that striations in C2C12 myotubes form only when cells are plated in a very small elastic modulus range on polyacrylamide gels, centered at 12 kPa (Engler et al., 2004), while a later study confirmed that C2C12 cells on alginate gels with an elastic modulus below 10 kPa do not differentiate and form myotubes, but found no reduction in myotube formation or activity of the myogenic marker muscle creatine kinase (MCK) in cells grown on stiffer substrates, up to 50 kPa (Boontheekul et al., 2007). An additional study found that neonatal rat cardiomyocytes form aligned striations, have peak expression of SERCA2a, peak concentrations of stored calcium, peak calcium transient concentrations and peak force generation when plated on substrates with elastic modulus of 10 kPa (Jacot et al., 2008). As a reference, the elastic modulus of the left ventricle of normal Lewis rats was measured as 18 ± 2 kPa, and this increased to 55 ± 15 kPa in infarcted areas (Berry et al., 2006).

5.7 Role of Structural Proteins and Related Signaling Pathways in Mechanotransduction and Heart Failure

Human genetic studies and mouse models of cardiomyopathy have emphasized the importance of the myocyte cytoskeleton in cardiac disease. In fact, genetic mutations in genes encoding for actin (ACTC) (Olson et al., 1998), titin (TTN) (Satoh et al., 1999; Gerull et al., 2002), myosin binding protein (MYBPC3) (Niimura et al., 1998), myosin heavy chain (MYH7, MYH6) (McKenna, 1993; Ching et al., 2005), troponin I (TNNI3) (Kimura et al., 1997), troponin T (TNNT2) (Thierfelder et al., 1994), and desmin (Goldfarb et al., 1998) have all been linked to the development of cardiomyopathy. Specifically, cytoskeletal proteins from the LIM domain family, and in particular Muscle LIM Protein (MLP), cypher and Four-and-a-Half LIM domain protein (FHL), have been shown to play an essential role in cardiac disease. Even though the main focus of this section will be to address the critical role of the myocyte cytoskeleton in mechanotransduction, evidence at the tissue level suggests that mechanotransduction may also take place at the level of the ECM, the sarcolemma, and the nucleus. Consequently, the role of these various tissue subsystems in mechanotransduction will also be addressed.

5.7.1 ECM in Mechanotransduction

The extracellular matrix (ECM) is the part of myocardial tissue that provides structural support to myocytes and serves as an anchorage of cells. In cardiac tissue, the ECM consists of collagens (primarily type 1 and 3), proteoglycans, noncollagenous glycoproteins such as fibronectin and laminin, growth factors and cytokines, and extracellular proteases (Goldsmith and Borg, 2002). The ECM binds to cell transmembrane receptors known as integrins (Ruoslahti, 1991). Mechanical forces of systole and diastole are transmitted through the ECM network where they are focused on these integrin receptors. It is through this mechanism that the ECM contributes to the induction of hypertrophic biochemical processes initiated at cell matrix adhesions (focal adhesions) via integrin mediated pathways (Wang et al., 1993; Katsumi et al., 2004). Changes in the organization of collagen, adhesion of myocytes to collagen, and the fluid properties of noncollagenous components of the ECM may effect mechanical force transmission between the ECM network and cardiac myocytes and may modulate mechanical load-induced biochemical signaling, cardiac remodeling, and growth (Goldsmith and Borg, 2002). For example, stretch of cardiac fibroblasts caused the activation of mitogen activated protein kinases (MAPK) and ERK2 in a matrix specific way. The ERK2 pathway was activated only in cells cultured on fibronectin and could be mediated by the α5β1 integrin that is expressed in cardiac cells and has been shown to exclusively bind to fibronectin (Ross, 2002). These results support the theory that changes in how cells adhere to the ECM may affect hypertrophic signaling and hence remodeling and growth.

5.7.2 Integrins

The most popular in vitro experimental model utilized to study mechanotransduction is one where cells adhered to elastic substrates are subject to either a single round of increased strain (static stretch) or repeated cycles of increased strain (cyclical stretch) (Sadoshima et al., 1992; Yamazaki et al., 1995; Komuro et al., 1996; van Wamel et al., 2000; Shyu et al., 2001; Gopalan et al., 2003; de Jonge et al., 2007). Externally applied mechanical loads to these types of in vitro systems are transmitted to intracellular structures via cell contacts, thereby implicating integrins in mechanotransduction. Integrins are transmembrane receptors composed of α and β subunit heterodimers, which bind to the ECM and to cytoskeletal proteins such as α-actinin, talin, tensin, and FAK (Hynes and Lander, 1992; Juliano and Haskill, 1993; Lewis and Schwartz, 1995; Kamm and Kaazempur-Mofrad, 2004). In myocytes the β1 integrin isoform (α1β1, α3β1, α5β1, α6β1, α7β1, α10β1, and α11β1) is predominantly expressed in the postnatal heart (Ross, 2002). Its role in the development of myocardial hypertrophy has been demonstrated. An increase in β1 integrin expression resulted in increased ANP levels and protein expression in the mouse heart (Ross et al., 1998). The inhibition of β1 function and signaling resulted in reduced adrenergically mediated hypertrophy. There are two likely mechanisms through which integrins act in mechanotransduction. First, the formation of focal adhesion complexes may not only contribute to strengthening cell adhesion to the ECM, it may also organize proteins in such a way that signals may be transmitted more efficiently. Second, physical forces acting on integrins, which are transmitted to focal adhesions, may alter the conformation of certain force or strain sensitive components of the focal adhesion, thereby activating new binding interactions between proteins or activating enzymatic sites. In support of this theory, computer simulations have predicted that force-induced conformational changes in specific regions of FAK influence its binding affinity to paxillin (Kamm and Kaazempur-Mofrad, 2004).

5.7.3 Sarcolemma in Mechanotransduction

The sarcolemma is the cell membrane of myocytes that is composed of two amphipathic phospholipid monolayers. This bilayer contains peripheral proteins attached to the inner surface of the sarcolemma and transmembrane proteins spanning through the thickness of the sarcolemma. Important sarcolemmal proteins that have been implicated in mechanotransduction are G-protein coupled receptors, ion channels, and effector enzymes (Omens, 1998).

The sarcolemma is a likely transmitter of external or internal loads. Mechanical stresses may cause deformation of the sarcolemma, which can induce conformational changes in membrane bound enzymes such as phospholipase C (Komuro et al., 1991; Sadoshima et al., 1993), mechanically activated ion channels such as stretch activated channels (as discussed earlier in Section 5.4.1), and the Na^+/H^+ exchanger, which functions to regulate intracellular pH and the concentration of Na^+ (Matsuda et al., 1996; Bers, 2001; Lyford et al., 2002).

5.7.3.1 Phospholipase C

Phospholipases are enzymes that catalyze the breakdown of phospholipids into fatty acids and other lipophilic substances, which may act as secondary messenger molecules. There are three members in the phospholipase C family: PLCγ, PLCδ, and PLCβ, all of which are membrane coupled and dependant upon calcium for optimal activity (Rhee and Choi, 1992). Mechanical stress has been evidenced to stimulate phospholipase C activity and increase intracellular calcium levels, which results in increased protein kinase C activity (PKC) (Komuro et al., 1991; Sadoshima et al., 1993). PKC has been suggested to play a biochemical role in mechanical stress induced hypertrophy. PKC translocation from the cytoplasm to cardiac particulate fraction has been evidenced in vitro and in vivo with hypertrophic agonist treatment (Paul et al., 1997; De Windt et al., 2000; Clerk and Sugden, 2001). Furthermore, it has been confirmed that the suppression of PKC activity is enough to blunt the development of stretch-induced hypertrophy in cultured neonatal cardiac myocytes (Yamazaki et al., 1995).

5.7.3.2 G-Protein Coupled Receptors

G-protein coupled receptors (GPCR) are transmembrane-spanning receptors that are coupled to G proteins (G_s, G_i, G_q, or $G_{11/12}$), which function as signal transducers. In the heart, many studies have implicated the GPCR mediated G_q pathway in the development of pathological cardiac hypertrophy (Salazar et al., 2007). Studies done on transgenic mice have shown that overexpression of the G_q protein in the heart was associated with the development of hypertrophy and heart failure (Heineke and Molkentin, 2006), while the inhibition of G_q signaling resulted in a blunted, but beneficial response to hypertrophy following transverse aortic constriction (Esposito et al., 2002; Suzuki et al., 2002). Endothelin 1 and Ang II are two agonists of the G_q GPCR pathway that have been implicated in the development of mechanical load induced hypertrophy. Endothelin 1 is induced by stretch in neonatal cardiac myocytes and has been evidenced to stimulate the activation of two kinases implicated in the development of hypertrophy: Raf1 and MAPK (Yamazaki et al., 1996). The secretion of Ang II from cultured cardiac myocytes was also induced by mechanical stretch (Sadoshima et al., 1993). Treatment of neonatal cardiac myocytes with Ang II resulted in the activation of Raf1 and MEK (two hypertrophic signaling molecules), which have been shown to activate ERKs in vitro (Yamazaki et al., 1998b). Furthermore, Ang II has been evidenced to induce cardiomyocyte hypertrophy directly, by increasing protein synthesis (Baker et al., 1992). Consequently, Ang II and endothelin 1 have been suggested to be involved in autocrine and paracrine mechanisms that mediate stretch induced hypertrophy. In addition to these findings, the Ang II type 1 receptor (AT1) has been shown to be activated by mechanical stretch alone. In vitro and in vivo experiments have shown that the hypertrophic response to an increase in mechanical load is not blunted in myocytes of angiotensinogen deficient mice (Zou et al., 2004). In support of these findings, mechanical stretch did not activate ERKs in HEK293 cells and Cos7 cells that did not express the AT1 receptor, but when the AT1 receptor was over-expressed in these cells, stretch activated ERK phosphorylation was detected. This response was inhibited with AT1 receptor blockers.

5.7.3.3 Stretch Activated Ion Channels

Stretch activated channels (SACs) are ion channels which open their 'pore' in response to mechanical deformation, thereby allowing the passage of ions like Na^+, K^+, and Ca^{2+}. Although the involvement of SACs in mechanotransduction is not fully understood (Sadoshima et al., 1992; Yamazaki et al., 1998a), there is some evidence in the literature supporting the role of SACs in mechanical load induced hypertrophy. In a study published by Cooper's group, stretch of contracting papillary muscles increased the rate of protein synthesis which was associated with an increase Na^+ uptake most likely mediated by SACs (Cooper et al., 1989). In support of this hypothesis, the load-induced effect on protein synthesis was inhibited with streptomycin treatment, a SAC blocker, with no effect on the systolic tension developed by these specimens. An increase in Ca^{2+} influx and intracellular Ca^{2+} concentration through SACs has also been reported in cultured cardiac myocytes (Tatsukawa et al., 1997). There are several mechanisms by which an increase in intracellular calcium concentration may contribute to the development of hypertrophy. First, PLC, endothelin-1, and calcineurin are three signaling molecules implicated in the development of hypertrophy, which depend upon calcium for optimal activity (Rhee and Choi, 1992; Zhu et al., 2000; Houser and Molkentin, 2008). Second, an increase in intracellular calcium will most likely have a positive effect on the systolic tension developed by cardiac muscle, resulting in greater mechanical loading of the tissue.

5.7.3.4 Na/H Exchanger

The Na^+/H^+ exchanger is a transmembrane glycoprotein that functions to regulate intracellular pH and the concentration of Na^+ by electroneutrally exchanging intracellular H^+ for extracellular Na^+ (Cingolani and Ennis, 2007). In vitro studies have shown that mechanical loading of cardiomyocytes resulted in the enhanced activity of the Na^+/H^+ exchanger, which is associated with the induction of hypertrophy at the molecular level (Yamazaki et al., 1998a; Cingolani and Ennis, 2007). Furthermore, treatment of cultured cardiomyocytes with HOE 694, a specific inhibitor of the Na^+/H^+ exchanger, notably attenuated stretch induced activation of the ERK pathway and stimulation of protein synthesis (Yamazaki et al., 1998a). In vivo studies showed enhanced activity of the Na^+/H^+ exchanger in the myocardium of spontaneously hypertensive rats (Schussheim and Radda, 1995; Kusumoto et al., 2001; Hardt and Sadoshima, 2002). Additionally, cardiac hypertrophy in rats induced by the continuous treatment with isoproterenol was prevented by the inhibition of the Na^+/H^+ exchanger (Ennis et al., 2003). Taken together these results suggest a vital role for the Na^+/H^+ exchanger in the development of mechanical stress-induced hypertrophy.

5.7.4 Cytoskeleton in Mechanotransduction

The cytoskeleton is an interconnected structure of proteins that contributes to the stabilization of cell shape and cell structure (Hein et al., 2000) and is an obvious candidate for mechanotransduction (Forgacs, 1995; Maniotis et al., 1997). A mechanical signal applied to the cell membrane can traverse the cytoskeletal network of proteins, by inducing molecular deformations and changes in protein conformation, and can reach the nucleus relatively undiminished (Shafrir and Forgacs, 2002). The cytoskeleton of cardiac myocytes can be divided into three parts: the force generating sarcomeric cytoskeleton composed mainly of actin and myosin, the intrasarcomeric cytoskeleton containing titin and α-actinin, and the extrasarcomeric cytoskeleton which includes desmin, microtubules, and intermediate filaments (Chen and Chien, 1999). Deficiencies in proteins belonging to any one of these cytoskeletal parts, can lead to significant changes in cytoskeletal architecture. For example, a deletion of MLP or CapZ from the intrasarcomeric cytoskeleton leads to defects in Z-disc alignment, causes changes in Z-disc width and/ or length, and contributes to the disorganization of thin filaments and the contractile apparatus which may affect the contractile function of the heart (Arber et al., 1997; Hart and Cooper, 1999; Clark et al., 2002). Dilated cardiomyopathy, resulting in both hypertrophy and heart failure, can develop from deletion of the MLP (Arber et al., 1997) protein. Desmin, an extrasarcomeric protein, functions as a lateral linker of adjacent z-discs and connects myofibrils with the sarcolemma. Removal of desmin results in lateral misalignment of myofibrils as well as decreased attachment of myofibrils to the sarcolemma (Clark et al., 2002), and desmin mutations have been associated with cardiomyopathies (Goldfarb et al., 1998). Effective diastolic and systolic force transmission from the inside of myocytes (for example) to z-discs, the sarcolemma, the basal lamina, and to the extracellular matrix depends on the structure of the cytoskeleton. The cytoskeleton may also play a crucial role in load sensing and cellular signaling by acting as a scaffold for signaling cascades, providing sites where multiple signaling molecules can localize and attach (Pyle et al., 2002).

5.7.4.1 Titin

Titin is the most abundant protein of the intrasarcomeric cytoskeleton (Clark et al., 2002) and has multiple functions that contribute to myocardial mechanics and signaling. Titin spans half the sarcomere from z-line to m-line. When sarcomeres are stretched, the extension of titin's I-band gives rise to diastolic force (Granzier and Labeit, 2004). Titin's extensible region is composed of proximal and distal immunoglobulin (Ig) like domains, the PEVK segment, and the N2B region. When stretched, these different titin regions, having distinct stiffness values, extend sequentially, where the Ig segments extend first, followed by the extension of PEVK and N2B segments. When sarcomeres are shortened below slack length, during systolic contraction for example, the compression of titin's I-band gives rise to the restoring force that pushes z-discs toward their slack positions. Titin also functions as a scaffold for signaling molecules, which traffic between the cytoskeleton and the nucleus. Titin is bound to the z-disc by Tcap, which mechanically connects titin to signaling molecules: PKC, MLP, and calcineurin (Hoshijima, 2006). At the I-band, titin interacts by way of FHL2 with hypertrophic signaling molecule ERK2 (Purcell et al., 2004). Titin's I-band region is also associated with FHL1, which binds to signaling molecules: Raf, MEK, and ERK (see Section 5.7.4.3). At the C-terminal end of titin, which is mechanically bound to the M-line, titin contains a catalytic serine-threonine kinase domain, which indirectly interacts with MURF2, a signaling molecule involved in myogenesis that can translocate between the cytosol and nucleus under atrophic conditions (Lange et al., 2005). Changes in the deformation of specific titin regions, which are accompanied by conformational changes in titin, may occur when titin associated structural proteins such as MLP or FHL1 are eliminated from the cytoarchitecture and may affect titin function as a contributor to myocardial mechanics and/ or mechanosensing. It may also affect force transmission in the fiber direction of cardiac tissue to 'mechanosensors' embedded within the cytoskeleton, leading to alterations in the biochemical response of the tissue to an increase in mechanical load.

In excised, isometrically contracting papillary muscles diastolic stress and/or strain has been proposed as the primary mechanical stimulus for myocardial remodeling and load induced hypertrophy (Guterl et al., 2007; Raskin et al., 2008). This evidence supports the claim that titin may play a major role in mechanotransduction, since within the physiological sarcomere length range (1.85–2.4 μm) titin is the primary contributor to the diastolic tension of cardiac muscle (Granzier and Irving, 1995; Li et al., 2002).

5.7.4.2 Muscle LIM Protein

Muscle LIM protein, MLP, is a member of a unique subclass of LIM proteins that serve several functions in the cytosol as well as in the nucleus (Arber and Caroni, 1996). Cytoplasmic MLP is localized at the z-disc and functions as a mechanical stabilizer. At the z-disc MLP binds to Tcap, α-actinin, and N-Rap (another LIM domain protein that binds to actin, talin, and vinculin) (Hoshijima, 2006). In the nucleus MLP may have signaling roles, where it has been shown to interact with transcriptional regulators of myogenesis: MyoD and Murf4. MLP knock out mice develop dilated cardiomyopathy with progressive heart failure (Arber et al., 1997). The primary effects of MLP deficiency in mice did not alter the development of systolic stresses in the heart (Lorenzen-Schmidt et al., 2005), but did contribute to disorganized and widened z-disc structure, diastolic myocardial dysfunction, affecting the diastolic stiffness of heart tissue and dysfunction in the development of cardiac hypertrophy, possibly due to impairment of mechanotransduction (Knöll et al., 2002).

5.7.4.3 Four-and-a-Half LIM Domain Protein

FHL1 is a member of the four and a half LIM domain protein subfamily (Lee et al., 1998). There are four identified members of this protein family: FHL1, FHL2, FHL3, and FHL4. FHL2 and FHL3 are highly expressed in striated muscle (Chu et al., 2000; Johannessen et al., 2006). FHL1 is expressed in most tissues. FHL1 expression is significantly upregulated in hearts of mice treated with hypertrophic agonist and in hearts of human patients exhibiting dilated and hypertrophic cardiomyopathy (Chu et al., 2000; Lim et al., 2001; Gaussin et al., 2003). Recent studies have shown that FHL1 is localized to titin's N2B region; and the direct interaction of FHL1 with titin was supported with co-purification of FHL1 with titin N2B unique sequence (Sheikh and Chen, 2008). The interaction of FHL1 with signaling molecules: Raf1, MEK1/2, and ERK2, was also determined. These studies further revealed that at baseline (i.e. normal loading conditions) FHL1 deficient cardiac muscle was not phenotypically different than controls. However, with pressure overload FHL1 deficient cardiac muscle had a blunted response to hypertrophic growth, coupled with reduced ANP gene expression, lower ERK2 phosphorylation levels, and preserved cardiac function. Furthermore, when compared to wild-type controls FHL1 deficient right ventricular papillary muscles exhibit greater myocardial compliance and a reduced hypertrophic response, characterized by lower ANP gene expression levels, to a chronic increase in mechanical load. These studies revealed that FHL1 plays an essential role in the development of mechanical stress induced myocardial hypertrophy. Furthermore, these studies have suggested that FHL1 may play some role in mechanotransduction by linking the passive stretch domain of titin to the ERK pathway to modulate cardiac hypertrophy.

In contrast to MLP knockout mice, impairment of mechanotransduction in FHL1 deficient mice did not result in the development of heart failure (Sheikh and Chen, 2008). This may be a result of differences in biochemical signaling between FHL1 and MLP knockout mice. The Gq mediated ERK kinase pathway, which is blunted in FHL1 knockout mice, is among the best established pathways for the development of pathological hypertrophy (Esposito et al., 2002; Minamino et al., 2002; Sheikh and Chen, 2008). Unlike in MLP knockout mice, z-disc structure was not affected with a deletion in FHL1. Disruption of z-disc structure is commonly associated with the development of heart failure (Zhou et al., 2001; Epstein and Davis,

2003; Hoshijima, 2006). The z-disc functions as a junctional complex where various cytoskeletal structures are concentrated.

5.7.4.4 Troponin C

Troponin C (TnC) is a very highly conserved protein (Roher et al., 1986) that couples the concentration of Ca^{2+} in the cytoplasm with the generation of force. Upon association of TnC with calcium, it effects a conformational change in the troponin-tropomysin complex that results in opening of actin-myosin binding sites. Mutations in TnC have been implicated in a hereditable form of dilated cardiomyopathy (Mogensen et al., 2004).

Experiments in skinned cardiac muscle preparations have shown that rapid length changes result in dissociation of Ca^{2+} from the myofilaments (Allen and Kentish, 1988). Ca^{2+} transients measured in intact preparations have also been shown to change when the muscle is allowed to shorten (Janssen and de Tombe, 1997). While some have interpreted these results as evidence for intrinsic force sensing by TnC (Hunter et al., 1998; Niederer et al., 2006), soluble S1 myosin fragments have been demonstrated to increase affinity of TnC for Ca^{2+} in the absence of force (Robinson et al., 2004; Davis et al., 2007; Dong et al., 2007). These most recent results suggest that stretch affects Ca^{2+} binding to TnC indirectly by increasing myosin crossbridge detachment.

5.7.5 The Nucleus in Mechanotransduction

The nucleus is the largest and stiffest organelle within the cell, contains the genome, and is the site of transcriptional regulation. Intra- and extra-cellular forces alter nuclear shape and structure. For these reasons, the nucleus has been implicated in the processes of mechanotransduction.

The main mechanism by which the nucleus is believed to act as a mechanosensor is mediated by chromatin. Chromatin is the major component of the nuclear interior and functions to package DNA into the small volume of the nucleus. Forces applied onto the nucleus induce changes in nuclear shape, which are associated with changes in chromatin shape and organization. It is these changes in chromatin that are believed to affect transcription and regulate gene expression (Dahl et al., 2008).

5.7.6 Influence of the RhoA/ROCK Pathway in Cardiomyocyte Mechanotransduction

Many studies of mechanotransduction in attachment-dependent cells, especially mechanotransduction impinging on reorganization of the actin cytoskeleton, have shown involvement of a signaling pathway initiating with activation of the small GTPase Ras homolog gene family member A (RhoA) and Rho-associated coiled-coil containing kinases (ROCK I and II), which are activated upon binding to RhoA-GTP (Riento and Ridley, 2003). Mechanical stress acting upon integrin connections to the ECM can lead to activation of RhoA, which binds ROCK and leads to a signaling cascade that prevents actin depolymerization (Peters and Michel, 2007). In addition, ROCK inhibits myosin light chain phosphatase and may enhance the activity of myosin light chain kinase, both of which result in increased phosphorylation of myosin regulatory light chain (MLC) and increased stress generation in the actin/myosin skeleton as well as increased contractility of smooth muscle cells (Kimura et al., 1996). These effects result in an increase in the number and size of focal adhesions and lead to the development of stress fibers (Nobes and Hall, 1995). RhoA can be specifically inhibited by botulinum ADP-ribosyltransferase C3, or C3 toxin (Aktories et al., 1989). ROCK is necessary for cardiac fibrosis in models of heart failure and can be specifically inhibited by either fasudil or Y-27632, each of which are equipotent (Noma et al., 2006).

Studies have shown that cyclic stretch can increase the activity of RhoA in striated muscle (Zhang et al., 2007). Studies using magnetic collagen-coated beads have shown that applying a physiologically relevant tensile force of approximately 480 pN to Rat-2 cardiac fibroblasts results in a transient activation of RhoA and downstream effects of LIMK-1 phosphorylation, MRTF-A translocation to the nucleus, a transient increased expression (~15 min) of total actin and a longer-term increased expression (~180 min) of smooth muscle α-actin, a marker of myofibroblast differentiation. Each of these effects was blocked by the addition of the ROCK inhibitor Y27632. Interestingly, the applied stress resulted in slightly decreased levels of MLC phosphorylation, even though MLC is canonically activated downstream of ROCK (Zhao et al., 2007). Another study showed that neonatal rat cardiomyocytes cultured on gels of varying modulus for 7 days developed stress fibers on gels with a modulus stiffer than 10 kPa. Along with stress fiber formation, these cardiomyocytes did not form aligned sarcomeres and a typical rod-like myocytes shape, and had lower expression of SERCA2a, reduced calcium storage, smaller calcium transients and generated less force when contracting. However, when RhoA activity was inhibited with C3 toxin or when ROCK activity was inhibited with hydroxyfasudil, these cells formed aligned sarcomeres and generated forces higher than those generated by cells on softer substrates, as would be predicted from the more isometric contractions due to the stiffer substrate (Jacot et al., 2008).

One study has also found that cell spread area can influence the activation of RhoA and ROCK in human mesenchymal stem cells, and in turn influence the cells' decisions to differentiate into adipocytes or osteoblasts (McBeath et al., 2004).

5.8 Summary and Conclusion

In summary, researchers have used many techniques to apply mechanical stimuli to in vitro cultures of cardiac myocytes, including direct application of stretch, indentation and shear stress as well as altering the mechanics of the material supporting beating myocytes. Because the structure of myocyte cultures heavily influences function, and because the mechanical responses of myocytes are highly directional, techniques have also been developed for patterning and aligning in vitro cultures of cardiomyocytes. Directional stretch, indentation and shear stress in cardiomyocytes can all induce beating through putative stretch-activated ion channels, as well as reduce membrane potential, shorten action potential duration, modulate gap junction communication between myocytes, and alter conduction velocity. In addition, direct stretch and shear stress can induce a hyperplasic response in fetal and neonatal cardiac cells and a hypertrophic response in mature cardiac cells. Furthermore, the change in the stiffness of the culture substrate can affect myocyte behavior in a bell-shaped manner, with observations of highest myocyte contractile force in cells in environments with approximately native myocardial stiffness and loss of function in softer or stiffer environments. The application of static or dynamic external force also appears to enhance the differentiation of stem cells into cardiac cells, as measured both by cell function and by the expression of protein markers specific to cardiomyocytes.

The mechanotransduction of these signals appears to initiate through stretch-sensitive ion channels in the membrane or sarcolemma, cell–cell connections or connections between the ECM-binding proteins (specifically, integrins) and the internal cytoskeleton. Signaling can progress through several pathways, including pathways mediated by G-coupled protein receptors, phospholipase C/PKC, MAPK/ERK2/JNK2 and RhoA/ROCK. Proteins initially thought to have only a structural role, linking together portions of the sarcomere structures and the cytoskeleton, including MLP, the FHL proteins, and titin, have been found to play key roles in mechanotransductive signaling. Additionally, nuclear deformation may play an important role in mechanotransduction.

In conclusion, mechanical signals can greatly affect cardiomyocyte function, both in short-term behavior (including beating frequencies, conduction velocities and contractile forces), in long term behaviors such as the hypertrophic response, and in development. These mechanical signals and their sensing by the cell are critical for normal function in the myocyte, and defects in mechanotransduction pathways can be important mechanisms of cardiac disease.

References

Aktories K, Braun U, Rosener S, Just I and Hall A (1989) The rho gene product expressed in E. coli is a substrate of botulinum ADP-ribosyltransferase C3. Biochem Biophys Res Commun 158: 209–13.

Allen D and Kentish J (1985) The cellular basis of the length-tension relation in cardiac muscle. J Mol Cell Cardiol 17: 821–40.

Allen DG and Kentish JC (1988) Calcium concentration in the myoplasm of skinned ferret ventricular muscle following changes in muscle length. J Physiol 407: 489–503.

Antonitsis P, Ioannidou-Papagiannaki E, Kaidoglou A, Charokopos N, Kalogeridis A, Kouzi-Koliakou K, Kyriakopoulou I, Klonizakis I and Papakonstantinou C (2008) Cardiomyogenic potential of human adult bone marrow mesenchymal stem cells in vitro. Thorac Cardiovasc Surg 56: 77–82.

Arber S and Caroni P (1996) Specificity of single LIM motifs in targeting and LIM/LIM interactions in situ. Genes Dev 10: 289–300.
Arber S, Hunter JJ, Ross J, Hongo M Jr., Sansig G, Borg J, Perriard J-C, Chien KR and Caroni P (1997) MLP-deficient mice exhibit a disruption of cardiac cytoarchitectural organization, dilated cardiomyopathy, and heart failure. Cell 88: 393–403.
Arts T, Prinzen FW, Snoeckx LH and Reneman RS (1995) A model approach to the adaptation of cardiac structure by mechanical feedback in the environment of the cell. Adv Exp Med Biol 382: 217–28.
Asher CR and Klein AL (2002) Diastolic heart failure: Restrictive cardiomyopathy, constrictive pericarditis, and cardiac tamponade: clinical and echocardiographic evaluation. Cardiol Rev 10: 218–29.
Baker KM, Booz GW and Dostal DE (1992) Cardiac actions of angiotensin II: role of an intracardiac renin angiotensin system. Annu Rev Physiol 54: 227–41.
Balgude AP, Yu X, Szymanski A and Bellamkonda RV (2001) Agarose gel stiffness determines rate of DRG neurite extension in 3D cultures. Biomaterials 22: 1077–84.
Banes AJ, Gilbert J, Taylor D and Monbureau O (1985) A new vacuum-operated stress-providing instrument that applies static or variable duration cyclic tension or compression to cells in vitro. J Cell Sci 75: 35–42.
Barbee KA (2002) Role of subcellular shear-stress distributions in endothelial cell mechanotransduction. Ann Biomed Eng 30: 472–82.
Barbosa ME, Alenina N and Bader M (2005) Induction and analysis of cardiac hypertrophy in transgenic animal models. Methods Mol Med 112: 339–52.
Baumgarten CM and Clemo HF (2003) Swelling-activated chloride channels in cardiac physiology and pathophysiology. Prog Biophys Mol Biol 82: 25–42.
Bearzi C, Rota M, Hosoda T, Tillmanns J, Nascimbene A, De Angelis A, Yasuzawa-Amano S, Trofimova I, Siggins RW, Lecapitaine N, Cascapera S, Beltrami AP, D'Alessandro DA, Zias E, Quaini F, Urbanek K, Michler RE, Bolli R, Kajstura J, Leri A and Anversa P (2007) Human cardiac stem cells. Proc Natl Acad Sci U S A 104: 14068–73.
Berry MF, Engler AJ, Woo YJ, Pirolli TJ, Bish LT, Jayasankar V, Morine KJ, Gardner TJ, Discher DE and Sweeney HL (2006) Mesenchymal stem cell injection after myocardial infarction improves myocardial compliance. Am J Physiol Heart Circ Physiol 290: H2196–203.
Bers DM (2001). *Excitation Contraction Coupling and Cardiac Contractile Force*. Kluwer Academic, Norwell, MA.
Bett GCL and Sachs F (1997) Cardiac mechanosensitivity and stretch-activated ion channels. Trends Cardiovasc Med 7: 4–8.
Bian W and Tung L (2006) Structure-related initiation of reentry by rapid pacing in monolayers of cardiac cells. Circ Res 98: e29–38.
Boheler KR, Czyz J, Tweedie D, Yang HT, Anisimov SV and Wobus AM (2002) Differentiation of pluripotent embryonic stem cells into cardiomyocytes. Circ Res 91: 189–201.
Bolognese L, Dellavesa P, Rossi L, Sarasso G, Bongo AS and Scianaro MC (1994) Prognostic value of left ventricular mass in uncomplicated acute myocardial infarction and one-vessel coronary artery disease. Am J Cardiol 73: 1–5.
Boontheekul T, Hill EE, Kong HJ and Mooney DJ (2007) Regulating myoblast phenotype through controlled gel stiffness and degradation. Tissue Eng 13: 1431–42.
Brown TD (2000) Techniques for mechanical stimulation of cells in vitro: a review. J Biomech 33: 3–14.
Brown XQ, Ookawa K and Wong JY (2005) Evaluation of polydimethylsiloxane scaffolds with physiologically-relevant elastic moduli: interplay of substrate mechanics and surface chemistry effects on vascular smooth muscle cell response. Biomaterials 26: 3123–9.
Bupha-Intr T, Holmes JW and Janssen PM (2007) Induction of hypertrophy in vitro by mechanical loading in adult rabbit myocardium. Am J Physiol Heart Circ Physiol 293: H3759–67.
Bursac N, Parker KK, Iravanian S and Tung L (2002) Cardiomyocyte cultures with controlled macroscopic anisotropy: a model for functional electrophysiological studies of cardiac muscle. Circ Res 91: e45–54.

Calaghan SC, Belus A and White E (2003) Do stretch-induced changes in intracellular calcium modify the electrical activity of cardiac muscle? Prog Biophys Mol Biol 82: 81–95.
Calaghan SC, Le Guennec JY and White E (2004) Cytoskeletal modulation of electrical and mechanical activity in cardiac myocytes. Prog Biophys Mol Biol 84: 29–59.
Camelliti P, Gallagher J, Kohl P and McCulloch A (2006) Micropatterned cell cultures on elastic membranes as an in vitro model of myocardium. Nat Protoc 1: 1379–91.
Cazorla O, Pascarel C, Brette F and Le Guennec JY (1999) Modulation of ions channels and membrane receptors activities by mechanical interventions in cardiomyocytes: possible mechanisms for mechanosensitivity. Prog Biophys Mol Biol 71: 29–58.
Chen J and Chien KR (1999) Complexity in simplicity: monogenic disorders and complex cardiomyopathies. J Clin Invest 103: 1483–5.
Ching YH, Ghosh TK, Cross SJ, Packham EA, Honeyman L, Loughna S, Robinson TE, Dearlove AM, Ribas G, Bonser AJ, Thomas NR, Scotter AJ, Caves LS, Tyrrell GP, Newbury-Ecob RA, Munnich A, Bonnet D and Brook JD (2005) Mutation in myosin heavy chain 6 causes atrial septal defect. Nat Genet 37: 423–8.
Chiu J, Chen L, Lee P, Lee C, Lo L, Usami S and Chien S (2003) Shear stress inhibits adhesion molecule expression in vascular endothelial cells induced by coculture with smooth muscle cells. Blood 101: 2667–74.
Chu PH, Ruiz-Lozano P, Zhou Q, Cai C and Chen J (2000) Expression patterns of FHL/SLIM family members suggest important functional roles in skeletal muscle and cardiovascular system. Mech Dev 95: 259–65.
Cingolani HE and Ennis IL (2007) Sodium-hydrogen exchanger, cardiac overload, and myocardial hypertrophy. Circulation 115: 1090–100.
Clark KA, McElhinny AS, Beckerle MC and Gregorio CC (2002) Striated muscle cytoarchitecture: an intricate web of form and function. Annu Rev Cell Dev Biol 18: 637–706.
Clerk A and Sugden PH (2001) Untangling the web: specific signaling from PKC isoforms to MAPK cascades. Circ Res 89: 847–9.
Cooper G IV, Kent RL and Mann DL (1989) Load induction of cardiac hypertrophy. J Mol Cell Cardiol 21: 11–30.
Cooper Gt (2006) Cytoskeletal networks and the regulation of cardiac contractility: microtubules, hypertrophy, and cardiac dysfunction. Am J Physiol Heart Circ Physiol 291: H1003–14.
Dahl KN, Ribeiro AJ and Lammerding J (2008) Nuclear shape, mechanics, and mechanotransduction. Circ Res 102: 1307–18.
Davis JP, Norman C, Kobayashi T, Solaro RJ, Swartz DR and Tikunova SB (2007) Effects of thin and thick filament proteins on calcium binding and exchange with cardiac troponin C. Biophys J 92: 3195–206.
de Jonge HW, Dekkers DH, Houtsmuller AB, Sharma HS and Lamers JM (2007) Differential signaling and hypertrophic responses in cyclically stretched vs endothelin-1 stimulated neonatal rat cardiomyocytes. Cell Biochem Biophys 47: 21–32.
De Windt LJ, Lim HW, Haq S, Force T and Molkentin JD (2000) Calcineurin promotes protein kinase C and c-jun NH2-terminal kinase activation in the heart. cross-talk between cardiac hypertrophic signaling pathways. J Biol Chem 275: 13571–9.
Deutsch J, Motlagh D, Russell B and Desai TA (2000) Fabrication of microtextured membranes for cardiac myocyte attachment and orientation. J Biomed Mater Res 53: 267–75.
Dewey CF Jr., Bussolari SR, Gimbrone MA Jr., and Davies PF (1981) The dynamic response of vascular endothelial cells to fluid shear stress. J Biomech Eng 103: 177–85.
Ding B, Price RL, Goldsmith EC, Borg TK, Yan X, Douglas PS, Weinberg EO, Bartunek J, Thielen T, Didenko V and Lorell BH (2000) Left ventricular hypertrophy in ascending aortic stenosis mice: anoikis and the progression to early failure. Circulation 101: 2854–62.
Discher DE, Janmey P and Wang YL (2005) Tissue cells feel and respond to the stiffness of their substrate. Science 310: 1139–43.
Dong WJ, Jayasundar JJ, An J, Xing J and Cheung HC (2007) Effects of PKA phosphorylation of cardiac troponin I and strong crossbridge on conformational transitions of the N-domain of cardiac troponin C in regulated thin filaments. Biochemistry 46: 9752–61.

Engelmayr GC Jr, Sales VL, Mayer JE Jr, and Sacks MS (2006) Cyclic flexure and laminar flow synergistically accelerate mesenchymal stem cell-mediated engineered tissue formation: implications for engineered heart valve tissues. Biomaterials 27: 6083–95.

Engler AJ, Griffin MA, Sen S, Bonnemann CG, Sweeney HL and Discher DE (2004) Myotubes differentiate optimally on substrates with tissue-like stiffness: pathological implications for soft or stiff microenvironments. J Cell Biol 166: 877–87.

Engler AJ, Sen S, Sweeney HL and Discher DE (2006) Matrix elasticity directs stem cell lineage specification. Cell 126: 677–89.

Ennis IL, Escudero EM, Console GM, Camihort G, Dumm CG, Seidler RW, Camilión de Hurtado MC and Cingolani HE (2003) Regression of isoproterenol-induced cardiac hypertrophy by Na^{+}/H^{+} exchanger inhibition. Hypertension 41: 1324–9.

Entcheva E and Bien H (2005) Acoustic micromachining of three-dimensional surfaces for biological applications. Lab Chip 5: 179–83.

Epstein ND and Davis JS (2003) Sensing stretch is fundamental. Cell Biochem Biophys 112: 147–50.

Eschenhagen T, Fink C, Remmers U, Scholz H, Wattchow J, Weil J, Zimmermann W, Dohmen HH, Schafer H, Bishopric N, Wakatsuki T and Elson EL (1997) Three-dimensional reconstitution of embryonic cardiomyocytes in a collagen matrix: a new heart muscle model system. FASEB J 11: 683–94.

Esposito G, Rapacciuolo A, Naga Prasad SV, Takaoka H, Thomas SA, Koch WJ and Rockman HA (2002) Genetic alterations that inhibit in vivo pressure-overload hypertrophy prevent cardiac dysfunction despite increased wall stress. Circulation 105: 85–92.

Folch A and Toner M (1998) Cellular micropatterns on biocompatible materials. Biotechnol Prog 14: 388–92.

Forgacs G (1995) On the possible role of cytoskeletal filamentous networks in intracellular signaling: an approach based on percolation. J Cell Sci 108: 2131–43.

Frey N and Olson EN (2003) Cardiac hypertrophy: the good, the bad, and the ugly. Annu Rev Physiol 65: 45–79.

Fukuda K (2003) Use of adult marrow mesenchymal stem cells for regeneration of cardiomyocytes. Bone Marrow Transplant 32(1): S25–7.

Garnier D (1994) Attachment procedures for mechanical manipulation of isolated cardiac myocytes: a challenge. Cardiovasc Res 28: 1758–64.

Gaussin V, Tomlinson JE, Depre C, Engelhardt S, Antos CL, Takagi G, Hein L, Topper JN, Liggett SB, Olson EN, Lohse MJ, Vatner SF and Vatner DE (2003) Common genomic response in different mouse models of beta-adrenergic-induced cardiomyopathy. Circulation 108: 2926–33.

Genes NG, Rowley JA, Mooney DJ and Bonassar LJ (2004) Effect of substrate mechanics on chondrocyte adhesion to modified alginate surfaces. Arch Biochem Biophys 422: 161–7.

Gerull B, Gramlich M, Atherton J, McNabb M, Trombitas K, Sasse-Klaassen S, Seidman JG, Seidman C, Granzier H, Labeit S, Frenneaux M and Thierfelder L (2002) Mutations of TTN, encoding the giant muscle filament titin, cause familial dilated cardiomyopathy. Nat Genet 30: 201–4.

Goldfarb LG, Park KY, Cervenakova L, Gorokhova S, Lee HS, Vasconcelos O, Nagle JW, Semino-Mora C, Sivakumar K and Dalakas MC (1998) Missense mutations in desmin associated with familial cardiac and skeletal myopathy. Nat Genet 19: 402–3.

Goldsmith EC and Borg TK (2002) The dynamic interaction of the extracellular matrix in cardiac remodeling. J Card Fail 8: S314–8.

Gopalan SM, Flaim C, Bhatia S, Hoshijima M, Knöll R, Chien KR, Omens JH and McCulloch AD (2003) Anisotropic stretch-induced hypertrophy in neonatal ventricular myocytes micropatterned on deformable elastomers. Biotechnol Bioeng 81: 578–87.

Granzier HL and Irving TC (1995) Passive tension in cardiac muscle: contribution of collagen, titin, microtubules, and intermediate filaments. Biophys J 68: 1027–41.

Granzier HL and Labeit S (2004) The giant protein titin: a major player in myocardial mechanics, signaling, and disease. Circ Res 94: 284–95.

Groban ES, Narayanan A and Jacobson MP (2006) Conformational changes in protein loops and helices induced by post-translational phosphorylation. PLoS Comput Biol 2: e32.

Grossman W, Jones D and McLaurin LP (1975) Wall stress and patterns of hypertrophy in the human left ventricle. J Clin Invest 56: 56–64.

Guo XM, Zhao YS, Chang HX, Wang CY, E LL, Zhang XA, Duan CM, Dong LZ, Jiang H, Li J, Song Y and Yang XJ (2006) Creation of engineered cardiac tissue in vitro from mouse embryonic stem cells. Circulation 113: 2229–37.

Gupta S, Das B and Sen S (2007) Cardiac hypertrophy: mechanisms and therapeutic opportunities. Antioxid Redox Signal 9: 623–52.

Guterl KA, Haggart CR, Janssen PM and Holmes JW (2007) Isometric contraction induces rapid myocyte remodeling in cultured rat right ventricular papillary muscles. Am J Physiol Heart Circ Physiol 293: H3707–12.

Gwak SJ, Bhang SH, Kim IK, Kim SS, Cho SW, Jeon O, Yoo KJ, Putnam AJ and Kim BS (2008) The effect of cyclic strain on embryonic stem cell-derived cardiomyocytes. Biomaterials 29: 844–56.

Hagiwara N, Masuda H, Shoda M and Irisawa H (1992) Stretch-activated anion currents of rabbit cardiac myocytes. J Physiol 456: 285–302.

Hardt SE and Sadoshima J (2002) Glycogen synthase kinase-3beta: a novel regulator of cardiac hypertrophy and development. Circ Res 90: 1055–106.

Hart MC and Cooper JA (1999) Vertebrate isoforms of actin capping protein beta have distinct functions in vivo. J Cell Biol 147: 1287–98.

Hein S, Kostin S, Heling A, Maeno Y and Schaper J (2000) The role of the cytoskeleton in heart failure. Cardvasc Res 45: 273–8.

Heineke J and Molkentin JD (2006) Regulation of cardiac hypertrophy by intracellular signalling pathways. Nat Rev Mol Cell Biol 7: 589–600.

Hidalgo-Bastida LA, Barry JJ, Everitt NM, Rose FR, Buttery LD, Hall IP, Claycomb WC and Shakesheff KM (2007) Cell adhesion and mechanical properties of a flexible scaffold for cardiac tissue engineering. Acta Biomater 3: 457–62.

Hongo K, White E, Le Guennec JY and Orchard CH (1996) Changes in [Ca^{2+}]i, [Na^+]i and Ca^{2+} current in isolated rat ventricular myocytes following an increase in cell length. J Physiol 491 (Pt 3): 609–19.

Hoshijima M (2006) Mechanical stress-strain sensors embedded in cardiac cytoskeleton: Z disk, titin, and associated structures. Am J Physiol Heart Circ Physiol 290: H1313–25.

Houser SR and Molkentin JD (2008) Does contractile Ca^{2+} control calcineurin-NFAT signaling and pathological hypertrophy in cardiac myocytes. Sci Signal 1: pe31.

Hu H and Sachs F (1996) Mechanically activated currents in chick heart cells. J Membr Biol 154: 205–16.

Hu H and Sachs F (1997) Stretch-activated ion channels in the heart. J Mol Cell Cardiol 29: 1511–23.

Hunter PJ, McCulloch AD and ter Keurs HE (1998) Modelling the mechanical properties of cardiac muscle. Prog Biophys Mol Biol 69: 289–331.

Hynes RO and Lander AD (1992) Contact and adhesive specificities in the associations, migrations, and targeting of cells and axons. Cell Biochem Biophys 68: 303–22.

Iijima Y, Nagai T, Mizukami M, Matsuura K, Ogura T, Wada H, Toko H, Akazawa H, Takano H, Nakaya H and Komuro I (2003) Beating is necessary for transdifferentiation of skeletal muscle-derived cells into cardiomyocytes. FASEB J 17: 1361–3.

Illi B, Scopece A, Nanni S, Farsetti A, Morgante L, Biglioli P, Capogrossi MC and Gaetano C (2005) Epigenetic histone modification and cardiovascular lineage programming in mouse embryonic stem cells exposed to laminar shear stress. Circ Res 96: 501–8.

Isenberg G, Kazanski V, Kondratev D, Gallitelli MF, Kiseleva I and Kamkin A (2003) Differential effects of stretch and compression on membrane currents and [Na^+]c in ventricular myocytes. Prog Biophys Mol Biol 82: 43–56.

Jacot JG, McCulloch AD and Omens JH (2008) Substrate Stiffness Affects the Functional Maturation of Neonatal Rat Ventricular Myocytes. Biophys J 95: 3479–87.
Janmey PA and McCulloch CA (2007) Cell mechanics: integrating cell responses to mechanical stimuli. Annu Rev Biomed Eng 9: 1–34.
Janssen PM and de Tombe PP (1997) Uncontrolled sarcomere shortening increases intracellular Ca^{2+} transient in rat cardiac trabeculae. Am J Physiol 272: H1892–7.
Jarygin C, Hanze J and Lang RE (1994) Gene expression of atrial natriuretic peptide in rat papillary muscle. rapid induction by mechanical loading. FEBS Lett 346: 185–8.
Johannessen M, Moller S, Hansen T, Moens U and Van Ghelue M (2006) The multifunctional roles of the four-and-a-half-LIM only protein FHL2. Cell Mol Life Sci 63: 268–84.
Juliano RL and Haskill S (1993) Signal transduction from the extracellular matrix. J Cell Biol 120: 577–85.
Kamkin A, Kiseleva I and Isenberg G (2000) Stretch-activated currents in ventricular myocytes: amplitude and arrhythmogenic effects increase with hypertrophy. Cardiovasc Res 48: 409–20.
Kamkin A, Kiseleva I, Wagner KD, Bohm J, Theres H, Gunther J and Scholz H (2003) Characterization of stretch-activated ion currents in isolated atrial myocytes from human hearts. Pflugers Arch 446: 339–46.
Kamm RD and Kaazempur-Mofrad MR (2004) On the molecular basis for mechanotransduction. Mech Chem Biosyst 1: 201–9.
Kaplan S, Gard J, Carvajal-Huerta L, Ruiz-Cabezas J, Thiene G and Saffitz J (2004a) Structural and molecular pathology of the heart in Carvajal syndrome. Cardiovasc Pathol 13: 26–32.
Kaplan S, Gard J, Protonotarios N, Tsatsopoulou A, Spiliopoulou C, Anastasakis A, Squarcioni C, McKenna W, Thiene G, Basso C, Brousse N, Fontaine G and Saffitz J (2004b) Remodeling of myocyte gap junctions in arrhythmogenic right ventricular cardiomyopathy due to a deletion in plakoglobin (Naxos disease). Heart Rhythm 1: 3–11.
Kaprielian RR, Gunning M, Dupont E, Sheppard MN, Rothery SM, Underwood R, Pennell DJ, Fox K, Pepper J, Poole-Wilson PA and Severs NJ (1998) Downregulation of immunodetectable connexin43 and decreased gap junction size in the pathogenesis of chronic hibernation in the human left ventricle. Circulation 97: 651–60.
Katsumi A, Orr AW, Tzima E and Schwartz MA (2004) Integrins in mechanotransduction. J Biol Chem 279: 12001–4.
Katz A (2002) Ernest Henry Starling, his predecessors, and the "Law of the Heart". Circulation 106: 2986–92.
Kimura A, Harada H, Park JE, Nishi H, Satoh M, Takahashi M, Hiroi S, Sasaoka T, Ohbuchi N, Nakamura T, Koyanagi T, Hwang TH, Choo JA, Chung KS, Hasegawa A, Nagai R, Okazaki O, Nakamura H, Matsuzaki M, Sakamoto T, Toshima H, Koga Y, Imaizumi T and Sasazuki T (1997) Mutations in the cardiac troponin I gene associated with hypertrophic cardiomyopathy. Nat Genet 16: 379–82.
Kimura K, Ito M, Amano M, Chihara K, Fukata Y, Nakafuku M, Yamamori B, Feng J, Nakano T, Okawa K, Iwamatsu A and Kaibuchi K (1996) Regulation of myosin phosphatase by Rho and Rho-associated kinase (Rho-kinase). Science 273: 245–8.
Knöll R, Hoshijima M, Hoffman HM, Person V, Lorenzen-Schmidt I, Bang ML, Hayashi T, Shiga N, Yasukawa H, Schaper W, McKenna W, Yokoyama M, Schork NJ, Omens JH, McCulloch AD, Kimura A, Gregorio CC, Poller W, Schaper J, Schultheiss HP and Chien KR (2002) The cardiac mechanical stretch sensor machinery involves a Z disk complex that is defective in a subset of human dilated cardiomyopathy. Cell 111: 943–56.
Kohl P and Sachs F (2001). Mechanoelectric feedback in cardiac cells. Phil Trans R Soc Lond A 359: 1173–85.
Komuro I, Kaida T, Shibazaki Y, Kurabayashi M, Katoh Y, Hoh E, Takaku F and Yazaki Y (1990) Stretching cardiac myocytes stimulates protooncogene expression. J Biol Chem 265: 3595–8.
Komuro I, Katoh Y, Kaida T, Shibazaki Y, Kurabayashi M, Hoh E, Takaku F and Yazaki Y (1991) Mechanical loading stimulates cell hypertrophy and specific gene expression in cultured rat cardiac myocytes. J Biol Chem 266: 1265–8.

Komuro I, Kudo S, Yamazaki T, Zou Y, Shiojima I and Yazaki Y (1996) Mechanical stretch activates the stress-activated protein kinases in cardiac myocytes. FASEB J 10: 631–6.

Kong CR, Bursac N and Tung L (2005) Mechanoelectrical excitation by fluid jets in monolayers of cultured cardiac myocytes. J Appl Physiol 98: 2328–36; discussion 2320.

Koren MJ, Devereux RB, Casale PN, Savage DD and Laragh JH (1991) Relation of left ventricular mass and geometry to morbidity and mortality in uncomplicated essential hypertension. Ann Intern Med 114: 345–52.

Kostin S, Dammer S, Hein S, Klovekorn WP, Bauer EP and Schaper J (2004) Connexin 43 expression and distribution in compensated and decompensated cardiac hypertrophy in patients with aortic stenosis. Cardiovasc Res 62: 426–36.

Kostin S, Rieger M, Dammer S, Hein S, Richter M, Klovekorn WP, Bauer EP and Schaper J (2003) Gap junction remodeling and altered connexin43 expression in the failing human heart. Mol Cell Biochem 242: 135–44.

Kusumoto K, Haist JV and Karmazyn M (2001) Na(+)/H(+) exchange inhibition reduces hypertrophy and heart failure after myocardial infarction in rats. Am J Physiol Heart Circ Physiol 280: H738–45.

Laks MM, Morady F, Adomian GE and Swan HJ (1970) Presence of widened and multiple intercalated discs in the hypertrophied canine heart. Circ Res 27: 391–402.

Lammerding J, Kamm RD and Lee RT (2004) Mechanotransduction in cardiac myocytes. Ann N Y Acad Sci 1015: 53–70.

Lange S, Xiang F, Yakovenko A, Vihola A, Hackman P, Rostkova E, Kristensen J, Brandmeier B, Franzen G, Hedberg B, Gunnarsson LG, Hughes SM, Marchand S, Sejersen T, Richard I, Edstrom L, Ehler E, Udd B and Gautel M (2005) The kinase domain of titin controls muscle gene expression and protein turnover. Science 308: 1599–603.

Le Guennec JY, Peineau N, Argibay JA, Mongo KG and Garnier D (1990) A new method of attachment of isolated mammalian ventricular myocytes for tension recording: length dependence of passive and active tension. J Mol Cell Cardiol 22: 1083–93.

Lee AA, Delhaas T, Waldman LK, MacKenna DA, Villarreal FJ and McCulloch AD (1996) An equibiaxial strain system for cultured cells. Am J Physiol 271: C1400–8.

Lee S, Kim J, Li Y, Son M and Woo S (2008) Fluid pressure modulates L-type Ca^{2+} channel via enhancement of Ca^{2+}-induced Ca^{2+} release in rat ventricular myocytes. Am J Physiol Cell Physiol 294: C966–76.

Lee SM, Tsui SK, Chan KK, Garcia-Barcelo M, Waye MM, Fung K, Liew CC and Lee CY (1998) Chromosomal mapping, tissue distribution and cDNA sequence of four-and-a-half LIM domain protein 1 (FHL1). Gene 216: 163–70.

Levy D, Garrison RJ, Savage DD, Kannel WB and Castelli WP (1990) Prognostic implications of echocardiographically determined left ventricular mass in the framingham heart study. N Engl J Med 322: 1561–6.

Lewis JM and Schwartz MA (1995) Mapping in vivo associations of cytoplasmic proteins with integrin beta 1 cytoplasmic domain mutants. Mol Biol Cell 6: 151–60.

Li H, Linke WA, Oberhauser AF, Carrion-Vazquez M, Kerkvliet J, Lu H, Marszalek PE and Fernandez JM (2002) Reverse engineering of the giant muscle protein titin. Nature 418: 998–1002.

Lim DS, Roberts R and Marian AJ (2001) Expression profiling of cardiac genes in human hypertrophic cardiomyopathy: insight into the pathogenesis of phenotypes. J Am Coll Cardiol 38: 1175–80.

Lin NJ, Drzal PL and Lin-Gibson S (2007a) Two-dimensional gradient platforms for rapid assessment of dental polymers: a chemical, mechanical and biological evaluation. Dent Mater 23: 1211–20.

Lin RZ, Ho CT, Liu CH and Chang HY (2006) Dielectrophoresis based-cell patterning for tissue engineering. Biotechnol J 1: 949–57.

Lin W, Laitko U, Juranka PF and Morris CE (2007b) Dual stretch responses of mHCN2 pacemaker channels: accelerated activation, accelerated deactivation. Biophys J 92: 1559–72.

Lorenzen-Schmidt I, Schmid-Schonbein GW, Giles WR, McCulloch AD, Chien S and Omens JH (2006) Chronotropic response of cultured neonatal rat ventricular myocytes to short-term fluid shear. Cell Biochem Biophys 46: 113–22.
Lorenzen-Schmidt I, Stuyvers BD, ter Keurs HEDJ, Date M, Hoshijima M, Chien K, McCulloch AD and Omens JH (2005) Young MLP deficient mice show diastolic dysfunction before the onset of dilated cardiomyopathy. J Mol Cell Cardiol 39: 241–50.
Lyford GL, Strege PR, Shepard A, Ou Y, Ermilov L, Miller SM, Gibbons SJ, Rae J, Szurszewski JH and Farrugia G (2002) alpha(1C) (Ca(V)1.2) L-type calcium channel mediates mechanosensitive calcium regulation. Am J Physiol Cell Physiol 283: C1001–8.
Makino S, Fukuda K, Miyoshi S, Konishi F, Kodama H, Pan J, Sano M, Takahashi T, Hori S, Abe H, Hata J, Umezawa A and Ogawa S (1999) Cardiomyocytes can be generated from marrow stromal cells in vitro. J Clin Invest 103: 697–705.
Maniotis AJ, Chen CS and Ingber DE (1997) Demonstration of mechanical connections between integrins, cytoskeletal filaments, and nucleoplasm that stabilize nuclear structure. Proc Nat Acad Sci 94: 849–54.
Maron B and Ferrans V (1973) Significance of multiple intercalated discs in hypertrophied human myocardium. Am J Pathol 73: 81–96.
Matsuda N, Hagiwara N, Shoda M, Kasanuki H and Hosoda S (1996) Enhancement of the L-type Ca^{2+} current by mechanical stimulation in single rabbit cardiac myocytes. Circ Res 78: 650–9.
Matsuda T, Fujio Y, Nariai T, Ito T, Yamane M, Takatani T, Takahashi K and Azuma J (2006) N-cadherin signals through Rac1 determine the localization of connexin 43 in cardiac myocytes. J Mol Cell Cardiol 40: 495–502.
Mauritz C, Schwanke K, Reppel M, Neef S, Katsirntaki K, Maier LS, Nguemo F, Menke S, Haustein M, Hescheler J, Hasenfuss G and Martin U (2008) Generation of functional murine cardiac myocytes from induced pluripotent stem cells. Circulation 118: 507–17.
McBeath R, Pirone DM, Nelson CM, Bhadriraju K and Chen CS (2004) Cell shape, cytoskeletal tension, and RhoA regulate stem cell lineage commitment. Dev Cell 6: 483–95.
McDevitt TC, Angello JC, Whitney ML, Reinecke H, Hauschka SD, Murry CE and Stayton PS (2002) In vitro generation of differentiated cardiac myofibers on micropatterned laminin surfaces. J Biomed Mater Res 60: 472–9.
McKenna WJ (1993) Hypertrophic cardiomyopathy: an update. Cardiologia 38: 277–81.
Miller CE, Donlon KJ, Toia L, Wong CL and Chess PR (2000) Cyclic strain induces proliferation of cultured embryonic heart cells. In Vitro Cell Dev Biol Anim 36: 633–9.
Minamino T, Yujiri T, Terada N, Taffet GE, Michael LH, Johnson GL and Schneider MD (2002) MEKK1 is essential for cardiac hypertrophy and dysfunction induced by Gq. Proc Natl Acad Sci U S A 99: 3866–71.
Mogensen J, Murphy RT, Shaw T, Bahl A, Redwood C, Watkins H, Burke M, Elliott PM and McKenna WJ (2004) Severe disease expression of cardiac troponin C and T mutations in patients with idiopathic dilated cardiomyopathy. J Am Coll Cardiol 44: 2033–40.
Narazaki G, Uosaki H, Teranishi M, Okita K, Kim B, Matsuoka S, Yamanaka S and Yamashita JK (2008) Directed and systematic differentiation of cardiovascular cells from mouse induced pluripotent stem cells. Circulation 118: 498–506.
Niederer SA, Hunter PJ and Smith NP (2006) A quantitative analysis of cardiac myocyte relaxation: a simulation study. Biophys J 90: 1697–722.
Niimura H, Bachinski LL, Sangwatanaroj S, Watkins H, Chudley AE, McKenna W, Kristinsson A, Roberts R, Sole M, Maron BJ, Seidman JG and Seidman CE (1998) Mutations in the gene for cardiac myosin-binding protein C and late-onset familial hypertrophic cardiomyopathy. N Engl J Med 338: 1248–57.
Nobes CD and Hall A (1995) Rho, rac, and cdc42 GTPases regulate the assembly of multimolecular focal complexes associated with actin stress fibers, lamellipodia, and filopodia. Cell 81: 53–62.
Noma K, Oyama N and Liao JK (2006) Physiological role of ROCKs in the cardiovascular system. Am J Physiol Cell Physiol 290: C661–8.

Nygren JM, Jovinge S, Breitbach M, Sawen P, Roll W, Hescheler J, Taneera J, Fleischmann BK and Jacobsen SE (2004) Bone marrow-derived hematopoietic cells generate cardiomyocytes at a low frequency through cell fusion, but not transdifferentiation. Nat Med 10: 494–501.

Olson TM, Michels VV, Thibodeau SN, Tai YS and Keating MT (1998) Actin mutations in dilated cardiomyopathy, a heritable form of heart failure. Science 280: 750–2.

Omens JH (1998) Stress and strain as regulators of myocardial growth. Prog Biophys Mol Biol 69: 559–72.

Orr W, Helmke B, Blackman B and Schwartz MA (2006) Mechanisms of mechanotransduction. Cell Develop 10: 11–20.

Paul K, Ball NA, Dorn GW IInd and Walsh RA (1997) Left ventricular stretch stimulates angiotensin II–mediated phosphatidylinositol hydrolysis and protein kinase C epsilon isoform translocation in adult guinea pig hearts. Circ Res 81: 643–50.

Paul S (2003) Ventricular remodeling. Crit Car Nurse Clin N Am 15: 407–11.

Pelham RJ Jr. and Wang YL (1998) Cell locomotion and focal adhesions are regulated by the mechanical properties of the substrate. Biol Bull 194: 348–9; discussion 349–50.

Pereira WC, Khushnooma I, Madkaikar M and Ghosh K (2008) Reproducible methodology for the isolation of mesenchymal stem cells from human umbilical cord and its potential for cardiomyocyte generation. J Tissue Eng Regen Med 2: 394–9.

Peters NS, Green CR, Poole-Wilson PA and Severs NJ (1993) Reduced content of connexin43 gap junctions in ventricular myocardium from hypertrophied and ischemic human hearts. Circulation 88: 864–75.

Peters SLM and Michel MC (2007) The RhoA/Rho kinase pathway in the myocardium. Cardiovasc Res 75: 3–4.

Peterson MB and Lesch M (1972) Protein synthesis and amino acid transport in the isolated rabbit right ventricular papillary muscle. effect of isometric tension development. Circ Res 31: 317–27.

Peyton SR, Ghajar CM, Khatiwala CB and Putnam AJ (2007) The emergence of ECM mechanics and cytoskeletal tension as important regulators of cell function. Cell Biochem Biophys 47: 300–20.

Peyton SR, Raub CB, Keschrumrus VP and Putnam AJ (2006) The use of poly(ethylene glycol) hydrogels to investigate the impact of ECM chemistry and mechanics on smooth muscle cells. Biomaterials 27: 4881–93.

Pimentel RC, Yamada KA, Kleber AG and Saffitz JE (2002) Autocrine regulation of myocyte Cx43 expression by VEGF. Circ Res 90: 671–7.

Pleumsamran A and Kim D (1995) Membrane stretch augments the cardiac muscarinic K^+ channel activity. J Membr Biol 148: 287–97.

Purcell NH, Darwis D, Bueno OF, Müller JM, Schüle R and Molkentin JD (2004) Extracellular signal-regulated kinase 2 interacts with and is negatively regulated by the LIM-only protein FHL2 in cardiomyocytes. Mol Cell Biol 24: 1081–95.

Pyle WG, Hart MC, Cooper JA, Sumandea MP, de Tombe PP and Solaro RJ (2002) Actin capping protein: an essential element in protein kinase signaling to the myofilaments. Circ Res 90: 1299–306.

Rakusan K (1984). Cardiac growth, maturation, and aging. In: Zak R (eds.). *Growth of the Heart and Disease*. Raven Press, New York, pp. 131–64.

Rana OR, Zobel C, Saygili E, Brixius K, Gramley F, Schimpf T, Mischke K, Frechen D, Knackstedt C, Schwinger RH and Schauerte P (2008) A simple device to apply equibiaxial strain to cells cultured on flexible membranes. Am J Physiol Heart Circ Physiol 294: H532–40.

Raskin AM, Hoshijima M, Swanson E, McCulloch AD and Omens JH (2009) Hypertrophic expression induced by chronic stretch of excised mouse heart muscle. Mol Cell Biomech: 6: 145–160.

Rhee SG and Choi KD (1992) Regulation of inositol phospholipid-specific phospholipase C isozymes. J Biol Chem 267: 12393–6.

Riemer TL and Tung L (2003) Stretch-induced excitation and action potential changes of single cardiac cells. Prog Biophys Mol Biol 82: 97–110.

Riento K and Ridley AJ (2003) Rocks: multifunctional kinases in cell behaviour. Nat Rev Mol Cell Biol 4: 446–56.
Riha GM, Wang X, Wang H, Chai H, Mu H, Lin PH, Lumsden AB, Yao Q and Chen C (2007) Cyclic strain induces vascular smooth muscle cell differentiation from murine embryonic mesenchymal progenitor cells. Surgery 141: 394–402.
Robinson JM, Dong WJ, Xing J and Cheung HC (2004) Switching of troponin I: Ca(2+) and myosin-induced activation of heart muscle. J Mol Biol 340: 295–305.
Rockman HA, Ross RS, Harris AN, Knowlton KU, Steinhelper ME, Field LJ, Ross J Jr. and Chien KR (1991) Segregation of atrial-specific and inducible expression of an atrial natriuretic factor in an in vivo murine model of cardiac hypertrophy. Proc Natl Acad Sci 88: 8277–81.
Rodic N, Rutenberg MS and Terada N (2004) Cell fusion and reprogramming: resolving our transdifferences. Trends Mol Med 10: 93–6.
Roher A, Lieska N and Spitz W (1986) The amino acid sequence of human cardiac troponin-C. Muscle Nerve 9: 73–7.
Rohr S, Fluckiger-Labrada R and Kucera JP (2003) Photolithographically defined deposition of attachment factors as a versatile method for patterning the growth of different cell types in culture. Pflugers Arch 446: 125–32.
Rohr S, Kleber AG and Kucera JP (1999) Optical recording of impulse propagation in designer cultures. Cardiac tissue architectures inducing ultra-slow conduction. Trends Cardiovasc Med 9: 173–9.
Ross RS (2002) The extracellular connections: the role of integrins in myocardial remodeling. J Card Fail 8: S326–31.
Ross RS, Pham C, Shai SY, Goldhaber JI, Fenczik C, Glembotski CC, Ginsberg MH and Loftus JC (1998) Beta1 integrins participate in the hypertrophic response of rat ventricular myocytes. Circ Res 82: 1160–72.
Ruoslahti E (1991) Integrins. J Clin Invest 87: 1–5.
Ruwhof C and van der Laarse A (2000) Mechanical stress-induced cardiac hypertrophy: mechanisms and signal transduction pathways. Cardiovasc Res 47: 23–37.
Sadoshima J, Jahn L, Takahashi T, Kulik TJ and Izumo S (1992) Molecular characterization of the stretch-induced adaptation of cultured cardiac cells. An in vitro model of load-induced cardiac hypertrophy. J Biol Chem 267: 10551–60.
Sadoshima J, Xu Y, Slayter HS and Izumo S (1993) Autocrine release of angiotensin II mediates stretch-induced hypertrophy of cardiac myocytes in vitro. Cell 75: 977–84.
Saha S, Ji L, de Pablo JJ and Palecek SP (2006) Inhibition of human embryonic stem cell differentiation by mechanical strain. J Cell Physiol 206: 126–37.
Salazar NC, Chen J and Rockman HA (2007) Cardiac GPCRs: GPCR signaling in healthy and failing hearts. Biochim Biophys Acta 1768: 1006–18.
Sasaki N, Mitsuiye T and Noma A (1992) Effects of mechanical stretch on membrane currents of single ventricular myocytes of guinea-pig heart. Jpn J Physiol 42: 957–70.
Satoh M, Takahashi M, Sakamoto T, Hiroe M, Marumo F and Kimura A (1999) Structural analysis of the titin gene in hypertrophic cardiomyopathy: identification of a novel disease gene. Biochem Biophys Res Commun 262: 411–7.
Saygili E, Rana OR, Reuter H, Frank K, Schwinger RH, Muller-Ehmsen J and Zobel C (2007) Losartan prevents stretch-induced electrical remodeling in cultured atrial neonatal myocytes. Am J Physiol Heart Circ Physiol 292: H2898–905.
Schmelter M, Ateghang B, Helmig S, Wartenberg M and Sauer H (2006) Embryonic stem cells utilize reactive oxygen species as transducers of mechanical strain-induced cardiovascular differentiation. FASEB J 20: 1182–4.
Schussheim AE and Radda GK (1995) Altered Na(+)- H(+)- exchange activity in the spontaneously hypertensive perfused rat heart. J Mol Cell Cardiol 27: 1475–81.
Shafrir Y and Forgacs G (2002) Mechanotransduction through the cytoskeleton. Am J Physiol Cell Physiol 282: C479–86.

Shanker AJ, Yamada K, Green KG, Yamada KA and Saffitz JE (2005) Matrix-protein-specific regulation of Cx43 expression in cardiac myocytes subjected to mechanical load. Circ Res 96: 558–66.

Shapira-Schweitzer K and Seliktar D (2007) Matrix stiffness affects spontaneous contraction of cardiomyocytes cultured within a PEGylated fibrinogen biomaterial. Acta Biomater 3: 33–41.

Sheikh F, Raskin A, Chu PH, Lange S, Domenighetti AA, Zheng M, Liang X, Zhang T, YajimaT, Gu Y, Dalton ND, Mahata SK, Dorn GW II, Heller-Brown J, Peterson KL, Omens JH, McCulloch DA and Chen J (2008) An FHL1-containing complex within the cardiomyocyte sarcomere mediates hypertrophic biomechanical stress responses in mice. J Clin Invest. 118: 3870–3880.

Shimko VF and Claycomb WC (2008) Effect of mechanical loading on three-dimensional cultures of embryonic stem cell-derived cardiomyocytes. Tissue Eng Part A 14: 49–58.

Shyu KG, Chen CC, Wang BW and Kuan P (2001) Angiotensin II receptor antagonist blocks the expression of connexin43 induced by cyclical mechanical stretch in cultured neonatal rat cardiac myocytes. J Mol Cell Cardiol 33: 691–8.

Sigurdson W, Ruknudin A and Sachs F (1992) Calcium imaging of mechanically induced fluxes in tissue-cultured chick heart: role of stretch-activated ion channels. Am J Physiol 262: H1110–5.

Simpson DG, Majeski M, Borg TK and Terracio L (1999) Regulation of cardiac myocyte protein turnover and myofibrillar structure in vitro by specific directions of stretch. Circ Res 85: e59–69.

Simpson DG, Terracio L, Terracio M, Price RL, Turner DC and Borg TK (1994) Modulation of cardiac myocyte phenotype in vitro by the composition and orientation of the extracellular matrix. J Cell Physiol 161: 89–105.

Smith RR, Barile L, Cho HC, Leppo MK, Hare JM, Messina E, Giacomello A, Abraham MR and Marban E (2007) Regenerative potential of cardiosphere-derived cells expanded from percutaneous endomyocardial biopsy specimens. Circulation 115: 896–908.

Sung D, Mills RW, Schettler J, Narayan SM, Omens JH and McCulloch AD (2003) Ventricular filling slows epicardial conduction and increases action potential duration in an optical mapping study of the isolated rabbit heart. J Cardiovasc Electrophysiol 14: 739–49.

Suzuki M, Carlson KM, Marchuk DA and Rockman HA (2002) Genetic modifier loci affecting survival and cardiac function in murine dilated cardiomyopathy. Circulation 105: 1824–9.

Takahashi K, Tanabe K, Ohnuki M, Narita M, Ichisaka T, Tomoda K and Yamanaka S (2007) Induction of pluripotent stem cells from adult human fibroblasts by defined factors. Cell 131: 861–72.

Takahashi K and Yamanaka S (2006) Induction of pluripotent stem cells from mouse embryonic and adult fibroblast cultures by defined factors. Cell 126: 663–76.

Tan JL, Liu W, Nelson CM, Raghavan S and Chen CS (2004) Simple approach to micropattern cells on common culture substrates by tuning substrate wettability. Tissue Eng 10: 865–72.

Tanaka N, Mao L, DeLano FA, Sentianin EM, Chien KR, Schmid-Schönbein GW and Ross J Jr (1997) Left ventricular volumes and function in the embryonic mouse heart. Am J Physiol 273: H1368–H1376.

Tatsukawa Y, Kiyosue T and Arita M (1997) Mechanical stretch increases intracellular calcium concentration in cultured ventricular cells from neonatal rats. Heart Vessels 12: 128–135.

Terracio L, Miller B and Borg TK (1988) Effects of cyclic mechanical stimulation of the cellular components of the heart: in vitro. In Vitro Cell Dev Biol 24: 53–8.

Terracio L, Tingstrom A, Peters WH IIIrd and Borg TK (1990) A potential role for mechanical stimulation in cardiac development. Ann N Y Acad Sci 588: 48–60.

Thierfelder L, Watkins H, MacRae C, Lamas R, McKenna W, Vosberg HP, Seidman JG and Seidman CE (1994) Alpha-tropomyosin and cardiac troponin T mutations cause familial hypertrophic cardiomyopathy: a disease of the sarcomere. Cell 77: 701–12.

Tobita K, Liu LJ, Janczewski AM, Tinney JP, Nonemaker JM, Augustine S, Stolz DB, Shroff SG and Keller BB (2006) Engineered early embryonic cardiac tissue retains proliferative and contractile properties of developing embryonic myocardium. Am J Physiol Heart Circ Physiol 291: H1829–37.

Tomanek R and Cooper Gt (1981) Morphological changes in the mechanically unloaded myocardial cell. Anat Rec 200: 271–80.

Vakili BA, Okin PM and Devereux RB (2001) Prognostic implications of left ventricular hypertrophy. Am Heart J 141: 334–41.

Van Wagoner DR (1993) Mechanosensitive gating of atrial ATP-sensitive potassium channels. Circ Res 72: 973–83.

van Wamel AJ, Ruwhof C, van der Valk-Kokshoorn LJ, Schrier PI and van der Laarse A (2000) Rapid effects of stretched myocardial and vascular cells on gene expression of neonatal rat cardiomyocytes with emphasis on autocrine and paracrine mechanisms. Arch Biochem Biophys 381: 67–73.

Vandenburgh HH, Solerssi R, Shansky J, Adams JW, Henderson SA and Lemaire J (1995) Response of neonatal rat cardiomyocytes to repetitive mechanical stimulation in vitro. Ann N Y Acad Sci 752: 19–29.

Vogel V (2006) Mechanotransduction involving multimodular proteins: converting force into biochemical signals. Annu Rev Biophys Biomol Struct 35: 459–88.

Wang N, Bulter JP and Ingber DE (1993) Mechanotransduction across the cell surface and through the cytoskeleton. Science 260: 1124–7.

Wang TL, Tseng YZ and Chang H (2000) Regulation of connexin 43 gene expression by cyclical mechanical stretch in neonatal rat cardiomyocytes. Biochem Biophys Res Commun 267: 551–7.

Ward DF Jr., Salasznyk RM, Klees RF, Backiel J, Agius P, Bennett K, Boskey A and Plopper GE (2007) Mechanical strain enhances extracellular matrix-induced gene focusing and promotes osteogenic differentiation of human mesenchymal stem cells through an extracellular-related kinase-dependent pathway. Stem Cells Dev 16: 467–80.

White E, Le Guennec JY, Nigretto JM, Gannier F, Argibay JA and Garnier D (1993) The effects of increasing cell length on auxotonic contractions; membrane potential and intracellular calcium transients in single guinea-pig ventricular myocytes. Exp Physiol 78: 65–78.

Wong JY, Leach JB and Brown XQ (2004) Balance of chemistry, topography and mechanics at the cell-biomaterial interface: issues and challenges for assessing the role of substrate mechanics on cell response. Surf Sci 570: 119–133.

Wong JY, Velasco A, Rajagopalan P and Pham Q (2003) Directed movement of vascular smooth muscle cells on gradient-compliant hydrogels. Langmuir 19: 1908–13.

Woo SH, Risius T and Morad M (2007) Modulation of local Ca^{2+} release sites by rapid fluid puffing in rat atrial myocytes. Cell Calcium 41: 397–403.

Xu W, Zhang X, Qian H, Zhu W, Sun X, Hu J, Zhou H and Chen Y (2004) Mesenchymal stem cells from adult human bone marrow differentiate into a cardiomyocyte phenotype in vitro. Exp Biol Med (Maywood) 229: 623–31.

Yamada K, Green K and Saffitz J (2005) Distinct pathways regulate expression of cardiac electrical and mechanical junctional proteins in response to stretch. Circ Res 97: 346–53.

Yamada Y, Sakurada K, Takeda Y, Gojo S and Umezawa A (2007) Single-cell-derived mesenchymal stem cells overexpressing Csx/Nkx2.5 and GATA4 undergo the stochastic cardiomyogenic fate and behave like transient amplifying cells. Exp Cell Res 313: 698–706.

Yamazaki T, Komuro I, Kudoh S, Zou Y, Nagai R, Aikawa R, Uozumi H and Yazaki Y (1998a) Role of ion channels and exchangers in mechanical stretch-induced cardiomyocyte hypertrophy. Circ Res 82: 430–7.

Yamazaki T, Komuro I, Kudoh S, Zou Y, Shiojima I, Hiroi Y, Mizuno T, Maemura K, Kurihara H, Aikawa R, Takano H and Yazaki Y (1996) Endothelin-1 is involved in mechanical stress-induced cardiomyocyte hypertrophy. J Biol Chem 271: 3221–8.

Yamazaki T, Komuro I, Kudoh S, Zou Y, Shiojima I, Mizuno T, Takano H, Hiroi Y, Ueki K, Tobe K et al. (1995) Mechanical stress activates protein kinase cascade of phosphorylation in neonatal rat cardiac myocytes. J Clin Invest 96: 438–46.

Yamazaki T, Komuro I and Yazaki Y (1998b) Signalling pathways for cardiac hypertrophy. Cell Signal 10: 693–8.

Yang M, Lim CC, Liao R and Zhang X (2007) A novel microfluidic impedance assay for monitoring endothelin-induced cardiomyocyte hypertrophy. Biosens Bioelectron 22: 1688–93.

Zaari N, Rajagopalan P, Kim SK, Engler AJ and Wong JY (2004) Photopolymerization in microfluidic gradient generators: microscale control of substrate compliance to manipulate cell response. Adv Mater 16: 2133–7.

Zeng T, Bett G and Sachs F (2000) Stretch-activated whole cell currents in adult rat cardiac myocytes. Am J Physiol Heart Circ Physiol 278: H548–57.

Zhang SJ, Truskey GA and Kraus WE (2007) Effect of cyclic stretch on beta1D-integrin expression and activation of FAK and RhoA. Am J Physiol Cell Physiol 292: C2057–69.

Zhang Y, Sekar RB, McCulloch AD and Tung L (2008) Cell cultures as models of cardiac mechanoelectric feedback. Prog Biophys Mol Biol 97: 367–82.

Zhang YH, Youm JB, Sung HK, Lee SH, Ryu SY, Ho WK and Earm YE (2000) Stretch-activated and background non-selective cation channels in rat atrial myocytes. J Physiol 523 Pt 3: 607–19.

Zhao XH, Laschinger C, Arora P, Szaszi K, Kapus A and McCulloch CA (2007) Force activates smooth muscle alpha-actin promoter activity through the Rho signaling pathway. J Cell Sci 120: 1801–9.

Zhao YS, Wang CY, Li DX, Zhang XZ, Qiao Y, Guo XM, Wang XL, Dun CM, Dong LZ and Song Y (2005) Construction of a unidirectionally beating 3-dimensional cardiac muscle construct. J Heart Lung Transplant 24: 1091–7.

Zhou Q, Chu PH, Huang C, Cheng CF, Martone ME, Knoll G, Shelton GD, Evans S and Chen J (2001) Ablation of Cypher, a PDZ-LIM domain Z-line protein, causes a severe form of congenital myopathy. J Cell Biol 155: 605–12.

Zhu W, Zou Y, Shiojima I, Kudoh S, Aikawa R, Hayashi D, Mizukami M, Toko H, Shibasaki F, Yazaki Y, Nagai R and Komuro I (2000) Ca^{2+}/calmodulin-dependent kinase II and calcineurin play critical roles in endothelin-1-induced cardiomyocyte hypertrophy. J Biol Chem 275: 15239–45.

Zhuang J, Yamada KA, Saffitz JE and Kléber AG (2000) Pulsatile stretch remodels cell-to-cell communication in cultured myocytes. Circ Res 87: 316–22.

Zimmermann WH, Schneiderbanger K, Schubert P, Didie M, Munzel F, Heubach JF, Kostin S, Neuhuber WL and Eschenhagen T (2002) Tissue engineering of a differentiated cardiac muscle construct. Circ Res 90: 223–30.

Zou Y, Akazawa H, Qin Y, Sano M, Takano H, Minamino T, Makita N, Iwanaga K, Zhu W, Kudoh S, Toko H, Tamura K, Kihara M, Nagai T, Fukamizu A, Umemura S, Iiri T, Fujita T and Komuro I (2004) Mechanical stress activates angiotensin II type 1 receptor without the involvement of angiotensin II. Nat Cell Biol 6: 499–506.

Chapter 6
Stretch-Activated Channels in the Heart: Contribution to Cardiac Performance

Marie-Louise Ward and David G. Allen

Abstract Stretch-activated ion channels are widely expressed in most cell types and play an important role in a variety of normal cell functions, including volume regulation and length detection. In the heart, transduction of mechanical energy into cellular responses is an essential component of cardiac function. The heart is passively stretched, and actively shortens in every cardiac cycle; in addition, longer-term changes in volume occur during exercise, and in diseases such as heart failure. In this article, we discuss the importance of stretch-activated ion channels as mechano-transducers in the heart, with emphasis on their contribution to the regulation of contractile performance. As well, the role of stretch-activated channels in modifying the electrical activity of the heart is also discussed.

Keywords Stretch-activated channels · Non-selective stretch-activated cation channels · Mechano-sensitive ion currents · Slow force response · TRP channels

6.1 Introduction

During each cardiac cycle individual myocytes are stretched passively during diastole and actively shorten during systole as part of the normal pump cycle. In addition to these normal beat-to-beat variations in cell length, longer-term events also expose the cardiac myocytes to mechanical deformation. For instance, at the start of exercise, skeletal muscle activity increases venous return, raises central venous pressure and stretches the heart. On a longer timescale, diseases such as heart failure are characterised by increased central venous pressure, which causes chronic stretch of the heart. As well as these volume changes, fibre orientation, extracellular matrix composition, and hypertrophy of the heart will all impact mechanically on

M-L. Ward (✉)
Department of Physiology, Faculty of Medical and Health Sciences,
University of Auckland, Auckland, New Zealand
e-mail: m.ward@auckland.ac.nz

A. Kamkin, I. Kiseleva (eds.), *Mechanosensitivity of the Heart*,
Mechanosensitivity in Cells and Tissues 3, DOI 10.1007/978-90-481-2850-1_6,

individual cardiac myocytes. Although the performance of the heart is continuously influenced by nervous and hormonal input, it has long been understood that regulatory mechanisms intrinsic to the myocytes themselves also exist. Frank, Starling and colleagues demonstrated that an increase in ventricular filling (i.e. "end-diastolic volume") brought about an immediate increase in the force of contraction. This property of cardiac muscle is better known as the Frank-Starling relation, or "Starling's Law of the Heart" (Frank, 1895; Patterson and Starling, 1914), and is important in ensuring the heart adjusts its output to match venous return. In 1912, Von Anrep reported on another stretch-dependent regulatory mechanism intrinsic to cardiac muscle. He observed that when the resistance to aortic outflow was increased, there was an abrupt increase in ventricular end-diastolic volume that then gradually returned back to normal over a period of 10–15 min. Although the observations of Frank, Starling, and Von Anrep were made almost 100 years ago, the mechanisms that give rise to their observations are still debated. We also know that mechanical deformation, or stretch, of cardiac muscle can increase heart rate (Bainbridge, 1915; Blinks, 1956; Cooper and Kohl, 2005), cause diastolic depolarisation (Penefsky and Hoffman, 1963), alter action potential duration and shape (Allen, 1977; Lab, 1978b; Franz et al., 1989; Calkins et al., 1991; Zabel et al., 1996) and induce arrhythmias (Franz et al., 1989; Hansen et al., 1990), but exactly how the cardiac myocyte acts as a transducer of mechanical signals is not fully understood.

Membrane channels sensitive to mechanical stimulation (mechano-sensitive channels, or MSCs) are present in most cell types, and play an essential role in how cells and higher organisms interact with their immediate environment (for a recent review see Blount et al., 2008). The MSC super family is made up of many different channel types, with a diverse range of primary structures, indicating that they have evolved independently on a number of occasions during evolutionary history (Martinac and Kloda, 2003). These channels show a wide variety of mechanical means of activation, or inactivation, particularly so for those from specialized cells serving as sensory transducers. The superfamily also includes channels whose main mechanism of regulation is not mechanical, yet are sensitive to mechanical input, such as the voltage-gated channels (for review see Morris and Laitko, 2005). Functionally, the MSCs appear to separate into those that are stimulated by stress in the cytoskeleton (such as the stereocilia of cells in the cochlea (Corey, 2003), for example) and those that are stimulated by stress in the lipid bilayer (Hamill and Martinac, 2001; Chiang et al., 2004; Suchyna et al., 2004; Maroto et al., 2005). More recently, a group of channels have been identified that are stimulated indirectly via stretch-dependent signalling cascades (Browe and Baumgarten, 2004; Caldiz et al., 2007; Dyachenko et al., 2008a, b), although there is some debate as to whether these should be labelled as mechano-sensitive. MSCs can be selective, or non-selective, for cations, or conduct anions, as well as being either activated or inactivated by mechanical stimulation.

A subgroup of MSC are the stretch-activated channels (SACs) first described by Guharay and Sachs in 1984 in patch-clamped embryonic skeletal muscle cells. As their name suggests, these channels open in response to mechanical stretch (as opposed to those MSCs that are *inactivated* by stretch (Morris and Sigurdson,

1988)). Whilst making measurements of nicotinic ion channels in excised patches from skeletal muscle, Guharay and Sachs noticed that when suction was applied to the patch, channel activity increased, even in the absence of agonists. The stretch-activated channels described were cation selective, but discriminated poorly between Na^+ and K^+ ions (Guharay and Sachs, 1984). Since that time, SAC have been found to be present in many different cell types, including cardiac myocytes (Craelius et al., 1988; Kim, 1992; Ruknudin et al., 1993; Hu and Sachs, 1996, 1997; Zeng et al., 2000). Here we review stretch-activated channels in the heart, and discuss their contribution to contractile performance. We also consider channels that are indirectly activated by stretch, for instance by signalling pathways, since it is becoming increasingly evident that they also contribute to the stretch-dependent increased cardiac performance.

6.2 The Cardiac Response to Stretch

It is thought that at least 3 different cellular mechanisms are involved in the heart's response to stretch (for review see Allen and Kentish, 1985). (i) Increased overlap between the thick and thin filaments (Gordon et al., 1966; Fabiato and Fabiato, 1975). (ii) Increased Ca^{2+} sensitivity of the contractile machinery (Hibberd and Jewell, 1982; Fukuda and Granzier, 2005). (iii) Increased Ca^{2+} transients (the systolic rise in Ca^{2+} which activates the contractile proteins) which gradually become larger over some minutes after a stretch (Allen and Kurihara, 1982; Kentish and Wrzosek, 1998). The first two of these mechanisms are rapid responses that occur immediately on stretching cardiac muscle. The third mechanism is commonly known as the slow force response, or SFR, and was first recognised by Parmley and Chuck (1973) who noted a slow increase in force following a stretch that took 5–10 min to reach completion. They argued that the increases in overlap of the thick and thin filaments should occur instantaneously with the stretch, and that the slow phenomenon was likely to be caused by changes in the degree of activation of the contractile proteins. These ideas were confirmed when it was shown that, following a stretch, there was a slow increase in the magnitude of the Ca^{2+} transients which caused the slow increase in force (Allen and Kurihara, 1982; Kentish and Wrzosek, 1998). Figure 6.1 shows an example of the biphasic response to stretch obtained when ventricular muscle is lengthened. Ventricular trabeculae isolated from mouse hearts were loaded with fura-2 as an indicator of intracellular [Ca^{2+}], and subjected to step increases in muscle length, as described in Ward et al., (2008). Figure 6.1a shows intracellular [Ca^{2+}] (340/380 fura-2 ratio, top trace), isometric force (middle trace), and muscle length (lower trace), before, during, and after a 2 min stretch in a representative trabecula. Note that an increase in twitch force is apparent immediately following the stretch (the *rapid* response), with no change in the intracellular Ca^{2+} transient amplitude. This is followed by a slower increase in the amplitude of both the Ca^{2+} transient, and the twitch, that develops over a period of a few minutes (the *slow force response,* or SFR). Figure 6.1b shows individual transients and twitches taken at the points labelled in 1a. It is well established that the

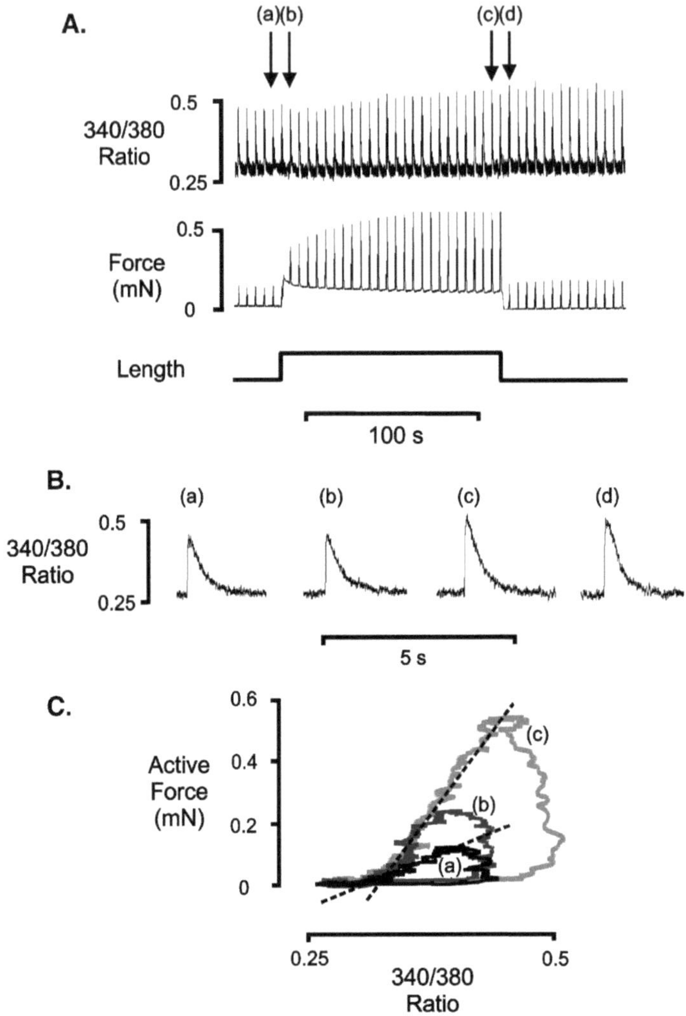

Fig. 6.1 The slow force response to longitudinal stretch. **Panel A** shows the fura-2 340/380 ratio (*top traces*), isometric force (*middle*), and muscle length (*bottom*) for a representative muscle subjected to a 2 min stretch. **Panel B** shows individual calcium transients with an expanded timescale taken from the time points labelled in Panel A. Phase plots of isometric force against 340/380 ratio are shown in **Panel C** for the same time points. *Dotted lines* fitted to the relaxation component of the phase plots indicate a change in the myofilament Ca^{2+} sensitivity immediately after the stretch (comparison between (a) and (b)), that does not change further during the SFR (comparison between (b) and (c)). (Panels A and B from Ward et al., 2008, with permission.)

rapid response is largely due to a change in the sensitivity of the contractile proteins to Ca^{2+}, whereas no further change in Ca^{2+} sensitivity occurs after the stretch. Figure 6.1c shows phase plots of force versus Ca^{2+} taken from the points labelled in 1a above. The dashed lines show the change in slope of the relaxation component of each example phase plot, and are representative of myofilament Ca^{2+} sensitivity. The leftward shift in the force-$[Ca^{2+}]_i$ relationship during the relaxation phase following stretch represents a stretch-dependent increase in myofilament Ca^{2+} sensitivity that then remains constant throughout the SFR.

6.3 Possible Mechanisms of the Slow Force Response

The SFR has been observed in vivo (Lew, 1993), in isolated perfused hearts (Tucci et al., 1984; Burkhoff et al., 1991), in isolated multicellular preparations (Parmley and Chuck, 1973; Kentish and Wrzosek, 1998; Alvarez et al., 1999; Pérez et al., 2001), and in isolated myocytes (White et al., 1995), suggesting that the underlying mechanisms are, at least in part, intrinsic to the cardiac myocytes themselves. Ca^{2+} transients could increase if there was additional Ca^{2+} entry into the muscle, either associated with excitation-contraction coupling (for example, via a stretch-dependent increase in L-type Ca^{2+} channel influx) or via some other pathway that is continuously active. Both mechanisms would lead to a greater Ca^{2+} load within the intracellular store, or sarcoplasmic reticulum (SR), with a correspondingly greater release. Alternatively, Ca^{2+} transients would increase if stretch induced greater SR Ca^{2+} release, although mechanisms of this sort are only short-lived because the increased Ca^{2+} transient causes greater extrusion of Ca^{2+} from the cell and leads to a reduction of the store Ca^{2+} (for review see Eisner et al., 2000).

6.3.1 Role of the Sarcoplasmic Reticulum in the SFR

A number of studies have shown that a functional SR is *not* a requirement for the development of the SFR since it is observed in the presence of specific SR inhibitors (Bluhm and Lew, 1995; Kentish and Wrzosek, 1998), although the magnitude of the SFR response is diminished in some (Chuck and Parmley, 1980; von Lewinski et al., 2004), but not all (Kentish and Wrzosek, 1998; Calaghan and White, 2004) species. This would suggest that the SFR is mediated by a mechanism that increases SR Ca^{2+} load, but is dependent on some additional mechanism(s). One suggestion is that stretch enhances nitric oxide (NO) release, which increases Ca^{2+} spark frequency and triggers more SR Ca^{2+} release in response to the action potential (Vila Petroff et al., 2001). However, a subsequent study in which NO production was inhibited with L-NAME failed to support this mechanism (Calaghan and White, 2004). Another difficulty with this idea is that agents which enhance Ca^{2+} release deplete the SR store and so have no permanent effect (Trafford et al., 1998); thus stretch-dependent NO release would need to be combined with some other mechanism to enhance Ca^{2+} release persistently and cause the SFR.

A steady increase in diastolic Ca^{2+} has been reported during the SFR (Steele and Smith, 1993; White et al., 1993; Alvarez et al., 1999; Ward et al., 2008), suggesting a process not directly connected with excitation-contraction coupling per se. Figure 6.2 shows the stretch-dependent increase in intracellular $[Ca^{2+}]$ recorded from quiescent RV trabeculae from mouse hearts subjected to axial stretch (modified

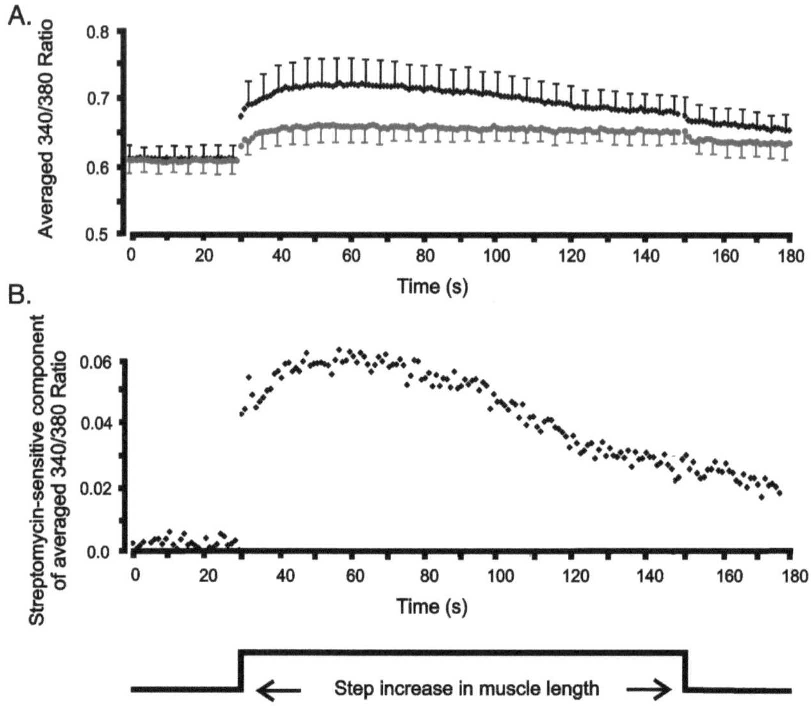

Fig. 6.2 Stretch-dependent Ca^{2+} influx in quiescent muscles. **Panel A** shows the averaged fura-2 340/380 ratio in response to a step increase in length (schematically shown below) for 5 muscles in the absence of stimulation (quiescent), before (*black*), and during (*grey*), exposure to 400 μM streptomycin. **Panel B** shows the streptomycin-sensitive component of the fluorescence ratio during stretch obtained by subtracting the averaged fluorescence of five muscles during exposure to streptomycin from the same muscles stretched in the absence of streptomycin. For each muscle, the filtered fluorescence was averaged over 1 s time intervals, for 30 s before, 120 s immediately after, and 30 s following, the length change. Note the change in the ordinate scale between Panel A and Panel B. (Panel B from (Ward et al., 2008), with permission.)

from Ward et al., 2008) where stimulated L-type Ca^{2+} channel openings and subsequent SR Ca^{2+} release are absent. Here, the averaged fura-2 340/380 ratio (Fig. 6.2a) is shown for 5 quiescent trabeculae from mouse hearts subjected to a 2 min stretch before (black symbols), and during (grey symbols), exposure to 400 μM streptomycin. Data were obtained by averaging the filtered fluorescence from each trabecula over 1 s intervals for 30 s before, 120 s immediately following, and 30 s after the step increase in muscle length. Representative error bars only are shown, for clarity. Stretching in the absence of streptomycin (black symbols) produced an apparent rapid rise in intracellular Ca^{2+} that was followed by a slow increase in Ca^{2+} that reached a plateau ~ 30 s after the stretch. Returning to short length initially produced only a small decrease in fluorescence, with a slow return to the baseline, pre-stretch, level over the next few minutes. A similar time course is seen for the SFR shown in a representative muscle in Fig. 6.1b during stimulation at 0.1 Hz. Exposure to streptomycin substantially reduced, but did not eliminate, the increase in the fura-2 ratio with stretch. Rapid changes in length can produce artefactual changes in fluorescence signals, and these can be minimised by subtracting the streptomycin data from the control. Thus Fig. 6.2b shows the mean, streptomycin-sensitive, component of the stretch-induced increase in fluorescence for five quiescent muscles. These observations suggest that a Ca^{2+} entry pathway is invoked by stretch that is active during diastole, and with a streptomycin-sensitive component. Further support for this mechanism is that stretch timed to occur only during diastole causes both the SFR and the increase in Ca^{2+} transients (Allen et al., 1988).

6.3.2 *Stretch-Activation of Na^+ /H^+ Exchanger*

One class of mechanism that could explain the slow force increase is based on the release of angiotensin II and endothelin by stretched cardiac muscle (Sadoshima and Izumo, 1993). These autocrine agents bind to their receptors and increase the activity of the cardiac Na^+/H^+ exchanger (NHE1, Cingolani et al., 1998). Cingolani and colleagues have demonstrated that stretched rabbit muscle shows an intracellular alkalosis, consistent with activation of NHE1, which was blocked by the NHE1 inhibitor EIPA, as well as by angiotensin and endothelin receptor blockers (Cingolani et al., 1998; Alvarez et al., 1999). Increased NHE1 activity also increased $[Na^+]_i$ and this would increase $[Ca^{2+}]_i$ through activity of the cardiac Na^+/Ca^{2+} exchanger. Other studies have confirmed some, but not all, aspects of this pathway. For instance, a study of failing human myocardium exhibited a slow force response (SFR) to stretch, but no pH_i change (von Lewinski et al., 2004). In a later study on rabbit myocardium, the same group observed a pH_i change, but occurring after the force had stabilized (Luers et al., 2005). Nevertheless, in both these studies, inhibitors of NHE1 reduced the magnitude of the SFR and the rise in $[Na^+]_i$, though blockers of angiotensin and endothelin receptors were ineffective.

In contrast, we (Ward et al., 2008), and others (Kondratev et al., 2005), have shown that the slow force response is *independent* of NHE1 activation. Although an increase in $[Na^+]_i$ was observed during the SFR, this was not eliminated by applying HOE 642, an inhibitor of NHE1. To further explore our finding that NHE1 activation was not essential to the SFR, we utilized a cardiac model of the SFR (Niederer and Smith, 2007). Modelling confirmed our experimental finding that the SFR can occur in the absence of NHE1 activation, provided sarcolemmal Ca^{2+} permeability was increased during stretch. Such increased Ca^{2+} permeability would occur if non-specific stretch-activated cation channels (SAC_{NSC}) were opened during stretch, allowing Ca^{2+}, and/or Na^+ (with subsequent Ca^{2+} increase through the Na^+/Ca^{2+} exchanger) influx.

6.3.3 *Stretch-Activated Channels and the SFR*

A plausible explanation of the SFR is that stretch activates channels that allow either Na^+ entry, which would then permit increased $[Ca^{2+}]_i$ via Na^+/Ca^{2+}-exchanger, or

Ca^{2+} entry, which would directly enhance SR Ca^{2+} load and Ca^{2+} release. Stretch-activated channels have been described in both atria and ventricles of different species (Craelius et al., 1988; Hu and Sachs, 1996; Zhang et al., 2000; Belus and White, 2003; Isenberg et al., 2003). Channels are permeable to monovalent cations and to Ca^{2+}, and can therefore act as a source of intracellular Ca^{2+} either directly, or indirectly, via the Na^+/Ca^{2+}-exchanger in response to an increase in intracellular $[Na^+]$. The involvement of such stretch-activated channels that are non-selective for cations (SAC_{NSC}) in the SFR will be discussed in more detail in the sections that follow.

6.4 Stretch-Sensitive Channels in the Heart

Stretch-sensitive channels were first discovered by applying negative pressure to the membrane within a patch pipette (Guharay and Sachs, 1984), but it has always been appreciated that the relation between this stimulus and the stretch applied to the heart by raised intra-chamber pressure is tenuous. Multicellular cardiac preparations, such as papillary muscle, or trabeculae, in which the myocytes are aligned along the length of the preparation have been successfully used to investigate the cardiac response to axial stretch. However, in order to assess cellular responses during stretch, such as ion channel activity, it is necessary to work with isolated cells. White et al. (1993) successfully attached flexible carbon fibres to isolated ventricular myocytes and subjected the myocytes to stretch (White et al., 1993). They were thus able to combine stretch with the simultaneous measurement of force development, intracellular Ca^{2+}, and membrane potential.

During normal (linear, or axial) stretch the surface area of a ventricular myocyte must increase and it is thought that, at least in skeletal muscle, the additional membrane is provided, in part, by the opening of folds in the membrane (Dulhunty and Franzini-Armstrong, 1975). Thus the linear strain on the membrane may be much less than suggested by the fractional length change. Given that the volume of cells is constant, when ventricular muscle is stretched, the cross-sectional area must fall and this suggests that the length of the T-tubules will become smaller. However, the T-tubular system is geometrically complex, and consists not only of tubules at right angles (i.e. "transverse") to the external sarcolemma, but also of a sub-population of interconnecting longitudinal tubules (Soeller and Cannell, 1999). Thus it is quite difficult to envisage the way in which these dimensional changes will affect channels situated in the T-tubular membrane. Equally, the actin cytoskeleton is thought to make numerous connections to lipid rafts and binding proteins within the lipid rafts, and the caveolae, but exactly how linear stretch affects the cytoskeleton or channels embedded in the membrane is unclear.

In contrast to linear (also referred to as "longitudinal" or "axial") stretch, three types of stimuli have been regularly used to activate SACs in ventricular myocytes. (i) Increase in cell volume by hypotonic solutions (Tseng, 1992; Matsuda et al., 1996; Vandenberg et al., 1996; Baumgarten and Clemo, 2003), or by cell inflation using positive pressure applied to the patch pipette (Hagiwara et al., 1992; Matsuda et al., 1996), (ii) radial stretch as applied to magnetic beads fixed to integrins by antibodies (Browe and Baumgarten, 2003, 2004), and (iii) the shear stress applied to an isolated cell by means of two tools attached to the upper surface while the lower surface is attached to the glass (Kamkin et al., 2000; Dyachenko et al., 2008a). While each of these may provide some linear stretch of the membrane, it is also clear that they will lead to stretching of the T-tubules so that, if SACs are located in the T-tubules, they will see a very different stimulus under these circumstances compared to linear stretch.

6.4.1 Stretch Sensitivity of Voltage-Gated Ion Channels

There is some evidence that voltage-gated ion channels are responsive to mechanical deformation of the lipid bilayer in isolated atrial myocytes (Matsuda et al., 1996), and when expressed in oocytes (Taberean et al., 1999; Gu et al., 2001). Typically, these studies use either inflation of the cell by applying positive pressure via the pipette, or osmotic cell swelling, as the mechanical stimulus in the whole-cell configuration. Pressure changes can also be applied directly to the cell membrane via the recording electrode in patch clamp mode. Morris and Juranka (2007) expressed the α-subunit of human heart voltage gated Na^+ channels in oocytes and found that increasing pipette pressure in cell-attached patches reversibly accelerated the time course of the recorded Na^+ current. Since voltage-dependent Na^+ channels are responsible for initiation and conduction of action potentials in nerve and muscle, a mechano-sensitive change in gating kinetics might be a contributing factor to stretch-induced cardiac arrhythmias (see below). Alternatively, in some circumstances, stretch might indirectly influence voltage-gated channel gating subsequent to the activation of other mechano-sensitive channels that modify membrane potential (Zabel et al., 1996).

L-type Ca^{2+} channels have been shown to be mechano-sensitive in a variety of tissues, including vascular smooth muscle (Langton, 1993), gastrointestinal smooth muscle (Farrugia et al., 1999), and cardiac atrial myocytes (Matsuda et al., 1996). Matsuda et al. (1996) found a stretch-dependent increase in the L-type Ca^{2+} channel open probability using whole-cell voltage clamp of isolated sino-atrial and atrial myocytes. At each membrane potential tested, the L-type Ca^{2+} current was reversibly increased by osmotic cell swelling and by cell inflation when a positive pressure of 10–15 cm H_2O was applied via the patch pipette. Lyford et al. (2002) investigated the mechano-sensitive regulation of smooth muscle Ca^{2+} entry using the pore-forming protein of the L-type Ca^{2+} channel, the α_{1C} subunit, cloned from human intestinal muscle and expressed in a heterologous system. They found that the cloned channel exhibited the same mechano-sensitive behaviour as the native channel at both the single-channel level and at the whole-cell current level. These authors also heterologously expressed human cardiac α_{1C}-subunit with a jejunal β2 Ca^{2+} channel subunit and found that the same mechano-sensitivity exhibited. They concluded that the components necessary for L-type Ca^{2+} channel sensitivity were contained within the α_{1C}-subunit (Lyford et al., 2002), and were not peculiar to the intestinal smooth muscle splice variant. It would seem reasonable to assume, therefore, that mechano-sensitivity of the L-type Ca^{2+} channel might contribute to the stretch-dependent increase in contractility of ventricular tissue by enhancing the Ca^{2+} current. Moreover, Calaghan et al. (1999) reported that axial stretch increased cellular levels of cAMP in isolated ferrit papillary muscles, which should also increase L-type Ca^{2+} current. In light of these findings, it is perhaps surprising that no evidence of L-type Ca^{2+} mechano-sensitivity has been found in isolated ventricular myocytes (Sasaki et al., 1992; Hongo et al., 1996; Kamkin et al., 2000; Belus and White, 2003) or human atrial myocytes (Kamkin et al., 2003b). It is interesting to note, however, that these studies used axial stretch rather than osmotically induced changes in cell volume (Matsuda et al., 1996), whole cell inflation (Matsuda et al., 1996), or increased patch pipette pressure (Morris and Juranka, 2007). It would therefore seem reasonable to believe that the nature of the mechanical stimulus applied determines the observed responses, to some degree at least.

6.4.2 Stretch-Activated Channels Non-selective for Cations (SAC_{NSC})

Evidence of the presence of stretch-activated channels non-selective for cations (SAC_{NSC}) in cardiac myocytes comes from a number of studies that have recorded whole cell currents during stretch in isolated cells. These include cultured (Hu and Sachs, 1996), neo-natal (Craelius et al., 1988), and adult (Sasaki et al., 1992; Bett and Sachs, 2000; Kamkin et al., 2000; Zeng et al., 2000; Belus and White, 2003; Isenberg et al., 2003; Kamkin et al., 2003a; Dyachenko et al., 2008a, b) ventricular myocytes, as well as atrial (Zhang et al., 2000; Kamkin et al., 2003b) and sino-atrial node (Cooper et al., 2000) myocytes. The resulting stretch-activated currents reverse at around –10 to 0 mV, and are not selective between cations (SAC_{NSC}). Thus they are permeable to Na^+, K^+ and possibly to Ca^{2+}, although the permeability to Ca^{2+} is low, and Ca^{2+} can also act as a partial blocker of the channel when monovalent cations are the main charge carrier (Kamkin et al., 2003a). Stretch-activated channels were first discovered by patch clamping in embryonic skeletal muscle (Guharay and Sachs, 1984), but, surprisingly, these channels have never been patch clamped in adult ventricular myocytes, despite repeated attempts (Zeng et al., 2000). Craelius et al. (1988) is the only study to identify SACs by patch clamping in ventricular cells, and they used neonatal cells kept in storage for up to 4 days. Either aspect might mean the absence of T-tubules, with channels normally found in T-tubules expressed in the surface membrane. Consistent with this idea, Zeng et al. (2000) speculated that SACs were located in the T-tubules of adult ventricular myocytes, and therefore unavailable to the patch electrode. However, SAC_{NSC} have been patch clamped in atrial cells (Kim, 1993; Zhang et al., 2000) where they were found to have Ca^{2+} permeability only slightly smaller than Na^+ (Kim, 1993). Interestingly, atrial myocytes show greater inhomogeneity in T-tubule distribution than found in ventricular myocardium (Avettev and Navaratnam, 1978; Tidball et al., 1991), with some atrial cells being devoid of T-tubules altogether (Leeson, 1980).

Many different mechanical stimulation techniques have been used in studies of isolated myocytes, mostly quantified by measured changes in sarcomere length. Longitudinal stretch has been successfully applied by attaching carbon fibres to the myocytes (Le Guennec et al., 1991; White et al., 1993; Belus and White, 2003), or by attaching patch electrodes to each end of the cell (Zhang et al., 2000). Another popular technique is to apply suction using the micro-pipette in cell-attached patch-clamp experiments, as used on neonatal ventricular myocytes by Craelius et al. (1988). Hu and Sachs (1996) also mechanically evoked currents in isolated ventricular myocytes from neonatal chick hearts by pressing on the cells with a micropipette. A second pipette was used in their experiments for measurement of currents through a perforated patch (Hu and Sachs, 1996). They identified currents carried by Na^+ and K^+, but not Cl^-, that were independent of extracellular $[Ca^{2+}]$. Hu and Sachs (1996) also carried out single channel studies using standard techniques (Hamill et al., 1981) where they identified two types of stretch-activated ion channels: a 21 pS nonspecific cation-selective reversing at 2 mV; and a 90 pS K^+ selective reversing at –70 mV in normal saline (Hu and Sachs, 1996).

6.4.3 Stretch-Activated Channels Selective for K^+

Stretch-activated K^+ channels in the heart were first described in rat atrial cells by Kim (1992). These channels were activated by negative pressure from a patch pipette, and were also activated by arachadonic acid, and by intracellular acidosis. As a result, a new class of K^+ channels were identified, characterized by 4 trans-membrane domains, and 2 pore domains (K_{2P} channels, for review see Patel and Honoré, 2001). Initially these channels were found in brain tissue, and were not thought to be expressed in the heart; subsequently one of this class, TREK-1, has been observed in rat ventricular myocytes by several groups. It appears to be the same channel as originally described by Kim (Li et al., 2006). The channel is expressed in longitudinal stripes on the myocyte surface membrane and can also be activated by longitudinal stretch (Li et al., 2006). Increasing K^+ permeability will shorten the action potential duration and tend to cause hyperpolarization of the resting membrane potential. Activation of this channel may therefore help to counterbalance the arrhythmogenic potential of SAC_{NSC} channels. Alternatively, activation of the K_{2P} channels could shorten the atrial refractory period increasing the likelihood of early after depolarizations and arrhythmias (Ninio and Saint, 2008). Application of blockers of SAC_{NSC} have been shown to reduce the stretch-induced vulnerability to atrial fibrillation (Bode et al., 2000; Bode and Franz, 2001), without a change in the stretch-related drop in refractoriness (Bode et al., 2000), suggesting that atrial stretch activates both K_{2P} channels (which were not affected by the concentration of SAC_{NSC} blockers used) and SAC_{NSC}.

6.4.4 Volume Sensitive Chloride Channels

It has long been recognized that swelling cardiac cells, and many other cell types, activates a Cl⁻ current ($I_{Cl,swell}$) which assists in the recovery of cell volume (Tseng, 1992). This current has been described in atrial and ventricular myocytes of many mammalian species, and seems to be chronically activated in some diseases, including heart failure (for review see Baumgarten and Clemo, 2003). The chloride current can be activated by extracellular hypo-osmotic solutions, or by the use of hyper-osmotic intracellular solutions in a patch electrode, or by physically swelling the cells by applying positive pressure to a whole-cell attached patch pipette. Unitary Cl^- channels can also be observed by applying negative pressure to inside-out patches (Sato and Koumi, 1998). The channel can also be activated by anionic amphipaths, which enter the outer leaflet of the lipid bilayer and induce increased curvature. Given these properties, $I_{Cl,swell}$ may also be activated by membrane deformation and this current is sometimes observed during stretch and other mechanical interventions (Browe and Baumgarten, 2003, 2004).

In a recent development, Browe and Baumgarten (2003) have shown that magnetic beads coated with antibodies to integrins will attach to the surface of ventricular myocytes. When a magnetic field is applied, causing outward stretch on the beads, the $I_{Cl,swell}$ is activated. They have extensively investigated the activation pathway of this current, and proposed a stretch-activated signalling cascade similar to that shown in Fig. 6.3. The integrin signalling pathway involves focal adhesion kinases (FAK) and/or src kinases, that contributes to the stretch-induced release of angiotensin (AII) (Sadoshima et al., 1993). Activation of the AII Type 1 receptor

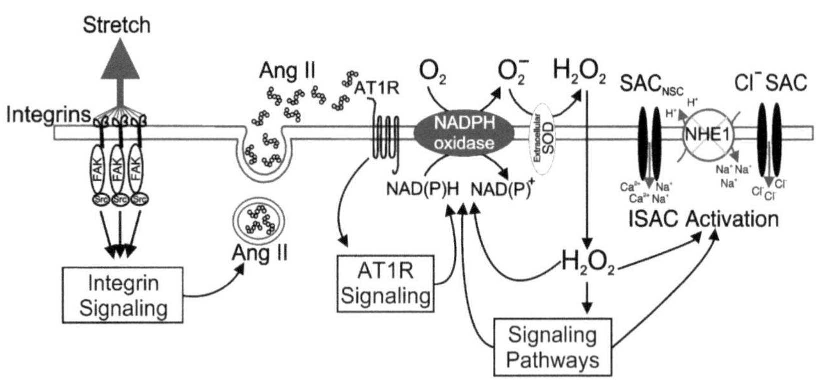

Fig. 6.3 Proposed model of stretch-activation in cardiac myocytes. A simplified model of the mechano-transduction process coupling β1 integrin stretch to activation of Cl^- SAC, NHE1, and SAC_{NSC} in ventricular myocytes. It is proposed that stretch triggers a mechanism that leads to the paracrine/autocrine release of angiotensin II (Ang II) from secretory vesicles. Ang II then binds to the AT1 receptor (AT1R) and activates a signaling cascade that, in turn, activates NADPH oxidase and production of ROS. H_2O_2 then crosses the cell membrane and activates I_{SAC} either directly, or via some ROS-sensitive signaling pathway. (Modified, and redrawn from Browe and Baumgarten, 2004, with permission.)

(AT1R) stimulates the activity of NADPH oxidase, which produces extracellular superoxide (SOD). Extracellular superoxide dismutase converts the extracellular, and cell impermeant, superoxide to H_2O_2 which then directly or indirectly activates $I_{Cl,swell}$. Whereas Browe and Baumgarten (2004) propose this pathway as a model of activation of the chloride SAC, we have modified the pathway to include stretch activation of NHE1 and SAC_{NSC} in our Fig. 6.3. It should be emphasized, though, that it is not yet clear just how stretch activates SACs. Whereas there is evidence that stretch activates signalling pathways via integrin receptors that results in SAC opening (Wang et al., 1993; Browe and Baumgarten, 2003, 2004), it does not seem likely that this is the *only* means of their activation in cardiac myocytes, since SACs are also present in excised, and cell attached, patches (Bustamante et al., 1991; Ruknudin et al., 1993). This suggests that SAC are also gated by tension developed in cytoskeleton structures coupled to the channel proteins, as well as by deformation developed in the lipid bilayer itself.

6.4.5 Pharmacological Agents that Block Stretch-Activated Channels

In the heart, lanthanides and aminoglycosidic antibiotics have been used to block SACs, but these pharmacological agents, although readily available, are non-specific. SACs are generally blocked by 5–10 μM Gd^{3+} (Yang and Sachs, 1989), the most commonly used lanthanide for blocking cardiac SACs (Kamkin et al., 2000; Zeng et al., 2000; Zhang et al., 2000; Kamkin et al., 2003a). It is not always recognized that Gd^{3+} binds extremely tightly to $CO_3{}^{2-}$, $PO_3{}^{3-}$ and many proteins (Caldwell et al., 1998), effectively reducing the concentration of free [Gd^{3+}] available (see discussion in White, 2006). Thus, to test the efficacy of Gd^{3+} for blocking of SAC_{NSC} it is desirable to use solutions lacking $PO_3{}^{3-}$, $HCO_3{}^-/CO_2$ buffering, and proteins, which is often impractical. Despite this issue, Gd^{3+} has been found to be effective blocker of SAC_{NSC} in physiological solutions (Yeung et al., 2003), and even in vivo (Takagi et al., 1999). Gd^{3+} is non-specific and can block L-type Ca^{2+} currents (Sadoshima et al., 1992; Lacampagne et al., 1994), Na^+ currents (Li and Baumgarten, 2001), K^+ currents (Hongo et al., 1997), the Na^+/Ca^{2+} exchanger (Zhang and Hancox, 2000), and the stretch-sensitive K^+ channels, TREK-1 (Patel and Honoré, 2001).

The aminoglycoside antibiotics (streptomycin, neomycin, kanamycin, and gentamicin) are also commonly used to block SAC_{NSC}. Streptomycin is an effective blocker at 100–200 μM (Winegar et al., 1996), though streptomycin can also block L-type Ca^{2+} and the delayed rectifier currents in isolated myocytes from guinea-pig (Belus and White, 2002). However, we found that 400 μM streptomycin had no affect on the amplitude of the Ca^{2+} transient in un-stretched mouse trabeculae (Ward et al., 2008), perhaps suggesting species differences in the response to streptomycin. A recent discovery is that a peptide isolated from a spider venom (GsMTx-4) is a more potent, and specific, blocker of SAC_{NSC} (Suchyna et al., 2000). GsMTx-4 blocks SAC_{NSC} in astocytes and chick hearts with a KD of 500 nM, and appears to have no other effect on the cardiac action potential (Suchyna et al., 2000). While initially thought to be specific for the SAC_{NSC} encoded by TRPC1 (see Section 6.5, below), there are now several reports that GsMTx-4 also blocks the SAC_{NSC} encoded by TRPC6 (Spassova et al., 2006; Dyachenko et al., 2008a). It has also been shown that the peptide synthesized entirely from D amino acids blocks the channel as effectively as the normal L form (Suchyna et al., 2004). Thus it appears that channel blockade is not stereospecific, suggesting that GsMTx-4 is acting at the lipid–protein boundary and disturbing the gating mechanism rather than blocking the pore. Further, it appears that it may have this effect on different proteins that encode SAC_{NSC}.

Beech and his group have shown that it is possible to design antibodies to an extracellular epitope near the putative pore on the TRPC protein sequence (Xu and Beech, 2001). If such an antibody blocks the channel, it provides strong evidence that the particular TRPC protein to which the antibody was designed is a functional channel (Benham, 2005). The existing antibodies for TRPC1 and TRPC6 bind to intracellular sites, and Dyachenko et al. (2008a) have shown that an existing commercial antibody to TRPC6 when applied via the patch pipette (i.e. intracellularly) effectively blocks SAC_{NSC} in mouse ventricular myocytes.

6.4.6 Effect of Stretch-Activated Channel Blockers on the SFR

The contribution of stretch-activated channels to the SFR was investigated in mouse trabeculae subjected to a 2 min stretch in the presence of three different blockers of SAC (Ward et al., 2008). Following a pre-drug, control stretch, muscles were returned to short length, and allowed to recover. They were then exposed to either 400 μM streptomycin sulphate, 10 μM $GdCl_3$, or 10 μM GsMTx-4, and equilibrated for 15 min at short length to allow for diffusion of the blockers throughout the multicellular preparation. Figure 6.4 shows averaged data before, during, and after

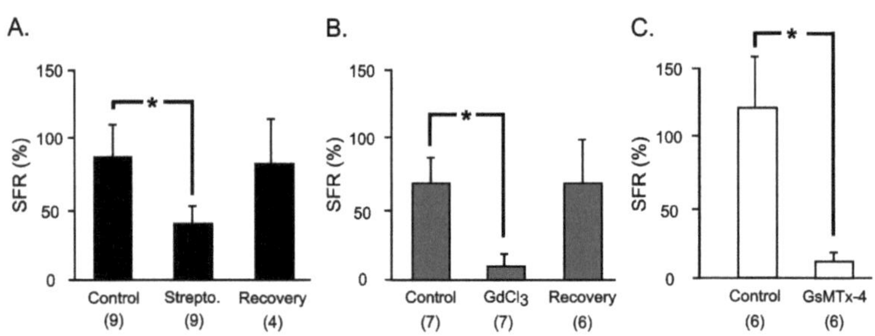

Fig. 6.4 Effect of stretch-activated channel blockers on the slow force response. Panels show pooled data from mouse ventricular trabeculae (number of muscles shown in *parenthesis*) subjected to a 2 min stretch before, during, and after exposure to 400 μM streptomycin sulphate (**Panel A**), 10 μM gadolinium chloride (**Panel B**), 10 μM GsMTx-4 (**Panel C**). *$P < 0.05$. (From Ward et al., 2008, with permission.)

exposure to each of the SAC blockers. The magnitude of the SFR was reversibly decreased by both streptomycin (Fig. 6.4a), and Gd^{3+} (Fig. 6.4b). Incubation in 10 μM GsMTx–4 for 15 min also reduced the magnitude of the SFR (Fig. 6.4c), with a further reduction after 30 min of incubation (data not shown). Recovery of the SFR following washout of GsMTx-4 was poor. Previously, Callaghan and White (2004) showed that streptomycin, used at 40–80 μM concentration reduced the amplitude of the SFR to stretch in both rat papillary muscles and isolated myocytes. In contrast, neither Gd^{3+} nor streptomycin reduced the SFR in failing human myocardium (von Lewinski et al., 2004), perhaps suggesting that the role of SAC in the SFR differs between species.

6.5 Molecular Candidates for Cardiac SAC_{NSC}

A critical issue is the molecular identification of SAC_{NSC}. Maroto et al. (2005) showed that the purified SAC_{NSC} from frog oocytes had the correct molecular weight for TRPC1 and bound the TRPC1 antibody. These amphibian SAC_{NSC} have similar properties (single channel conductance, ion selectivity, and voltage insensitivity) to the SAC_{NSC} observed in mammalian skeletal muscle. Maroto et al. expressed human TRPC1 in COS cells and observed a 10-fold increase in SAC_{NSC} expression. Subsequently, it has been shown that the SAC_{NSC} blocker, GsMTx-4, blocks the expressed channel (Bowman et al., 2007). Further support for the idea that TRPC proteins might encode SAC_{NSC} derives from work on dystrophic muscle that over-expresses a SAC_{NSC} (Franco-Obregon and Lansman, 2002). Vandebrouck et al., (2002) showed that *mdx* muscle fibres contained a store-operated channel and expressed TRPC1, TRPC4 and TRPC6 in the surface membrane. Using a silencing strategy, they successfully knocked down expression of TRPC1 and TRPC6, and showed that this reduced the expression of the store-operated channel. Later the same group (Ducret et al., 2006) showed that the store operated channel and SAC_{NSC} have virtually identical electrophysiological and pharmacological properties, and suggested they might be properties of a single channel.

More recently, however, it was found that over expression of TRPC1 in either COS or CHO cells did not reliably increase the endogenous SAC_{NSC} activity (Gottlieb et al., 2008). They also showed that, although TRPC1 was extensively expressed, the protein was mainly intracellular and showed no obvious membrane expression. This suggests that the problem may be in the trafficking of the expressed protein to the membrane. In support of this view, we recently expressed TRPC1 in C2 myoblasts (an established cell line of skeletal muscle origin) and observed that the expression was mainly intracellular (Gervásio et al., 2008). Furthermore, H_2O_2, which was able to activate SAC_{NSC} in dystrophic skeletal muscle cells, failed to generate Ca^{2+} influx when the transfected TRPC1 was intracellular only. However, when TRPC1 was co-expressed with caveolin-3, some of the TRPC1 was expressed in the membrane, and Ca^{2+} influx was stimulated by H_2O_2. Thus our interpretation is that TRPC1 requires caveolin-3 to assist its trafficking to the membrane, and when appropriately inserted in the membrane, it can be activated by ROS and presumably also by membrane stretch.

The above experiments suggest that TRPC1 may contribute to the SAC_{NSC} observed in skeletal muscle. What is the molecular candidate for SAC_{NSC} in ventricular muscle? As noted earlier, direct activation by pressure of non-specific cation channels has not been observed in adult ventricular muscle. However, stretch-activated currents have been observed by several groups when isolated myocytes are either stretched (Kamkin et al., 2000; Zeng et al., 2000) or subject to shearing strain (Hu and Sachs, 1996; Bett and Sachs, 2000; Dyachenko and Isenberg, 2007; Dyachenko et al., 2008a, b). The recent studies by Dyachenko et al. utilize mouse ventricular cells fixed to a coverslip and held down by a patch-clamp electrode at one end in whole cell mode. The upper surface of the other end of the cell is stretched by attachment of a second glass tool, so that the cell is subject to a shearing strain. Robust currents can be elicited by this mechanism that are similar, but not identical, to stretching rat ventricular myocytes (Zeng et al., 2000). These currents are abolished by de-tubulation, suggesting that they arise from channels within the T-tubules. Two main channels appeared to contribute to the measured current: a non-specific cation current (reversal potential ~ –10 mV) activated by stretch, and blocked by streptomycin and GsMTx-4 (denoted I_{NS}); secondly, an inwardly rectifying K^+ current inhibited by stretch and blocked by replacing K^+ by Cs^+ and thought to be the inward rectifier (denoted I_{K1}). I_{NS} was blocked by an antibody to TRPC6 applied intracellularly, but unaffected when the antibody was applied extracellularly. TRPC6 was expressed only within the T-tubules, consistent with elimination of the current by de-tubulation. In the second paper the authors explored the activation of I_{NS}. Thus, at present, TRPC1 and TRPC6 seem the strongest contenders for the genes encoding SAC_{NSC} with the evidence favouring TRPC1 in skeletal muscle, and TRPC6 in ventricular muscle of at least some species.

6.6 Stretch-Induced Arrhythmias

Mechanical stress or strain can also lead to changes in the electrical activity of the heart (Bainbridge, 1915). Depending on the type, and timing, of the mechanical stimulus, this can be seen as either changes in action potential duration (Lab, 1978b; Franz et al., 1989; Calkins et al., 1991; Zabel et al., 1996; Kamkin et al., 2000), or as transient membrane repolarization, or depolarization (Lab, 1978a; Zabel et al., 1996). Since SAC are implicated in these changes in electrical activity, it can be easily understood that the timing of the mechanical stimulus in relation to the cardiac action potential is crucial. As discussed previously, non-selective cationic SACs are permeant to Ca^{2+} ions as well as to Na^+ and K^+, whereas other SACs are selective for K^+ and possibly Cl^- ions (Hagiwara et al., 1992; Ruknudin et al., 1993). Influx, or efflux, of individual ion species through open channels will therefore be dependent on their reversal potential in relation to the membrane potential.

Computer simulations of action potentials obtained for different degrees of steady-state stretch showed that AP duration at negative membrane potentials was increased with stretch in guinea-pig ventricular myocytes, whereas, at positive potentials, stretch shortened the AP duration (Zabel et al., 1996). It was noted also in this study that the crossover of action potentials occurred close to the predicted reversal potential for the SAC_{NSC} stretch-activated currents, at approximately –10 mV (Sasaki et al., 1992; Hu and Sachs, 1996). Experimental results obtained from guinea-pig, and failing human, ventricular myocytes yielded similar results when cells were subjected to a localized stretch between a stylus and a patch pipette (Kamkin et al., 2000), as did an earlier study in which monophasic AP recordings were made from the surface of isolated canine hearts (Franz et al., 1989). In this study, AP duration was decreased with stretch at 20% repolarization, but increased at 90% repolarization, when the ventricles were subjected to an increase in LV volume. Kamkin et al. (2000) in voltage-clamped myocytes showed that stretch activated several ionic currents, the most prominent being a stretch-activated current through non-selective cation channels (I_{SAC}). They showed that I_{SAC} became steady within 200 ms of stretch, and remained steady throughout the stretch. An interesting feature of their study was that the stretch sensitivity, determined from the slope of the I_{SAC} versus the amplitude of the stretch (pA/μm), was increased with age (myocytes isolated from 3 month, in comparison to 15 month, old animals), and with hypertrophy (myocytes isolated from 15 month old spontaneously hypertensive animals). It would therefore appear that age, as well as hypertrophy, increases the risk of stretch-induced arrhythmias (Kamkin et al., 2000).

Zabel et al. (1996) also showed that, in isolated beating hearts, the timing of the ventricular volume increase determined the nature of the electrophysiological response. This was consistent with their patch-clamp studies in isolated cells in which only stretch pulses applied at the end of the AP, or during diastole, produced ectopic beats as a result of transient depolarizations. An example of stretch-induced spontaneous activity in a multicellular preparation, a mouse RV trabecula, is shown in Fig. 6.5. Figure 6.5a shows isometric force (top), fura-2 340/380 ratio (upper middle), stimulus voltage (lower middle), and trabecula length (bottom), before during and after a 2 min stretch. A dramatic increase in the resting Ca^{2+} accompanied the stretch-induced spontaneous activity in this muscle, with some full sized Ca^{2+} transients in response to stimulation occurring initially. As the duration of the stretch continued, the spontaneous activity increased with no return to resting levels of $[Ca^{2+}]_i$ between electrical stimuli applied at 0.2 Hz. On returning the muscle again to short length the spontaneous activity was very much reduced with a return to stimulated responses only after ~30 s at short length. Interestingly, stretch-induced spontaneous activity occurs in only ~2% of isolated trabeculae (Ward et al., 2008), whereas stretching the ventricle of isolated, perfused, hearts by balloon inflation is commonly used to induce ventricular arrhythmia (Franz et al., 1989; Hansen et al., 1990; Franz et al., 1992; Zabel et al., 1996).

In humans, a moderate mechanical impact to the pre-cordial region of the chest can give rise to cardiac arrhythmias that are frequently fatal (*commotio cordis*), in the absence of any morphological damage to the heart (Kohl et al., 1999; Link et al.,

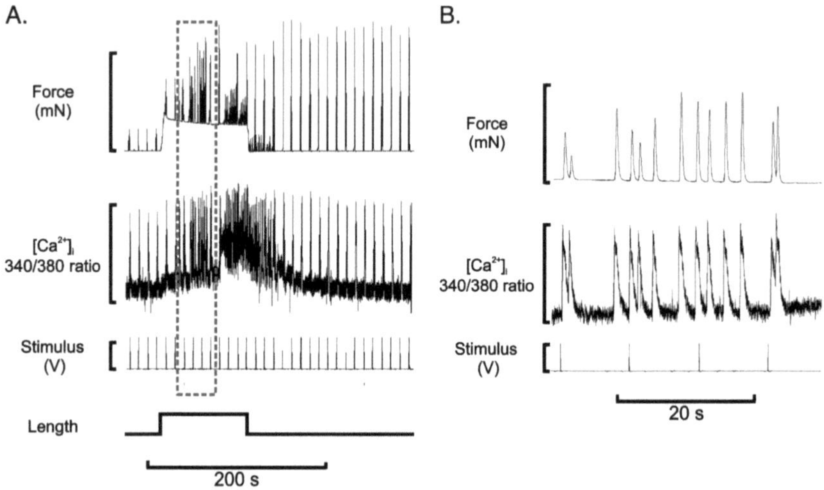

Fig. 6.5 Stretch-induced arrhythmia. Shown here is an example of spontaneous arrhythmia developed following stretch in an RV trabecula from mouse. **Panel A** shows the 340/380 ratio (*top*), isometric force (*middle*), and stimulus voltage (0.1.Hz, lower) before, during, and after a step change in muscle length (*bottom*). Similar spontaneous activity on stretching occurred in <2% of mouse trabeculae. On returning to short length again the spontaneous activity ceased within ~30s, with both the Ca^{2+} transient amplitude, and twitch force being very much potentiated. **Panel B** shows a section of data taken from **Panel A** (*rectangle* within *dotted grey lines*) with the time axis expanded

1999; Nesbitt et al., 2001). Initiation of these arrhythmias has been attributed to mechano-electric feedback due to the recruitment of SACs (Kohl et al., 1999). Modelling of *commotio cordis*-like events at the level of a single cell has shown that SAC_{NSC} openings induced by an impact delivered during diastole can cause sufficient depolarization to trigger an extra AP, while outward currents through stretch-activated K^+ channels change AP duration, depending on their precise timing (Li et al., 2004). Furthermore, the modelling of Li et al. strongly suggests that, although initiation of the *commotio cordis* arrhythmias was via SAC_{NSC}, the interaction of other SAC populations (including K^+-selective channels) was crucial in determining whether the arrhythmias were sustained, or not.

6.7 Overview of the Contribution of Stretch to Cardiac Performance: SFR

For a simple observation, the SFR seems to have generated a relatively rich range of proposed mechanisms. It is widely accepted that the SFR is quite variable, even in the hands of a single laboratory and species. This variability has led some to describe it as an epiphenomenon (Blinks and Endoh, 1986), of little fundamental importance. However the observation of a robust SFR in intact hearts (Tucci et al., 1984; Lew, 1993) and the physiological importance of increasing cardiac output in the face of a rise in filling pressure suggest to us that its physiological importance is considerable (Allen and Kentish, 1985). There are also many suggestions that the phenomenon may have different mechanisms in different tissues. For instance the variable involvement of NHE1 has been noted above. TRPC6, thought to encode SAC_{NSC} in mouse ventricular muscle, is not expressed in rat ventricular muscle (Dyachenko et al., 2008a), although rat muscle does exhibit SAC_{NSC} (Kamkin et al., 2000). I_{K1} is reportedly activated by stretch in some species (Kamkin et al., 2000), but inactivated by stretch in mouse (Dyachenko et al., 2008a). Permeability of SAC_{NSC} to Ca^{2+} was found to be substantial in rat, guinea pig and human ventricular myocytes (Kamkin et al., 2000), but low in mouse ventricular myocytes (Kamkin et al., 2003a). After treatment with cytochalasin, which inhibits cytoskeletal connections, SAC_{NSC} in skeletal muscle became more sensitive to stretch (Guharay and Sachs, 1984), whereas SAC_{NSC} in mouse ventricular cells became completely insensitive to stretch (Kamkin et al., 2003a).

The idea that activation of NHE1 has a role in SFR was first suggested by Cingolani et al. (1998). They proposed a pathway in which endogenous AII release was stimulated by stretch, which subsequently stimulated activity of NHE1. In a more recent paper, Caldiz et al. (2007) have extended these ideas by showing that ROS production increases during SFR, and that this is blocked by two non-specific blockers of NADPH oxidase, DPI and apocyanin. They also proposed that ROS production was amplified by mitochondria on the basis that glybenclamide and 5HT, blockers of mitochondrial K_{ATP} channel, both minimised the SFR. The final step of their proposed pathway is that ROS activated NHE1 by means of the ERK 1/2 kinases. What is clear from this mechanism is that any other ROS sensitive mechanism might be activated and also contribute. Caldiz et al. argue the importance of NHE1 on the basis that NHE1 inhibitors reduce the size of the SFR and also reduce the increase in $[Na^+]_i$. However these arguments are far from overwhelming. The effectiveness of NHE1 inhibitors in preventing SFR are variable in the literature, and, more important, Niederer and Smith (2007) have shown in a modeling study that inhibiting NHE1 can reduce the size of the SFR even when NHE1 is not affected by muscle length. In other words, NHE1 modulates the magnitude of the SFR irrespective of whether it is length-dependent or not. Furthermore, the rise in $[Na^+]_i$ could arise from activity of a SAC_{NSC} though one would not expect the rise to be prevented by NHE1 blockers if that were the case, unless their specificity is called into question. (A recent publication does suggest that this is the case (Garciarena et al., 2008).) We propose that the length-dependent activation of AII release, and subsequent increase in activity of NADPH oxidase and in ROS production, could be the starting elements in the SFR. Caldiz et al. demonstrate the role of this pathway in NHE1 activation, Browe and Baumgarten have shown a similar pathway in relation to $I_{Cl,swell}$, and Dyachneko et al. have demonstrated its importance in relation to SAC_{NSC}. The classic NADPH oxidase (NOX2) was first described in phagocytes, and produces superoxide in the extracellular space or into vacuoles. Superoxide is charged, and membrane impermeant, and Browe and Baumgarten proposed that extracellular superoxide dismutase converted it to H_2O_2, which is membrane permeant and may be the signaling molecule that subsequently affects channels or exchangers. Thus it is likely that the H_2O_2 signalling pathway is spatially restricted to near membrane targets. One possibility is src kinase, which is activated by H_2O_2 (Chen et al., 2005) and may phosphorylate channels or exchangers. Alternatively, H_2O_2, or its products such as the hydroxyl radical, may produce oxidative modifications of proteins that modify their function. As an example, Gervásio et al. (2008) expressed TRPC1 and caveolin-3 in myoblasts and found that this combination could produce Ca^{2+} entry when the cells were exposed to H_2O_2. This stimulation of Ca^{2+} entry was prevented by PP2, an inhibitor of src kinase, suggesting that phosphorylation of some component was essential to TRPC1 activation. Obviously the downstream events will depend on the magnitude of expression of NHE1, SAC_{NS}, $I_{Cl,swell}$ and perhaps also their distribution in relation to the generation of ROS.

6.8 Conclusion

Transduction of mechanical forces via a variety of stretch-activated channels provides an intrinsic means of regulating both the contractility, and the electrical activity of the heart. Although the emphasis in this Review has been on the role of stretch-activated channels and their relatively short-term effects on myocytes, it is recognized that cardiac mechano-transduction is complex, occurring at many different levels, and over different timescales. Never-the-less, for all its complexity, mechano-transduction in the heart must begin with the conversion of a mechanical signal into some biochemical or electrical response that then alters cellular function in some way. Although it seems likely that SACs might provide the initial response to mechanical stimuli in the heart, since they typically respond within tens of milliseconds, there is still much debate as to their role. We think much of the variability in the literature arises from the differential expression of channels, and the likelihood that different types of stretch (linear stretch, swelling, distorting membrane in a patch pipette, shear stress) activate different classes of channels. It also seems likely that, although stretch-activated channels were first discovered by their response to pressure-induced deformation of the membrane in a patch-pipette, this mechanical stimulus is not their normal means of activation. Instead, we think the pathways which have been described for the activation of $I_{Cl,swell}$, I_{NSC} and NHE1 may be their physiological activating mechanism. The common elements of this pathway involve integrin activation, AII release, activation of NADPH oxidase, and release of superoxide. A complex array of channels and exchangers may then be activated depending on their expression in a particular tissue, and spatial factors. These complex factors may help explain the variability between tissues, animals and laboratories in investigating the role of stretch-activated channels in the heart.

Acknowledgments Our work was supported by the National Health and Medical Research Council of Australia. M-L Ward gratefully acknowledges financial support from the P. J. Smith NZ Freemason's Travelling Fellowship. The authors would also like to thank Dr Patricia Cooper for comments and discussion on this manuscript.

References

Allen DG. (1977). On the relationship between action potential duration and tension in cat papillary muscle. Cardiovascular Research 11, 210–218.

Allen DG and Kentish JC. (1985). The cellular basis of the length-tension relation in cardiac muscle. Journal of Molecular and Cellular Cardiology 17, 821–840.

Allen DG and Kurihara S. (1982). The effects of muscle length on intracellular calcium transients in mammalian cardiac muscle. Journal of Physiology 327, 79–94.

Allen DG, Nichols CG and Smith GL. (1988). The effects of changes in muscle length during diastole on the calcium transient in ferret ventricular muscle. Journal of Physiology 406, 359–370.

Alvarez BV, Perez NG, Ennis IL, Camilion de Hurtado MC and Cingolani HE. (1999). Mechanisms underlying the increase in force and Ca^{2+} transient that follow stretch of cardiac muscle: a possible explanation of the Anrep effect. Circulation Research 85, 716–722.

Avettev AS and Navaratnam V. (1978). The T-tubule system in the specialized and general myocardium of the rat. Journal of Anatomy 127, 125–140.

Bainbridge FA. (1915). The influence of venous filling upon the rate of the heart. Journal of Physiology 50, 65–84.

Baumgarten CM and Clemo HF. (2003). Swelling-activated chloride channels in cardiac physiology and pathophysiology. Progress in Biophysics and Molecular Biology 82, 25–42.

Belus A and White E. (2002). Effects of streptomycin sulphate on I_{CaL}, I_{Kr} and I_{Ks} in guinea-pig ventricular myocytes. European Journal of Pharmacology 445, 171–178.

Belus A and White E. (2003). Streptomycin and intracellular calcium modulate the response of single guinea-pig ventricular myocytes to axial stretch. Journal of Physiology 546, 501–509.

Benham CD. (2005). Simple recipe for blocking ion channels. Nature Biotechnology 10, 1234–1235.

Bett GC and Sachs F. (2000). Whole-cell mechanosensitive currents in rat ventricular myocytes activated by direct stimulation. Journal of Membrane Biology 173, 255–263.

Blinks JR. (1956). Positive chronotropic effect of increasing right atrial pressure in the isolated mammalian heart. American Journal of Physiology 186, 299–303.

Blinks JR, Endoh M. (1986). Modification of myofibrillar responsiveness to Ca++ as an inotropic mechanism. Circulation 73, 11185–11198.

Blount P, Li Y, Moe PC and Iscla I. (2008). Mechanosensitive channels gated by membrane tension: Bacteria and beyond, Vol. 1. Springer, New York, 71–101.

Bluhm WF and Lew WYW. (1995). Sarcoplasmic reticulum in cardiac length-dependent activation in rabbits. American Journal of Physiology 269, H965–H972.

Bode F and Franz MR. (2001). Tarantula peptide inhibits atrial fibrillation. Nature 409, 35.

Bode F, Katchman A, Woosley RL and Franz MR. (2000). Gadolinium decreases stretch-induced vulnerability to atrial fibrillation. Circulation 101, 2200–2205.

Bowman CL, Gottlieb PA, Suchyna TM, Murphy YK and Sachs F. (2007). Mechanosensitive ion channels and the peptide inhibitor GsMTx-4: history, properties, mechanisms and pharmacology. Toxicon 49, 249–270.

Browe DM and Baumgarten CM. (2003). Stretch of β1 integrin activates an outwardly-rectifying chloride current via FAK and Src in rabbit ventricular myocytes. Journal of General Physiology 122, 689–702.

Browe DM and Baumgarten CM. (2004). Angiotensin II (AT1) receptors and NADPH oxidase regulate Cl^- current elicited by β1 integrin stretch in rabbit ventricular myocytes. Journal of General Physiology 124, 273–287.

Burkhoff D, de Tombe PP, Hunter WC and Kass DA. (1991). Contractile strength and mechanical efficiency of left ventricle are enhanced by physiological afterload. American Journal of Physiology 260, H569–H578.

Bustamante JO, Ruknudin A and Sachs F. (1991). Stretch-activated channels in heart cells: relevance to cardiac hypertrophy. Journal of Cardiovascular Pharmacology 17, S110–S113.

Calaghan S and White E. (2004). Activation of Na^+-H^+ exchange and stretch-activated channels underlies the slow inotropic response to stretch in myocytes and muscle from the rat heart. Journal of Physiology 559, 205–214.
Calaghan SC, Colyer J and White E. (1999). Cyclic AMP but not phosphorylation of phospholamban contributes to the slow inotropic response to stretch in isolated ferrit papillary muscle. Pflügers Archive 437, 780–782.
Caldiz CI, Garciarena CD, Dulce RA, Novaretto LP, Yeves, Alejandra M., Ennis IL, Cingolani HE, Chiappe de Cingolani G and Perez GN. (2007). Mitochondrial reactice oxygen species activate the slow force response to stretch in feline myocardium. Journal of Physiology 584, 895–905.
Caldwell RA, Clemo HF and Baumgarten CM. (1998). Using gadolinium to identify stretch-activated channels: technical considerations. American Journal of Physiology 275, C619–C621.
Calkins H, Levine JH and Kass DA. (1991). Electrophysiological effect of varied rate and extent of acute in vivo left ventricular load increase. Cardiovascular Research 25, 637–644.
Chen DB, Li SM, Qian XX, Moon C and Zheng J. (2005). Tyrosine phosphorylation of caveolan 1 by oxidative stress is reversible and dependent on the c-src tyrosine kinase but not mitogen-activated protein kinase pathways in placental artery endothelial cells. Biology of Reproduction 73, 761–772.
Chiang C-S, Anishkin A and Sukharev S. (2004). Gating of the large mechanosensitive channel in situ: Estimation of the spatial scale of the transition from channel population responses. Biophysical Journal 86, 2846–2861.
Chuck LH and Parmley WW. (1980). Caffeine reversal of length-dependent changes in myocardial contractile state in the cat. Circulation Research 47, 592–598.
Cingolani HE, Alvarez BV, Ennis IL and Camili6n de Hurtado MC. (1998). Stretch-induced akalanization of feline papillary muscle: an autocrine-paracrine system. Circulation Research 83, 775–780.
Cooper P, Lei M, Cheng L-X and Kohl P. (2000). Axial stretch increases spontaneous pacemaker activity in rabbit isolated sinoatrial node cells. Journal of Applied Physiology 89, 2099–2104.
Cooper PJ and Kohl P. (2005). Species- and preparation-dependence of stretch effects on sino-atrial node pacemaking. Annals of the New York Academy of Sciences 1047, 324–335.
Corey DP. (2003). New TRP channels in hearing and mechanosensation. Neuron 39, 585–588.
Craelius W, Chen V and El-Sherif N. (1988). Stretch-activated ion channels in ventricular myocytes. Bioscience Reports 8, 407–414.
Ducret T, Vandebrouck C, Cao ML, Lebacq J and Gailly P. (2006). Functional role of store-operated and stretch-activated channels in murine adult skeletal muscle fibres. Journal of Physiology 575, 913–924.
Dulhunty AF and Franzini-Armstrong C. (1975). The relative contributions of the folds and caveolae to the surface membrane of skeletal muscle fibres at different lengths. Journal of Physiology 250, 513–539.
Dyachenko V and Isenberg G. (2007). Ventricular myocytes: deformation induced sarcomee misalignment modulates TRPC6 and KIR2.3 channels. In Proceedings of the 4th International Conference on Cardiac Mechano-Electric Feedback, pp. 84.
Dyachenko V, Husse B, Rueckschloss U, Isenberg G. (2009a). Mechanical deformation of ventricular myocytes modulates both TRPC6 and Kir2.3 channels. Cell Calcium 45, 38–54.
Dyachenko V, Rueckschlos U, Isenberg G. (2009b). Modulation of cardiac mechanosensitive ion channels involves superoxide, nitric oxide and peroxynitrite. Cell Calcium 45, 55–64.
Eisner DA, Choi HS, Díaz ME, O'Neill SC and Trafford AW. (2000). Integrative analysis of calcium cycling in cardiac muscle. Circulation Research 87, 1087–1094.
Fabiato A and Fabiato F. (1975). Dependence of the contractile activation of skinned cardiac cells on sarcomere length. Nature 256, 54–56.
Farrugia G, Holm AN, Rich A, Sarr MG, Szurszewski JH and Rae JL. (1999). A mechanosensitive calcium channels in human intestinal smooth muscle cells. Gastroenterology 117, 900–905.
Franco-Obregon A and Lansman JB. (2002). Changes in mechanosensitive channel gating following mechanical stimulation in skeletal muscle myotubes from the mdx mouse. Journal of Physiology 539, 391–407.

Frank O. (1895). Zur dynamik des hersmuskels. Zeitschrift für Biologie 32, 370–447.
Franz MR, Burkhoff D, Yue DT and Sagawa K. (1989). Mechanically induced action potential changes and arrhythmia in isolated and in situ canine hearts. Cardiovascular Research 23, 213–223.
Franz MR, Cima R, Wang D, Profitt D and Kurz R. (1992). Electrophysiological effects of myocardial stretch and mechanical determinants of stretch-activated arrhythmias. Circulation 86, 968–978.
Fukuda N and Granzier HL. (2005). Titan/connectin-based modulation of the Frank-Starling mechanism of the heart. Journal of Muscle Research and Cell Motility 26, 319–323.
Garciarena CD, Caldiz CI, Correa MV, Schinella GR, Mosca SM, Chiappe de Cinglani GE, Cingolani HE, Ennis IL. (2008) Na+/H+ exchanger-1 inhibitors decrease myocardial superoxide production via direct mitochondrial action. Journal of Applied Physiology 105, 1706–1713.
Gervásio OL, Whitehead NP, Yeung EW, Phillips WD and Allen DG. (2008). TRPC1 binds to caveolan-3 and is regulated by Src kinase -role in duchenne muscular dystrophy. Journal of Cell Science 121, 2246–2255.
Gordon AM, Huxley AF and Julian FJ. (1966). The variation in isometric tension with sarcomere length in vertebrate muscle fibres. Journal of Physiology 184, 170–192.
Gottlieb P, Folgering J, Maroto R, Rso A, Wood TG, Kurosky A, Bowman C, Bichet D, Patel A, Sachs F, Martinac B, Hamill OP and Honoré E. (2008). Revisiting TRPC1 and TRPC6 mechanosensitivity. Pflügers Archive 455, 1097–1103.
Gu CX, Juranka PF and Morris CE. (2001). Stretch-activation and stretch-inactivation of Shaker-IR, a voltage-gated K^+ channel. Biophysical Journal 80, 2678–2693.
Guharay F and Sachs F. (1984). Stretch-activated single ion channel currents in tissue-cultured embryonic chick skeletal muscle. Journal of Physiology 352, 685–701.
Hagiwara N, Masuda H, Shoda M and Irisawa H. (1992). Stretch-activated anion currents of rabbit cardiac myocytes. Journal of Physiology 456, 285–302.
Hamill OP and Martinac B. (2001). Molecular basis of mechanotransduction in living cells. Physiological Reviews 81, 685–740.
Hamill OP, Neher ME, Sakmann B and Sigworth FJ. (1981). Improved patch-clamp techniques for high-resolution current recording from cells and cell-free membrane patches. Pflügers Archive 391, 85–100.
Hansen DE, Craig CS and Hondeghem LM. (1990). Stretch-induced arrhythmias in the isolated canine ventricle. Evidence for the importance of mechanelectrical feedback. Circulation 81, 1094–1105.
Hibberd MG and Jewell BR. (1982). Calcium- and length-dependent force production in rat ventricular muscle. Journal of Physiology 329, 527–540.
Hongo K, Pascarel C, Cazorla O, Gannier F, Le Guennec J-Y and White E. (1997). Gadolinium blocks the delayed rectifier potassium current in isolated guinea-pig ventricular myocytes. Experimental Physiology 82, 647–656.
Hongo K, White E, Le Guennec J-Y and Orchard CH. (1996). Changes in $[Ca^{2+}]_i$, $[Na^+]_i$ and Ca^{2+} current in isolated rat ventricular myocytes following an increase in cell length. Journal of Physiology 491, 609–619.
Hu H and Sach F. (1997). Stretch-activated ion channels in the heart. Journal of Molecular and Cellular Cardiology 29, 1511–1523.
Hu H and Sachs F. (1996). Mechanically activated currents in chick heart cells. Journal of Membrane Biology 154, 205–216.
Isenberg G, Kazanski V, Kondratev D, Gallitelli MF, Kiseleva I and Kamkin A. (2003). Differential effects of stretch and compression on membrane currents and $[Na^+]$c in ventricular myocytes. Progress in Biophysics and Biophysical Chemistry 82, 43–56.
Kamkin A, Kiseleva G and Isenberg G. (2000). Stretch-activated currents in ventricular myocytes: amplitude and arrhythmogenic effects increase with hypertrophy. Cardiovascular Research 48, 409–420.
Kamkin A, Kiseleva I and Isenberg G. (2003a). Ion selectivity of stretch-activated cation currents in mouse ventricular myocytes. Pflügers Archive 446, 220–231.

Kamkin A, Kiseleva I, Wagner K-D, Bohm J, Theres H, Günther J and Scholz H. (2003b). Characterization of stretch-activated ion currents in isolated atrial myocytes from human hearts. Pflugers Archive 446, 339–346.
Kentish JC and Wrzosek A. (1998). Changes in force and cytosolic Ca^{2+} concentration after length changes in isolated rat ventricular trabeculae. Journal of Physiology 506, 431–444.
Kim D. (1992). A mechanosensitive K^+ channel in heart cells. Activation by arachidonic acid. Journal of General Physiology 100, 1021–1040.
Kim D. (1993). Novel cation-selective mechanosensitive ion channel in the atrial cell membrane. Circulation Research 72, 225–231.
Kohl P, Hunter P and Noble D. (1999). Stretch-induced changes in heart rate and rhythm: clinical observations, experiments and mathematical models. Progress in Biophysics and Molecular Biology 71, 91–138.
Kondratev D, Christ A and Gallitelli MF. (2005). Inhibition of the Na^+-H^+ exchanger with cariporide abolishes stretch-induced calcium but not sodium accumulation in mouse ventricular myocytes. Cell Calcium 37, 69–80.
Lab MJ. (1978a). Depolarization produced by mechanical changes in normal and abnormal myocardium. Journal of Physiology 284, 143P–144P.
Lab MJ. (1978b). Mechanically dependent changes in action potentials recorded from the intact frog ventricle. Circulation Research 42, 519–528.
Lacampagne A, Gannier F, Argibay JA, Garnier D and Le Guennec JY. (1994). The stretch-activated channel blocker gadolinium also blocks L-type calcium channels in isolated ventricular myocytes of the guinea-pig. Biochimica et Biophysica Acta 1191, 205–208.
Langton PD. (1993). Calcium currents recorded from isolated myocytes of rat basilar artery are stretch sensitive. Journal of Physiology 471, 1–11.
Le Guennec JY, White E, Gannier F, Argibay JA and Garnier D. (1991). Stretch-induced increase of resting intracellular calcium concentration in single guinea-pig ventricular myocytes. Experimental Physiology 76, 975–978.
Leeson TS. (1980). T-tubules, couplings and myofibrillar arrangements in rat atrial myocardium. Acta Anatomica (Basel) 108, 374–388.
Lew WYW. (1993). Mechanisms of volume-induced increase in left ventricular contractility. American Journal of Physiology 265, H1778–H1786.
Li GR and Baumgarten CM. (2001). Modulation of cardiac Na^+ current by gadolinium, a blocker of stretch-induced arrhythmias. American Journal of Physiology 280, H272–H279.
Li W, Kohl P and Trayanova N. (2004). Induction of ventricular arrhythmias following mechanical impact: a simulation study in 3D. Journal of Molecular Histology 35, 679–686.
Li XT, Dyachenko V, Zuzarte M, Putzke C, Preisig-Müller R, Isenberg G and Daut J. (2006). The stretch-activated potassium channel TREK-1 in rat cardiac ventricular muscle. Cardiovascular Research 69, 86–97.
Link M, Wang P, VanderBrink BA, Avelar E, Pandian NG, Maron BJ and Estes NA. (1999). Selective activation of the K^+-ATP channels is a mechanism by which sudden death is produced by low-energy chest-wall impact (commotio cordis). Circulation 100, 413–418.
Luers C, Fialka F, Elgner A, Zhu D, Kochskamper J, von Lewinski D, Pieske, B. (2005). Stretch-dependent modulation of [Na+]i, [Ca2+]i, and pHi in rabbit myocardium: -a mechanism for the slow force response. Cardiovascular Research 68, 454–463.
Lyford GL, Strege PR, Shepard A, Ou Y, Ermilov L, Miller SM, Gibbons SJ, Rae JL, Szurszewski JZ and Farrugia G. (2002). $\alpha_{1C}(Ca_v1.2)$ L-type calcium channel mediates mechanosensitive calcium regulation. American Journal of Physiology 283, C1001–C1008.
Maroto R, Raso A, Wood TG, Kurosky A, Martinac B and Hamill OP. (2005). TRPC1 forms the stretch-activated cation channels in vertebrate cells. Nature Cell Biology 7, 179–185.
Martinac B and Kloda A. (2003). Evolutionary origins of mechanosensitive ion channels. Progress in Biophysics and Molecular Biology 82, 11–24.
Matsuda N, Hagiwara N, Shoda M, Kasanuki H and Hosoda S. (1996). Enhancement of the L-type Ca^{2+} current by mechanical stimulation in single rabbit cardiac myocytes. Circulation Research 78, 650–659.

Morris C and Laitko U. (2005). The mechanosensitivity of voltage-gated channels may contribute to cardiacmechano-electric feedback. Elsevier Saunders, Philadelphia, pp. 42–52.

Morris CE and Juranka PF. (2007). Nav channel mechanosensitivity: Activation and inactivation accelerate reversibly with stretch. Biophysical Journal 93, 822–833.

Morris CE and Sigurdson WJ. (1988). Stretch-inactivated ion channels coexist with stretch-activated ion channels. Science 243, 807–809.

Nesbitt A, Cooper PJ and Kohl P. (2001). Rediscovering commotio cordis. Lancet 357, 1195–1197.

Niederer SA and Smith NP. (2007). A mathematical model of the slow force response to stretch in rat ventricular myocytes. Biophysical Journal 92, 4030–4044.

Ninio DM and Saint DA. (2008). The role of stretch-activated channels in atrial fibrillation and the impact of intracellular acidosis. Progress in Biophysics and Molecular Biology 97, 401–416.

Parmley WW and Chuck L. (1973). Length-dependent changes in myocardial contractile state. American Journal of Physiology 224, 1195–1199.

Patel AJ and Honoré E. (2001). Properties and modulation of mammalian 2P domain K^+ channels. Trends in Neuroscience 24, 339–346.

Patterson S and Starling EH. (1914). On the mechanical factors which determine the output of the ventricles. Journal of Physiology 48, 357–379.

Penefsky ZA and Hoffman BF. (1963). Effects of stretch on mechanical and electrical properties of cardiac muscle. American Journal of Physiology 204, 433–438.

Pérez NG, de Hurtado MC and Cingolani HE. (2001). Reverse mode of the Na^+-Ca^{2+} exchange after myocardial stretch: underlying mechanism of the slow force response. Circulation Research 88, 376–382.

Ruknudin A, Sachs F and Bustamante JO. (1993). Stretch-activated ion channels in tissue-cultured chick heart. American Journal of Physiology – Heart Circulation Physiology 264, H960-H972.

Sadoshima J and Izumo S. (1993). Mechanical stretch rapidly activates multiple signal transduction pathways in cardiac myocytes: potential involvement of an autocrine/paracrine mechanism. The EMBO Journal 12, 1681–1692.

Sadoshima J, Takahashi T, Jahn L and Izumo S. (1992). Roles of mechano-sensitive ion channels, cytoskeleton, and contractile activity in stretch-induced immediate-early gene expression and hypertrophy of cardiac myocytes. In Proceedings of the National Academy of Science USA, pp. 9905–9909.

Sadoshima J, Xu Y, Slayter HS and Izumo S. (1993). Autocrine release of angiotensin II mediates stretch-induced hypertrophy of cardiac myocytes in vitro. Cell 75, 977–984.

Sasaki N, Mitsuiye T and Noma A. (1992). Effects of mechanical stretch on membrane currents of single ventricular myocytes of guinea-pig heart. Japanese Journal of Physiology 42, 957–970.

Sato R and Koumi S-I. (1998). Characterization of the stretch-activated chloride channel in isolated human atrial myocytes. Journal of Membrane Biology 163, 67–76.

Soeller C and Cannell MB. (1999). Examination of the transverse tubular system in living cardiac rat myocytes by 2-photon microscopy and digital image-processing techniques. Circulation Research 84, 266–275.

Spassova MA, Hewavitharana T, Xu W, Soboloff J and Gill DL. (2006). A common mechanism underlies the stretch activation and receptor activation of TRPC6 channels. Proceedings of the National Academy of Science USA 103, 16586–16591.

Steele DS and Smith GL. (1993). Effects of muscle length on diastolic $[Ca^{2+}]_i$ in isolated guinea-pig ventricular trabeculae. Journal of Physiology 467, 328P.

Suchyna TM, Johnson JH, Hamer K, Leykam JF, Gage DA, Clemo HF, Baumgarten CM and Sachs F. (2000). Identification of a peptide toxin from grammostola spatulata spider venom that blocks cation-selective stretch-activated channels. Journal of General Physiology 115, 583–598.

Suchyna TM, Tape SE, Koeppe II RE, Andersen OS, Sachs F and Gottlied PA. (2004). Bilayer-dependent inhibition of mechanosensitive channels by neuroactive peptide enantiomers. Nature 430, 235–240.

Taberean IV, Juranka P and Morris CE. (1999). Membrane stretch affects gating modes of a skeletal muscle sodium channel. Biophysical Journal 77, 758–774.

Takagi S, Miyazaki T, Moritani K, Miyoshi S, Furukawa Y, Ito S and Ogawa S. (1999). Gadolinium suppresses stretch-induced increases in the differences in epicardial and endocardial monophasic action potential durations and ventricular arrhythmias in dogs. Japanese Circulation Journal 63, 296–302.

Tidball JG, Cederdahl JE and Bers DM. (1991). Quantitative analysis of regional variability in the distribution of transverse tubules in rabbit myocardium. Cell Tissue Research 264, 293–298.

Trafford AW, Díaz ME and Eisner DA. (1998). Stimulation of Ca-induced Ca release only transiently increases the systolic Ca transient: measurements of Ca fluxes and sarcoplasmic reticulum Ca. Cardiovascular Research 37, 710–717.

Tseng G. (1992). Cell swelling increases membrane conductance of canine cardiac cells: evidence for a volume-sensitive Cl channel. American Journal of Physiology 262(31), C1056–C1068.

Tucci PJ, Bregagnollo EA, Spadaro J, Cicogna AC and Riberiro MC. (1984). Length dependence of activation studied in the isovolumic blood-perfused dog heart. Circulation Research 55, 59–66.

Vandebrouck C, Martin D, Colson-Van Schoor M, Debaix H and Gailly P. (2002). Involvement of TRPC in the abnormal calcium influx observed in dystrophic (mdx) mouse skeletal muscle fibers. Journal of Cell Biology 158, 1089–1096.

Vandenberg JI, Rees SA, Wright AR and Powell T. (1996). Cell swelling and ion transport pathways in cardiac myocytes. Cardiovascular Research 32, 85–97.

Vila Petroff MG, Kim SH, Pepe S, Dessy C, Marban E, Balligand J-L and Sollott SJ. (2001). Endogenous nitric oxide mechanisms mediate the stretch dependence of Ca^{2+} release in cardiomyocytes. Nature Cell Biology 3, 867–873.

von Anrep G. (1912). On the part played by the suprarenals in the normal vascular reactions of the body. Journal of Physiology 45, 307–317.

von Lewinski D, Stumme B, Fialka F, Luers C and Pieske B. (2004). Functional relevance of the stretch-dependent slow force response in failing human myocardium. Circulation Research 94, 1392–1398.

Wang N, Butler JP and Ingber DE. (1993). Mechanotransduction across the cell surface and through the cytoskeleton. Science 260, 1124–1127.

Ward M-L, Williams IA, Cooper PJ, Chu Y, Ju Y-K and Allen DG. (2008). Stretch-activated channels in the heart: contributions to length-dependence and to cardiomyopathy. Progress in Biophysics and Molecular Biology 97, 232–249.

White E. (2006). Mechanosensitive channels: therapeutic targets in the myocardium? Current Pharmaceutical Design 12, 3645–3663.

White E, Boyett MR and Orchard CH. (1995). The effects of mechanical loading and changes of length on single guinea-pig ventricular myocytes. Journal of Physiology 482, 93–107.

White E, Le Guennec J, Nigretto J, Gannier F, Argibay J and Garnier D. (1993). The effects of increasing cell length on auxotonic contractions; membrane potential and intracellular calcium transients in single guinea-pig ventricular myocytes. Experimental Physiology 78, 65–78.

Winegar BD, Haws CM and Lansman JB. (1996). Subconductance block of single mechanosensitive ion channels in skeletal fibres by aminoglycoside antibiotics. Journal of General Physiology 107, 433–443.

Xu SZ and Beech DJ. (2001). TrpC1 is a membrane-spanning subunit of store-operated Ca^{2+} channels in native vascular smooth muscle cells. Circulation Research 88, 84–87.

Yang XC and Sachs F. (1989). Block of stretch-activated ion channels in Xenopus oocytes by gadolinium and calcium ions. Science 243, 1068–1071.

Yeung EW, Head SI and Allen DG. (2003). Gadolinium reduces the short-term stretch-induced muscle damage in isolated mdx mouse muscle fibres. Journal of Physiology 552, 449–458.

Zabel M, Koller BS, Sachs F and Franz MR. (1996). Stretch-induced voltage changes in the isolated beating heart: importance of the timing of stretch and implications for stretch-activated ion channels. Cardiovascular Research 32, 120–130.

Zeng T, Bett GC and Sachs F. (2000). Stretch-activated whole cell currents in adult rat cardiac myocytes. American Journal of Physiology 278, H548–H557.

Zhang YH and Hancox JC. (2000). Gadolinium inhibits Na^+-Ca^{2+} exchanger current in guinea-pig isolated ventricular myocytes. British Journal of Pharmacology 130, 485–488.
Zhang YH, Youm JB, Sung HK, Lee SH, Ryu SY, Ho WK and Earm YE. (2000). Stretch-activated and background non-selective cation channels in rat atrial myocytes. Journal of Physiology 523, 607–619.

Chapter 7
Effects of Applied Stretch on Native and Recombinant Cardiac Na^+ Currents

Umberto Banderali, Robert B. Clark, Catherine E. Morris, Martin Fink and Wayne R. Giles

Abstract In the mammalian heart, electrical activity triggers and strongly modulates the contractions. In addition, under both physiological and pathophysiological conditions the mechanical activity of the heart may change tissue excitability, the action potential waveform and/or the pattern of conduction. In some cases, this mechanoelectrical feedback can alter the myocardium such that extrasystoles or rhythm disturbances are observed. It is thought that this sensitivity to mechanical perturbations is due to stretch-induced activation or alteration of ion channels which are expressed in the sarcolemma of cardiac myocytes. In the present manuscript, we describe studies on the effects of membrane stretch on the Na^+ channel alpha subunit, $Na_V1.5$ (which is predominant in the adult mammalian heart). Three different approaches have been utilized: (i) recordings of Na^+ current from adult rat ventricular myocytes, (ii) studies of currents due to this Na^+ channel transcript expressed in a *Xenopus laevis* oocyte preparation, and (iii) integration of these findings, following appropriate alterations of the descriptors for this Na^+ current in a mathematical model of the human ventricular action potential. The results demonstrate that in both native mammalian myocytes and in the heterologous expression system, applied stretch causes the Na^+ current to activate at more negative membrane potentials. Stretch also significantly increases the Na^+ current density. When these effects are incorporated into a mathematical model of the human ventricular action potential, myocyte excitability is enhanced, and there is also a significant increase in the maximum rate of rise in the action potential. Thus, in the mammalian heart the effects of stretch on conventional time- and voltage-dependent intrinsic Na^+ currents need to be taken into account when attempting to understand either the basis for, or the consequences of mechanoelectrical feedback.

U. Banderali (✉)
Faculty of Kinesiology, University of Calgary, Calgary, AB, Canada
e-mail: ubanderali@kin.ucalgary.ca

W.R. Giles (✉)
Faculty of Kinesiology, University of Calgary, Calgary, AB, Canada
e-mail: wgiles@ucalgary.ca

A. Kamkin, I. Kiseleva (eds.), *Mechanosensitivity of the Heart*, Mechanosensitivity in Cells and Tissues 3, DOI 10.1007/978-90-481-2850-1_7,

Keywords Na^+ current · Membrane stretch · Nav1.5 · Mathematical modeling · Human ventricle

7.1 Introduction

Cyclic contraction of the heart is a fundamental physiological event. Under most circumstances this excitation-contraction process is thought of as being unidirectional. That is, the contraction of the heart is triggered and modulated by the properties of the action potential, and the Ca^{2+} influx during the action potential initiates intracellular Ca^{2+} release, followed by myocyte shortening and/or tissue contraction. During relaxation, and in diastole, the myocardium essentially resets itself so that this excitation-contraction coupling cycle can be repeated. However, selected observations in the classical literature, and a number of detailed investigations within the last 20 years, have drawn attention to the fact that the mechanical activity of the heart itself can alter fundamental aspects of cardiac electrophysiology (Lab, 1998; Kohl et al., 2006). Such alterations can occur both during the relatively long cardiac action potential and in the following resting or diastolic period (Horner et al., 1996). These electrophysiological changes are thought to be due to mechanical effects being transmitted to ion channels. Collectively, they have been termed "contraction-excitation coupling". Here, these effects will be referred to as "mechanotransduction" or "stretch-induced effects" (Horner et al., 1996; Isenberg et al., 2003; Eijsbouts et al., 2004; Nishimura et al., 2006; Taggart and Lab, 2008). Not surprisingly, although these effects can be recorded under physiological conditions in the adult heart they are more prominent in the embryonic and developing heart. Contraction-excitation coupling is also thought to be important in pathophysiological settings including heart failure, selected cardiomyopathies, and atrial dilation (Manios et al., 2006; Otway et al., 2007; Kuijpers et al., 2007; Taggart and Lab, 2008).

Results published within the past decade have helped to define some aspects of the mechanisms and the consequences of mechanotransduction in the mammalian heart (Lab, 1998; Zhang et al., 2008a). Much of the initial focus and a significant amount of present effort is directed toward identification and characterization of so-called "stretch activated ion channels" (Hamill, 2006; Yaum et al., 2006). In general, these are thought to be nonselective cation channels having ion transfer functions (current-voltage curves) which are approximately linear (Kohl et al., 2006; Zhang et al., 2008b; Nishimura et al., 2008). However, in the fields of epithelial transport, neurophysiology, and cardiovascular sciences, it is now known that conventional ion selective time- and voltage-dependant channels undergo significant and selective biophysical changes in response to applied stretch (c.f. Tabarean and Morris, 2002; Laitko and Morris, 2004; Morris and Juranka, 2007a; Kalifa et al., 2007).

In the present manuscript, our main goal is to present and evaluate the possibility that the main type of Na^+ channel which is expressed in the mammalian heart can be modulated by applied stretch. To carry out this work conventional cell-attached patch clamp recordings have been made from adult myocytes and from a heterologous expression system, the *Xenopus* oocyte in which the predominant mammalian cardiac Na^+ channel alpha subunit isoform ($Na_V1.5$) was expressed. Effects of stretch were identified by developing and applying three part protocols in which parts 1 and 3 served as control recordings while part 2 involved selected and fixed amounts of applied negative pressure. This pressure was maintained for the entire duration of the data acquisition that was needed for obtaining an activation curve, or for delineation of the time course of the recovery of this transient inward Na^+ current (Morris and Juranka, 2007b). Data analysis and integration, and an illustration of the observed effects on excitability of the action potential of the mammalian ventricular myocyte was achieved by adapting the ten Tusscher model of the action potential of the human ventricular myocyte for this purpose (Ten Tusscher et al., 2004).

7.2 Effects of Stretch on Na^+ Currents

7.2.1 Modification by Stretch of Na^+ Current Density and Kinetics

The experimental work presented in this manuscript, and many of the other papers in this volume, provide evidence that either transient or steady-state stretch applied to biological cells can elicit significant electrophysiological responses. In our study the focus has been the human cardiac Na^+ channel. Our results demonstrate significant, reversible stretch-induced changes on the gating of this conductance. Specifically, stretch causes changes in the activation and inactivation processes which regulate opening and closing of this Na^+ channel. These changes are such that within a narrow window of time the Na^+ current is enhanced. The effects of membrane stretch on the biophysical properties of mammalian cardiac Na^+ channels, $Na_V1.5$, are illustrated in Fig. 7.1. Panels B and C in this figure show significant and reversible effects of stretch on both the kinetics and the peak size of Na^+ currents. In Panel B records were obtained from an adult rat ventricular myocyte. Very similar results were obtained from so-called recombinant Na^+ current records from *Xenopus* oocyte experiments (Panel C). In both sets of experiments the myocyte or oocyte was clamped at a holding potential approximately 75 mV more hyperpolarized than the resting potential. From this negative potential 50 ms depolarizing voltage clamp steps of either 95 (Panel B) or 90 (Panel C) mV were applied to activate macroscopic Na^+ currents. To assess reversibility of stretch effects on I_{Na}, this voltage clamp protocol was applied under control conditions, during stretch, and then after a recovery period in which no stretch was applied. The results consistently showed reversible increases in peak current as well as acceleration of the kinetics of activation and inactivation.

Our results also demonstrate that this enhancement in current is substantial only in a quite narrow range of membrane voltages. This range is very close to the negative slope region of the current-voltage relationship, and/or the voltage threshold for initiation of the action potential. In fact, the increase in peak current is

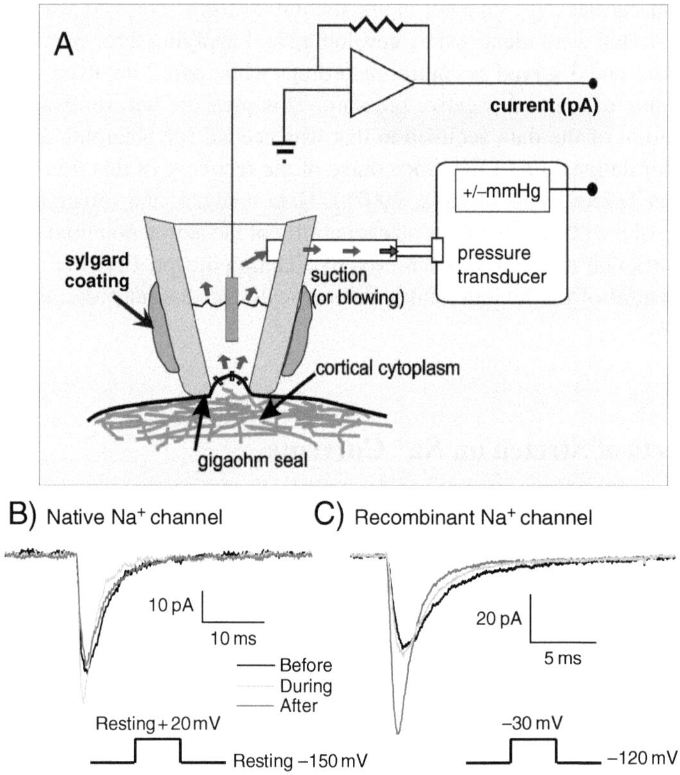

Fig. 7.1 **Panel A** Experimental setup for patch-clamp experiments involving applied stretch. The diagram illustrates a patch pipette sealed to the plasma membrane of a *Xenopus laevis* oocyte or a cardiac myocyte. The pipettes used in the experiments were pulled from thick-walled glass tubes and pipette tips were coated with sylgard to reduce electrical noise. A pressure transducer connected to the patch pipette allowed control and monitoring of the positive or negative pressure applied to the membrane patch. With the oocyte preparation, a high-K^+ solution containing (in mM) 89 KCl, 0.4 $CaCl_2$, 5 HEPES, 0.8 $MgCl_2$ (pH 7.4 with KOH) was used as extracellular medium. In the recording pipette, the KCl was substituted with 89 mM NaCl (pH adjusted to 7.4 by adding NaOH). $LaCl_3$ (250 μM) was added to the pipette solution to block the endogenous stretch activated cation channels. The La^{3+} ion at this concentration completely abolished the endogenous stretch activated currents and had only a small blocking effect of Nav1.5 (Bustamante 1987; Nathan et al., 1988). The human heart wild type Nav1.5 plasmid (pSP64T-hH1) was kindly provided by Dr. Al George (Makita et al., 1996). In the experiments in which native cardiac Na^+ currents were studied, adult rat ventricular myocytes were prepared for patch recordings as described previously (Ward and Giles, 1997). The same solution containing (in mM) 150 NaCl, 5 KCl, 2 $MgCl_2$, 1 $CaCl_2$, 10 HEPES, 5.5 glucose (pH 7.4 with NaOH) was used both in the bath and in the pipette. No ion channel inhibitors were added. All other experimental conditions were identical to those in the experiments on recombinant channels. (For more details on experimental methods, see Morris and Juranka, 2007b) **Panel B** Cell-attached patch-clamp recording of sodium currents from an adult rat cardiac myocyte. The membrane was held at a voltage 75 mV more hyperpolarized than its resting potential (approx. –80 mV). A rectangular voltage clamp step to 20 mV more depolarized than resting potential elicited a transient inward current. Three current records are superposed: before application of negative pressure to the pipette (*black*), during application of –30 mmHg suction

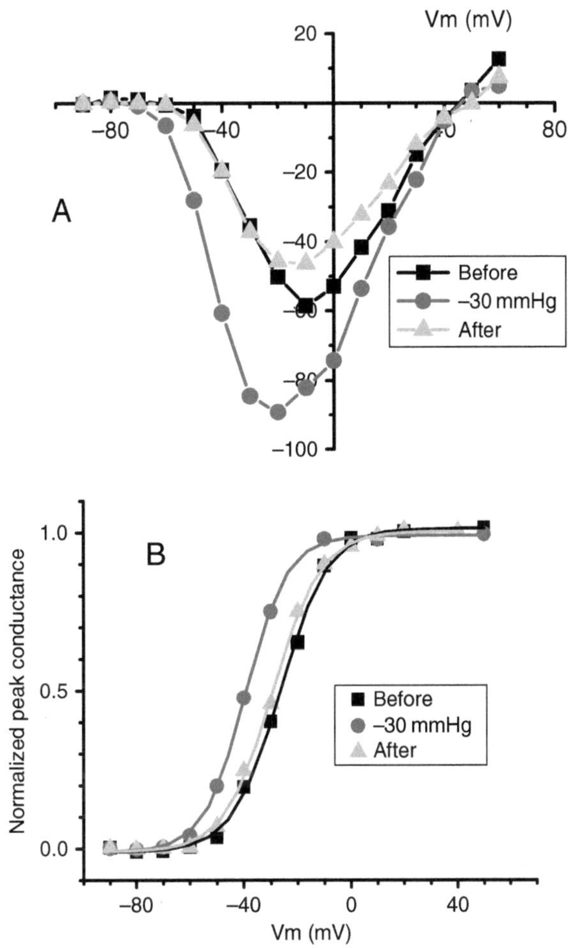

Fig. 7.2 **Panel A** Peak current vs. voltage (I/V) relations for the recombinant Nav1.5 channel recorded (i) before application of negative pressure to the patch pipette, (ii) during application of –30 mmHg suction and (iii) after return to the non stretch conditions. Note that membrane stretch induces a reversible increase in peak currents which is more evident at the "foot" of the I/V curve, and throughout the voltage range of increasing conductance. **Panel B** Steady state activation curve for I_{Na}. Note that stretch shifts this conductance vs. voltage relation approx 11 mV in the hyperpolarizing direction. These recordings were made using 0.25 mM La^{3+} ion in the patch pipette to inhibit the oocyte endogenous stretch activated channels. The conductance values in the steady-state activation curve were obtained dividing the peak currents in the I–V curve by the Na^+ ion driving force and normalizing the results to the maximum

Fig. 7.1 (*dark grey*) and after return to non stretch conditions (*grey*). The purpose of Before-During-After protocols was to ensure that stretch effects on I_{Na} were reversible. The color code will be the same in all figures. **Panel C** Cell attached patch clamp recording of sodium currents on a *Xenopus laevis* oocyte which expressed the human cardiac Nav1.5. Na^+ current holding potential was –120 mV and a rectangular voltage pulse to –30 mV elicited the transient current. The same experimental protocol as in **Panel B** was performed. The results obtained on the recombinant Nav1.5 channel were very similar to those observed when studying native sodium currents in cardiac myocytes. Specifically, a reversible increase in peak currents accompanied by reversible acceleration of both activation and inactivation kinetics of the channel in response to stretch

caused by changes in the absolute rates of activation and inactivation. The consequence of these changes in Na^+ kinetics is most pronounced at membrane voltages near the foot of the I–V relation. This effect can be seen in the I–V relation presented in Fig. 7.2a. As shown near the foot of the I–V relation, stretch produces an approximately three fold increase in peak current. In contrast, there is very little effect at more depolarized voltages when the Na^+ conductance is near maximum even though stretch still accelerates the current kinetics (as explained in Morris and Juranka, 2007b). This effect, an increase in transient current which is limited to the membrane potential range where the current is first activated and which saturates when conductance is maximal, is more likely explained by changes in kinetics than by a stretch-induced increase in the number of Na^+ channels in the patch. Consistent with this hypothesis, and as shown in Fig. 7.2b, the steady-state activation curve for I_{Na} is shifted approximately 10 mV toward more hyperpolarized membrane voltages during stretch.

7.2.2 Changes in Biophysical Properties of I_{Na} Depend on Amount of Stretch

Figure 7.3 consists of a family of I_{Na} recordings from a *Xenopus* oocyte in which six different amounts of negative pressure were applied to determine whether the effects of stretch shown in Fig. 7.2 were graded. The voltage clamp protocol is described in the figure legend. All of these changes were reversible with the first significant alterations being observed at approximately –5 mmHg pressure, and maximal effects obtained at approximately –30 mmHg. In this figure all current traces have been normalized to the value of peak current with no stretch. The superimposed records illustrate the changes in both activation and inactivation kinetics. Note that

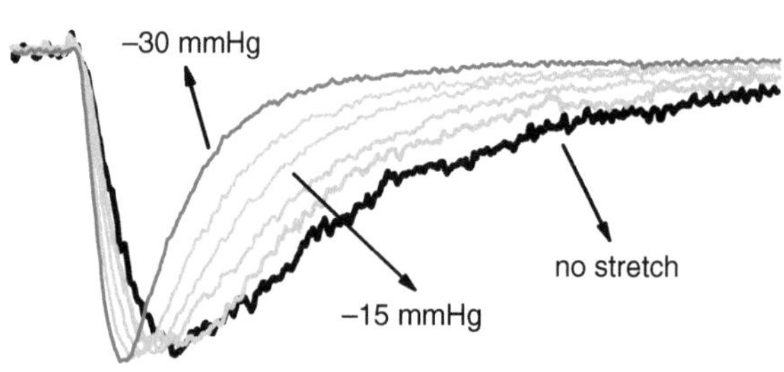

Fig. 7.3 The stretch-induced effects on recombinant Nav1.5 currents appear to be proportional to the negative pressure which is applied. Six recordings from the same patch are plotted each different applied negative pressures, obtained under 0 (*black*), –5, –10, –15, –20 and –30 mmHg respectively: (colors from *light grey* to *grey*). The holding potential was –120 mV and a rectangular voltage clamp step to –30 mV elicited each transient increased current. The amplitude of each current peak has been normalized to the control (no stretch) current in order to illustrate the effects of progressively increasing stretch on the Na^+ channel kinetics. Note that gradual increases in negative pressure accelerate both activation and inactivation of Nav1.5 current

both are accelerated by stretch. In most experiments, negative pressures larger than –40 mmHg caused seal rupture.

7.3 Pathophysiological Implications

7.3.1 Sodium Channels and Stretch-Induced Arrhythmias

Previous results have demonstrated that the antiarrhythmic compound, flecainide, can alter the response of isolated ventricular myocytes to stretch (Kalifa et al., 2007). This effect can derive from its ability to block Na^+ current. Similarly, gadolinium has been reported to be a blocker of stretch-induced arrhythmias (Li and Baumgarten 2001). In this case too, the antiarrhythmic effect may be ascribed to the inhibitory action of gadolinium on Na^+ currents.

Our voltage clamp findings suggest that as a consequence of stretch the threshold for initiation of the action potential moves in the hyperpolarizing direction, becoming closer to the resting potential. Preliminary computational work, which is illustrated in the next section, is consistent with this. A second prediction is that the larger net inward Na^+ current could result in increased conduction velocity, at least within tissue segments experiencing relatively large stretch or mechanical displacement. However, this supposition remains to be documented.

7.3.2 Mathematical Simulations of the Human Ventricular Action Potential

The pattern of results illustrated in Figs 7.2–7.3 was observed consistently in both of the types of experimental preparations utilized in this study. These findings raise fundamental questions concerning the functional significance of stretch-induced changes in the biophysical properties of I_{Na}. Mathematical modeling can, in principle, be useful in integrating these findings and illustrating the consequences of either a set of assumptions, or in this case a subset of experimental findings. Figure 7.4 shows the output of simulations of the human ventricular action potential waveform carried out using the ten Tusscher model of the human ventricular action potential, as described in our recent papers (Fink et al., 2006, 2008). When the mathematical descriptors of the Na^+ current in this model are changed from control values to values consistent with our experimental findings, significant changes in excitability were observed. Specifically as shown in Fig. 7.4b the maximum rate of rise of the action potential increased substantially and the latency between the applied stimulus and the regenerative depolarization was reduced. Consistent with this effect, and as expected, the strength duration curve for the membrane action potential was altered in such a way that the simulated stretch-induced changes would be expected to reduce the threshold for excitation of the action potential and therefore increase the excitability of the ventricular myocardium.

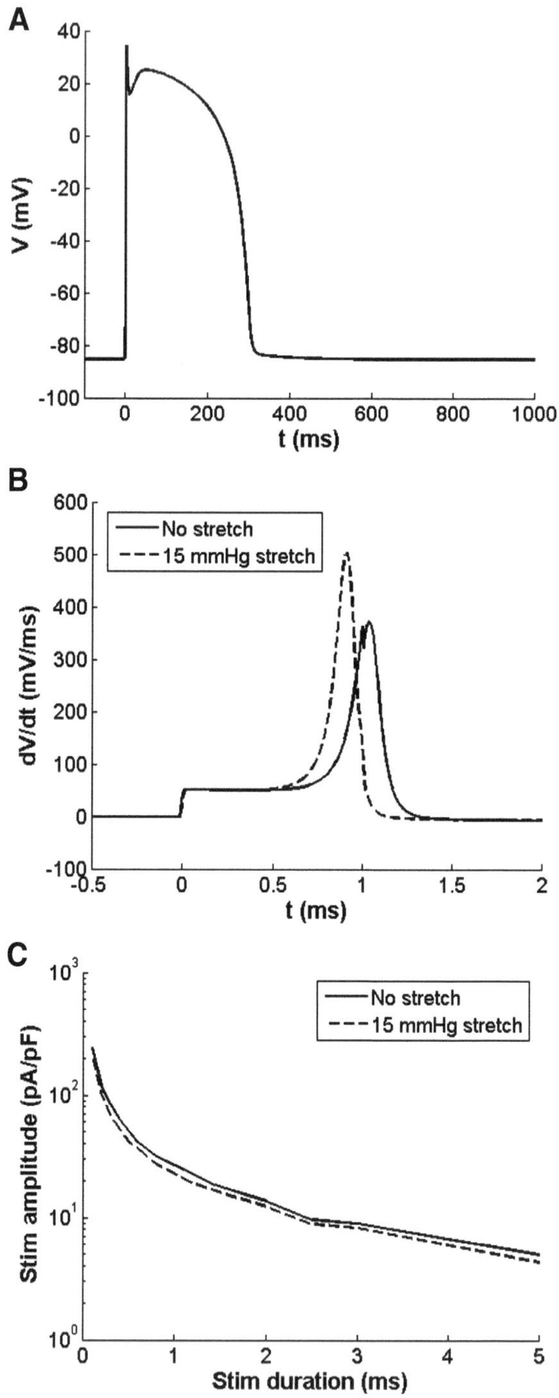

Fig. 7.4 Mathematical simulation of the changes in excitability and/or action potential waveform in a human ventricle myocyte. The changes in the size and kinetics of I_{Na} (identified in Figs. 7.2–7.3) were incorporated in the ten Tusscher model of the human ventricular action potential and action potentials were elicited at 1 Hz. These changes included: (i) a hyperpolarizing shift of the m-gate (i.e., activation) kinetics by 0.5 mV per 1 mmHg of stretch and (ii) acceleration of all gating kinetics by a factor of 1.0. In **Panel A,** two action potentials are superimposed. The control action potential and the action potential computed assuming –15 mmHg negative pressure are virtually identical. **Panel B** Plot of the first derivative of the voltage with respect to time, i.e., the maximum "upstroke velocity". As expected from the underlying increase in I_{Na} the dV/dt max increases with stretch, which is 371mV/ms for the control and 503 mV/ms in the presence of stretch. Note also that the activation starts earlier for stretch. **Panel C** Plot of the excitability thresholds: the stimulus amplitudes necessary to elicit an action potential vs. the stimulus duration on a semilogarithmic scale. The two curves are quite distinct. Note that in the control case an approx. 1.16 times larger stimulus is required than during applied stretch

7.3.3 Experimental and Computational Limitations

Our computations are made on the basis of non conducted or membrane action potentials. Moreover, the expression levels of the Na^+ current may not be uniform within the ventricle either in the base-to-apex axis, or in the transmural aspect of the left ventricle (Ashamalla et al., 2001; Antzelevitch and Belardinelli 2006; Brette and Orchard 2006; Stones et al., 2007). It is also known that the stress-strain relationships in both the left ventricle and the right ventricle vary with time during the cardiac cycle (Nishimura et al., 2006; Hoshijima 2006).

7.3.4 Possible Chamber-Specific Effects of Stretch on I_{Na}, in the Heart

Identification of significant electrophysiological effects of stretch on the biophysical properties of the Na^+ channel isoform which is predominant in the mammalian heart should also be considered in the context of chamber-specific effects within the heart. Although chamber pressure differentials are much smaller in the right ventricle than in the left, the relatively thin wall of the right ventricle may result in it being more susceptible to stretch under physiological conditions including exercise onset (Reddy et al., 2007), or in the setting of pulmonary hypertension. Somewhat similarly both the right and the left atria are highly deformable compared to the left ventricle. This property and the fact that the resting potential in the atrium is relatively depolarized, may combine to make the stretch-sensitive effects on the Na^+ current very important in atrial physiology/pathophysiology (c.f. Kneller et al., 2005). The relatively depolarized resting potential of atrial tissue (or ventricular tissue in the setting of hypoxic or ischaemic challenge) could also result in the effects of stretch on the time course of reactivation of the Na^+ current becoming relevant.

7.3.5 Stretch Alters the Kinetics of Reactivation of Cardiac Na^+ Current

A well known property of mammalian Na^+ channels is that both steady state inactivation and the kinetics of reactivation (recovery) show strong voltage-dependence within the range of membrane potentials of physiological interest. These properties plays an important role in modulating the refractory period in the mammalian heart. It therefore was of interest to explore the effects of stretch on reactivation or recovery kinetics of I_{Na}. Figure 7.5 consists of data obtained from the *Xenopus* oocyte preparation. In these experiments the membrane potential was held at either –120 mV, or –80 mV and two short (15 ms) depolarizing voltage clamp pulses were applied at selected inter-pulse intervals. The first pulse served to activate the Na^+ current and allow it to inactivate. As a result, the peak amplitude of the current during the second pulse, taken as a ratio to the current elicited by the first pulse provides a measurement of the relative amount of recovery from inactivation. Raw data illustrating the current records at a holding potential of –120 and –80 mV are shown in Panels 5A and 5B respectively. Panel C in Fig. 7.5 consists of aggregate

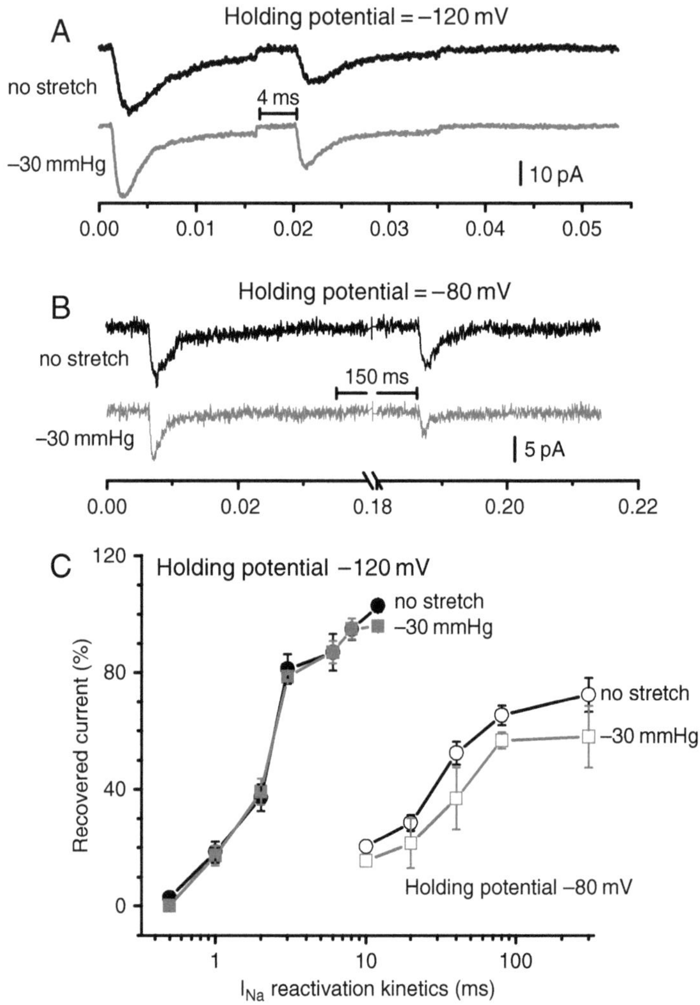

Fig. 7.5 Effects of stretch on I_{Na} reactivation. **Panel A** Recordings from a reactivation protocol when Nav1.5 was studied at a holding potential of –120 mV. Two successive rectangular voltage clamp steps to –40 mV separated by a time interval which varied between 1 and 15 ms were applied. In the example shown here, an interval of 4 ms was used. The black trace is the recording before application of negative pressure. Maximum current recovery was approx. 53%. The grey trace shows the results when the experiment was repeated with a –30 mmHg stretch. Maximum current recovery was approx 54%. **Panel B** Recordings of a reactivation protocol of Nav1.5 from a holding potential of –80 mV. Two successive rectangular voltage clamp steps to –20 mV were separated by a time interval which varied between 10 and 300 ms. In the example shown here, the time interval is 150 ms. Black denotes the recording before application of negative pressure. Maximum current recovery was 82%. Grey shows results from an experiment repeated with a –30 mmHg stimulus. Peak current recovery was 40%. **Panel C** Semilog plot of the percentage of current recovered vs. recovery time for Nav1.5 at holding potentials of –120 (*filled symbols*) and –80 (*open symbols*) mV. Note that at holding potential –120 mV, recovery is much faster than at –80 mV and at –120 mV stretch has no effect ($n = 9$). In contrast, at –80 mV stretch (–30 mmHg) appears to slow the recovery from inactivation ($n = 2$). In this graph, the stretch data points are averages of the data before and after stretch

data which illustrates the entire time-course of reactivation at these two holding potentials in the absence, and in the presence of significant (30 mmHg) stretch. Note that at the very hyperpolarized holding potential there is little effect of stretch on recovery kinetics; however, the preliminary data obtained at –80 mV suggest that stretch may slow the recovery process.

Previous findings agree with our own in showing that extreme hyperpolarized membrane voltages stretch has no effect on the time-course of reactivation of the Na^+ current (Li and Baumgarten, 2001). However, and in contrast, when the holding potential is in the range of that of the resting potential of the atrium, applied stretch may alter reactivation (and thus the refractory period) quite substantially.

7.4 What is the Basis for this Mechanotransduction?

Our results, and previous work, raise questions concerning the microanatomical features of the mechanotransduction system or mechanism which is responsible for the biophysical changes which we have observed. It is known that at least one of the classical components of mechanotransduction sensors, the integrin family of multifunctional proteins, are expressed in surface indentations or cavaeolae in mammalian hearts (Hoshijima, 2006; Dyachenko et al., 2008). It is also known that a significant fraction or subpopulation of cardiac Na^+ channels are expressed or targeted to these cavaeolae (Yarbrough et al., 2002; Vatta et al., 2006). This microdomain signaling complex may serve as an important element in mechanotransduction in the myocytes of mammalian heart. The innervation of the heart includes intracardiac ganglia which may also serve a role in mechanotransduction. Interestingly, the nerve cell bodies in these ganglia appear to express low levels of the sodium channel isoform, $Na_V1.5$ (Scornik et al., 2006).

A significant missing data set needed for forming a working hypothesis concerning mechanotransduction in the ventricle could be obtained if semiquantitative mechanical measurements and manipulations could be combined with measurements of membrane potential and/or underlying ionic currents. Recently, studies from which some of these results have been obtained have been published. In general, it would appear that a 15–20% lengthening of an isolated myocyte results in a significant depolarization of the resting potential in the ventricular myocardium (Nishimura et al. 2008). It has even been suggested that this amount of applied stretch can alter the permeability properties of the stretch-activated, nonselective cation channels which are presumed to be activated. However, this interpretation assumes that nonselective cation channels, as opposed to specific alterations in conventional time- and voltage-dependent conductances, are the main targets for applied stretch in excitable membranes (von Lewinski et al., 2003; Lin et al. 2007). Table 7.1 summarizes some of the effects of stretch and/or shear stress on different types of recombinant voltage gated channels.

Table 7.1 Mechanic stimuli can affect voltage gated channels (VCGs). Studies on recombinant VGCs expressed in heterologous systems have shown that both stretch (**A**) and shear stress (**B**) can modify the behaviour of these channels

Channel type	Details about channel	Effects of stretch	Comments on stimulus	References
A. Reversible effects of membrane stretch on recombinant VGCs				
Kv1	Shaker-IR (Shaker, fast inactivation removed)	Increased NP_{open} (macroscopic and single channel I_K), accelerated activation	Oocyte patch stretch	Gu et al. (2001)
Kv1-5aa	Shaker-IR-5aa: rate limiting step = independent activation motions of V-sensors	Accelerated activation and inactivation (same-fold), left-shifted g/V	Oocyte patch stretch	(Laitko and Morris 2004)
Kv-ILT	Shaker-IR-ILT: rate limiting step = concerted pre-pore opening	Slower activation, smaller steady-state I_K, right-shifted g/V	Oocyte patch stretch	Laitko et al. (2006)
Kv3.2	Shaw2 mutant F335A	Steady-state I_K increases, no change of activation speed	Oocyte patch stretch	Laitko et al. (2006)
Cav2.2	N-type (α_{1B} subunit co-expressed with several auxiliaries)	Peak I_{Ba} increases transiently, no change of activation speed, inactivation accelerates	Whole-cell inflation by positive pressure HEK cell	Calabrese et al. (2002)
Cav2.2	N-type (α_{1B} subunit only)	Unitary current NP_{open} increases	Patch stretch HEK cell	Calabrese et al. (2002)
Cav1.2	L-type	Peak I_{Ba} increases	Increased intracell. Pressure	Farrugia et al. (1999)

Table 7.1 (continued)

Channel type	Details about channel	Effects of stretch	Comments on stimulus	References
HCN2	Homotetramers	Depolarization-induced deactivation and hyperpolarization-induced activation both accelerate	Oocyte patch stretch	Lin et al. (2007)
Nav1.5	α subunit only	Activation and inactivation accelerate to the same extent	Oocyte patch stretch	Morris and Juranka (2007b)
B. Effects of shear flow and osmotic swelling/shrinking on VGCs				
Kv7.1 KCNQ1	Co-expressed in oocytes with KCNE β-subunit and aquaporin AQP1	Reversible increase of steady-state current.	Two electrode clamp, hypo-osmotic swelling	Grunnet et al. (2003)
Kv7.1 KCNQ1	LQT1-related mutant compared to WT expressed in HEK cells	Faster activation, left-shift of g/V	Hypotonic swelling	Otway et al. (2007
Nav1.5	SCN5A α subunit in human interstitial intestinal cells	Overall increase in transient currents elicited by flow	Whole-cell recording Stimulus: no flow-flow- no flow	Strege et al. (2003)
Cav1.2	L-type	Overall increase in transient currents elicited by flow	Whole-cell recording Stimulus: no flow-flow- no flow	Farrugia et al. (1999)
HCN2	Homotetramers coexpressed in oocytes with aquaporin AQP1	Increase in currents with no modification of kinetics	Two electrode clamp, hypoosmotic or isosmotic swelling	Calloe et al. (2005)

Our results, and many of the other findings in this volume, provide meaningful starting points for beginning to attempt to understand the roles of the human cardiac Na^+ current in triggering or maintaining cardiac rhythm disorders. These include some types of genetic mutations known to target cardiac Na^+ channels (Bennett et al., 1995; Dumaine et al., 1996; Clancy and Rudy 1999; Ruan et al., 2007).

Acknowledgements The Giles laboratory is funded by the Canadian Institutes of Health Research and the Heart and Stroke Foundation of Canada. In addition, W. Giles holds a Medical Scientist Award from the Alberta Heritage Foundation for Medical Research. The Morris laboratory is funded by the Canadian Institutes of Health Research and the Heart and Stroke Foundation of Canada.

References

Antzelevitch C, Belardinelli L (2006) The role of sodium channel currents in modulating transmural dispersion of repolarization and arrhythmogenesis. J. Cardiovasc Electrophysiol 17: 579–585.

Ashamalla SM, Navarro D, Ward CA (2001) Gradient of sodium current across the left ventricular wall of adult rat hearts. J Physiol 536:439–443.

Bennett PB, Yazawa K, Makita N, George AL Jr (1995) Molecular mechanism for an inherited cardiac arrhythmia. Nature 376:683–685.

Brette F, Orchard CH (2006) Density and sub-cellular distribution of cardiac and neuronal sodium channel isoforms in rat ventricular myocytes. Biochem Biophys Res Commun 348:1163–1166.

Bustamante JO (1987) Modification of sodium currents by lanthanum and lanthanide ions in human heart cells. Can J Physiol Pharmacol 65:591–597.

Clancy CE, Rudy Y (1999) Linking a genetic defect to its cellular phenotype in a cardiac arrhythmia. Nature 400:566–569.

Calabrese B, Tabarean IV, Juranka PF, Morris CE (2002) Mechanosensitivity of N-type calcium channel current. Biophys J 83:2560–2574

Calloe K, Elmedyb P, Olesen SP, Jorgensen NK, Grunnet M (2005) Hyposmotic cell swelling as a novel mechanism for modulation of cloned HCN2 channels. Biophys J 89:2159–2169.

Dumaine R, Wang Q, Keating MT, Hartmann HA, Schwartz PJ, Brown AM, Kirsch GE (1996) Multiple mechanisms of Na^+ channel – linked long-QT syndrome. Circ Res 78: 916–924.

Dyachenko V, Christ A, Gubanov R, Isenberg G (2008) Bending of z-lines by mechanical stimuli: An input signal for integrin dependent modulation of ion channels? Prog Biophys Mol Biol 97:196–216.

Eijsbouts SC, Houben RP, Blaauw Y, Schotten U, Allessie MA (2004) Synergistic action of atrial dilation and sodium channel blockade on conduction in rabbit atria. J Cardiovasc Electrophysiol 15:1453–1461.

Farrugia G, Holm AN, Rich A, Sarr MG, Szurszewski JH, Rae JL (1999) A mechanosensitive calcium channel in human intestinal smooth muscle cells. Gastroenterology 117:900–905

Fink M, Giles WR, Noble D (2006). Contributions of inwardly-rectifying K^+ currents to repolarization assessed using mathematical models of human ventricular myocytes. Philos Transact A Math Phys Eng Sci 364:1207–1222.

Fink M, Noble D, Giles WR (2008). Contributions of HERG K^+ currents to repolarization of the human ventricular action potential. Prog Biophys Mol Biol 96:357–376.

Grunnet M, Jespersen T, MacAulay N, Jorgensen NK, Schmitt N, Pongs O, Olesen SP, Klaerke DA (2003) KCNQ1 channels sense small changes in cell volume. J Physiol 549:419–427.

Gu CX, Juranka PF, Morris CE (2001) Stretch-activation and stretch-inactivation of Shaker-IR, a voltage-gated K^+ channel. Biophys J 80:2678–2693.

Hamill OP (2006) Twenty odd years of stretch-sensitive channels. Pflugers Arch 453:333–351.

Horner SM, Dick DJ, Murphy CF, Lab MJ (1996) Cycle length dependence of the electrophysiological effects of increased load on the myocardium. Circulation 94:1131–1136.

Hoshijima M (2006) Mechanical stress-strain sensors embedded in cardiac cytoskeleton: Z disk, titin, and associated structures. Am J Physiol Heart Circ Physiol 290: H1313–H1325.

Isenberg G, Kazansky V, Kondratev D, Gallitelli MF, Kiseleva I, Kamkin A (2003) Differential effects of stretch and compression on membrane currents and $[Na^+]_c$ in ventricular myocytes. Prog Biophys Mol Biol 82:43–56.

Kalifa J, Bernard M, Gout B, Bril A, Cozma D, Laurent P, Chalvidan T, Deharo JC, Djiane P, Cozzone P, Maixent JM (2007) Anti-arrhythmic effects of I(Na), I(Kr), and combined I(Kr-I(CaL) blockade in an experimental model of acute stretch-related atrial fibrillation. Cardiovasc Drugs Ther 21:47–53.

Kneller J, Kalifa J, Zou R, Zaitsev AV, Warren M, Berenfeld O, Vigmond EJ, Leon LJ, Nattel S, Jalife J (2005) Mechanisms of atrial fibrillation termination by pure sodium channel blockade in an ionically-realistic mathematical model. Circ Res 96:e35–e47.

Kohl P, Bollensdorff C, Garny A. (2006) Effects of mechanosensitive ion channels on ventricular electrophysiology: Experimental and theoretical models. Exp Physiol 91:307–321.

Kuijpers NH, ten Eikelder HM, Bovendeerd PH, Verheule S, Hilberts PA (2007) Mechanoelectric feedback leads to conduction slowing and block in acutely dilated atria: a modeling study of cardiac electromechanics. Am J Physiol Heart Circ Physiol 292:H2832–H2853.

Lab MJ (1998) Mechanosensitivity as an integrative system in heart: an audit. Prog Biophys Mol Biol 71:7–27.

Laitko U, Juranka PF, Morris CE (2006) Membrane stretch slows the concerted step prior to opening in a Kv channel. J Gen Physiol 1237:687–701

Laitko U, Morris CE (2004) Membrane tension accelerates rate-limiting voltage-dependent activation and slow inactivation steps in a Shaker channel. J Gen Physiol 123:135–154.

Li G-R, Baumgarten CM (2001) Modulation of cardiac Na^+ current by gadolinium, a blocker of stretch-induced arrhythmias. Am J Physiol Heart Circ Physiol 280:H272–279.

Lin W, Laitko U, Juranka PF, Morris CE (2007) Dual stretch responses of mHCN2 pacemaker channels: accelerated activation, accelerated deactivation. Biophys J 92:1559–1572.

Makita N, Bennett PB Jr, George AL Jr (1996) Multiple domains contribute to the distinct inactivation properties of human heart and skeletal muscle Na^+ channels. Circ Res 78: 244–252.

Manios EG, Mavrakis HE, Kanoupakis EM, Kallergis EM, Kafarakis PK, Vardas PE (2006) Evidence of mechanoelectric feedback in the atria of patients with atrioventricular nodal reentrant tachycardia. J Interv Card Electrophysiol 16:51–57.

Morris CE, Juranka PF (2007a) Lipid stress at play: mechanosensitivity of voltage-gated channels. Curr Top Membr 59:298–338.

Morris CE, Juranka PF (2007b) Nav channel mechanosensitivity: activation and inactivation accelerate reversibly with stretch. Biophys J 93:822–833.

Nathan RD, Kanai K, Clark RB Giles WR (1988) Selective block of calcium current by lanthanum in single bullfrog atrial cells. J Gen Physiol 91:549–572.

Nishimura S, Kawai Y, Nakajima T, Hosoya Y, Fujita H, Katoh M, Yamashita H, Nagai R, Sugiura S (2006) Membrane potential of rat ventricular myocytes responds to axial stretch in phase, amplitude and speed-dependent manners. Cardiovasc Res 72:403–411.

Nishimura S, Seo K, Nagasaki M, Hosoya Y, Yamashita H, Fujita H, Nagai R, Sugiura S (2008) Responses of single-ventricular myocytes to dynamic axial stretching. Prog Biophys Mol Biol 97:282–297.

Otway R, Vandenberg JI, Guo G, Varghese A, Castro ML, Liu J, Zhao J, Bursill JA, Wyse KR, Crotty H, Baddeley O, Walker B, Kuchar D, Thorburn C, Fatkin D (2007) Stretch-sensitive KCNQ1 mutation A link between genetic and environmental factors in the pathogenesis of atrial fibrillation? J Am Coll Cardiol 49:578–586.

Reddy PR, Reinier K, Singh T, Mariani R, Gunson K, Jui J, Chugh SS (2007) Physical activity as a trigger of sudden cardiac arrest: The Oregon Sudden Unexpected Death Study. Int J Cardiol doi:10.1016/j.ijcard.2007.10.024.

Ruan Y, Liu N, Bloise R, Napolitano C, Priori SG (2007) Gating properties of SCN5A mutations and the response to mexiletine in long-QT syndrome type 3 patients. Circulation 116: 1137–1144.

Scornik FS, Desai M, Brugada R, Guerchicoff A, Pollevick GD, Antzelevitch C, Perez GJ (2006) Functional expression of "cardiac-type" Nav1.5 sodium channel in canine intracardiac ganglia. Heart Rhythm 3:842–840.

Stones R, Calaghan SC, Billeter R, Harrison SM, White E (2007) Transmural variations in gene expression of stretch modulated proteins in the left ventricle. Pfluegers Arch 454:545–549

Strege PR, Ou Y, Sha L, Rich A, Gibbons SJ, Szurszewki JH, Sarr MG, Farrugia G (2003) Sodium current in human intestinal interstitial cells of Cajal. Am J Physiol Gastrointest Liver Physiol 285:G1111–G1121.

Tabarean IV, Morris CE (2002) Membrane stretch accelerates activation and slow inactivation in Shaker channels with S3-S4 linker deletions. Biophys J 82:2982–2994.

Taggart P, Lab M (2008) Cardiac mechano-electric feedback and electrical restitution in humans. Prog Biophys Mol Biol 97:452–460.

Ten Tusscher KHWJ, Noble D, Noble PJ, Panfilov AV (2004) A model for human ventricular tissue. Am J Physiol (Heart Circ Physiol) 286:H1573–H1589.

Vatta M, Ackerman MJ, Ye B, Makielski JC, Ughanze EE, Taylor EW, Tester DJ, Balijepalli RC, Foell JD, Li Z, Kamp TJ, Towbin JA (2006) Mutant caveolin-3 induces persistent late sodium current and is associated with long-QT syndrome. Circulation 114:2104–2112.

von Lewinski D, Stumme B, Maier LS, Luers C, Bers DM, Pieske B (2003) Stretch-dependent slow force response in isolated rabbit myocardium is Na^+ dependent. Cardiovasc Res 15:1052–1061.

Ward CA, Giles WR (1997) Ionic mechanism of the effects of hydrogen peroxide in rat ventricular myocytes. J Physiol 500:631–642.

Yarbrough TL, Lu T, Lee HC, Shibata EF (2002) Localization of cardiac sodium channels in caveolin-rich membrane domains: Regulation of sodium current amplitude. Circ Res 90: 443–449.

Yaum JB, Han J, Kim N, Zhang YH, Kim E, Joo H, Hun Leem C, Joon Kim S, Cha KA, Earm YE (2006) Role of stretch-activated channels on the stretch-induced changes of rat atrial myocytes. Prog Biophys Mol Biol 90:186–206.

Zhang Y, Sekar RB, McCulloch D, Tung L (2008a) Cell cultures as models of cardiac mechano-electric feedback. Prog Biophys Mol Biol 97:367–382.

Zhang Y, Youm JB, Earm YE (2008b) Stretch-activated non-selective cation channels: a casual link between mechanical stretch and atrial natriuretic peptide secretion. Prog Biophys Mol Biol 98:1–9.

Chapter 8
Mechanosensitive Alterations of Action Potentials and Membrane Currents in Healthy and Diseased Cardiomyocytes: Cardiac Tissue and Isolated Cell

Ilya Lozinsky and Andre Kamkin

Abstract Several electrophysiological alterations in the heart, which were ascribed to mechanoelectric feedback have been reported. First of all, they include changes in mechano-gated channels, mechanosensitive whole-cell currents which lead to membrane depolarization which is equivalent to a decrease in the resting membrane potential and elicited stretch-induced depolarizations, that appear during repolarization phase of cardiomyocyte action potential. Stretch-induced depolarization during action potentials provoke extra-action potentials when the stretch-induced depolarizations reach a threshold potential. Mechano-gated channels and mechanosensitive whole-cell currents are the cellular meachanisms underlying this phenomenon. In this review we discuss some open questions about mechanosensitive ionic currents in freshly isolated single cardiomyocytes. We will demonstrate certain methods of direct mechanical deformation of isolated cardiomyocytes for the purpose of electrophysiological investigation, including different experimental approaches to application of stretch and compression to pressure the cardiomyocytes. It is necessary to note that brick-like isolated cardiomyocytes stick to the bottom of the perfusion chamber in two different positions: edgewise, staying on the narrow side, or broad-wise. Partly these different positions of cells define the cell reaction to deformation. The reaction to stretch is identical in cardiomyocytes, occupying both positions (edgewise and broad-wise). However, the reaction to compression is different and is determined by the position of a cell. We demonstrate the possibility of simultaneous recording of mechano-gated single channels (in cell-attached mode) and mechanosensitive whole-cell currents during direct deformation of the whole cell. We discuss the results of stretch and compression of freshly isolated atrial cardiomyocytes from healthy and diseased animals and humans. Isolated cardiomyocytes respond to stretch with membrane depolarization, prolongation of their action potential (AP) and extra-APs that correlated with the amplitude of a non-selective stretch-activated current (I_{SAC}). At negative potentials, I_{SAC} is negative

A. Kamkin (✉)
Department of Fundamental and Applied Physiology, Russian State Medical University, Moscow, Russia
e-mail: Kamkin.A@g23.relcom.ru;andre.kamkin@gmail.com

A. Kamkin, I. Kiseleva (eds.), *Mechanosensitivity of the Heart*,
Mechanosensitivity in Cells and Tissues 3, DOI 10.1007/978-90-481-2850-1_8,

and carried by a transmembrane influx of Na^+ ions. In this review we discuss some of the recent advances from intracellular recordings of the bioelectrical activity of cardiomyocytes during mechanical stretch of healthy and diseased tissues from animals and humans. The sensitivity of the AP to mechanical stretch was significantly increased in hypertrophied myocardium, and this could be related to the expression of SACs. We suppose that they are the basic findings that may explain mechanism of some arrhythmias and fibrillation.

Keywords Cardiomyocytes · Mechano-gated channel · Mechanosensitive whole-cell currents · Action potential · Resting potential · Active force · Resting force · Myocardial hypertrophy

8.1 Introduction

Mechanoelectric feedback (Lab, 1968, 1996, 1998) describes the situation in which a mechanical stimulus is transduced into an electrical signal. Several electrophysiological alterations in the heart, which were ascribed to mechano-electric feedback have been reported. They include: (1) changes of the monophasic action potential duration, (2) a decrease in the resting membrane potential, (3) a decrease in monophasic action potential amplitude, (4) development of stretch-induced depolarizations, (5) ectopic beats originating from stretch induced depolarization (which reached the threshold potential for depolarization) in myocardium sustaining the stretch, (6) other changes of the cardiac potentials (Franz, 2000), and changes in intercellular interaction through gap junctions (Kamkin et al., 2005a). However, in most studies of mechano-electric feedback, the monophasic action potential recording method was used (Franz, 2000), since it is extremely difficult to maintain a glass-microelectrode positioned stably within a cell of a beating heart and even harder to ascertain the effects of simultaneously imposed mechanical perturbations (Franz, 1996). Dudel and Trautwein (1954) stretched papillary muscle while measuring the microelectrode action potential (AP).

In this review we will discuss some of the recent advances from intracellular recordings of the bioelectrical activity of cardiomyocytes during mechanical stretch of healthy and diseased tissues from animals and human. Furthermore, we will discuss some open questions about mechano-sensitive ionic currents in isolated cardiomyocytes.

8.2 Methods of Direct Mechanical Deformation of Isolated Cardiomyocytes for Electrophysiological Registration

Direct mechanical deforming of a cell is the simplest form of mechanical stimulation. However from methodological point of view stretching/compressing of a cell is a very challenging and complicated procedure. Several approaches exist up to date. Choice of any of them depends on the cell type, which is used as experimental model for investigation. Stretching/compressing freshly isolated cardiomyocytes is the most challenging among them.

8.2.1 *Cell Stretch*

8.2.1.1 Axial Stretch by Two Patch Pipettes

Most commonly used method of stretching of single cells is employing two patch pipettes, one of which is used for recording the whole cell current, while the other patch-pipette with a wide opening lifts and stretches the cell (Fig. 8.1). This approach is difficult to implement because, in most preparations, attachment of pulling probes to the cell without producing local stress is impossible and often irreversible damage terminates the experiment. Stretching of cells by two patch pipettes was successfully used in several cell types, for example in studies conducted on vascular smooth muscle cells (Davis et al. 1992), single ventricular myocytes (Sasaki et al., 1992), smooth muscle cells from blood vessels and the urinary bladder (Wellner and Isenberg, 1993, 1994, 1995) and on cardiac fibroblasts (Kamkin et al., 2003a, b, 2005c, 2008a). It was shown that in some cases auricle cardiomyocytes can be effectively stretched using this method (Zhang et al., 2000). Other cells require different stretch methods (Sachs and Morris 1998).

In general, stretching of cells by two patch pipettes is analyzed in detail by Sachs and Morris (1998), membrane mechanics under various types of mechanical interference with the cell was studied in detail by Morris (1990), and membrane tension is analyzed in the work by Morris and Homann (2001). It can be suggestedthat for

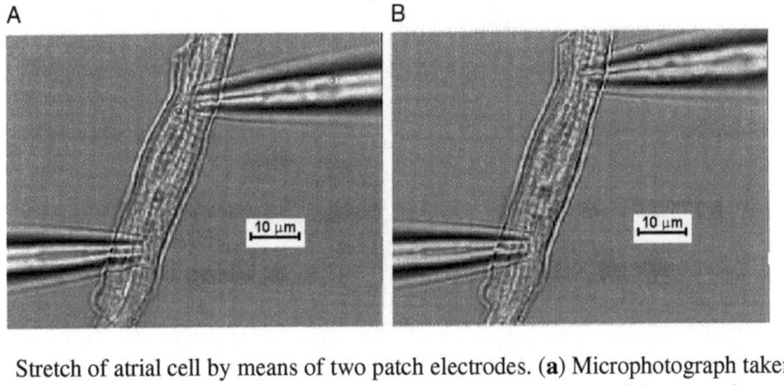

Fig. 8.1 Stretch of atrial cell by means of two patch electrodes. (**a**) Microphotograph taken when patch electrodes were just attached to the cell. (**b**) After displacement of the second patch electrode from the first one in longitudinal direction (by 15% of initial distance between electrodes). Reproduced from Zhang et al. (2000) with copyright permission of the Blackwell Publishing and J Physiol (Lond)

certain cell types like, for example, smooth muscle cells and fibroblasts, homogeneous axial stretch can be obtained by using two patch-pipettes. At that in most preparations it is relatively easy to fix pipettes to cells that are stretched without deformation.

8.2.1.2 Axial Stretch by Two Glass Capillaries

Axial stretch of the sarcomeres was achieved by sucking both cell ends into the openings of two glass pipettes, distance between which was varied by piezo devices (Zeng et al., 2000). Isolated ventricular cells with clear sarcomeres were held by two concentric glass pipettes, with the inner pipette serving as a stop to prevent the cell from being sucked up by the outer pipette. The outer pipette was pulled from a glass capillary with the inner tip diameter of ~15 μm. The inner pipette was made from a glass capillary with the outer tip diameter of ~12 μm. The inner pipette was inserted into the outer pipette by a manipulator, leaving a gap of ~8 μm to the tip of the outer pipette. The tip of the outside pipette was then cut by fusion of the tip to the filament of a micro forge. The cut end was lightly fire-polished so that the tips of the inside and outside pipettes were forged together and formed a cup to hold the cell. A third manipulator was used to attach a patch-pipette for whole cell electrophysiology (Fig. 8.2). Probably this method can be considered as homogeneous axial cell stretching.

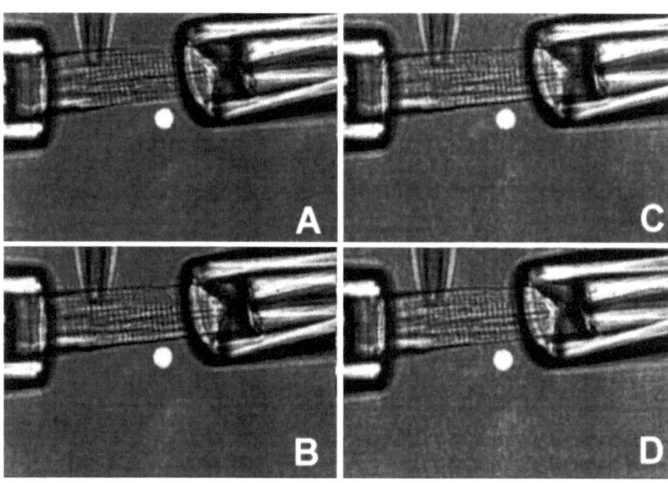

Fig. 8.2 Axial stretch. Cell was attached to two concentric pipettes (*left* and *right*) and voltage-clamped by the third (*middle*). Right concentric pipette moved right to stretch the cell, and patch pipette moved with local strain to reduce stress around tip. Reproduced from Zeng et al. (2000) with copyright permission of the Am Physiol Soc and Am J Physiol

8.2.1.3 Local Axial Stretch by Glass Stylus and Axial Stretch by Two Glass Styluses

This method is very similar to the previous one, but instead of SP uses fire-polished glass stylus (S) that adheres to the cell surface. This method of local stretch uses displacement of the glass stylus together with part of the cell surface and underlying sarcomeres (Fig. 8.3). It does not stretch the bottom of the cell that was attached to the glass bottom of the chamber where attachment was facilitated by coating it with poly-L-lysine. This method applies a component of shear stress to some components of the cell as well as axial and non axial stress (Isenberg et al., 2003, 2005; Kamkin et al., 2000a, 2003c, 2005b).

By positioning the stylus and patch-pipette ~40 μm apart before attaching them to the cell the area of stretched membrane was restricted to approximately one third of the cell length. To monitor changes of the surface membrane, micro beads of 4 μm diameter were added 20 min in advance. The cell attached glass stylus (S) is displaced from the patch pipette (P) (Fig. 8.3: from 31 to 38 μm or by 22%). The resulting stretch increased the distance between S and the micro bead (B) and the distance between B and P to the same extent (22% in Fig. 8.3b), suggesting

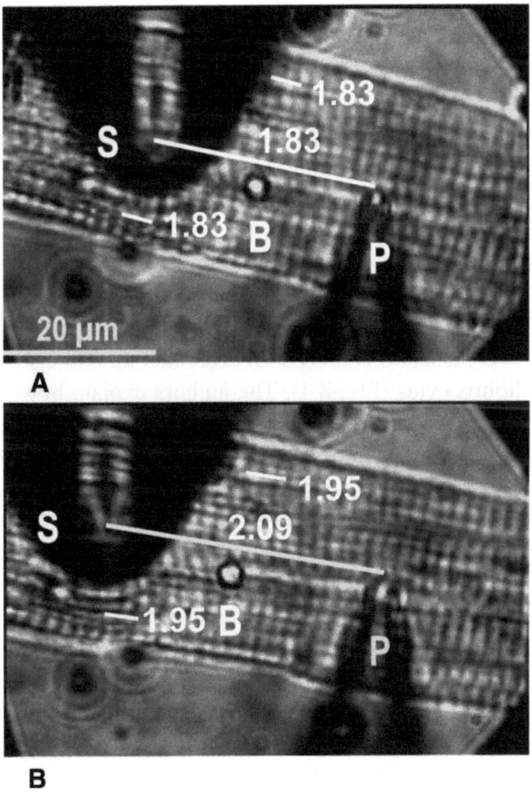

Fig. 8.3 Local axial stretch. Mechanical stimulation: local stretch of a guinea pig ventricular myocyte. (**a**) Before, (**b**) during stretch. Original microphotographs. Labels for glass stylus S, micro bead B attached to the cell surface, patch pipette P, and a line connecting S over B to P. Increasing the distance S–P by 7 μm (connecting line from 31 to 38 mm or by 22%) increases distance S–B and B–P by the same extent of 22%. Reproduced from Kamkin et al. (2000a) with copyright permission from Elsevier, Oxford University Press and Cardiovasc Res

that the stylus did not slip on the surface of the membrane. The sarcomere length (SL), however, did not follow the glass stylus as expected. On average, SL before stretch was 1.83±0.01 μm (Fig. 8.3a evaluations in the area on top or below the line S–B–P). During local stretch, the average SL increased to 1.92±0.08 μm or by 5%. As indicated by numbers, sarcomere length increased up to 2.09 μm between S and P in the center of the cell and only to 1.95 μm at the peripheral regions. Left to S and on the right to P, the sarcomere length did not change significantly (Fig. 8.3). The different lengthening, 5% SL versus 22% surface (micro bead) and the increased S.D. of the SL are the first indication that the stylus induced surface deformation, which spread with spatial decrement.

Nevertheless this method has certain limitations (Kamkin et al., 2008a). Stretching the cell by one of the styluses causes local inhomogeneous stretch of cell area. Appearing stress gradients between stylus and pipette are different but homogeneous. They are highest and most homogeneous in the center of the cell (Fig. 8.3b) next to the drawn line (as next to the line the sarcomere lengths are equal). But the gradients of appearing stress decrease on the background of imaginary parallel lines drawn with Δl towards upper and lower cell edges. At the same time within the boundaries of each imaginary line they are homogeneous and sarcomere lengths are equal (Fig 8.3b). In any case they are enough for producing a different electro-physiological response formed by a different in volume stretching of a local cell zone. At that there was strict dependency registered of inward currents volumes depending on the stretch degree. The method has allowed to investigate mechanosensitive whole-cell currents of isolated ventricular and atrial cardiomyocites in healthy hearts of mice, rats, guinea-pigs, rabbits and humans, study the dependence of mechanosensitive whole-cell currents on the age of animals and humans and study those currents on the background of various heart failures, e.g., hypertrophy aft, hypertrophy following MI (Kamkin et al., 2005b). The method of stretching the cell with adhered glass stylus can be used for axial stretch of cells if the investigator uses two glass styluses (one at each end of the cell), while positioning patch electrode at the center of the cell (Kamkin et al., 2008a).

Dyachenko et al. (2008) described the effects of mechanical stimulation of cardiomyocytes (Fig. 8.4). The authors explain how mechanical stimuli may interact with cardiomyocytes. Membranes of cell surface and T-tubules were labeled with ANEPPS (0.5 μmol/L). Application of stretch in x-direction and shear in y-direction displaced the upper cell part versus the cell bottom that remained attached to the cover slip. Dyachenko et al. (2008) hypothesized that such mechanical stimulation may modulate channel activity, or else would trigger a mechanoinduced signalling with ion channels located downstream. Displacement of S changed the orientation of the T-tubular stripes, which authors termed as ''z-line deviation'', in experiments with axial displacement of S, in contrast to y-shear or z-press, which did not produce such effect. Though z-line deviations were detectable in xz- and in xy-sections (Fig. 8.4a and b). Authors hypothesized that this force (deflected connections in Fig. 8.4b) can activate integrin dependent signaling cascades leading to channel modulation (Dyachenko et al., 2008).

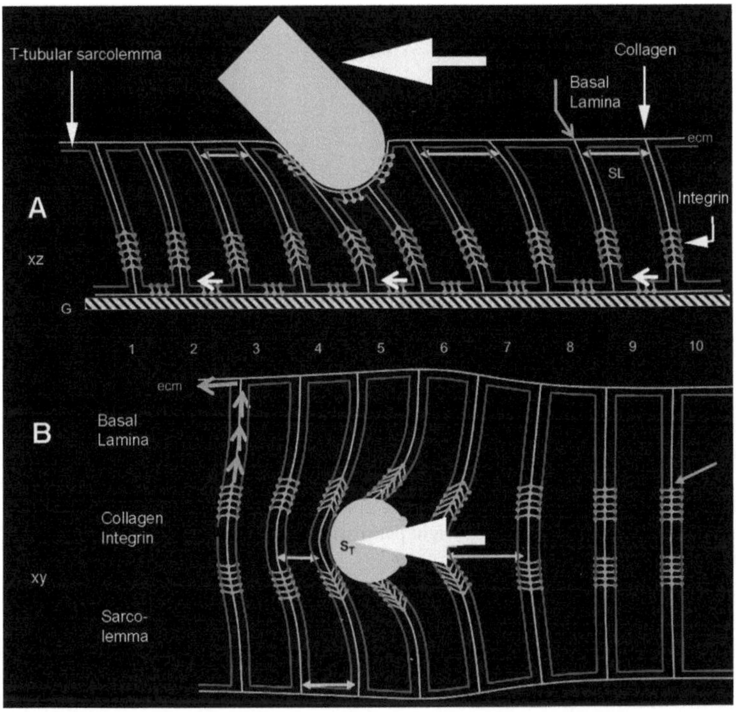

Fig. 8.4 Transmission of mechanical stimuli from glass stylus in cardiomyocytes. T-tubules are enlarged 10 times in diameter in order to show binding of collagen (in basal lamina) to integrins in T-tubular sarcolemma.← → local SL. *Top right*: Transmission of shear forces (*arrow*) from bilyer to integrins, located inside the T-tubules. (**a**) xz-section. Stylus displaces bilayer under stylus versus that at the bottom of the cell, which is attached to the glass cover slip (G, hatched area). Shear forces may activate integrin signaling at the adhesion plaques indicated as contacts between collagen and glass (*white arrows* directed to the *left*). These signals should extend up to the cell edge (here out of scale). (**b**) xy-section. Assuming that collagen is less compliant than lipid bilayer membrane imbedding integrins, z-line bending will shear collagen versus integrins. Resulting force (symbolized by deflected connections) may activate integrin dependent signaling. Sarcomere 8, distant from S_T without lengthening or z line bending. 4–6: on the right of ST SL are longer and z-lines are bent. 2–3: on left of S_T SL are shorter and z-lines are bent. z-Line bending increases length of T-tubules by 3% (3), 14% (4, 6) or by 22% (5). Reproduced from Dyachenko et al. (2008) with copyright permission from Elsevier and Prog Biophys Mol Biol

8.2.1.4 Local Axial Stretch by Two Thin Carbon Fibers

Several groups succeeded in attaching a thin carbon fiber to the cell end (Fig. 8.5). When the fiber was moved together with the cell end, an almost homogenous lengthening of the sarcomeres that were far from the probes could be achieved (Le Guennec J-Y et al., 1991; White et al., 1993, 1995). However it is difficult to position the cell in such way that both of its ends are in contact with a carbon fiber, and to achieve a good contact simultaneously between both ends of the cell and the

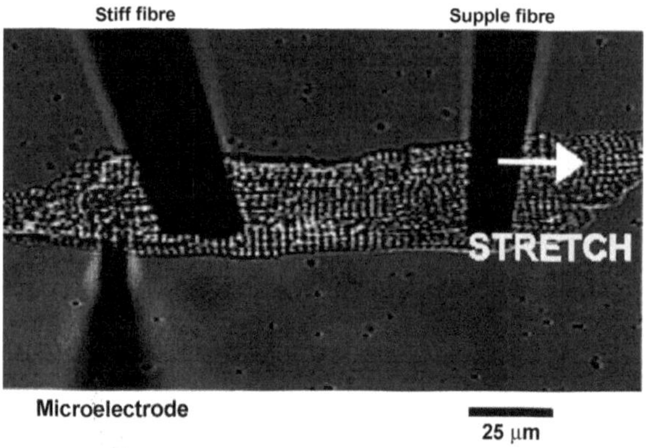

Fig. 8.5 Carbon fiber method. The microelectrode is used to record membrane potentials or currents and is placed behind the stiff fiber to protect the site of impalement during the stretch. The supple fiber is used to stretch the cell and its displacement during stimulation is a measure of active force development. Increased sarcomere spacing is used as index of stretch. Reproduced from Belus and White (2003) with copyright permission of the Blackwell Publishing and J Physiol (London)

two fibers. This method is useful for registering the contractile activity of the cell when bending one of the fibers is used as strain gauge.

8.2.2 Brick-Like Isolated Cardiomyocyte has Two Different Surfaces

It is necessary to note that brick-like isolated cardiomyocytes stuck to the bottom of the perfusion chamber in two different positions: edgewise, staying on the narrow side or broad-wise (Kamkin et al., 2005b). In some experiments after seal formation and whole cell access, cardiomyocytes were rolled using the patch-pipette to attach to the glass bottom from edgewise, staying on the narrow side (Fig. 8.6a) to broad-wise (Fig. 8.6d). This allows analyzing the whole-cell current during stretch and compression at different positions on the same cardiomyocyte.

As we have shown – a proportion of the data will be presented below – the reaction to stretch was identical in cardiomyocytes, occupying both positions (edgewise and broad-wise). However, the reaction to compression was different and determined by the position of a cell.

Thus, the brick-like cardiomyocytes have narrow and wide sides. Therefore, we considered two variants of compressing cardiomyocytes – a compression on the narrow side and a compression on the wide side. It was not difficult to perform these experiments, because brick-like isolated cardiomyocytes, are stuck to a bottom of perfusion chamber in two positions: edgewise, staying on the narrow side or broadwise. In some experiments after seal formation and whole cell access,

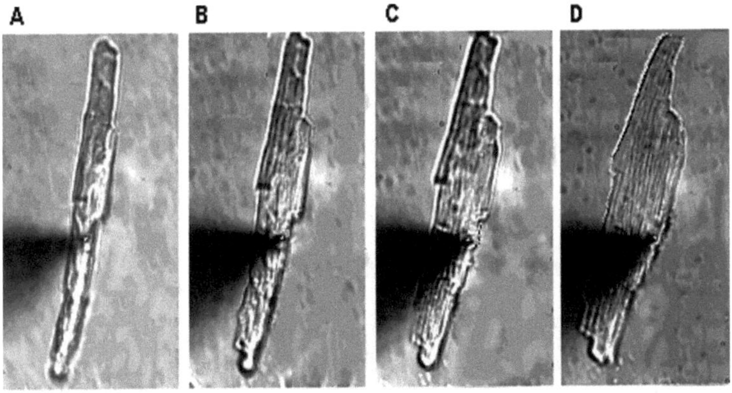

Fig. 8.6 Brick-like myocytes were rolled to attach to the glass bottom in two different positions. (**a**) Edgewise, staying on the narrow side, (**b**) and (**c**) turning of the cell by patch-pipette, (**d**) broadwise, staying on the broad side. Reproduced from Kamkin et al. (2005b) with copyright permission of the Academia Publishing House Ltd

cardiomyocytes were rolled by the patch-pipette to attach to the glass bottom from edgewise, staying on the narrow side to broadwise (Fig. 8.6). It allowed on the same cardiomyocyte, placed in different positions, to study whole-cell current during stretch and compression. Myocytes were compressed by pressing the cell surface with the glass stylus towards the glass bottom of the chamber. Reaction to stretch of cells, which were lying in two different positions, was the same. The effects of compression depended on the orientation of the cardiomyocyte, and was mediated by different mechanisms (Kamkin et al., 2005b).

When we examine a situation under patch-pipette, this situation today is clear – MSCs respond to mechanical forces in the plane of the cell membrane (membrane tension), not to hydrostatic pressure perpendicular to it (Gustin et al., 1988; Sokabe and Sachs, 1990; Sokabe et al., 1991). Membrane tension is changed by application of negative or positive pressure to the patch pipette. Honoré demonstrated that buckling of the patch increases the local curvature in both directions, but always leads to channel closing. Thus, membrane curvature, per se, cannot account for channel activation, and we favor the traditional interpretation that the primary stimulus for activation of MGCs is tension, not curvature (Honoré et al., 2006).

In whole cells, stress is distributed among various components associated with the cortex of the cell (Akinlaja and Sachs, 1998). The lipid bilayer is far from homogeneous in content (Lillemeier et al., 2006), let alone stress. The cytoskeleton applies forces parallel and normal to the bilayer (Suchyna and Sachs, 2007). The forces in the extracellular matrix are unknown.

Cell stretching by glass stylus causes local homogeneous axial stretch of the cell, which leads to activation of certain MGCs. If we perform local compression with a glass stylus a ball segment shaped dent is formed on the cell surface. It is related to the fact that that the acting tip of stylus ball shaped with the diameter of approximately 20 μm. The stylus is brought to the cell nearly vertically which forms a spherical dent of 6 μm for this experiment. The result is that vertical pressing creates a dent of about 900 μm^3 on the cell surface. In these conditions will start changes of all vectors of forces of various components associated with the cell, such as cytoskeleton, membrane, extracellular matrix.

8.2.3 Compression of the Cardiomyocytes

The compressing of cells causes certain questions. Partially they were considered above. In a different approach, whole-cell mechanosensitive currents have been evoked by pressing on a spherical cell with the tip of one pipette while voltage clamping with another (Hu and Sachs 1994, 1995, 1996). It is known that some MGCs respond when bowed toward the nucleus. Examples are found in glial cells (Bowman et al., 1992; Bowman and Lohr, 1996), smooth muscle cells (Kawahara, 1993) and endothelial cells (Marchenko and Sage, 1996).

Bett and Sachs reported currents from an approximately normal stimulus by pressing the top of a cell with the side of a patch pipette with a sinusoidal stimulus and the observed currents were phase locked to the stimulus (Bett and Sachs, 2000). This is probably the least invasive of any experiments on MGCs. Most striking in this work was the data showing that mechanical sensitivity was not visible in resting cells but only became visible after exercising the cell. And the sensitivity developed as an abrupt shift as though the channels were like a mechanoprotective boundary that had ruptured. This boundary would slowly recover over time with gentle stimulation causing the current to disappear but it would be awakened with a temporary increase in the stimulus. This emphasizes that the cell is not in equilibrium and that the cytoskeleton and other components are dynamically active (Mizuno et al., 2007) so stresses are never under control. There is stress softening (Chaudhuri et al., 2007) as well as hardening and fluctuations far from equilibrium.

Freshly isolated cardiac fibroblasts respond differently (mechanosensitive whole-cell currents) to stretching and tip pressing, two patch-pipettes were used (Kamkin et al., 2003a, b, 2005c). In these studies it was demonstrated that stretch and local stress around the tips in the form of compression have different effects on fibroblast SACs. Probably cytoskeleton transmits the stress energy to the channel in different ways depending on which the channel either opens or closes.

The role of cytoskeleton can be found in different reactions of the cell towards stretch and compression of the cardiomyocytes lying edgewise, staying on the narrow side and lying broadwise, staying on the broad side. If we press one part of a cell as shown in Fig. 8.7 for the cardiomyocyte the other parts of the cell will be stretched. In this case the reaction of the cell to local pressing may be the same or different from the cellular reaction to stretch, but it should be happening simultaneously with it and would be partially masked by it.

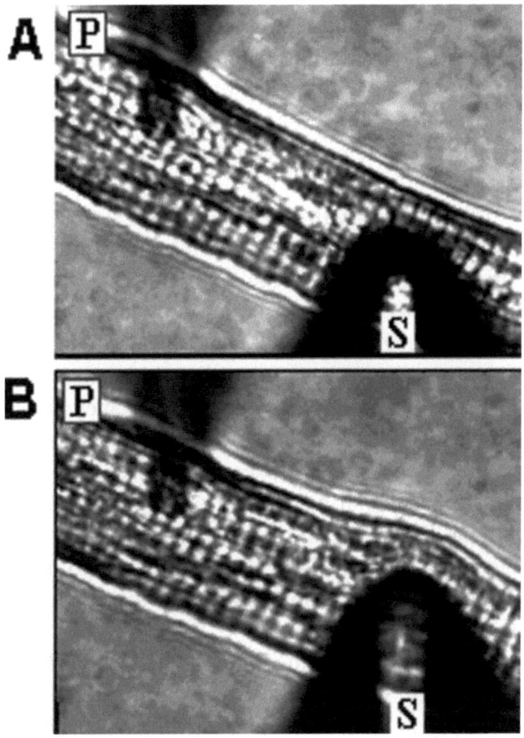

Fig. 8.7 Normal force deformation of ventricular cardiomyocyte with two glass tools. P: patch pipette fixing the myocyte to the glass cover slip. S: fire polished glass stylus. Stress caused by vertical movement of S towards the cover slip by 6 μm (**b**). Note: Under these conditions pressing the cell surface forms a dent in the shape of ball segment. It is due to the fact that the acting form of stylus tip is ball shaped with the diameter of about 20 μm. The stylus is brought to the cell nearly vertically which results in spherical dent in its surface 6 μm deep for this experiment. The result is that vertical pressing creates a dent on the cell surface about 900 μm^3. Reproduced from Kamkin et al. (2005b) with copyright permission of the Academia Publishing House Ltd

8.2.4 Simultaneous Recording of Single Channels and Whole-Cell Currents During Direct Deformation of the Whole Cell

The only relevant measurement would be observing singles in a whole cell recording and that requires a very small cell. So far there are no reports of successful single channels recordings during stretch of cardiac myocytes. However even if it were possible to perform such recordings the issue with such experiment would be that during stretch intracellular concentration of calcium is rising, which in turn would activate Ca^{2+}-sensitive K^+ channels and/or other pathways. For the time it was presumed that it is much worse that the seal of a patch pipette isolates most of the stress in the patch from that in the cell.

However our group managed to record simultaneously single channels activity and whole-cell currents during axial stretch in freshly isolated cardiac fibroblasts (See A. Kamkin, I. Kiseleva and I. Lozinsky in this volume). Single MGCs and mechanosensitive (MS) whole-cell currents were recorded with the delay of less then 1 s from isolated atrial fibroblasts using the cell-attached and whole-cell patch-clamp configurations, respectively. Mechanical stress was applied using the patch pipette used for the whole-cell recordings. Two types of MGCs with conductances of 43 and 87 pS were observed. Both MGCs displayed linear current-voltage relationships with the reversal potential around 0 mV. The mixture of cytochalasin D and colchicines completely blocked both the whole-cell MG currents and MGCs.

The situation with research of MGCs of cardiomyocytes is similar. Till recently it was considered to be impossible to record single MGCs in cell attached mode during direct stretch of the cardiomyocyte since on one hand the membrane seals patch electrode and stretching does not affect the membrane patch in the electrode opening, and on the other hand stretch triggers changes in cellular signaling, primarily in calcium mediated signaling, which in turn modulates all other ion channels, which are located in the membrane patch in the electrode opening. We managed to overcome those difficulties by means of employment of one patch pipette in cell attach mode, which is used for registration of singles MGCs, and use of the second patch electrode in whole cell mode, which we used for intracellular BAPTA dialysis (Fig. 8.8). The tip of the patch electrode was dipped into 2 μM of di-8-ANEPPS before experiment, which allowed to visualize it. Cofocal setup was mounted on an inverted microscope using a water immersion objective.

Labeled tips of both patch pipettes are visible on the microphotograph as small white spots. Stretch was applied via fire polished tip of the pipette, which is present as a big grey spot on the photograph. We managed to stretch the cell (Fig. 8.9) and to register single MGCs in cell attached (Fig. 8.10) and mechanosensitive whole cell currents (Kamkin and Kiseleva, 2008b).

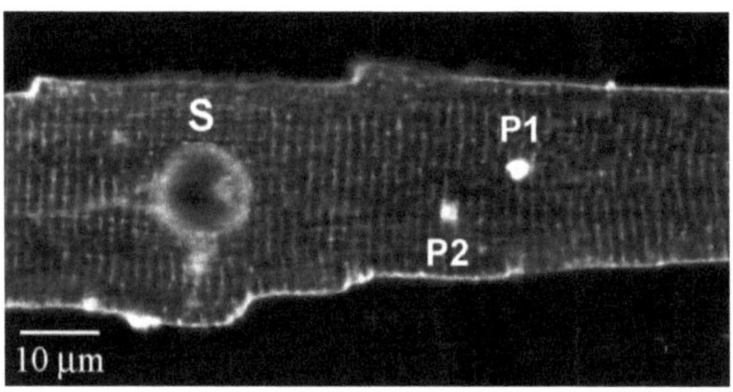

Fig. 8.8 Stretch of the cardiomyocyte between one patch electrode, registering in cell attach mode (P1) single MSC activity, and second electrode, recording in whole cell mode and used for BAPTA dialysis (P2). Labeled tips of both patch pipettes are visible on the microphotograph as small white spots. Axial stretch of the cardiomyocyte and its release (*double arrow*) was performed by means of displacement of the fire polished patch pipette (S), the tip of which is visible as a big *grey* spot. Reproduced from Kamkin and Kiseleva (2008b) with copyright permission of the Academia Publishing House Ltd

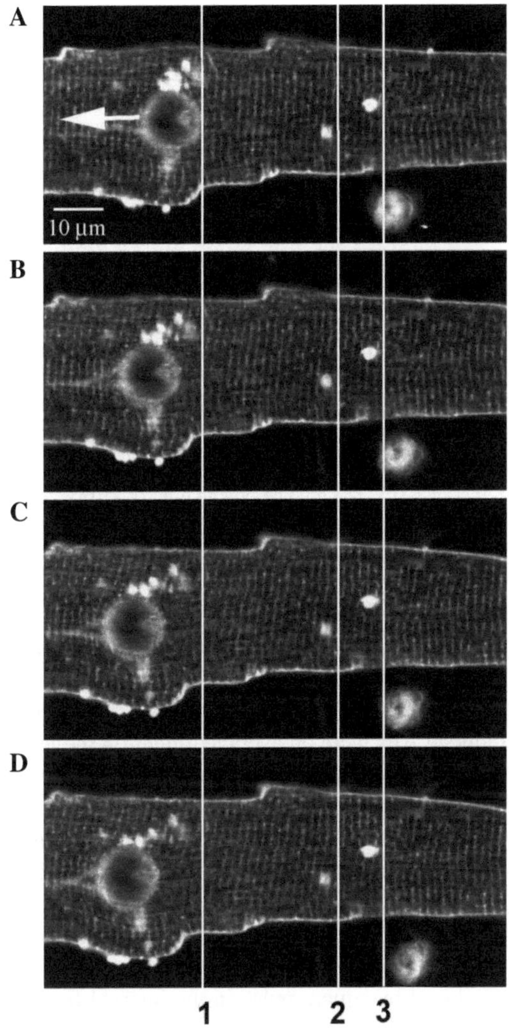

Fig. 8.9 Microphotographs of the cardiomyocyte during stretching (from resting condition on «a» till «d»). Reproduced from Kamkin et al. (2008b) with copyright permission of the Academia Publishing House Ltd

8.3 Stretch of Freshly Isolated Atrial Cardiomyocytes

8.3.1 Freshly Isolated Atrial Cardiomyocytes from Healthy Animals

The following data was acquired from investigation of rat atrial cardiomyocytes by means of nistating perforated patch recording in whole cell configuration during cellular stretch by two patch electrodes (Fig. 8.1a, b). In normal Tyrode solution

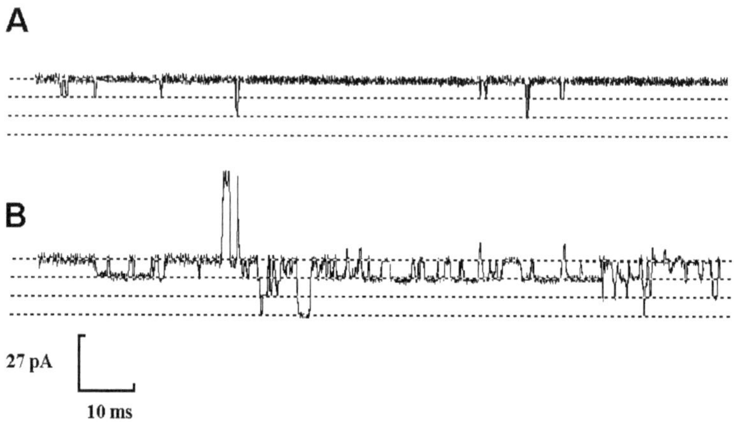

Fig. 8.10 Single MGC activity of cardiomyocyte recorded in the cell-attached mode. On-line pen records. Sample traces show the activity of single MGCs in control (**a**) and during a 8 μm cell stretch (**b**). The holding potential was –45 mV. Reproduced from Kamkin et al. (2008b) with copyright permission of the Academia Publishing House Ltd

resting membrane potential was –63.6±0.58 mV, APD90 was 108.8±11.9 ms. During cellular stretch by 20% by two patch electrodes cellular membrane depolarized to –54.6±4.40 mV. In several cases we observed spontaneous extra AP. Stretch induced APs were on average longer on the level of APD90 by 32.2±8.8 MC. Their amplitude was slightly reduced. All of those changes were completely reversible (Zhang et al., 2000).

In this study authors tested the hypothesis that I_{SAC} is a non selective cation current by means of voltage-clamp recordings with different ionic composition of bathing and pipette solutions (Zhang et al., 2000). In one of experimental series authors used physiological Tyrode as a bathing solution in combination with physiological pipette solution. The membrane potential was adjusted to –30 mV and was displaced by square pulse in the interval from +60 to –120 mV. Acquired current-voltage (*I-V*) relationship showed strong inward rectification, which is typical for cardiac atrial myocytes. Cellular stretch by 20% by means of two patch pipettes increased inward and outward currents. I_{SAC} was calculated as a difference current between current before and after application of stretch at a given potential step. The reversal potential was –6.1±3.7 mV. Observed changes were completely reversible. Such value of the reversal potential of I_{SAC} means that I_{SAC} is a nonselective ionic current.

In symmetrical sodium solution Na_i/Na_o (in which Na^+ concentration in the pipette and the bathing solution is the same and in absence of K^+ and Ca^{2+}) I-V curve becomes linear with $E_{rev} = 0$ mV (Fig. 8.11a). In symmetrical cesium solution Cs_i/Cs_o (in absence of K^+ and Ca^{2+}) after application of 100 μM of E-4031 for blockade of Cs^+ permeability through the rapidly activating delayed rectifier K^+ channels cellular stretch activated I_{SAC}.

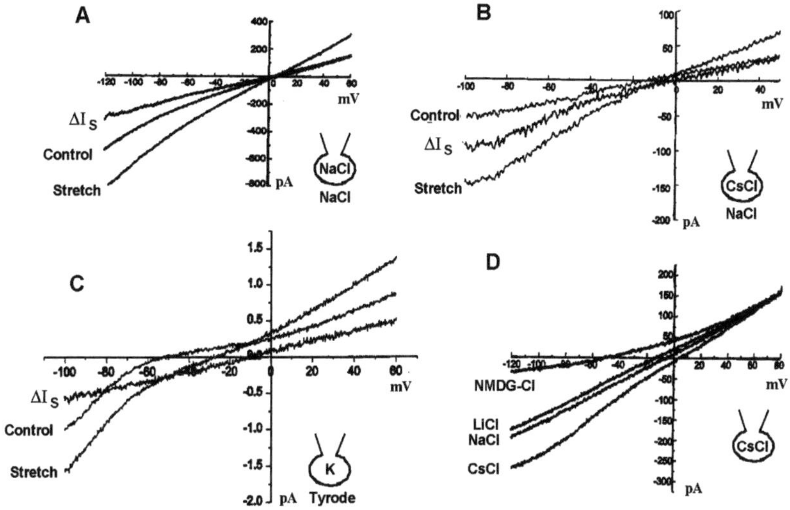

Fig. 8.11 Ionic nature of I_{SAC}. (**a**) *I-V* relationships of I_{SAC} recorded in symmetrical NaCl solutions. Direct stretch of the cell by 20% still elicited linear I_{SAC} (ΔI_S). (**b**) *I-V* relationships of $I_{SAC,NC}$ obtained in 140 mM NaCl external and 140 mM CsCl internal solutions. Mechanical stretch shifted the reversal potential of the current (stretch) slightly to the right. The difference current (SAC) reversed near 0 mV, suggesting that the permeability of Cs^+ and Na^+ for $I_{SAC,NC}$ were almost identical. (**c**) *I-V* relationships of I_{SAC} in normal Tyrode external and high-K^+ internal solutions. Currents were elicited by ramp pulses from +60 to –120 mV with d*V*/d*t* of –225 mV s^{-1}. The amplitude of the current was normalized to the current measured at –100 mV in control, and normalized *I-V* curves obtained from 7 cells were averaged. Stretch of the cell by 20% caused an increase in both the inward and outward currents (from $I_{control}$ to $I_{stretch}$). The difference between $I_{stretch}$ and $I_{control}$ represents I_{SAC}. (**d**) *I-V* relationships of $I_{NSC,b}$ obtained in 140 mM CsCl internal solution with the external solution being changed from solution containing CsCl to solutions containing NaCl, LiCl or NMDG-Cl. Current amplitude was reduced in the inward direction without change in the outward current, and the reversal potential was shifted in a negative direction. Modified from Zhang et al. (2000) with copyright permission of the Blackwell Publishing and J Physiol (London)

In order to compare the coefficients of relative permeability of different cations through SAC authors registered I_{SAC} and measured the reversal potential. Figure 8.11b represents I-V curve recorded in isotonic solution Cs_i/Na_o. Before the stretch E_{rev} was –13 mV. During stretch current increased and E_{rev} shifted to the right. I-V curve of I_{SAC} reversed at –2 mV. E_{rev} of I_{SAC} was also measured in different solutions such as Cs_i/Li_o, Na_i/Cs_o, Na_i/Li_o, which yielded similar values of it. P_{Cs}:P_{Na}:P_{Li} ratio was 1.05:1:0.98. This data shows that permeability for Na^+, Li^+ and Cs^+ through SAC is practically identical.

Authors calculated R_K/P_{Na} from E_{rev} of I_{SAC}, acquired at physiological ionic concentrations (Fig. 8.11c); to be equal 1.32 mV. Their results demonstrate that I_{SAC}, which is triggered by direct cellular stretch, is a non selective cationic current. Authors named this current $I_{SAC,NS}$.

Application of DIDS (0.1 mM), which is a known blocker of chloride channels, or replacement of Cl^- to aspartate did not affect I_{SAC}, which shows that Cl^- component is not present in I_{SAC}, triggered by stretching of the cell. Since this current is not affected by elimination of Cl^- or by application of DIDS, this background nonselective cation current authors named as $I_{NSC,b}$. Sift of E_{rev} as a result of cellular stretch (Fig. 8.10b) implies differences in ionic permeability between $I_{SAC,NS}$ and $I_{NSC,b}$.

For determination of ionic permeability of $I_{NSC,b}$ currents were triggered similarly to those, measured for $I_{SAC,NS}$ which allowed to plot the I-V curve for $I_{NSC,b}$ (Fig. 8.11d). Pipette solution for those experiments contained 140 mM of CsCl. In isotonic solution with Cs^+ I-V curve was linier with E_{rev} of 0 mV. The replacement of external Cs^+ to Na^+ or Li^+ decreased the inward current and shifted the E_{rev} to negative direction. The replacement by $NMDG^+$ caused significant depression of the current. This suggests that most of the inward current is carried by cations. Authors calculated the permeability ratio P_{Cs}:P_{Na}:P_{Li} to be 1.49:1:0.70.

Those results suggest that background nonselective cationic channels (NSC_b) can affect the form of action potential via generation of inward ionic currents at the level of resting membrane potential and during late depolarization phase of the AP. However other possible mechanisms can not be out ruled, since Gd^{3+} effects are not selective. On the other hand mechanical stretch in any way depolarizes the membrane and causes spontaneous APs even in presence of Gd^{3+} (up to 0.1 mM). Those results prove observations that $I_{SAC,NC}$ is not sensitive to Gd^{3+}. Difference in pharmacological profiles of $I_{SAC,NC}$ and $I_{NSC,b}$ demonstrates that those currents are originating from different populations of non selective cationic channels (Zhang et al., 2000; Youm et al., 2005).

8.3.2 Stretch of Freshly Isolated Atrial Cardiomyocytes from Patients with Cardiac Hypertrophy

Authors performed electrophysiological recordings on cardiac myocytes that were freshly isolated from patients with various heart diseases (Kamkin et al., 2003d). It was shown recently that application of mechanical stretch to atrial cardiac tissue and isolated atrial cardiomyocytes (Kamkin et al., 2000b; Nazir and Lab, 1996; Zhang et al., 2000; Youm et al., 2005) increased the AP duration, induced diastolic depolarization, and caused extra-APs. These phenomena were at least in part due to activation of stretch-operated ion channels. Authors demonstrated that stretch-activated non-selective cation channels are also contained in human atrial myocytes and potentially are related to the pathophysiology of the diseased heart.

In the current-clamp mode, human atrial cardiomyocytes responded to stretch with membrane potential depolarization, and increasing stretch by 2 μm (S1 in

Fig. 8.12a) and 3 μm (S2 in Fig. 8.12a) induced stretch-induced depolarizations at APD90. Remarkably, extra-APs were observed upon stretching of atrial cardiac myocytes by 4 μm (S3 in Fig. 8.12b) and 6 μm (S4 in Fig. 8.12b) under current-clamp conditions.

Those findings show, that human atrial myocytes indeed contain stretch-activated ion channels. It was demonstrated that transmembrane influx, presumably of sodium ions, through non-selective cation channels may have a role in cardiac arrythmogenesis.

Replacement of K^+ by Cs^+ ions in the bathing and electrode solutions for suppression of inwardly rectifying K^+ currents allowed the separation of net currents into current components (Fig. 8.13a). A typical current-voltage relationship for the late current I_L and $I_{Ca\text{-}L}$ is shown in Fig. 8.13b.

The example in Fig. 8.14 illustrates the voltage-dependence of the membrane late currents before and during application of mechanical stretch. Stretch-induced currents increased with the intensity of stretch. Increasing stretch at –45 mV by 2 μm, 4 μm, and 6 μm caused stretch-induced late currents of –38 ± 6 pA, –254 ± 17 pA, and –595 ± 22 pA, respectively. These effects were reversible upon release of stretch (Fig. 8.14a).

Upon stretching the atrial myocytes, the stretch-activated difference currents I_{SAC} followed an almost linear voltage-dependence and crossed the voltage axis at a reversal potential E_{rev} of 0 mV (Fig. 8.14b). At –45 mV, I_{SAC} was –54 ± 18 pA with 2 μm of stretch, –272 ± 22 pA with 4 μm of stretch and –613 ± 22 pA with 6 μm of stretch.

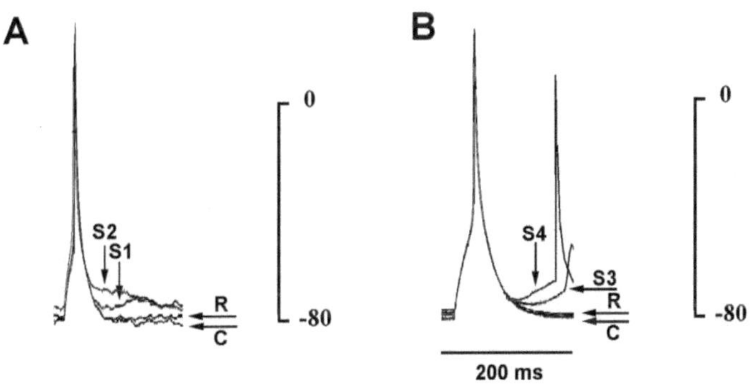

Fig. 8.12 Membrane depolarization, prolongation of the action potential and extra-action potentials in response to mechanical stretch. Right atrial cardiomyocytes from patients undergoing open-heart surgery were stretched from the control value (**c**) by 2 μm (S1), 3 μm (S2), 4 μm (S3), and 6 μm (S4). (**a**) Application of 2 μm and 3 μm of stretch induced membrane depolarization and prolongation of the action potential duration measured at 90% (APD90) of depolarization. Relaxation of the membrane (R) completely reversed the stretch-induced membrane depolarization. (**b**) Increasing stretch by 4 μm and 6 μm increased the amplitude of the after depolarization near APD90 and caused extra-action potentials when reaching the threshold level. Note: the cells were stimulated at a frequency of 1 Hz with 100 pA pulses for 10 ms. Reproduced from Kamkin et al. (2003d) with copyright permission of the Springer and Pflüg Arch - Europ J Physiol

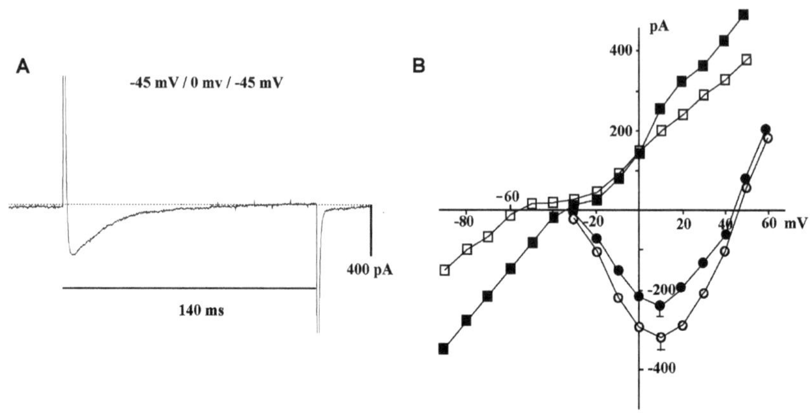

Fig. 8.13 Example of L-type Ca^{2+} channel current, $I_{Ca\text{-}L}$, and late current, I_L, measured at the end of the depolarizing 140 ms pulse in a human atrial myocyte during suppression of I_K. (**a**) Original current registration in the whole-cell configuration. The depolarizing current pulses had durations of 140 ms and were applied with the holding potential set to –45 mV. (**b**) Current-voltage relation of the currents measured at the end of the pulse (late current I_L, *squares*) and peak Ca^{2+} channel currents (*circles*). Current-voltage relations were recorded in the absence (*open symbols*) and presence (*filled symbols*) of stretch. On average, stretching the isolated cells by 2 μm decreased $I_{Ca\text{-}L}$ from –255±77 pA to –181±56 pA at a membrane potential of +10 mV. This value of stretch changed I_L from +16±9 pA to –56±9 pA (at –45 mV). Modified from Kamkin et al. (2003d) with copyright permission of the Springer and Pflüg Arch - Europ J Physiol

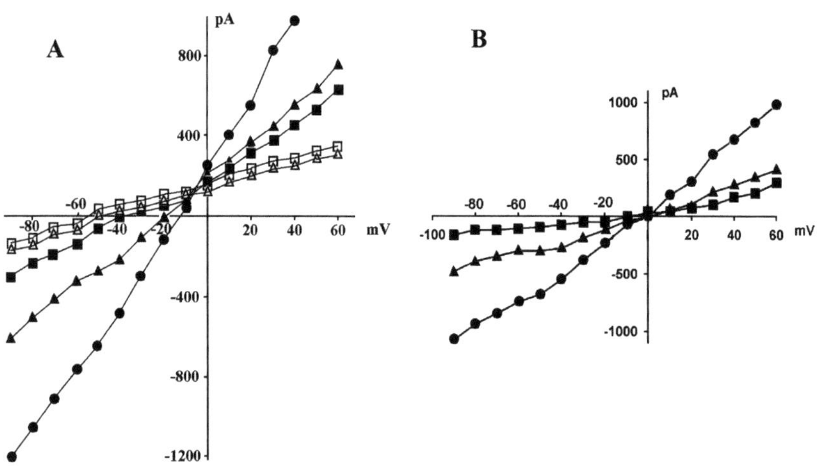

Fig. 8.14 Modulation of net membrane currents by stretch during suppression of I_K. (**a**) Current-voltage relation of the late current I_L measured at the end of the test pulse (*empty squares* – before stretch; *empty triangles* – after removal of stretch; *filled squares* – during 2 μm stretch; *filled triangles* – during 4 μm stretch; *filled circles* – during 6 μm stretch). (**b**) Voltage dependence and reversal potential of the stretch-activated difference currents (I_{SAC}). *Squares* – 2 μm of stretch (E_{rev} = –10 mV); *triangles* – 4 μm of stretch (E_{rev} = 0 mV); *circles* – 6 μm of stretch (E_{rev} = 0 mV). Modified from Kamkin et al. (2003d), with copyright permission of the Springer and Pflüg Arch - Europ J Physiol

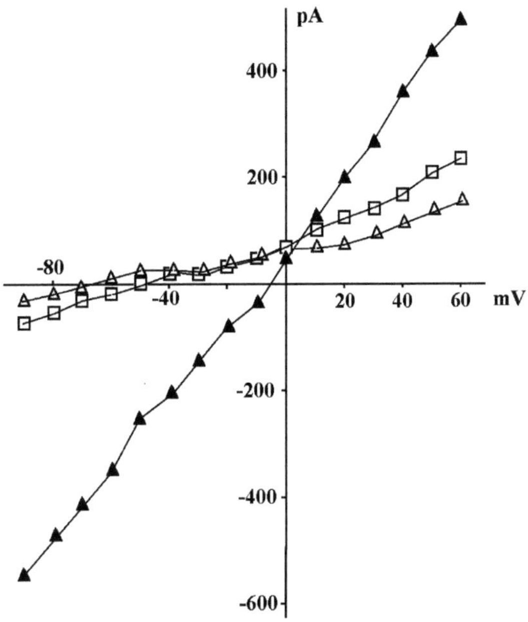

Fig. 8.15 Gd^{3+}-sensitivity of stretch activated currents during suppression of I_K. Stretch-induced late currents were calculated from the current-voltage relation at a holding potential of –45 mV. Voltage-dependence of stretch-induced late currents before application of stretch (*empty squares*; pre-stretch I_L was +12 pA); at 4 μm of stretch (*filled triangles*; I_L was –228 pA); and during stretch application in the presence at 5 μM Gd^{3+} (*empty triangles*; I_L was +23 pA). Modified from Kamkin et al. (2003d), with copyright permission of the Springer and Pflüg Arch - Europ J Physiol

Stretch-induced late currents were suppressed within 8–10 min after application of 5 μM Gd^{3+} (Fig. 8.15). I_{SAC} was insensitive to substitution of Cl^- by aspartate ions suggesting that I_{SAC} is carried by cations rather than by Cl^-. Thus, the linear voltage-dependence, the value of E_{rev}, Gd^{3+} sensitivity, and Cl^- insensitivity suggest that stretch-activated I_{SAC} flows through mechanically sensitive non-selective cation channels

During application of stretch, the sustained inward current I_{SAC} of approximately –500 pA is expected to cause intracellular accumulation of Na^+ and Ca^{2+} ions, which may affect other current components. Indeed, our findings made with Cs^+ dialyzed cells (suppression of I_K) showed that mechanical stretch suppressed the L-type Ca^{2+} channel current $I_{Ca\text{-}L}$ (Fig. 8.13b). Inhibition of $I_{Ca\text{-}L}$ was not observed when the cells were dialyzed with 5 mM BAPTA for 5 min prior to application of stretch, whilst I_{SAC} remained unchanged by this maneuver. For example, imposing a stretch of 3 μm at a holding potential of –45 mV produced I_{SAC} of –140 ± 10 and –148 ± 20 pA in the presence or absence of 5 mM BAPTA, respectively (not significant). The effect of BAPTA is likely due to chelating of Ca^{2+} ions. Hence, the observed stretch-induced reduction of $I_{Ca\text{-}L}$ can be explained as "Ca^{2+} inactivation" due to intracellular Ca^{2+} accumulation (Zühlke et al., 1999; Wellner and Isenberg, 1994). Thus, the negative I_{SAC} that was presumably carried by influx of Na^+ ions at negative membrane potentials may cause depolarization and extra-APs, if the depolarization reaches threshold.

8.4 Stretch of Freshly Isolated Ventricular Cardiomyocytes

8.4.1 Freshly Isolated Ventricular Cardiomyocytes from Healthy Animals

For studying whole cell currents, ventricular cardiomyocytes for the first time have been stretched by Kamkin, Kiseleva, Isenberg (Kamkin et al., 2000a) and Zeng, Bett, Sachs (Zeng et al., 2000) and further whole cell currents have been studied in detail (Isenberg et al., 2003; Kamkin et al., 2003c). Also other effects of stretch on isolated cardiomyocytes have been reported (Belus and White, 2003; Calaghan and White, 2004; Dyachenko et al., 2006, 2008, 2009a, b; Isenberg et al., 2004; Kondratev and Gallitelli, 2003; Kondratev et al., 2005; Xian Tao Li et al., 2006).

8.4.1.1 Modulation of the Resting Membrane Potential and Action Potential

In physiological Tyrode in case of employment of patch-clamp method in whole-cell configuration freshly isolated guinea pig and mouse ventricular cardiomyocytes (Fig. 8.16) respond to stretching by 6 μm to 10 μm by depolarization of their membrane, by prolongation of their AP duration and by generation of extra-APs. Stretch release completely reverses those changes.

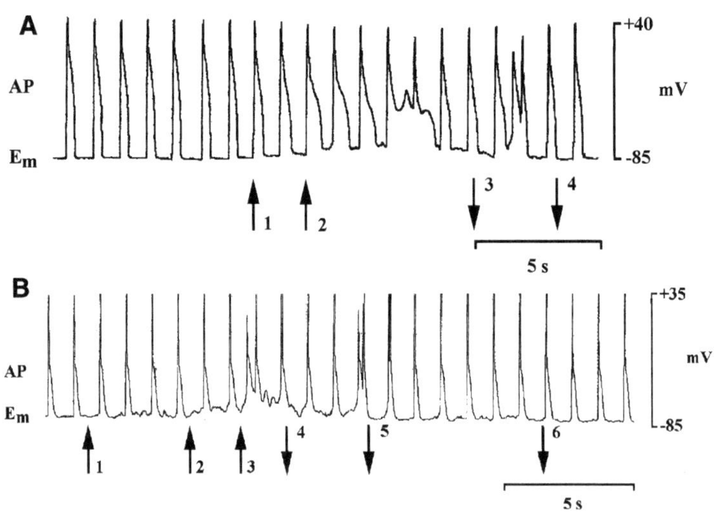

Fig. 8.16 Effects of stretching of freshly isolated ventricular guinea pig (**a**) and mouse (**b**) ventricular cardiomyocyte on resting membrane potential and action potential properties. ↑ indicates beginning of stretch application, while ↓ indicates the release of stretch to the control levels. AP – action potential, E_m – resting membrane potential. Modified from Kamkin et al. (2003c), with copyright permission of the Springer and Pflüg Arch - Europ J Physiol

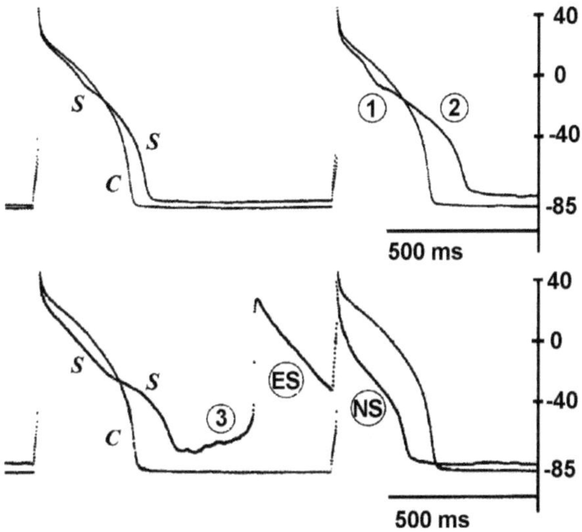

Fig. 8.17 Mechanically induced depolarization of the membrane of freshly isolated guinea pig ventricular cardiomyocyte, prolongation of the action potential and generation of extra action potentials. (**a**) Ventricular cardiomyocyte, stretched by 4 μm. Changes of resting membrane potential and AP form. (**b**) Ventricular cardiomyocyte, stretched by 8 μm. Changes of resting membrane potential, AP form and generation of extra action potential. (**c**) Control, S – potentials of the stretched cell, ES – Extra-AP, 1 – early depolarization phase, 2 – late depolarization phase, 3 – diastolic depolarization. Modified from Kamkin et al. (2000a), with copyright permission from Elsevier, Oxford University Press and Cardiovasc Res

Figure 8.17a, b demonstrates comparison of APs of freshly isolated ventricular guinea pig cardiomyocyte before and during stretch, applied when the cell was perfused with normal physiological Tyrode, recorded by means of patch-clamp method in whole cell configuration. Freshly isolated ventricular cardiomyocytes of healthy guinea pigs respond to stretch by modulation of the resting membrane potential and of duration of the action potential (Kamkin et al., 2003c). It was shown that stretching by 2 μm does not affect the resting membrane potential and the form of the action potential, while stretch by 4 μm depolarizes resting membrane potential by 3±1 mV (Fig. 8.17a), and stretch by 8 μm depolarizes it by 6±2 mV (Fig. 8.17b).

Stretch by 8 μm prolongs APD90 from 360 ±14 ms to 502 ±27 ms. Besides that stretch by 8 and 10 μm triggers extra-APs, which begin at the level of diastolic depolarization (mark 3 on Fig. 8.17b). Similar changes of depolarization (prolongation of MAPD90) were recorded via surface electrodes from intact hearts of animals (Zabel et al., 1996) and caused early after depolarization and arrhythmias in humans (Taggart et al., 1992). Similar data was acquired by means of microelectrode recordings in tissue fragments of healthy hearts (Kamkin et al., 2000b; Kiseleva et al., 2000).

Stretch of ventricular cardiomyocytes from healthy rats in two capillaries led to depolarization of the membrane and prolongation of APD90. Stretch by 5 μm during 20 s shifted resting membrane potential from –62.4±1.1 mV to –59.7±1.2 mV.

Similar stretch led to prolongation of APD90 to 764±50 ms in comparison to control APD90 of 621±12 ms. APD50 also increased from 516±19 ms to 597±22 ms (Zeng et al., 2000). However in those experiments authors used low concentrations of calcium in the perfusion solution.

Thus all studies, which employed direct stretching of cell up to date report two major reversible reactions to stretch of atrial and ventricular cardiomyocytes – membrane depolarization and APD90 prolongation. Both of those processes lead to generation of extra APs, if the depolarization of the membrane (as a sum of change of the resting membrane potential and mechanically induced depolarization at the level of APD90) reaches threshold for action potential generation. Therefore single freshly isolated ventricular cardiomyocytes can generate extra APs, which could lead to extrasysltolic beats in response to stretch. We noted earlier that extrasystolic beats as the reaction to stretch of atrial cells is a known phenomena. At the same time extrasystolic excitation of isolated ventricular cardiomyocytes does not fit Starling Effect, where stretch of ventricular cardiomyocytes does not produce extrasysltolic beats. (Extrasystolic beats of ventricles in experiments employing inflatable balloon or catheters in our opinion most likely originate from local stimulus). It is possible to assume that the mismatch between the response of a single cell and the whole heart to stretching has the following explanation. Stepwise stretching of the isolated cardiomyocyte via SAC activation translates to stepwise depolarization of the membrane, which after reaching the threshold at a certain level of stretching leads to generation of extra APs. In following sections of our manuscript we will describe a causal relationship between the amount of current flowing through SACs and the extent of stretch, which is applied to the cell. Stepwise stretching of ventricles in Starling experiments can not lead to similar stretching of all cardiac cells, due to inhomogeneity of cardiac tissue. In other words all cardiac cells experience different amount of mechanical stimulus. Some cells probably are not stretched at all. In this case during stretching of ventricles action potentials of different cardiac cells should be different. Due to leveling of potentials through low resistance contacts between neighbouring cells (through gap junctions) single cells would not reach threshold for firing and therefore there would be no extra APs.

8.4.1.2 Modulation of Basal Membrane Current by Stretch

We studied the reaction to stretch of whole-cell currents and it was shown that the reaction to stretch was identical in cardiomyocytes, occupying both positions (Fig. 8.6 : edgewise and broadwise). However, reaction to compression was different and was determined by the position of a cell. Figure 8.18 shows membrane currents by means of on-line pen records.

A 6 μm stretch shifted the holding current at –45 mV by –0.16 nA to more negative values (Fig. 8.18a). The stretch induced current change completed during the time of mechanical movement (usually 200 ms), i.e. an activation time course could not be observed. Stretches by 2 μm and 4 μm did not change the currents (not illustrated). During stretches of 6, 8, 10, and 12 μm, the amount of negative cur-

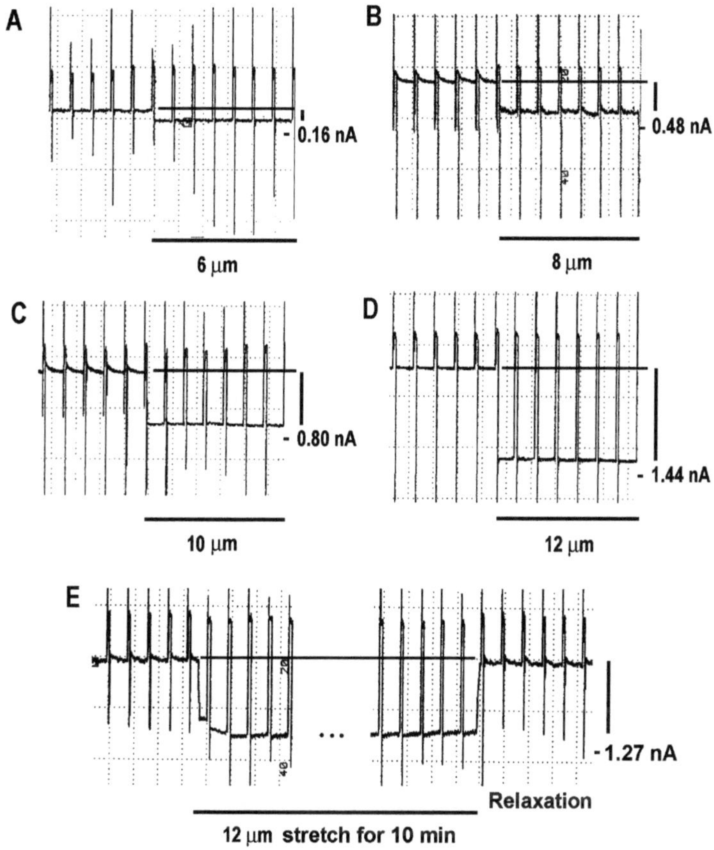

Fig. 8.18 Induction of net inward currents by local stretch is graded and reversible. On-line pen records of membrane current; K^+ currents are not suppressed. Membrane potential clamped to a holding potential of –45 mV, 140 ms pulses to 0 mV at 1 Hz. Amplitude of stretch and amount of stretch-induced inward current at –45 mV are indicated. (**a**) to (**d**): Amount of inward current increases with the extent of stretch. (**e**): During 10 min of stretch, inward current remains constant meaning no inactivation occurs. After stretch, current relaxes back to the value before stretch. Note: Stretches by 2 μm and 4 μm did not change the current (not illustrated). Reproduced from Kamkin et al. (2003c), with copyright permission of the Springer and Pflüg Arch - Europ J Physiol

rent increased with the extent of stretch (Fig. 8.18 from a to d). During continuous stretch, the inward current remained constant (tested up to 15 min), i.e. inactivation with time was not observed (Fig. 8.18e). The effect of stretch on the current was reversible, i.e. the current returned to the value before stretch when the stretch was relaxed by moving the stylus to its position before stretch (Fig. 8.18e). Figure 8.18 shows furthermore that the amplitude of the stretch induced inward current gradually increased with the extent of stretch.

Figure 8.19 demonstrates that the value of the late current and stretch induced inward current gradually increased with the extent of stretch (Kamkin et al., 2003c).

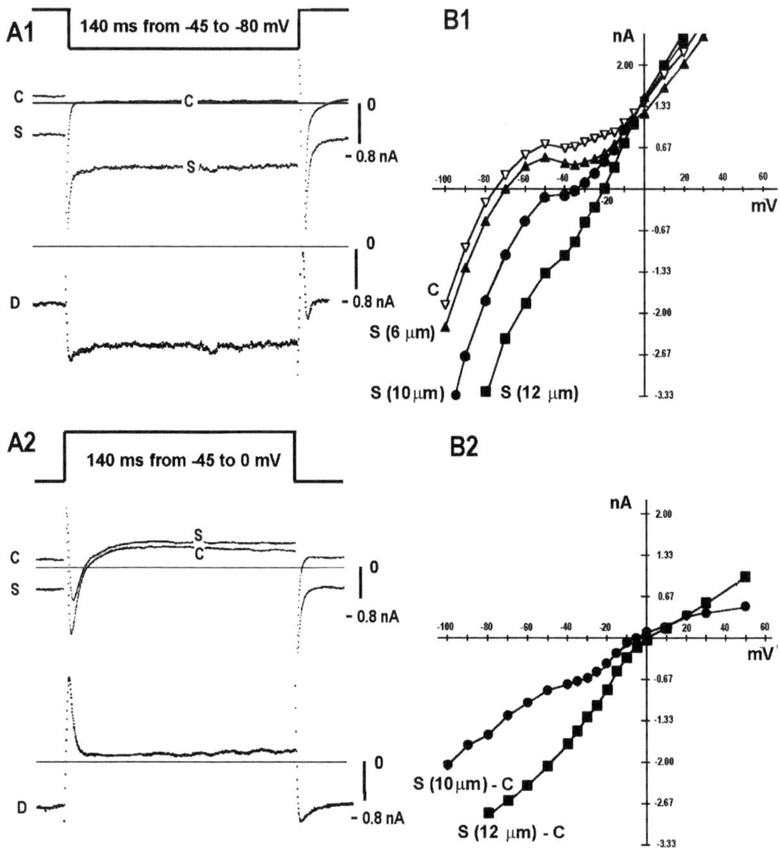

Fig. 8.19 Voltage-dependence of stretch-induced membrane currents, K^+ currents not suppressed. (**a**) Holding potential –45 mV, 140 ms pulses to –80 mV (**a1**) or 0 mV (**a2**). Traces of currents before (label C) and during 12 µm stretch (label S), and stretch induced difference currents (label D). Note: during-stretch current is more negative than pre-stretch current at –80 and –45 mV, however, more positive at 0 mV. (**b1**) Current voltage relation (*I-V* curve) of late current I_L measured at end of pulse before stretch (*empty triangles*), during 6 µm stretch (*filled triangles*), 10 µm stretch (*filled circles*) or 12 µm stretch (*filled squares*). (**b2**) Difference currents activated by 10 µm stretch (*filled circles*, $E_{rev} = -5$ mV) or by 12 µm stretch (*filled squares*, $E_{rev} = 0$ mV). Reproduced from Kamkin et al. (2003c) with copyright permission of the Springer and Pflüg Arch - Europ J Physiol

As an example, we depicted only the effect of stretch on the time-dependent net membrane currents in K_{in}/K_{out} configuration, which is shown in Fig. 8.19a.

The holding potential of freshly isolated guinea pig cardiomyocytes was set at –45 mV and then was shifted by square pulses of 140 ms duration with amplitude in the range from –100 mV to +100 mV.

After shifting of potential from the holding potential to –80 mV (Fig. 8.19a1) in absence of stretch current was equal to +0.15 nA ("C" at the beginning of the curve). In presence of stretch by 12 μm at the same potential current was equal to –0.65 nA ("S" at the beginning of the curve). At the holding potential of –45 mV, differential mechanically induced current was equal to –80 nA ("D" at the beginning of the curve). At –80 mV current in absence of stretch was close to zero and practically did not depend on the duration of the pulse (Fig. 8.19a1, curve C). Stretch changed this current from 0 to –1.30 nA (Fig. 8.19a1, curve S). Therefore differential mechanically induced current was equal to –1.30 nA (Fig. 8.19a1 , curve D) and practically did not depend on the impulse duration.

At a step to 0 mV (Fig. 8.19a2) stretch led to a shift of the late current, measured at the end of the voltage step (Late current: I_L) from 0.30 to 0.45 nA. Thus mechanically induced current was positive (+0.15 nA). Steps to –45 mV from до 0 мВ induced L-type Ca^{2+}-current ($I_{ca\text{-}L}$). Stretch decreased $I_{ca\text{-}L}$. Mechanoinduced difference L type Ca^{2+} current $I_{Ca\text{-}L}$, is shown as a positive peak on Fig. 8. 19a2 – curve D.

Figure 8.19b1 compares I-V curves of I_L currents before stretch (empty symbols) and during stretch (filled symbols). I-V curve before stretch has N-shape and crosses potential axis at –74 mV (potential at current which equals zero: E_0), which is equivalent to the resting membrane potential of non-clamped cell. Stretch by 6 μm shifts basal current to more negative values (filled triangles) and E_0 to –70 mV. I-V curves before and during application of stretch cross approximately at –10 mV, while at more positive potentials mechanically induced late current is increasing during stretch. Stretching by 10 μm shifts basal current to even more negative values (filled circles) and depolarizes E_0 to –35 mV. Significant stretching by 12 μm eliminates N-shape of the I-V curve and depolarizes E_0 to –15 mV (filled squares).

Difference between late currents before and during stretch are plotted as I-V curves on Fig. 8.19b2. Those two curves, corresponding to stretching of the cell by 10 μm (filled circles) and by 12 μm (filled squares) demonstrate practically linear dependency and reverse at – 5 mV and 0 mV accordingly.

8.4.1.3 I_{SAC}, The Stretch-Activated Current Through Non-selective Cation Channels

Investigation of I_{SAC} requires blockade of other current components in ventricular myocytes. Such blockade of K^+ currents is readily achievable after substitution of K^+ ions by Cs^+ ions in perfusion and intracellular solutions. In order to remove potassium from the cell interior investigators dialyze cells with cesium based solution (Isenberg, 1976).

Voltage Dependence of I_{SAC}

Late currents, resembling I_{SAC}, which are measured at the end of depolarization step (Fig. 8.20a), are used for plotting of current voltage relationships (Fig. 8.20b). Before the mechanical stimulation (open triangles), current-voltage dependance was

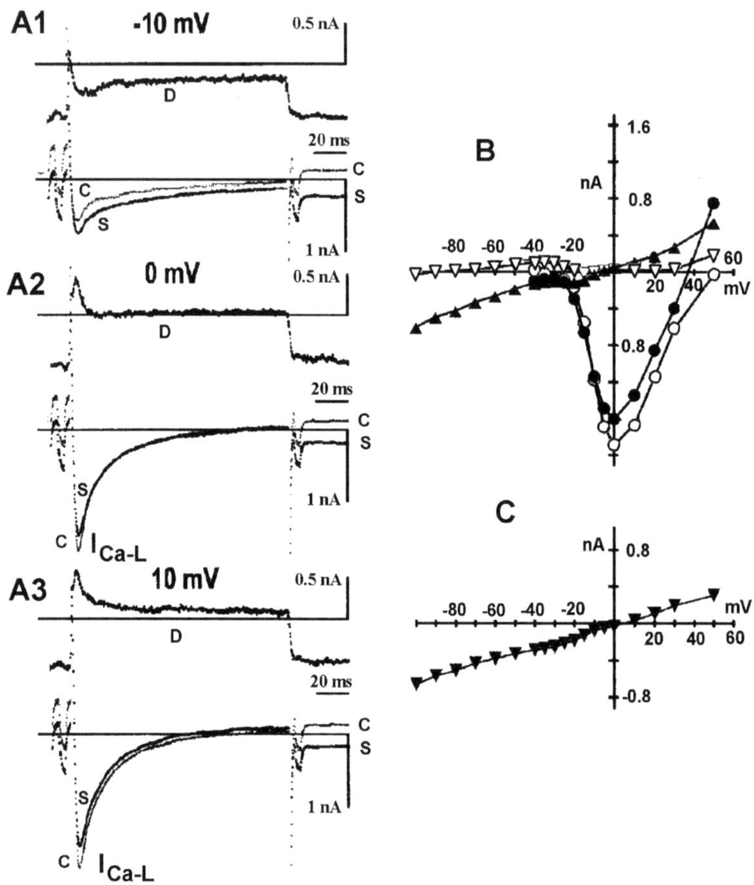

Fig. 8.20 Voltage dependence of the stretch induced current I_{SAC} through non-selective cation channels. Currents at the end of 140 ms pulses were plotted versus the potential of the clamp pulse. Superimposed K^+ currents were blocked by replacing K^+ by Cs^+ ions in both extracellular Tyrode solution and electrode solution. (**a**) Holding potential –45 mV, 140 ms pulses to –10 mV (**a1**) or 0 mV (**a2**) or 10 mV (**a3**). Traces of currents before (label C) and during 8 μm stretch (label S), and stretch induced difference currents (label D). (**b**) Current voltage relation (*I-V* curve) of late current I_L measured at end of pulse before stretch (*empty triangles*), during 8 μm stretch (*filled triangles*). (**c**) Definition of I_{SAC} as stretch induced difference current. Reproduced from Kamkin et al. (2000a) with copyright permission of the Elsevier, Oxford University Press and Cardiovasc Res

flat small because Cs^+ blocks the K^+ current components. The slope at negative potentials is used for determination of input resistance of the cell, which for this study was 500 MΩ. Application of a 8 μm stretch shifts the I-V curve (filled triangles) and lowers the input resistance to 62.5 MΩ. Stretch, flattens the I-V curve, which intersects intersected voltage axis at –5 mV. Stretch induced difference current I_{SAC} (Fig. 8.20c) has almost linear voltage-dependence, i.e. its conductance G_{SAC} remains constant regardless of the potential.

Linear I-V curves with reversal potentials between 0 and –15 mV were reported for both axial stretch (Belus and White, 2003; Zeng et al., 2000) and for local stretch (Kamkin et al., 2000a, 2003c). Since at potentials negative to –10 mV I_{SAC} is inward, its activation can depolarize the diastolic membrane potential, and it can retard the repolarization phase 2, from the plateau back to the resting potential. On the controrary at positive potentials I_{SAC} been outward should shorten the initial repolarization phase 1. Thus, the properties of the mechanically activated current I_{SAC} can explain the stretch induced modification of the resting and action potentials in non-clamped cardiomyocytes.

I_{SAC} is a Non-selective Cation Current

Investigation of permeability properties of I_{SAC} demonstrated that chloride ions do not contribute to it, which essentially means that ion channels, which are responsible for I_{SAC} generation, belong to a cation selective channel (Kamkin et al., 2000a, 2003c; Zeng et al., 2000). SACs are best permeable for Cs^+ ions (Kamkin et al., 2000a; Zhang et al., 2000). The reversal potential of I_{SAC} is laying in the range of –10 mV. This indicates that K^+ and Na^+ ions pass through the channel with the same permeability. For mouse ventricular myocytes superfused with a Ca^{2+} free medium, the permeability ratio is P_{Cs}:P_K:P_{Na}:P_{Li}:P_{TMA}:P_{NMG} = 1.3:1.1:1.0:0.8:0.1:0.03 (Kamkin et al., 2000a). Even large organic cations like tetramethyl ammonium (TMA^+) or N-methyl glucosamine (NMG^+) can carry a small inward current through SACs. In another study of rat atrial myocytes (different extracellular Ca^{2+} concentrations: 0 mM in Kamkin et al., 2000a, 0.2 mM in Zeng et al., 2000 and 1.8 mM in Zhang et al., 2000 and the patch electrode: low or high concentrations of EGTA or BAPTA) the permeability ratio is P_{Cs} :P_{Na} :P_{Li} =1.05:1:0.98 for the stretch activated current and P_{Cs} :P_{Na} :P_{Li}= 1.49:1.0:0.79 for the non-selective background current.

In guinea pig ventricular cells authors report Ca^{2+} carried inward current (150 mM Na^+ ions were replaced by 90 mM Ca^{2+}). At –100 mV, there was a Ca^{2+} carried inward current that reversed polarity at +15 mV. This study reported similar permeability of SACs for Ca^{2+} ions (passing inwardly) and K^+ ions (flowing outwardly) ions. However, for calcium carried I_{SAC} the I-V curve was no longer linear. Negative to –20 mV it bended to become nearly independent of membrane potential (Kamkin et al., 2000a, 2003c). In summary, SACs are Ca^{2+} permeable, albeit with low permeability (Wellner and Isenberg, 1993; Hamill and Martinac, 2001).

Cations La^{3+} or Gd^{3+} (3-valent) block I_{SAC}, while Ca^{2+} is known to hinder the permeation of the Cs^+ ions throught SACs (Kamkin et al., 2000a). 8 μM concentration, which is reported by this group to be sufficient for saturation of the blocking effects, is much lower than 100 μM concentrations used in other reports. The difference between those concentrations can result from the fact that the on-rate of the block increases with the concentration. Gd^{3+} block of I_{SAC} is not completely reversible due to high affinity to Gd^{3+} (Yang and Sachs, 1989).

SACs and non-selective background channels are both expressed in rat atrial myocytes. One study reported a Gadolinium block of SACs with IC_{50} of 46 μM.

Still it is important to note that even 100 μM of Gd^{3+} is insufficient to completely abolish stretch-induced I_{SAC} and diastolic depolarization (Zhang et al., 2000), though this report was not confirmed by other groups yet. Gd^{3+} is not a selective blocker of SACs since it also inhibits potassium currents I_{K1} and I_{Ko} (Hongo et al., 1997). Amiloride is also used to block SACs (Hamill et al., 1992), though it also has a low selectivity to SACs.

8.5 Compression of Isolated Ventricular Cardiomyocytes

It is obvious, that each cardiomyocyte in multicellular preparations is disposed to different extent of stretch or compression even in the pause between contraction and relaxation of the myocardium. Most cardiomyocytes are compressed during each contraction (during a systole). The brick-like cardiomyocytes have narrow and wide sides. Therefore, we considered two variants of a compression of cardiomyocytes – a compression on the narrow side and a compression on the wide side. It was not difficult to perform these experiments, because brick-like isolated cardiomyocytes, stuck to a bottom of perfusion chamber in two positions: edgewise, staying on the narrow side or broadwise. In some experiments after seal formation and whole cell access, cardiomyocytes were rolled by the patch-pipette to attach to the glass bottom from edgewise, staying on the narrow side (Fig. 8.6a) to broadwise (Fig. 8.6d). It allowed on the same cardiomyocyte, placed in different positions, to the study of whole-cell current during stretch and compression.

Cardiomyocytes were compressed by pressing the cell surface with the glass stylus towards the glass bottom of the chamber (Fig. 8.21). The effects of compression depended on the orientation of the myocyte (Fig. 8.21a or b).

Figure 8.22 demonstrated the effect of local pressure on the cardiomyocyte, with K^+ ions in both, bath and electrode solution, which is in edgewise position (Kamkin et al., 2005b). During compression, we registered depolarization of the resting membrane and reduction of I_{K1}. Voltage dependence of compression-induced currents reminded voltage dependence of stretch-induced currents (see Fig. 8.19b1). This reduction of I_{K1} was blocked by 8 μM of non-selective blocker Gd^{3+}. Gadolin-

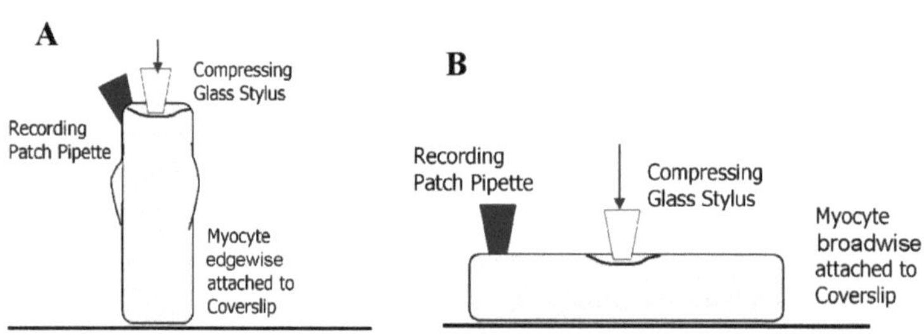

Fig. 8.21 Schematic drawing for the two variants of compression of cardiomyocytes – compression on the narrow side (**a**) and compression on the wide side (**b**)

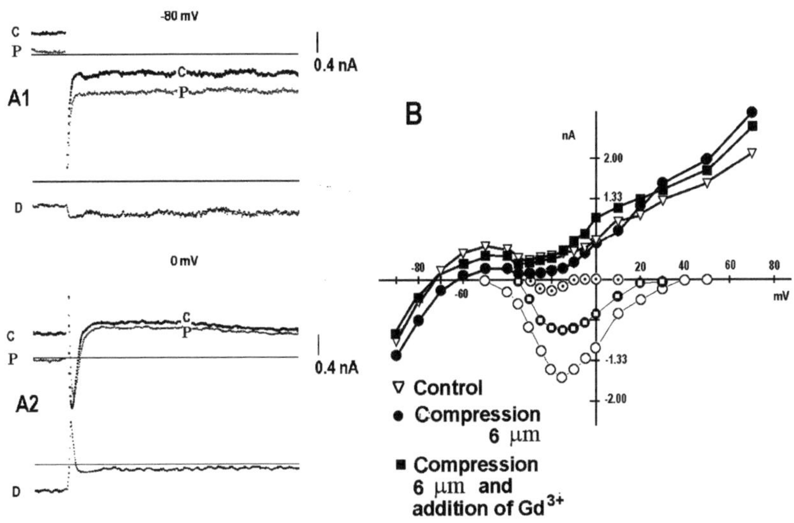

Fig. 8.22 Pressure on the cell by 6 μm, which is in edgewise position with K^+_{in}/K^+_{out} solutions. (**a**) Starting from –45 mV, the membrane potential was adjusted for 140 ms to –80 mV (**a1**) and 0 mV (**a2**), respectively. Shown are the net membrane currents before (labeled as C) and during application of mechanical pressure (labeled as P), as well as the compression-induced difference currents (labeled as D). Note, that the compression-activated current is more negative than the pre-compression current at –80, –45 mV and 0 mV. The current is inward at both –45 and –80 mV. (**b**) Current-voltage relation of the late current I_L measured at the end of the test pulse (*empty triangles* – before compression; *filled circles* – during 6 μm compression; *filled squares* – during compression by 6 μm after addition of Gd^{3+}. Note, that $I_{L\text{-}Ca}$ decreased during compression of the cell (*empty circles* in comparison to empty *fat circles*). Gd^{3+} blocked $I_{L\text{-}Ca}$ (*empty circles* with point). Reproduced from Kamkin and Kiseleva (2005b) with copyright permission from Academia Publishing House Ltd

ium also blocks L-type calcium channels in isolated ventricular myocytes of the guinea-pig (Lacampagne et al., 1994), in mouse ventricular myocytes (Kamkin et al., 2003c), and ventricular myocytes from human heart (Kamkin et al., 2000a). Moreover, gadolinium blocks the delayed rectifier potassium current in isolated guinea-pig ventricular myocytes (Hongo et al., 1997). Nevertheless, it was possible to assume the activation of I_{PA} (pressure activated current – PA) through non-selective cation channels. Negative I_{PA} appeared and decayed rapidly upon exposure/release of compression. I_{PA} was –0.12 ± 0.04 nA at 2 μm, –0.28 ± 0.05 nA at 4 μm; –0.32 ± 0.02 nA at 6 μm, and –0.70 ± 0.1 nA at 8 μm compression (at –45 mV). Compression also reduced I_{Ca}. Since this effect was prevented by cell dialysis with 5 mM BAPTA, the reduced I_{Ca} may be caused by elevated $[Ca^{2+}]_c$ (Kamkin et al., 2000a, 2003c). Thus, edgewise attached cells responded to compression with depolarization of the resting membrane, and reduction of I_{K1} through inwardly-rectifying K^+-channels.

Later Dyachenko et al. (2008) reported that edgewise attached cells respond in the same way to stretch and compression (Fig. 8.23) due to x-stretch/x-compression. Stretch shifted the I-V curve downwardly. Assuming I_{K1} to be zero at E_K, $I(E_K)$

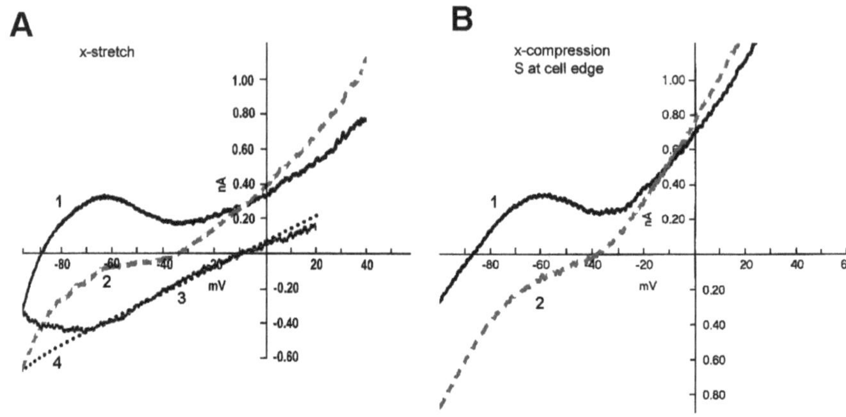

Fig. 8.23 Modulation of ion currents by mechanical deformation. I–V curves before 1 and during mechanical stimulus 2. (**a**) 10 μm axial x-stretch. Difference current 3 deviates from linear voltage dependence 4. (**b**) 10 μm axial x-compression without sarcomere lengthening (stylus attached to left cell edge) changed I–V curve similar as stretch. Reproduced from Dyachenko et al. (2008) with copyright permission from Elsevier and Prog Biophys Mol Biol

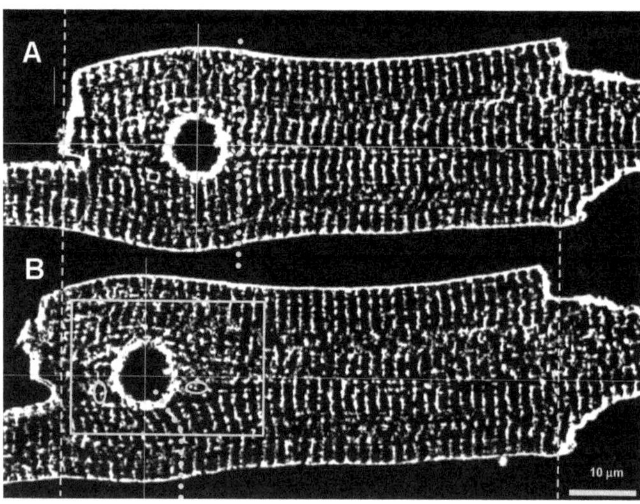

Fig. 8.24 Reversible z-line bending by 10 μm x-stretch – xy images before (**a**) and during stretch (**b**). *Arrow* marks $\Delta x = 8.3$ μm. *Dashed lines* for easier recognition of displaced cell edges. Leftward movement of S bends T-tubular stripes to the left, dotted lines mark angle of z-line bending in reference to y-axis and to z-axis. Reproduced from Dyachenko et al. (2008) with copyright permission from Elsevier and Prog Biophys Mol Biol

is interpreted to indicate I_{ns}. The stretch-induced difference current (Fig. 8.23a) reversed polarity at –9±3 mV. Figure 8.24 demonstrate reversible z-line bending by 10 μm x-stretch – xy images before (a) and during stretch (b). Axial compression of edgewise attached cells changed the I-V curves similar as axial stretch did (Fig. 8.23b)

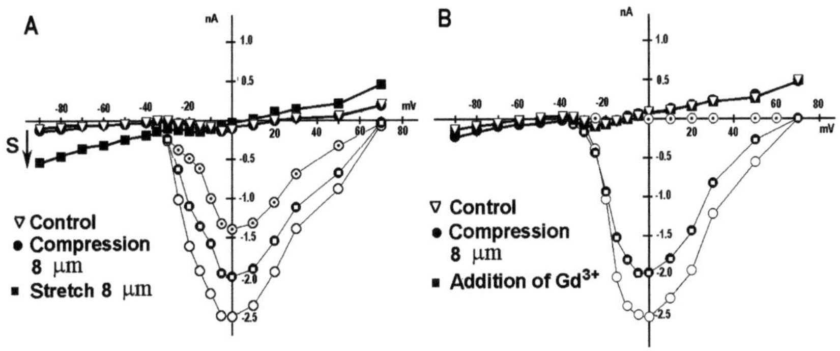

Fig. 8.25 Mechanosensitive currents in edgewise attached cells after block of K-currents by substitution of both extracellular and intracellular K^+-ions with Cs^+-ions. (**a**) Compression by 8 μm does not induce I_{PA} for example through non-selective cation channels – I_{NS}, (*empty triangles* before compression in comparison to *filled circles* during compression) but reduces $I_{Ca\text{-}L}$. Following compression, 8 μm stretch was applied; stretch induced I_{SAC} (*filled squares*). Note: $I_{L\text{-}Ca}$ before (*empty circles*), during (*empty fat circles*) compression and during stretch (*empty circles* with point). (**b**) Another cell in Cs_{in}/Cs_{out} solutions. Compression by 8 μm did not induce an inward I_{NS}, (*empty triangles* before compression in comparison to *filled circles* during compression), and addition of Gd^{3+} did not change the late currents either (*filled squares*). The L-type calcium current, $I_{L\text{-}Ca}$, however, is suppressed suggesting that the compression was effective (*empty circles* before compression in comparison to empty fat circles). Gd^{3+} blocked the $I_{L\text{-}Ca}$ (*empty circles* with point). Reproduced from Kamkin and Kiseleva (2005b) with copyright permission from Academia Publishing House Ltd

After block of K^+-currents by substitution of both extracellular and intracellular K^+-ions with Cs^+-ions (Fig. 8. 25a) even compression by 8 μm does not induce I_{PA} (Kamkin et al., 2005b). However, the stretch by 8 μm of the same cell in all cases caused the appearance of I_{SAC} through non-selective cation channels (Fig. 8.25a). It is impossible to tell that the compression as against a stretch does not influence to the cell, because in absence of BAPTA in both cases (at a compression and a stretch) L-type calcium current decreases (Fig. 8.25). Gadolinium does not render influence on a late current I_L of the cell subject to a compression even by 8 μm (Fig. 8.25b). All three curves – a control curve of a I_L, that is before compression of a cell, I_L curve on a background of a compression and I_L curve after addition of gadolinium are similar in all of experiments. Thus, negative to –45 mV, I_{K1} rectified inwardly, and was blocked by replacing extracellular and intracellular K^+ by Cs^+ ions.

As a whole, pressure on the cell, which is in edgewise position with K^+_{in}/K^+_{out} solutions modulated I_{K1} through inwardly-rectifying K-channels. The effect of compression of edgewise cells is not shown in Cs^+_{in}/Cs^+_{out}, though the cell reacted to deformation as $I_{L\text{-}Ca}$ decreased and reacted to stretch (Kamkin et al., 2005b).

Similar observations were reported in following studies (Dyachenko et al., 2008). The deviation of the difference current from the straight line was attributed to the stretch-deactivation of G_{K1} because it was abolished by substituting extracellular K^+ by Cs^+ ions or by intracellular dialysis of Kir2.3 antibodies (Dyachenko et al.,

2009a). Stretch reduced maximal G_{K1} by 25±11% without changing the gating parameter. At positive potentials, stretch increased the outward currents, mostly due to outward current through G_{ns}. The changes in G_{ns} and G_{K1} increased with the amplitude of stretch and were reversible, i.e. they disappeared upon relaxation from stretch (Dyachenko et al., 2009a). For the simultaneous activation of non-selective cation currents and deactivation of K1-currents we have introduced the term "stretch modulation of ion currents" (SMIC) (Dyachenko et al., 2009a).

Figure 8.26 depicts the effect of local pressure on the cardiomyocyte, with K^+ ions in both bath and electrode solution, which is in broad-wise position. Hyperpolarization is seen as shift of the zero current potential from –75 to –80 mV. When pressure was applied to broad-wise attached myocytes, it increased I_{K1} through inwardly-rectifying K^+-channels. Nevertheless, it was possible to assume the inactivation of I_{PA} (pressure activated current – PA) through non-selective cation channels (Kamkin et al., 2005b).

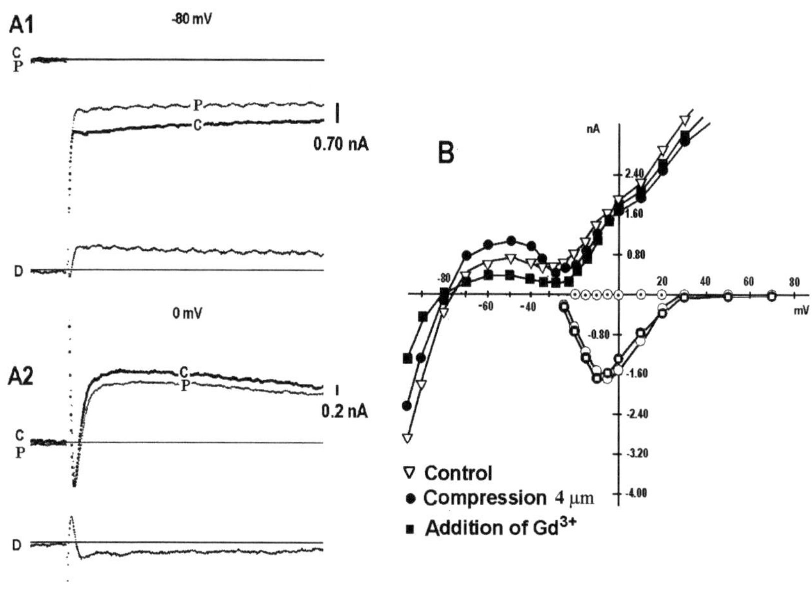

Fig. 8.26 Pressure on the cell, which is in broad-wise position with K^+_{in}/K^+_{out} solutions. (**a**) Starting from –45 mV, the membrane potential was adjusted for 140 ms to –80 mV (**a1**) and 0 mV (**a2**), respectively. Shown are the net membrane currents before (labeled as C) and during application of mechanical pressure (labeled as P), as well as the compression-induced difference currents (labeled as D). Note, that the compression-activated current is more positive than the pre-compression current at –80, but more negative at 0 mV. (**b**) current-voltage relation of the late current I_L measured at the end of the test pulse (*empty triangles* – before compression; *filled circles* – during 4 μm compression; *filled squares* – during compression) by 4 μm after addition of Gd^{3+}. Note, that there was no change in the L-type calcium current $I_{L\text{-}Ca}$ (*empty circles*– before compression, *empty fat circles* –during compression). Gadolinium blocks $I_{L\text{-}Ca}$ (*empty circles* with point). Reproduced from Kamkin and Kiseleva (2005b) with copyright permission from Academia Publishing House Ltd

Thus, pressure on the cell, which is in broadwise position with $K^+{}_{in}/K^+{}_{out}$ solutions increased I_{K1} through inwardly-rectifying K-channels. It could be possibly also deactivation of a background I_{NS}. Compression did not reduce $I_{Ca}6$

When broadwise-attached cells with blocked K^+ currents were studied (Kamkin et al., 2005b), compression by 4 μm decreased late current I_L. It could be deactivation of I_{NS} (Fig. 8.27b). However, the stretch by 4 μm of the same cell in all cases causes the appearance of I_{SAC} through non-selective cation channels (Fig. 8.27b).

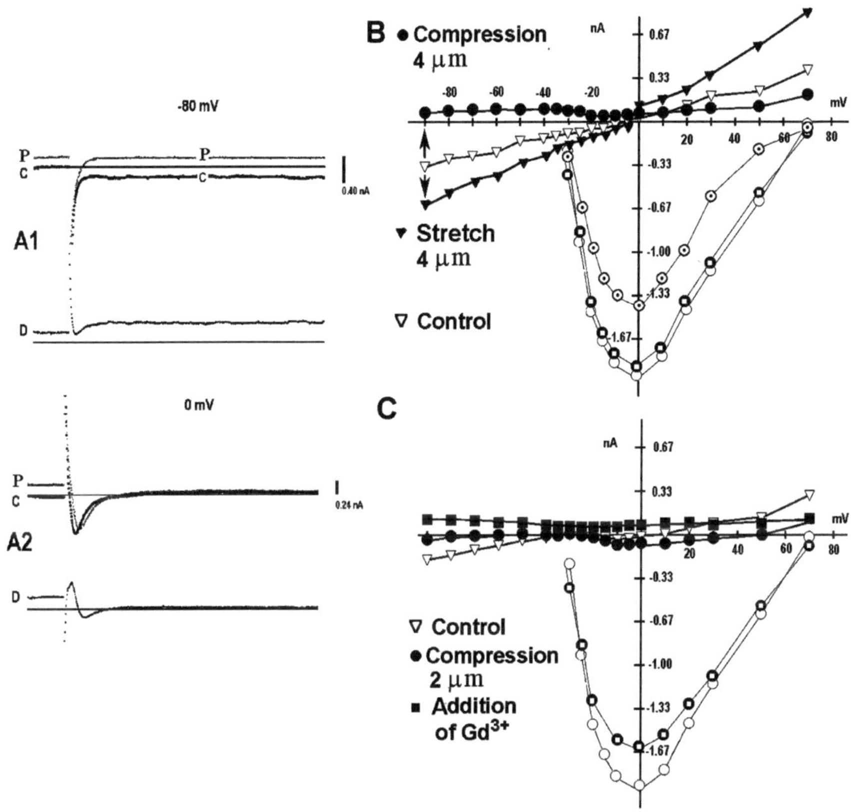

Fig. 8.27 Broad-wise attached cell with blocked K-currents (Cs_{in}/Cs_{out} solution). (**a**) Starting from –45 mV, the membrane potential was adjusted for 140 ms to –80 mV (**a1**) and 0 mV (**a2**). Shown are the net membrane currents before (labeled as C) and during application of mechanical pressure (labeled as P), as well as the compression-induced difference currents (labeled as D). (**b**) Compression by 4 μm does not induce I_{PA} for example through non-selective cation channels – $I_{NS,}$ (*empty triangles* before compression in comparison to *filled circles* during compression) and eliminates inward component of late current I_L. Following compression, 4 μm stretch was applied and demonstrates stretch induced I_{SAC} (*filled triangles*). Note: $I_{L\text{-}Ca}$ before (*empty circles*), during (*empty fat circles*) compression, and during stretch (*empty circles* with point). (**c**) Compression by 2 μm shifts the I_L (*empty triangles* before compression in comparison to *filled circles* during compression) in the positive direction. Addition of Gd^{3+} during compression (*filled squares*). Reproduced from Kamkin and Kiseleva (2005b) with copyright permission from Academia Publishing House Ltd

After compression by 4 μm, the curve of late current I_L takes a position, which is inconveniently to explain for us. Stronger compression does not result in change of the I_L curve. If on the cell to apply a smaller compression by 2 μm, typical reduction of a late current is registered (Fig. 8.27c). However, addition of gadolinium results in a curve that takes the same position as the one at a compression by 4 μm (Fig. 8.27c).

Pressure upon the outer cell surface can modify several ionic currents. It is known, for example, that some mechanosensitive channels are activated, when force of pressure is directed in relation to the nucleus of a cell. This has been reported for glial cells (Bowman et al., 1992), smooth muscle cells (Kawahara, 1993), and endothelial cells (Marchenko and Sage, 1996).

Those results suggested that compression modulates inwardly-rectifying K^+-channels. Whether the conductance is activated or deactivated depends on the orientation of the compression in regard to the long or short diameter of the cell. Compression has different effects, dependent on the angle to the width and height of the cell (edge- and broadwise attached cells) (Kamkin et al., 2005b).

Later it was shown (Isenberg et al., 2003) by analyzing compression with a ramp command, that edgewise attached myocytes responded to compression with a decrease of the K^+ currents I_{Ko} and I_{K1}. When pressure was applied to broad-wise attached myocytes, it increased the two potassium currents I_{K1} and I_{Ko} (Fig. 8.28). The opportunity of activation of a I_{NS} during a local compression of cardiomyocytes in different position is not clear and demands further experiments.

In broad-wise attached cells, the deformation has a lower working space of e.g. 6 μm only, i.e. high amplitudes of stretch damage more easily. The stress

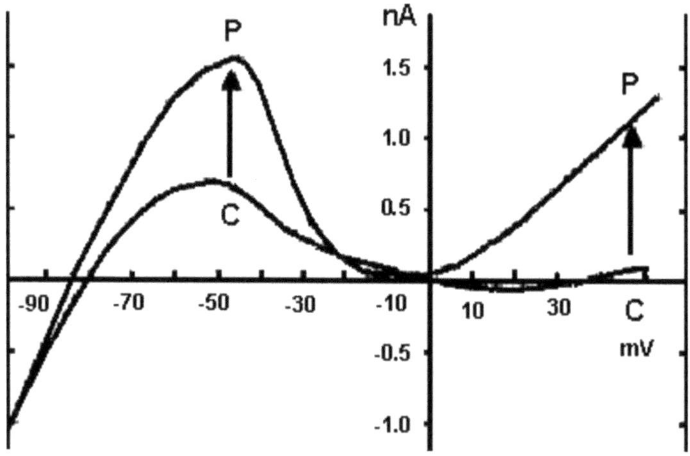

Fig. 8.28 Currents before and during 3 μm compression, broad-wise attached myocytes. Note: hyperpolarisation is seen as shift of the zero current potential from −80 to −84 mV. Note: Currents measured before and during compression with a ramp command from +60 to −100 mV at 100 mV/s. Modified from Isenberg et al. (2003), with copyright permission from Elsevier and Prog Biophys Mol Biol

may be translated up to the interface between the cell button and the coverslip. There might be less bulging at the lateral side because the distance from glass stylus to outer membrane is longer, or it might be more energy dissipated inside the cell.

The dependence on the direction of the effects of local deformation can be explained as follows: a similar standard movement of the stylus will reduce the height in broad-wise attached cells more than in edgewise attached cell, however, it will increase the local diameter that stretches the cell surface more in the edgewise than in the broad-wise attached cells (Isenberg et al., 2003).

The asymmetry of the response is difficult to explain with the current models that attribute channel gating to changes in the surface tension of the lipid bilayer. The responses are easier to understand on the assumption that the energy of local compression is transferred by cytoskeletal elements to the channel protein (Kamkin et al., 2003c).

Treatment of the cells with cytochalasin D, which is thought to disrupt F-actin reduced the amplitude of I_{SAC} during continuous stretch (Kamkin et al., 2003c). It attenuated the activation of I_{K1} by pressure of broad-wise attached cells. When the cells were pre-treated before application of stretch, the mechanosensitivity was reduced or abolished, i.e. the mechanical stimuli (stretch or pressure) became ineffective (Kamkin et al., 2003c). Nearly identical results were obtained by cell dialysis with 5 μM colchicin, i.e. depolymerisation of tubulin reduced or abolished the mechanosensitivity of I_{NS} and I_{K1} (Isenberg et al., 2003). From these observations, we concluded that an intact cytoskeleton is necessary for the mechanosensitive gating of ion channels and K^+ gating. We might see the cytoskeleton as part of a pathway that transforms the exogenous mechanical energy into activation energy of membrane channels. This hypothesis is in line with the increased mechanosensitivity of I_{NS} in hypertrophied cells (Kamkin et al., 2000a, b; Kiseleva et al., 2000) with enhanced stiffness resulting from a pathological high expression of tubulin in the cytoskeletal cortex (Watson et al., 1996).

8.6 Freshly Isolated Ventricular Cardiomyocytes from Hearts with Pathology

8.6.1 Modulation of Resting Membrane Potential and Action Potential Properties

Investigation of effects of stretch on freshly isolated ventricular cardiomyocytes of healthy young rats (3 months old), performed by means of patch clamp method in whole cell configuration when cells were perfused by normal physiological Tyrode, showed that stretching them by 2 and 4 μm did not induce depolarization and did not affect APD duration. Further increase of the extent of stretch applied was sufficient to induce depolarization of the membrane and prolongation of the APD90 by

32% in all cells under investigation. So stretching cardiomyocyte by 8 μm caused depolarization by 6 mV and prolongation of APD by 32%. Stretching by same 8 μm of healthy old rat ventricular cardiomyocytes (15 months old rats were used) caused much greater depolarization by 11 mV and APD90 prolongation by 43%. At this level of stretching authors observed appearance of extra action potentials. Stretching of old rat ventricular cardiomyocytes (15 months old rats were used), which suffered from spontaneous hypertension, by only 2 μm already caused depolarization by 8 mV and APD90 prolongation by 39%. Stretching cardiomyocytes of those rats by 4 and 6 μm in all cases induced extra action potentials (Kamkin et al., 2000a).

Freshly isolated ventricular cardiomyocytes from human patients suffering from infarction respond to stretch in the same manner as guinea pig cardiomyocytes (Kamkin et al., 2000a). However in order to achieve comparable magnitude of changes of resting membrane potential and action potential from cardiomyocytes of pathological human hearts one needs mechanical stimulus, which is from twice to four times smaller in magnitude than the one, which is sufficient for cardiomyocytes from healthy guinea pigs. This high sensitivity to stretch of freshly isolated cardiomyocytes from hearts of patients, who suffer from cardiac disease, goes in hand with data, acquired by means of microelectrode method from cardiac tissue fragments of animals with infarction (Kamkin et al., 2000b; Kiseleva et al., 2000). Stretching of isolated cells by just 2 μm depolarizes cardiomyocyte by 10 ± 3 mV. Besides that APD90, which was 341 ± 15 ms before application of stretch, increases to 571 ± 68 ms during stretch by 2 μm. Stretching speeds up repolarization during initial phase 1 of AP, but prolongs repolarization during late phase 2 of AP (Kamkin et al., 2000a).

Confirmation of increased sensitivity of single enzymatic isolated hypertrophied cardiac cells to stretch (Kamkin et al., 2000a) is of high importance since it complements previous studies, which reported increased sensitivity to stretch of hypertrophied ventricular (Kiseleva et al., 2000) and atrial tissue (Kamkin et al., 2000b).

During investigation of stretch induced current stretching of cells to predefined levels (for example by 10 μm) took only 200 ms, and this time was defined by the frequency of switching between steps of the motor of the micromanipulator. Activation of I_{SAC} (at –45 mV) was observed during whole 200 ms of stretching (similar observation was reported in a study of effects of stretch on cardiomyocytes from hearts of human patients with pathology, guinea pigs and rats) (Kamkin et al., 2000a). Fast activation of I_{SAC} confirms earlier findings from investigation of effects of stretch of cardiomyocytes from healthy rats (Zeng et al., 2000). During stretch I_{SAC} (at –45 mV) was constant for the whole duration of the recording, which lasted for minutes, without showing any signs of inactivation or adaptation.

Absence of inactivation in the end of long lasting application of stretch was also found in ventricular cardiomyocytes from human patients with infarction, in ventricular cardiomyocytes of healthy guinea pigs, healthy and spontaneously hypertensive rats (Kamkin et al., 2000a).

8.6.2 Dependence of Mechanosensitivity of Cardiomyocytes from Age and Presence of Hypertension

I_{SAC} amplitude of freshly isolated cardiomyocytes is increasing with increase of stretch. Stretching of cardiomyocytes by 2 μm from patients with depressed calcium current at –45 mV, triggered I_{SAC} of –116±25 pA, while stretch by 4 μm produced I_{SAC} of –483±28 pA (Fig. 8.29) (Kamkin et al., 2000a).

At holding potential of –45 mV stretch by 2 μm of cardiomyocytes of healthy guinea pigs was not sufficient to trigger I_{SAC}. Stretching by 4 μm was sufficient to trigger I_{SAC} of –65±21 pA only in 2 out of 19 cells. It is important to note that stretching by values exceeding 4 μm produced I_{SAC} in all cells. Stretching by 6 μm triggered I_{SAC} of –300 ± 89 pA, while stretching by 8 μm lead to I_{SAC} of –557 ± 78 pA, and stretch by 10 μm triggered I_{SAC} of –1050 ± 190 pA (Fig. 8.29) (Kamkin et al., 2000a).

Thus cardiomyocytes from human patients after infarction have much greater sensitivity to stretch then cardiomyocytes of healthy guinea pigs. This high sensitivity can be explained by species differences, which is in our opinion less likely then explanation that it originating from hypertrophy, which results from pathological cardiac tissue remodeling (Boluyt et al., 1995). In described study (Kamkin et al.,

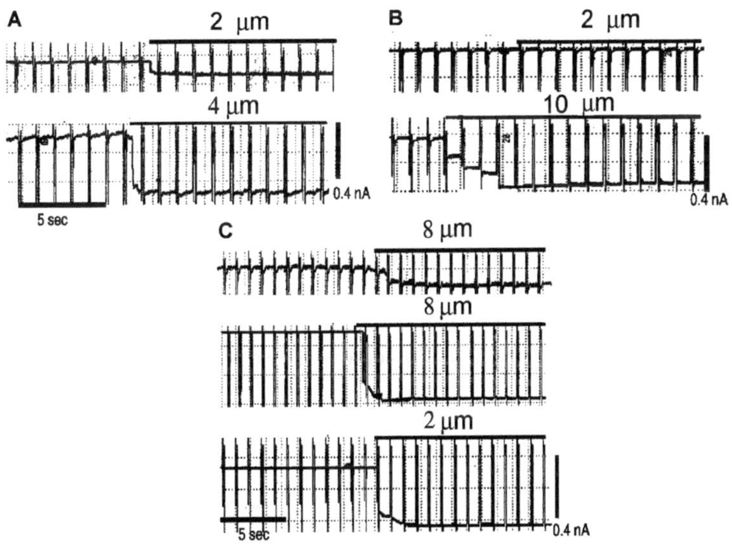

Fig. 8.29 Current recordings before and after application of stretch of the cardiomyocyte (pen recordings). I_K was suppressed. (**a**) Human ventricular cardiomyocyte from patients after infarction. (**b**) 3 months old guinea pig ventricular cardiomyocyte, (**c**) rat ventricular cardiomyocyte. Top trace – healthy 3 months old rat, middle trace – healthy 15 months old rat, bottom trace – spontaneously hypertensive 15 months old rat. The line above each current recording indicates application of stretch. Reproduced from Kamkin et al. (2000a) with copyright permission from Elsevier, Oxford University Press and Cardiovasc Res

2000a) cellular hypertrophy was determined by morphological and electrophysiological methods. Measurements of membrane capacity showed that it was twice as big for cardiomyocytes of human patients in comparison with cardiomyocytes from healthy guinea pigs.

In order to test the hypothesis that this high sensitivity to stretch is linked with ventricular hypertrophy one group investigated the effects of stretch on cardiomyocytes of healthy young rats and spontaneously hypertensive rats. Hypertensive rats did not show any signs of heart failure, however their hearts were hypertrophied, which was determined by heart-to-body weight measurements. This parameter of hearts of spontaneously hypertensive rats was twice of that of healthy young rats (Fig. 8.29). Besides that the capacity of the membrane and the surface of the cellular slice was also two times as big. During blockade of potassium currents rat ventricular cardiomyocytes responded to stretch by I_{SAC}, which was very similar to I_{SAC} of cells of guinea pigs, i.e. had a linear I-V dependence, reversal potential of -4 ± 3 mV, and was suppressible by 5 μM of Gd^{3+}.

Figure 8.30 summarizes data from animal and human experiments (Kamkin et al., 2000a; Kamkin and Kiseleva, 2008b). Stretching of ventricular cardiomyocytes of young 3 months old rats by 2 and 4 μm did not trigger I_{SAC}. Further increase of stretch produced I_{SAC} in all cells. Stretch of the cardiomyocyte by 8 μm induced inward current of -269 ± 40 pA (at holding potential of -45 mV). Similar stretch of ventricular cardiomyocytes of healthy 15 months old rats by 8 μm triggered much greater I_{SAC} of -460 ± 55 pA, while stretching of ventricular cardiomyocytes of spontaneously hypertensive 15 months old rats by just 2 μm triggered I_{SAC}, of -420 ± 110 pA. Stretch of cardiomyocytes from hypertensive rats by 4 μm triggered I_{SAC} of -1205 ± 110 pA, and by 6 μm – I_{SAC} of -1500 pA.

Sensitivity of cells to stretch is characterized by the angle of curves, which describe I_{SAC} as a function of stretch. Approximation of sensitivity to stretch was assessed as velocity of change of the curve during stretch by more then 4 μm (linear regression). For rat cardiomyocytes this sensitivity was 77 pA/μm for young (3 months old) rats and 132 pA/μm for old (15 months old) rats, and it was 270 pA/μm for cardiomyocytes of old (15 months old) spontaneously hypertensive rats. This demonstrates that cellular sensitivity to stretch is increasing with hypertrophy, which is developing with age and to much greater extent with hypertension.

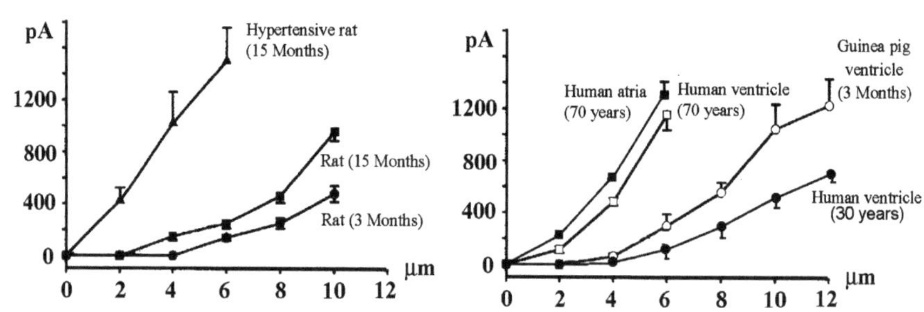

Fig. 8.30 Dependence of I_{SAC} from the extent of stretch, which was applied to the cardiomyocyte

Stretching of human atrial and ventricular cardiomyocytes (Fig. 8.30) from patients after myocardial infarction by 2 and 4 μm triggered I_{SAC} in the range from –300 pA to –600 pA, respectively. Stretching by 6 μm induces inward current of approximately –1200 pA. Similar extent of stretching by 6 μm of human ventricular cardiomyocytes from healthy humans initiates much smaller I_{SAC} of –200 pA.

Sensitivity to stretching of cardimyocites from sick human patients was 260 pA/μm, which is similar to that of myocytes from spontaneously hypertensive old rats. In general Fig. 8.29 demonstrates that sensitivity of cardiomyocytes to stretch is increasing with age and is extremely high in case of ventricular hypertrophy (Kamkin et al., 2000a).

8.7 From Isolated Cell with Patch-Pipette to Heart Tissue with Microelectrodes

The above information completely explains the earlier data obtained from the heart tissue by intracellular microelectrodes. More than that, these data obtained from heart tissue fragments or whole heart now acquire a totally different meaning. Early results, received from stretching tissues by various methods with biopotentials registered as the background, have been completely confirmed. More than that, during recent years at the isolated cell level were shown practically all the basic mechanisms of the processes registered by microelectrodes in the conditions of tissue stretching. Further on we show how the mechano-induced effects manifest in healthy and sick heart tissue, at that the cited works were performed by microelectrode technique, i.e. with high precision. This is especially important because in heart tissue each cell is naturally surrounded by other cells and therefore it creates maximal closeness to natural conditions. This opens up considerable opportunities for potential application of scientific data by practicing doctors and for creating new medicines. In other words the present information makes possible the "back to future" transition, i.e. using older methods for receiving new data.

8.8 Mechano-Electric Feedback in Atrium from Healthy and Diseased Animals and Human

8.8.1 Mechano-Electric Feedback in Right Atrial Tissue in Healthy Rats

The intact atrial tissue shows mechano-electric transduction, as has been reviewed from experiments with extracellular registration (see for example: Lab, 1998; Nazir and Lab, 1996). Mechano-electric feedback in atrial cardiomyocytes, studied by the intracellular microelectrode technique has been shown in rat hearts (Kamkin et al., 1998, 2000b). Figure 8.31 demonstrates one representative example of the effect of sustained stretch (1.75 ± 0.04 mN) in control healthy (sham operated) rats on action potential duration (APD) at three different time points of repolarization. APD25 was

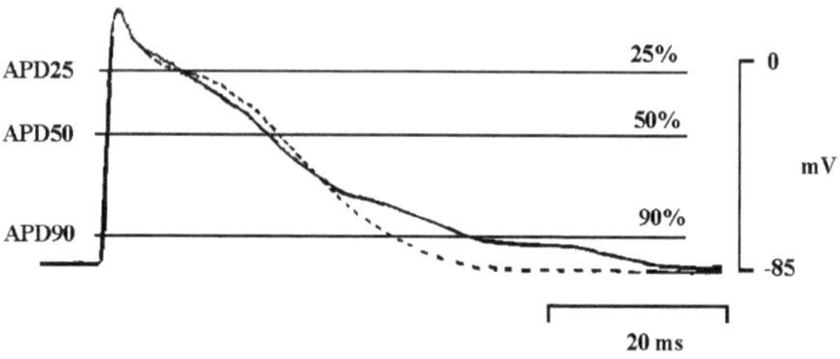

Fig. 8.31 Superimposed action potentials from right atrial cardiomyocytes in sham operated rats at a preload of 1 mN (*dotted line*) and 1.7 mN (*solid line*) of stretch. Stretch produces depolarization (SID) near APD90. Reprinted from Kamkin et al. (2000b), with permission from Elsevier and J Mol Cell Cardiol

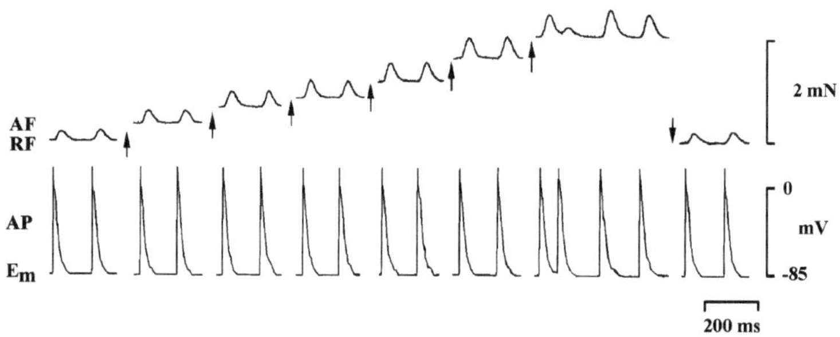

Fig. 8.32 Response of the resting membrane potential (E_m) and action potential (AP) of an atrial cardiomyocyte from a sham operated rat to mechanical stretch. Stepwise increases in resting force due to physical stretch (indicated by ↑) up to 1.7 mN significantly increased APD90. A further increase in stretch to 2.0 mN produced premature extra-action potentials (extra-AP). Removal of stretch (indicated by ↓) reversed these effects. Reprinted from Kamkin et al. (2000b), with permission from Elsevier and J Mol Cell Cardiol

unaffected, APD50 was shortened, whereas APD90 was increased significantly. The increase in APD90 was due to delayed repolarization resulting from stretch-induced depolarization (SID).

A further increase in stretch to 2 mN evoked extra-APs (Fig. 8.32). These effects were completely reversible upon release of stretch. Resting membrane potentials (E_m) and AP amplitudes were not significantly changed by mechanical stretch (only a depolarization of approximately 5 mV was observed). These findings demonstrate that the electrical properties of right atrial tissue from normal heart are sensitive to mechanical stretch.

8.8.2 Mechano-Electric Feedback in Right Atrial Tissue from Animals with Cardiac Hypertrophy After Infarction and Human Atria

Whilst myocardial infarction (MI) is a regional pathology, it can lead to alterations in global hemodynamic load of the heart, thereby inducing cardiac hypertrophy.

Consequently, changes in systemic hemodynamics may cause phenotype modulation of right atrial cardiomyocytes after left ventricular MI (Pfeffer et al., 1991).

There is considerable evidence to suggest electrophysiological abnormalities in hypertrophied cardiac myocytes after MI. Data from other models of cardiac hypertrophy indicate that arrhythmia develops more readily in hypertrophied ventricular and atrial myocardium than in healthy tissue (Aronson and Ming, 1993; Hart, 1994; Pye and Cobbe, 1992).

After phenotype modulation of atrial myocardium, a close correlation between atrial size and the occurrence of atrial fibrillations has been observed (for review see: Nazir and Lab, 1996). Moreover, congestive heart failure is frequently associated with atrial fibrillation, which has been reported recently as an important predictor for the mortality of patients with MI (Crenshaw et al., 1997). Therefore, we investigated AP characteristics after MI and a potential role for atrial fibrillation.

Two types of APs could be monitored in the atria from animals with MI. The first type of AP had a similar time course like APs at APD25 and APD50 in sham operated rats, but was considerably lengthened near APD90 (Fig. 8.33a). The second type of AP showed substantial enlargement during APD25, APD50, and APD90 (Fig. 8.33b) (Kamkin et al., 2000b; Kiseleva et al., 1998).

These two types of APs responded differently to application of stretch (Fig. 8.34). Thus, the first type of AP was associated with SID, which arose at APD90 (AP-SID90, Fig. 8.34a), whereas the second type of AP developed SIDs at APD50 (AP-SID50, Fig. 8.34b).

Although compensatory hypertrophy of the surviving myocardium is considered as an important adaptive response of the heart (Pfeffer et al., 1991), it can have potentially adverse effects. Ultrastructural alterations, such as swelling and destruc-

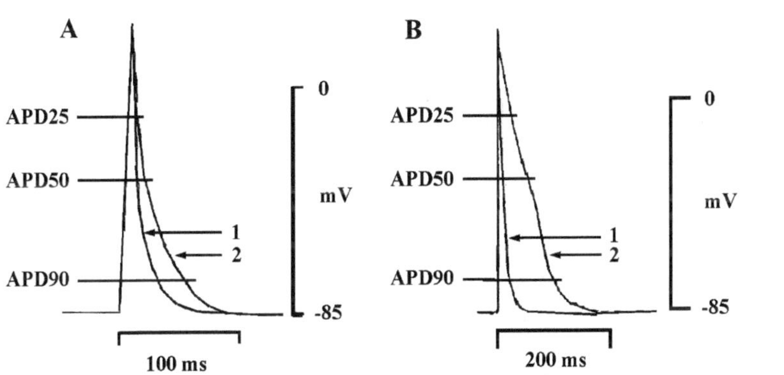

Fig. 8.33 Superimposed action potentials in a right atrial cardiomyocyte from a sham operated rat (1) and a rat after MI (2). The preload was set to 1 mN. Remodeling after MI revealed two types of action potentials (AR). (**a**) The first type of action potential (AR) is similar at APD25 and APD50 compared to action potentials of the control group; but it was considerably lengthened at APD90. (**b**) The second type showed substantial prolongation in APD25, APD50, and APD90 in comparison to control. Modified from Kiseleva et al. (1998) with permission from Elsevier and J Mol Cell Cardiol

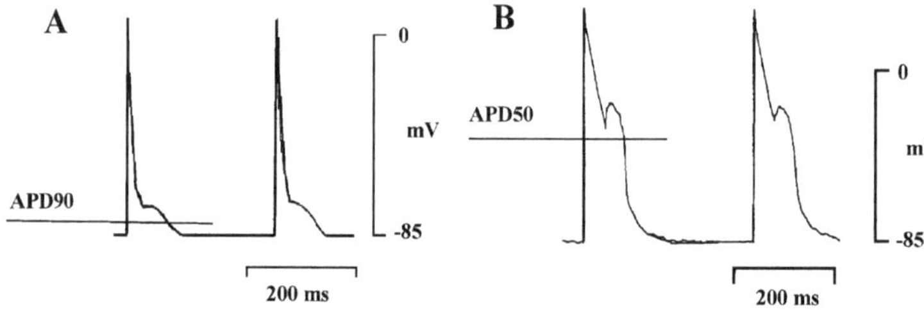

Fig. 8.34 Action potentials from right atrial cardiomyocytes at a preload of 1 mN in rats after MI. Two types of stretch-induced depolarization (SID), arising either at APD90 (**a**) or at APD50 (**b**) could be monitored. Reproduced from Kamkin and Kiseleva (2005b) with copyright permission from Academia Publishing House Ltd

tion of mitochondria and the sarcoplasmic reticulum, cellular edema, and loss of clear structure and enlargement of myofilaments have been described for hypertrophied myocardium after MI (Guski et al., 1991; Weitbrecht et al., 1983). The lowered frequency of spontaneous contractions in the MI group might reflect a specific phenomenon of the interaction of cardiomyocytes through connexons with cardiac fibroblasts, which significantly hyperpolarize after MI (see review Kamkin at al., 2009: in this book).

Consistent with data obtained in other models of cardiac hypertrophy (Aronson and Nordin, 1984; Hart et al., 1997), we found prolongation of the APD. Hypertrophy has been reported to increase APD dispersion possibly due to structural heterogeneity (Qin et al., 1996). The two types of APs may result from non-uniform remodelling causing heterogeneous APs in cardiomyocytes. This mechanism would provide an explanation for the moderate increase in APD90, which was seen in the type AP-SID90s only, and the more general increase in APD25, APD50, and APD90 in the type AP-SID50s. That is, the AP-SID90 types may be recorded from myocytes being affected by remodelling in a different way than cells showing the AP-SID50 type characteristics.

Most importantly, SIDs could be elicited at a much lower level of stretch (0.19 $\pm$ 0.02 mN) in tissue preparations that were obtained from rats with MI compared to specimens from sham operated animals (1.75 ± 0.04 mN) (Kamkin et al., 2000b).

After MI, a significantly lower level of stretch (0.19 ± 0.02 mN) was sufficient to elicit SIDs in both types of APs. Figure 8.35 shows the effect of increasing stretch (MI group) on SIDs near APD90 (AP-SID90 type). Figure 8.35A was recorded at the standard preload of 1 mN. Whilst application of a weak stretch (0.15 mN) did not produce SIDs in sham-operated rats, the same amount of stretch induced a SID after MI (Fig. 8.35b). Stretch did not change APD25. APD50 was minimally shortened, whereas APD90 was significantly increased. The increase in APD90 was due to the SID. Moreover, only a small increase in stretch (up to 0.2 mN) produced SIDs, which were accompanied by extra-APs (Fig. 8.35c). These premature excitations, taking a paired or bigeminal form, had reduced amplitudes as would be expected in the case of a partially depolarized cell membrane. However, these effects on AP

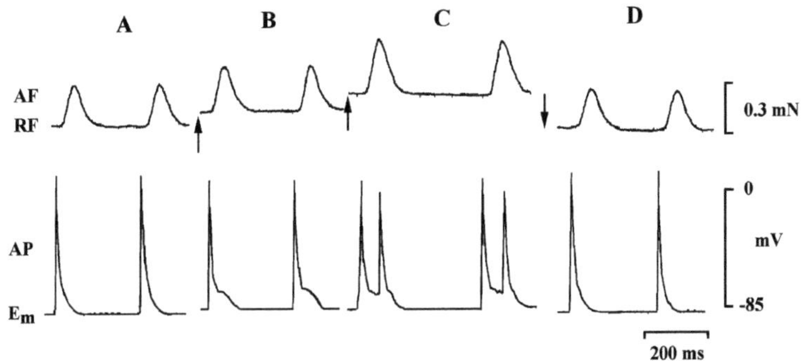

Fig. 8.35 Example of one type of response of the membrane potential of a cardiomyocyte to stretch applied by the micrometer to right atrial tissue in the post-infarction group. Stepwise increases in resting force due to stretch (indicated by ↑) up to 0.15 mN (**a** to **b**) produced stretch-activated depolarization (SID) near APD90 (cell is of type SID90). Further increase in stretch to 0.2 mN resulted in extra action potentials (**c**). Release of the stretch (↓) resulted in complete reversibility of stretch-induced effects (**d**). Reprinted from Kamkin et al. (2000b) with permission from Elsevier and J Mol Cell Cardiol

configuration were seen at an approx. 10-fold lower degree of stretch as compared to sham operated rats. The effects were completely reversible upon cessation of stretch (Fig. 8.35d).

Increasing stretch to more than 0.2 mN prolonged APD90 due to SID and produced extra-APs in the SID90 types. Although we did not record an electrogram, these extra-APs seemed to even elicit atrial tachyarrhythmia (Fig. 8.36 – same preparation as in Fig. 8.35). The irregular amplitude of the potentials and the barely visible and irregular global contractions suggested atrial fibrillation. Note that the two large AF records were associated with an initial large AP, but the following three or so small rapid APs occurred within each force trace. Apparent tachyarrhythmia as well as atrial fibrillation was observed in the preparations after MI. Tach-

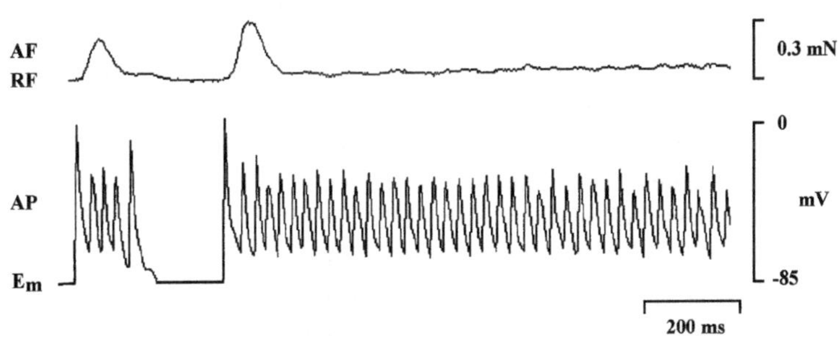

Fig. 8.36 Example of stretch-induced atrial fibrillation in the group with MI. The terminal action potential of the first brief episode of tachyarrhythmia shows a clear SID. The next episode of tachyarrhythmia is maintained. Initially, it shows action potentials of small and moderately irregular amplitude, which deteriorate to very irregular amplitudes (same preparation as in Fig. 8.8). Reprinted from Kamkin et al. (2000b) with permission from Elsevier and J Mol Cell Cardiol

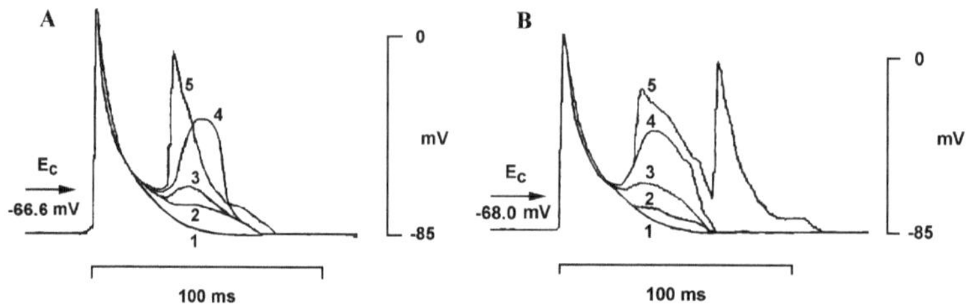

Fig. 8.37 Superimposed action potentials during stepwise increases in stretch. AP SID90 types (**a**) Trace 1 was recorded at a preload of 1 mN. Stepwise increases in length produced SIDs (2–4). The highest stretch SID induced an extra action potential (5) coinciding with the onset of the SID. (**b**) Trace 1 was recorded at a preload of 1 mN. Stepwise increases in length produced SIDs (2–4). The highest stretch SID induced an extra action potential (5) coinciding with the onset of the SID. Reprinted from Kamkin et al. (2000b) with permission from Elsevier and J Mol Cell Cardiol

yarrhythmia and atrial fibrillations were never observed in sham operated rats. That is, in addition to the greater sensitivity of the atrial AP configuration to stretch, the appearance of fibrillations is a particular characteristic of this preparation of post-infarction atrial myocardium. These effects were also reversible upon release of stretch.

Figure 8.37a demonstrates the dynamics of development of one extra-AP in the cardiomyocyte of the rat right atrium after MI. The preparation was exposed to increasing stretch (2–4) up to 0.2 mN. The second potential arises, when SID reaches the critical level of depolarization ($E_C = -66.6$ mV) (Kamkin et al., 2000b). A similar effect but with two additional APs is demonstrated in Fig. 8.37b.

Exposure to 40 μmol/l of Gd^{3+}, a dose, which is normally used to suppress stretch-activated events in tissue (Hu and Sachs, 1997; Ward and White, 1994), had only little effect on contractile activity in all preparations. Developed peak force decreased to 95 % of control, but, importantly, gadolinium had already exerted its suppressing effect on SIDs approximately 5 min before.

In the AP-SID50 types, stretch induced a membrane potential response near APD50 (Fig. 8.38). Very small amounts of stretch, less than 0.2 mN, increased developed peak force and produced SIDs near APD50, which significantly prolonged APD50. Stretch also increased APD90, whereas APD25 remained unaffected. The increases in APD50 and APD90 may result from the SIDs. The APD50 possibly represents another type of response to stretch because these very early SIDs were never followed by extra-APs. However, premature APs were also observed most likely after the refractory period and near to the threshold potential (Fig. 8.39). Release of stretch completely reversed the effects.

In the AP-SID50 type, increases in stretch produced increments in SID (steps 2–4 in Fig. 8.39). We also observed generation of extra-APs not at APD50, but with some delay. These additional APs occurred immediately upon mechanical stimulation, i.e. as soon as the threshold potential was reached (steps 2–4 in Fig. 8.39). Also in this case, exposure to 40 μmol/l of Gd^{3+} blocked the stretch-induced effects.

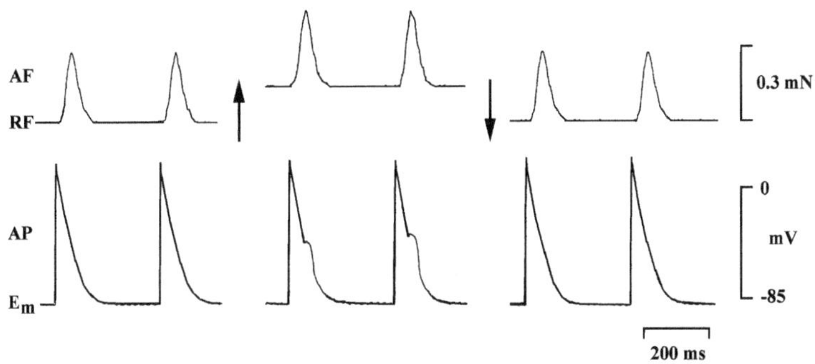

Fig. 8.38 Example of another type of response of cardiomyocyte membrane potential to stretch applied by the micrometer to right atrial tissue in the post-ventricular infarction group. Stretch (↑) induced a depolarization (SID) during the plateau of action potential (i.e. near APD50 – cell is of type SID50). Release of stretch (↓) completely reversed the effects. Further stretch did not produce premature (extra) action potentials or atrial arrhythmia Reprinted from Kamkin et al. (2000b) with permission from Elsevier and J Mol Cell Cardiol

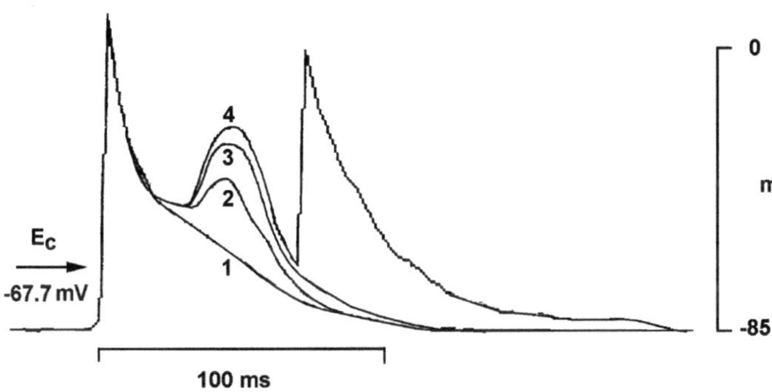

Fig. 8.39 Superimposed action potentials during stepwise increases in stretch. APs were of the SID50 type. Recordings were performed at a preload of 1 mN (1), and during stepwise increases of stretch, which produced a stepwise rise of the SIDs (2–4). The highest stretch, producing SID (4), induced an extra action potential some 40 ms after the onset of the SID when the threshold potential (E_C – critical membrane potential) was reached Reprinted from Kamkin et al. (2000b) with permission from Elsevier and J Mol Cell Cardiol

In human atrial myocardium, baseline electrophysiological parameters recorded in cardiomyocytes are as follows: AP amplitude 71 ± 8 mV, resting potential –63 ± 10 mV, overshoot 8 ± 6 mV, duration 452 ± 65 ms (Kamkin et al., 1999).

Figure 8.40 demonstrates a representative example of the effect of sustained stretch on AP configuration at three different levels of repolarization in human atrial tissue. APD25 and APD50 were unaffected, whereas APD90 was significantly increased. The increase in APD90 was associated with stretch-activated depolarization (SID).

Long-lasting stretch increased active force development. In human atrial tissue, lengthening increased the isometric peak force from 1.9±0.3 mN (curve 1) to 2.1

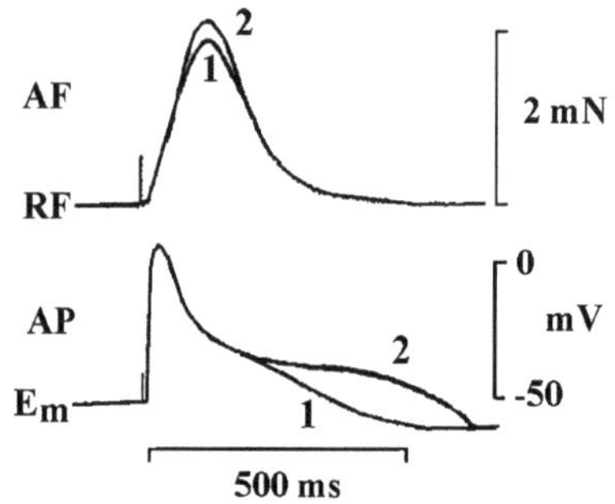

Fig. 8.40 Simultaneous recording of the isometric force development – resting force (RF), active force (AF) of a human atrial trabecula (*upper trace*) and action potential (AP) of a cardiomyocyte (*lower trace*). Superimposed action potentials at a preload of 1 mN (1) and active force (AF) amplitude of 1.9 mN and during the increase of isometric force amplitude (AF) to 2.1 mN by stretch of the tissue. Note that mechanical stretch produces SID (curve 2) near APD90 in comparison with the control (curve 1). Reproduced from Kamkin and Kiseleva (2005b) with copyright permission from Academia Publishing House Ltd

± 0.2 mN (curve 2). These effects were completely reversible upon removal of stretch. The resting membrane potential and the AP amplitude were not affected by stretch. The SIDs at APD90 were suppressed completely after 10 min of application of 40 μM Gd^{3+}. These experiments with Gd^{3+}, which blocked the stretch-induced electrophysiological changes, strongly suggest the involvement of SACs.

8.9 Mechano-Electric Feedback in Ventricle from Healthy and Diseased Animals and Human

8.9.1 Mechano-Electric Feedback in Left Ventricular Tissue in Healthy Rats

Mechanoelectric feedback in ventricular cardiomyocytes, studied by the microelectrode technique has been shown in rat heart (Kamkin et al., 1998; Kiseleva et al., 1998, 2000). Figure 8.41demonstrates the typical APs, which were registered in left ventricular tissue of rats.

In normal hearts, AP configurations of different cardiomyocytes in the endocardial region of the ventricle were relatively uniform. Long-lasting stretch of the tissue led to an increase in active force development, which could modulate the electrophysiological function of the cardiac myocytes. The resting membrane potential and the AP amplitude were not affected by stretch.

Figure 8.42 illustrates the typical effect of prolonged stretch of the ventricular preparations from a sham-operated rat. Application of physical stretch resulted in an increase in active force. APD25 and APD50 remained constant, whereas APD90 was significantly increased during stretch. The rise in APD90 was associated with stretch-activated depolarization (SID). Application of 40 μM Gd^{3+} reduced the

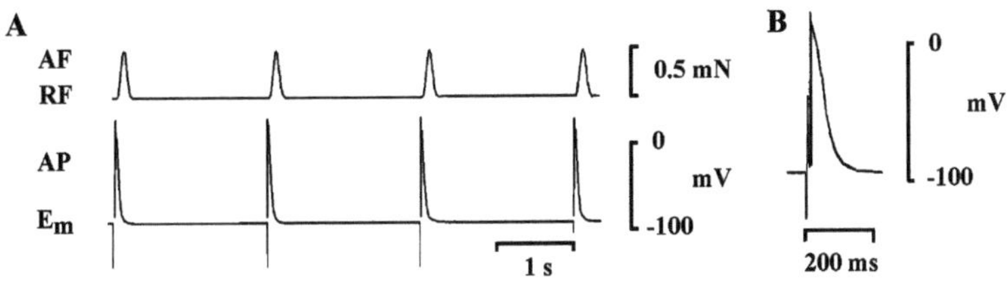

Fig. 8.41 Continuous recordings of active force (AF), resting force (RF) and action potentials (AP) from the sham operated group (**a**). A single AP from a sham-operated rat is shown at higher resolution (**b**). Note that the ventricular preparations from sham-operated rats were stimulated electrically. Modified from Kiseleva et al. (2000), with permission from Elsevier, Oxford University Press and Cardiovasc Res

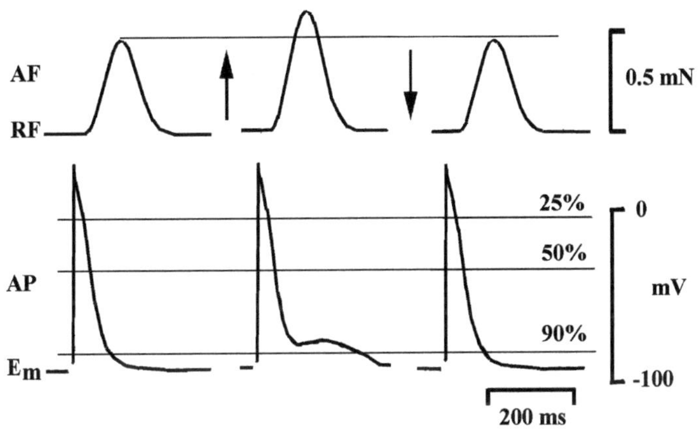

Fig. 8.42 Response of the membrane potential of a cardiomyocyte to long-lasting mechanical stretch of the left ventricular preparation from a sham-operated rat. Increasing stretch (indicated by ↑) caused a membrane depolarisation, which was completely reversible upon removal of stretch (indicated by ↓). Single registrations, which are shown, were obtained from continuous recordings 15 s after application or removal of stretch. Modified from Kiseleva et al. (2000), with permission from Elsevier, Oxford University Press and Cardiovasc Res

active force only insignificantly by 5% but completely suppressed SIDs near to APD90. The resting membrane potential and the AP amplitude were unaffected by Gd^{3+}.

8.9.2 *Mechano-Electric Feedback in Ventricular Tissue from Animals with Cardiac Hypertrophy After Infarction*

Remodelling of the heart after MI is associated with phenotypic changes and hypertrophy of the surviving myocardium (Qin et al., 1996). Structural remodelling may include an increase in myocyte cross-sectional area and length of the left ventricular myocytes (Anversa et al., 1984, 1986).

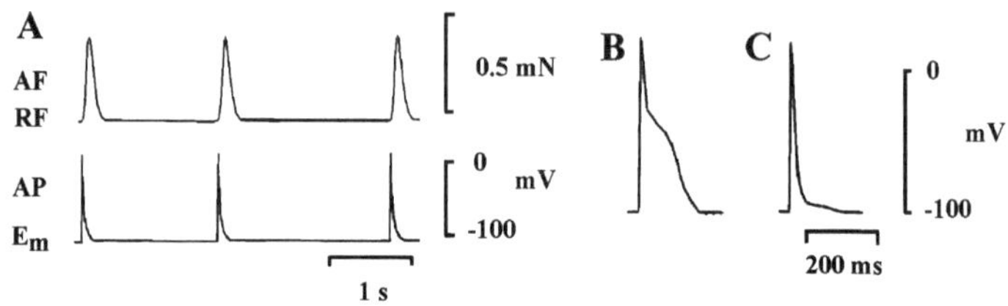

Fig. 8.43 Continuous recordings of active force (AF), resting force (RF), and action potentials (AP) in ventricular cardiomyocytes from rats with MI (**a**). Single APs are shown in Figs. B and C. Note that the ventricular preparations after sham operation were stimulated electrically whereas preparations after MI generated spontaneous electrical and contractile activity. The different action potential configuration in (**b**) and (**c**) points to the increased heterogeneity after MI. Modified from Kiseleva et al. (2000), with permission from Elsevier, Oxford University Press and Cardiovasc Res

Approximately 90% of the preparations from rats with MI (16.5 ± 0.6 % of the left ventricular endocardial surface) generated spontaneous APs and contractile activity, which was never observed in the tissue of sham operated rats. Representative examples of APs in myocardial preparations after MI are depicted in Fig. 8.43. The preload was adjusted to a level, which allowed active force development of 0.5 mN. Post-infarct remodelling was associated with typical changes of AP configuration in ventricular cardiomyocytes. In comparison to sham-operated rats, cardiomyocytes from animals with MI had a similar AP amplitude, a more negative resting membrane potential (–95.2 ± 1.3 mV vs. –88.6 ± 0.8 mV, $P < 0.005$), and a prolonged AP duration at progressing levels of repolarization (APD90: 129 ± 15 ms vs. 86±3 ms, $P < 0.05$).

The configuration of APs was more heterogeneous in the cardiac myocytes from rats with MI than in cells that were recorded from sham-operated rats. Consistent with previous findings, the duration of APs was increased and the time course of repolarization showed marked heterogeneity in left ventricular myocytes in the remodelling heart after MI (Qin et al., 1996). Prolongation of the APD was explained by decreased K^+ outward currents, rather than by changes in Ca^{2+} inward currents (Qin et al., 1996). The electrophysiological heterogeneity may result from differences in remodelling between individual cells. The scar tissue and the adjacent myocardium are less distensible and more resistant to mechanical stretch than the cells at a distance from the infarcted area. It is known from the classical work by Tennant and Wiggers (1935) that mechanical forces can deform the non-contracting scar region. These forces act differently on both, the necrotic and the non-infarcted myocardium, which may account for the observed electrophysiological heterogeneity in the myocardium adjacent to the scar tissue.

Application of mechanical stretch led to an increase in active force development, which could provide a trigger for changes in the electrical function of cardiac myocytes. In comparison to the extensive stretch that was required to change the configuration of APs in the myocardium of sham-operated rats (150 μm; increase in active force from 0.48 to 0.88 mN), a far weaker mechanical stimulation was sufficient to produce increases in APD90 in tissue preparations from rats with MI

(20 μm; increase in active force from 0.5 to 0.57 mN). These effects were completely reversible upon removal of stretch. The resting membrane potential and the AP amplitudes were not affected by stretch.

In the cases of relatively short APD, the SIDs appeared near to APD90 (Fig. 8.44). In the cases of extremely prolonged APs, SID was already observed close to APD50 (Fig. 8.45) at similar intensities of mechanical stretch as in the experiments with SIDs near to APD90. Application of weak stretch caused SIDs with small amplitudes. Increasing stretch led to a further rise in the SID. Removal of stretch demonstrated the complete reversibility of alterations in AP configuration. SID at APD90 and APD50 were completely suppressed after application of 40 μM Gd^{3+}.

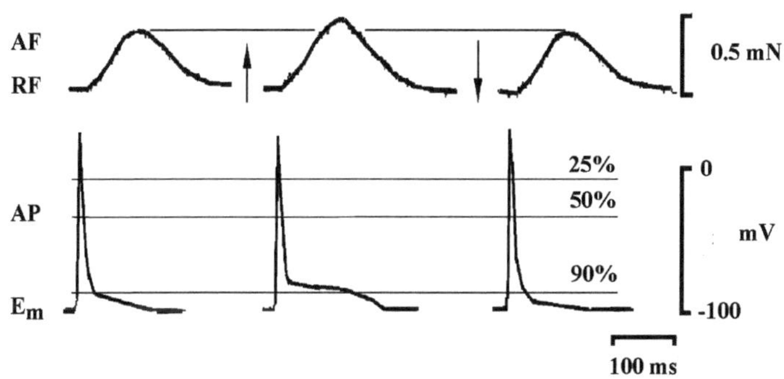

Fig. 8.44 Representative example of the response of the cardiomyocyte membrane potential to long-lasting stretch in rats with MI. The increase in stretch (indicated by ↑) led to depolarization near to APD90. This effect was completely reversible upon release of stretch (↓). The individual registrations were obtained from continuous recordings 15 s after each step of stretch or release of stretch. Reprint from Kiseleva et al. (2000), with permission from Elseiver, Oxford University Press and Cardiovasc Res

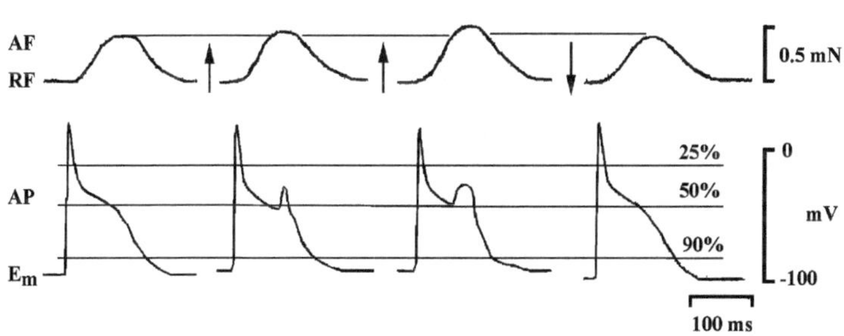

Fig. 8.45 Response of the membrane potential of a cardiomyocyte to long-lasting mechanical stretch in a left ventricular preparation from a rat with MI. The increase in stretch (indicated by ↑) caused depolarization near to APD50. Removal of stretch (↓) resulted in a complete reversibility of this effect. The single registrations were obtained from continuous recordings 15 s after each step of stretch or relaxation. Reprint from Kiseleva et al. (2000), with permission from Elsevier, Oxford University Press and Cardiovasc Res

One major finding of this study is that in remodelled myocardium adjacent to the scar, stretch elicited SIDs more readily compared to ventricular myocardium from sham-operated rats.

8.10 Conclusion

In summary, our findings indicate that (i) stretch of the tissue leads to appearance of SID of cardiomyocytes in tissue fragments, which could be in APD50 and APD90 in atrial cardiomyocytes as well as in ventricular cardiomyocytes; (ii) increment of stretch causes proportional and reversible changes in the AP associated with SID; (iii) in the level of APD90 appear extra-APs when SID amplitude reaches threshold; (iv) isolated cardiomyocytes respond to mechanical stimulation with membrane depolarization, prolongation of the action potential and extra-APs; (v) I_{SAC} is the major cause of these stretch-induced events. At negative potentials, I_{SAC} was negative and carried by influx of Na^+ ions, and induced diastolic depolarization or SID; (vi) reaction to a stretch was identical in cardiomyocytes, occupying both positions (edgewise and broad-wise). However, reaction to compression was different and was determined by the position of a cell; (vii) the sensitivity of the AP to mechanical stretch is significantly increased in hypertrophied myocardium; (viii) this increase in the sensitivity to stretch could be related to expression of SACs.

Further work is required to elucidate signalling pathways, which regulate MGCs function and intercellular interactions. Special interest represents interaction of cardiomyocytes with cardiac fibroblasts and signaling pathways of its regulation.

References

Akinlaja J, Sachs F (1998) The breakdown of cell membranes by electrical and mechanical stress. Biophys J 75: 247–254.

Anversa P, Beghi C, Kikkawa Y, Olivetti G (1986) Myocardial infarction in rats: infarct size, myocyte hypertrophy, and capillary growth. Circ Res 58:26–37.

Anversa P, Beghi C, McDonald SL, Levicky V, Kikkawa Y, Olivetti G (1984) Morphometry of right ventricular hypertrophy induced by myocardial infarction in the rats. Am J Pathol 116:504–513.

Aronson RS, Ming Z (1993) Cellular mechanisms of arrhythmias in hypertrophied and failing myocardium. Circulation 83 (VII):76–83.

Aronson RS, Nordin C (1984) Electrophysiologic properties of hypertrophied myo-cytes isolated from rats with renal hypertension. Eur Heart J 5 (F):339–345.

Belus A, White E (2003) Streptomycin and intracellular calcium modulate the response of single guinea-pig ventricular myocytes to axial stretch. J Physiol (London) 546:501–509.

Bett GCL, Sachs F (2000) Whole-cell mechanosensitive currents in rat ventricular myocytes activated by direct stimulation. J Membrane Biol 173:255–263.

Boluyt MO, Bing OHL, Lakatta EG (1995) The ageing spontaneously hypertensive rat as a model of the transition from stable compensated hypertrophy to heart failure. Eur Heart J 16:19–30.

Bowman CL, Lohr JW (1996) Mechanotransducing ion channels in C6 glioma cells. Glia 18(3):161–176.

Bowman CL, Ding JP, Sachs F, Sokabe M (1992) Mechanotransducing ion channels in astrocytes. Brain Res 584:272–286.

Calaghan S, White E (2004) Activation of Na^{+}-H^{+} exchange and stretch-activated channels underlies the slow inotropic response to stretch in myocytes and muscle from the rat heart. J Physiol (London) 559:205–214.

Chaudhuri O, Parekh SH, Fletcher DA (2007) Reversible stress softening of actin networks. Nature 445(7125):295–298.

Crenshaw BS, Ward SR, Granger CB, Stebbins AL, Topol EJ, Califf RM (1997) Atrial fibrillation in the setting of acute myocardial infarction: the GUSTO-I experience. Global utilization of streptokinase and TPA for occluded coronary arteries. J Am Coll Cardiol 30:406–413.

Davis MJ, Donovitz JA, Hood JD (1992) Stretch-activated single-channel and whole cell currents in vascular smooth muscle cells. Am J Physiol 262:C1083–C1088.

Dudel J, Trautwein W (1954) Das Aktionspotential und Mechanogramm des Herzmuskels unter dem Einfluss der Dehnung. Cardiologie 25:344.

Dyachenko V, Christ A, Gubanov R, Isenberg G (2008) Bending of z-lines by mechanical stimuli: an input signal for integrin dependent modulation of ion channels? Prog Biophys Mol Biol 97(2–3):196–216.

Dyachenko V, Husse B, Rueckschloss U, Isenberg G (2009a) Mechanical deformation of ventricular myocytes modulates both TRPC6 and Kir2.3 channels. Cell Calcium Jan; 45(1):38–54.

Dyachenko V, Rueckschloss U, Isenberg G (2006) Aging aggravates heterogeneities in cell-size and stress-intolerance of cardiac ventricular myocytes. Exp Gerontol 41(5):489–496.

Dyachenko V, Rueckschloss U, Isenberg G (2009b) Modulation of cardiac mechanosensitive ion channels involves superoxide, nitric oxide and peroxynitrite. Cell Calcium Jan; 45(1):55–64.

Franz MR (1996) Mechano-electrical feedback in ventricular myocardium. Cardiovasc Res 32: 15–24.

Franz MR (2000) Mechano-electrical feedback. Cardiovasc Res 45:263–266.

Guski H, Meyer R, Fernandez-Britto JE (1991) Morphometric, histochemical and autoradiographic studies on myocardial cells in experimental cardiac hypertrophy and ischemia. Exp Pathol 41:79–97.

Gustin MC, Zhou XL, Martinac B, Kung C (1988) A mechanosensitive ion channel in the yeast plasma membrane. Science 242:762–765.

Hamill OP, Lane JW, McBride DW Jr. (1992) Amiloride: A molecular probe for mechanosensitive channels. Trends Pharmacol Sci 13:373–376.

Hamill OP, Martinac B (2001) Molecular basis of mechanotransduction in living cells. Physiol Rev 81:685–740.

Hart G (1994) Cellular electrophysiology in cardiac hypertrophy and failure. Cardiovasc Res 28:933–946.

Hart G, Bryant SM, Shipsey SJ (1997) Regional differences in electrical and mechanical properties of myocytes from guinea-pig hearts with mild left ventricular hypertrophy. Cardiovasc Res 35(2):315–323.

Hongo K, Pascarel C, Cazorla O, Gannier F, LeGuennec JY, White E (1997) Gadolinium blocks the delayed rectifier potassium current in isolated guinea-pig ventricular myocytes. Exp Physiol 82:647–656.

Honoré E, Patel AJ, Chemin J, Suchyna T, Sachs F (2006) Desensitization of mechano-gated K2P channels. Proc Natl Acad Sci USA 103(18):6859–6864.

Hu H, Sachs F (1994) Effects of mechanical stimulation on embryonic chick heart cells. Biophys J 66:A170.

Hu H, Sachs F (1995) Whole cell mechanosensitive currents in acutely isolated chick heart cells: correlation with mechanosensitive channels. Biophys J 68:A393.

Hu H, Sachs F (1996) Mechanically activated currents in chick heart cells. J Membr Biol 154:205–216.

Hu H, Sachs F (1997) Stretch-activated ion channels in the heart. J Mol Cell Cardiol 29:1511–1523.
Isenberg G (1976) Cardiac Purkinje fibres. Cesium as a tool to block inward rectifying potassium currents. Pflügers Arch – Eur J Physiol 365:99–106.
Isenberg G, Kazanski V, Dyachenko V (2004) How does shear stress modulate ion conductances of ventricular myocytes? J Mol Cell Cardiol 36:738.
Isenberg G, Kazanski V, Kondratev D, Gallitelli MF, Kiseleva I, Kamkin A (2003) Differential effects of stretch and compression on membrane currents and $[Na^+]_C$ in ventricular myocytes. Progr Biophys Mol Biol 82:43–56 (Review).
Isenberg G, Kondratev D, Dyachenko V, Kazanski V, Gallitelli MF (2005) Isolated cardiomyocytes: Mechanosensitivity of action potential, membrane current and ion concentration. In: Kamkin A and Kiseleva I (eds.) Mechanosensitivity in Cells and Tissues. Academia Publishing House Ltd., Moscow, pp. 126–164 (Review).
Kamkin A, Kiseleva I (2008b) Physiology and Molecular Biology of Cell Membranes. Monograph. Academia Publishing House Ltd., Moscow, 586 pp. (Russian)
Kamkin A, Kiseleva I, Isenberg G (2000a) Stretch-activated currents in ventricular myocytes: amplitude and arrhythmogenic effects increase with hypertrophy. Cardiovasc Res 48: 409–420.
Kamkin A, Kiseleva I, Isenberg G (2003c) Ion selectivity of stretch-activated cation currents in mouse ventricular myocytes. Pflügers Arch – Europ J Physiol 446(2):220–231.
Kamkin A, Kiseleva I, Lozinsky I (2008a) Experimental methods of studying mechanosensitive channels and possible errors in data interpretation. In: Kamkin A and Kiseleva I (eds.) Mechanosensitivity in Cells and Tissues 1. Mechanosensitive Ion Channels. Springer, pp. 3–35 (Review).
Kamkin A, Kiseleva I, Lozinsky I (2009) The role of mechanosensitive fibroblasts in the heart: Evidence from acutely isolated single cells, cultured cells and from intracellular microelectrode recordings on multicellular preparations from healthy and diseased cardiac tissue. In: Kamkin A and Kiseleva I (eds) Mechanosensitivity in cells and Tissues 3. Mechanosensitivity of the Heart. Springer, pp. 239–266 (Review).
Kamkin A, Kiseleva I, Lozinsky I, Scholz H (2005a) Electrical interaction of mechanosensitive fibroblasts and myocytes in the heart. Basic Res Cardiol 100(4):337–345.
Kamkin A, Kiseleva I, Wagner KD, Lammerich A, Bohm J, Persson PB, Gunther J (1999) Mechanically induced potentials in fibroblasts from human right atrium. Exp Physiol 84:347–356.
Kamkin A, Kiseleva I, Wagner KD, Bohm J, Theres H, Günther J, Scholz H (2003d) Characterization of stretch-activated ion currents in isolated atrial myocytes from human hearts. Pflügers Arch – Europ J Physiol 446(3):339–346.
Kamkin A, Kiseleva I, Wagner KD, Leiterer KP, Theres H, Scholz H, Günther J, Lab MJ (2000b) Mechanoelectric feedback in right atrium after left ventricular infarction in rats. J Mol Cell Cardiol 32:465–477.
Kamkin A, Kiseleva I, Wagner KD, Theres H, Leiterer KP, Gunther J, Lab MJ (1998) Mechanoelectric feedback in right atrium after ventricular infarction in rats. Pflügers Arch – Europ J Physiol 435(6):O20–4/R79.
Kamkin A, Kiseleva I, Isenberg G (2003a) Activation and inactivation of a non-selective cation conductance by local mechanical deformation of acutely isolated cardiac fibroblasts. Cardiovasc Res 57:793–803.
Kamkin A, Kiseleva I, Isenberg G, Wagner KD, Günther J, Theres H, Scholz H (2003b) Cardiac fbroblasts and the mechano-electric feedback mechanism in healthy and diseased hearts. Prog Biophys Mol Biol 82:111–120.
Kamkin A, Kiseleva I, Lozinsky I, Wagner KD, Isenberg G, Scholz H (2005c) The role of mechanosensitive fibroblasts in the heart. In: Kamkin A and Kiseleva I (ed.) Mechanosensitivity in Cells and Tissues. Academia Publishing House Ltd., pp. 203–229 (Review).
Kamkin A, Kiseleva I, Wagner KD, Scholz H (2005b) Mechano-electric feedback in the heart: Evidence from intracellular microelectrode recordings on multicellular preparations and single

cells from healthy and diseased tissue. In: Kamkin A and Kiseleva I (eds.) Mechanosensitivity in Cells and Tissues. Academia Publishing House Ltd., pp. 165–202 (Review).
Kawahara K (1993) Stretch-activated channels in renal tubule. Nippon Rinsho 51:2201–2208.
Kiseleva I, Kamkin A, Pylaev A, Kondratjev D, Leiterer KP, Theres H, Wagner KD, Persson PB, Günther J (1998) Electrophysiological properties of mechanosensitive atrial fibroblasts from chronic infarcted rat hearts. J Mol Cell Cardiol 30:1083–1093.
Kiseleva I, Kamkin A, Wagner KD, Theres H, Leiterer KP, Gunther J (1998) Stretch-induced alterations in action potential configuration of ventricular myocytes in the chronic phase following myocardial infarction. Pflügers Arch – Europ J Physiol 435(6):P43–10, R205.
Kiseleva I, Kamkin A, Wagner KD, Theres H, Ladhoff A, Scholz H, Günther J, Lab MJ (2000) Mechano-electric feedback after left ventricular infarction in rats. Cardiovasc Res 45:370–378.
Kondratev D, Christ A, Gallitelli MF (2005) Inhibition of the Na^+-H^+ exchanger with cariporide abolishes stret-induced calcium but not sodium accumulation in mouse ventricular myocytes. Cell Calcium 37:69–80.
Kondratev D, Gallitelli MF (2003) Increments in the concentration of sodium and calcium in cell compartments of stretched mouse ventricular myocytes. Cell Calcium 34:193–203.
Lab MJ (1968) Is there mechanoelectric transduction in cardiac muscle? The mono-phasic action potential of the frog ventricle during isometric and isotonic contraction with calcium deficient perfusions. S Afr J Med Sci 33:60 (Abstract).
Lab MJ (1996) Mechanoelectric feedback (transduction) in heart: concepts and implications. Cardiovasc Res 32:3–14.
Lab MJ (1998) Mechanosensitivity as an integrative system in heart: an audit. Prog Biophys Mol Biol 71:7–27.
Lacampagne A, Gannier F, Argibay J, Garnier D, Le Guennec JC (1994) The stretch-activated ion channel blocker gadolinium also blocks L-type calcium channels in isolated ventricular myocytes of the guinea-pig. Biochim Biophys Acta 1191:205–208.
Le Guennec J-Y, White E, Gannier F, Argibay JA, Garnier D (1991) Stretch-induced increase of resting intracellular calcium concentration in single guinea-pig ventricular myocytes. Exp Physiol 76:975–978.
Lillemeier BF, Pfeiffer JR, Surviladze Z, Wilson BS, Davis MM (2006) Plasma membrane-associated proteins are clustered into islands attached to the cytoskeleton. Proc Natl Acad Sci USA 103(50):18992–18997.
Marchenko SM, Sage SO (1996) Mechanosensitive ion channels from endothelium of excised rat aorta. Biophys J 70:A365.
Mizuno D, Tardin C, Schmidt CF, Mackintosh FC (2007) Nonequilibrium mechanics of active cytoskeletal networks. Science 315(5810):370–373.
Morris CE (1990) Mechanosensitive ion channels. J Membr Biol 113:93–107.
Morris CE, Homann U (2001) Cell surface area regulation and membrane tension. J Membr Biol 179(2):79–102.
Nazir SA, Lab MJ (1996) Mechanoelectric feedback and atrial arrhythmias. Cardiovasc Res 32: 52–61.
Pfeffer JM, Pfeffer MA, Fletcher PJ, Braunwald E (1991) Progressive ventricular remodeling in rat with myocardial infarction. Am J Physiol 260:H1406–H1414.
Pye PM, Cobbe SM (1992) Mechanisms of ventricular arrhythmias in cardiac failure and hypertrophy. Cardiovasc Res 26:740–750.
Qin D, Zhang ZH, Caref EB, Boutjdir M, Jain P, El-Sherif N (1996) Cellular and ionic basis of arrhythmias in postinfarction remodeled ventricular myocardium. Circ Res 79:461–473.
Sachs F, Morris CE (1998) Mechanosensitive ion channels in nonspecialized cells. Rev Physiol Biochem Pharmacol 132:1–77 (Review).
Sasaki N, Mitsuiye T, Noma A (1992) Effects of mechanical stretch on membrane currents of single ventricular myocytes of guinea-pig heart. Jpn J Physiol 42:957–970.
Sokabe M, Sachs F (1990) The structure and dynamics of patch clamped membrane, a study using differential interference contrast microscopy. J Cell Biol 111, 599–606.

Sokabe M, Sachs F, Jing Z (1991) Quantitative video microscopy of patch clamped membranes, stress, strain, capacitance and stretch channel activation. Biophys J 59, 722–728.
Suchyna TM, Sachs F (2007) Mechanosensitive channel properties and membrane mechanics in mouse dystrophic myotubes. J Physiol (London) 581(1):369–387.
Taggart P, Sutton P, Lab M, Runnalls M, O′Brien W, Treasure T (1992) Effect of abrupt changes in ventricular loading on repolarization induced by transient aortic occlusion in humans. Am J Physiol 263(3 Pt 2):H816–H823.
Tennant R, Wiggers CJ (1935) The effect of coronary occlusion on myocardial contraction. Am J Physiol 112:351–361.
Ward H, White E (1994) Reduction in the contraction and intracellular calcium transient of single rat ventricular myocytes by gadolinium and the attenuation of these effects by extracellular NaH_2PO_4. Exp Physiol 79:107–110.
Watson PA, Hannan R, Carl LL, Giger KE (1996) Contractile activity and passive stretch regulate tubulin mRNA and protein content in cardiac myocytes. Am J Physiol – Cell Physiol 271:C684–C689.
Weitbrecht M, Schaper J, Zanker K, Blumel G, Mathes P (1983) Morphology and mitochondrial function of the surviving myocardium following myocardial infarction in the cat. Basic Res Cardiol 78:423–434.
Wellner MC, Isenberg G (1993) Properties of stretch activated channels in myocytes from the guinea-pig urinary bladder. J Physiol (London) 466:412–425.
Wellner MC, Isenberg G (1994) Stretch effects on whole-cell currents of guinea-pig urinary bladder myocytes. J Physiol (London) 480(3):439–448.
Wellner MC, Isenberg G (1995) cAMP accelerates the decay of stretch-activated inward currents in guinea-pig urinary bladder myocytes. J Physiol (London) 482:141–156.
White E, Boyett MR, Orchard CH (1995) The effects of mechanical loading and changes of length on single guinea-pig ventricular myocytes. J Physiol (London) 482:93–107.
White E, Le Guennec IY, Nigretto JM, Gannier F, Argibay JA, Gamier D (1993) The effects of increasing cell length on auxotonic contractions: membrane potential and intracellular calcium transients in single guinea-pig ventricular myocytes. Exp Physiol 78:65–78.
Xian Tao Li, Dyachenko V, Zuzarte M, Putzke C, Preisig-Müller R, Isenberg G, Daut J (2006) The stretch-activated potassium channel TREK-1 in rat cardiac ventricular muscle. Cardiovasc Res 69(1):86–97.
Yang XC, Sachs F (1989) Block of stretch-activated ion channels in Xenopus oocytes by gadolinium and calcium ions. Science 243:1068–1071.
Youm JB, Han J, Kim N, Zhang YH, Kim E, Leem CH, Kim SJ, Earm YE (2005) Role of stretch-activated channels in the heart: Action potential and Ca^{2+} transients. In: Kamkin A and Kiseleva I (eds.) Mechanosensitivity in Cells and Tissues. Academia Publishing House Ltd., pp. 103–125 (Review).
Zabel M, Koller BS, Sachs F, Franz MR (1996) Stretch-induced voltage changes in the isolated beating heart: importance of the timing of stretch and implications for stretch-activated ion channels. Cardiovasc Res 32:120–130.
Zeng T, Bett GCL, Sachs F (2000) Stretch-activated whole cell currents in adult rat cardiac myocytes. Am J Physiol 278:H548–H557.
Zhang YH, Youm JB, Sung HK, Lee SH, Ryu SY, Lee S-H, Ho W-K, Earm YE (2000) Stretch-activated and background non-selective cation channels in rat atrial myocytes. J Physiol (London) 523:607–619.
Zühlke RD, Pitt GS, Deisseroth K, Tsien RW, Reuter H (1999) Calmodulin supports both inactivation and facilitation of L type calcium channels. Nature 399:159–162.

Chapter 9
The Role of Mechanosensitive Fibroblasts in the Heart: Evidence from Acutely Isolated Single Cells, Cultured Cells and from Intracellular Microelectrode Recordings on Multicellular Preparations from Healthy and Diseased Cardiac Tissue

Andre Kamkin, Irina Kiseleva, and Ilya Lozinsky

Abstract Cardiac fibroblasts are electrically non-excitable cells that respond to mechanical deformation of the cells with typical changes of their membrane potential. These changes of fibroblasts membrane potential are determined by operation of mechano-gated channel (MGCs). Two types of MGCs with conductances of 43 pS and 87 pS were observed during direct deformation of the fresh isolated cells. Cell compression augment the whole-cell MG currents and increase the frequency and duration of single MGC openings. Both MGCs (with conductance levels of 43 pS and 87 pS) displayed linear current-voltage relationships with the reversal potential around 0 mv. Cell stretch inactivated the whole-cell MG currents and abolished the activity of single MGCs. These channels, mainly permeable for sodium ions, are activated by compression of the cell leading to depolarization, and are inactivated by stretch, which in turn leads to hyperpolarization. Cultured cardiac fibroblasts preserve MGCs up to 5 days without special flexible substrates and possess electrophysiological properties of freshly isolated cells. Thus, cardiac fibroblasts function as mechano-electric transducers in the heart and represent the cellular substrate for a cardiac mechano-electrical feedback mechanism. Cardiac fibroblasts respond to spontaneous contractions of the myocardium with rhythmical changes of their resting membrane potential. This phenomenon is referred to as mechanically induced potential (MIP) and is thought to participate in the mechano-electric feedback mechanism of the heart. Enhanced sensitivity of the cardiac fibroblasts to mechanical deformation is increasing with age and during hypertrophy. It is contributing to electrical instability and arrhythmia after myocardial infarction. Recent findings indicate that these processes involve the transfer of electrical signals via gap junctions. In this

A. Kamkin (✉)
Department of Fundamental and Applied Physiology, Russian State Medical University, Moscow, Russia
e-mail: Kamkin.A@g23.relcom.ru; andre.kamkin@gmail.com

A. Kamkin, I. Kiseleva (eds.), *Mechanosensitivity of the Heart*,
Mechanosensitivity in Cells and Tissues 3, DOI 10.1007/978-90-481-2850-1_9,

article we will discuss recent progress in the electrophysiology of cardiac fibroblasts and their role in mechano-electric feedback in healthy and diseased hearts.

Keywords Cardiac fibroblasts · Mechano-gated channel · Mechanosensitive whole-cell currents · Action potential · Resting potential · Active force · Resting force · Myocardial hypertrophy · Gap junction · Connexon

9.1 Introduction

Investigation of mechanosensitivity of cardiac fibroblasts is a new rapidly developing field of research. Its importance originates from the modern perception that fibroblasts play an important role in regulatory mechanisms of mechano-electric feedback in heart. In addition to cardiomyocytes, cardiac fibroblasts have been reported recently to function as mechano-electric transducers and to serve as a potential source for arrhythmias.

Fibroblasts comprise a volume fraction of approximately 5–10% of the cardiac tissue (Anversa et al., 1980). Fibroblasts constitute the most numerous non-myocyte cell population (more than 90% of the non-myocyte cells) in the heart (Eghbali et al., 1998). Other cells, including vascular smooth muscle and endothelial cells represent smaller fractions of this population (Adler et al., 1981). In the atrial sinus node region, fibroblasts and the connective tissue occupy between 45% (Shiraishi et al., 1992) and 75% (Davies and Pomerance, 1972) of the volume. Fibroblasts are highly heterogeneous mesenchymal cells (Camelliti et al., 2005; Chang et al., 2002; Kamkin et al., 2005a, b).

Fibroblasts are the major producers of collagen in connective tissues. All investigated sinoatrial nodes (in rabbit, guinea pig, cat, and pig) contain large amounts (45% or more) of collagen (Opthof et al., 1987). In contrast to cardiomyocytes that synthesize collagen type IV exclusively, fibroblasts also produce types I, III and VI (Camelliti et al., 2005). Cardiac fibroblasts form a network of cells that are connected to each other via specific cadherins and connexins, to the extracellular matrix via integrins, and to myocytes by a variety of receptors, including connexins (Banerjee et al., 2006). Various growth factors (MacKenna et al., 2000), such as transforming growth factor-β1 (Leask, 2007), platelet-derived growth factor-β (Burstein et al., 2008), fibroblast growth factor-2 (Virag et al., 2007), insulin-like growth factor-1 (Horio et al., 2005), and cytokines, such as interleukin 1 (Bujak et al., 2008), tumor necrosis factor (Turner et al., 2007) are secreted by fibroblasts. Cytokines regulate matrix metalloproteinases and migration in cardiac fibroblasts (Brown et al., 2007). They are responsible for maintenance of balance between synthesis and degradation of growth factors, cytokines, matrix metalloproteinases, and tissue inhibitors of metalloproteinases. Fibroblasts can acquire an activated phenotype in which they are usually referred to as "myofibroblasts" which express α-smooth muscle actin (Tomasek et al., 2002). In the heart cardiac fibroblasts influence cardiomyocyte phenotype (LaFramboise, et al., 2007). The activated phenotype is associated with increased proliferative activity, and enhanced secretion of extracellular matrix components such as type I collagen, fibronectin, and tenascin-C.

The basic mechanisms of mechanotransduction have been intensively investigated in the past few years because mechanical loading signals critically influence the development, remodeling, and pathogenesis of tissues. As a result, various mechanisms have been proposed for mechanotransduction in a variety of cell types such as fibroblasts of different lines (for review see Thampatty and Wang, 2008) and cardiac fibroblasts (for review see Kamkin et al., 2003b). Their sensitivity to mechanical stimulus has been found also in tendon fibroblasts, ligament fibroblasts and dermal fibroblasts (Banes et al., 1995; Lee et al., 1996; Wang and Thampatty, 2006; Yost et al., 2000).

Cells are equipped with numerous receptors that allow them to detect and respond to mechanical forces. In addition, the cytoskeleton and other structural components are able to transmit and modulate cellular tension via integrins and focal adhesion sites that link with extracellular matrix. Mechano-gated channel (MGCs), receptor tyrosine kinases, and G proteins are all implicated in cellular mechanotransduction. However, the precise mechanotransduction mechanisms remain yet to be understood completely.

In general, the earliest responses to mechanical stimuli include opening of MGCs, release of soluble mediators, phosphorylation of focal adhesion kinases, and activation of small guanosine triphosphatases (Sadoshima and Izumo, 1997). Subsequent to these responses, numerous intracellular signaling pathways are triggered such as mitogen activated protein kinase, protein kinase C, and nuclear factor-κB (Chiquet and Fluck, 2001) to regulate extracellular matrix gene transcription (MacKenna et al., 2000).

Firstly, back in the end of 1980s and in the beginning of 1990s it has been shown that cardiac fibroblasts synthesize and excrete different bio-active compounds and therefore participate in regulation of cardiac function (Brilla et al., 1993; Eghbali et al., 1991; Sadoshimaet and Izumo, 1993; Villarreal and Dillmann, 1992; Weber and Brilla, 1991; Weber et al., 1993, 1994). A little bit earlier electrophysiological properties of cardiac fibroblasts were investigated. In 1986 A. Kamkin and I. Kiseleva published a description of electrophysiological properties of cardiac fibroblasts in the whole beating heart, in isolated whole heart and in cardiac tissue fragments. Electrotonical interaction of those cells with each other was also shown (Kiseleva et al, 1987; Kamkin et al., 1988). Since no histological studies were performed at this point of time initially those cells were termed as "atypical" nonmuscular cardiac cells (for example, Kamkin et al., 1988). We suggested that these cells are fibroblasts of the heart. A typical feature of the cells was the sensitivity of their resting potential to mechanical deformation of the plasma membrane, i.e. during the spontaneous atrial contractions (Kohl et al. 1992).

Later by means of employment of microelectrode technique and histological approaches those "atypical" nonmuscular cardiac cells were shown to be cardiac fibroblast (Kondratjev et al., 1993; Kohl et al., 1994), which have a pronounced response to mechanical stimulation (Kamkin et al., 1995; Kiseleva et al., 1996, 1998). Finally lately several groups published reports of studies of cell-to-cell fibroblasts-to-fibroblast and fibroblast-to-cardiomyocyte interaction in cardiac tissue (Camelliti et al., 2004a, b, 2005; Kamkin et al., 2005a).

Recent study (Camelliti et al., 2004b) of the rabbit sinoatrial node region, revealed two spatially distinct fibroblast populations, which expressed different types of connexins, the gap junction forming proteins. Fibroblasts are coupled via connexin40 (Cx40) in fibroblast-rich areas devoid of myocytes, and by Cx45 in regions of the node where fibroblasts intermingle with myocytes. In contrast to the cardiomyocytes, fibroblasts are electrically non-excitable cells, and their electrophysiological function remained unexplored until recently. Novel studies demonstrated the existence of MGCs in the plasma membrane of cardiac fibroblasts (Kamkin et al., 2003a, b). These channels were similar to the MGCs in other cell types (for review see reference (Sachs and Morris, 1998), suggesting that cardiac fibroblasts can transform mechanical stimuli into electrical signals (Kamkin et al., 2003a, b), which has been postulated in earlier studies (Kohl et at. 1992). If one assumes a role for cardiac fibroblasts in the mechano-electric feedback mechanism of the heart, the question arises as to how membrane potential changes that occur in cardiac fibroblasts upon mechanical stimulation, are transmitted to the adjacent cardiomyocytes. Over decades, fibroblasts in the heart were thought to be electrically isolated from the cardiomyocytes. However, novel results testify that fibroblasts form a coupled network of cells, which is structurally and functionally interconnected to the myocytes (Camelliti et al., 2004a, b; Kohl et al., 1994). Thus, the formation of single gap junction channels between cardiac myocytes and fibroblasts may facilitate the transfer of electrical signals between the two cell types (Maziere de et al., 1992; Kiseleva et al., 1996).

In this article we will review some of the recent advances in the electrophysiology of cardiac fibroblasts. We will also discuss the role of cardiac fibroblasts in the mechano-electric feedback in healthy and diseased hearts.

9.2 Electrical Properties of Single Cardiac Fibroblasts

Low values of the resting membrane potential (E_m) are typical to cultured fibroblasts in different cell lines. E_m of cultured mouse fibroblasts in different cell lines is reported to be in the range between –10 mV ÷ –20 mV and –20 mV ÷ –25 mV (Nelson et al., 1972; Okada et al., 1977, 1984; Tsuchiya et al., 1981), –13 mV ÷ –17 mV (Ince et al., 1984), –15 mV ÷ –40 mV (Okada et al., 1981).

More recently, fibroblasts were isolated enzymatically from atria and ventricles. The membrane currents of these freshly isolated cells were studied by means of the patch-clamp technique (Kamkin et al., 2003a, b). Isolated fibroblasts were used for voltage-clamp analysis of ionic currents generating mechanically-induced potentials. These cells had E_m of –37±3 mV, an input resistance of 514±11 MΩ and a membrane capacity of 18±3 pF.

Examples, typical for the mechanosensitive (MS) currents of cardiac fibroblasts are presented in Figs. 9.1a and 9.2a. In the absence of electrical pulses, the membrane was clamped to a holding potential of –45 mV, a value slightly more

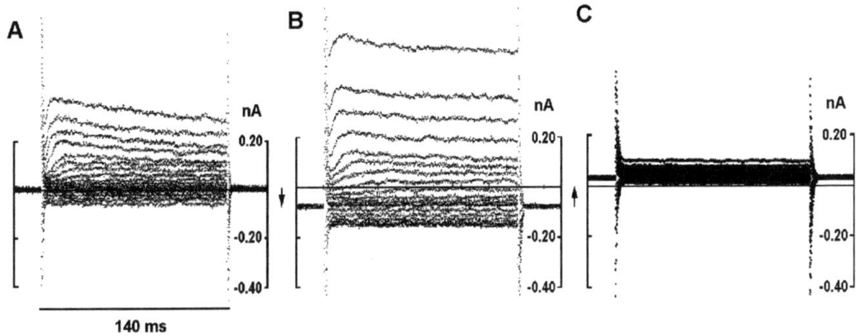

Fig. 9.1 Whole-cell currents from cardiac fibroblasts in the absence (**a**) and presence (**b**) of mechanical compression (2 μm). (**c**) Effect of 8 μM Gd^{3+} on the whole-cell currents during sustained compression. Reproduced from Kamkin et al. (2003a), with permission from Elsevier, Oxford University Press and Cardiovasc Res

negative than the normal resting potential of the fibroblasts. From this holding potential, application of test pulses with 140 ms duration adjusted the membrane potential to values between –100 and +50 mV. Addition of 8 μM Gd^{3+} shifted E_0 to –90±5 mV during the first 7 min. Since these values did not differ from those measured with two patch-pipettes it has been postulated that activation of G_{ns} by the second cell attached patch-pipette was negligible. Those results suggested that a Gd^{3+} -sensitive non-selective membrane conductance G_{ns} is active in atrial fibroblasts under "normal recording conditions", and that this G_{ns} moves the resting potential away from the E_K (Kamkin et al., 2003a).

Cultured fibroblasts in this study had resting potential of –32±3 mV, a membrane resistance of 531±32 MΩ, and a membrane capacity of 18±3 pF (Lozinsky I, Kiseleva I, Kamkin A, 2009. Unpublished data).

9.2.1 Mechanosensitive Whole-Cell Currents During Compression of the Isolated and Cultured Fibroblasts

Figure 9.1 shows how net membrane currents are modulated by compression. Mechanical compression of the fresh isolated fibroblasts shifted the holding current at –45 mV to more negative values (compare the initial traces in Fig. 9.1b and a). Compression increased the current amplitudes during the depolarizing clamp steps without changing their time course. At negative potentials, the currents were more negative than under control conditions without compression. Hence, mechanical compression increased the membrane conductance of the atrial fibroblasts. Finally, when 8 μM Gd^{3+} was added to the bath solution, not only the compression-induced currents but also a large portion of the currents in the absence of mechanical compression was blocked (Fig. 9.1c). These findings strongly suggest that mechanical compression activates a MS conductance in the cardiac fibroblasts (Kamkin et al., 2003a).

Cultured up to 5 days atrial fibroblasts were used for voltage-clamp analysis of ionic generating mechanically-induced potentials in different bathing solutions and of patch-pipette solutions with different pCa. Fibroblasts were mechanically deformed by two patch-pipettes. Cardiac fibroblasts cultured during 5 days specifically respond to mechanical deformation. Axial compression of the cultured cardiac fibroblasts by 2, 3 and 4 μm caused increase in depolarization by activating inward currents through a non-selective cation conductance. Compression-induced depolarization and I_{ci} are independent of the pCa and this is a non-selective Gd^{3+}-sensitive cation conductance. The MS currents were carried by Na^+, K^+ and Cs^+, and were blocked by application of Gd^{3+}. (Lozinsky I, Kiseleva I, Kamkin A, 2009. Unpublished data). These data suggest that cultured cardiac fibroblasts preserve MGCs up to 5 days without special flexible substrates, which allow application of the stretch and compression, and does not change their conductance during direct axial compression or stretch of the cells.

9.2.2 Mechanosensitive Whole-Cell Currents During Stretch of the Isolated and Cultured Fibroblasts

Figure 9.2 shows membrane currents of fresh isolated fibroblasts recorded under control conditions and during 2 μm lateral stretch. Stretch shifted the holding current at –45 mV to more positive values (beginning of the traces in Fig. 9.2b, marked by arrow). Stretch almost blocked the inward currents at negative potentials, and lowered the outward currents at positive potentials. These results suggest that stretch reduced the membrane conductance without significant changes of the time course of the currents (Fig. 9.2b). The experiment was completed by addition of 8 μM Gd^{3+} to the bathing solution. Gd^{3+} further reduced currents during sustained stretch (Fig. 9.2c). Since Gd^{3+} is known to block MGCs in a variety of cell types, these results suggest that cardiac fibroblasts contain MGCs that are inactivated by stretch and can be blocked by Gd^{3+} (Kamkin et al., 2003a).

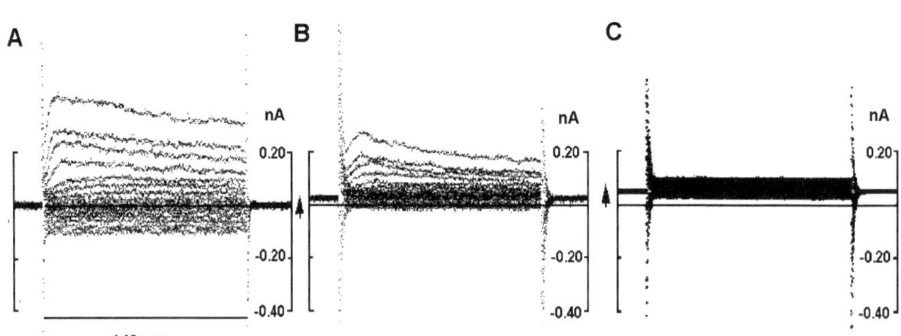

Fig. 9.2 Series of whole-cell currents from cardiac fibroblasts in the absence (**a**) and presence (**b**) of stretch (2 μm). (**c**) Effect of 8 μM Gd^{3+} on the whole-cell cur-rents during sustained stretch Changes of the holding current are marked by *arrows*. Reproduced from Kamkin et al. (2003a), with permission from Elsevier, Oxford University Press and Cardiovasc Res

Axial stretch of the cultured up to 5 days atrial fibroblasts by 2, 3 and 4 μm depressed inward currents by means of stretch-induced suppression of non-selective cation conductance. Stretch-induced hyperpolarization and I_{str} are independent of the pCa. The changes in membrane currents continued as long as stretch (or compression) was sustained, i.e. there were no signs of adaptation (tested up to 15 min) (Lozinsky I, Kiseleva I, Kamkin A, 2009. Unpublished data). From the data of earlier experiments on acutely isolated rat cardiac fibroblasts together with similar findings from studies of cultures we can conclude that the value of E_0 and MS current have a strict dependence on the extent of the cellular deformation.

9.2.3 *Other Ion Currents of Cardiac Fibroblasts*

In addition to MGCs, cardiac fibroblasts were reported to express voltage-dependent potassium K_v channels. It was shown that neonatal cardiac fibroblasts express at least three different K_v channels, outward K^+ currents of which were measured in cultured cells. The majority of cells express a transient outward K^+ current (I_{to}) that activates at potentials positive to –40 mV and partially inactivate during depolarizing voltage steps. A smaller number of cells express one of two types of kinetically distinct, delayed-rectifier K^+ currents (I_K fast I_{Kf} and I_K slow I_{Ks}). Immunoblot analysis reveals the presence of $K_v1.4$, $K_v1.2$, $K_v1.5$, and $K_v2.1$ alpha-subunits but not $K_v4.2$ or $K_v1.6$ alpha-subunits in the fibroblasts. Thus neonatal cardiac fibroblasts express at least three different K_v channels (Walsh and Zhang, 2008).

Ca^{2+}-activated K^+ currents ($I_{K[Ca]}$) were reported to be found in cultured human cardiac fibroblasts. In this study, in whole-cell configuration, depolarizing pulses evoked $I_{K[Ca]}$ in an outward rectification in these cells. A large-conductance, Ca^{2+}-activated K^+ currents (BK_{Ca}) channel with single-channel conductance of 162 ± 8 pS was also found in human cardiac fibroblasts. Western blot analysis revealed the presence of α-subunit of BK_{Ca} channels (Wang et al., 2006).

9.2.4 *Current-Voltage Relations During Compression and Stretch of the Isolated and Cultured Fibroblasts*

The currents at the end of the 140 ms long voltage-pulses were assembled in current-voltage relations (*I-V* curves, Fig. 9.3). Without mechanical stimulation, the *I-V* curves intersected the voltage axis at –37 mV. This zero-current potential corresponds to the normal resting membrane potential V_m ($=E_m$) of the non-clamped fibroblasts. Mechanical compression of isolated fibroblasts caused downward flection of the *I-V* curve and shifted V_m to more positive values (Fig. 9.3a). In contrast, when freshly isolated atrial fibroblasts were stretched, the *I-V* curve was bended upwardly, and intersected the voltage axis at a more negative V_m (Fig. 9.3c). Both, the compression-stimulated and the stretch-reduced currents reversed their polarity close to 0 mV, as would be expected for a non-selective cation MGCs that can conduct Na^+, K^+ and Cs^+ ions. The compression-induced current was inhibited by

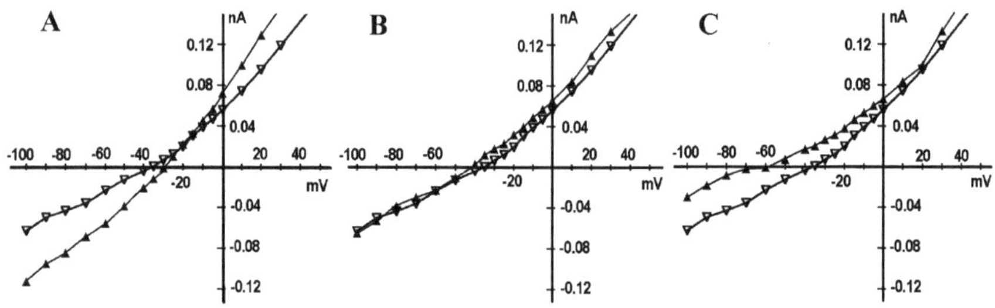

Fig. 9.3 Mechanosensitivity of membrane currents in fibroblasts that were freshly isolated from rat atria. Current-voltage relations (*I-V* curves, current is measured at the end of the 140 ms clamp pulses). (**a**) *I-V* curves before (*empty triangles*) and during 2 μm compression (*filled triangles*). Note the shift of the zero-current potential from V_m –37 to –28 mV. (**b**) Reversibility of the compression-induced changes (*empty triangles* before, *filled triangles* 2 min after stretch). (**c**) *I-V* curves before (*empty triangles*) and during 2 μm stretch (*filled triangles*). Note the shift of V_m from –37 to –60 mV. *I-V* curves A, B, C were recorded from the same cell. Modified from Kamkin et al. (2003a), with permission from Elsevier, Oxford University Press and Cardiovasc Res

low concentrations of Gd^{3+} (8 μM; Fig. 9.4), a compound, which is commonly used to block current flow through MGCs (Kamkin et al., 2003a, b).

Similar to the effect of mechanical stretch, Gd^{3+} hyperpolarized the membrane potential of the fibroblasts, albeit to a larger extent (Fig. 9.4). The

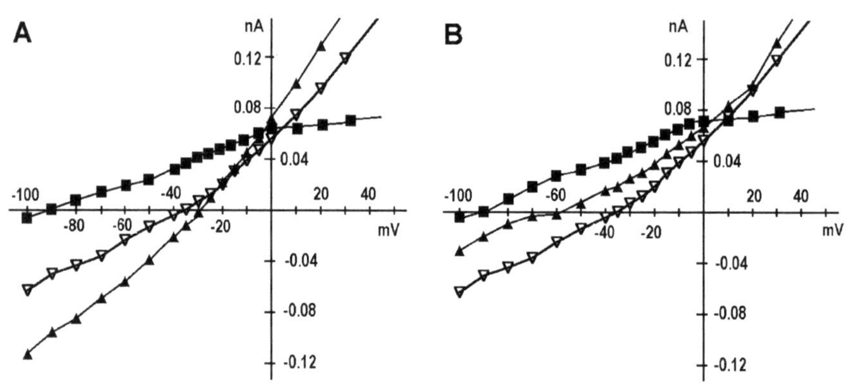

Fig. 9.4 Inhibition of compression-induced currents in freshly isolated fibroblasts from rat atria with Gd^{3+} (8 μM). Current-voltage relations (*I-V* curves, the current was measured at the end of 140 ms clamp pulses). (**a**) *I-V* curve before (*empty triangles*) and during 2 μm compression (*filled triangles*). The zero-current potential V_m was shifted from –37 to –28 mV during mechanical compression. The effect of mechanical compression was abolished in the presence of Gd^{3+} (*filled squares*). Note the shift of V_m to –90 mV and the decrease in negative currents with Gd^{3+}. (**b**) *I-V* curve before (*empty triangles*) and during 2 μm stretch (*filled triangles*). The zero-current potential V_m was shifted from –37 to –60 mV during mechanical stretch. In the presence of Gd^{3+} (*filled squares*) V_m shift to –90 mV and the decrease in negative currents with Gd^{3+}. Modified from Kamkin et al. (2003a), with permission from Elsevier, Oxford University Press and Cardiovasc Res

voltage-dependence and the Gd^{3+}-sensitivity of the compression-induced currents resembled the MS currents in cardiomyocytes (Isenberg et al., 2003; Kamkin et al., 2000a, 2003c, d; Zeng et al., 2000; Zhang et al., 2000) and other cell types (for review about stretch-activated, non-selective cation channels (SACs) see Sachs F, Morris CE, 1998). Hence, we conclude that modulation of MGCs is responsible at least in part for the mechanosensitivity of cardiac fibroblasts.

9.2.5 Single Mechano-Gated Channels in Cardiac Fibroblasts

Single MGCs are activated by mechanical deformation of acutely isolated cardiac fibroblasts. Single MGCs and MS whole-cell currents were practically simultaneously recorded from isolated atrial fibroblasts using the cell-attached and whole-cell patch-clamp configurations or cell-attached pipette (first pipette) and whole-cell patch-clamp pipette (second pipette), respectively. Under resting conditions occasional short openings of two types of MGCs with conductances of 43 pS and 87 pS were observed (Fig. 9.5a). Small (1 μm) mechanical deformations affected neither MGCs nor whole-cell MG currents. Cell compression (2, 3 and 4 μm) augmented the whole-cell MG currents and increased the frequency and duration of single MGC openings (Fig. 9.5b : 2 μm of compression). Both MGCs (with conductance levels of 43 and 87 pS) displayed linear current-voltage relationships with the reversal potential around 0 mV. Cell stretch (2, 3 and 4 μm) inactivated the whole-cell MG currents and abolished the activity of single MGCs (Fig. 9.5c : 2 μm of stretch). Thus, we conclude that fibroblasts sense the directionality of the applied stress. Hence, MGCs, and their (likely) cytoskeletal attachments, represent a vectorial sensor in cardiac fibroblasts. Gd^{3+} (8 μM) blocked the whole-cell MG currents within 5 min after the beginning of application, MGCs recorded in the cell mode were blocked immediately when Gd^{3+} was added to the intrapipette solution. The mixture of cytochalasin D and colchicines (100 μM each) completely blocked both the whole-cell MS currents and MGCs. Thus, rat atrial fibroblasts express two types of MGCs whose activity is governed by cell deformation (Kamkin A, Kiseleva I, Kirischuk S. 2009. Unpublished data).

To our knowledge, this is the first time when single MGCs where activated by mechanical deformation of the whole cell and recorded in the cell-attached mode. In the vast majority of experiments single MGCs were activated by changing the pressure in the patch pipette (see for reviews Hamill, 2006; Hamill and Martinac, 2001; Sachs and Morris, 1998). Therefore, it is interesting to compare two types of stimuli used to activate MGCs (pressure in the pipette and mechanical deformation of the cell). Probability of MGC opening (P_0) appears to be a suitable parameter for this. Using the data presented by Bustamante et al. (1991), we can slightly speculate that a 2 μm compression is equivalent to 20 mmHg in the pipette, a 3 μm compression-induced effects are comparable with those produced by 40 mmHg, and a 4 μm compression may correspond to 60–80 mm Hg in the patch pipette.

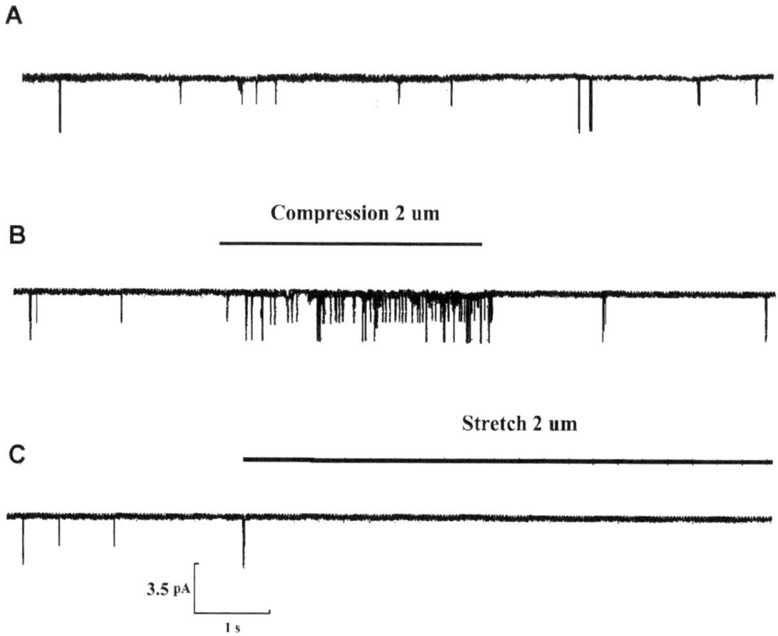

Fig. 9.5 Single MGC activity recorded in the cell-attached mode during mechanical deformation of acutely isolated cardiac fibroblasts. On-line pen records. Sample traces show the activity of single MGCs in control ("resting compression") (**a**), during a 2 μm cell compression (**b**) and during a 2 μm cell stretch (**c**). The holding potential was –45 mV. (Kamkin A, Kiseleva I, Kirischuk S. 2009. Unpublished data)

9.3 Electrical Properties of Cardiac Fibroblasts in Heart Tissue

The electrical properties of cardiac fibroblasts were studied for the first time in 1986 (Kiseleva et al., 1987; Kamkin et al., 1988). The intracellular recordings were accomplished on multicellular tissue preparations with the use of fine microelectrodes (Kamkin et al., 1999, 2003e, 2002, 2001; Kiseleva et al., 1996, 1998; Kohl et al., 1992, 1994; Kohl and Noble, 1996). Fibroblasts in rat atrial tissue had a resting membrane potential (E_m) of approximately –22 mV and input resistances of ≈0.5 GΩ (Kamkin et al., 2002; Kiseleva et al., 1998). A typical feature of the cells was the sensitivity of their E_m to mechanical deformation of the plasma membrane, i.e. during the spontaneous atrial contractions.

The findings obtained with freshly isolated single fibroblasts were similar to those reported for multicellular preparations. However, fibroblasts that were embedded in their normal three-dimensional tissue environment differed in some aspects from isolated cells. For example, fibroblasts in multi-cellular preparations had a lower E_m of approximately –22 mV with a wide distribution range between –5 and –70 mV (Fig. 9.6).

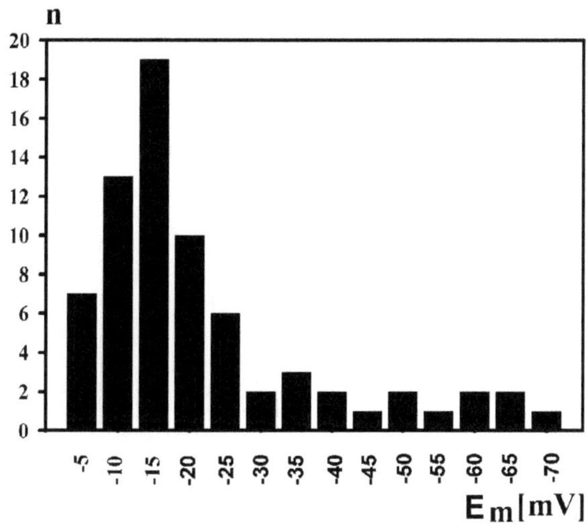

Fig. 9.6 Distribution of the membrane potential of mechanosensitive fibroblasts in the rat sinus node region. Reproduced from Kiseleva et al. (1998) with permission from Elsevier and J Mol Cell Physiol

The wide distribution range of E_m suggests that fibroblast in their normal environment are exposed to variable mechanical stretch or compression, even in the pause between the contractions and relaxations of the myocardium. A similar distribution of E_m of atrial fibroblasts was found in non-contracting tissue. A step-by-step increase of the amount of stretch gradually hyperpolarized E_m only in 80% of the fibroblasts (Fig. 9.7) and reduced the frequency of spontaneous contractions of the myocardial tissue preparations (Kamkin et al., 1999, 2003e).

At the same time, a step-by-step increase of stretch of the tissue depolarized $\mathrm{E_m}$ in 20% of the fibroblasts (Fig. 9.8). This latter reaction to artificial stretch results from compression of fibroblasts, which could explain why in situ during tissue contraction (systole) not all of the fibroblasts are depolarized, and during tissue relaxation (diastole) not all of fibroblast are hyperpolarized. In our review we will discuss the typical response of fibroblasts – depolarization during compression and hyperpolarization during stretch.

Most fibroblasts are probably compressed during the contractions, which will depolarize their membrane potential, thereby eliciting so-called "mechanically induced potentials" (MIPs) (Kiseleva et al., 1996). MIPs in cardiac fibroblasts differ from action potentials in cardiomyocytes by their lack of a fast upstroke velocity and overshoot. They usually start with a delay of several milliseconds after the onset of action potentials (Kamkin et al., 2003e; Kiseleva et al., 1996, 1987; Kohl et al., 1992). MIPs can be detected only in contracting atrial tissue and strictly follow the rhythm of contractions. Upon hyperpolarization of the resting potential, the amplitude of the MIPs increases, whereas membrane depolarization will decrease the MIP amplitudes.

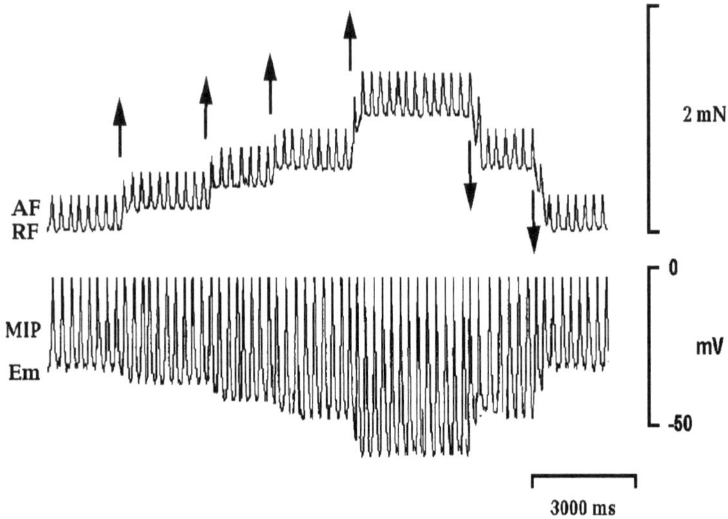

Fig. 9.7 Specific reaction of the membrane potential of a rat atrial fibroblast to mechanical stretch of the tissue. Synchronous registration of the mechanogram (*top curve*) and mechanically induced potentials (MIP) of the fibroblast (*bottom curve*). The symbol (↑) indicates the time point of stretch application, whereas ↓ marks the release of applied stretch. AF – active force, RF – resting force, MIP – mechanically induced potential of the fibroblast, E_m – resting potential of the fibroblast. Modified from Kiseleva et al. (1998) with permission from Elsevier and J Mol Cell Physiol

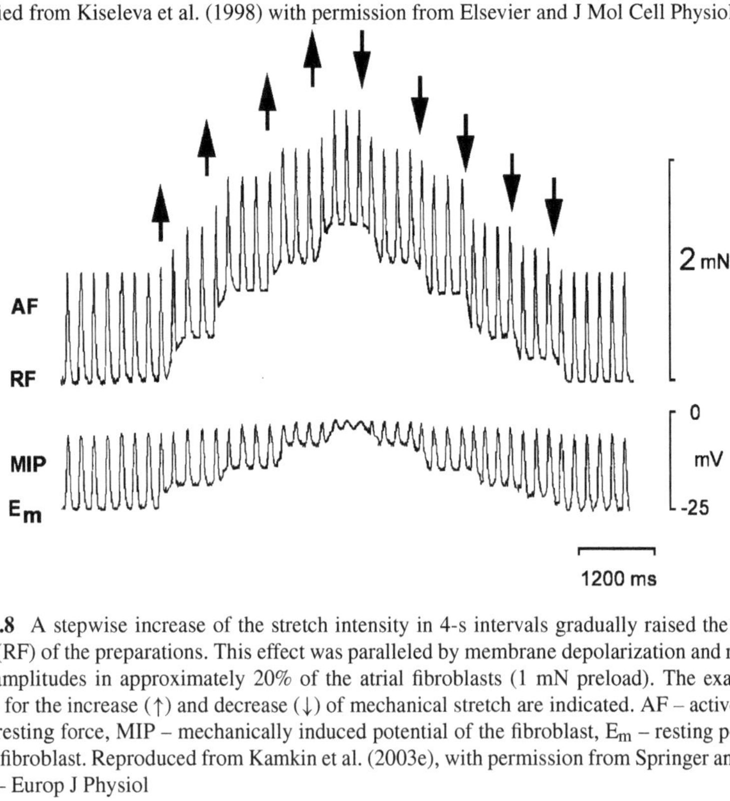

Fig. 9.8 A stepwise increase of the stretch intensity in 4-s intervals gradually raised the resting force (RF) of the preparations. This effect was paralleled by membrane depolarization and reduced MIP amplitudes in approximately 20% of the atrial fibroblasts (1 mN preload). The exact time points for the increase (↑) and decrease (↓) of mechanical stretch are indicated. AF – active force, RF – resting force, MIP – mechanically induced potential of the fibroblast, E_m – resting potential of the fibroblast. Reproduced from Kamkin et al. (2003e), with permission from Springer and Pflüg Arch – Europ J Physiol

It is a major challenge to analyze the pathways, through which mechanical compression may activate ion channels in the plasma membrane of atrial fibroblasts. One possibility would be that the mechanical energy is transmitted through changed tension of the lipid bilayer (Sachs and Morris, 1998). Alternatively, the cytoskeleton may conduct mechanical stress from the site of physical deformation to the channel protein (Kamkin et al., 2001; Sachs and Morris, 1998). The latter possibility is supported by recent observations with multicellular preparations, where disruption of cytoskeletal proteins – through depolymerisation of F-actin with cytochalasin D and tubulin degradation with colchicine – decreased the MIP amplitudes (Kamkin et al., 2001). This effect of cytochalasin D on the MIP amplitudes could be attenuated by intracellular application of Mg-ATP to promote actin polymerisation (Chhabra et al., 2000). Thus, similar to the findings made with other stretch-activated ion channels and different types of voltage-gated channels (Galli and DeFelice, 1994; Johnson and Byerly, 1993; Maltsev and Undrovinas, 1997), the cytoskeleton appears to play a major role in the signal transfer from the site of mechanical deformation to the ion channels in the plasma membrane of atrial fibroblasts. Cytochalasin D and colchicine did not fully suppress the MIPs in atrial fibroblasts suggesting that cytoskeletal disruption by these drugs was either incomplete, or that additional, cytoskeleton-independent mechanisms are involved in the regulation of mechanically operated ion channels in atrial fibroblasts (Sachs and Morris, 1998).

9.4 Age Dependence of the Electrical Properties of Cardiac Fibroblasts

Cardiac arrhythmia is a clinical disposition that affects mainly the older population. Since fibroblasts in the heart have been implicated in the pathophysiology of cardiac arrhythmia, it is important to explore whether the mechano-electrical properties of the cells would change with increasing age of the individuals. In a first attempt to resolve this issue, we analyzed the electrophysiological characteristics of fibroblasts from the right atria of young (3 months) and old (15 months) healthy rats. In both groups, the membrane potential of individual fibroblasts in multicellular right atrial preparations varied between –5 and –70 mV. However, several interesting differences in the electrophysiological properties between cardiac fibroblasts from 3 to 15 months old animals were detected. For example, the distribution of resting membrane potential values varied remarkably between young (Fig. 9.9a) and old (Fig. 9.9b) rats. The majority of atrial fibroblasts in young rats had resting membrane potentials between –10 and –15 mV (n=124). For comparison, E_m of cardiac fibroblasts from old rats was mostly in the range from –20 to –25 mV (n=162). The mean values of the fibroblast membrane potentials were 19.23 ± 1.02 mV and 29.41 ± 1.23 mV in young and old animals, respectively.

The membrane resistance of atrial fibroblasts from young rats was usually between 0.4 and 0.6 GΩ, whilst older animals had membrane resistances between 0.7 and 0.9 GΩ. Thus, the higher membrane potential of fibroblasts from old rats was associated with increased membrane resistances.

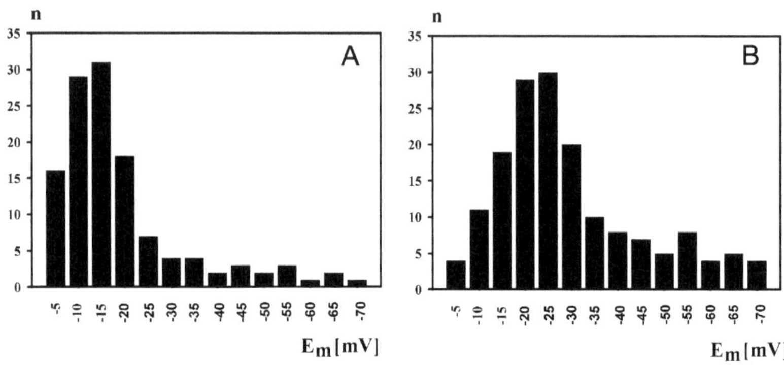

Fig. 9.9 Distribution of the resting membrane potentials of right atrial fibroblasts from 3 months old (**a**) and 15 months old (**b**) rats. Reproduced from Kamkin et al. (2005b) with permission from Academia Publishing House Ltd

The following experiments were performed on fibroblasts from human right atrial trabeculae. The tissue biopsies were obtained from patients between 65 and 78 years of age, who underwent aorto-coronary bypass surgery. The right atrial function of the patients was clinically normal. The resting membrane potential of these fibroblasts was -15.9 ± 2.1 mV and the membrane resistance 4.1 ± 0.1 GΩ (Kamkin et al., 1999).

9.5 Altered Electrical Function of Cardiac Fibroblasts After Myocardial Infarction

Contractile failure of the heart due to cardiac arrhythmia is a major cause for cardiovascular mortality. The risk for arrhythmia is particularly high after myocardial infarction (D'Alonzo yet al., 1995; Kleber and Fast, 1997; Naccarella et al., 2000; Podzuweit et al., 1987; Pogwizd and Corr, 1992; Underwood et al., 1997). The mechanisms that underlie the electrical instability of the ischemic heart are not well understood. Recent findings suggest that a phenomenon, which is commonly referred to as mechano-electric feedback, might play an important role in cardiac arrhythmia (Dean and Lab, 1989a,b; Franz, 1996; Nazir and Lab, 1996; Ninio and Saint, 2008; Ravens, 2003; Taggart, 1996; Taggart and Lab, 2008). Mechano-electric feedback refers to the situation that mechanical changes in the myocardium, which may result from variable filling pressures, can modulate the electrical function of the heart. This, in turn, may affect the spontaneous electrical activity of the myocardium (Lab, 1968).

The use of microelectrode recordings and the patch-clamp technology made it possible to analyze in detail membrane potential changes of cardiac cells in response to mechanical stretch. Electrophysiological studies revealed that mechano-electric coupling, i.e. the electrical response of cardiomyocytes to stretch, involves the transmembrane influx of cations through stretch-activated channels (Isenberg et al., 2003; Kamkin et al., 2000a, 2003c, d; Zeng et al., 2000; Zhang et al., 2000). Correspondingly, the influx of cations into cardiomyocytes through stretch-activated channels was enhanced and extra action potentials occurred under ischemia, which could possibly elicit cardiac arrhythmia (Dean and Lab 1989, 1990; Dilly and Lab, 1988; Kamkin et al., 2000b, 2003d; Murphy et al., 1996, 1994; Nazir and Lab, 1996). The sensitivity of the cardiac myocytes to mechanical stretch was enhanced after infarction, suggesting increased expression levels of stretch-activated ion channels during myocardial remodeling after infarction (Kamkin et al., 2000a, b; Kiseleva et al., 2000).

In addition to the electrophysiological changes of cardiomyocytes, fibroblasts in the heart showed pronounced hyperpolarization of their resting membrane potential (Fig. 9.10) and increased sensitivity to mechanical stretch after infarction. Modulation of the electrical function of the atrial fibroblasts may destabilize the rhythmical activity of the ischemic myocardium. This idea is supported by the observation that mechanical stretch hyperpolarized E_m of atrial fibroblasts to a higher extent in tissue specimens from infarcted (Fig. 9.11) than from healthy rat hearts (Kamkin et al., 2002). Furthermore, the susceptibility of the membrane potential of the fibroblasts to mechanical stimulation correlated positively with the infarct size, and artificial stretching of the cells reduced the frequency of spontaneous contractions (Kamkin et al., 2002). While these findings document a role of atrial fibroblasts in the mechano-electric function of normal and diseased hearts, it remains to be clarified as to how electrical signals that are generated by the atrial fibroblasts can modulate the spontaneous contractile activity of the myocardium. An important clue to this issue may come from a detailed analysis of the pathways, through which fibroblasts and myocytes in the heart communicate with each other.

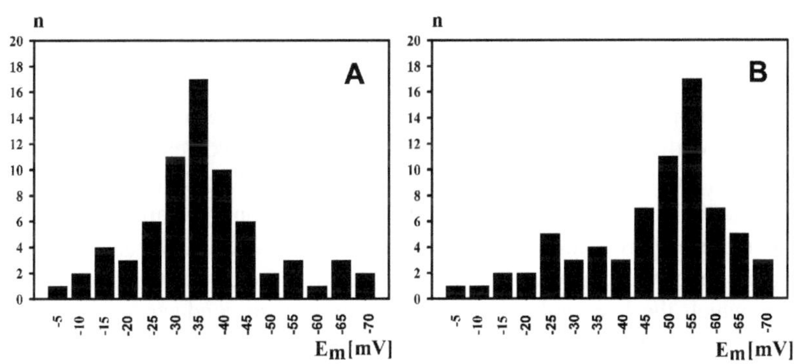

Fig. 9.10 Frequency distribution of the resting membrane potential of fibroblasts from rat atria 20 days after experimental myocardial infarction. (**a**) Shows the data from rats with small infarct sizes of 16.5 ± 0.6 % of the left endocardial circumference (**a**). The data shown in (**b**) were obtained from rats with extensive myocardial infarction including 40 ± 1.3 % of the left endocardial circumference. Reproduced from Kiseleva et al. (1998), with permission from Elsevier and J Mol Cell Physiol

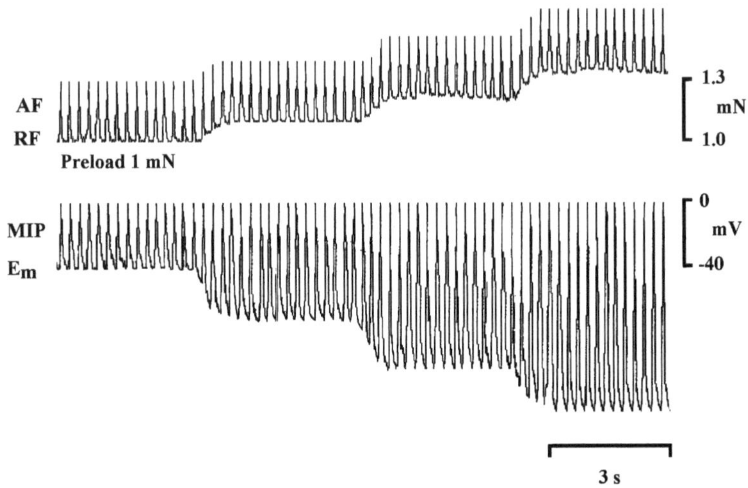

Fig. 9.11 Response of the membrane potential of a right atrial fibroblast to long lasting artificial stretch of the tissue 20 days after experimental myocardial infarction (approx. 40% of the left endocardial circumference). Note the excessive hyperpolarization of E_m in response to artificial stretch of the tissue. AF – active force, RF – resting force, E_m – resting membrane potential, MIP – mechanically induced potential of the fibroblast. Reproduced from Kamkin et al. (2005b), with permission from Academia Publishing House Ltd

9.6 Electrical Coupling Between Fibroblasts and Myocytes

9.6.1 Electrical Communication Among Cardiac Fibroblasts

Electrical communication of fibroblasts in the heart was studied in right atrial tissue preparations from various species with the use of microelectrode recording techniques. A bidirectional transcellular electrical current flow could be monitored in 65% of the fibroblasts from frog hearts when two microelectrodes were randomly inserted into different cells at a distance of $\approx$40 μm (Kamkin et al., 1988; Kiseleva et al., 1987; Kohl et al., 1992, 1994; Kohl and Noble, 1996; Kohl, 2003). Artificial hyperpolarization of one fibroblast shifted the membrane potential of the second cell to more negative values and vice versa (Fig. 9.12a, c). By consequence, the MIP amplitudes were increased not only in the polarized cell, but also in the fibroblast, to which the hyperpolarizing current pulses were conducted (Fig. 9.12a, c). Conversely, depolarization of one fibrobroblast shifted the membrane potential of the second cell to less negative values, and this effect was also bidirectional (Fig. 9.12b, and d). The transmission coefficient was 0.11 ± 0.02 mV and did not depend on the magnitude of artificial intracellular polarization (Kiseleva et al., 1987).

Intracellular injection of Lucifer Yellow into a cardiac fibroblast resulted in spreading of the dye to other fibroblasts through gap junctions (Fig. 9.13).

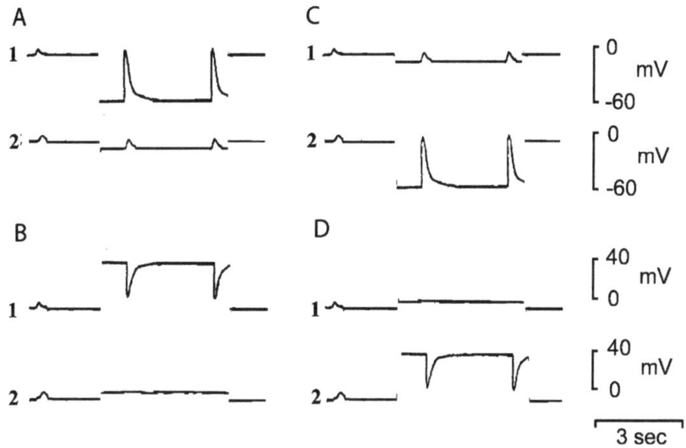

Fig. 9.12 Bidirectional electrical coupling of two fibroblasts in the right atrium of a frog. The two microelectrodes were located 40 μm apart from each other. (**a**) Artificial intracellular hyperpolarization of one cell (1) shifted the membrane potential of the second fibroblast to more negative values (2). (**b**) Injection of depolarizing current pulses into the first fibroblast (1) shifted the membrane potential of the other cell to less negative values (2). Electrical interaction between the two fibroblasts was bi-directional in that artificial intracellular hyperpolarization (**c**) or depolarization (**d**) of the second fibroblast caused similar changes of the membrane potential in the first cell. Modified from Kamkin et al. (1988) and Kiseleva et al. (1987)

Furthermore, recent data demonstrated the existence of two populations of fibroblasts in the sinoatrial node region of rabbit hearts; one, which expresses the gap junction protein connexin (Cx) 45, and a second type of fibroblast, which is positive for Cx40. Additionally, some fibroblasts in the heart were reported to contain Cx43 (Camelliti, 2004b).

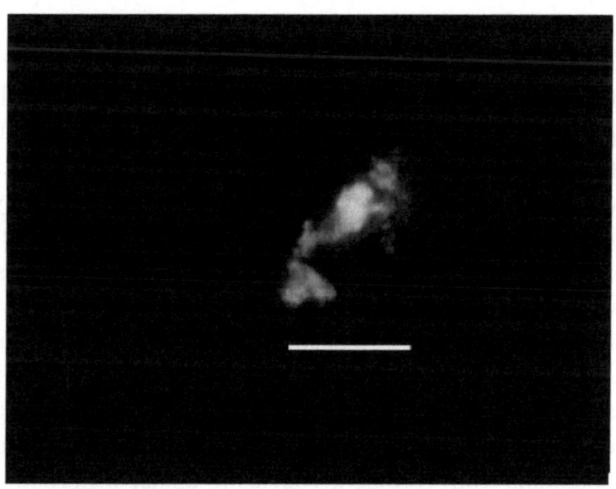

Fig. 9.13 Light photomicrograph demonstrating the intercellular transfer of Lucifer Yellow between to fibroblasts through gap junctions. Scale bar indicates 100 μm. Reproduced from Kiseleva et al. (1996), with permission from Elsevier, Oxford University Press and Cardiovasc Res

In approximately 20% of the cells, monodirectional coupling between two fibroblasts at a distance of ≈40 μm was observed (Kiseleva et al., 1993). In this case, artificial polarization of one cell caused a shift of the membrane potential of the second cell, but not vice versa. Monodirectional current flow between two cardiac fibroblasts can be explained on the basis of variable input resistances of the cells. For example, in two adjacent cells with different input resistances, any change of the membrane potential of the cell with low input resistance would cause a similar membrane potential shift of the cell with high input resistance. However, due to its lower input resistance, the first cell would be less susceptible to membrane potential changes, which are transmitted from the other fibroblasts. Notably, the input resistance of a cell is dependent not only on its size, but also on its three-dimensional shape and, in particular, on its electrical coupling to neighbouring cells. Monodirectional current flow between two cardiac fibroblasts can therefore result from different input resistances due to asymmetrical coupling of their cell membrane with adjacent cardiomyocytes.

9.6.2 Electrical Interaction Between Fibroblasts and Myocytes in the Heart

A detailed knowledge of the intercellular routes, through which fibroblasts and myocytes in the heart communicate to each other is necessary. Several attempts to demonstrate electrotonic interaction of atrial fibroblasts and cardiomyocytes via gap junctions were not successful. Electron microscopic studies showed clearly, that the plasma membranes of fibroblasts and myocytes in the sinoatrial node region form close contacts (Kamkin et al., 2002; Kiseleva et al., 1998; Maziere de et al., 1992), but without typical gap junction structures visible (Maziere de et al., 1992). We therefore assumed that single gap junction channels rather than clusters of cell–cell contacts might exist in the cell membrane of atrial fibroblasts and cardiac myocytes (Kohl et al., 1994). Unfortunately, single gap junction channels were difficult to detect by means of standard morphological methods (Kamkin et al., 2002; Kiseleva et al., 1998; Maziere de et al., 1992). More recently, the gap junction protein Cx45 was identified in the sinus node region of rabbit hearts at sites where the plasma membranes of cardiomyocytes and atrial fibroblasts formed close contacts (Camelliti et al., 2004b). Furthermore, intracellular injection of Lucifer Yellow into cardiac fibroblasts resulted in lateral spreading of the dye along extended threads of interconnected fibroblasts and, in particular, between neighbouring fibroblasts and cardiomyocytes in the sinoatrial node. These findings strongly suggest that electrical signals, which may arise in cardiac fibroblasts upon mechanical stimulation, can be transmitted to the adjacent cardiomyocytes (Camelliti et al., 2004b).

Electrical interaction of myocytes and fibroblasts in the heart via gap junctions is also supported by the results of intracellular microelectrode studies. As discussed earlier, gap junctions constitute an important pathway for the transfer of action potentials from cardiac myocytes to adjacent fibroblasts (Rook et al., 1988, 1989,

1990). Isolated fibroblasts had a E_m of approximately –37 mV, whereas the resting membrane potential of isolated cardiomyocytes was normally in the range of –80 mV. Cultures that were enriched in cardiac fibroblasts had E_m between –10 and –20 mV, while the membrane potential of isolated myoblasts was approximately –60 mV. Electrical coupling of the two cell types in mixed cultures is indicated by a depolarization of the membrane potential of the myoblasts to –50 mV and a shift of E_m of the co-cultured fibroblasts to more negative values between –40 and –60 mV.

Due to extensive gap junctional coupling, fibroblasts in the myocardial tissue had a E_m, which is close to the resting membrane potential of the cardiac myocytes (approx. –80 mV). Intracellular microelectrode recordings demonstrated that membrane potential changes in fibroblasts in response to the action potentials in cardiac myocytes occurred with a much slower up-stroke velocity (delayed time course) in the myocardial tissue (Fig. 9.14).

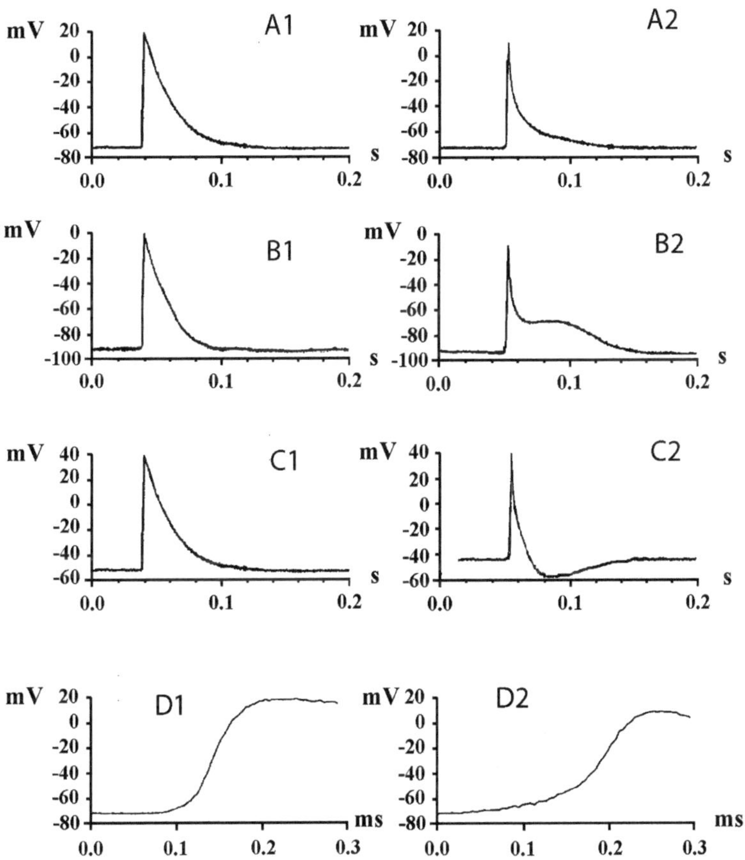

Fig. 9.14 Effect of intracellular de- and hyperpolarization on the action potential in a cardiomyocyte (**a1**, **b1**, **c1** and **d1**) and the membrane potential of a rat cardiac fibroblast (**a2**, **b2**, **c2** and **d2**). See text for more details Modified from Kohl et al. (1994) with permission of the Blackwell Publishing and Exp Physiol

In the experiment, which is shown in Fig. 9.14, both cell types could be distinguished from each other by their different responses to de- and hyperpolarizing intracellular current injections (10^{-9} A). Intracellular polarization of this small current does not lead to changes in action potential (AP) con-figuration of the cardiomyocytes, because all cardiomyocytes have input resistances, which are below the microelectrode resistance. In contrast, the input resistance of fibroblasts is much higher than the microelectrode resistance, and therefore polarization of a fibroblast will lead to changes in MIP amplitude. A typical AP registered from a cardiomyocyte is shown in Fig. 9.14a1, whereas Fig. 9.14a2 depicts the AP-like potential recorded in a fibroblast. Intracellular hyperpolarization of the cardiac myocyte did not change the shape of its AP due to the low input resistance of the cell (Fig. 9.14b1). Hyperpolarization of the fibroblast membrane also had no impact on the amplitude of the AP-like potential, but resulted in a delayed repolarization of the membrane potential instead (Fig. 9.14b2). This "plateau phase" corresponds to the mechanically induced potential (MIP), which is unmasked from the AP-like potential upon hyperpolarization of the membrane potential. Depolarizing current injections caused a transient hyperpolarization of the membrane potential of the cardiac fibroblast right after the AP transmitted from the cardiac myocyte (Fig. 9.14c2). Depolarization of the cardiac myocyte did not affect the time course of the AP in this cell (Fig. 9.14c1). Moreover, the upstroke velocity of the AP (Fig. 9.14d1) is typical for cardio-myocytes. The AP-like potential (Fig. 9.14d2) never reaches the upstroke velocity of the AP depolarization since it is much slower than that.

Intracellular injection of Lucifer Yellow into a cell with AP-like potentials (Fig. 9.14 a2, b2, c2, d2) demonstrates that it is cardiac fibroblast (Kamkin et al., 2005b). These results provide further evidence that myocytes and fibroblasts in the heart are electrically coupled through gap junctions (Kohl et al., 1994).

9.7 Role of Cardiomyocyte – Fibroblast Coupling in Healthy and Diseased Hearts

Cardiac fibroblasts and myocytes are cellular components of the mechano-electric feedback mechanism in the heart. Fibroblasts can change their membrane potential in response to mechanical deformation (compression) of the plasma membrane such that contraction of the myocardium compresses the fibroblasts and depolarizes their membrane potential. These contraction-evoked membrane depolarizations are referred to as mechanically induced potentials (MIPs) and result from a transmembrane influx of cations, mainly of Na^+, through MGCs in the plasma membrane of these cells (Kamkin et al, 2003a, b). On the contrary, hyperpolarization of the fibroblasts occurs in response to channel inactivation during mechanical stretch of the cells (Kamkin et al, 2003a, b).

In addition to MGCs, cardiac fibroblasts may also contain voltage-dependent potassium K_V channels in their plasma membrane (Kamkin et al., 2003a, b). Voltage-sensitive potassium currents were monitored in freshly isolated cardiac fibroblasts (Figs. 9.1a and 9.2a). Furthermore, the compression-activated conductance followed an outwardly rectifying voltage dependence.

Unlike the cardiac fibroblasts, cardiomyocytes become depolarized by mechanical stretch. Stretch-induced depolarization of their membrane potential is determined by the influx of cations through SACs (Isenberg et al., 2003; Kamkin et al., 2000a, 2003c,2003d; Zeng et al., 2000; Zhang et al., 2000). When the membrane depolarization reaches threshold, extra action potentials can occur, which potentially give rise to arrhythmic myocardial contractions (Kamkin et al., 2000b). Thus, mechanical stretch exerts opposite effects on the membrane potential of the cardiomyocytes and fibroblasts: it depolarizes the membrane potential of the cardiomyocytes and hyperpolarizes the fibroblast membrane.

Electrical communication between cardiac myocytes and fibroblasts is based on the presence of Cx45-positive gap junction channels. Camelliti et al. estimated that approximately 10% of the overall Cx45 protein content in the heart is localized to the sites of contact formation between fibroblasts and myocytes (Camelliti et al., 2004b). Since the input membrane resistance is significantly higher in fibroblasts than in the cardiomyocytes, even a relatively small number of Cx45 molecules would be sufficient to maintain the electrical interaction between myocytes and fibroblasts in the heart.

Mathematical modeling was performed for a cell pair consisting of a single myocyte from the sinoatrial node weakly coupled by 30 gap junction channels to a mechanosensitive fibroblast. Theoretical analysis predicted that depolarization of E_m of the fibroblast would decrease the duration of slowly depolarizing voltage trajectory of the pacemaker potential (between action potentials) of adjacent sinoatrial node myocytes, thereby increasing the frequency of spontaneous atrial contractions (Kohl et al., 1994; Kohl and Noble, 1996). On the contrary, hyperpolarization of E_m of the fibroblast would increase the time of slowly depolarizing voltage trajectory of the pacemaker potential of adjacent sinoatrial node myocyte. This, in turn, would decrease the frequency of spontaneous atrial contractions.

It is therefore likely that MIPs have a role in the mechano-electric feed-back in the heart. The depolarization phase of the MIP does not influence the AP, because it occurs during the absolute refractory period of the cardiomyocytes. However, repolarization of the MIP can possibly affect the duration of the repolarization phase of the AP at the level of APD50 or APD90 (action potential duration at 50 and 90% of repolarization). Repolarization of the MIP together with stretch-induced depolarization of AP can increase APD50 or APD90. As a consequence, increases of APD50 and APD90 may provide a risk factor for the occurrence of arrhythmias.

Tissue remodeling after myocardial infarction is associated with profound changes in both, the number and the electrical properties of cardiac fibroblasts. For example, the resting membrane potential of rat atrial fibroblasts increased in parallel to the infarct size after coronary artery occlusion. Furthermore, atrial fibroblasts from infarcted rat hearts were approximately 10-times more sensitive to mechanical deformation of their plasma membrane and responded with a stronger hyperpolarization of their membrane potential than fibroblasts from healthy hearts (Kamkin et al., 2002).

Furthermore, remodeling after myocardial infarction also increased the sensitivity of the cardiomyocytes to mechanical stretch. Thus, very small amounts of stretch were sufficient to evoke membrane depolarization of the cardiomyocytes in the postischemic heart. Mechanically induced membrane depolarization resulted from activation of mechanosensitive ion channels in the plasma membrane of the cardiomyocytes. These channels exhibit a selective permeability for cations, mainly for sodium ions (Isenberg et al., 2003; Kamkin et al., 2000a, 2003c, d; Zeng et al., 2000; Zhang et al., 2000). Extra action potentials appeared, when the mechanically induced depolarizations reached a threshold. As a consequence, extrasystoles and, in the worst case, fibrillations can occur.

Focal disorganization of gap junctions and down-regulation of Cx43, a major cardiac gap junction protein, are typical features of myocardial remodeling. These processes may also play a permissive role for the development of arrhythmia in human cardiomyopathies (Kostin et al., 2003, 2004). Abnormal localization of Cx43 was often observed on lateral surfaces of surviving cardiomyocytes after myocardial infarction in rats (Matsushita et al., 1999). Cx43 was redistributed from intercalated discs to the lateral surface of structurally compromised myocytes (Camelliti et al., 2004a). These findings suggest that ischemia caused ectopic expression of Cx43, which could be responsible for abnormal electrical conductivity around the infarcted area (Matsushita et al., 1999). A changing pattern of connexin expression in sheep was caused by the invasion of Cx45-positive fibroblasts into the ischemic zone within 24 h after infarction. The number of Cx45-expressing fibroblasts reached a maximum after 6 days. It is important to note that the sensitivity of the cardiac fibroblasts to mechanical stretch was highest 7 days after myocardial infarction due to coronary artery occlusion (Kamkin et al., 2002). From thereon, the expression of Cx43 increased, whereas Cx45 expression was reduced (Camelliti et al., 2004a).

The rapid infiltration of the damaged tissue with highly coupled fibro-blasts likely interferes with the excitation control and the spreading of electrical signals from the infarction border zone to the surrounding tissue. In addition to the enhanced mechanosensitivity of cardiac fibroblasts and myocytes after infarction, altered electrical interaction between the two cell types may increase the risk for contractile dysfunction. This possibility is in agreement with the finding that the nonspecific gap junction channel uncoupler heptanol limited necrosis and infarct size (Saltman et al., 2002; Garcia-Dorado et al., 2004).

9.8 Conclusions and Perspectives

A detailed knowledge of the trafficking routes, through which cardiac myocytes and fibroblasts communicate with each other is necessary for the understanding of normal and impaired cardiac function. Interaction of fibroblasts and myocytes involves the intercellular exchange of electrical signals via gap junctions (Camelliti et al., 2004a, b). Gap junction channels were demonstrated with the use of electrophysiological and morphological techniques at the sites of contact formation between cardiomyocytes and fibroblasts. Gap junctions between atrial fibroblasts and cells of the sinoatrial node may provide the structural correlate for the transmission of membrane potential changes from the fibroblasts to the cardiac pacemaker cells. Through this pathway, an increase in wall distension of the right atrium, i.e. due to enhanced venous re-turn, may elicit a chronotropic response of the heart (Kohl et al., 1994). This may have serious implications for the contractile activity of the heart resulting in a decrease of heart rates, when hyperpolarization of the fibroblasts exceeds depolarization of the cardiomyocytes. Furthermore, extra systoles and cardiac arrhythmia may occur, when mechanically induced membrane depolarization of the cardiac myocytes predominates.

The gap junctional conductivity can be regulated by protein phosphorylation involving different protein kinases, i.e. protein kinase A, C, and G as well as tyrosine kinases and mitogen-activated protein kinase. Some studies have demonstrated that activation of cAMP-dependent protein kinase A can enhance gap junction coupling in Cx43-positive cardiomyocytes. The conductance of gap junctions can also be regulated by small ions like H^+, Ca^{2+}, Mg^{2+} and Na^+. An increase in the intracellular concentration of Na^+ may reduce the gap junction conductance. A drop of the intracellular pH can also reduce gap junctional conductivity, with the histidine residue 95 acting as a pH sensor in Cx43 channels (Dhein, 1998).

In addition to gap junction regulation by extracellular mechanical forces, there is a close relation between gap junctions and adhesion junctions and their linkage to the cytoskeleton. This can be inferred from experiments on neoformation of cell-to-cell coupling, concomitant upregulation of adherens and gap junctions after mechanical stretch, and human cardiomyopathies caused by genetic defects in cell-cell adhesion junction proteins. The molecular mechanisms responsible for the interaction between mechanical and functional cell-to-cell coupling remain to be elucidated (for review see Saffitz and Kléber, 2004).

It is likely that stretch-activated changes in cardiac myocyte structure and function are mediated by signaling pathways that are initiated by interactions between integrins and extracellular matrix proteins. Indeed, overexpression of β_1 integrin was sufficient to induce a hypertrophic response in cultured neonatal rat ventricular myocytes and enhancedthe effects of β_1 adrenergic stimulation (Ross et al., 1998).

Future research will need to elucidate the complete signaling pathway by which mechanical stimulation of cardiac fibroblasts alters the gap junctional communication of cardiac cells. A detailed analysis of the molecular signaling mechanisms between cardiac myocytes and fibroblasts may eventually allow one to develop new strategies for the treatment of cardiac arrhythmias.

Acknowledgements This study was supported by grants from the Russian Foundation for Basic Research (09-04-01277-a)

References

Adler CP, Ringlage WP, Bohm N (1981) DNS-Gehalt und Zellzahl in Herz und Leber von Kindern. Pathol Res Pract 172:25–41 (German).

Anversa P, Olivetti G, Melissari M, Loud AV (1980) Stereological measurement of cellular and subcellular hypertrophy and hyperplasia in the papillary muscle of adult rat. J Mol Cell Cardiol 12:781–795.

Banerjee I, Yekkala K, Borg TK, Baudino TA (2006) Dynamic interactions between myocytes, fibroblasts, and extracellular matrix. Ann N Y Acad Sci. 1080:76–84 (Review).

Banes AJ, Tsuzaki M, Hu P, Brigman B, Brown T, Almekinders L, Lawrence WT, Fischer T (1995). PDGF-BB, IGF-I and mechanical load stimulate DNA synthesis in avian tendon fibroblasts in vitro. J Biomech 28:1505–1513.

Brilla CG, Maisch B, Weber KT (1993) Renin-angiotensin system and myocardial collagen matrix remodeling in hypertensive heart disease: in vivo and in vitro studies on collagen matrix regulation. Clin Invest 71:S35–S41.

Brown RD, Jones GM, Laird RE, Hudson P, Long CS (2007) Cytokines regulate matrix metalloproteinases and migration in cardiac fibroblasts. Biochem Biophys Res Commun 362(1):200–205.

Bujak M, Dobaczewski M, Chatila K, Mendoza LH, Li N, Reddy A, Frangogiannis NG (2008) Interleukin-1 receptor type I signaling critically regulates infarct healing and cardiac remodeling. Am J Pathol 173(1):57–67.

Burstein B, Libby E, Calderone A, Nattel S (2008) Differential behaviors of atrial versus ventricular fibroblasts: a potential role for platelet-derived growth factor in atrial-ventricular remodeling differences. Circulation 117(13):1630–1641.

Bustamante JO, Ruknudin A, Sachs F (1991) Stretch-activated channels in heart cells: relevance to cardiac hypertrophy. J Cardiovasc Pharmacol 17(2):S110–S113.

Camelliti P, Borg TK and Kohl P (2005) Structural and functional characterisation of cardiac fibroblasts. Cardiovasc Res 65:40–51.

Camelliti P, Devlin GP, Matthews KG, Kohl P (2004a) Spatially and temporally distinct expression of fibroblast connexins after sheep ventricular infarction. Cardiovasc Res 62(2): 415–425.

Camelliti P, Green CR, LeGrice I, Kohl P (2004b) Fibroblast network in rabbit sinoatrial node: structural and functional identification of homogeneous and heterogeneous cell coupling. Circ Res 94(6):828–835.

Chang HY, Chi JT, Dudoit S, Bondre C, van de Rijn M, Botstein D, Brown PO (2002) Diversity, topographic differentiation, and positional memory in human fibroblasts. Proc Natl Acad Sci USA 99:12877–12882.

Chhabra D, Nosworthy NJ, dos Remedios CG (2000) The role of ATP, ADP and divalent cations in the formation of binary and ternary complexes of actin, cofilin and DNase I. Electrophoresis 21:3863–3869.

Chiquet M, and Fluck M (2001) Early Response to Mechanical Stress: Form Signals at the Surface to Altered Gene Expression. Elsevier, Amsterdam, pp. 97–110.

D'Alonzo AJ, Sewter JC, Darbenzio RB, Hess TA (1995) Effects of dofetilide on electrical dispersion and arrhythmias in post-infarcted anesthetized dogs. Basic Res Cardiol 90(5):424–434.

Davies MJ, Pomerance A (1972) Quantitative study of ageing changes in the human sinoatrial node and internodal tracts. Br Heart J 34:150–160.

Dean JW, Lab MJ (1989a) Arrhythmia in heart failure: role of mechanically induced changes in electrophysiology. Lancet 1(8650):1309–1312 (Review).

Dean JW, Lab MJ (1989b) Effect of changes in load on monophasic action potential and segment length of pig heart in situ. Cardiovasc Res 23:887–896.

Dean JW, Lab MJ (1990) Regional changes in myocardial refractoriness during load manipulation in the in-situ pig heart. J Physiol 429:387–400.

Dhein S (1998) Gap junction channels in the cardiovascular system: pharmacological and physiological modulation. Trends Pharmacol Sci 19(6):229–241.

Dilly SG, Lab MJ (1988) Electrophysiological alternans and restitution during acute regional ischaemia in myocardium of anaesthetized pig. J Physiol (London) 402:315–333.

Eghbali M, Czaja MJ, Zeydel M, Weiner FR, Zern MA, Seifter S, Blumenfeld OO. (1988) Collagen chain mRNAs in isolated heart cells from young and adult rats. J Mol Cell Cardiol 20: 267–276.

Eghbali M, Tomek R, Sukhatme VP, Woods C, Bhambi B (1991) Differential effects of transforming growth factor-β1 and phorbol myristate acetate on cardiac fibroblasts: regulation of fibrillar collagen mRNAs and expression of early transcription factors. Circ Res 69:483–490.

Franz MR. (1996) Mechanoelectrical feedback in ventricular myocardium. Cardiovasc Res 32(1):15–24 (Review).

Galli A, DeFelice LJ (1994) Inactivation of l-type Ca channels in embryonic chick ventricular cells: dependence on the cytoskeletal agents colchicine and taxol. Biophys J 67:2296–2304.

Garcia-Dorado D, Rodriguez-Sinovas A, Ruiz-Meana M (2004) Gap junction-mediated spread of cell injury and death during myocardial ischemia-reperfusion. Cardiovasc Res 61:386–401.

Hamill OP (2006) Twenty odd years of stretch-sensitive channels Pflügers Arch – Eur J Physiol 453:333–351 (Review).

Hamill OP and Martinac B (2001) Molecular basis of mechanotransduction in living cells Physiol Rev 81(2):685–740 (Review).

Horio T, Maki T, Kishimoto I, Tokudome T, Okumura H, Yoshihara F, Suga S, Takeo S, Kawano Y, Kangawa K (2005) Production and autocrine/paracrine effects of endogenous insulin-like growth factor-1 in rat cardiac fibroblasts. Regul Pept 124(1–3):65–72.

Ince C, Leijh PCJ, Meijer J, van Bavel E, Yrey DL. 1984. Oscillatory hyperpolarizations and resting membrane potentials of mouse fibroblast and macrophage cell lines. J Physiol (London) 352:625–635.

Isenberg G, Kazanski V, Kondratev D, Gallitelli MF, Kiseleva I, Kamkin A (2003) Differential effects of stretch and compression on membrane currents and $[Na^+]_c$ in ventricular myocytes. Prog Biophys Mol Biol 82(1–3):43–56 (Review).

Johnson BD, Byerly L (1993) A cytoskeletal mechanism for Ca^{2+} channel metabolic dependence and inactivation by intracellular Ca^{2+}. Neuron 10:797–804.

Kamkin A, Kiseleva I, Isenberg G (2000a) Stretch-activated currents in ventricular myocytes: amplitude and arrhythmogenic effects increase with hypertrophy. Cardiovasc Res 48:409–420.

Kamkin A, Kiseleva I, Isenberg G (2003a) Activation and inactivation of a non-selective cation conductance by local mechanical deformation of acutely isolated cardiac fibroblasts. Cardiovasc Res 57(3):793–803.

Kamkin A, Kiseleva I, Isenberg G (2003c) Ion selectivity of stretch-activated cation currents in mouse ventricular myocytes. Pflügers Arch 446(2):220–231.

Kamkin A, Kiseleva I, Isenberg G, Wagner KD, Günther J, Theres H, Scholz H (2003b) Cardiac fibroblasts and the mechanoelectric feedback mechanism in healthy and diseased hearts. Prog Biophys Mol Biol 82(1–3):111–120.

Kamkin A, Kiseleva I, Kircheis R, Kositzky G (1988) Bioelectric activity of frog atrium cells with non-typical impulse activity. Abhandlungen der Akademie der Wissenschaften der DDR (Abteilung Mathematik – Naturwissenschaft – Technik) 1:103–106.

Kamkin A, Kiseleva I, Lozinsky I, Scholz H (2005a) Electrical interaction of mechanosensitive fibroblasts and myocytes in the heart. Basic Res Cardiol 100(4):337–345.

Kamkin A, Kiseleva I, Lozinsky I, Wagner KD, Isenberg G, Scholz H (2005b) The role of mechanosensitive fibroblasts in the heart. In: Kamkin A and Kiseleva I (eds.) Mechanosensitivity in Cells and Tissues. Academia Publishing House Ltd, pp. 203–229. (See also PubMed, Book)

Kamkin A, Kiseleva I, Pylaev A, Lab MJ, Kohl P (1995) Cardiac fibroblasts – a cellular substrate for mechanosensitivity in frog sinus venosus. J Physiol (London) 483:24P.

Kamkin A, Kiseleva I, Wagner KD, Bohm J, Theres H, Günther J, Scholz H (2003d) Characterization of stretch-activated ion currents in isolated atrial myocytes from human hearts. Pflügers Arch 446(3):339–346.

Kamkin A, Kiseleva I, Wagner KD, Lammerich A, Bohm J, Persson PB, Günther J (1999) Mechanically induced potentials in fibroblasts from human right atrium. Exp Physiol 84:347–356.

Kamkin A, Kiseleva I, Wagner KD, Leiterer KP, Theres H, Scholz H, Günther J, Lab MJ (2000b) Mechanoelectric feedback in right atrium after left ventricular infarction in rats. J Mol Cell Cardiol 32:465–477.

Kamkin A, Kiseleva I, Wagner KD, Lozinsky I Günther J, Scholz H (2003e) Mechanically induced potentials in atrial fibroblasts from rat hearts are sensitive to hypoxia/reoxygenation. Pflügers Arch 446(2):169–174.

Kamkin A, Kiseleva I, Wagner KD, Pylaev A, Leiterer KP, Theres H, Scholz H, Günther J, Isenberg G (2002) A possible role for atrial fibroblasts in postinfarction bradycardia. Am J Physiol 282:H842–H849.

Kamkin A, Kiseleva I, Wagner KD, Scholz H, Theres H, Kazanski V, Lozinsky I, Günther J, Isenberg G (2001) Mechanically induced potentials in rat atrial fibroblasts depend on actin and tubulin polymerisation. Pflügers Arch 442:487–497.

Kiseleva I, Kamkin A, Kohl P, Lab M (1996) Calcium and mechanically induced potentials in fibroblasts of rat atrium. Cardiovasc Res 32:98–111.

Kiseleva I, Kamkin A, Leiterer KP, Kohl P (1993) Interaction of mechanosensitive cells with surrounding cells in the right atrium of the rat heart. J Mol Cell Cardiol 25(1):S76 (Abstract)

Kiseleva I, Kamkin A, Pylaev A, Kondratjev D, Leiterer KP, Theres H, Wagner KD, Persson PB, Günther J (1998) Electrophysiological properties of mechanosensitive atrial fibroblasts from chronic infarcted rat hearts. J Mol Cell Cardiol 30:1083–1093.

Kiseleva I, Kamkin A, Wagner KD, Theres H, Ladhoff A, Scholz H, Günther J, Lab MJ (2000) Mechanoelectric feedback after left ventricular infarction in rats. Cardiovasc Res 45(2): 370–378.

Kiseleva IS, Kamkin AG, Kircheis R, Kositski GI (1987) Intercellular electrotonical interaction in the cardiac sinus node in the frog. Rep Acad Sci USSR 292(6):1502–1505 (Russian).

Kleber AG, Fast V (1997) Molecular and cellular aspects of re-entrant arrhythmias. Basic Res Cardiol 92 (1):111–119 (Review).

Kohl P (2003) Heterogeneous cell coupling in the heart an electrophysiological role for fibroblasts. Circ Res 93:381–383.

Kohl P, Kamkin A, Kiseleva I, Streubel T (1992) Mechanosensitive cells in the atrium of frog heart. Exp Physiol 77:213–216.

Kohl P, Kamkin AG, Kiseleva IS, Noble D (1994) Mechanosensitive fibroblasts in the sinoatrial node region of rat heart: interaction with cardiomyocytes and possible role. Exp Physiol 79:943–956.

Kohl P, Noble D (1996) Mechanosensitive connective tissue: potential influence on heart rhythm. Cardiovasc Res 32:62–68.

Kondratjev D, Pylaev A, Kiseleva I, Kohl P, Kamkin A. 1993. Mechanosensitive electrically nonexcitable cells in the atria of the frog and rat heart. J Physiol (London) 473:258P.

Kostin S, Dammer S, Hein S, Klovekorn WP, Bauer EP, Schaper J (2004) Connexin 43 expression and distribution in compensated and decompensated cardiac hypertrophy in patients with aortic stenosis. Cardiovasc Res 62(2):426–436.

Kostin S, Rieger M, Dammer S, Hein S, Richter M, Klovekorn WP, Bauer EP, Schaper J (2003) Gap junction remodeling and altered connexin43 expression in the failing human heart. Mol Cell Biochem 242(1–2):135–144.

Lab MJ (1968) Is there mechanoelectric transduction in cardiac muscle? The mono-phasic action potential of the frog ventricle during isometric and isotonic contraction with calcium deficient perfusions. S Afr J Med Sci 33:60 (Abstract).

LaFramboise WA, Scalise D, Stoodley P, Graner SR, Guthrie RD, Magovern JA, Becich MJ (2007) Cardiac fibroblasts influence cardiomyocyte phenotype in vitro. Am J Physiol Cell Physiol 292(5):C1799–C1808.

Leask A (2007) TGFbeta, cardiac fibroblasts, and the fibrotic response. Cardiovasc Res 74(2): 207–212 (Review).

Lee AA, Delhaas T, Waldman LK, MacKenna DA, Villarreal FJ, McCulloch AD (1996) An equibiaxial strain system for cultured cells. Am J Physiol 271:C1400–C1408.

MacKenna D, Summerour SR, and Villarreal FJ (2000) Role of mechanical factors in modulating cardiac fibroblast function and extracellular matrix synthesis. Cardiovasc Res 46:257–263.

Maltsev VA, Undrovinas AI (1997) Cytokleleton modulates coupling between availability and activation of cardiac sodium channel. Am J Physiol 273:H1832–H1840.

Matsushita T, Oyamada M, Fujimoto K, Yasuda Y, Masuda S, Wada Y, Oka T, Takamatsu T (1999) Remodeling of cell-cell and cell-extracellular matrix interactions at the border zone of rat myocardial infarcts. Circ Res 85(11):1046–1055.

Maziere de AMGL, Ginneken van ACG, Wilders R, Jongsma HJ, Bouman LN (1992) Spatial and functional relationship between myocytes and fibroblasts in the rabbit sinoatrial node. J Mol Cell Cardiol 24:567–578.

Murphy CF, Horner SM, Dick DJ, Coen B, Lab MJ (1996) Electrical altemans and the onset of rate-induced pulsus alternans during acute regional ischaemia in the anaesthetised pig heart. Cardiovasc Res 32:138–147.

Murphy CF, Lab MJ, Horner SM, Dick DJ, Harrison FG (1994) Regional electromechanical alternans in anesthetized pig hearts: modulation by mechanoelectric feedback. Am J Physiol 267:H1726–1735.

Naccarella F, Lepera G, Rolli A. (2000) Arrhythmic risk stratification of post-myocardial infarction patients. Curr Opin Cardiol 15:1–6 (Review).

Nazir SA, Lab MJ (1996) Mechanoelectric feedback and atrial arrhythmias. Cardiovasc Res 32(1):52–61 (Review).

Nelson PG, Peacock J, Minna J. 1972. An active electrical response in fibroblasts. J Gen Physiol 60:58–71.

Ninio DM and Saint DA (2008) The role of stretch-activated channels in atrial fibrillation and impact of intracellular acidosis. Prog Biophys Mol Biol 97(2–3):401–416 (Review).

Okada Y, Doida Y, Roy G, Tsuchiya W, Inouye K, Inouye A. 1977. Oscillation of membrane potential in L cells. I. Basic characteristics. J Membr Biol 35:319–335.

Okada Y, Tsuchiya W, Yada T, Yano J, Yawo H. 1981. Phagocytic activity and hyperpolarizing responses in L-strain mouse fibroblasts. J Physiol (London) 313:101–119.

Okada Y, Yada T, Ohno-Shosaku T, Oiki S, Ueda S, Machida K. 1984. Exogenous ATP induces electrical membrane responses in fibroblasts. Exp Cell Res 152:552–557.

Opthof T, deJonge B, Jongsma HJ, Bouman LN (1987) Functional morphology of the mammalian sinuatrial node. Eur Heart J 8:1249–1259.

Podzuweit T, Braun W, Muller A, Schaper W (1987) Arrhythmias and infarction in the ischemic pig heart are not mediated by xanthine oxidase-derived free oxygen radicals. Basic Res Cardiol 82(5):493–505.

Pogwizd SM, Corr B (1992) The contribution of nonreentrant mechanisms to malign-nant ventricular arrhythmias. Basic Res Cardiol 87(2):115–129 (Review).

Ravens U (2003) Mechanoelectric feedback and arrhythmias. Prog Biophys Mol Biol 82 (1–3):255–266 (Review).

Rook MB, de Jonge B, Jongsma HJ, Masson-Pevet MA (1990) Gap junction formation and functional interaction between neonatal rat cardiocytes in culture: a correlative physiological and ultrastructural study. J Membrane Biol 118:179–192.

Rook MB, Jongsma HJ, de Jonge B (1989) Single channel currents of homo- and heterologous gap junctions between cardiac fibroblasts and myocytes. Pflügers Arch 414:95–98.

Rook MB, Jongsma HJ, van Ginneken ACG (1988) Properties of single gap junctional channels between isolated neonatal rat heart cells. Am J Physiol 255:H770–H782.

Ross RS, Pham C, Shai SY, Goldhaber JI, Fenczik C, Glembotski CC, Ginsberg MH, Loftus JC (1998) $_1$ integrins participate in the hypertrophic response of rat ventricular myocytes. Circ Res 82:1160–1172.

Sachs F, Morris CE (1998) Mechanosensitive ion channels in nonspecialized cells. Rev Physiol Biochem Pharmacol 132:1–77 (Review).

Sadoshima J, Izumo S (1997) The cellular and molecular response of cardiac myocytes to mechanical stress. Annu Rev Physiol 59:551–571.

Sadoshima J, Izumo S (1993) Molecular characterization of angiotensin II-induced hypertrophy of cardiac myocytes and hyperplasia of cardiac fibroblasts: critical role of the AT1 receptor subtype. Circ Res 73:413–423.

Saffitz JE, Kléber AG (2004) Effects of mechanical forces and mediators of hypertro-phy on remodeling of gap junctions in the heart. Circ Res 94:585–591 (Review).

Saltman AE, Aksehirli TO, Valiunas V, Gaudette GR, Matsuyama N, Brink P, Krukenkamp IB (2002) Gap junction uncoupling protects the heart against ischemia. J Thorac Cardiovasc Surg 124:371–376.
Shiraishi I, Takamatsu T, Mimikawa T, Onouchi Z, Fujita S (1992) Quantitative histological analysis of the human sinoatrial node during growth and aging. Circulation 85:2176–2184.
Taggart P (1996) Mechanoelectric feedback in the human heart. Cardiovasc Res 32(1):38–43 (Review).
Taggart P and Lab MJ (2008) Cardiac mechano-electric feedback and electrical restitution in humans. Prog Biophys Mol Biol 97(2–3):452–460 (Review).
Thampatty BP and Wang JH-C (2008) Mechanobiology of Fibroblasts. In: Kamkin A and Kiseleva I (eds.) Mechanosensitivity in Cells and Tissues 1. Mechanosensitive Ion Channels. Springer, pp. 351–378.
Tomasek JJ, Gabbiani G, Hinz B, Chaponnier C, Brown RA (2002) Myofibroblasts and mechanoregulation of connective tissue remodelling. Nat Rev Mol Cell Biol 3:349–363.
Tsuchiya W, Okada Y, Yano J, Inouye A, Sasaki S, Doida Y (1981) Effects of cytochalasin B and local anesthetics on electrical and morphological properties in L cells. ExpCell Res 133:83–92.
Turner NA, Mughal RS, Warburton P, O'Regan DJ, Ball SG, Porter KE (2007) Mechanism of TNFalpha-induced IL-1alpha, IL-1beta and IL-6 expression in human cardiac fibroblasts: effects of statins and thiazolidinediones. Cardiovasc Res 76(1):81–90.
Underwood RD, Sra J, Akhtar M (1997) Evaluation and treatment strategies in patients at high risk of sudden death post myocardial infarction. Clin Cardiol 20:753–758 (Review).
Villarreal FJ, Dillmann WH (1992) Cardiac hypertrophy induced chances in mRNA levels for TGF-β1 fibronectin, and collagen. Am J Physiol 262:H1861–H1866.
Virag JA, Rolle ML, Reece J, Hardouin S, Feigl EO, Murry CE (2007) Fibroblast growth factor-2 regulates myocardial infarct repair: effects on cell proliferation, scar contraction, and ventricular function. Am J Pathol 171(5):1431–1440.
Walsh KB, Zhang J (2008) Neonatal rat cardiac fibroblasts express three types of voltage-gated K^+ channels: regulation of a transient outward current by protein kinase C. Am J Physiol Heart Circ Physiol 294(2):H1010–H1017.
Wang J-HC and Thampatty BP (2006) Mechanoregulation of Fibroblast Function. In: Bowlin GL and Wnek G (eds.) Encyclopedia of Biomaterials and Biomedical Engineering. Taylor and Francis.
Wang YJ, Sung RJ, Lin MW, Wu SN (2006) Contribution of BK(Ca)-channel activity in human cardiac fibroblasts to electrical coupling of cardiomyocytes-fibroblasts. J Membr Biol 213(3):175–185.
Weber KT, Brilla CG (1991) Pathological hypertrophy and cardiac interstitium: fibrosis and renin-anglotensin-aldosterone system. Circulation 83:1849–1865.
Weber KT, Brilla CG, Janicki JS (1993) Myocardial fibrosis: functional significance and regulatory factors. Cardiovasc Res 27:341–348.
Weber KT, Sun Y, Tyagi SC, Cleutjens JPM. (1994) Collagen network of the myocardium: function, structural remodeling and regulatory mechanisms. J Mol Cell Cardiol 26:279–292.
Yost MJ, Simpson D, Wrona K, Ridley S, Ploehn HJ, Borg TK, Terracio L (2000) Design and construction of a uniaxial cell stretcher. Am J Physiol Heart Circ Physiol 279:H3124–H3130.
Zeng T, Bett GCL, Sachs F (2000) Stretch-activated whole cell currents in adult rat cardiac myocytes. Am J Physiol 278:H548–H557.
Zhang YH, Youm JB, Sung HK, Lee SH, Ryu SY, Ho WK, Earm YE (2000) Stretch-activated and background non-selective cation channels in rat atrial myocytes. J Physiol (London) 523(3):607–619.

Chapter 10
Scanning Ion Conductance Microscopy for Imaging and Mechanosensitive Activation of Selected Areas of Live Cells

Max J. Lab

Abstract Biological mechanosensitivity is increasing in importance, and means of studying it at the cell or membrane level ever demanding. In consequence we have developed a non-contact nanoscale method for applying force to selected areas on the surface of living cells. The method applies hydrostatic pressure through a nanopipette, the operative probe of a scanning ion conductance microscope (SICM). The pipette is kept above the cell surface using distance feedback. This prevents surface contact, and promotes non-invasive mechanical probing. First the microscope scans and images a living cell surface at high resolution – no applied pressure. Subsequently we apply pressure to areas selected from the scanned image for mechanosensitive studies, as well as studies of their nanomechanical properties.

Keywords Scanning microscopy · Scanning ion conductance microscopy · Mechanosensitivity · Mechanosensitive activation

10.1 Introduction

Mechanosensitivity is highly conserved in biology and investigative tools for study in living soft biological tissue at high resolution could be improved. In addition, cell structure is heterogeneous even at high resolution, and the nano-area investigated would be heterogeneous over the cell's surface – a means of coping with this nano-heterogeneity is also a requirement.

The mechanical properties of cells have been previously measured (Hochmuth, 2000; Svoboda and Block, 1994; Dai and Sheetz, 1995) and the atomic force microscope (AFM) using similar principles to ours, has been used so far for mapping local mechanical properties on a nanoscale (Stolz et al., 2004; Haga et al., 2000; Dulinska et al., 2006; Hofmann et al., 1997). Recently it has been used to study

M.J. Lab (✉)
National Heart and Lung Institute, Imperial College London, London, UK
e-mail: m.lab@imperial.ac.uk

A. Kamkin, I. Kiseleva (eds.), *Mechanosensitivity of the Heart*,
Mechanosensitivity in Cells and Tissues 3, DOI 10.1007/978-90-481-2850-1_10,

mechanosensitivity (Bacabac et al., 2008). However, the AFM directly contacts the cell surface, affecting what one measures. Here we present a method to non-invasively probe mechanosensitive channels and receptors (surface and subsurface) and also measure the cell's mechanical properties, and do so on the nanoscale.

10.2 The Scanning Ion Conductance Microscope

A scanning ion conductance microscope (SICM) that uses a nanopipette as the probe (Gorelik et al., 2003, 2006; Shevchuk et al., 2001; Zhang et al., 2005) is combined with a pneumatic and hydraulic pressure system to push or pull on cell surfaces (Fig.10.1). The microscope uses the ion current flowing between an electrode in the bath and an electrode in the nanopipette, which is also in the bath, to control the pipette-surface separation (Korchev et al., 1997a, b). The ion current through the pipette is determined by the resistance of the pipette and the solution in the bath (between the pipette and cell). However as the pipette approaches the cell's surface, narrowing the pipette-surface gap, the ion current reduces further. This reduction is used for distance feed-back control, and is set at about 1% – i.e. the current is 99% of its maximum value, and this means the pipette is controlled over the surface at a distance of a pipette inner radius. This system has provided high resolution non-contact imaging of a number of different living cells, and has followed structural re-arrangements of the cell membrane at high resolution (Gorelik et al., 2003; Shevchuk et al., 2006).

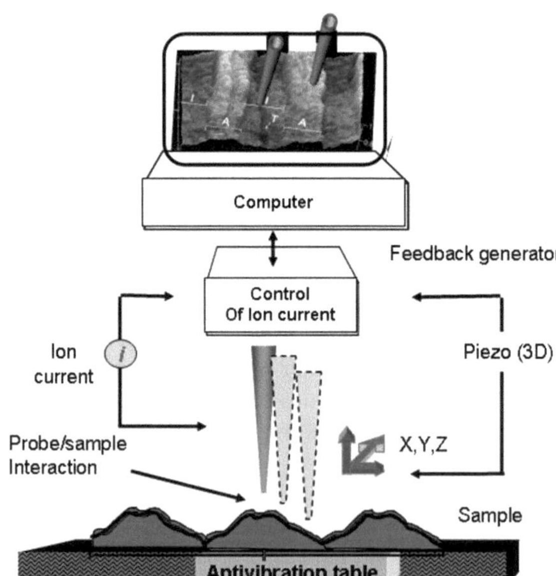

Fig. 10.1 Schematic diagram of the scanning ion conductance microscope (SICM) setup. During scanning the cell membrane with the nanopipette probe, the SICM feedback control system uses the ion current between the bath and nanopipette electrodes to keep a constant the pipette-surface distance. A resultant scanned image of a cardiomyocyte is at *top*, also diagramming the scanning nanopipette – 2 positions of the pipette are indicated (In image, A & I = sarcomeric bands, T = t tubule.)

10.2.1 Applied Pressure

During the scans for obtaining images, no hydrostatic pressure is applied through the nanopipette, but we modified the system for pressure to be applied through the same scanning pipette probe (Sanchez et al., 2008). This is to activate mechanosensitive parts of the surface membrane. Figure 10.2, schematically outlines the pressure system and depicts the membrane deformation during hydrostatic pressure applied via the pressure port and pipette to the cell surface. The deformation of the cell surface invokes the distance feedback control (see Fig. 10.1), adjusting the pipette position to keep the ion current constant, and importantly to maintain non-contact. Because the distance is also kept constant, and the pipette moves down and follows the indentation, the applied pressure determines the pipette's position, and we can non-invasively and quantitatively, locally indent the cell surface.

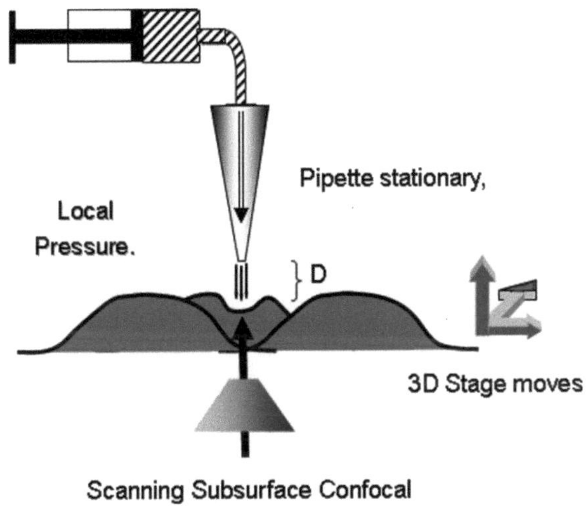

Fig. 10.2 Localised Mechanical Stimulation via pressure system, and Scanning Sub – Surface Confocal Microscope. Pressure is non-invasively applied via the nanopipette. Consequent deformation of the cell surface invokes the distance feedback control to keep a constant pipette-surface distance (D), providing a record of nanopipette position. Confocal volume is submembranar, so follows submembranar space to detect amplified 2nd messengers e.g. calcium

10.2.2 Calibration of Force Exerted Via the Nanopipette

We calibrated the SICM using an AFM cantilever – increasing pipette pressure bends the cantilever, and we measure the change in pipette position. The relationship between the cantilever deformation and applied pressure was linear, with no hysteresis. Since we know the nominal spring constant of the AFM cantilever, the distance moved by the cantilever can be directly converted into force allowing a direct conversion of applied pressure into applied force. The force exerted on the cantilever depends on the applied pressure and pipette radius. The linear plot obtained was in close agreement with the values obtained from the measured displacement and shows that the pressure applied at the pressure port is developed at the pipette tip.

10.3 Example Application

10.3.1 Probing (Activating) Mechanosensitive Receptors

We have previously produced SICM images (no applied pressure) showing that cardiomyocytes have repetitive scalloped topographic features (Korchev et al., 1997) – see for example image top of Fig. 10.1. This facilitates the selection of an area of interest, e.g. t tubule as for targeting kATP channels (Korchev et al., 2000).

As mechanically induced beats have been previously produced by a variety of techniques in a variety of preparations (Lab and Dean, 1991; Lab, 1996) – probably by activating mechanosensors, we tried to produce similar activations with our technique. We find that pressure applied to the cardiomyocytes via the nanopipette can mechanically stimulate contraction of a cardiomyocyte (Fig. 10.3).

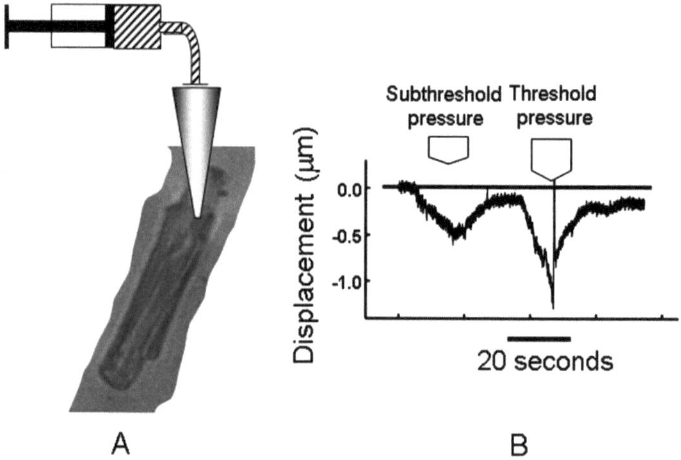

Fig. 10.3 Pressure induced activation of contraction in a cardiomyocyte. (**a**) Adult cardiomyocyte with diagrammed pressure application system. (**b**) Subthreshold pressure stimulation produces a small indentation of about 0.5 µm, whereas a higher pressure not only produces a larger indentation, but ilicits a contraction

This method has been applied to the body of a sensory neuron (Sanchez et al., 2007). This time we monitored intracellular calcium – the applied pressure indented the surface of the cell, producing a rise in intracellular calcium.

10.3.2 Mechanical Properties of Cells

Pressure applied through the pipette probe, can target a variety of cells to determine their mechanical properties, for example cardiomyocytes and erythrocytes (Sanchez et al., 2008). In erythrocytes the applied pressure deforms the cell linearly, and the gradient of pressure plotted against indentation distance can be used to calculate the cell's modulus of elasticity (Young modulus). The average Young's modulus was in good agreement with the literature (Lekka et al., 2005) validating the method.

10.4 Concluding Remarks

This chapter presents a non-contact method of applying force to soft living cell surfaces. It is based on the application of pressure via a nanopipette, and measuring the resulting surface deformation. It has the potential to probe surface mechanosensitivity, without cell contact. It also has the potential to broadly map the locations of mechanosensitive ion channels, in a similar fashion to that used by this laboratory to map kATP channels. (Korchev et al., 2000) In our case the probe would apply pressure (instead of K+) while scanning the cell surface, with the second, patch pipette measuring currents as the probe scans over and activates a mechanosensitive channel. The activated mechanosensitive receptor should produce an electrophysiological current as the scanning pipette passes over it.

The method can probe the mechanical properties of living cells on the nanoscale at defined positions on the cell surface, and seems well suited to activate mechanosensors on the surface and subsurface membrane, and quantitatively probe and map the nanomechanical properties of living cells.

References

Bacabac RG, Mizuno D, Schmidt CF, MacKintosh FC, Van Loon JJ, Klein-Nulend J, and Smit TH (2008) Round versus flat: bone cell morphology, elasticity, and mechanosensing. J Biomech 41: 1590–1598.

Dai J, and Sheetz MP (1995) Mechanical properties of neuronal growth cone membranes studied by tether formation with laser optical tweezers. Biophys J 68: 988–996.

Dulinska I, Targosz M, Strojny W, Lekka M, Czuba P, Balwierz W, and Szymonski M (2006) Stiffness of normal and pathological erythrocytes studied by means of atomic force microscopy. J Biochem Biophys Methods 66: 1–11.

Gorelik J, Ali NN, Shevchuk AI, Lab M, Williamson C, Harding SE, and Korchev YE (2006) Functional characterization of embryonic stem cell-derived cardiomyocytes using scanning ion conductance microscopy. Tissue Eng 12: 657–664.

Gorelik J, Shevchuk AI, Frolenkov GI, Diakonov IA, Lab MJ, Kros CJ, Richardson GP, Vodyanoy I, Edwards CR, Klenerman D, and Korchev YE (2003) Dynamic assembly of surface structures in living cells. Proc Natl Acad Sci USA 100: 5819–5822.

Haga H, Sasaki S, Kawabata K, Ito E, Ushiki T, and Sambongi T (2000) Elasticity mapping of living fibroblasts by AFM and immunofluorescence observation of the cytoskeleton. Ultramicroscopy 82: 253–258.

Hochmuth RM (2000) Micropipette aspiration of living cells. J Biomech 33: 15–22.

Hofmann UG, Rotsch C, Parak WJ, and Radmacher M (1997) Investigating the cytoskeleton of chicken cardiocytes with the atomic force microscope. J Struct Biol 119: 84–91.

Korchev YE, Bashford CL, Milovanovic M, Vodyanoy I, and Lab MJ (1997a) Scanning ion conductance microscopy of living cells. Biophys J 73: 653–658.

Korchev YE, Milovanovic M, Bashford CL, Bennett DC, Sviderskaya EV, Vodyanoy I, and Lab MJ (1997b) Specialized scanning ion-conductance microscope for imaging of living cells. J Microsc 188: 17–23.

Korchev YE, Negulyaev YA, Edwards CR, Vodyanoy I, and Lab MJ (2000) Functional localization of single active ion channels on the surface of a living cell. Nat Cell Biol 2: 616–619.

Lab MJ (1996) Mechanoelectric feedback (transduction) in heart: concepts and implications. Cardiovasc Res 32: 3–14.

Lab MJ and Dean J (1991) Myocardial mechanics and arrhythmia. J Cardiovasc Pharmacol 18(2): S72–S79.

Lekka M, Fornal M, Pyka-Fosciak G, Lebed K, Wizner B, Grodzicki T, and Styczen J (2005) Erythrocyte stiffness probed using atomic force microscope. Biorheology 42: 307–317.

Sanchez D, Anand U, Gorelik J, Benham CD, Bountra C, Lab M, Klenerman D, Birch R, Anand P, and Korchev Y (2007) Localized and non-contact mechanical stimulation of dorsal root ganglion sensory neurons using scanning ion conductance microscopy. J Neurosci Methods 159: 26–34.

Sanchez D, Johnson N, Li C, Novak P, Rheinlaender J, Zhang Y, Anand U, Anand P, Gorelik J, Frolenkov G, Benham C, Lab M, Ostanin V, Schaffer TE, Klenerman D, and Korchev Y (2008) Non-contact measurement of the local mechanical properties of living cells using pressure applied via a pipette. Biophys J 95(6): 3017–3027.

Shevchuk AI, Frolenkov GI, Sanchez D, James PS, Freedman N, Lab MJ, Jones R, Klenerman D, and Korchev YE (2006) Imaging proteins in membranes of living cells by high-resolution scanning ion conductance microscopy. Angew Chem Int Ed Engl 45: 2212–2216.

Shevchuk AI, Gorelik J, Harding SE, Lab MJ, Klenerman D, and Korchev YE (2001) Simultaneous measurement of Ca^{2+} and cellular dynamics: combined scanning ion conductance and optical microscopy to study contracting cardiac myocytes. Biophys J 81: 1759–1764.

Stolz M, Raiteri R, Daniels AU, VanLandingham MR, Baschong W, and Aebi U (2004) Dynamic elastic modulus of porcine articular cartilage determined at two different levels of tissue organization by indentation-type atomic force microscopy. Biophys J 86: 3269–3283.

Svoboda K, and Block SM (1994) Biological applications of optical forces. Annu Rev Biophys Biomol Struct 23: 247–285.

Zhang Y, Gorelik J, Sanchez D, Shevchuk A, Lab M, Vodyanoy I, Klenerman D, Edwards C, and Korchev Y (2005) Scanning ion conductance microscopy reveals how a functional renal epithelial monolayer maintains its integrity. Kidney Int 68: 1071–1077.

Part III
Mechano-Electric Feedback in the Whole Heart and a Computer Simulation Study

Chapter 11
The Contribution of MEF to Electrical Heterogeneity and Arrhythmogenesis

David A. Saint, Douglas Kelly, and Lorraine Mackenzie

Abstract Much progress has been made in understanding the mechanisms underlying arrhythmias. It is now clear that electrophysiological properties of the myocardium at a cellular level are very different in different parts of the heart, even over fairly small distances (for example across the thickness of the left ventricular wall). However, most models of arrhythmogenesis do not include the role that mechanical forces on the myocardium play in altering its electrophysiology (a process called mechanoelectric feedback, or MEF). Stretch of the myocardium can alter action potential morphology, propagation velocity and intracellular calcium handling, all of which can contribute to arrhythmogenesis. In particular, it is now becoming clear that MEF is not homogeneous in the heart. It is also clear that MEF is altered in some diseases such as hypertrophy, where it may explain the propensity to arrhythmias in these diseases. Here, we discuss the evidence that MEF is heterogeneous in the heart, in the same way that other electrophysiological properties are heterogeneous. The reasons why this may be are discussed. The possible role that this heterogeneity, and its modulation in some common cardiac diseases, plays in the induction of arrhythmias is also explored.

Keywords Mechanoelectric feedback · Stretch activated channels · Arrhythmias · Atrial fibrillation · Cardiac electrophysiology · Cardiac hypertrophy

11.1 Introduction

The electrophysiological properties of cells and tissues are different in different parts of the heart. The differences between SA node, atrial, and ventricular action potentials are obvious and well understood. These differences arise because of the expression of different types of ion channels in the different areas (for example,

D.A. Saint (✉)
The School of Molecular and Biomedical Science, University of Adelaide, Adelaide, SA, Australia
e-mail: david.saint@adelaide.edu.au

A. Kamkin, I. Kiseleva (eds.), *Mechanosensitivity of the Heart*, Mechanosensitivity in Cells and Tissues 3, DOI 10.1007/978-90-481-2850-1_11,

in SA node cells there is little expression of voltage dependent sodium current, no inward rectifier and a high level of Ih (If) compared to the currents in ventricular cells, which have a high level of sodium current, high level of inward rectifier, and low expression of Ih (Liu et al., 2008). The functional role of these gross differences is obvious; the need for pacemaking in the SA node vs a stable membrane potential but rapid conduction of an action potential once it is triggered in the ventricles. It is less apparent that there are electrophysiological differences at a smaller scale: for example, action potential morphology is different in different parts of the ventricle, even down to the level of transmural differences between epicardial, mid-myocardial and endocardial layers at the same level in the ventricle. The functional role of these small-scale differences is also less apparent, although it has been hypothesised that the differences in action potential duration between epi- and endocardial layers serves to reduce the dispersion (in space and time) of action potential repolarisation, and hence reduces transmural potential gradients. Such transmural gradients are arrhythmogenic; an increased dispersion, whether drug-induced or due to genetic mutations in ion channels, leads to the genesis of arrhythmias, typically of the re-entrant form (Barr et al., 1994; Han and Moe, 1964; Pye et al., 1994). Electrophysiological studies have lead to great insights into the mechanism and potential treatment of many arrhythmias. However, almost all current models of arrhythmogenesis neglect the influence of mechanical forces on the electrophysiology of the atria and ventricles, despite growing evidence that at least some arrhythmias have this as a major contributing mechanism.

A commonly held view of cardiac biomechanics is that electrical activation of the myocardium triggers and controls contraction and that this process is essentially divorced from the variation in mechanical forces on the myocardium during the cardiac cycle. This is an over-simplification of the true picture; stresses (or, more likely, strains)[1] in the myocardium can and do influence its electrical properties, and they do this on a beat- by- beat timescale. This process has been termed mechano-electric feedback (MEF). It has been hypothesised that MEF is important in the regulation of contractile force in the heart, and that it plays a role in minimising both mechanical and electrical heterogeneities which may arise from regional physiological disturbances such as mild ischaemia. Since mechanical forces at the cellular level in the myocardium are certainly heterogeneous, it seems reasonable to suppose that a system which senses forces and consequently influences electrophysiology (i.e. MEF) will also be heterogeneous. Indeed, there is now good evidence to suggest that this is so, but the detailed distribution of MEF in the normal myocardium is not known. It has also become apparent that both the magnitude and the distribution of MEF can be altered in many cardiac disease states, and that these changes may predispose the myocardium to arrhythmias. For example, the clinical link between mitral incompetence and an increased incidence of atrial fibrillation, or between cardiac

[1] Here, stress and strain are used in their engineering context: Stress is defined as force per unit area, while strain is the deformation produced by the application of such a force.

hypertrophy and an increased predisposition to ventricular arrhythmias, can plausibly be attributed to mechanical effects.

In this review we first briefly highlight the now well described heterogeneity in what might be called "conventional" electrophysiology in the heart, and how this is thought to be altered in some disease states, leading to arrhythmias. Then we discuss the evidence that mechanical forces can influence cardiac electrophysiology, and the mechanisms that underlie this. We then review recent evidence that MEF is heterogeneous in the heart, and the mechanisms by which this heterogeneity may arise. Evidence that MEF is altered in some disease states is then presented, along with mechanisms by which this may lead to arrhythmogenesis in these diseases. Finally, the possibility that MEF may provide a new therapeutic avenue for treatment of some arrhythmias is briefly explored.

11.2 Electrical Heterogeneity and Arrhythmias

11.2.1 Atrial Arrhythmias

The most obvious heterogeneity in the atria is, of course, anatomical. Structures such as the ostia of the pulmonary veins provide obvious interruptions to action potential propagation, and hence can promote re-entrant arrhythmias around the anatomical obstruction (Cherry et al., 2007). There is also growing evidence that the pulmonary veins themselves can act as sites of enhanced automaticity, or rapid repetitive generation of ectopic action potentials (Chen et al., 2006; Chou et al., 2005). Similar effects have been seen at the superior vena cava (Shah et al., 2002). This involvement of the pulmonary veins has led to the recent increase in the use of left atrial ablation, and in particular isolation of the pulmonary veins, as a treatment for atrial fibrillation (AF) (Wright et al., 2008). However, there are smaller scale heterogeneities in the atria; even in apparently uniform tissue, electrophysiological anisotropy is evident as, for example, anisotropic conduction velocities. In addition, effective refractory periods can vary over small distances. The role that these heterogeneities play in atrial arrhythmias is unclear, but consideration of them has led to more sophisticated models of arrhythmogenesis (Jalife, 2003; Pertsov et al., 1993; Ulphani et al., 2007)

11.2.2 Ventricular Arrhythmias

Myocytes isolated from different regions of the left ventricle (i.e. epicardial vs endocardial, apex vs base) display differences in electrophysiological properties. The most obvious of these differences is in the repolarising currents, such as the transient outward K^+ current (I_{to}) (Patel and Campbell, 2005). I_{to} is the major repolarising current in some species (e.g. rat), and in humans is responsible for early (Phase 1) repolarisation. In rat, the major channel responsible for I_{to} is Kv4.2, with

Kv4.3 being the major channel sub-type in some other species. Kv4.2 and KvLQT1 are expressed differentially across the ventricular wall (Dixon et al., 1996; Pereon et al., 2000), with the expression of Kv4.2 being more than eight times higher in epicardial muscle cells compared to papillary muscle cells in rat hearts (Dixon and McKinnon, 1994). Similar results have been noted in mouse (Fiset and Giles, 2006). There are also pronounced transmural differences in the expression levels of other ion channels (Gaborit et al., 2007). These cellular differences result in macroscopic electrical gradients within the ventricular wall (Antzelevitch, 2004; Antzelevitch et al., 1991). Evidence also exists for gradients running from the base to the apex of the heart (Wan et al., 2003; Sengupta et al., 2006). However, it should be noted that transmural gradients of action potential durations in whole heart are often not as pronounced as these results in isolated cells would suggest (Taggart et al., 2001), possibly due to the electrotonic coupling of myocytes across the wall.

It is thought that a beneficial consequence of these differences in action potential duration in the normal myocardium is that repolarisation tends to be synchronised across the ventricular wall despite later activation of epicardial myocytes as the action potential propagates through the myocardium. This reduction in dispersion of repolarisation reduces potential arrhythmogenic local circuits (Cowan et al., 1988; Volders et al., 2000). This idea of a protective effect is borne out by the observation that transmural electrophysiological gradients can be accentuated in many disease states, such as cardiac hypertrophy or inherited gene defects, and this accentuation is correlated with increases the propensity to arrhythmias (Antzelevitch et al., 1999; Stilli et al., 2004; Yan et al., 2001).

However, regional differences in other electrophysiological parameters are also apparent. The structure of the ventricle wall is complex, with muscle layers arranged in sheets with connective tissue between them and this structural anisotropy leads to a corresponding electrical anisotropy, particularly in conduction velocity (Hooks et al., 2007). Superimposed on this anatomical complexity, transmural differences in conduction velocity arise as a result of differences in connexin-43 expression, which can be accentuated in hypertrophy and failure (Wiegerinck et al., 2008). These effects on conduction velocity within the ventricular wall also have important consequences for the mechanism of arrhythmias (Poelzing and Rosenbaum, 2004; Valderrabano, 2007).

As a further complication, there are transmural gradients of calcium handling mechanisms (Cordeiro et al., 2004). Many arrhythmias, particularly early afterdepolarisations (EADs) and delayed afterdepolarisations (DADs) arise through defects in calcium handling mechanisms, and these gradients may affect these also.

11.2.3 Summary

Much effort has been devoted to mapping the heterogeneity of electrophysiology in the heart and understanding how this either protects the normal heart from arrhythmias or, conversely, predisposes the heart to arrhythmias when the normal pattern of heterogeneity is disturbed in disease states such as cardiac hypertrophy. Although there has been evidence for many decades that mechanical forces on the myocardium can influence its electrophysiology (MEF), the incorporation of MEF into models of arrhythmogenesis has only recently been considered.

11.3 A Brief Look at MEF

11.3.1 MEF at the Cellular and Whole Heart Level

It is obvious that the electrical activity of the heart drives its mechanical activity. It is less obvious, but perhaps no less important, that the reverse occurs; i.e. that mechanical stimuli can cause electrical changes in the myocardium. As early as 1915, Bainbridge reported that heart rate of anaesthetised dogs was increased when the right atrium was distended by injecting fluids into the jugular vein (Bainbridge, 1915). These chronotropic responses to stretch are intrinsic to the SA node itself, since they can be demonstrated in isolated tissues (Cooper et al., 2000). It is possible that this mechanism contributes to respiratory sinus arrhythmia: – the phasic alteration of heart rate with respiration (Zhang et al., 2002), and it is possible that it is part of the compensatory increase in cardiac output which accompanies increased venous return.

In the atria, early studies of the electrophysiological effects of stretch sometimes produced conflicting results, but the consensus gained from more recent studies using improved methods is that acute atrial stretch produces a shortening of the action potential at early repolarisation and a shortening of refractory periods. Consistent with a shortening of refractory period, acute dilation of the atria greatly increases the propensity to atrial tachyarrhythmias and fibrillation both in animal models (Bode et al., 2001; Ninio et al., 2005; Ravelli and Allessie, 1997; Saint, 2002; Satoh and Zipes, 1996), and in humans (Antoniou et al., 1997; Tse et al., 2001).

At the cellular level, stretch-activated (and stretch-inactivated) currents have been demonstrated directly in isolated atrial myocytes (Kamkin et al., 2003b; Niu and Sachs, 2003) and in fibroblasts (Kamkin et al., 2003a), which are electrically coupled to the myocytes and can influence their electrophysiology.

MEF has also been demonstrated in the ventricles of all species examined (including man) (Franz et al., 1992; Hansen et al., 1990; Stacy et al., 1992; Taggart and Sutton, 1999). In 1954, Dudel and Trautwein were the first to study the effects of stretch on the action potential of cat papillary muscle and dog Purkinje fibre (Dudel and Trautwein, 1954). They found that the resting potential and action potentials were not affected by stretch up to a tension of 1,000 g/cm^2 (at higher tensions, injury of the preparation occurred and the amplitude of the action potential decreased). However, subsequent studies have been able to demonstrate electrophysiological changes at levels of stretch (or strain) that are within physiological ranges. Passive stretch produces a rapid hyperpolarisation if applied during systole, but a depolarisation if applied during diastole (Zabel et al., 1996a). Later studies have also shown that the effects of stretch are phase- and speed-dependent (Nishimura et al., 2006). In humans, mechanoelectric feedback alters action potential duration within one beat of an abrupt change in load (Taggart et al., 1992). In whole heart, if the stretch is large enough or applied quickly enough, or occurs during a vulnerable period of the cardiac cycle, it can trigger ectopic beats. It is thought that this response in humans is responsible for "Commotio cordis" – the sudden death of an otherwise healthy individual due to a blow to the chest. Again, since they can be demonstrated in isolated tissues (Franz et al, 1992) and at the level of single myocytes (White et al., 1995; Zeng et al., 2000), these responses must be intrinsic to the myocardium itself, rather than of reflex origin (Fig. 11.1).

Mechanical forces on cardiac tissue may produce other electrophysiological changes, as well as changes in action potential morphology. For example, it has been shown that conduction velocity is slowed by tissue stretch.[2] In 1963, Penefsky

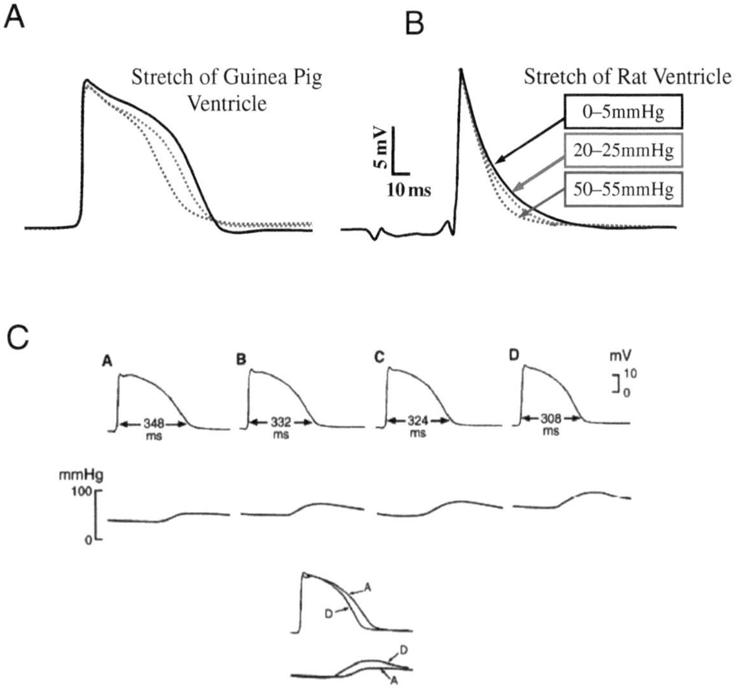

Fig. 11.1 Changes in action potential duration with stretch of the ventricle. **Panel A**: Superimposed recordings of epicardial MAPs in an isolated Guinea pig heart at intraventricular pressures of 0–5 mmHg (*solid line*), 20–25 mmHg (*inner dotted line*) and 50–55 mmHg (*innermost dotted line*). **Panel B**: Same plots, but for an isolated rat heart (Kelly, Mackenzie and Saint, unpublished data). **Panel C**: Upper traces: Sequential ventricular epicardial monophasic action potentials in a patient coming off bypass and, lower traces, radial artery pressure. As the workload on the heart increases, the action potential shortens. Lowermost figure: MAPs at the start of coming off bypass (**A**) and at the end of bypass (**D**) (from Taggart and Sutton, 1999)

[2] When discussing the effects of stretch on conduction velocity, one must distinguish between changes in conduction time between two points separated by a constant distance (apparent con-

and Hoffman studied the effects of stretch on conduction in atrial and ventricular myocardium (Penefsky and Hoffman, 1963). With mild stretch, the true velocity of conduction remained constant and the transmembrane potential did not change. During high levels of stretch, the membrane resting potential, upstroke velocity, and overshoot of the action potential declined, which they concluded indicated a decrease in the rapid sodium current. (There are very few subsequent reports that the voltage dependent sodium current is itself stretch sensitive (Morris and Juranka, 2007)). It is more likely that changes in the passive properties of the tissue with stretch are responsible for changes in conduction velocity. Such effects of stretch on the cable properties of cardiac muscle was investigated by Deck (1964) and Dominguez and Fozzard (1979). Deck concluded that changes in membrane resistance (increase) and capacitance (decrease), together with a decrease in core resistance, could be responsible for the change in the apparent CV. Dominguez and Fozzard argued that geometric alterations, such as unfolding of the membrane and uncoiling of the muscle, could partly explain the effects of stretch on CV. A recent study in isolated rabbit hearts concluded that conduction slowing during ventricular volume loading was not attributable to stretch-activated currents or altered resting membrane potential, but a reduction of intercellular resistance with a concurrent increase of effective membrane capacitance (Mills et al., 2008). In general, acute stretch results in conduction slowing (Kuijpers et al., 2007; Sung et al., 2003) and this may exacerbate the arrhythmogenic effect of stretch, particularly for re-entrant type arrhythmias.

11.3.2 Mechanisms Responsible for MEF

It has been suspected for some time that stretch sensitive, or mechanosensitive, ion channels are present in the heart of both animals and humans, as shown by the electrophysiological responses of the myocardium to stretch (Ravelli et al., 1994; Taggart et al., 1992). Two main types of stretch sensitive channels have been described in cardiac tissues: (1) a non-selective cation channel (SAC), and (2) some members of the tandem pore family of potassium selective channels (Hu and Sachs, 1997). There have been sporadic reports that some other ion channels are stretch-sensitive (e.g. the K_{ATP} channel (Van Wagoner, 1993) muscarinic K^+ channel (Ji et al., 1998) and sodium channels (Morris and Juranka, 2007)) but none of these other channel types appears to have the specialised gating structures to sense membrane stretch, and their response to membrane stretch is generally less striking than the tandem pore or SAC channels. The contribution of direct effects of stretch on other channel types to MEF is therefore probably minor. On the other hand, it should be noted that calcium handling mechanisms in cardiac cells are stretch-sensitive,

duction velocity) and changes in conduction between two landmarks on the surface of the muscle (true conduction velocity). The distinction can have important implications for mechanisms of arrhythmogenesis.

and these effects may also produce electrophysiological changes without invoking the contribution of stretch sensitive channels (see below).

11.3.2.1 The Potassium Selective Stretch-Sensitive Channels:

The stretch-sensitive potassium selective channels that have been recorded in cardiac tissue are members of the tandem pore family of channels. All tandem pore potassium channels have the signature K^+ pore sequence TXGYG or TXGFG, in the same way as the Kv family of channels, but there are two such sequences on a single subunit, hence the name "two pore" or "tandem pore". The subunits have four transmembrane segments, and lack a voltage sensing sequence analogous to the S4 segment of Kv family of channels. Hence, none of the members of the family have obvious intrinsic voltage dependence.

The tandem pore channels exhibit a range of properties, including sensitivity to extracellular pH, intracellular pH, fatty acids, membrane stretch, etc. Since the cloning of the first tandem pore channel (Lesage et al., 1996), about a dozen members have been identified and sequenced. At least some members are present in all tissues, although their function is not yet clear (Lesage and Lazdunski, 2000). Of the tandem pore family, three are activated by membrane stretch: TREK-1 (Twik-RElated K^+ channel), TREK-2 and TRAAK (Twik-Related Arachidonic Acid-stimulated K^+ channel) (Aimond et al., 2000). They have been cloned in mouse and human (Fink et al., 1996; Fink et al., 1998) and the properties of the expressed channels approximate very closely with the properties observed in cardiac tissues (Terrenoire et al., 2001). The mechanosensitivity of TREK-1, TREK-2 and TRAAK resides in the carboxy terminus: – truncation of the carboxyterminus abolishes stretch sensitive gating (Honore et al., 2002; Maingret et al., 2002) Studies using site-directed mutagenesis have pinpointed the amino acids involved in this gating (Honore et al., 2002). TREK-1 and TREK-2 have been shown to be expressed at moderately high levels in animal cardiac tissue (Kelly et al., 2006; Liu and Saint, 2004; Tan et al., 2002, 2004; Xian Tao et al., 2006) although the (somewhat sparse) literature on human heart suggests that expression is low (Lesage and Lazdunski, 2000; Lesage et al., 2000). This is surprising, since electrophysiological evidence consistent with the presecnce of mechanosensitive K^+ channels is well documented in human heart (Eckardt et al., 2001).

11.3.2.2 Non-selective Cation Channels (SACs)

SACs have been recorded in heart cells using patch clamp techniques (Hu and Sachs, 1997), and block of SACs with Gd^{3+} or streptomycin affects stretch induced changes in myocardial action potentials (Takagi et al., 1999). Frustratingly, the gene coding for SACs was until very recently unknown, so expression studies similar to those on TREK and TRAAK have not been possible. Recently, it has been shown that one of the TRP family of channels, TRPC1, is stretch sensitive and has the biophysical characteristics of SAC channels (Maroto et al., 2005). All TRP channels have six transmembrane domains, are thought to assemble as homo- or hetero-tetramers and can be divided into seven sub-families (Pedersen et al., 2005). Of these, TRPC channels are the most likely candidates for being the SAC in heart. TRPC channels are expressed in human heart, particularly TRPC1/4/5 and 6 (Riccio et al., 2002), and in rat heart (mainly TRPC3), and it has been shown that TRPC3 can mediate stretch-induced hypertrophic signalling in rat myocytes (Bush et al., 2006). Patch clamp studies have shown that SACs are blocked by gadolinium (Bowman et al., 2007), as is TRPC3 (Shlykov et al., 2003). Certainly there is evidence that TRP channels (of various families) can be mechanosensitive (Barritt and Rychkov, 2005; O'neil and Heller, 2005, and this, together with the demonstration that TRPC1 is stretch sensitive (Maroto et al., 2005), makes it highly likely that the stretch sensitive cation channel (SAC) in the heart is a member of the TRPC family. However, direct evidence that TRPC-1 or TRPC-3 plays such a role in cardiac muscle is currently less abundant (but see Williams and Allen, 2007).

11.3.2.3 Other Mechanisms Contributing to MEF

Stretch-dependent electrophysiological changes can be induced by mechanisms other than by the activation of stretch sensitive ion channels in the sarcolemma. One of the main ways in which stretch may affect electrophysiology is via the well known change in myofilament calcium affinity, coupled with the action of the electrogenic sodium-calcium exchanger (NCX). When cardiac muscle is stretched there is a rapid increase in force that is associated with an increase in the myofilament sensitivity to Ca^{2+}. Developed force increases in the absence of an increase in the intracellular Ca^{2+} transient and there is a leftward shift in the force-pCa relationship (Fabiato and Fabiato, 1978). This modulation of calcium sensitivity is thought to be due to decreased lattice spacing between the filaments, (Cazorla et al., 2001; Fukuda et al., 2005). Hence, any alterations in the force on the myocytes alters the "calcium buffering" effect of the myofilaments, and leads to an increase or decrease in the amount of calcium bound to the myofilaments. A rapid application of force can thus reduce free Ca^{2+} in the cytoplasm, while a rapid release of tension would lead to a rapid rise in cytosolic Ca^{2+} (Calaghan et al., 2003). These sorts of effects of stretch have been shown in both atrial (Tavi et al., 1998) and ventricular (Housmans et al., 1983) tissue.

Other calcium handling mechanisms may be sensitive to stretch, or to shear stress on the myocytes. In atrial myocytes, "puffing" solutions onto the myocytes activated internal calcium release (Woo et al., 2007). Puffing of pressurized (200 mm H_2O) solutions triggered or enhanced spontaneously occurring peripheral sparks by five- to six-fold and central Ca^{2+} sparks by two- to three-fold, without altering the unitary spark properties. Exposure to higher pressure flows (400 mm H_2O) often triggered longitudinally spreading Ca^{2+} waves. The authors suggested that pressurized flows may directly modulate Ca^{2+} signalling by activating the intracellular Ca^{2+} release sites. It has been suggested that these calcium signals are due to release of calcium from a pool distinct from the SR, possibly the mitochondria (Belmonte and Morad, 2008).

Since the sodium-calcium exchanger (NCX) is electrogenic, any process which changes intracellular calcium will alter the activity of this exchanger, and hence alter the membrane potential. This is thought to be the underlying mechanism of

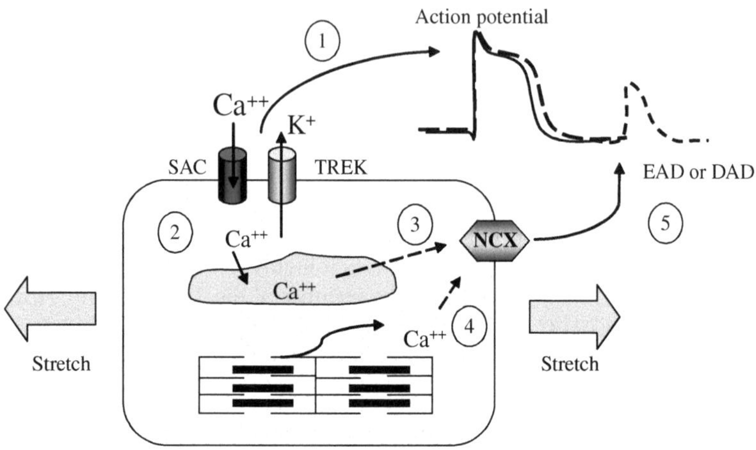

Fig. 11.2 Mechanisms of MEF. Stretch of myocytes (or fibroblasts) directly opens non-selective cation channels (SACs) and/or K^+ selective channels (TREK) which can directly affect the action potential (1). Ca^{2+} influx through SACs contributes to SR store load (2). If this is large enough, the SR releases Ca^{2+} (3) which is pumped out through the NCX, producing a current large enough to produce EADs or DADs (5). Stretch also increases myofilament Ca^{2+} affinity. Ca^{2+} dissociated from the myofilaments on release of stretch is pumped out by NCX (4), again producing EADs or DADs

EADs and DADs in cardiac tissue as a consequence of calcium overload, but the idea that stretch induced changes in calcium handling may give rise to arrhythmias by a similar mechanism has not been well investigated (Fig. 11.2).

11.4 The Role of MEF in Normal Heart and Cardiac Diseases

The normal physiological role of MEF in the heart is not clear. It may be that MEF arises as a consequence of other mechanisms that are physiologically beneficial, such as the change in myofilament sensitivity to calcium that contributes to an enhanced Frank-Starling relation but that indirectly alters calcium dependent conductances and exchangers. However, the presence of ion channels that appear to be evolved to respond to stretch directly, such as TREK-1 and TRPC-1 or -3, implies that there is a useful role for MEF. It has been hypothesised that MEF is important in controlling the beat to beat force of contraction in the ventricle in response to fluctuations in load, but most data suggests that changes in action potential morphology or conduction velocity with physiological levels of stretch in the normal heart are small, and do not produce changes in cardiac function (i.e. do not contribute to the Frank-Starling relation). Alternatively, MEF may play a role in controlling the dispersion of repolarisation in areas of the myocardium where action potential conduction velocity is locally slowed (Lab, 1999). In this model, MEF would act to counter the pro-arrhythmic alterations in electrophysiology that might occur with, for example, localised regions of relative ischaemia at high levels of myocardial oxygen demand (Fig. 11.3).

If MEF does play an important role in the heart, the question arises as to why it is so hard to demonstrate, and when it is demonstrated, why is it so small? It may be that the conditions in normally functioning myocardium suppress (or do not activate) MEF, whereas conditions in ischaemia or other myocardial stressors enhance MEF: – for example, TREK activity is enhanced by internal acidosis, and lysophospholipids, and these agents increase its stretch sensitivity. Hence, stretch-dependent effects of TREK which are small in normal tissue may be greatly enhanced in ischemic myocardium. Similar studies on SACs have not been done.

Here, we also propose an alternative hypothesis which is, as yet, untested. In humans, the foetal heart begins to beat at about week five of gestation but is not innervated by the sympathetic nervous system until week 22 (Huang et al., 1996). It may be that MEF is part of a foetal mechanism contributing to regulation of cardiac output, and it is small in the adult because it is down-regulated once the sympathetic nervous system establishes innervation and adopts this role. This idea would fit with the changes in gene expression seen in the heart during hypertrophy and failure, which has been described as a re-expression of the foetal genotype. Tentative recent reports suggest that some of the changes seen in hypertrophy include an up-regulation of stretch-sensitive channels TREK-1 (Zhao et al., 2007) and TRPC (Ohba et al., 2007). Consistent with this, stretch activated currents are enhanced in myocytes from hypertrophic hearts (Kamkin et al., 2000). MEF at the cellular level

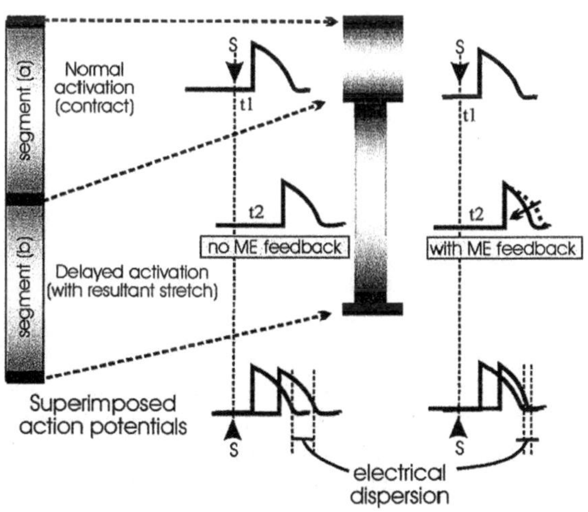

Fig. 11.3 MEF reduces dispersion of repolarisation. Diagram shows action potentials of two segments of myocardium in series during mechanical interaction. Segment (**a**), stimulated at (S), produces an action potential after a delay of t1 ms (top action potential). If segment (**b**), has a conduction delay (t2), it will experience delayed mechanical activation. Hence segment (**a**) stretches segment (**b**). MEF will then shorten the action potential in segment (**b**), reducing the arrhythmogenic dispersion in repolarisation which would otherwise occur. Diagram from Lab (1999)

has been shown to be increased in hypertrophic myocardium and in post infarct myocardium (Kamkin et al., 2000; Kiseleva et al., 2000). It seems likely that this upregulation of stretch activated channels contributes to the propensity to arrhythmias seen in the hypertrophic and failing heart.

11.4.1 The Heterogeneity of MEF in the Heart

While it is clear that MEF exists in the heart, information on the *distribution* of MEF (i.e. regional differences in the response of the myocardium to stretch) is sparse. Since the electrophysiology of the heart is obviously heterogeneous in other ways (distribution of voltage dependent channels, etc), it is reasonable to propose that MEF is similarly heterogeneous. In addition to its proposed role in protecting against increased dispersion of repolarisation in the face of regional alterations in conduction velocity, heterogeneous MEF may serve to protect against the electrophysiological consequences of unequal distributions in stress in the myocardium and the resulting uneven changes in myofilament Ca^{2+} sensitivity. The idea that MEF is heterogeneous in the heart, and the consequences of this for its contractile function has recently been reviewed by Stones et al. (2008). However, study of the electrophysiological consequences of heterogeneous MEF, both in normal heart and for the generation of arrhythmias, is still in its infancy.

11.4.1.1 Atria

As noted above, because the atria are anatomically complex, with ostia of the major veins in both atria, and the anatomical discontinuities presented by the crista terminalis and the coronary sinus, there are large regional differences in electrophysiology. These are reflected in regional differences in MEF; for example, the pulmonary veins themselves appear to have enhanced responses to stretch and this can lead to the generation of atrial arrhythmias (Chang et al., 2007; Kalifa et al., 2003; Seol et al., 2008). Acute stretch of the atria produces shortening of the action potential and reduction of effective refractory periods, but this has generally been measured at only single or a few sites in the atria, so that information on the relative changes in the different parts of the atria is sparse. A recent study using isolated dog hearts in which effective refractory period was measured at four sites showed that ERP dispersion increased significantly during high atrial pressure and that AF inducibility significantly correlated with the ERP dispersion, suggesting that an increased inhomogeneity in atrial electrophysiological properties during atrial dilatation contributed to the inducibility of AF (Huang et al., 2003). In addition, one of the effects of stretch in the atria is slowing of conduction velocity, as roughly measured between two recording points (Sideris et al., 1994). Later studies have shown similar results (Chorro et al., 1998; Kuijpers et al., 2007; Sung et al., 2003) Importantly, the changes in conduction velocity are heterogeneous. The effects of acute atrial dilatation on the occurrence of local conduction delays and block was investigated by Eijsbouts et al. in the isolated rabbit heart model. Areas with slow conduction

(10–20 cm/sec) and lines of conduction block (10 cm/sec) were identified during pacing from four different directions. It was shown that acute atrial dilatation not only depressed atrial conduction, but promoted spatial heterogeneity in conduction by causing conduction blocks which occurred parallel to the boundaries of large trabeculae (Eijsbouts et al., 2003).

As a further consideration, mechanosensitive currents have been demonstrated in atrial myocytes, but fibroblasts are also mechanosensitive. The coupling between fibroblasts and myocytes is heterogeneous (Camelliti et al., 2004) and hence the effect of stretch on electrophysiological changes may also be heterogeneous because of this.

11.4.1.2 Ventricles

On one hand, the heterogeneity of MEF in the ventricles is simpler than in the atria in that the anatomy of ventricles is somewhat simpler, but on the other hand it is complicated by the fact that stresses in the ventricular wall vary between sub-endocardial and sub-epicardial layers (Tendulkar and Harken, 2006) (Fig. 11.4).

These regional differences in strain would be expected to produce differences in MEF, but this depends on the distribution of the "gain" of MEF at a cellular level (i.e. how much electrophysiological change is produced at the cellular level for the same level of strain). This different cellular "gain" of MEF could in turn be due to different gene expression levels of stretch-sensitive channels. The latter idea, that gene expression of stretch sensitive proteins is heterogeneous in the heart, has recently been reviewed (Stones et al., 2007; Stones et al., 2008), but, to date, very

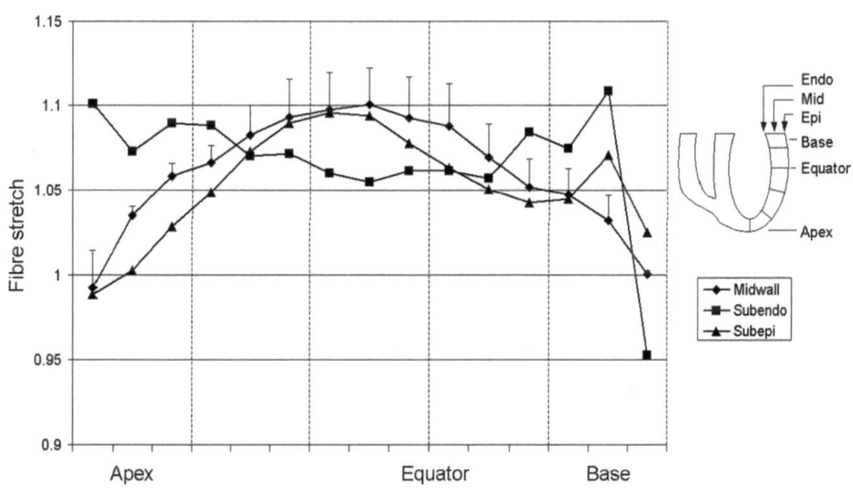

Fig. 11.4 Plot of stresses calculated in the model of pig heart at different points of the myocardium plotted from base to apex for sub-endocardial (*squares*), mid-myocardial (*circles*) and sub-epicardial (*triangles*) layers. At the apex and the base, stresses are highest in the subendocardium and lowest in subepicardium and midmyocardium, while at the equator the situation is reversed (from Kelly et al., 2006)

few studies have been done on whether MEF is heterogeneous at the cellular level in the ventricular myocardium.

11.4.1.3 Distribution of Stretch Sensitive Channels in Ventricle

Potassium Selective Channels TREK-1, TREK-2 and TRAAK

In contrast to the voltage dependent potassium channels, the regional distribution of the tandem pore potassium channels within the heart is almost unknown. We have shown the distribution of nine members of the family between chambers in rat heart, and that the expression of TREK-1 is heterogeneous between epicardial and endocardial cells in the rat left ventricle (Kelly et al., 2006; Tan et al., 2004). The expression levels correlated with the magnitude of chloroform-activated currents (presumed to be carried by TREK-1) in epicardial and endocardial cells. A study by a different group, using Northern Blot techniques, has shown that TBAK-1 and TASK-1 (non-mechanosensitive members of the tandem pore potassium channels), are not differentially expressed in epicardial and endocardial cells (Kim et al., 1999), reinforcing the idea that perhaps the distribution of the stretch-sensitive tandem pore channel is a response to distribution of strain on the myocardium.

The Stretch Sensitive Cation Channel (SACs)

There is almost no data on the distribution of SACs in the ventricles. Stones et al. reported recently that the gene expression level of TRPC-1 (the gene thought to code for SACs) was not different in epi- and endo-cardial rat ventricular myocytes (Stones et al., 2007).

Genes Involved in Other Possible Mechanisms of MEF

Stones et al. also measured the level of mRNA of 12 targets that are associated with events related to mechanical stimulation, such as stress-induced hypertrophy, stretch regulation of contractility and mechanosensitive ion channels. They showed that six of the 12 targets had significantly higher levels of mRNA in ENDO than EPI. They hypothesised that the differential distribution was related to the different stresses in the epi- and endocardial myocardium.

Regional Differences in MEF in the Ventricle

Dutertre et al. reported in 1972 that mechanically induced changes in action potential duration were dissimilar in different parts of the intact left ventricle (Dutertre et al., 1972). In other studies, the decrease in epicardial effective refractory period in the in situ pig heart in response to aortic clamping (Dean and Lab, 1990) and in the rabbit heart in response to left ventricular balloon inflation (Reiter et al., 1988) has been reported to be greatest in the apical region. Zabel, Portnoy and Franz reported results from a study in which recordings were made from six epicardial MAP electrodes in isolated rabbit hearts. Sustained stretch was applied by inflating an intra-ventricular balloon. Upon stretch, differential shortening of MAP occurred, so that dispersion of $APD_{(90)}$ was increased during steady-state pacing and during premature extrastimulation (Zabel et al., 1996b). There are transmural differences in MEF also. In human heart, it was shown that an abrupt change in ventricular filling, within the physiological range, increased QT dispersion in subjects with abnormal ventricular function but not in subjects with normal ventricles (James et al., 2002). In in situ dog hearts, rapid increase in left ventricular pressure, induced by aortic occlusion, produced a rapid (within five beats) effect on monophasic action potential duration (Takagi et al., 1999). Epicardial MAP duration was decreased but no change in endocardial MAP duration was seen, so the transmural gradient in MAP duration at 90% repolarisation rose from approximately 22 to 28 ms. This effect, and the generation of arrhythmias, was prevented by treatment with Gd^{3+}, a blocker of non-specific cationic mechanosensitive ion channels. Given that other studies have shown (in rats) a greater expression of TREK-1 in endocardial cells compared to epicardial, one might have predicted greater effects of occlusion on endocardium, if TREK-1 were a major contributor to this effect. Since this was not seen, and given the block of the effect by Gd^{3+}, it seems that SACs were the mechanism behind this effect in dogs. However, this does not preclude the possibility that TREK also contributes to MEF in these circumstances. It is also possible, indeed likely, that there are inter-species differences in the relative importance of the different mechanosensitive channels.

We have recently shown that differential transmural electrical responses to stretch can be observed in rat heart (Fig. 11.5).

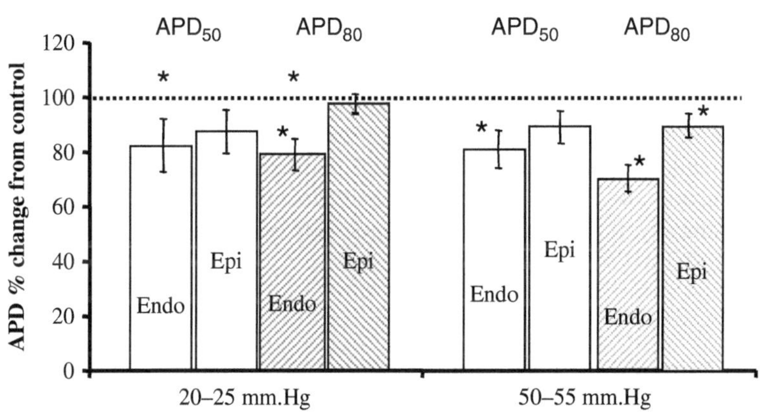

Fig. 11.5 Effect of stretch on epicardial and endocardial action potentials in rat heart. Isolated adult rat hearts were perfused in a Langendorff setup, and had a balloon in the left ventricle to allow recording and manipulation of intraventricular pressure. MAPs were recorded from the subepicardium and the subendocardial layers simultaneously. Leftmost bars: Change in epicardial and endocardial APD_{50} (*plain bars*) and APD_{80} (*hatched bars*) at intraventricular pressures between 20 and 25 mmHg. Rightmost bars, data for intraventricular pressure of 50–55 mmHg. The value for "no stretch" condition is taken as 100% (*dotted line*). *Stars* indicate a significant change from the "no stretch" condition ($P < 0.05$. $n = 8$) (Kelly, Mackenzie and Saint, unpublished data)

11.5 The Contribution of MEF to Arrhythmogenesis

The relevance of MEF to clinical arrhythmias was reviewed almost ten years ago by Kohl et al. (1999). They concluded from data then available that MEF was likely to be involved in the induction of premature excitation in lower order pacemaker cells, the initiation of ectopic beats and the triggering and maintenance of tachyarrhythmias or cardiac arrest. However, they also noted that the precise cellular and subcellular correlates for those effects were uncertain, and they raised several issues that needed to be resolved regarding the nature of the stretch induced "signal" and its pathophysiological role. Many subsequent studies have tackled these issues. Overall, the weight of evidence seems to be growing that MEF is involved in many types of clinically relevant arrhythmias, both atrial and ventricular.

11.5.1 Atrial Arrhythmias

Although the precise mechanisms are still being elucidated, there is no doubt that stretch is a contributing factor in the genesis and maintenance of many types of AF (Allessie et al., 2001). In animal studies, both acute and chronic atrial dilation have been shown to increase vulnerability to AF (Boyden and Hoffman, 1981; Ninio et al., 2005; Ravelli and Allessie, 1997; Saint, 2002; Solti et al., 1989a, b). Clinically, atrial dilation is known to be an important independent risk factor for AF (Psaty et al., 1997; Vaziri et al., 1994). Some time ago clinical studies showed a strong relationship between increased left atrial size and the development of AF (Henry et al., 1976; Takahashi et al., 1982), a relationship that has been confirmed in later studies (e.g.) (Tanabe et al., 2001). Early studies suggested that atrial size is a predictor of successful cardioversion and maintenance of sinus rhythm (Brodsky et al., 1989; Hoglund and Rosenhamer, 1985) although some later studies have been equivocal (Lin et al., 2002). However, it is a common clinical observation that p-wave duration is a good predictor of atrial fibrillation (Dogan et al., 2004), and p-wave duration or dispersion is thought to reflect atrial stretch (Mitro and Spegar, 2006) Clinically, there is a good correlation between intra-atrial pressure (which reduces ERP) and AF (Manios et al., 2006).

11.5.2 Ventricular Arrhythmias

It has been hypothesised that the increased risk of arrhythmia in post-infarct patients may be due to the abnormal wall movement of the infarcted zone producing stretch-induced ectopic beats in the ventricular wall (Taggart, 1996; Babuty and Lab, 2001). Mechanical effects on the heart have also been implicated in some instances of sudden cardiac death in otherwise health individuals. Over a century ago there were reports of cardiac rhythm disturbances and sudden death caused by non-penetrating mechanical impacts to the chest. Nélaton (1876) reported a case of instant death of a manual labourer due to precordial impact. Post mortem examination showed no signs of internal structural damage. This condition became known as "Commotio cordis". A sudden impact to the heart, especially if it is applied during the vulnerable period of the cardiac cycle, can initiate ventricular tachyarrhythmias, including sustained VF (Babuty and Lab, 2001; Kohl et al., 1999; Link and Estes, 2007). This is likely due to the initiation of ectopic beats. In isolated animal hearts, it is relatively easy to initiate such beats. Occasionally, it is possible to demonstrate the initiation of an ectopic beat in one layer of the myocardium but not the other (Fig. 11.6). This type of extreme transmural difference would be expected to be highly arrhythmogenic (although it does not generally result in sustained arrhythmia in rat heart in

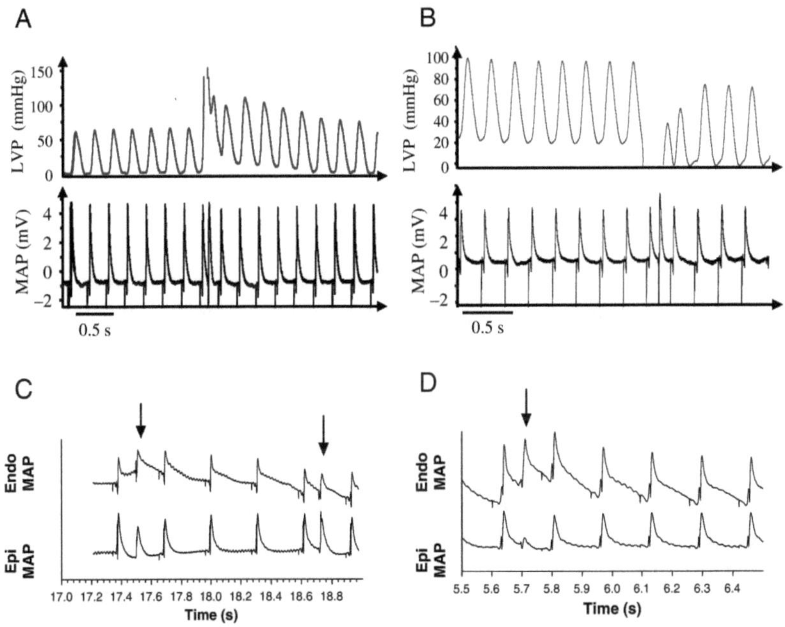

Fig. 11.6 Generation of ectopic beats by stretch. Isolated adult rat hearts were perfused in a Langendorff setup. A balloon was placed in the left ventricle for recording of intraventricular pressure and to allow rapid changes in ventricular volume via a servo-driven syringe triggered by the stimulator. A suction electrode recorded epicardial surface MAPs. **Panel A**: top trace is intraventricular pressure, lower trace is MAP recording. A sudden increase in intraventricular pressure to about 30 mmHg produced an ectopic beat. **Panel B**. Same setup as panel A. A rapid reduction of intraventricular pressure also produced an ectopic beat. **Panel C**. A dual recording electrode was implanted in the ventricular wall to allow simultaneous recording of epicardial and endocardial MAPs. Top trace shows endocardial recording and lower trace epicardial (pressure trace not shown for clarity). On rapid increase in intraventricular pressure (*at arrow*), an ectopic beat was generated in both epi- and endo-cardial layers. Second *arrow* shows the end of the pressure pulse (rapid deflation), which also triggered an ectopic beat in both layers. **Panel D**: In some instances, rapid inflation (*at arrow*) triggered an ectopic beat in only the endocardial layer (Kelly, Mackenzie and Saint, unpublished data)

our hands-this may be due to lack of a suitable arrhythmogenic substrate-see later (Fig. 11.7))

The generation of ectopic beats by sudden stretch is likely due to activation of stretch-activated channels. However, evidence for this is sparse, perhaps due to a lack of good pharmacological tools for manipulation of either SACs or TREK channels. Gadolinium (a blocker of SACs) inhibits stretch induced ventricular arrhythmias (Hansen et al., 1991) although the use of Gd^{3+} is problematical in physiological solutions containing phosphate and bicarbonate (White, 2006). It's likely that the generation of an ectopic beat on rapid release is due to a different mechanism-rapid dissociation of Ca^{2+} from the myofilaments on release could greatly increase the current through the electrogenic transporter NCX, leading to EAD or DAD type ectopic beats (Fig. 11.2).

More complex effects of stretch can be observed: Taggart and Lab (2008) discuss the idea that the rate of action potential electrical restitution can be altered in an inhomogeneous way by stretch, and that this may be important in arrhythmogenesis.

11.6 Conclusion

In studies of stretch induced AF, it has been shown that the blockers of SACs Gd^{3+}, GSmTx-4 and streptomycin can all reduce AF (Franz and Bode, 2003; Ninio et al., 2005; Ninio and Saint, 2008). However, they do not modify the reduction in ERP that occurs with stretch. This raises the question as to how they stop or ameliorate AF in these models. The explanation is likely to lie in the idea that fibrillation needs not only a substrate that can sustain it (a reduced ERP and/or conduction velocity) but it also requires a trigger. It may be that activation of TREK reduces ERP and provides a substrate, while activation of SACs provides the trigger (Fig. 11.7). In contrast to the atria, there are much fewer studies on stretch-induced ventricular arrhythmias, but it seems likely that the same mechanisms will apply.

The therapeutic possibilities of targeting stretch activated channels has been reviewed by several authors in recent years (Bowman et al., 2007; Saint, 2002; Savelieva and Camm, 2008; White, 2006). For some types of arrhythmias, this is an appealing prospect: Atrial fibrillation is often associated with atrial distension, and some ventricular arrhythmias (apart from commotio cordis) may be triggered mechanically, for example in post-MI patients. However, progress has been slow. This may be due to a lack of detailed understanding of the role of stretch-activated channels in the heart, and the lack of good pharmacological tools to investigate them. This lack of selective pharmacological tools impedes our ability to conduct experiments, which in turn impedes our understanding of the mechanisms – a difficult "catch 22". However, there is no doubt that this will eventually be overcome, and that agents targeting stretch-activated channels (or other mechanisms of MEF) will assume a greater profile in the future (Fig. 11.7).

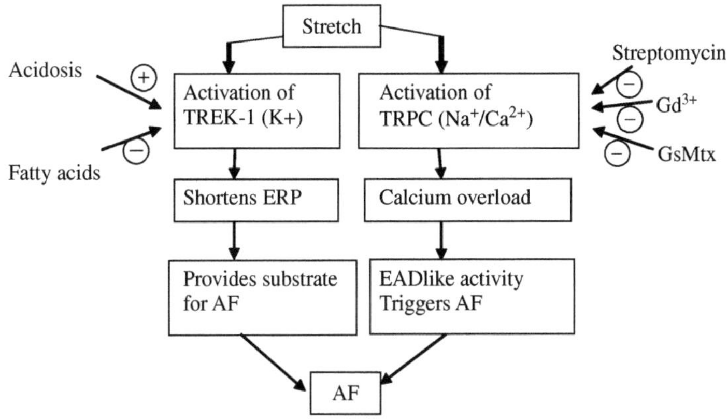

Fig. 11.7 "Trigger and substrate model". We postulate that stretch of the myocardium activates TREK channels, which causes a shortening of action potential duration and ERP, and so provides the substrate for AF. This effect is exacerbated by acidosis and ameliorated by fatty acids (*left hand side of figure*). Stretch also activates SACs which allow Ca^{2+} influx, contributing to SR Ca^{2+} overload and triggering ectopic beats via EADs or DADs. This effect can be blocked be Gd3+, Streptomycin and GsMtx-4 (from Ninio and Saint, 2008)

References

Aimond F, Rauzier JM, Bony C and Vassort G (2000) Simultaneous activation of p38 MAPK and p42/44 MAPK by ATP stimulates the K^+ current ITREK in cardiomyocytes. J Biol Chem 275, 39110–6.

Allessie MA, Boyden PA, Camm AJ, Kleber AG, Lab MJ, Legato MJ, Rosen MR, Schwartz PJ, Spooner PM, Van Wagoner DR and Waldo AL (2001) Pathophysiology and prevention of atrial fibrillation. Circulation 103, 769–77.

Antoniou A, Milonas D, Kanakakis J, Rokas S and Sideris DA (1997) Contraction-excitation feedback in human atrial fibrillation. Clin Cardiol 20, 473–6.

Antzelevitch C (2004) Cellular basis and mechanism underlying normal and abnormal myocardial repolarization and arrhythmogenesis. Ann Med 36 Suppl 1, 5–14.

Antzelevitch C, Sicouri S, Litovsky SH, Lukas A, Krishnan SC, Di Diego JM, Gintant GA and Liu DW (1991) Heterogeneity within the ventricular wall. Electrophysiology and pharmacology of epicardial, endocardial, and M cells. Circ Res 69, 1427–49.

Antzelevitch C, Yan GX and Shimizu W (1999) Transmural dispersion of repolarization and arrhythmogenicity: the Brugada syndrome versus the long QT syndrome. J Electrocardiol 32 Suppl, 158–65.

Babuty D and Lab MJ (2001) Mechanoelectric contributions to sudden cardiac death. Cardiovasc Res 50, 270–9.

Bainbridge FA (1915) The influence of venous filling upon the rate of the heart. J Physiol 50, 65–84.

Barr CS, Naas A, Freeman M, Lang CC and Struthers AD (1994) QT dispersion and sudden unexpected death in chronic heart failure. Lancet 343, 327–9.

Barritt, G. and Rychkov, G. (2005) TRPs as mechanosensitive channels. *Nat Cell Biol* 7, 105–7.

Belmonte S and Morad M (2008) 'Pressure-flow'-triggered intracellular Ca^{2+} transients in rat cardiac myocytes: possible mechanisms and role of mitochondria. J Physiol 586, 1379–97.

Bode F, Sachs F and Franz MR (2001) Tarantula peptide inhibits atrial fibrillation. Nature 409, 35–6.

Bowman CL, Gottlieb PA, Suchyna TM, Murphy YK and Sachs F (2007) Mechanosensitive ion channels and the peptide inhibitor GsMTx-4: history, properties, mechanisms and pharmacology. Toxicon 49, 249–70.

Boyden PA and Hoffman BF (1981) The effects on atrial electrophysiology and structure of surgically induced right atrial enlargement in dogs. Circ Res 49, 1319–31.

Brodsky MA, Allen BJ, Capparelli EV, Luckett CR, Morton R and Henry WL (1989) Factors determining maintenance of sinus rhythm after chronic atrial fibrillation with left atrial dilatation. Am J Cardiol 63, 1065–8.

Bush EW, Hood DB, Papst PJ, Chapo JA, Minobe W, Bristow MR, Olson EN and McKinsey TA (2006) Canonical transient receptor potential channels promote cardiomyocyte hypertrophy through activation of calcineurin signaling. J Biol Chem 281, 33487–96.

Calaghan SC, Belus A and White E (2003) Do stretch-induced changes in intracellular calcium modify the electrical activity of cardiac muscle? Prog Biophys Mol Biol 82, 81–95.

Camelliti P, Green CR, LeGrice I and Kohl P (2004) Fibroblast network in rabbit sinoatrial node: structural and functional identification of homogeneous and heterogeneous cell coupling. Circ Res 94, 828–35.

Cazorla O, Wu Y, Irving TC and Granzier H (2001) Titin-based modulation of calcium sensitivity of active tension in mouse skinned cardiac myocytes. Circ Res 88, 1028–35.

Chang SL, Chen YC, Chen YJ, Wangcharoen W, Lee SH, Lin CI and Chen SA (2007) Mechanoelectrical feedback regulates the arrhythmogenic activity of pulmonary veins. Heart 93, 82–8.

Chen PS, Chou CC, Tan AY, Zhou S, Fishbein MC, Hwang C, Karagueuzian HS and Lin SF (2006) The mechanisms of atrial fibrillation. J Cardiovasc Electrophysiol 17 Suppl 3, S2–7.

Cherry EM, Ehrlich JR, Nattel S and Fenton FH (2007) Pulmonary vein reentry – properties and size matter: insights from a computational analysis. Heart Rhythm 4, 1553–62.

Chorro FJ, Egea S, Mainar L, Canoves J, Sanchis J, Llavador E, Lopez-Merino V and Such L (1998) [Acute changes in wavelength of the process of auricular activation induced by stretching. Experimental study]. Rev Esp Cardiol 51, 874–83.

Chou CC, Nihei M, Zhou S, Tan A, Kawase A, Macias ES, Fishbein MC, Lin SF and Chen PS (2005) Intracellular calcium dynamics and anisotropic reentry in isolated canine pulmonary veins and left atrium. Circulation 111, 2889–97.

Cooper PJ, Lei M, Cheng LX and Kohl P (2000) Selected contribution: axial stretch increases spontaneous pacemaker activity in rabbit isolated sinoatrial node cells. J Appl Physiol 89, 2099–104.

Cordeiro JM, Greene L, Heilmann C, Antzelevitch D and Antzelevitch C (2004) Transmural heterogeneity of calcium activity and mechanical function in the canine left ventricle. Am J Physiol Heart Circ Physiol 286, H1471–9.

Cowan JC, Hilton CJ, Griffiths CJ, Tansuphaswadikul S, Bourke JP, Murray A and Campbell RW (1988) Sequence of epicardial repolarisation and configuration of the T wave. Br Heart J 60, 424–33.

Dean, J. W. and Lab, M. J. (1990) Regional changes in ventricular excitability during load manipulation of the in situ pig heart. *J Physiol* 429, 387–400.

Deck KA (1964) [Changes in the resting potential and the cable properties of Purkinje fibers during stretch]. Pflugers Arch Gesamte Physiol Menschen Tiere 280, 131–40.

Dixon JE and McKinnon D (1994) Quantitative analysis of potassium channel mRNA expression in atrial and ventricular muscle of rats. Circ Res 75, 252–60.

Dixon JE, Shi W, Wang HS, McDonald C, Yu H, Wymore RS, Cohen IS and McKinnon D (1996) Role of the Kv4.3 K^+ channel in ventricular muscle. A molecular correlate for the transient outward current. Circ Res 79, 659–68.

Dogan A, Avsar A and Ozturk M (2004) P-wave dispersion for predicting maintenance of sinus rhythm after cardioversion of atrial fibrillation. Am J Cardiol 93, 368–71.

Dominguez G and Fozzard HA (1979) Effect of stretch on conduction velocity and cable properties of cardiac Purkinje fibers. Am J Physiol 237, C119–24.

Dudel J and Trautwein W (1954) [Effect of stretch on action potentials and mechanogram of the heart muscle.]. Cardiologia 25, 344–62.

Dutertre J, Jean CF, Cartier R and Dieudonne JM (1972) Measurement of tissular strain with a tripod-like transducer – I. Med Biol Eng 10, 277–81.

Eckardt L, Kirchhof P, Breithardt G and Haverkamp W (2001) Load-induced changes in repolarization: evidence from experimental and clinical data. Basic Res Cardiol 96, 369–80.

Eijsbouts SC, Majidi M, van Zandvoort M and Allessie MA (2003) Effects of acute atrial dilation on heterogeneity in conduction in the isolated rabbit heart. J Cardiovasc Electrophysiol 14, 269–78.

Fabiato A and Fabiato F (1978) Myofilament-generated tension oscillations during partial calcium activation and activation dependence of the sarcomere length-tension relation of skinned cardiac cells. J Gen Physiol 72, 667–99.

Fink M, Duprat F, Lesage F, Reyes R, Romey G, Heurteaux C and Lazdunski M (1996) Cloning, functional expression and brain localization of a novel unconventional outward rectifier K^+ channel. Embo J 15, 6854–62.

Fink M, Lesage F, Duprat F, Heurteaux C, Reyes R, Fosset M and Lazdunski M (1998) A neuronal two P domain K^+ channel stimulated by arachidonic acid and polyunsaturated fatty acids. Embo J 17, 3297–308.

Fiset C and Giles WR (2006) Transmural gradients of repolarization and excitation-contraction coupling in mouse ventricle. Circ Res 98, 1237–9.

Franz MR and Bode F (2003) Mechano-electrical feedback underlying arrhythmias: the atrial fibrillation case. Prog Biophys Mol Biol 82, 163–74.

Franz MR, Cima R, Wang D, Profitt D and Kurz R (1992) Electrophysiological effects of myocardial stretch and mechanical determinants of stretch-activated arrhythmias. Circulation 86, 968–78.

Fukuda N, Wu Y, Farman G, Irving TC and Granzier H (2005) Titin-based modulation of active tension and interfilament lattice spacing in skinned rat cardiac muscle. Pflugers Arch 449, 449–57.

Gaborit N, Le Bouter S, Szuts V, Varro A, Escande D, Nattel S and Demolombe S (2007) Regional and tissue specific transcript signatures of ion channel genes in the non-diseased human heart. J Physiol 582, 675–93.

Han J and Moe GK (1964) Nonuniform recovery of excitability in ventricular muscle. Circ Res 14, 44–60.

Hansen DE, Borganelli M, Stacy GP Jr. and Taylor LK (1991) Dose-dependent inhibition of stretch-induced arrhythmias by gadolinium in isolated canine ventricles. Evidence for a unique mode of antiarrhythmic action. Circ Res 69, 820–31.

Hansen DE, Craig CS and Hondeghem LM (1990) Stretch-induced arrhythmias in the isolated canine ventricle. Evidence for the importance of mechanoelectrical feedback. Circulation 81, 1094–105.

Henry WL, Morganroth J, Pearlman AS, Clark CE, Redwood DR, Itscoitz SB and Epstein SE (1976) Relation between echocardiographically determined left atrial size and atrial fibrillation. Circulation 53, 273–9.

Hoglund C and Rosenhamer G (1985) Echocardiographic left atrial dimension as a predictor of maintaining sinus rhythm after conversion of atrial fibrillation. Acta Med Scand 217, 411–5.

Honore E, Maingret F, Lazdunski M and Patez AJ (2002) An intracellular proton sensor commands lipid- and mechano-gating of the K^+ channel TREK-1. Embo J 21, 2968–76.

Hooks DA, Trew ML, Caldwell BJ, Sands GB, LeGrice IJ and Smaill BH (2007) Laminar arrangement of ventricular myocytes influences electrical behavior of the heart. Circ Res 101, e103–12.

Housmans PR, Lee NK and Blinks JR (1983) Active shortening retards the decline of the intracellular calcium transient in mammalian heart muscle. Science 221, 159–61.

Hu H and Sachs F (1997) Stretch-activated ion channels in the heart. J Mol Cell Cardiol 29, 1511–23.

Huang JL, Tai CT, Chen JT, Ting CT, Chen YT, Chang MS and Chen SA (2003) Effect of atrial dilatation on electrophysiologic properties and inducibility of atrial fibrillation. Basic Res Cardiol 98, 16–24.

Huang MH, Friend DS, Sunday ME, Singh K, Haley K, Austen KF, Kelly RA and Smith TW (1996) An intrinsic adrenergic system in mammalian heart. J Clin Invest 98, 1298–303.
Jalife J (2003) Experimental and clinical AF mechanisms: bridging the divide. J Interv Card Electrophysiol 9, 85–92.
James PR, Hardman SM and Taggart P (2002) Physiological changes in ventricular filling alter cardiac electrophysiology in patients with abnormal ventricular function. Heart 88, 149–52.
Ji S, John SA, Lu Y and Weiss JN (1998) Mechanosensitivity of the cardiac muscarinic potassium channel. A novel property conferred by Kir3.4 subunit. J Biol Chem 273, 1324–8.
Kalifa J, Jalife J, Zaitsev AV, Bagwe S, Warren M, Moreno J, Berenfeld O and Nattel S (2003) Intra-atrial pressure increases rate and organization of waves emanating from the superior pulmonary veins during atrial fibrillation. Circulation 108, 668–71.
Kamkin A, Kiseleva I and Isenberg G (2000) Stretch-activated currents in ventricular myocytes: amplitude and arrhythmogenic effects increase with hypertrophy. Cardiovasc Res 48, 409–20.
Kamkin A, Kiseleva I and Isenberg G (2003a) Activation and inactivation of a non-selective cation conductance by local mechanical deformation of acutely isolated cardiac fibroblasts. Cardiovasc Res 57, 793–803.
Kamkin A, Kiseleva I, Wagner KD, Bohm J, Theres H, Gunther J and Scholz H (2003b) Characterization of stretch-activated ion currents in isolated atrial myocytes from human hearts. Pflugers Arch 446, 339–46.
Kelly D, Mackenzie L, Hunter P, Smaill B and Saint DA (2006) Gene expression of stretch-activated channels and mechanoelectric feedback in the heart. Clin Exp Pharmacol Physiol 33, 642–8.
Kim Y, Bang H and Kim D (1999) TBAK-1 and TASK-1, two-pore K(+) channel subunits: kinetic properties and expression in rat heart. Am J Physiol 277, H1669–78.
Kiseleva I, Kamkin A, Wagner KD, Theres H, Ladhoff A, Scholz H, Gunther J and Lab MJ (2000) Mechanoelectric feedback after left ventricular infarction in rats. Cardiovasc Res 45, 370–8.
Kohl P, Hunter P and Noble D (1999) Stretch-induced changes in heart rate and rhythm: clinical observations, experiments and mathematical models. Prog Biophys Mol Biol 71, 91–138.
Kuijpers NH, ten Eikelder HM, Bovendeerd PH, Verheule S, Arts T and Hilbers PA (2007) Mechanoelectric feedback leads to conduction slowing and block in acutely dilated atria: a modeling study of cardiac electromechanics. Am J Physiol Heart Circ Physiol 292, H2832–53.
Lab MJ (1999) Mechanosensitivity as an integrative system in heart: an audit. Prog Biophys Mol Biol 71, 7–27.
Lesage F, Guillemare E, Fink M, Duprat F, Lazdunski M, Romey G and Barhanin J (1996) TWIK-1, a ubiquitous human weakly inward rectifying K^+ channel with a novel structure. Embo J 15, 1004–11.
Lesage F and Lazdunski M (2000) Molecular and functional properties of two-pore-domain potassium channels. Am J Physiol Renal Physiol 279, F793–801.
Lesage F, Terrenoire C, Romey G and Lazdunski M (2000) Human TREK2, a 2P domain mechanosensitive K^+ channel with multiple regulations by polyunsaturated fatty acids, lysophospholipids, and Gs, Gi, and Gq protein-coupled receptors. J Biol Chem 275, 28398–405.
Lin JM, Lin JL, Lai LP, Tseng YZ and Stephen Huang SK (2002) Predictors of clinical recurrence after successful electrical cardioversion of chronic persistent atrial fibrillation: clinical and electrophysiological observations. Cardiology 97, 133–7.
Link MS and Estes NA 3rd (2007) Mechanically induced ventricular fibrillation (commotio cordis). Heart Rhythm 4, 529–32.
Liu J, Noble PJ, Xiao G, Abdelrahman M, Dobrzynski H, Boyett MR, Lei M and Noble D (2008) Role of pacemaking current in cardiac nodes: insights from a comparative study of sinoatrial node and atrioventricular node. Prog Biophys Mol Biol 96, 294–304.
Liu W and Saint DA (2004) Heterogeneous expression of tandem-pore K^+ channel genes in adult and embryonic rat heart quantified by real-time polymerase chain reaction. Clin Exp Pharmacol Physiol 31, 174–8.
Maingret F, Honore E, Lazdunski M and Patel AJ (2002) Molecular basis of the voltage-dependent gating of TREK-1, a mechano-sensitive K(+) channel. Biochem Biophys Res Commun 292, 339–46.

Manios EG, Mavrakis HE, Kanoupakis EM, Kallergis EM, Kafarakis PK and Vardas PE (2006) Evidence of mechanoelectric feedback in the atria of patients with atrioventricular nodal reentrant tachycardia. J Interv Card Electrophysiol 16, 51–7.
Maroto R, Raso A, Wood TG, Kurosky A, Martinac B and Hamill OP (2005) TRPC1 forms the stretch-activated cation channel in vertebrate cells. Nat Cell Biol 7, 179–85.
Mills RW, Narayan SM and McCulloch AD (2008) Mechanisms of conduction slowing during myocardial stretch by ventricular volume loading in the rabbit. Am J Physiol Heart Circ Physiol 295, H1270–78.
Mitro P and Spegar J (2006) Dynamic changes of P-wave duration and P-wave axis during head-up tilt test in patients with vasovagal syncope. Pacing Clin Electrophysiol 29, 742–6.
Morris CE and Juranka PF (2007) Nav channel mechanosensitivity: activation and inactivation accelerate reversibly with stretch. Biophys J 93, 822–33.
Nélaton, A., 1876. In: Pean, J., Editor, 1876. *Elements de pathologie chirurgicale* Vol. 4, Librairie Germer Bateliere et Co., Paris.
Ninio DM, Murphy KJ, Howe PR and Saint DA (2005) Dietary fish oil protects against stretch-induced vulnerability to atrial fibrillation in a rabbit model. J Cardiovasc Electrophysiol 16, 1189–94.
Ninio DM and Saint DA (2008) The role of stretch-activated channels in atrial fibrillation and the impact of intracellular acidosis. Prog Biophys Mol Biol 97, 401–16.
Nishimura S, Kawai Y, Nakajima T, Hosoya Y, Fujita H, Katoh M, Yamashita H, Nagai R and Sugiura S (2006) Membrane potential of rat ventricular myocytes responds to axial stretch in phase, amplitude and speed-dependent manners. Cardiovasc Res 72, 403–11.
Niu W and Sachs F (2003) Dynamic properties of stretch-activated K^+ channels in adult rat atrial myocytes. Prog Biophys Mol Biol 82, 121–35.
Ohba T, Watanabe H, Murakami M, Takahashi Y, Iino K, Kuromitsu S, Mori Y, Ono K, Iijima T and Ito H (2007) Upregulation of TRPC1 in the development of cardiac hypertrophy. J Mol Cell Cardiol 42, 498–507.
O'Neil, R. G. and Heller, S. (2005) The mechanosensitive nature of TRPV channels. *Pflugers Arch* 451, 193–203.
Patel SP and Campbell DL (2005) Transient outward potassium current, 'Ito', phenotypes in the mammalian left ventricle: underlying molecular, cellular and biophysical mechanisms. J Physiol 569, 7–39.
Pedersen SF, Owsianik G and Nilius B (2005) TRP channels: an overview. Cell Calcium 38, 233–52.
Penefsky Z and Hoffman B (1963) Effect of stretch on mechanical and electrical properties of cardiac muscle. Am J Physiol 204, 433–38.
Pereon Y, Demolombe S, Baro I, Drouin E, Charpentier F and Escande D (2000) Differential expression of KvLQT1 isoforms across the human ventricular wall. Am J Physiol Heart Circ Physiol 278, H1908–15.
Pertsov AM, Davidenko JM, Salomonsz R, Baxter WT and Jalife J (1993) Spiral waves of excitation underlie reentrant activity in isolated cardiac muscle. Circ Res 72, 631–50.
Poelzing S and Rosenbaum DS (2004) Altered connexin43 expression produces arrhythmia substrate in heart failure. Am J Physiol Heart Circ Physiol 287, H1762–70.
Psaty BM, Manolio TA, Kuller LH, Kronmal RA, Cushman M, Fried LP, White R, Furberg CD and Rautaharju PM (1997) Incidence of and risk factors for atrial fibrillation in older adults. Circulation 96, 2455–61.
Pye M, Quinn AC and Cobbe SM (1994) QT interval dispersion: a non-invasive marker of susceptibility to arrhythmia in patients with sustained ventricular arrhythmias? Br Heart J 71, 511–4.
Ravelli F and Allessie M (1997) Effects of atrial dilatation on refractory period and vulnerability to atrial fibrillation in the isolated Langendorff-perfused rabbit heart. Circulation 96, 1686–95.
Ravelli F, Disertori M, Cozzi F, Antolini R and Allessie MA (1994) Ventricular beats induce variations in cycle length of rapid (type II) atrial flutter in humans. Evidence of leading circle reentry. Circulation 89, 2107–16.

Reiter, M. J., Synhorst, D. P. and Mann, D. E. (1988) Electrophysiological effects of acute ventricular dilatation in the isolated rabbit heart. *Circ Res* 62, 554–62.
Riccio A, Medhurst AD, Mattei C, Kelsell RE, Calver AR, Randall AD, Benham CD and Pangalos MN (2002) mRNA distribution analysis of human TRPC family in CNS and peripheral tissues. Brain Res Mol Brain Res 109, 95–104.
Saint DA (2002) Stretch-activated channels in the heart: their role in arrhythmias and potential as antiarrhythmic drug targets. Drug Dev Res 55, 53–8.
Satoh T and Zipes DP (1996) Unequal atrial stretch in dogs increases dispersion of refractoriness conducive to developing atrial fibrillation. J Cardiovasc Electrophysiol 7, 833–42.
Savelieva I and Camm J (2008) Anti-arrhythmic drug therapy for atrial fibrillation: current anti-arrhythmic drugs, investigational agents, and innovative approaches. Europace 10, 647–65.
Sengupta, P. P., Khandheria, B. K., Korinek, J., Wang, J., Jahangir, A., Seward, J. B. and Belohlavek, M. (2006) Apex-to-base dispersion in regional timing of left ventricular shortening and lengthening. *J Am Coll Cardiol* 47, 163–72.
Seol CA, Kim WT, Ha JM, Choe H, Jang YJ, Youm JB, Earm YE and Leem CH (2008) Stretch-activated currents in cardiomyocytes isolated from rabbit pulmonary veins. Prog Biophys Mol Biol 97, 217–31.
Shah DC, Haissaguerre M, Jais P and Clementy J (2002) High-resolution mapping of tachycardia originating from the superior vena cava: evidence of electrical heterogeneity, slow conduction, and possible circus movement reentry. J Cardiovasc Electrophysiol 13, 388–92.
Shlykov SG, Yang M, Alcorn JL and Sanborn BM (2003) Capacitative cation entry in human myometrial cells and augmentation by hTrpC3 overexpression. Biol Reprod 69, 647–55.
Sideris DA, Toumanidis ST, Thodorakis M, Kostopoulos K, Tselepatiotis E, Langoura C, Stringli T and Moulopoulos SD (1994) Some observations on the mechanism of pressure related atrial fibrillation. Eur Heart J 15, 1585–9.
Solti F, Vecsey T and Kekesi V (1989a) Effect of atrial dilatation on the tendency of atrial arrhythmias. Acta Physiol Hung 74, 49–55.
Solti F, Vecsey T, Kekesi V and Juhasz-Nagy A (1989b) The effect of atrial dilatation on the genesis of atrial arrhythmias. Cardiovasc Res 23, 882–6.
Stacy GP Jr., Jobe RL, Taylor LK and Hansen DE (1992) Stretch-induced depolarizations as a trigger of arrhythmias in isolated canine left ventricles. Am J Physiol 263, H613–21.
Stilli D, Berni R, Bocchi L, Zaniboni M, Cacciani F, Sgoifo A and Musso E (2004) Vulnerability to ventricular arrhythmias [corrected] and heterogeneity of action potential duration in normal rats. Exp Physiol 89, 387–96.
Stones R, Calaghan SC, Billeter R, Harrison SM and White E (2007) Transmural variations in gene expression of stretch-modulated proteins in the rat left ventricle. Pflugers Arch 454, 545–9.
Stones R, Gilbert SH, Benoist D and White E (2008) Inhomogeneity in the response to mechanical stimulation: cardiac muscle function and gene expression. Prog Biophys Mol Biol 97, 268–81.
Sung D, Mills RW, Schettler J, Narayan SM, Omens JH and McCulloch AD (2003) Ventricular filling slows epicardial conduction and increases action potential duration in an optical mapping study of the isolated rabbit heart. J Cardiovasc Electrophysiol 14, 739–49.
Taggart, P. (1996) Mechano-electric feedback in the human heart. *Cardiovasc Res* 32, 38–43.
Taggart P and Lab M (2008) Cardiac mechano-electric feedback and electrical restitution in humans. Prog Biophys Mol Biol 97, 452–60.
Taggart P, Sutton P, Lab M, Runnalls M, O'Brien W and Treasure T (1992) Effect of abrupt changes in ventricular loading on repolarization induced by transient aortic occlusion in humans. Am J Physiol 263, H816–23.
Taggart P and Sutton PM (1999) Cardiac mechano-electric feedback in man: clinical relevance. Prog Biophys Mol Biol 71, 139–54.
Taggart P, Sutton PM, Opthof T, Coronel R, Trimlett R, Pugsley W and Kallis P (2001) Transmural repolarisation in the left ventricle in humans during normoxia and ischaemia. Cardiovasc Res 50, 454–62.

Takagi S, Miyazaki T, Moritani K, Miyoshi S, Furukawa Y, Ito S and Ogawa S (1999) Gadolinium suppresses stretch-induced increases in the differences in epicardial and endocardial monophasic action potential durations and ventricular arrhythmias in dogs. Jpn Circ J 63, 296–302.
Takahashi N, Imataka K, Seki A and Fujii J (1982) Left atrial enlargement in patients with paroxysmal atrial fibrillation. Jpn Heart J 23, 677–83.
Tan JH, Liu W and Saint DA (2002) Trek-like potassium channels in rat cardiac ventricular myocytes are activated by intracellular ATP. J Membr Biol 185, 201–7.
Tan JH, Liu W and Saint DA (2004) Differential expression of the mechanosensitive potassium channel TREK-1 in epicardial and endocardial myocytes in rat ventricle. Exp Physiol 89, 237–42.
Tanabe T, Deguchi Y, Handa S, Takahashi A and Fukushi H (2001) Longer longitudinal atrial dimension in patients with idiopathic paroxysmal atrial fibrillation: a possible cause of atrial fibrillation. Am Heart J 142, 669–78.
Tavi P, Han C and Weckstrom M (1998) Mechanisms of stretch-induced changes in $[Ca^{2+}]i$ in rat atrial myocytes: role of increased troponin C affinity and stretch-activated ion channels. Circ Res 83, 1165–77.
Tendulkar AP and Harken AH (2006) Mechanics of the normal heart. J Card Surg 21, 615–20.
Terrenoire C, Lauritzen I, Lesage F, Romey G and Lazdunski M (2001) A TREK-1-like potassium channel in atrial cells inhibited by beta-adrenergic stimulation and activated by volatile anesthetics. Circ Res 89, 336–42.
Tse HF, Pelosi F, Oral H, Knight BP, Strickberger SA and Morady F (2001) Effects of simultaneous atrioventricular pacing on atrial refractoriness and atrial fibrillation inducibility: role of atrial mechanoelectrical feedback. J Cardiovasc Electrophysiol 12, 43–50.
Ulphani JS, Ng J, Aggarwal R, Cain JH, Gordon D, Yang E, Morris AR, Arora R, Goldberger JJ and Kadish AH (2007) Frequency gradients during two different forms of fibrillation in the canine atria. Heart Rhythm 4, 1315–23.
Valderrabano M (2007) Influence of anisotropic conduction properties in the propagation of the cardiac action potential. Prog Biophys Mol Biol 94, 144–68.
Van Wagoner DR (1993) Mechanosensitive gating of atrial ATP-sensitive potassium channels. Circ Res 72, 973–83.
Vaziri SM, Larson MG, Benjamin EJ and Levy D (1994) Echocardiographic predictors of nonrheumatic atrial fibrillation. The Framingham Heart Study. Circulation 89, 724–30.
Volders PG, Vos MA, Szabo B, Sipido KR, de Groot SH, Gorgels AP, Wellens HJ and Lazzara R (2000) Progress in the understanding of cardiac early afterdepolarizations and torsades de pointes: time to revise current concepts. Cardiovasc Res 46, 376–92.
Wan, X., Bryant, S. M. and Hart, G. (2003) A topographical study of mechanical and electrical properties of single myocytes isolated from normal guinea-pig ventricular muscle. *J Anat* 202, 525–36.
White E (2006) Mechanosensitive channels: therapeutic targets in the myocardium? Curr Pharm Des 12, 3645–63.
White E, Boyett MR and Orchard CH (1995) The effects of mechanical loading and changes of length on single guinea-pig ventricular myocytes. J Physiol 482 (Pt 1), 93–107.
Wiegerinck RF, van Veen TA, Belterman CN, Schumacher CA, Noorman M, de Bakker JM and Coronel R (2008) Transmural dispersion of refractoriness and conduction velocity is associated with heterogeneously reduced connexin43 in a rabbit model of heart failure. Heart Rhythm 5, 1178–85.
Williams, I. A. and Allen, D. G. (2007) Intracellular calcium handling in ventricular myocytes from mdx mice. *Am J Physiol Heart Circ Physiol* 292, H846–55.
Woo SH, Risius T and Morad M (2007) Modulation of local Ca^{2+} release sites by rapid fluid puffing in rat atrial myocytes. Cell Calcium 41, 397–403.
Wright M, Haissaguerre M, Knecht S, Matsuo S, O'Neill MD, Nault I, Lellouche N, Hocini M, Sacher F and Jais P (2008) State of the art: catheter ablation of atrial fibrillation. J Cardiovasc Electrophysiol 19, 583–92.

Xian Tao L, Dyachenko V, Zuzarte M, Putzke C, Preisig-Muller R, Isenberg G and Daut J (2006) The stretch-activated potassium channel TREK-1 in rat cardiac ventricular muscle. Cardiovasc Res 69, 86–97.

Yan GX, Rials SJ, Wu Y, Liu T, Xu X, Marinchak RA and Kowey PR (2001) Ventricular hypertrophy amplifies transmural repolarization dispersion and induces early afterdepolarization. Am J Physiol Heart Circ Physiol 281, H1968–75.

Zabel M, Koller BS, Sachs F and Franz MR (1996a) Stretch-induced voltage changes in the isolated beating heart: importance of the timing of stretch and implications for stretch-activated ion channels. Cardiovasc Res 32, 120–30.

Zabel M, Portnoy S and Franz MR (1996b) Effect of sustained load on dispersion of ventricular repolarization and conduction time in the isolated intact rabbit heart. J Cardiovasc Electrophysiol 7, 9–16.

Zeng T, Bett GC and Sachs F (2000) Stretch-activated whole cell currents in adult rat cardiac myocytes. Am J Physiol Heart Circ Physiol 278, H548–57.

Zhang R, Iwasaki K, Zuckerman JH, Behbehani K, Crandall CG and Levine BD (2002) Mechanism of blood pressure and R-R variability: insights from ganglion blockade in humans. J Physiol 543, 337–48.

Zhao F, Dong L, Cheng L, Zeng Q and Su F (2007) Effects of acute mechanical stretch on the expression of mechanosensitive potassium channel TREK-1 in rat left ventricle. J Huazhong Univ Sci Technolog Med Sci 27, 385–7.

Chapter 12
Mechanical Modulation of a Reentrant Arrhythmia: The Atrial Flutter Case

Flavia Ravelli and Michela Masè

Abstract Atrial flutter (AFL) is a supraventricular arrhythmia, based on a reentrant mechanism, which presents small fluctuations in cycle length. We report on studies in humans and animals which disclosed the nature of these variations and supported their mechanical origin. The sources of the spontaneous variability of atrial flutter cycle length have been identified in ventricular contraction and respiration, which cause phasic variations in atrial interval. The phase-response curves have been shown to be closely related to atrial volume changes during ventricular and respiratory activities and oscillations in cycle length have been reported to be independent of autonomic tone. All this evidence has led to the formulation of the mechano-electrical feedback (MEF) paradigm, which suggests that changes in atrial volume directly affect atrial flutter cycle length variability via direct alteration of the reentrant circuit size and mechano-electrical modulation of conduction velocity. Theoretical predictions of experimental variability patterns by a closed-loop mathematical model of AFL variability, including a MEF branch, provided additional evidence in favour of a mechanically-mediated mechanism at the basis of atrial flutter cycle length variability.

Keywords Atrial flutter · Cycle length variability · Mechano-electrical feedback · Reentry · Mathematical modelling

12.1 The Beat-to-Beat Variability of Atrial Flutter Cycle Length

Atrial flutter (AFL) is a common supraventricular arrhythmia, characterized by a very rapid, highly regular rhythm of the atria (240–350 beats/min) and by the presence of some degree of atrioventricular (AV) block including 2:1, 3:1 or 4:1 ratios and advanced AV block. Studies in animal models (see citations in Waldo (1998))

F. Ravelli (✉)
Laboratory of Biophysics and Biosignals, Department of Physics, Faculty of Science, University of Trento, Trento, Italy
e-mail: ravelli@science.unitn.it

A. Kamkin, I. Kiseleva (eds.), *Mechanosensitivity of the Heart*,
Mechanosensitivity in Cells and Tissues 3, DOI 10.1007/978-90-481-2850-1_12,

and more recent studies in patients, principally using catheter electrode mapping and pacing techniques (Cosio et al., 1990, 1996; Olshansky et al., 1990; Olgin et al., 1995; Kalman et al., 1996; Nakagawa et al., 1996; Waldo, 2000), have helped to clarify the underlying mechanisms of atrial flutter, revealing the anatomical and/or functional barriers which determine the reentrant arrhythmia. Typical atrial flutter has been shown to be caused by a reentrant circuit in the right atrium, in which the impulse travels up the atrial septum, emerges epicardially in the superior right atrium, travels inferiorly down the right atrial free wall, and reenters the right atrial septum, passing through an isthmus bounded by the inferior vena cava, Eustachian ridge, the coronary sinus ostium on one side and the tricuspid valve annulus on the other side (Fig. 12.1, panel a). Differently atypical AFL may originate from the right or left atrium without involving the cavotricuspid isthmus (Waldo, 2000).

Despite the stability of the reentrant circuit underlying atrial flutter, several observations have indicated that atrial flutter is not a strictly regular rhythm (Lewis, 1920; Wells et al., 1979; Lammers et al., 1991; Ravelli et al., 1994). In fact the cycle length of the arrhythmia, set by the revolution time of the reentry, has been documented to be subject to small beat-to-beat variations with standard deviation of about 5 ms (Wells et al., 1979; Lammers et al., 1991; Ravelli et al., 1994) (Fig. 12.1, panel b). The possibility of small variations in atrial flutter rate was first emphasized by Lewis in 1920 who showed that "variations in the length of intra-auricular cycles averaged less than 0.0009–0.0077 of a second" (Lewis, 1920). Lewis also observed that the "maximal variation of the cycles may amount to but usually does not exceed 0.02 second". Wells et al., (1979) made the first attempt to analyze the beat-to-beat cycle length in human atrial flutter. In postoperative type I atrial flutter, the authors measured a mean beat-to-beat variation of 4.3 ms and reported a bimodal distribution of flutter cycle lengths in some patients, which was related to the presence of alternans in beat-to-beat cycle length. However the components of the variability as well as the mechanisms of the variations remained uncleared for a long time.

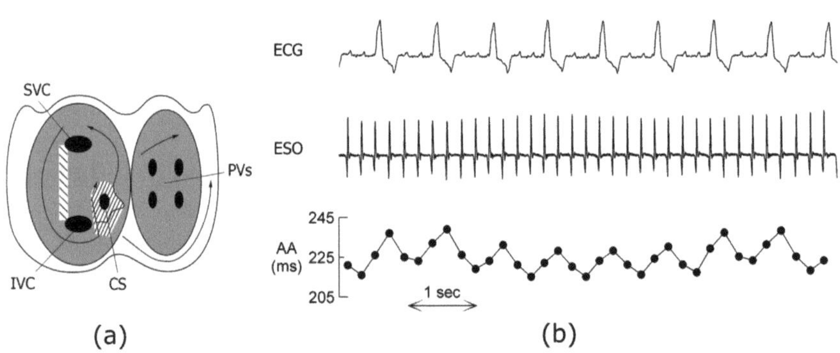

Fig. 12.1 Beat-to-beat variability of atrial flutter cycle length. (**a**) Schematic representation of the reentrant circuit responsible for typical atrial flutter. (**b**) Simultaneous recording of electrocardiogram (ECG) and atrial electrogram (ESO) during atrial flutter and corresponding atrial flutter interval variability series (AA). AA intervals represent time intervals between consecutive activation times in the atrial electrogram

In this review the findings supporting the mechanical origin of AFL cycle length variability will be summarized and the mechano-electrical (MEF) hypothesis for atrial flutter cycle length variability will be discussed. After introducing the first experimental evidence testifying the existence of a direct relation between atrial flutter rate and atrial volume (Section 12.2), we will go over the steps which have led to the formulation of the MEF paradigm (Section 12.3). Starting from the recognition of ventricular and respiratory activities as the main determinants of the spontaneous variability of AFL cycle length, the mechanical origin of the variability will be disclosed first through the experimental exclusion of concurrent hypothesis, specifically an autonomic mediation, and then by the observation of the strict relation between cycle length variability and atrial volume changes related to ventricular and respiratory activities. Finally in Section 12.4 experimental data will be complemented with a recent modelling study of atrial flutter cycle length variability, which has provided a mathematical ordering of the wide spectrum of cycle length variability patterns experimentally observed.

12.2 The Relation Between Atrial Size and Atrial Flutter Rate

The existence of a relationship between atrial size and atrial flutter rate was outlined by several studies many decades ago. In dogs, the rate of experimentally induced atrial flutter varied with atrial size and distension. Acute or chronic atrial enlargement slowed the flutter rate (Hayden et al., 1967; Boyden and Hoffman, 1981). Similarly, atrial flutter rate was lower in humans with markedly enlarged atria (Rytand et al., 1958).

Consistent experimental evidence concerning the role of atrial volume in the determination of atrial flutter rate was provided by the studies of Waxman et al., (1991, 1992). The first study analyzed the effects provoked on atrial flutter cycle length by manoeuvres which predictably affected venous return and thus cardiac volume. Specifically the mean atrial flutter cycle length was monitored in atrial flutter patients at rest, during passive upright tilting, Valsalva manoeuvre and along respiratory cycle. Since these interventions produced, together with hemodynamical changes, variations in autonomic tone, which could indirectly affect atrial flutter rate, measurements were repeated under autonomic blockade by atropine and propranolol. Independently of autonomic tone, the reduction in cardiac size produced by passive upright tilting, the strain phase of Valsalva manoeuvre and expiration significantly decreased atrial flutter cycle length, suggestive of a direct influence of cardiac volume on atrial flutter properties (Waxman et al., 1991). Corroborating results were obtained in the subsequent study (Waxman et al., 1992), which in turn examined the effects of 1:1 atrioventricular (AV) conduction on atrial flutter rate and atrial pressure in patients and dogs. The condition of 1:1 AV conduction was accomplished in patients by exercise, upright tilting, isoproterenol infusion or simulated by ventricular pacing. It determined a significant increase in atrial flutter cycle length, which returned to control values when 1:1 AV conduction ceased. The increase in atrial flutter cycle length was associated with the immediate rise in atrial pressure produced by 1:1 AV conduction. To disclose the mechanism of the phenomenon similar studies were carried out in dogs with experimentally induced atrial flutter. In addition to the condition of 1:1 AV conduction, simulated by ventricular pacing, the effects of vena cava occlusion were evaluated. Experiments were repeated after vagatomy and isoproterenol administration to evaluate the role of autonomic tone on cycle length variations. Similarly to the results obtained in patients, the development of 1:1 AV conduction was associated with a rise in atrial filling pressure and it consistently lengthened the atrial flutter cycle length during control conditions (Fig. 12.2) as well as following vagatomy and during the administration of isoproterenol. Inferior vena cava occlusion was instead associated with a fall in atrial filling pressure and significantly shortened the average atrial flutter cycle length, independently of vagus nerve activity or adrenergic agonist administration (Waxman et al., 1992). These results implied that rise and fall in atrial filling pressure directly affected the characteristics of atrial flutter circuit, and thus could modulate the atrial flutter rate.

The interdependence between electrophysiological and hemodynamical conditions in atrial flutter was remarked by the study of Vulliemin et al., (1994). To test the hypothesis that atrial cycle length was mainly determined by atrial dimensions, echocardiographic measurements of atrial size were performed in patients during atrial flutter, while inducing atrial volume and pressure changes by postural and pharmacologic means. An excellent correlation between changes in atrial size and induced changes in AFL cycle length was observed comparing data at rest and during head-up tilt manoeuvre. This directly demonstrated the existence of a strong

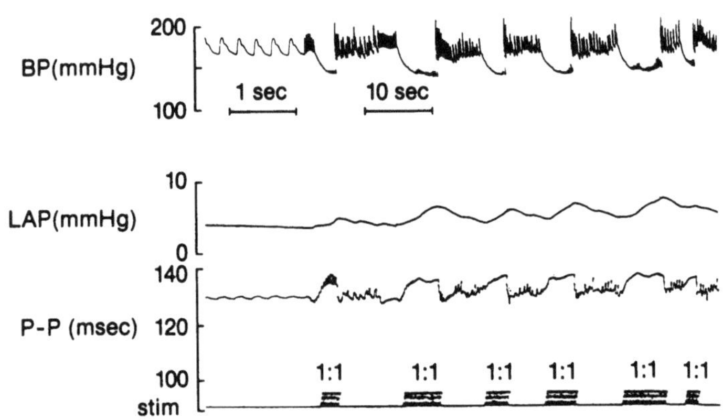

Fig. 12.2 Effects of 1:1 AV conduction on flutter rate and right atrial pressure in dogs. Simultaneous recording of blood pressure (BP), mean left atrial pressure (LAP), flutter cycle interval (P-P), and stimulus code marker (stim). The dog is in atrial flutter with a 2:1 ventricular response. In the segments indicated by symbol 1:1, ventricular pacing was carried out to simulate 1:1 AV conduction (VAT pacing). This immediately caused the mean LAP to rise and the flutter cycle interval to prolong by approximately 8 ms. At the same time, the systolic arterial blood pressure fell by 100 mmHg. This was observed during each of the six consecutive trials of 1:1 AV conduction. Modified from Waxman et al. (1992)

relation between atrial size and atrial cycle length in typical atrial flutter, underlining the importance of the right heart preload and atrial size in the determination of the electrophysiological characteristics of the arrhythmia.

12.3 The Mechanical Origin of Atrial Flutter Cycle Length Variability

Similarly to the documented variations in mean atrial flutter rate induced by atrial volume changes, experimental evidence has demonstrated that also the spontaneous variability of atrial flutter cycle length is correlated with atrial volume variations related to ventricular and respiratory cycles.

12.3.1 Ventricular Contraction and Respiration: The Sources of AFL Cycle Length Variability

The small spontaneous beat-to-beat variability of atrial flutter cycle length is not random, but periodic patterns have been observed. The nature of these fluctuactions has been investigated both in animals and humans (Lammers et al., 1991; Waxman et al., 1991; Ravelli et al., 1994; Yamashita et al., 1994; Stambler and Ellenbogen, 1996). Lammers et al., (1991) and Ravelli et al., (1994) showed the existence of fluctuations in atrial flutter interval related to ventricular activity. Waxman et al., (1991) observed variations in atrial flutter rate caused by respiration. In a recent study (Ravelli et al., 2008) all the oscillatory components of atrial flutter variability pattern were characterized in 30 patients with typical atrial flutter by application of spectral analysis. The study demonstrated that the spontaneous variability of atrial flutter cycle length was composed of a prevalent oscillation at the frequency of ventricular contraction (1.7 Hz), which constituted 54% of the total spectral power, and a second oscillation at the frequency of respiration (0.32 Hz), which represented 22% of the power (Fig. 12.3). Moreover the correlation of atrial flutter variability with ventricular and respiratory activities was dynamically evidenced in the study by performing driven manoeuvres. Ventricular pacing in patients with implanted pacemaker showed that changes in the ventricular frequency shifted the high-frequency oscillation peak in atrial cycle lengths. Similarly in patients under controlled respiration changes in the respiratory frequency shifted the low-frequency cycle length oscillation peak.

The relationship between the variations in AFL cycle length and the two perturbation sources was investigated both in humans and animals by phase-response curves (Lammers et al., 1991; Waxman et al., 1991; Yamashita et al., 1994). The relationship between AFL cycle length and ventricular activity in human atrial flutter was investigated by Lammers et al., (1991) and Ravelli et al., (1994) by constructing a phase-plot in which the flutter cycle length was displayed against the time after the previous QRS complex. This showed that atrial interval fluctuations were strictly coupled to the moment of ventricular activation. After the onset of the

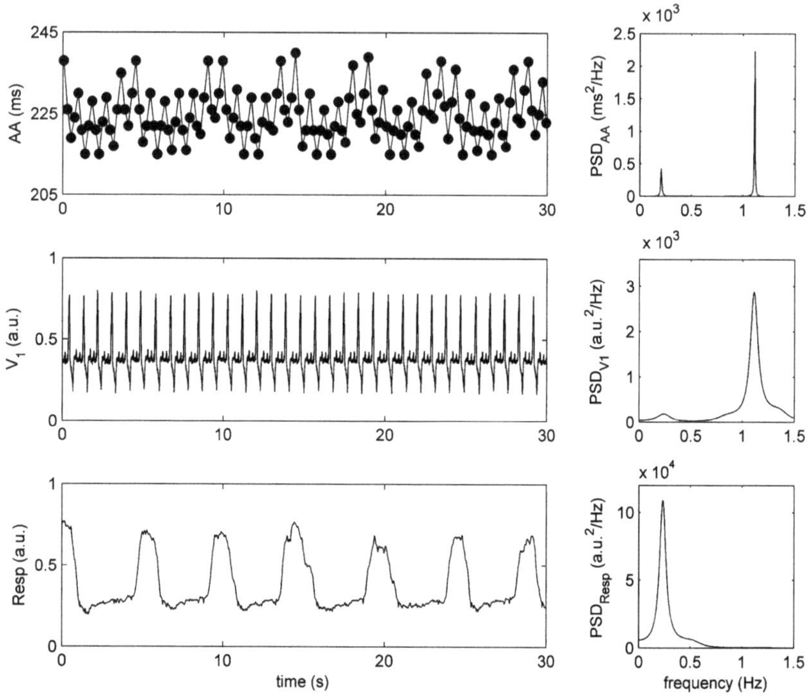

Fig. 12.3 Sources of atrial flutter cycle length variability. *Left panels*. Atrial cycle length series (*top*) and corresponding ECG, lead V_1 (*middle*), and respiratory signal (*bottom*) in a representative patient. *Right panels*. Power spectral density of atrial cycle length series (*top*), ECG (*middle*) and respiratory (*bottom*) signals. Note that the atrial cycle length spectrum shows two major spectral components in correspondence with the respiratory (0.21 Hz) and ECG (1.11 Hz) peaks, respectively. From Ravelli et al. (2008) with permission from Elsevier

QRS complex, the atrial flutter interval in typical atrial flutter gradually increased by an average of 1.8% and reached a maximum value after about 450 ms. Thereafter, the intervals decreased until the next ventricular beat occurred (Fig. 12.4). A similar phase response curve was found in a dog model of atrial flutter (Yamashita et al., 1994). The critical role of ventricular contraction in the modulation of atrial flutter interval variability was also evidenced by applying carotid sinus massage, a manoeuvre which temporally prevented ventricular activation. The ventricular asystole caused by the massage markedly reduced or even abolished the high-frequency variations in atrial flutter rate (Lammers et al., 1991).

The dependence of AFL cycle length changes on respiratory phase was investigated by Waxman et al., (1991) and Ravelli et al., (2008). Both studies evidenced the existence of a paradoxical modulation pattern, with longer cycle lengths during inspiration and shorter during expiration (Fig. 12.5). This is opposite to the effects of respiration during sinus rhythm (respiratory sinus arrhythmia), where heart rate increases during inspiration and decreases during expiration (Eckberg, 1983; Eckberg, 2003; Yasuma and Hayano, 2004).

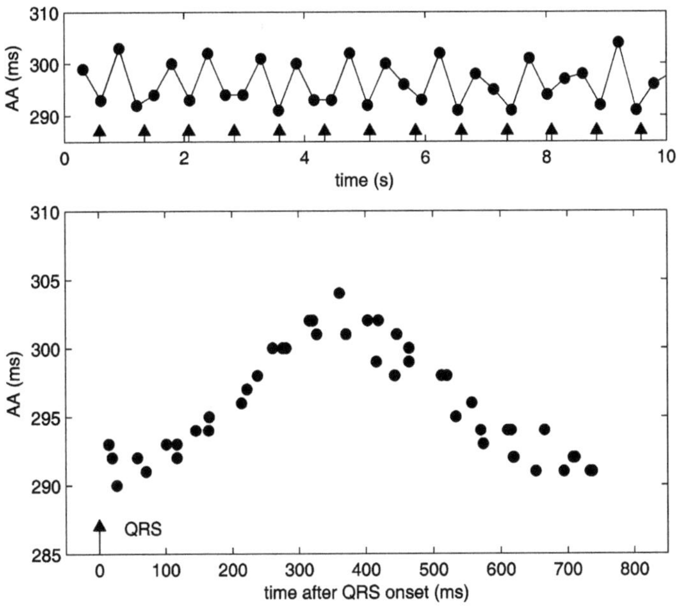

Fig. 12.4 Phasic variations of atrial flutter cycle length in relation to the onset of the QRS complex in a patient with typical AFL during ventricular pacing. In the *upper panel*, part of the sequential flutter intervals are plotted. Ventricular beats are marked by *vertical arrows*. In the *lower panel* the flutter cycle length are time aligned to the onset of QRS complexes. The phase-response curve shows a prolongation of AFL interval after the ventricular beat followed by a shortening

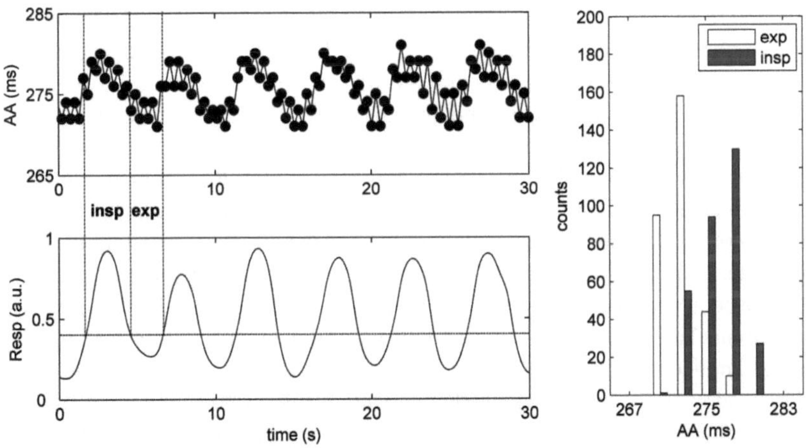

Fig. 12.5 Dependence of atrial flutter cycle length on respiratory phase. *Left panels*. Atrial cycle length series (*top*) and corresponding respiratory signal (*bottom*) in a representative patient. Inspiration (insp) and expiration (exp) phases are indicated and divided by the horizontal line in the respiratory signal. *Right panel*. Distributions of atrial cycle lengths occurring during expiration (*white bars*) and inspiration (*gray bars*). Note the paradoxical lengthening of atrial cycle lengths during inspiration with respect to expiration. From Ravelli et al. (2008) with permission from Elsevier

12.3.2 The Response to Pharmacological Denervation

The studies presented in the previous section identified in ventricular contraction and respiration the sources of AFL cycle length variations and characterized the features of the modulation. The following investigation pointed towards the disclosure of the mechanisms generating the modulation and in particular the evaluation of an autonomic mediation in the phenomenon.

Autonomic nervous reflexes caused by hemodynamic variations associated to ventricular and respiratory activities may hypothetically cause the changes in AFL cycle length. The rise in arterial pressure elicited by ventricular systole and the changes in arterial pressure due to the inspiratory increase in venous return to the heart result in stimulation of arterial baroreceptors, which in turn causes reflex responses (Triedman and Saul, 1994). Additional reflexes may originate from stimulation of atrial mechanoreceptors due to changes in atrial volume induced by both ventricular contraction and respiration (Paintal, 1973; Baertschi and Gann, 1977) and from stimulation of lung receptors by respiration (Taha et al., 1995). Changes in autonomic nervous activity are known to influence the electrophysiological properties of the atrial muscle, thus hypothetically affecting atrial flutter rate. Parasympatetic activity has been shown to shorten the atrial refractory period (Rensma et al., 1988; Liu and Nattel, 1997), while increased sympathomimetic activity has less clear-cut effects on atrial electrophysiology with both shortened (Farges et al., 1977; Liu and Nattel, 1997) and unchanged (Vargas et al., 1975; Rensma et al., 1988) refractoriness reported. Differently, conduction velocity has been reported to be either not changed (Rensma et al., 1988) or slightly increased (Liu and Nattel, 1997) by both sympathetic and parasympathetic activities.

Although neurally-mediated alterations in atrial electrophysiological properties may hypothetically cause the changes in AFL cycle length, sound evidence has been reported which excludes a role of neural mechanisms in the generation of AFL interval variability. Studies in humans and animals showed the persistence of AFL cycle length changes due to ventricular and respiratory activities after pharmacological denervation (Waxman et al., 1991; Yamashita et al., 1994; Ravelli et al., 2008) and in denervated heart (Stambler and Ellenbogen, 1996). Waxman et al., (1991) showed that autonomic blockade did not alter significantly the effects of respiration on atrial flutter rate. Phasic variations in AFL cycle length associated to ventricular contraction have been shown to persist in a dog model of atrial flutter after pharmacological denervation (Yamashita, 1994). In a recent study (Ravelli et al., 2008) spectral analysis was applied to atrial flutter interval series before and after combined muscarinic and beta-adrenergic receptor blockade to quantitatively evaluate the involvement of the autonomic nervous system in both ventricular and respiratory oscillations. The results of the analysis showed the variability to persist after autonomic blockade, since no significant decrease in atrial cycle length spectral powers and standard deviations was observed after administration of blocking agents (Fig. 12.6).

Other observations suggest that a neural mechanism is unlikely in the generation of AFL interval variability. First autonomic modulation on typical atrial flutter should be small or ineffective, since the modulation does not act on a pacemaker

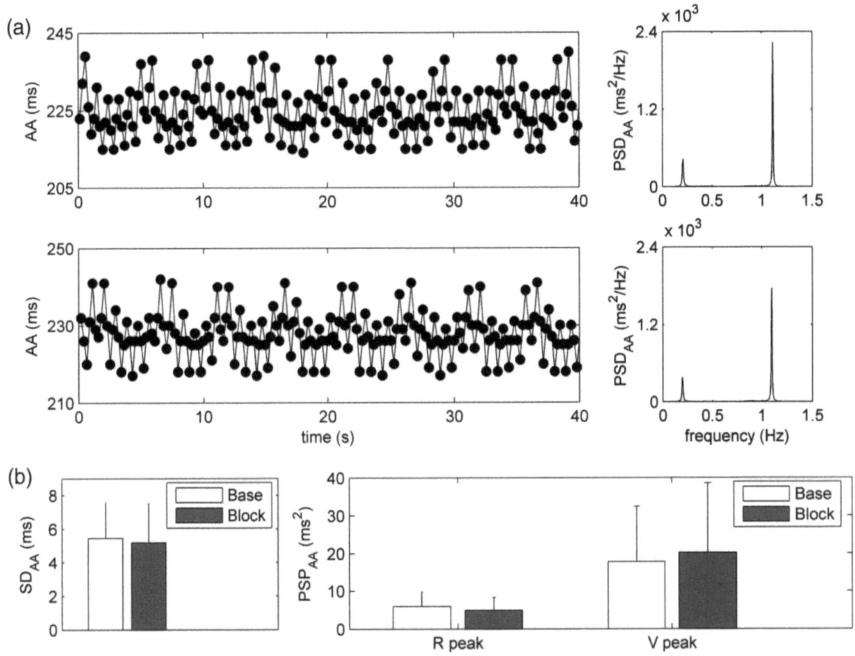

Fig. 12.6 Comparison of atrial flutter cycle length variability in basal condition and after autonomic blockade in a representative patient (**a**) and in the overall study group (b). (**a**) Atrial cycle length series (*left panels*) and corresponding power spectral densities (*right panels*) in basal condition (*upper panels*) and after autonomic blockade (*lower panels*) show that atrial variability patterns persist after autonomic blockade and the two main oscillation peaks are preserved. (**b**) The atrial cycle length standard deviation (SD) (*left panel*) and spectral powers (PSP) of the ventricular (V peak) and respiratory peaks (R peak) (*right panel*) remain almost unaffected after autonomic blockade (*gray bars*) with respect to basal condition (*white bars*). From Ravelli et al. (2008) with permission from Elsevier

activity, as during normal sinus rhythm, but on a reentrant mechanism mainly determined by conduction velocity, on which autonomic effects are limited. Second, the latency between the onset of ventricular contraction and the initial lengthening of the flutter cycle, estimated in about 50 ms (Lammers et al., 1991), is too short to involve nervous reflex mechanisms. Finally the presence of both respiratory and ventricular oscillations in AFL cycle length in heart transplant recipients (Stambler and Ellenbogen, 1996) also suggests the two oscillations to be independent of autonomic tone.

12.3.3 Correlation with Pressure/Volume Changes

The studies reported above point towards the existence of non-neural mechanisms which contribute to the origin of AFL cycle length variability. Evidence has been reported which supports a mechano-sensitive mechanism, intrinsic to the heart itself, which affects flutter rate in response to changes in atrial volume.

The phase-response curve of AFL cycle length variation has been shown to be closely related to changes in atrial pressure and/or volume following ventricular systole (Lammers et al., 1991; Ravelli et al., 1994; Yamashita et al., 1994). During one ventricular cycle atrial filling and atrial emptying phases alternate in concomitance with closure/opening of the AV valves, which causes consequent variations in atrial dimensions. With the onset of ventricular systole and the closure of the AV valves, the blood flows into the atria, producing an increase in volume accompanied by a continuous rise in pressure (v wave), while with the opening of the AV valves the atrial size decreases through atrial emptying (Jarvinen et al., 1996; Gaynor et al., 2005). In a study led in patients with typical atrial flutter the time course of AFL interval variations following the onset of QRS has been shown to be parallel to changes in atrial volume as determined from literature (Lammers et al., 1991). In fact, similarly to the AFL phase-response curve, described in Section 12.3.1, Matsuda et al. (1983) reported a gradual increase in atrial volume starting about 20–50 ms after the onset of the QRS complex, the maximal dimension of the atrial cavity being reached after 450–500 ms. The demonstration of a synchronism between the phasic variations in AFL cycle length and the hemodynamical changes in the atria was given in a subsequent study in patients (Ravelli et al., 1994) in which the intra-atrial pressure was monitored together with the atrial electrical activity. As shown in Fig. 12.7 the phase-response curve of AFL cycle length variations

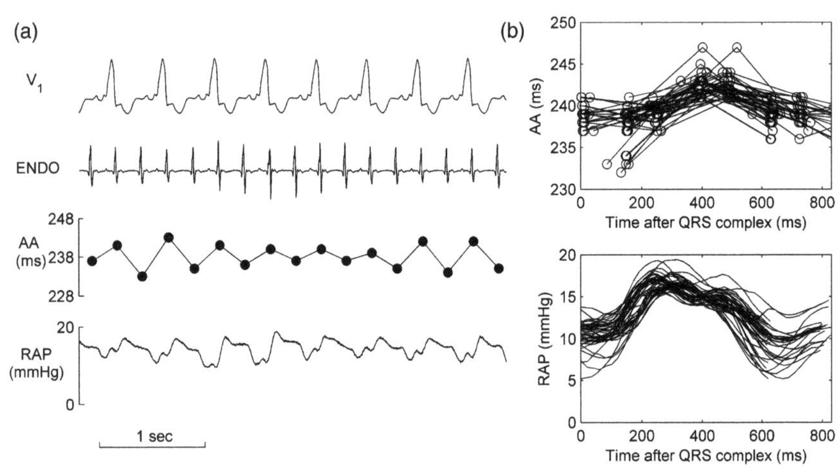

Fig. 12.7 Beat-to-beat fluctuations in atrial flutter cycle length and correlation with atrial pressure changes during ventricular cycle. (**a**) ECG (V_1), atrial electrogram (ENDO), atrial cycle length series (AA) and right atrial pressure (RAP) signal during atrial flutter in a patient with 2:1 AV conduction. (**b**) Comparison between atrial flutter cycle length variations and changes in atrial pressure after the onset of QRS in a patient with variable AV conduction. Note the synchronism between atrial flutter cycle length variations and atrial pressure changes. Modified from Ravelli et al. (1994)

was synchronous to atrial pressure changes associated with the onset of ventricular contraction. A correlation between phasic variations in AFL cycle length and atrial pressure changes was also found in a dog study, in which the phase-response curve of AFL cycle length variations was compared to both the intra-atrial pressure and its time derivative (Yamashita et al., 1994). The phase-response curve was shown to be more closely related to the curve of the velocity of atrial pressure changes, with no time lags, rather than with the curve of the atrial pressure itself.

Although they have not been directly correlated with actual measurements, also the AFL cycle length variations due to respiration can be associated with atrial volume changes. In fact inspiration increases both the venous return to the heart and the afterload of the ventricles by reducing the ventricular diastolic pressure relative to the extrathoracic arterial pressures, and expiration has the opposite effect (Robotham et al., 1978). These cyclic variations in venous return produce an increase in right atrial volume during inspiration and a decrease during expiration (Ferguson et al., 1989). The similar time course between the paradoxical increase of AFL interval during inspiration and decrease during expiration (see Fig. 12.5) and the respiratory-related atrial volume changes, as reported by Ferguson et al., (1989), suggests again a causal relationship between respiratory variability of AFL cycle length and volume variations.

12.3.4 The MEF Paradigm for Atrial Flutter Cycle Length Variability

The persistence of atrial flutter variability after pharamacological denervation and the striking correlation between changes in AFL cycle length and variations in atrial volume, presented in previous sections, suggest the direct influence of volume variations on atrial flutter reentry properties as the most likely mechanism underlying atrial flutter cycle length variability.

The specific mechanisms by which stretch may directly modulate atrial flutter reentrant circuit, inducing the described changes in atrial cycle length, are still hypothetical. Nevertheless a MEF paradigm for atrial flutter cycle length variability has been formulated (Ravelli et al., 2008), which aims to provide a reference framework for atrial flutter cycle length variability, relating the hemodynamic, electrophysiologic and microscopic perspectives involved in the issue of atrial flutter mechanical modulation. The structure of the MEF paradigm is schematized in Fig. 12.8. Ventricular contraction and respiration are assumed as the main sources of AFL cycle length variability and their modulation is exerted via hemodynamical changes, which affect atrial volume. In fact ventricular systole and inspiration involve an increase in atrial volume and stretch with corresponding increase of atrial flutter cycle length, while ventricular diastole and expiration have the opposite effect. Changes in the level of atrial stretch could modulate atrial flutter rate through modulation of different anatomical and/or electrophysiological factors, depending on the mechanism underlying atrial flutter. In the case of typical atrial flutter, the atrial cycle length is determined by the revolution time of a macroreentry with large excitable gap (Waldo,

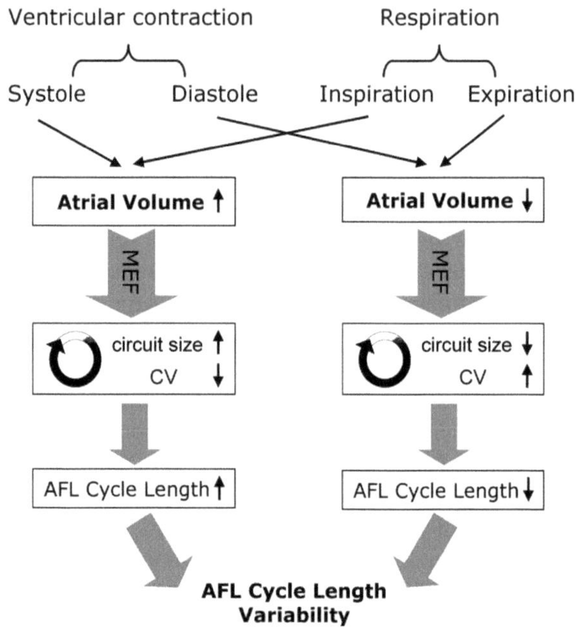

Fig. 12.8 The MEF paradigm for AFL cycle length variability. Schematic diagram showing the potential contribution of MEF to cycle length variability of typical atrial flutter. CV: conduction velocity. From Ravelli et al. (2008) with permission from Elsevier

2000) and thus is governed directly by the length of the circuit and inversely by conduction velocity (Allessie et al., 1987). Changes in refractory period, which determines the revolution time of functional reentries, instead, are not expected to produce rate variations in presence of a large excitable gap (Allessie et al., 1987). Thus, as indicated in Fig. 12.8, an increase/decrease in atrial volume by ventricular contraction and respiration is supposed to modulate atrial flutter cycle length through both a lengthening/shortening of the circuit size and a decrease/increase of conduction velocity.

Atrial volume and stretch changes could consistently produce changes in atrial flutter rate by an alteration of the geometrical properties of the reentrant circuit. In fact several observations, both in animal and human studies (Lammers et al., 1991; Waxman et al., 1991, 1992; Vulliemin et al., 1994) point to a relation between atrial size and cycle length. In particular, the study of Vulliemin et al., (1994) actually showed that atrial dimension variations induced by aimed manoeuvres had a direct influence on atrial flutter cycle length. In fact the concomitant measurement of right atrial cross-section by echocardiographic technique and of atrial flutter cycle length disclosed a strict correlation between the two variables suggesting an effect through lengthening of the circuit reentry pathway. Analogously the spontaneous variability of atrial flutter cycle length could be consistently explained by a geometrical deformation of the reentrant circuit associated with changes in atrial volume. Concerning the effects of ventricular contraction on atrial size, Jarvinen et al., (1996) found that during a cardiac cycle, the atrial volume varied by around 70%. If the atrial shape is assumed to be round, this increase should correspond to a lengthening of the circuit of 19% in a ventricular cycle, which may consistently produce the increase in cycle length associated to ventricular contraction. Similarly, changes in atrial volume observed during the respiratory phases (Ferguson et al., 1989) could involve sensitive deformations of the reentrant circuit and may account for the respiratory component of AFL cycle length variability.

Stretch may alter the atrial flutter rate also by changing the electrophysiological variables of the reentrant circuit. Experimental studies have demonstrated that stretch may modulate cardiac electrophysiological properties, such as action potential shape and duration (Lab, 1980; White et al., 1993; Riemer and Tung, 2003), tissue excitability (Tung and Zou, 1995; Riemer et al., 1998), passive and geometrical membrane properties (Deck, 1964; Dominguez and Fozzard, 1979), gap junction expression (Zhuang et al., 2000), and may provoke occurrence of ectopic activity (Franz, 1996). The modulation has been shown to operate on a beat-to-beat basis (Kaufmann et al., 1971; Lab, 1980, 1982) and to occur rapidly, involving a time lag of just 10–20 ms (Kaufmann et al., 1971). Stretch-activated ion channels (Sachs, 1991; Hu and Sachs, 1997) and modulation of intracellular calcium concentration by stretch-enhanced myofilament Ca^{2+} sensitivity or by stretch-activated calcium influx (Calaghan and White, 1999) have been suggested as the most likely candidates in the process of transduction of membrane tension in electrophysiological changes.

Focusing on atrial tissue, experimental and clinical studies have shown that changes in mechanical loading conditions may affect macroscopic electrophysiological properties of the atria (Nazir and Lab, 1996a; Franz and Bode, 2003; Ravelli, 2003), i.e. refractoriness (Kaseda and Zipes, 1988; Klein et al., 1990; Calkins et al., 1992; Ravelli and Allessie, 1997; Bode et al., 2000; Tse et al., 2001; Zarse et al., 2001; Ninio et al., 2005) and conduction velocity (Solti et al., 1989; Chorro et al., 1998; Eijsbouts et al., 2003). Consistent results regarding refractoriness were obtained in isolated preparations, which showed that atrial refractory period (Ravelli and Allessie, 1997; Bode et al., 2000; Zarse et al., 2001; Ninio et al., 2005) and action potential duration at early levels of repolarization (Nazir and Lab, 1996b; Tavi et al., 1998; Kamkin et al., 2000) were shortened by acute atrial dilatation. Although different effects of strain on conduction velocity have been observed experimentally and computationally for various species and tissue types, which include biphasic (Penefsky and Hoffman, 1963; Rosen et al., 1981; Sachse et al., 2004), constant (Spear and More, 1972; Zhu et al., 1997), increasing (Deck, 1964; Dominguez and Fozzard, 1979; Tavi et al., 1996; Reiter et al., 1997) and decreasing (Spear and More, 1972; Solti et al., 1989; Chorro et al., 1998; Eijsbouts et al., 2003; Sung et al., 2003) strain-velocity relationships, studies on atrial conduction in the isolated rabbit heart indicated a decreased conduction velocity in presence of atrial stretch. Specifically, high-density mapping studies by Chorro et al. (1998) showed a global decrease of conduction velocity of about 25% in the right atrium by balloon inflation. The effects of acute right atrial dilatation on conduction velocity were analyzed also

Eijsbouts et al. (2003), who showed that dilatation not only depressed atrial conduction, but also promoted spatial heterogeneity in conduction by causing conduction blocks parallel to the boundaries of large trabeculae. Consistently with experimental data, simulations by Kuijpers et al. (2007) using a cellular bidomain model, including stretch-activated channels and stretch influence on fiber conductivity, showed a stretch-dependent conduction slowing in homogenous and inhomogeneous atrial fibers. The study underlined the possible involvement of SACs in stretch-induced conduction disturbances, suggesting that a depolarized resting membrane potential produced by SAC currents could slow conduction through inactivation of Na^+ channels and lowering of the maximum upstroke velocity. Finally and interestingly for the atrial flutter case, the effect of atrial stretch by ventricular contraction on the intra-atrial conduction time were investigated by Yamashita et al. (1994) in a dog model during atrial pacing. The authors observed a consistent increase of the right atrial activation time following ventricular contraction, when the atrium was paced at a cycle length similar to atrial flutter. These variations were shown to persist after autonomic blockade, suggestive of an intrinsic mechanical mechanism. Thus both experimental and computational studies support the hypothesis that an increase in atrial flutter cycle length during ventricular contraction and inspiration could partially be explained by a stretch-induced decrease in conduction velocity.

The MEF paradigm offers an explanation, supported by sound experimental evidence, for the multilevel problem of atrial flutter cycle length variability. In the framework of mechano-electrical feedback it bridges the gap between the macroscopic changes observed in atrial flutter cycle length and the hemodynamical changes induced by ventricular contraction and respiration. In fact changes in atrial volume are supposed to modulate atrial flutter rate via direct alteration of circuit size and mechano-electrical modulation of conduction velocity. It also suggested a possible involvement of SACs as the microscopic determinants of the phenomenon.

12.4 A Mathematical Model for MEF Effects on Atrial Flutter Cycle Length Variability

The experimental studies on atrial flutter cycle length variability presented in previous sections have been recently complemented by a simulation study which has provided additional evidence to the MEF hypothesis and revealed an ordered structure in the patterns of atrial flutter variability.

Mathematical modelling is a useful technique for investigating the electrical activity of the heart, and several mechanisms of cardiac arrhythmias have been proposed on the basis of modelling studies (Moe et al., 1964; Panfilov and Pertsov, 2001). Modelling has also been used to determine the spatiotemporal properties of re-entrant arrhythmias by characterizing spiral wave dynamics emerging at changing models parameters (Zykov, 1987; Winfree, 1991, 1994; Pertsov et al., 1993 see references in Christini and Glass (2002)).

Since the simple schematization of the heart in terms of nonlinear relaxation oscillators by van der Pol and van der Mark in the 1920s (1928), the detail, complexity and predictive capability of mathematical models have sensitively evolved, leading to the modern biophysical models of cardiac tissue (Luo and Rudy, 1994; Courtemanche et al., 1998). Present ionic models of cardiac electrical activity are capable of reproducing a variety of experimental observations, and the biophysics of these models has been based on data from sound electrophysiological experiments.

More recently, efforts have been made to include cardiomechanics and mechano-electrical aspects in the modelling framework (Cherubini et al., 2008; Hunter et al., 1998; Kuijpers et al., 2007; Nash and Panfilov, 2004). In particular, concerning the second aspect, highly detailed ionic models of cardiac cells including stretch activated channels have been proposed (Sachs, 1994; Kohl et al., 1998; Rice et al., 1998; Riemer et al., 1998; Healy and McCulloch, 2005; Kuijpers et al., 2007), which incorporate linear, time independent, mechano-sensitive currents into single cell models, as well as into one- and two-dimensional cardiac network models. These models quantitatively reproduce the effects of maintained mechanical stretch on action potential characteristics such as amplitude, duration, maximum diastolic potential, peak upstroke velocity and conduction velocity, consistently with experimental measurements.

The great detail of ionic models presents the drawback of a high complexity and computational cost, which generally hinders the actual reproduction of clinical data. Therefore, together with the development of detailed ionic models, phenomenological models of cardiac arrhythmias have been developed (Glass et al., 1987, 1991; Shrier et al., 1987). These models make no assumptions on the specific microscopic events generating cardiac rhythms, schematizing the physiological key features of arrhythmias in terms of simple rules. Although mechanistic aspects can not be reliably tested by these models, they offer a means to identify and predict key structures of the observed clinical data.

Following a phenomenological approach, Masè et al., (2008) developed the first model of atrial and ventricular interval variability during atrial flutter. Based on the experimental evidence, presented in the previous sections, that oscillations in atrial cycle length during atrial flutter are cyclically related to ventricular contraction and respiration, the authors schematized the changes in mechanical environment correlated to ventricular and respiratory activities in terms of phase-dependent variability curves (see Section 12.4.1). This allowed the prediction of atrial and ventricular interval variability in atrial flutter patients during basal conditions and the determination of the dynamical structure of atrial variability in patients during ventricular pacing at changing frequency (see Section 12.4.2).

12.4.1 Model Formulation

The model was conceived to predict the variability of both atrial and ventricular interval series during atrial flutter in a closed-loop. Thus, as depicted in the schematic drawing in Fig. 12.9 (panel a), the model was composed of a MEF part consisting of a ventricular and a respiratory branch, which reproduced the effects

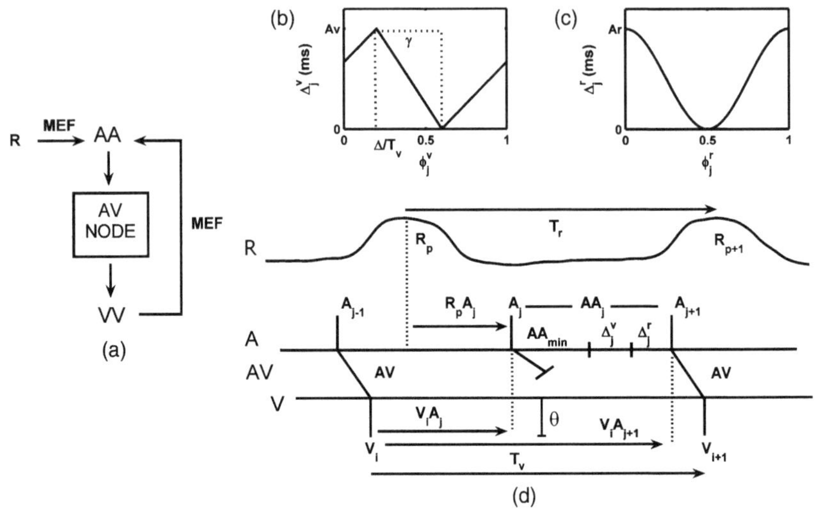

Fig. 12.9 Mathematical model of MEF effects on AFL cycle length variability (**a**) Electrophysiological principles of the model. Atrial activation determines ventricular activation after being filtered by the atrioventricular (AV) node. In turn ventricular contraction and respiration modulate atrial activity through mechano-electrical feedback (MEF). (**b**) and (**c**) Functional dependence of ventricular $\Delta^{v}{}_{j}$ and respiratory $\Delta^{r}{}_{j}$ modulations on the ventricular $\phi^{v}{}_{j} = V_iA_j/T_v$ and respiratory $\phi^{r}{}_{j} = R_pA_j/T_r$ phases of the atrial intervals in the ventricular and respiratory cycles. (**d**) Schematic representation of the mathematical model. An atrial input impinging the AV node at time A_{j-1} is conducted to the ventricle with a nodal conduction time *AV*, which leads to a ventricular beat at time V_i. In turn the ventricular beat V_i and the respiratory event R_p produce the phase-dependent modulations $\Delta^{v}{}_{j}$ and $\Delta^{r}{}_{j}$ on the duration of the following *AA* intervals, defining the occurrence of subsequent atrial beats. These are conducted to the ventricles if their recovery time V_iA_j is longer than the nodal refractory period θ and blocked otherwise. From Masè et al. (2008) with permission from Springer Science+Business Media

of ventricular and respiratory activities on atrial interval variability, and of an atrioventricular (AV) part, which mimicked the transmission of atrial beats through the AV node.

Concerning the MEF part of the model, the authors schematized the effects of ventricular and respiratory activities on atrial cycle length in terms of phase-dependent variability curves. Specifically, to calculate successive atrial activation times, atrial intervals were defined as (see Fig. 12.9, panel d):

$$AA_j = AA_{\min} + \Delta_j^{v} + \Delta_j^{r}$$

where $AA_{\min}$ was a constant minimal interval and Δ_j^{v} and Δ_j^{r} were the ventricular and respiratory modulation curves, respectively. The curves were assumed to depend solely on the phase of atrial intervals in ventricular and respiratory cycles. Specifically, to mimic the phase-dependences observed in patients (see Section 12.3.1), the ventricular modulation Δ_j^{v} was schematized by a piecewise linear function of the ventricular phase ϕ_j^{v} (see Fig. 12.9, panel b), while the respiratory modulation Δ_j^{r}

was expressed as an harmonic function of the respiratory phase ϕ_j^r (see Fig. 12.9, panel c).

The AV part of the model allowed the prediction of ventricular activation times by simulating the refractory and conductive properties of the AV node (Shrier et al., 1987). Specifically the authors assumed that atrial beats were conducted to the ventricle after a fixed conduction delay *AV*, provided that their recovery time V_iA_j from the previous ventricular beat (see Fig. 12.9, panel d) was longer than a constant refractory period, otherwise atrial beats were blocked.

Thus, as schematized in Fig. 12.9 (panel d), successive atrial and ventricular interval series could be predicted by iteration of a closed-loop procedure, given a set of model parameters and a sequence of respiratory events. The MEF part of the model determined successive atrial activation times by adding to the initial atrial event modulated AA intervals. Predicted atrial activations were in turn used as input of the AV part to compute ventricular activation times, which again were employed by the MEF part to compute ventricular modulations to atrial intervals, closing the loop structure of the model.

12.4.2 Model Predictions

The complete closed-loop model was validated in patients with atrial flutter and characterized by a certain degree of AV conduction block. Model parameters were estimated in each patient and the beat-to-beat agreement between model predictions and data was estimated by quantifying an average distance between recorded and predicted series (Jorgensen et al., 2002).

Representative examples of the prediction of atrial (upper panels) and ventricular (lower panels) interval series are displayed in Fig. 12.10 for patients with 2:1 (panel a) and 4:1 (panel b) AV conduction block. Simulated AA series closely predicted the pattern of atrial interval variability, composed of a fast ventricular oscillation and a slow respiratory oscillation. Results on the overall population of patients testified the ability of the model to follow on a beat-to-beat basis and over long time scales the time evolution of atrial variability. The model reproduced 96 $\pm$ 8% of atrial interval variability with small distance between recorded and predicted series (1.1 $\pm$ 0.2 ms, corresponding to 9 $\pm$ 2% of AA variability range), which quantitatively demonstrated the plausibility of a mechanical modulation of atrial flutter variability by ventricular contraction and respiration.

Through its AV branch the model accurately predicted the time course of ventricular interval variability (lower panels), which consisted of a slow oscillation at the frequency of respiration. On the whole, the model predicted 86 $\pm$ 21% of ventricular interval variability, with small distance between predicted and recorded series (3.2 $\pm$ 1.9 ms, corresponding to 11 $\pm$ 3% of VV variability range). This suggested that most part of ventricular variability during atrial flutter was originated by the respiratory mechanical modulation of atrial activity, filtered by the AV node, while other factors, as autonomic modulation (Warner et al., 1986; Warner and Loeb, 1986;

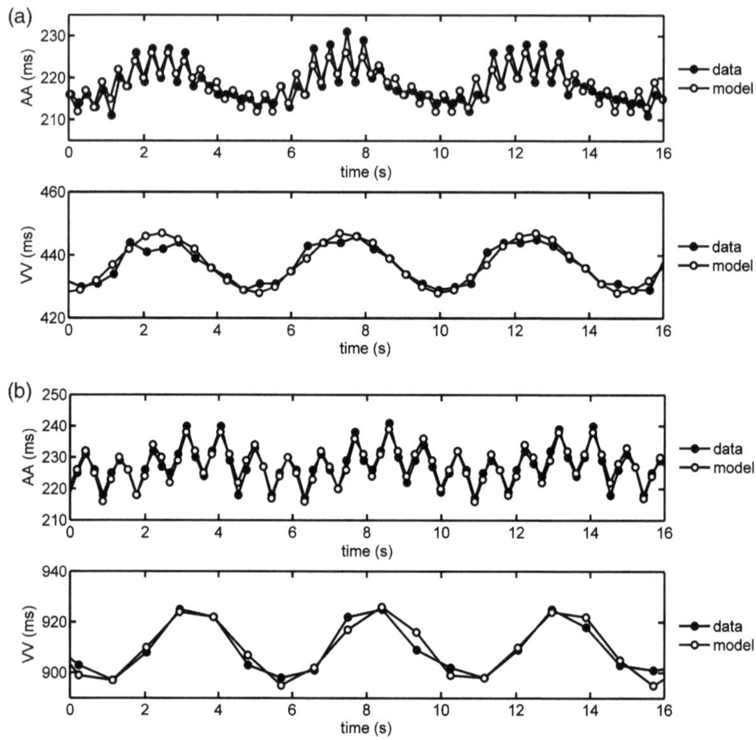

Fig. 12.10 Model predictions in patients with atrial flutter and spontaneous AV conduction. Atrial (AA) and ventricular (VV) intervals from data and model are shown in representative patients with 2:1 (**a**) and 4:1 (**b**) AV blocks, respectively. From Masè et al. (2008) with permission from Springer Science+Business Media

Nollo et al., 1994; Page et al., 1996; Kautzner et al., 2000), played a secondary role in heart rate variability during the arrhythmia.

Together with the validation of the complete model in closed-loop basal conditions, the authors performed a dynamical study of the properties of the MEF ventricular branch of the model, characterizing the patterns of atrial interval variability emerging at changing period T_v and amplitude A_v of the ventricular modulation. Theoretical predictions were compared with results obtained during ventricular pacing at increasing frequencies in patients with permanent pacemaker. The analysis of the model showed that atrial flutter variability could be sensitively influenced by periodic forcing and evidenced the possibility to entrain the variability. Specifically transitions between periodic and quasiperiodic AA series of different orders could be induced when the period and amplitude of ventricular modulation were changed. Periodic atrial rhythms of different orders were referred as N:M phase-locking patterns, since entrained atrial intervals occurred at N different phases in M ventricular cycles. The boundaries of the main phase-locking regions were numerically determined (see gray areas in the upper panel of Fig. 12.11) and resulted arranged according to a well-defined mathematical sequence, called typical Farey sequence (Allen, 1983; Bélair, 1986). In fact phase locking regions with decreasing

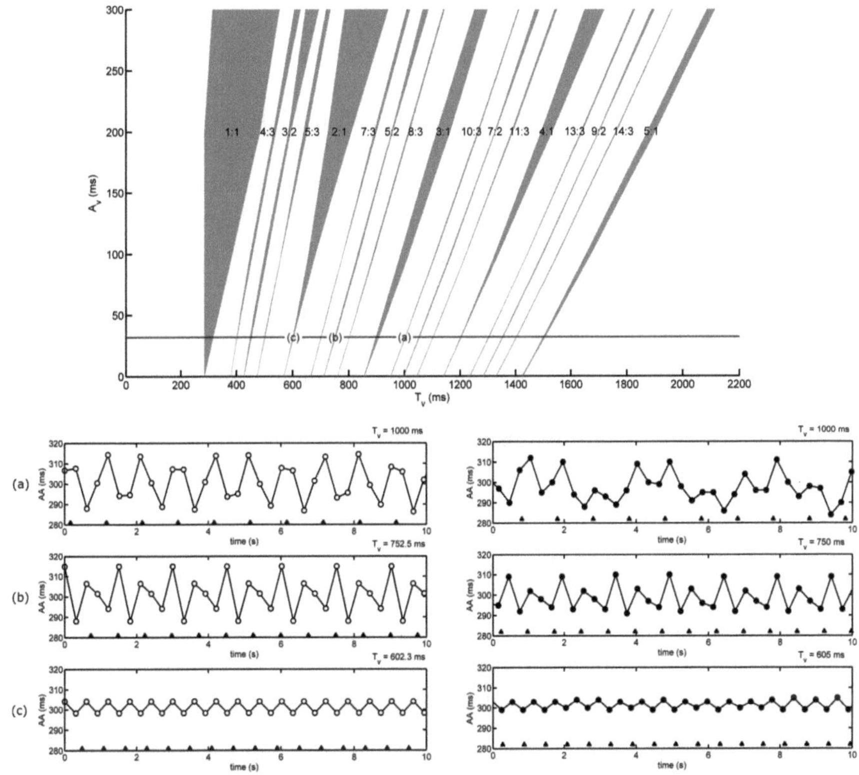

Fig. 12.11 Dynamical predictions of AFL cycle length variability in conditions of ventricular pacing. *Upper panel*. Boundaries of the phase-locking regions (*gray areas*) of the MEF ventricular branch of the model. Periodic AA series of *N:M* order are observed for parameter sets in the regions. *Lower panels*. Predicted (*left panels*) and real (*right panels*) AA series variability during ventricular pacing at increasing pacing frequencies T_V (indicated with corresponding letters in the *upper panel*). Triangles at the bottom of each panel represent successive ventricular activation times. Consistently with model predictions, no-locking (**a**), 5:2 (**b**) and 2:1 (**c**) phase locking conditions of atrial interval variability were observed in real *AA* series. Modified from Masè et al. (2008) with permission from Springer Science+Business Media

phase-locking ratios *M/N* were encountered at increasing forcing period T_V and transitions between *n:m* and *N:M* regions occurred through a *(n+N):(m+M)* region. The extension of the locking regions depended on the locking ratio and on the forcing amplitude A_V. For a given locking ratio, tongues got narrower in presence of weaker modulation (i.e., smaller A_V), which indicated that, in the physiological range of modulation of atrial flutter patients, the entrainment of atrial variability could occur just at well-defined pacing frequencies.

The possibility of transitions from unlocked to phase-locked variability patterns, predicted by the model, was actually verified during ventricular pacing at specific frequencies in patients with permanent pacemaker. Predicted atrial interval series obtained for pacing frequencies T_V in the unlocking region (panel a) and in the 5:2 (panel b) and 2:1 (panel c) phase-locking regions are displayed in the lower left panels of Fig. 12.11. In comparison with the unlocked state, where atrial intervals displayed an aperiodic behaviour and large variability (standard deviation = 9.2 ms, range = 32 ms), entrained conditions involved a reduction of interval variability, which was particularly marked in the 2:1 case (standard deviation = 3.0 ms, range = 6.0 ms). Recordings in patients (see Fig. 12.11, lower right panels) confirmed theoretical predictions, evidencing a transition from unlocked (panel a) to phase-locked conditions (panels b and c) at changing pacing frequency. In fact periodic atrial series of 5:2 and 2:1 orders were observed for pacing frequencies in proximity of the predicted 5:2 and 2:1 phase-locking regions. In agreement with model predictions, in the 2:1 entrainment condition the variability of the series decreased significantly (standard deviation = 2.1 ms, range = 6 ms).

Beyond confirming the goodness of the model, these results demonstrated the possibility of an external forcing of atrial flutter reentry and the existence of a well-defined structure of modulation patterns, which could involve sensitive effects on atrial interval variability.

12.5 Conclusions and Perspectives

In this chapter we have run through the intriguing research on atrial flutter cycle length variability, from its discovery in the 1920s through the experimental and theoretical findings which have led to the disclosure of its mechanism. The formulation of the MEF paradigm provides a consistent multi-level explanation for the issue of atrial flutter cycle length variability, which involves mechanisms of interaction at the hemodynamic, electrophysiologic and microscopic levels. Nevertheless to actually demonstrate how changes in atrial stretch are translated into changes in the conduction properties of the underlying reentrant circuit, further studies should be performed, involving beat-to-beat measurements of conduction velocity, refractory period and circuit pathway during atrial flutter. Such a challenging multi-measure study would complete the picture of atrial flutter cycle length variability.

Acknowledgements Michela Masè is recipient of a fellowship supported by Fondazione Cassa di Risparmio di Trento e Rovereto.

References

Allen T (1983) On the arithmetic of phase locking: Coupled neurons as a lattice on R^2. Physica D 6:305–320.
Allessie MA, Lammers WJEP, Rensma PL, Bonke FIM (1987) Flutter and fibrillation in experimental models: What has been learned that can be applied to humans? In: Brugada P, Wellens HJJ (eds.), Cardiac Arrhythmias: Where to Go From Here? Futura Publishing Co, Mount Kisco, NY, pp. 67–82.
Baertschi AJ and Gann DS (1977) Responses of atrial mechanoreceptors to pulsation of atrial volume. J. Physiol. 273:1–21.
Bélair J (1986) Periodic pulsatile stimulation of a nonlinear oscillator. J. Math. Biol. 24:217–232.
Bode F, Katchman A, Woosley RL, Franz MR (2000) Gadolinium decreases stretch-induced vulnerability to atrial fibrillation. Circulation 101:2200–2205.

Boyden PA and Hoffman BF (1981) The effects on atrial electrophysiology and structure of surgically induced right atrial enlargement in dogs. Circ. Res. 49:1319–1331.

Calaghan SC and White E (1999) The role of calcium in the response of cardiac muscle to stretch. Prog. Biophys. Mol. Biol. 71:59–90.

Calkins H, el Atassi R, Kalbfleisch S, Langberg J, Morady F (1992) Effects of an acute increase in atrial pressure on atrial refractoriness in humans. PACE 15:1674–1680.

Cherubini C, Filippi S, Nardinocchi P, Teresi L (2008) An electromechanical model of cardiac tissue: Constitutive issues and electrophysiological effects. Prog. Biophys. Mol. Biol. 97: 562–573.

Chorro FJ, Egea S, Mainar L, Canoves J, Sanchis J, Llavador E, Lopez-Merino V, Such L (1998) Acute changes in wavelength of the process of auricular activation induced by stretching. Experimental study. Rev. Esp. Cardiol. 51:874–883.

Christini DJ and Glass L (2002) Introduction: Mapping and control of complex cardiac arrhythmias. Chaos 12:732–739.

Cosio FG, Arribas F, Lopez-Gil M, Palacios J (1996) Atrial flutter mapping and ablation. I. Studying atrial flutter mechanisms by mapping and entrainment. PACE 19:841–853.

Cosio FG, Goicolea A, Lopez-Gil M, Arribas F, Barroso JL, Chicote R (1990) Atrial endocardial mapping in the rare form of atrial flutter. Am. J. Cardiol. 66:715–720.

Courtemanche M, Ramirez RJ, Nattel S (1998) Ionic mechanisms underlying human atrial action potential properties: Insights from a mathematical model. Am. J. Physiol. 275: H301–H321.

Deck KA (1964) Changes in the resting potential and the cable properties of Purkinje fibers during stretch. Pflugers Arch. Gesamte Physiol Menschen. Tiere. 280:131–140.

Dominguez G and Fozzard HA (1979) Effect of stretch on conduction velocity and cable properties of cardiac Purkinje fibers. Am. J. Physiol. 237:C119–C124.

Eckberg DL (1983) Human sinus arrhythmia as an index of vagal cardiac outflow. J. Appl. Physiol. 54:961–966.

Eckberg DL (2003) The human respiratory gate. J. Physiol. 548:339–352.

Eijsbouts S, Van Zandvoort M, Schotten U, Allessie M (2003) Effects of acute atrial dilation on heterogeneity in conduction in the isolated rabbit heart. J. Cardiovasc. Electrophysiol. 14: 269–278.

Farges JP, Ollagnier M, Faucon G (1977) Influence of acetylcholine, isoproterenol, quinidine and ouabain on effective refractory periods of atrial and ventricular myocardium in the dog. Arch. Int. Pharmacodyn. Ther. 227:206–219.

Ferguson JJ, Miller MJ, Aroesty JM, Sahagian P, Grossman W, McKay RG (1989) Assessment of right atrial pressure-volume relations in patients with and without an atrial septal defect. J. Am. Coll. Cardiol. 13:630–636.

Franz MR (1996) Mechano-electrical feedback in ventricular myocardium. Cardiovasc. Res. 32:15–24.

Franz MR and Bode F (2003) Mechano-electrical feedback underlying arrhythmias: The atrial fibrillation case. Prog. Biophys. Mol. Biol. 82:163–174.

Gaynor SL, Maniar HS, Prasad SM, Steendijk P, Moon MR (2005) Reservoir and conduit function of right atrium: Impact on right ventricular filling and cardiac output. Am. J. Physiol. Heart Circ. Physiol. 288:H2140–H2145.

Glass L (1991) Cardiac arrhythmias and circle maps – A classical problem. Chaos 1:13–19.

Glass L, Goldberger AL, Courtemanche M, Shrier A (1987) Nonlinear dynamics, chaos and complex cardiac arrhythmias. Proc. R. Soc. Lond. A 413:9–26.

Hayden WG, Hurley EJ, Rytand DA (1967) The mechanism of canine atrial flutter. Circ. Res. 20:496–505.

Healy SN and McCulloch AD (2005) An ionic model of stretch-activated and stretch-modulated currents in rabbit ventricular myocytes. Europace 7(2):128–134.

Hu H and Sachs F (1997) Stretch-activated ion channels in the heart. J. Mol. Cell Cardiol. 29: 1511–1523.

Hunter PJ, McCulloch AD, ter Keurs HE (1998) Modelling the mechanical properties of cardiac muscle. Prog. Biophys. Mol. Biol. 69:289–331.

Jarvinen VM, Kupari MM, Poutanen VP, Hekali PE (1996) Right and left atrial phasic volumetric function in mildly symptomatic dilated and hypertrophic cardiomyopathy: Cine MR imaging assessment. Radiology 198:487–495.

Jorgensen P, Schafer C, Guerra PG, Talajic M, Nattel S, Glass L (2002) A mathematical model of human atrioventricular nodal function incorporating concealed conduction. Bull. Math. Biol. 64:1083–1099.

Kalman JM, Olgin JE, Saxon LA, Fisher WG, Lee RJ, Lesh MD (1996) Activation and entrainment mapping defines the tricuspid annulus as the anterior barrier in typical atrial flutter. Circulation 94:398–406.

Kamkin A, Kiseleva I, Wagner KD, Leiterer KP, Theres H, Scholz H, Gunther J, Lab MJ (2000) Mechano-electric feedback in right atrium after left ventricular infarction in rats. J. Mol. Cell. Cardiol. 32:465–477.

Kaseda S and Zipes DP (1988) Contraction-excitation feedback in the atria: A cause of changes in refractoriness. J. Am. Coll. Cardiol. 11:1327–1336.

Kaufmann RL, Lab MJ, Hennekes R, Krause H (1971) Feedback interaction of mechanical and electrical events in the isolated mammalian ventricular myocardium (cat papillary muscle). Pflugers Arch. 324:100–123.

Kautzner J, Malik M, Camm AJ (2000) Autonomic modulation of AV nodal conduction. In: Mazgalev TN, Tchou PJ (eds.), Atrial-AV Nodal Electrophysiology: A View from the Millennium. Futura Publishing Company, New York, NY, pp. 237–250.

Klein LS, Miles WM, Zipes DP (1990) Effect of atrioventricular interval during pacing or reciprocating tachycardia on atrial size, pressure, and refractory period. Contraction-excitation feedback in human atrium. Circulation 82:60–68.

Kohl P, Day K, Noble D (1998) Cellular mechanisms of cardiac mechano-electric feedback in a mathematical model. Can. J. Cardiol. 14:111–119.

Kuijpers NH, ten Eikelder HM, Bovendeerd PH, Verheule S, Arts T, Hilbers PA (2007) Mechano-electric feedback leads to conduction slowing and block in acutely dilated atria: A modeling study of cardiac electromechanics. Am. J. Physiol. Heart Circ. Physiol. 292:H2832–H2853.

Lab MJ (1980) Transient depolarisation and action potential alterations following mechanical changes in isolated myocardium. Cardiovasc. Res. 14:624–637.

Lab MJ (1982) Contraction-excitation feedback in myocardium. Physiological basis and clinical relevance. Circ. Res. 50:757–766.

Lammers WJEP, Ravelli F, Disertori M, Antolini R, Furlanello F, Allessie MA (1991) Variations in human atrial flutter cycle length induced by ventricular beats: Evidence of a reentrant circuit with a partially excitable gap. J. Cardiovasc. Electrophysiol. 2:375–387.

Lewis T (1920) Observations upon flutter and fibrillation. I. The regularity of clinical auricolar flutter. Heart 7:127–130.

Liu L and Nattel S (1997) Differing sympathetic and vagal effects on atrial fibrillation in dogs: Role of refractoriness heterogeneity. Am. J. Physiol. 273:H805–H816.

Luo CH and Rudy Y (1994) A dynamic model of the cardiac ventricular action potential. I. Simulations of ionic currents and concentration changes. Circ. Res. 74:1071–1096.

Masè M, Glass L, Ravelli F (2008) A model for mechano-electrical feedback effects on atrial flutter interval variability. Bull. Math. Biol. 70:1326–1347.

Matsuda Y, Toma Y, Ogawa H, Matsuzaki M, Katayama K, Fujii T, Yoshino F, Moritani K, Kumada T, Kusukawa R (1983) Importance of left atrial function in patients with myocardial infarction. Circulation 67:566–571.

Moe GK, Rheinboldt WC, Abildskov JA (1964) A computer model of atrial fibrillation. Am. Heart J. 67:200–220.

Nakagawa H, Lazzara R, Khastgir T, Beckman KJ, McClelland JH, Imai S, Pitha JV, Becker AE, Arruda M, Gonzalez MD, Widman LE, Rome M, Neuhauser J, Wang X, Calame JD, Goudeau MD, Jackman WM (1996) Role of the tricuspid annulus and the eustachian valve/ridge on

atrial flutter. Relevance to catheter ablation of the septal isthmus and a new technique for rapid identification of ablation success. Circulation 94:407–424.
Nash MP and Panfilov AV (2004) Electromechanical model of excitable tissue to study reentrant cardiac arrhythmias. Prog. Biophys. Mol. Biol. 85:501–522.
Nazir SA and Lab MJ (1996a) Mechanoelectric feedback and atrial arrhythmias. Cardiovasc. Res. 32:52–61.
Nazir SA and Lab MJ (1996b) Mechanoelectric feedback in the atrium of the isolated guinea-pig heart. Cardiovasc. Res. 32:112–119.
Ninio DM, Murphy KJ, Howe PR, Saint DA (2005) Dietary fish oil protects against stretch-induced vulnerability to atrial fibrillation in a rabbit model. J. Cardiovasc. Electrophysiol. 16: 1189–1194.
Nollo G, Del Greco M, Ravelli F, Disertori M (1994) Evidence of low- and high-frequency oscillations in human AV interval variability: Evaluation with spectral analysis. Am. J. Physiol. 267:H1410–H1418.
Olgin JE, Kalman JM, Fitzpatrick AP, Lesh MD (1995) Role of right atrial endocardial structures as barriers to conduction during human type I atrial flutter. Activation and entrainment mapping guided by intracardiac echocardiography. Circulation 92:1839–1848.
Olshansky B, Okumura K, Hess PG, Waldo AL (1990) Demonstration of an area of slow conduction in human atrial flutter. J. Am. Coll. Cardiol. 16:1639–1648.
Page RL, Wharton JM, Prystowsky EN (1996) Effect of continuous vagal enhancement on concealed conduction and refractoriness within the atrioventricular node. Am. J. Cardiol. 77: 260–265.
Paintal AS (1973) Vagal sensory receptors and their reflex effects. Physiol. Rev. 53:159–227.
Panfilov AV and Pertsov AM (2001) Ventricular fibrillation: Evolution of the multiple wavelet hypothesis. Phil. Trans. Roy. Soc. Lond. 359:1315–1325.
Penefsky Z and Hoffman B (1963) Effect of stretch on mechanical and electrical properties of the cardiac muscle. Am. J. Physiol. 204:433–438.
Pertsov AM, Davidenko JM, Salomonsz R, Baxter WT, Jalife J (1993) Spiral waves of excitation underlie reentrant activity in isolated cardiac muscle. Circ. Res. 72:631–650.
Ravelli F (2003) Mechano-electric feedback and atrial fibrillation. Progr. Biophys. Mol. Biol. 82:137–149.
Ravelli F and Allessie M (1997) Effects of atrial dilatation on refractory period and vulnerability to atrial fibrillation in the isolated Langendorff-perfused rabbit heart. Circulation 96:1686–1695.
Ravelli F, Disertori M, Cozzi F, Antolini R, Allessie MA (1994) Ventricular beats induce variations in cycle length of rapid (type II) atrial flutter in humans. Evidence of leading circle reentry. Circulation 89:2107–2116.
Ravelli F, Masè M, Disertori M (2008) Mechanical modulation of atrial flutter cycle length. Prog. Biophys. Mol. Biol. 97:417–434.
Reiter MJ, Landers M, Zetelaki Z, Kirchhof CJ, Allessie MA (1997) Electrophysiological effects of acute dilatation in the isolated rabbit heart: Cycle length-dependent effects on ventricular refractoriness and conduction velocity. Circulation 96:4050–4056.
Rensma PL, Allessie MA, Lammers WJ, Bonke FI, Schalij MJ (1988) Length of excitation wave and susceptibility to reentrant atrial arrhythmias in normal conscious dogs. Circ. Res. 62: 395–410.
Rice JJ, Winslow RL, Dekanski J, McVeigh E (1998) Model studies of the role of mechano-sensitive currents in the generation of cardiac arrhythmias. J. Theor. Biol. 190:295–312.
Riemer TL, Sobie EA, Tung L (1998) Stretch-induced changes in arrhythmogenesis and excitability in experimentally based heart cell models. Am. J. Physiol 275:H431–H442.
Riemer TL and Tung L (2003) Stretch-induced excitation and action potential changes of single cardiac cells. Prog. Biophys. Mol. Biol. 82:97–110.
Robotham JL, Lixfeld W, Holland L, MacGregor D, Bryan AC, Rabson J (1978) Effects of respiration on cardiac performance. J. Appl. Physiol. 44:703–709.
Rosen MR, Legato MJ, Weiss RM (1981) Developmental changes in impulse conduction in the canine heart. Am. J. Physiol. 240:H546–H554.

Rytand DA, Onesti SJ, Bruns DL (1958) The atrial rate in patients with flutter; a relationship between atrial enlargement and slow rate. Stanford. Med. Bull. 16:169–202.
Sachs F (1991) Mechanical transduction by membrane ion channels: A mini review. Mol. Cell Biochem. 104:57–60.
Sachs F (1994) Modeling mechanical-electrical transduction in the heart. In: Mow VC, Guliak F, Tran-Son-Tray R, Hochmuth RM (eds.), Cell Mechanics and Cellular Engineering. Springer Verlag, New York, NY, pp. 308–328.
Sachse FB, Steadman BW, Bridge JHB, Punske BB, Taccardi B (2004) Conduction velocity in myocardium modulated by strain: Measurement instrumentation and initial results. Conf. Proc. IEEE Eng. Med. Biol. Soc. 5:3593–3596.
Shrier A, Dubarsky H, Rosengarten M, Guevara MR, Nattel S, Glass L (1987) Prediction of complex atrioventricular conduction rhythms in humans with use of the atrioventricular nodal recovery curve. Circulation 76:1196–1205.
Solti F, Vecsey T, Kekesi V, Juhasz-Nagy A (1989) The effect of atrial dilatation on the genesis of atrial arrhythmias. Cardiovasc. Res. 23:882–886.
Spear JF and More EN (1972) Stretch-induced excitation and conduction disturbances in the isolated rat myocardium. J. Electrocardiol. 5:15–24.
Stambler BS and Ellenbogen KA (1996) Elucidating the mechanisms of atrial flutter cycle length variability using power spectral analysis techniques. Circulation 94:2515–2525.
Sung D, Mills RW, Schettler J, Narayan SM, Omens JH, McCulloch AD (2003) Ventricular filling slows epicardial conduction and increases action potential duration in an optical mapping study of the isolated rabbit heart. J. Cardiovasc. Electrophysiol. 14:739–749.
Taha BH, Simon PM, Dempsey JA, Skatrud JB, Iber C (1995) Respiratory sinus arrhythmia in humans: An obligatory role for vagal feedback from the lungs. J. Appl. Physiol. 78:638–645.
Tavi P, Han C, Weckstrom M (1998) Mechanisms of stretch-induced changes in [Ca^{2+}]i in rat atrial myocytes: Role of increased troponin C affinity and stretch-activated ion channels. Circ. Res. 83:1165–1177.
Tavi P, Laine M, Weckstrom M (1996) Effect of gadolinium on stretch-induced changes in contraction and intracellularly recorded action- and afterpotentials of rat isolated atrium. Br. J. Pharmacol. 118:407–413.
Triedman JK and Saul JP (1994) Blood pressure modulation by central venous pressure and respiration. Buffering effects of the heart rate reflexes. Circulation 89:169–179.
Tse HF, Pelosi F, Oral H, Knight BP, Strickberger SA, Morady F (2001) Effects of simultaneous atrioventricular pacing on atrial refractoriness and atrial fibrillation inducibility: Role of atrial mechanoelectrical feedback. J. Cardiovasc. Electrophysiol. 12:43–50.
Tung L and Zou S (1995) Influence of stretch on excitation threshold of single frog ventricular cells. Exp. Physiol. 80:221–235.
van der Pol B and van der Mark J (1928) The heartbeat considered as a relaxation oscillation, and an electrical model of the heart. Phil. Mag. 6:763–775.
Vargas G, Akhtar M, Damato AN (1975) Electrophysiologic effects of isoproterenol on cardiac conduction system in man. Am. Heart J. 90:25–34.
Vulliemin P, Del Bufalo A, Schlaepfer J, Fromer M, Kappenberger L (1994) Relation between cycle length, volume, and pressure in type I atrial flutter. PACE 17:1391–1398.
Waldo AL (1998) Pathogenesis of atrial flutter. J. Cardiovasc. Electrophysiol. 9:S18–S25.
Waldo AL (2000) Treatment of atrial flutter. Heart 84:227–232.
Warner MR, deTarnowsky JM, Whitson CC, Loeb JM (1986) Beat-by-beat modulation of AV conduction. II. Autonomic neural mechanisms. Am. J. Physiol. 251:H1134–H1142.
Warner MR and Loeb JM (1986) Beat-by-beat modulation of AV conduction. I. Heart rate and respiratory influences. Am. J. Physiol. 251:H1126–H1133.
Waxman MB, Kirsh JA, Yao L, Cameron DA, Asta JA (1992) Slowing of the atrial flutter rate during 1:1 atrioventricular conduction in humans and dogs: An effect mediated through atrial pressure and volume. J. Cardiovasc. Electrophysiol. 3:544–557.

Waxman MB, Yao L, Cameron DA, Kirsh JA (1991) Effects of posture, Valsalva maneuver and respiration on atrial flutter rate: An effect mediated through cardiac volume. J. Am. Coll. Cardiol. 17:1545–1552.

Wells JL, MacLean WA, James TN, Waldo AL (1979) Characterization of atrial flutter. Studies in man after open heart surgery using fixed atrial electrodes. Circulation 60:665–673.

White E, Le Guennec JY, Nigretto JM, Gannier F, Argibay JA, Garnier D (1993) The effects of increasing cell length on auxotonic contractions; membrane potential and intracellular calcium transients in single guinea-pig ventricular myocytes. Exp. Physiol. 78:65–78.

Winfree AT (1991) Varieties of spiral wave behavior: An experimentalist's approach to the theory of excitable media. Chaos 1:303–334.

Winfree AT (1994) Electrical turbulence in three-dimensional heart muscle. Science 266: 1003–1006.

Yamashita T, Oikawa N, Murakawa Y, Nakajima T, Omata M, Inoue H (1994) Contraction-excitation feedback in atrial reentry: Role of velocity of mechanical stretch. Am. J. Physiol. 267:H1254–H1262.

Yasuma F and Hayano J (2004) Respiratory sinus arrhythmia: Why does the heartbeat synchronize with respiratory rhythm? Chest 125:683–690.

Zarse M, Stellbrink C, Athanatou E, Robert J, Schotten U, Hanrath P (2001) Verapamil prevents stretch-induced shortening of atrial effective refractory period in langendorff-perfused rabbit heart. J. Cardiovasc. Electrophysiol. 12:85–92.

Zhu WX, Johnson SB, Brandt R, Burnett J, Packer DL (1997) Impact of volume loading and load reduction on ventricular refractoriness and conduction properties in canine congestive heart failure. J. Am. Coll. Cardiol. 30:825–833.

Zhuang J, Yamada KA, Saffitz JE, Kleber AG (2000) Pulsatile stretch remodels cell-to-cell communication in cultured myocytes. Circ. Res. 87:316–322.

Zykov VS (1987) Simulation of Wave Processes in Excitable Media. Manchester University Press, Manchester.

Chapter 13
Early Hypertrophic Signals After Myocardial Stretch. Role of Reactive Oxygen Species and the Sodium/Hydrogen Exchanger

Horacio E. Cingolani, Néstor G. Pérez, Claudia I. Caldiz, Carolina D. Garciarena, Verónica C. De Giusti, María V. Correa, María C. Villa-Abrille, Alejandra M. Yeves, Irene L. Ennis, Gladys Chiappe de Cingolani, and Ernesto A. Aiello

Abstract In this chapter the enhanced activity of the cardiac Na^+/H^+ exchanger (NHE-1) after myocardial stretch is considered a key step of the intracellular signaling pathway leading to the slow force response to stretch as well as an early signal for the development of cardiac hypertrophy. We propose that the chain of events triggered by stretch begins with the release of small amounts of angiotensin II which in turn induce the release/formation of endothelin. The actions of these hormones trigger the production of mitochondrial reactive oxygen species that enhances NHE-1 activity, causing an increment in the intracellular Na^+ concentration which promotes the increase in intracellular Ca^{2+} concentration ($[Ca^{2+}]_i$) through the Na^+/Ca^{2+} exchanger. This $[Ca^{2+}]_i$ increase would trigger cardiac hypertrophy by activation of widely recognized Ca^{2+}-dependent intracellular signaling pathways.

Keywords Myocardium · Stretch · Sodium/hydrogen exchanger · Reactive oxygen species · Hypertrophy

13.1 Introduction

Adding electrons to oxygen produces sequentially: (1) superoxide anion (O_2^-), (2) hydrogen peroxide (H_2O_2), (3) hydroxil radical (OH^-) and finally water (H_2O) (Boveris, 1998). While H_2O_2 is not a free radical, this very reactive and membrane permeant molecule is included among the reactive oxygen species (ROS), together with the oxygen radicals O_2^- and OH^-. Mitochondria are the main source of ROS production, although NADPH oxidase and Xanthine Oxidase may also contribute to ROS formation (Giordano, 2005). The enzyme responsible for NO production

H.E. Cingolani (✉)
Facultad de Ciencias Médicas, Centro de Investigaciones Cardiovasculares, UNLP, La Plata, Argentina
e-mail: cicmes@infovia.com.ar

A. Kamkin, I. Kiseleva (eds.), *Mechanosensitivity of the Heart*, Mechanosensitivity in Cells and Tissues 3, DOI 10.1007/978-90-481-2850-1_13,

(Nitric Oxide Synthase, NOS) can also generate O_2^- under certain oxidative stress conditions (Takimoto et al., 2005).

During many years ROS were considered deleterious agents, but in the last years evidences of their effects as second messengers have emerged (D'Autreaux and Toledano, 2007). Furthermore, the concept that free radicals in the heart could be "friend or foe" depending on the magnitude, duration or timing of the redox signal has been recently suggested (Downey and Cohen, 2008). The cardiac Na^+/H^+ exchanger (NHE-1) is a target for ROS through the activation of kinases (Sabri et al., 1998; Snabaitis et al., 2002). ROS, kinases activation, and NHE-1 hyperactivity are three early hypertrtophic signals after myocardial stretch and/or stimulation by growth factors. Interestingly, inhibition of ROS, NHE-1 or growth factors results in regression of cardiac hypertrophy. The discussion of how these three factors are linked among them and how they are linked to other well known hypertrophic signals constitutes the aim of this chapter.

13.2 NHE-1 and Myocardial Stretch

In 1998 Bluhm et al. published the results obtained with an elegant theoretical ionic model of a ventricular myocyte used to analyze the changes in sarcolemmal ion fluxes following step changes in cardiac muscle length. They suggested that a sudden increase in muscle length might induce changes in sarcolemmal Na^+ influx leading to an increase in $[Na^+]_i$ and a concomitant increase in systolic Ca^{2+} entry through the Na^+/Ca^{2+} exchanger (NCX). However, the mechanism by which the increase in $[Na^+]_i$ takes place was not proposed. Since the NHE-1 is an important Na^+ entry pathway in cardiomyocytes, the possible role played by the exchanger will be analyzed in detail.

The finding of a stretch-induced myocardial alkalization in cat papillary muscles bathed with a bicarbonate-free medium was the first piece of evidence provided by our laboratory referent to NHE-1 activation by myocardial stretch and the main role played by this exchanger in the early signals leading to hypertrophy (Cingolani et al., 1998). The absence of bicarbonate in the medium allowed us to analyze the role of NHE-1 without the influence of bicarbonate-dependent intracellular pH (pH_i)-regulatory mechanisms. The stretch-induced myocardial alkalization was suppressed by either angiotensin II (Ang II) type 1 (AT_1) or endothelin (ET) type A (ET_A) receptors blockade, suggesting the involvement of these receptors in the stretch-induced activation of NHE-1 (Cingolani et al., 1998). In accordance with this, Sadoshima and co-authors (1993) have initially reported the release of Ang II after stretching cultured neonatal cardiomyocytes. They showed that the addition of the surrounding medium from stretched to non-stretched cardiomyocytes promoted hypertrophy, and that Ang II was the autocrine mediator of this effect. These authors also suggested that Ang II is stored in secretory vesicles in myocytes and released within 1 min by mechanical stretch (Sadoshima et al., 1993). Contemporarily, Ito et al. (1993) found in the same type of preparation that Ang II promotes the release/formation of ET-1, demonstrating that ET-1 is an autocrine factor in the mechanism of Ang II-induced cardiac hypertrophy. In addition, Yamazaki et al. (1996) found that, together with an increase in NHE-1 activity, stretch induced a rise in the concentration of ET-1 constitutively secreted from cardiomyocytes to the culture medium. The same authors showed that NHE-1 inhibition partially attenuated the stretch-induced mitogen-activated protein kinase (MAPK) activation. Our main contribution was to demonstrate the existence of a stretch-triggered autocrine/paracrine release of Ang II/ET leading to NHE-1 activation in an adult cardiac multicellular preparation (Cingolani et al., 1998; Alvarez et al., 1999; Perez et al., 2001). This finding allowed us to propose the hypothetical scheme depicted in Fig. 13.1. The proposed chain of events begins with the release of preformed Ang II and ends with an increase in the Ca^{2+} transient through reverse mode of NCX (NCX_{rev}) activation and/or forward mode of NCX ($NCX_{forward}$) inhibition secondary to the NHE-1 activation-mediated rise in $[Na^+]_i$. If we analyze the potential effects of NHE-1 activation on myocardial contractility, we should consider two different mechanisms: Na^+-triggered increase in the Ca^{2+} transient through NCX, and an increase in pH_i that would increase the contractile force by increasing myofilament Ca^{2+} responsiveness. Considering the latter possibility, it is important to emphasize that little or no change in pH_i is detected when the stimulating effect of stretch, exogenous Ang II or ET-1 on NHE-1 is studied in the presence of bicarbonate buffers (Cingolani et al., 1998; Alvarez et al., 1999; Perez et al., 2001, 2003; Aiello et al., 2005; Luers et al., 2005) The explanation for the lack of change in pH_i can be found in the fact that growth factors like Ang II and ET-1 simultaneously activate at least two opposing pH_i-regulatory mechanisms: the alkalinizing

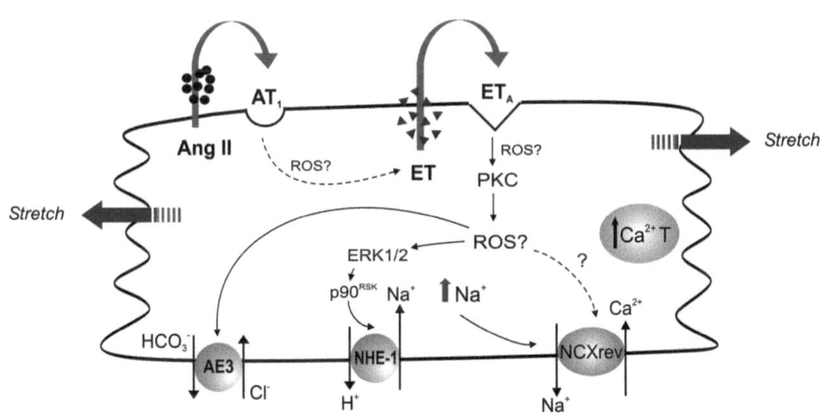

Fig. 13.1 A representation of the proposed autocrine/paracrine cascade of events following myocardial stretch. Endogenous Ang II is released from the myocytes activating AT_1 receptors in an autocrine fashion. Stimulation of AT_1 induces the release/formation of ET, which simultaneously activates NHE-1 and Cl^-–HCO_3^- exchanger through ET_A receptors. The activation of Cl^-– HCO_3^- exchanger prevents the expected intracellular alkalization due to NHE-1 activation but does not prevent the rise in $[Na^+]_i$. The increase in $[Na^+]_i$ drives the NCX in its reverse mode and this, together with a probable direct action on the exchanger, leads to the increase in Ca^{2+} transient ($Ca^{2+}T$)

NHE-1 and the acidifying Na^+–independent Cl^-–HCO_3^- anion exchanger (Ganz et al., 1988; Thomas, 1989; Camilion de Hurtado et al., 1998; de Hurtado et al., 2000; Alvarez et al., 2001; Cingolani et al., 2003a; Perez et al., 2003). The scheme in Fig. 13.1 illustrates the fact that Ang II through release/formation of ET-1, simultaneously stimulates NHE-1 and Cl^-–HCO_3^- exchanger, thus minimizing the changes in pH_i but without affecting the increase in $[Na^+]_i$ that follows NHE-1 activation. Therefore, NHE-1 activation can be detected as a pH_i increase only if bicarbonate is absent in the medium. We emphasize this point because the absence of changes in pH_i after growth factor stimulation in bicarbonate media is not widely recognized, though it was reported by Ganz et al. in 1988 in mesangial cells and a call for attention was published by Thomas (1989) in a letter to Nature one year later. More recently, Schafer et al. (2002) demonstrated that the hypertrophic response of cardiomyocytes to α- and β-adrenergic stimulation requires NHE-1 activation but not cellular alkalization. In summary, although there is enough evidence to suggest a direct correlation between activation of cellular acid extrusion mechanisms and proliferation, there is also enough evidence to state that proliferation can occur without changes in pH_i, and that changes in pH_i do not necessarily induce proliferation (Schafer et al., 2002; Ganz et al., 1988, 1990; Shrode et al., 1997). There is no agreement in the literature about the role played by the NHE-1 in growth and viability. While some authors report that NHE-1-deficient transgenic mice can grow at normal rate (Grinstein et al., 1989), others have shown that these animals exhibit growth retardation and are subject to slow-wave epilepsy (54–56).

The effects of myocardial stretch, exogenous Ang II and ET-1 on pH_i and $[Na^+]_i$ in cat papillary muscles are illustrated in Fig. 13.2. In these experiments, low doses of exogenous Ang II or ET-1 that probably reproduced those released after stretch

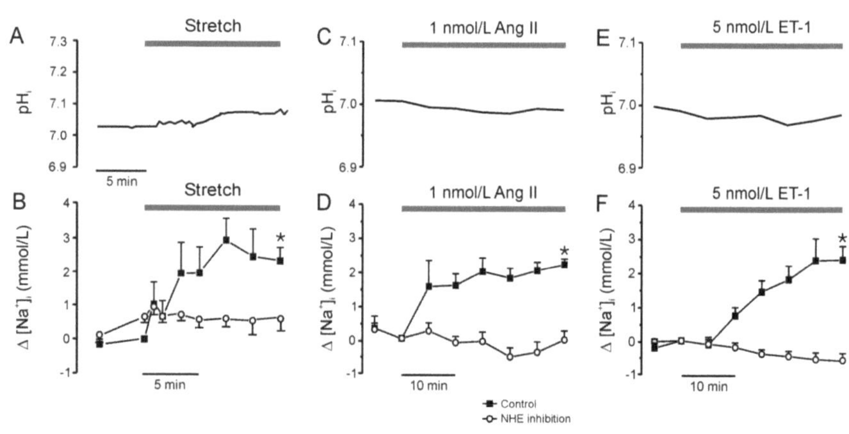

Fig. 13.2 Representative experiments showing that in the presence of bicarbonate, NHE-1 activation by stretch (**Panel A**), exogenous Ang II (**Panel C**) or ET-1 (**Panel E**) does not change pH_i. The same interventions promoted an increase in $[Na^+]_i$ that was prevented by NHE-1 blockade (pooled results of **Panels B, D** and **F**). * Indicates $P<0.05$ vs. NHE inhibition. Modified from Perez et al. (2003) with permission

did not affect pH_i but significantly increased $[Na^+]_i$. This rise in $[Na^+]_i$ was suppressed by NHE-1 inhibition. The ET receptor blockade exerted the same inhibitory effect after myocardial stretch and after the addition of exogenous Ang II or ET-1 (Perez et al., 2001, 2003). The role played by the Cl^-–HCO_3^- exchanger in preventing intracellular alkalization after myocardial stretch is better visualized by repeating the intervention in a bicarbonate medium before and after inhibition of the anion-exchanger with specific antibodies (see Fig. 13.3) (Cingolani et al., 2003a).

Under these conditions, an increase in pH_i takes place only after Cl^-–HCO_3^- exchanger inhibition. It is not clear whether changes in pH_i after the addition of growth factors or stretch stimulation localized to certain subcellular spaces within the myocyte may occur in the presence of bicarbonate-dependent mechanisms. The fact that an increase in pH_i stimulates protein synthesis (Fuller et al., 1990) does not necessarily mean that intracellular alkalization occurs after myocardial stretch, Ang II or ET-1 stimulation (Ganz et al., 1988; Schafer et al., 2002; Cingolani et al., 2005). We would like to emphasize that our proposal is valid for the concentration used by us. Higher concentrations of Ang II and/or ET-1 can trigger mechanisms other than those described herein.

It is known that the increase in $[Na^+]_i$ can induce an increase in $[Ca^{2+}]_i$ through the NCX as a result of a decrease in Ca^{2+} efflux (decreased forward mode) and/or an increase in Ca^{2+} entry (increased reverse mode). As mentioned before, the increase in $[Na^+]_i$ induced by stretch or by exogenous low doses of Ang II or ET-1 was prevented by blocking NHE-1 (Fig. 13.2) (Alvarez et al., 1999; Perez et al., 2001, 2003; Aiello et al., 2005). The increase in myocardial $[Na^+]_i$ detected in our experiments was ~3–6 mmol/L. In line with this, increases of similar magnitude were detected by Baartscheer et al. (2005) in the myocardium of rabbit failing hearts with enhanced activity of NHE-1 and by Luers et al. (2005) after stretching rabbit

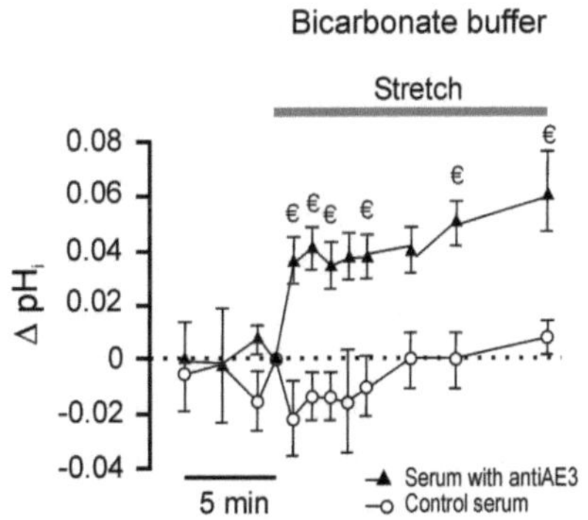

Fig. 13.3 When Cl^-– HCO_3^- exchanger activity is inhibited by a specific antibody against it, the slow increase in force after stretch is even greater than when the anion exchanger is operative, due to a rise in pH_i despite the presence of extracellular bicarbonate. Under this condition, the increase in myofilament responsiveness increases developed force in addition to the effect of the augmented Ca^{2+} transient. **€** $P < 0.05$ vs. Control serum. Modified from Cingolani et al. (2003a) with permission

myocardium. This increase in $[Na^+]_i$ shifts the reversal potential of NCX to a more negative voltage, thus allowing the NCX to operate in reverse mode for a longer period of time during the action potential and promoting Ca^{2+} influx to the cell which should be reflected by changes in contractility. As reported by Bers et al. (2003), cardiomyocytes have a limited capacity to buffer increases in $[Na^+]_i$ and the NCX is more sensitive than the Na^+/K^+ ATPase pump to a change in $[Na^+]_i$ of this magnitude.

Calculation of the estimated reversal potential of NCX in cat papillary muscles gives a value of –34 mV which is of the same order of magnitude as those estimated by other authors (Kusuoka et al., 1993; Bers, 2001), if we assume 10 mmol/L $[Na^+]_i$, 140 mmol/L extracellular Na^+, 1.5 mmol/L extracellular Ca^{2+} and a 150 nmol/L diastolic $[Ca^{2+}]_i$. The quick rise in sub-membrane $[Ca^{2+}]_i$ due to the Ca^{2+} transient that shifts the NCX reversal potential to even more positive voltages (Bers and Despa, 2006) would lead to a minimal contribution of the NCX_{rev} to basal contractility under normal conditions (Perez et al., 2001, 2003; Aiello et al., 2005). Accordingly, we have shown that NCX_{rev} inhibition with 5 μmol/L KB-R7943 did not affect basal contractility or the increase in contractility of ~20% promoted by rising extracellular Ca^{2+} from 1.35 to 1.9 mmol/L (Fig. 13.4) in cat papillary muscle. However, these results are in contrast to those obtained by Kurogouchi et al. (2000) in the dog myocardium that showed that KB-R7943 promoted a pronounced negative inotropic effect, discrepancy that might depend on the model and/or species used in each study. However, in isolated cat ventricular myocytes a decrease in basal inotropism of approximately 20% was detected after 1 μmol/L KB-R7943 (Cingolani et al., 2006). Therefore, this compound seems to exert greater negative inotropic effect in isolated myocytes.

The approximately 3–6 mmol/L increase in $[Na^+]_i$ induced by stretch (34), exogenous Ang II (Perez et al., 2003) or ET-1 (Aiello et al., 2005) in our experimental conditions certainly changes the scenario by shifting the reversal potential of NCX from –34 to –55 mV, allowing operation of the NCX reverse mode during a longer fraction of the action potential plateau. In line with the above-mentioned effect of stretch, Ang II and ET on $[Na^+]_i$, we detected a negative shift of the NCX reversal potential of –5 and –15 mV after treating isolated patch-clamped cat myocytes with 1 and 10 nmol/L ET-1, respectively (Aiello et al., 2005). Considering these experimental results, estimation of the ET-1-induced increase in $[Na^+]_i$ gives values of approximately 1.6 and 5.0 mmol/L for 1 and 10 nmol/L ET-1, respectively. These values are of the same order of magnitude as those measured in the bulk of the cytosol by epifluorescence in papillary muscles after addition of 5 nmol/L ET-1 (Perez et al., 2003). However, it is important to note that the increase in $[Na^+]_i$ in the isolated myocytes might reflect changes of this ion in a space in which intracellular dialysis with the solution of the patch pipette cannot maintain $[Na^+]$ at a constant level. The increase in $[Na^+]_i$ would tend to increase Ca^{2+} influx through reverse mode NCX during systole and to reduce Ca^{2+} extrusion via forward mode NCX during diastole that should necessarily end with an increase in the force of contraction as reported by us (Alvarez et al., 1999; Perez et al., 2001, 2003; Aiello et al., 2005).

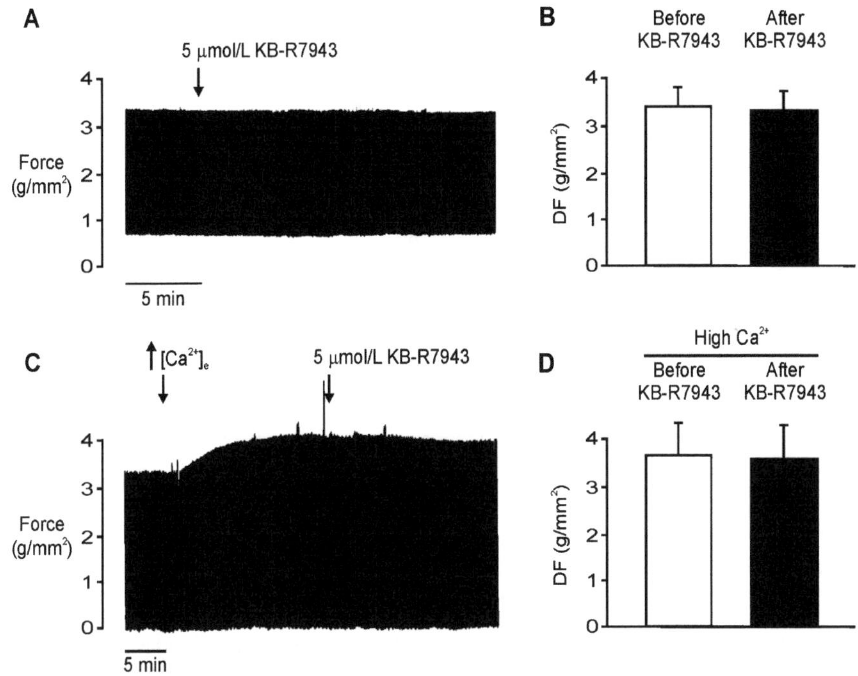

Fig. 13.4 Original force records showing the lack of effect of 5 μmol/L KB-R7943 (NCXrev blocker) on basal contractility (A, extracellular Ca^{2+}=1.35 mmol/L) and on the increase in contractility of ~20% promoted by increasing extracellular Ca^{2+} from 1.35 mmol/L to 1.9 mmol/L (C). Overall results of developed force (DF, in g/mm^2) for each type of experiments (B, *n*=6 and D, *n*=4). These results also strongly suggest that KB-R7943 at this concentration does not exert non-specific actions which may affect contractility. Reproduced from Perez et al. (Cingolani et al., 2003a) with permission

We have reported an increase in the Ca^{2+} transient amplitude of about 12% during the slow force response without changes in diastolic Ca^{2+} (Alvarez et al., 1999; Perez et al., 2001), result that coincides with that reported by Kentish and Wrzosek (1998). The reported lack of participation of the sarcoplasmic reticulum in this mechanism (Bluhm and Lew, 1995; Hongo et al., 1995; Kentish and Wrzosek, 1998) further supports the notion that the NCX_{rev} is one possible mechanism involved in the increase in Ca^{2+} transient.

The question that now arises is if this increase in $[Ca^{2+}]_i$ secondary to the increase in $[Na^+]_i$ is the only mechanism responsible for the positive inotropic effect when Ang II or ET are involved in the mechanism. Figure 13.5 shows that the developed force increases linearly with the increase in $[Na^+]_i$ caused by Na^+/K^+-ATPase inhibition, and that this increase is blunted by KB-R7943 (Fig. 13.5, inset). However, when $[Na^+]_i$ increases because of ET-1-induced activation of NHE-1 (Aiello et al., 2005), the increase in developed force lies above the linear relationship (Fig. 13.5). In addition, if ET-1 is applied when the rise in $[Na^+]_i$ caused by Na^+/K^+-ATPase inhibition reached a steady state in the presence

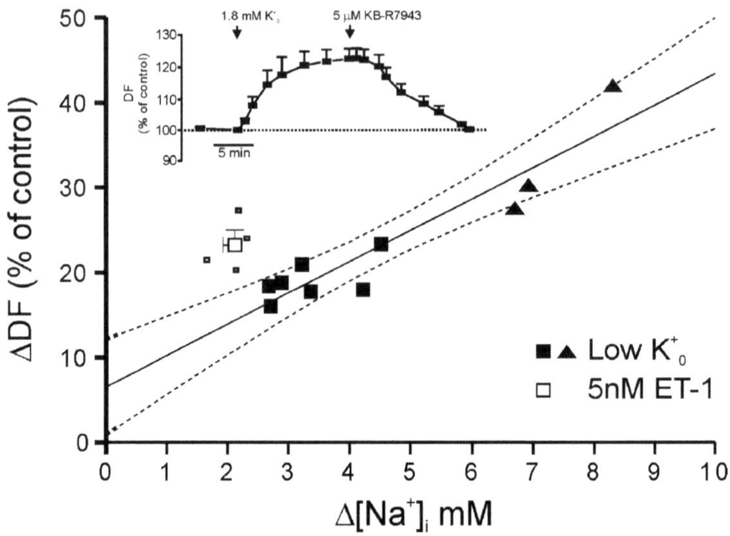

Fig. 13.5 The increase in $[Na^+]_i$ induced by partial inhibition of Na^+/K^+ ATPase by lowering extracellular K^+ (*squares*: 1.8 mmol/L; triangles: 0.9 mmol/L) increased developed force (DF) as a function of $[Na^+]_i$. This effect may be assigned to activation of NCXrev, because it was reverted by KB-R7943 (5 μmol/L; *inset*). However, when $[Na^+]_i$ levels were augmented by ET-1-induced NHE activation, the results lied above the relationship, suggesting that factors additional to the rise in $[Na^+]_i$ have taken place. Modified from Aiello et al. (2005) with permission

of NHE-1 inhibition, the peptide still produces a positive inotropic effect that is completely reversed by either inhibition of NCX_{rev} or protein kinase C (PKC) (Aiello et al., 2005). Patch-clamp experiments in isolated myocytes showed that ET-1 increases the NCX current and negatively shifts the NCX reversal potential (Aiello et al., 2005). Taken together, these data suggest that ET-1 is driving the reverse mode of the NCX by an NHE-1-mediated increase in $[Na^+]_i$ and by a direct stimulatory effect on the NCX, possibly by a PKC-dependent phosphorylation mechanism (Aiello et al., 2005). It is important to mention that PKC is a well known target of intracellular ROS (Juhaszova et al., 2004; Costa and Garlid, 2008). Thus, increased production of ROS by Ang II and/or ET-1 could stimulate PKC and might lead to the activation of both transporters, the NHE-1 and/or the NCX (Fig. 13.6).

Interestingly, experiments performed by Eigel et al. (2004) in guinea pig ventricular myocytes demonstrated that ROS activate NCX directly (Fig. 13.6). On the other hand, it was reported that Ang II or myocardial stretch, via AT_1 receptors stimulation, induces a ROS-mediated reduction of the transient outward potassium current (I_{to}) by a signaling pathway involving NADPH oxidase activation (Zhou et al., 2006). Moreover, Lu et al. (2008), recently reported that I_{to}, the slow delayed outward K^+ current (I_{Kslow}) and the steady-state K^+ current (I_{ss}) are phosphorylated and inhibited by p90RSK after ROS activation of this enzyme. Thus, decreased potassium currents would lead to a prolongation of action potential duration, which may eventually increase Ca^{2+} influx through NCX_{rev} (Fig. 13.6).

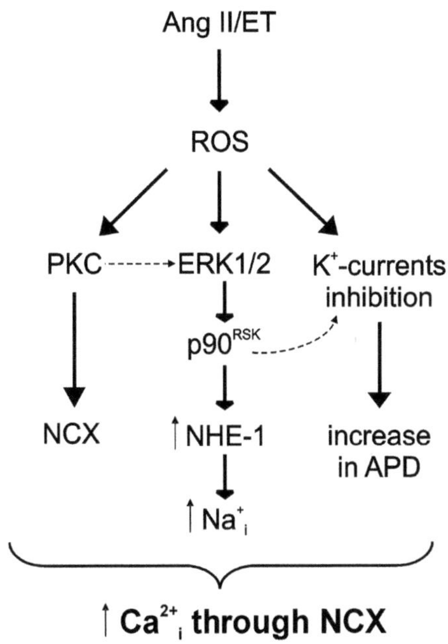

Fig. 13.6 Potential intracellular pathways mediated by ROS after stimulation of the autocrine Ang II/ET crosstalk. ROS generated by this autocrine mechanism might trigger the three different ROS-dependent pathways depicted in the Figure: (**a**) $[Na^+]_i$-independent and PKC-dependent pathway by direct stimulation of NCX; (**b**) $[Na^+]_i$-dependent pathway, consistent with a negative shift of the NCX reversal potential after a rise in $[Na^+]_i$ due to NHE-1 activation and (**c**) prolongation of the action potential duration due to inhibition of K^+ currents

In summary, it may be suggested that the reverse mode of cardiac NCX is modulated by myocardial stretch or, equivalently, by the Ang II/ET network, through the three different ROS-dependent pathways depicted in Fig. 13.6: (a) an $[Na^+]_i$-dependent pathway, consistent with a negative shift of the NCX reversal potential after a rise in $[Na^+]_i$ due to NHE-1 activation; (b) an $[Na^+]_i$–independent and protein kinase C-dependent pathway by direct stimulation of NCX; and (c) a prolongation of the action potential duration. All these intracellular pathways appear to be contributing in concert to the increase in Ca^{2+} after stretch.

The fact that Ang II triggers the beginning of the cascade of events leading to the slow force response has not been confirmed in all their steps. Activation of the NHE-1 after stretch has been confirmed in different species by several authors (Yamazaki et al., 1998; Alvarez et al., 1999; Calaghan and White, 2004; von Lewinski et al., 2004; Luers et al., 2005). However, the pathway leading to its activation is controversial. The release of Ang II and activation of the AT_1 receptors by stretch proposed by us in rat and cat myocardium (Cingolani et al., 1998; Alvarez et al., 1999; Perez et al., 2001), though reported in isolated rat myocytes (Sadoshima et al., 1993; Leri et al., 1998), was not confirmed by other investigators in ferret multicellular preparations (Calaghan and White, 2001). The role played by ET has been reported by Calaghan and White in ferret (Calaghan and White, 2001) and by us in rat (Alvarez et al., 1999) and cat myocardium (Cingolani et al., 1998; Perez et al., 2001), but it was not found in rabbit (Luers et al., 2005) or failing human myocardium (von Lewinski et al., 2004).

Whether the discrepancies are a matter of species differences is not apparent to us yet, but in any case, they leave open the possibility that under different experimental conditions some other mechanisms may be triggered by stretch. In this regard, another report by Calaghan and White (2004) shows activation of stretch-activated channels in addition to NHE-1 activation after myocardial stretch in rat ventricular myocytes and papillary muscles; Isenberg et al. (2005) proposed that myocardial stretch increases $[Na^+]_i$ and $[Ca^{2+}]_i$ in cell organelles partly by their influx through the stretch-activated channels, but they were unable to prevent the increase in $[Na^+]_i$ by cariporide. Interestingly, Hongo et al. (1996) demonstrated that the slow force response can also be detected in isolated cardiomyocytes, but they did not detect an increase in $[Na^+]_i$ during its development. In the same work, the authors also reported that L-type Ca^{2+} current is not involved in the slow force response. Vila Petroff et al. (2001) presented evidence that stretch activates the PI-3-kinase pathway to phosphorylate the endothelial isoform of nitric oxide synthase. Then nitric oxide stimulates Ca^{2+} release from the sarcoplasmic reticulum and promotes the slow force response. Unfortunately, the results of Vila-Petroff et al. (2001) could not be reproduced by other authors either in papillary muscle or isolated myocytes (Calaghan and White, 2004). This was certainly expected since the mechanism proposed by these authors requires a functional sarcoplasmic reticulum and the possible role of the sarcoplasmic reticulum in the slow force response has been clearly ruled out by several authors including Bluhm and Lew (1995), Hongo et al. (1995) and Kentish and Wrzosek (1998).

Another important aspect to consider in order to clarify the failure of detecting if ET is participating in the slow force response to stretch is to analyse the pharmacological intervention used to prove it. In this regard, Endoh et al. have clearly shown that high doses of the non-specific ET receptor antagonist TAK044 were necessary to prevent the inotropic effect of ET in the myocardium (Endoh et al., 1998). In our hands, either TAK044 or the selective ET_A receptor antagonist BQ123 (Fig. 13.7) blunted the slow force response (Alvarez et al., 1999; Perez et al., 2001). However, if based on the works of Calaghan and White (2001) and our own results (Cingolani et al., 1998; Alvarez et al., 1999; Perez et al., 2001) the role of ET after stretch is accepted in addition to the well known fact that Ang II induces release/formation of ET as shown in different studies by us (de Hurtado et al., 2000; Aiello et al., 2002; Perez et al., 2003; Cingolani et al., 2006) and others (Dohi et al., 1992; Imai et al., 1992; Chua et al., 1993; Ito et al., 1993; Fujisaki et al., 1995; Barton et al., 1997; Rajagopalan et al., 1997; Serneri et al., 1999; Muller et al., 2000; Ficai et al., 2001; Ortiz et al., 2001; Muller et al., 2002; Seccia et al., 2003), the rationale to accept our proposed chain of events seems to be plausible.

Regarding the identification of the ET isoform (s) that could be participating in the response to stretch, experiments in cat papillary muscles from our own laboratory showed an increase in ET-3 mRNA after stretch (Ennis et al., 2005). However, we should bear in mind that Tamamori et al. (1996) reported that, in cultured neonatal cardiomyocytes, ET-3 triggers the synthesis and release of ET-1, which in turn mediates a hypertrophic response. Therefore, though speculative, we should consider the possibility that the stretch of multicellular preparations triggers ET-3 release that might be responsible for the inotropic response and for the sequential release/formation of ET-1, which would induce cell growth. Moreover, while stretch could sequentially induce the release of ET-3 and ET-1, it is possible that exogenous

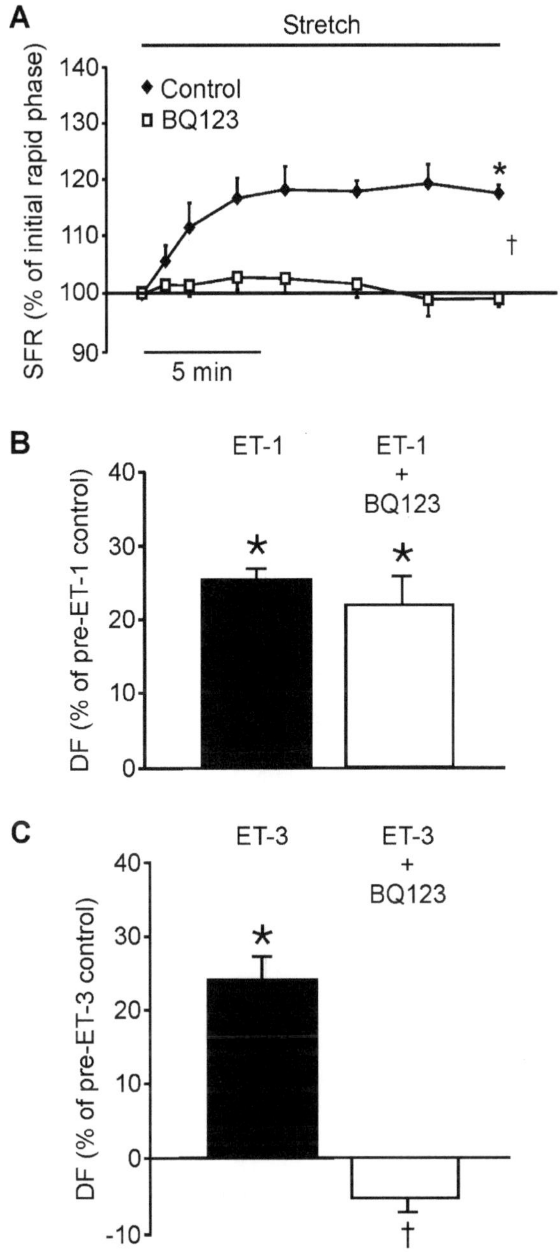

Fig. 13.7 **Panel A**: The stretch of rat papillary muscles promotes a slow force response (SFR) which stabilized after 10–15 min in a value ~20% greater initial phase. The ET_A blocker BQ123 canceled the SFR (the SFR was expressed as percent of the initial rapid phase). * Indicates $P<0.05$ vs. initial rapid phase, † indicates $P<0.05$ between curves. **Panel B** shows the lack of effect of BQ123 on the positive inotropic effect of 5 nmol/L ET-1. Comparative averaged results of developed force (DF) (expressed as percent of the pre-ET-1 value) after 30 min of incubation under both experimental conditions are shown. * Indicates $P<0.05$ vs. pre-ET-1 value. **Panel C**: Addition of 5 nmol/L ET-3 to a papillary muscle increased DF in a similar magnitude to the same dose of ET-1. This time, however, BQ123 canceled the increase in DF, suggesting that this may be the isoform involved in the SFR. DF was expressed as percent of the pre-ET-3 value after 30 min of incubation with the peptide. * Indicates $P<0.05$ vs. pre-ET-3 value, † indicates $P<0.05$ vs ET-3 alone. Modified from Ros et al. (2005) with permission

Ang II induces the release of ET-1 that in turn mediates, in this case, the increase in contractility. Supporting these speculations, we demonstrated, working with cat papillary muscles, that the same concentration of the ET_A blocker BQ123 (300 nmol/L) was able to cancel the slow force response to stretch and the inotropic effect induced by ET-3, but not that induced by ET-1 (Fig. 13.7) (Ros et al., 2005). However, we need to mention that we have recently demonstrated that the positive inotropic effect and the increase in ROS production induced by ET-1 in isolated cat ventricular myocytes were effectively blocked by 300 nmol/L BQ123 (De Giusti et al., 2008).

We can state that myocardial stretch-induced NHE-1 activation and the role of the NCX in increasing Ca^{2+} transient are confirmed facts. Considering the results of other investigators and our own (Cingolani et al., 1998; Alvarez et al., 1999; Perez et al., 2001; Calaghan and White, 2004; Cingolani et al., 2005; Luers et al., 2005) together with those from the experiments in isolated neonatal cardiomyocytes (Yamazaki et al., 1998), we can conclude that NHE-1 activation induced by myocardial stretch constitutes a relevant intracellular signal leading to myocardial hypertrophy. A recent publication support the idea that activation of NHE-1 is sufficient to generate Ca^{2+} signals that induce cardiac hypertrophy and failure (Nakamura et al., 2008). This signaling pathway can also be also evoked by equipotent doses of exogenous Ang II or ET-1 (Perez et al., 2003). Since it has been demonstrated that Ang II induces the release of ET-1 (see below), at least in some species, which in turn induces ROS formation and NHE-1 activation, the physiological chain of events depicted in Fig. 13.8 seems plausible.

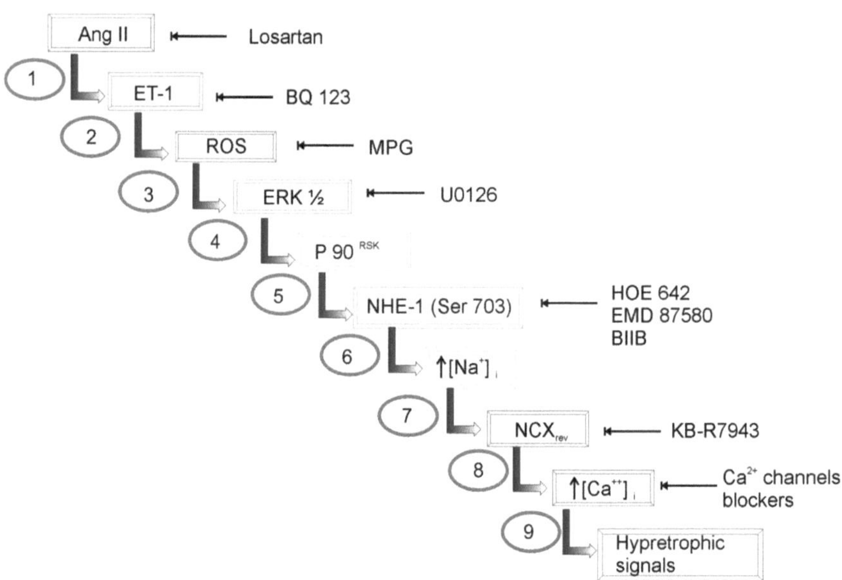

Fig. 13.8 Intracellular mechanisms triggered by 1 nmol/L Ang II. The figure schematizes the sequential steps that take place after activation of AT_1 receptors by Ang II, effect that can be blocked by the AT_1 blocker Losartan. Step 1: release of endogenous ET-1. Step 2: Increased ROS production after ET_A receptors activation, effect that can be blocked by the ET_A antagonist BQ123 and the ROS scavenger MPG. Step 3: Activation of the MAP kinase ERK 1/2 by ROS, effect that can be blocked by the MEK inhibitor U0126. Step 4: Phosphorylation and activation of P90RSK. Step 5: Phosphorylation and activation of the NHE-1, which can be blocked by the NHE-1 inhibitors HOE 642 (cariporide), EMD 87580 and BIIB. Step 6: Increase in the intracellular concentration of Na^+. Step 7: Activation of the reverse mode of the NCX, effect that can be inhibited by the blocker of the NCX_{rev}, KB-R7943. Step 8: Increase in the Ca^{2+} transient. Step 9: This increase in intracellular Ca^{2+} might lead to cardiac hypertrophy

13.3 Evidences for the ANG II-Induced Release of ET-1 Autocrine Mechanism

Many cardiovascular effects initially thought to be mediated by Ang II were in fact reported to be due to the paracrine/autocrine action of endogenous ET-1 released by the octapeptide (Ito et al., 1993; Rajagopalan et al., 1997; Liang and Gardner, 1998; Ortiz et al., 2001). The effects of stretch, which were mediated by the action of endogenous ET released by Ang II described above, were reflected by results obtained in cat papillary muscles. Since this is a multicellular preparation, it was not possible to elucidate if the action of ET-1 was paracrine or autocrine. However, working with isolated cat ventricular myocytes, we also reported that the increase in I_{NCX} induced by Ang II was blocked by ET receptors blockers, suggesting an autocrine interaction between these two hormones (Aiello et al., 2002). Accordingly, more recently we also showed that Ang II induced a concentration-dependent increase in sarcomere shortening of cat myocytes, which was downward shifted after ET receptors blockade (Fig. 13.9). This shift decreased the maximal effect of Ang II by approximately 30% and cancelled the effect of 1 nmol/L Ang II (Fig. 13.9). Therefore, these findings demonstrate that the increase in contractility induced by 1 nmol/L Ang II is entirely due to an autocrine pathway involving an ET isoform.

Further evidence that Ang II induces the release/production of ET from the myocyte was obtained in RT-PCR experiments performed in isolated cat myocytes

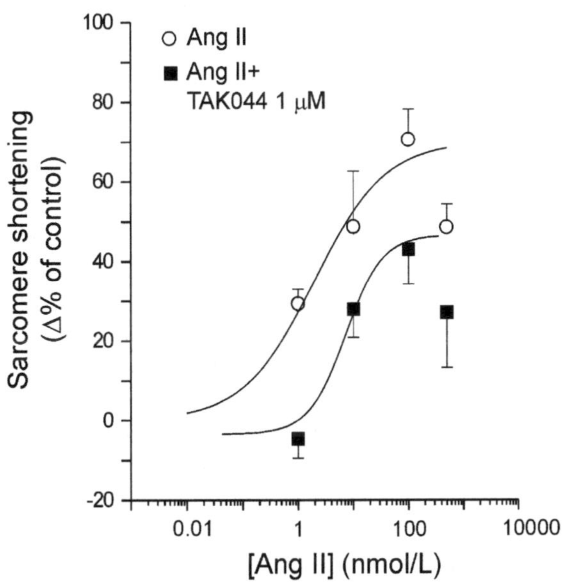

Fig. 13.9 Dose-response curve for different concentrations of Ang II, from 1 to 500 nmol/L, in the absence or presence of TAK044 (1 μmol/L). The maximal positive inotropic effect was obtained with 100 nmol/L Ang II. TAK044 shifted the dose-response curve to the right, and completely blocked the inotropic effect of 1 nmol/L Ang II indicating that this effect was entirely due to the action of the endogenous ET released/produced by Ang II. However, the data suggest that concentrations higher than 1 nmol/L are activating other mechanisms than the autocrine signal triggered by Ang II. Modified from Cingolani et al. (2006) with permission

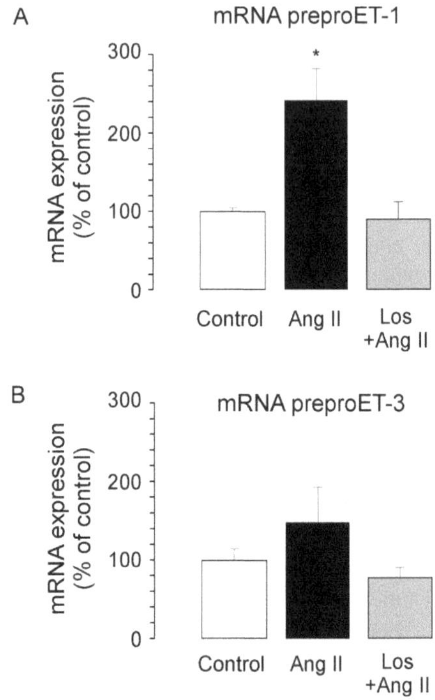

Fig. 13.10 Real-time RT-PCR. **Panel A** shows a significant increase in the expression of mRNA of preproET-1 induced by 1 nmol/L Ang II. This increase was prevented by losartan. The mRNA levels for preproET-3 (**Panel B**) does not change with 1 nmol/L Ang II. * Indicates $p<0.05$ vs. control. Modified from Cingolani et al. (2006) with permission

exposed to 1 nmol/L Ang II. Following 15 minutes of exposure to Ang II, isolated cat ventricular myocytes showed a significant increase in the expression of preproET-1 mRNA but not in that of preproET-3 mRNA (Fig. 13.10) (Cingolani et al., 2006). Although these RT-PCR experiments do not elucidate the mechanism by which Ang II induces the release/production of ET-1, they suggest that Ang II increases the "de novo" production of ET-1 in the isolated myocytes. It seems unlikely that the increase in preproET-1 mRNA levels in 15 min upon Ang II exposure could explain the acute positive inotropic effect induced by this peptide during that time frame, since production of mature ET-1 to be released by the myocyte would require a longer time period. However, it appears valid to assume, if the translation efficiency is not altered, that the Ang II-induced increase in ET-1 mRNA during this time reflects an increase in ET-1 synthesis secondary to the decrease in its intracellular pools due to its release, a mechanism probably leading to restore these intracellular pools.

13.4 The Slow Force Response as the Mechanical Counterpart of the Autocrine Mechanism Triggered by Stretch: the Anrep's Phenomenon

It is well known that two consecutive phases characterize the increase in force after myocardial stretch: A rapid and immediate one and the slow force response. The initial rapid change in force is induced by an increase in myofilament Ca^{2+}

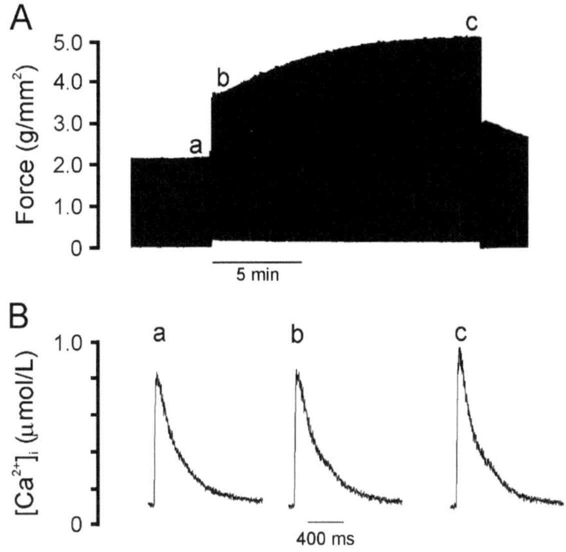

Fig. 13.11 After stretching a papillary muscle from 92 to 98% of Lmax, a sudden increase in force immediately occurs (a to b, **Panel A**), due to an increase in myofilament Ca^{2+} responsiveness. After that, a progressive increase in force develops during the next 10–15 min, the slow force response (SFR) (b to c), that is due to an increase in the Ca^{2+} transient (**Panel B**). Modified from Cingolani et al. (2001) with permission

responsiveness without changes in the Ca^{2+} transient whose underlying mechanisms are beyond the scope of this review (Fig. 13.11). The slow force response, in turn, is due to a progressive increase in the Ca^{2+} transient without changes in myofilament Ca^{2+} responsiveness during this phase (Fig. 13.11) (Allen and Kurihara, 1982; Kentish and Wrzosek, 1998; Alvarez et al., 1999). The increase in the Ca^{2+} level appears to result from the autocrine/paracrine mechanism described in the previous section. While the initial change in force after stretch seems to express the Frank-Starling mechanism, the slow force response may conceivably be the expression of Anrep's phenomenon.

In 1912, Von Anrep observed that when aortic pressure was elevated, ventricular volume initially increased and then declined to the starting volume. It appeared to him that an influence operating soon after myocardial dilatation caused an increase in myocardial contractility. His interpretation was that perhaps, the decrease in the flow to the adrenal glands induced the release of catecholamines and the consequent positive inotropic effect. In 1959, experiments by Rosenblueth et al. (1959) indicated that an increase in coronary perfusion pressure was not necessarily concomitant with the return of the heart to its initial volume. In 1960, Sarnoff et al. coined the term "pressure-induced homeometric autoregulation" to define the decrease in left ventricular end diastolic volume that follows an increase in diastolic volume due to a sudden increase in afterload. On the other hand, since the experiments of Sarnoff et al. (1960) were performed in isolated hearts, the study served to rule out the possibility of a role played by catecholamines in the described phenomenon. Interestingly, Sarnoff defined as "homeometric autoregulation" a phenomenon occurring in an organ which was not attributable to an influence by nerves or chemicals in its vicinity, paving the way for the idea of an autocrine/paracrine mechanism after cardiac stretching (Sarnoff et al., 1960). The existence of a real change in contractility during the homeometric autoregulation was challenged by the possibility of changes

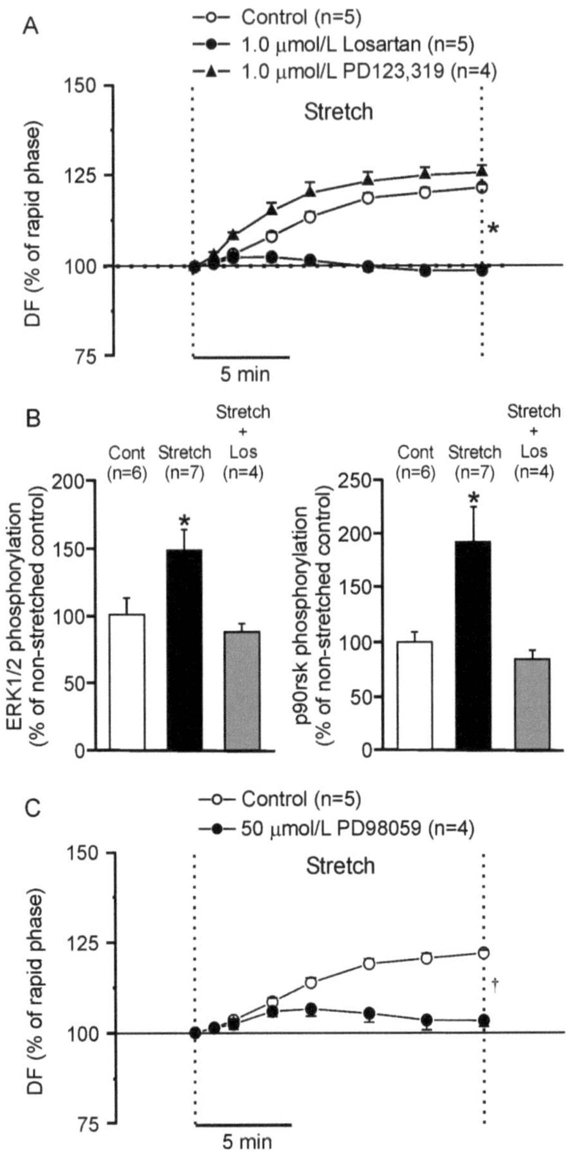

Fig. 13.12 Suppression of the slow force response (expressed as percent of initial rapid phase) after AT_1 but not AT_2 receptors blockade (Losartan and PD123, 319 respectively) (**Panel A**). Myocardial stretch significantly increased ERK1/2 and p90RSK phosphorylation, effect cancelled by losartan (Los) (**Panel B**). Inhibition of MEK (a kinase upstream ERK1/2 and downstream RAS) by PD98059 cancelled slow force response (expressed as percent of the initial rapid phase) (**Panel C**). *Indicates $P < 0.05$ vs. non-stretched control (cont); † indicates $P < 0.05$ control vs. PD98059. DF = developed force. Modified from Caldiz et al. (2007) with permission

in coronary blood flow distribution (Monroe et al., 1972). However, in 1973 Parmley and Chuck reproduced for the first time the contractile effect of stretch in isolated strips of ventricular myocardium. They showed that when the length of the muscle was increased, there were corresponding rapid and slow increases in the developed force. Since the slow force response to the change in length was still present in isolated muscles from animals treated with reserpine, those authors also ruled out the possibility of catecholamines released by nerve endings as having a role in the mechanism.

We and other authors have provided evidence that activation of NHE-1 after stretch play a key role in the development of the slow force response (Alvarez et al., 1999; Perez et al., 2001; Calaghan and White, 2004; von Lewinski et al., 2004; Luers et al., 2005), however, there is no agreement in the role played by Ang II and ET in NHE-1 activation (Sadoshima et al., 1993; Leri et al., 1998; Alvarez et al., 1999; Calaghan and White, 2001; Perez et al., 2001). Ang II is an octapeptide acting through its own G coupled receptors AT_1 and AT_2. $G_{\alpha}q$-$\beta\gamma$ activated by either Ang II or ET-1 targets the NHE through extracellular signal-regulated protein kinases 1/2 (ERK1/2)-p90 ribosomal S6 kinase (p90RSK). We showed that the slow force response was abolished by AT_1 receptors blockade (Alvarez et al., 1999; Perez et al., 2001) (Caldiz et al., 2007) but not by AT_2 receptors blockade (Caldiz et al., 2007) as shown in Fig. 13.12A. These results support the notion that Ang II is released after stretch and triggers the intracellular signaling pathways leading to slow force response. We should keep in mind that the release of Ang II from the cell after stretch and its link with ET-1 has been previously demonstrated (Ito et al., 1993; Sadoshima et al., 1993). Furthermore, a significant increase in ERK1/2 and p90RSK kinase phosphorylation can be detected after 15 minutes of stretch, effects that are both cancelled by AT_1 receptors blockade with losartan as shown in Fig. 13.12B (Caldiz et al., 2007). Finally, inhibition of MEK (a kinase that is upstream of ERK1/2 and downstream of RAS kinases) by PD98059 abolished the slow force response to stretch (Fig. 13.12C).

13.5 Role of ROS After Stretch, ANG II and ET-1

Ang II and ET-1 are well known activators of the NADPH oxidase (Lavigne et al., 2001; Giordano, 2005; Kimura et al., 2005b) and through this action it has been reported the phenomenon called "ROS-induced ROS-release", by which a small amount of ROS triggers a greater ROS production from the mitochondria (Fig. 13.13) (Zorov et al., 2000; Brandes, 2005; Kimura et al., 2005a). The possibility that this mechanism participates in the chain of events following stretch was examined. Figure 13.14A shows that stretch -in addition to its mechanical effect- induces an increase in intracellular ROS formation of approximately 30% above baseline levels. Furthermore, scavenging of ROS by N-(2-mercaptopropionyl)-glycine (MPG) or EUK8 inhibited both stretch-induced increase in ROS (Fig. 13.14A) and the slow force response (Fig. 13.14B). We

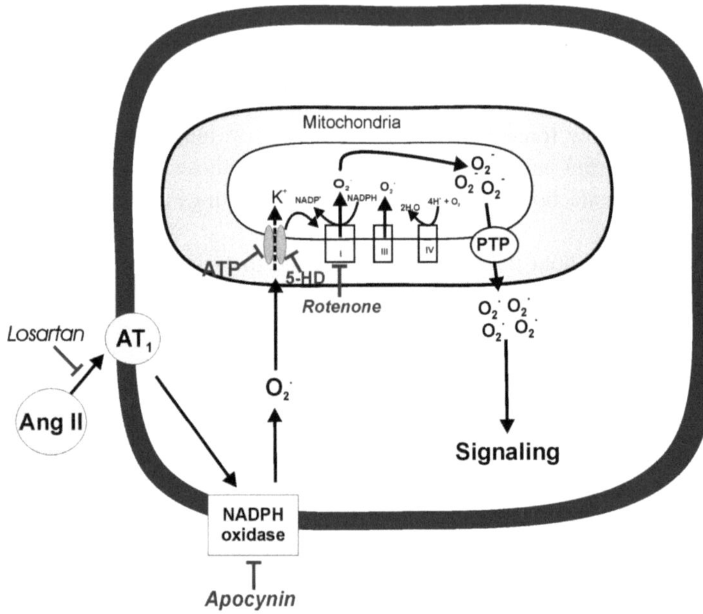

Fig. 13.13 The proposed "ROS-induced ROS-release mechanism". Stimulation of cardiac myocytes with Ang II leads via the action of AT_1 receptor to the assembly and activation of NADPH oxidase. The subsequently generated O_2^- stimulate mK_{ATP} channels, which augments the production of more O_2^- by the electron transport chain and allows the mitochondrial permeability transition pore (MPT) to open, facilitating the efflux of large amounts of O_2^- into the cytoplasm. O_2^- (or H_2O_2) can then act as signaling molecules in the cytosol (i.e. activating MAP kinases)

also found that the scavenging of ROS inhibited the increase in $[Na^+]_i$ that occurs in response to the stretch (Fig. 13.14C). We may hypothesize that activation of NAPDH oxidase after stretch would produce a small amount of O_2^-, which may open the ATP-sensitive mitochondrial potassium (mK_{ATP}) channels and produce a larger amount of O_2^- responsible for generating the slow force response. Therefore, if these assumptions were correct, the slow force response should be abolished by either NADPH oxidase inactivation or blockade of mK_{ATP} channels. As shown in Fig. 13.15A, slow force response was abolished after inhibition of NADPH oxidase inhibition (apocynin or diphenyleneiodonium chloride, DPI) or after blockade of mKATP channels (5-hydroxydecanoate, 5HD, or glibenclamide). The NHE-1-induced increase in $[Na^+]_i$ underlying the slow force response was also abolished by these interventions (Fig. 13.15B).

Ang II induced the production of O_2^- in a concentration-dependent manner in cat cardiac slices (Fig. 13.16A). Interestingly, the Ang II-induced concentration-dependent increase in O_2^- was very similar to the above shown (Fig. 13.9) concentration-dependent inotropic response curve (Fig. 13.16A), suggesting a potential correlation between Ang II-induced ROS production and positive

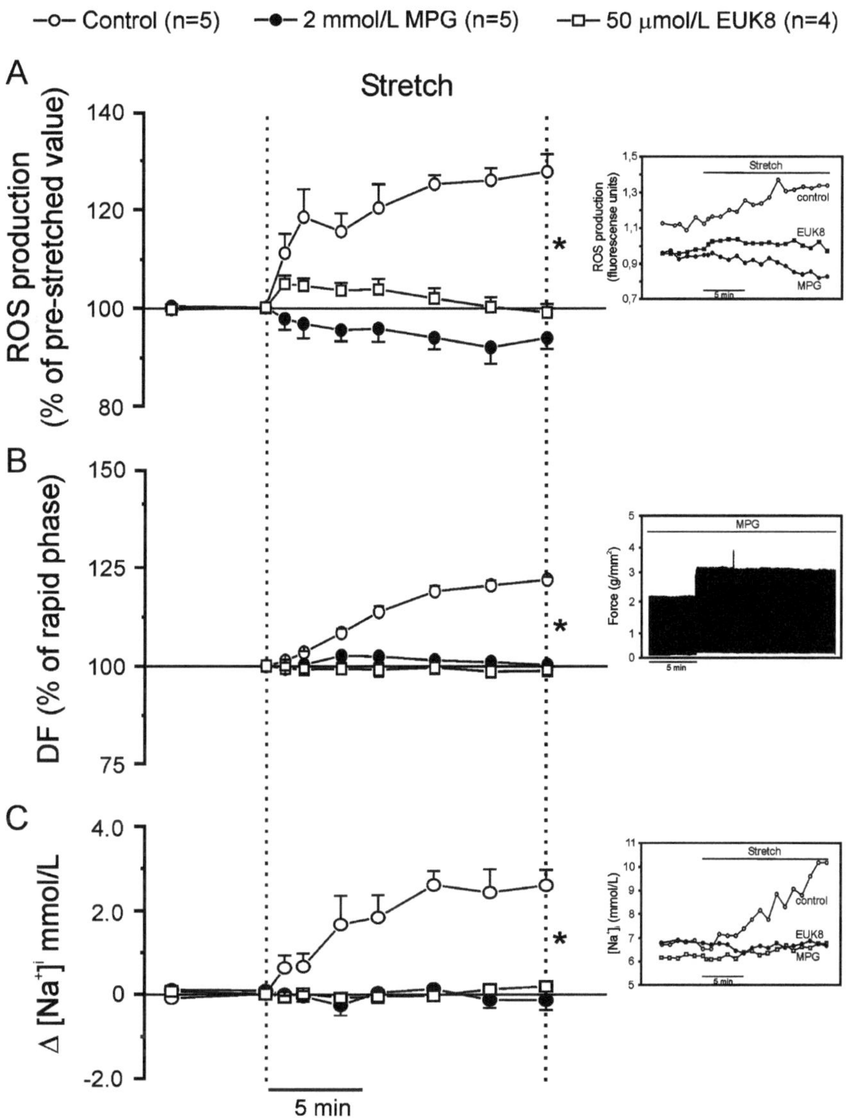

Fig. 13.14 Myocardial stretch induced an intracellular ROS increase of ~30% above the baseline levels that was cancelled by the ROS scavengers MPG and EUK8 (**Panel A**). MPG and EUK8 also cancelled the slow force response (expressed as percent of initial rapid phase) (**Panel B**). Furthermore, ROS scavenging also blunted stretch-induced increase in $(Na^+)_i$ (**Panel C**). *Insets* show original raw data. *Indicates $P < 0.05$ control vs. MPG and EUK8. DF = developed force. Modified from Caldiz et al. (2007) with permission

inotropy. The O_2^- production augmented by 1 nmol/L Ang II was abolished by AT_1 receptors blockade (losartan), ROS scavenging (MPG), NADPH oxidase inhibition (apocynin) and mK_{ATP} channels blockade (5HD or glibenclamide) as

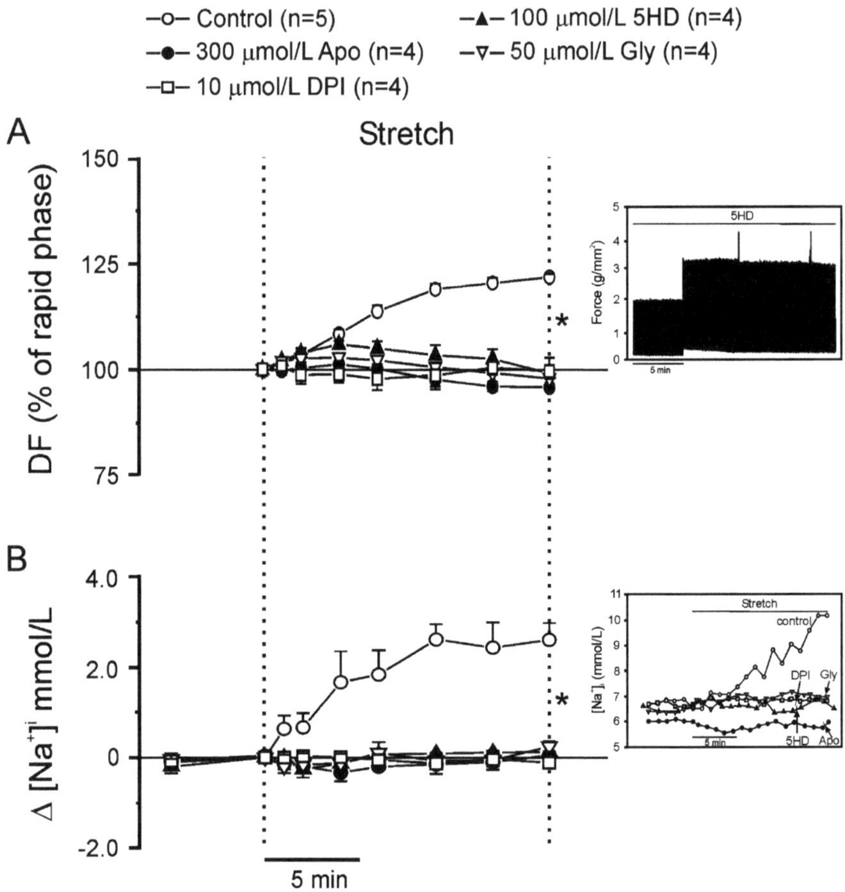

Fig. 13.15 NADPH oxidase inhibition by apocynin (Apo) or diphenyleneiodonium chloride (DPI) as well as mK_{ATP} channels blockade with 5-hydroxydecanoate (5HD) or glybenclamide (Gly) abolished slow force response (expressed as percent of initial rapid phase) (**Panel A**). All these interventions also cancelled NHE-1-mediated increase in $[Na^+]_i$ that accompanied the slow force response (**Panel B**). Insets show original raw data. * Indicates $P < 0.05$ control vs. all other groups. DF = developed force. Modified from Caldiz et al. (2007) with permission

shown in Fig. 13.16B. This Ang II-induced O_2^- production was also blunted by the non selective ET receptors blocker TAK044 and by the selective ET_A receptors antagonist BQ123 (unpublished observations), indicating that this effect is, in fact, mediated by endogenous ET released by Ang II. Consistently, MPG, apocynin, glybenclamide and 5HD also blocked the production of O_2^- induced by exogenous ET-1 in isolated cat ventricular myocytes (De Giusti et al., 2008) (Fig. 13.17). In line with these experiments, the ET-1-induced positive inotropic effect in cat ventricular myocytes was inhibited by these blockers (De Giusti et al., 2008) (Fig. 13.18), indicating that the "ROS-induced ROS-release" mechanism triggered

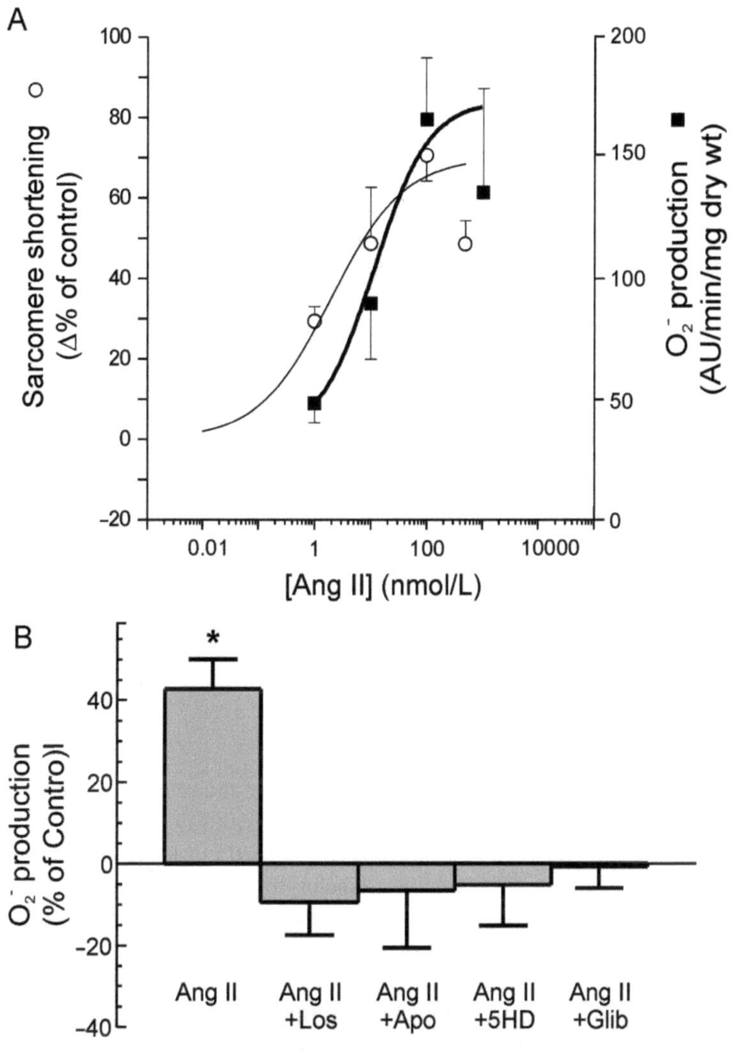

Fig. 13.16 **Panel A**: Ang II dose-response curves for the inotropic response and the production of O_2^-. The effect of different concentrations of Ang II on O_2^- production was assessed in cardiac tissue slices. Values of O_2^- production are expressed as the difference from control. The Ang II-induced concentration-dependent increase in O_2^- was very similar to the concentration-dependent inotropic response curve, suggesting a potential correlation between Ang II-induced ROS production and positive inotropy. **Panel B**: Superoxide production induced by 1 nmol/L Ang II (n = 34) in the absence and presence of 1 μmol/L losartan (Los, n = 8); 2 mmol/L MPG (n = 3); 300 μmol/L apocynin (Apo, n = 7); 100 μmol/L 5-hydroxydecanoate (5HD, n = 10) and 50 μmol/L glibenclamide (Gly, n = 6), expressed as percent of control values without additions and after 15 min of incubation. * Indicates $P < 0.05$ vs. control. Modified from Caldiz et al. (2007) and Garciarena et al. (2008) with permission

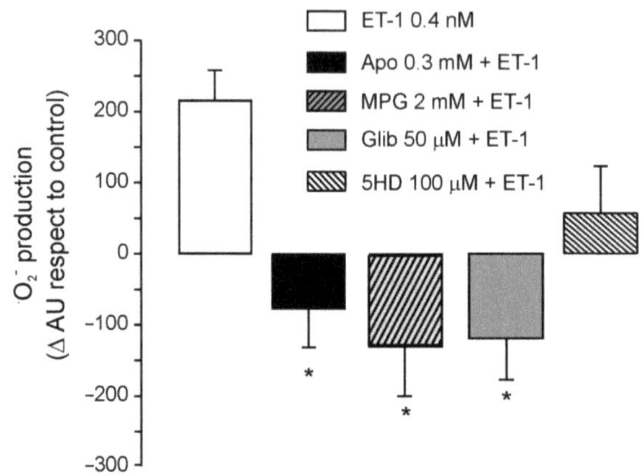

Fig. 13.17 Effect of MPG, the NADPH oxidase blocker, apocynin, and the mK_{ATP} blockers, glibenclamide and 5HD, on the ET-1-induced O_2^- production. Average increase in O_2^- after 15 min of 0.4 nmol/L ET-1 ($n = 12$). This increase was prevented by apocynin (Apo, 0.3 mmol/L; $n = 13$), MPG (2 mmol/L; $n = 12$) and glibenclamide (Glib, 50 μmol/L; $n = 14$) and attenuated by 5HD (100 μmol/L; $n = 15$) indicating that ET-1 is inducing the formation of O_2^- by activation of the NADPH oxidase, which in turn release O_2^- from the mitochondria after opening mK_{ATP} channels (ROS-induced-ROS-release). The results were expressed as the values in AU min $^{-1}$ 105 cells $^{-1}$ obtained in the presence of drugs minus control. *Indicates $p < 0.05$ vs. ET-1. Modified from De Giusti et al. (2008) with permission

by ET-1 participates in the inotropic response, being the release of mitochondrial ROS a step in the signaling cascade involved in this pathway.

The ET-1-induced positive inotropic effect observed in cat ventricular myocytes was also cancelled by the PKC inhibitor, chelerythrine, indicating that this kinase is involved in the intracellular pathway of this effect (De Giusti et al., 2008) (Fig. 13.19). However, the exact site of action of this enzyme in the chain of effects is unknown. One of these possible sites could be the activation of NADPH oxidase, since PKC activation is a critical step in the phosphorylation of the NADPH oxidase subunit p47phox and the subsequent assembly of this enzyme (Seshiah et al., 2002). However, the participation of PKC downstream NADPH oxidase activation can also be responsible for the effects of ET-1 on contractility. In relation to this matter, it is important to mention that the stimulation of the NHE-1 after PKC activation by ROS has been previously reported (Snabaitis et al., 2002). In addition, PKC can act upstream or downstream mK_{ATP} channels since PKC stimulation of mK_{ATP} channels (Sato et al., 1998) and PKC activation by mitochondrial ROS produced after mK_{ATP} channels opening (Juhaszova et al., 2004) have been reported. Moreover, a feed-forward mechanism in which mitochondrial swelling leads to activation of PKC, which stimulates mK_{ATP} channels and further increases mitochondrial swelling, has been also proposed (Juhaszova et al., 2004). Finally, the possibility that different PKC isoforms are acting upstream and downstream the production of ROS and/or the activation of mK_{ATP} channels might also be considered.

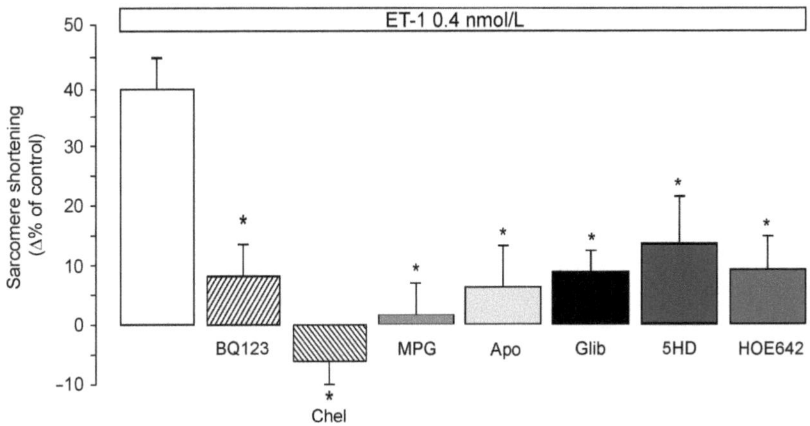

Fig. 13.18 The ET-1-induced positive inotropic effect is blunted by ET_A receptors, PKC and NADPH oxidase blockade, ROS scavenging, mK_{ATP} blockers and NHE inhibition. The average changes in SL shortening, expressed as delta percent of the control, with 0.4 nmol/L ET-1 ($n = 10$), and with the same concentration of ET-1 but in the presence of 0.3 μmol/L BQ123 ($n = 9$), 2 μmol/L chelerythrine (Chel, $n = 6$), 2 mmol/L MPG ($n = 8$), 0.3 mmol/L apocynin (Apo, $n = 13$), glibenclamide (Glib, 50 μmol/L, $n = 6$), 5HD (500 μmol/L, $n = 9$) and 5 μmol/L HOE642 (cariporide, $n = 7$) are shown. The positive inotropic effect induced by ET-1 was inhibited by BQ123, indicating that this effect is due to stimulation of the ET_A receptor. Prevention of the ET-1-induced increase in contractility with Chel suggests the participation of PKC in the intracellular pathway. Since MPG, Apo, Glib and 5HD also abolished this positive inotropic effect, the results suggest the participation of ET-1-induced ROS production by NADPH oxidase and the participation of mitochondrial ROS in this effect. Furthermore, the positive inotropic effect induced by ET-1 was inhibited by HOE642, indicating that this effect is mediated by NHE stimulation. *Indicates $p < 0.05$ vs. ET-1. Modified from De Giusti et al. (2008) with permission

The ET-1-induced positive inotropic effect was inhibited by NHE blockade with HOE642 (Fig. 13.18). Additionally, ET-1 was able to increase the proton flux (J_H) carried by the NHE during the recovery of intracellular acidosis induced by ammonium pulses and this effect was inhibited by scavenging ROS with MPG (De Giusti et al., 2008). These data are in line with previous results that have shown activation of the NHE after exogenous addition of H_2O_2 and stimulation of the MAPK ERK 1/2 pathway (Snabaitis et al., 2002) (Sabri et al., 1998). Consistently, ERK 1/2 phosphorylation induced by 1 nmol/L Ang II was cancelled by MPG, apocynin, glibenclamide, 5HD and the inhibitor of the complex I of the electron transport chain, rotenone (Fig. 13.19) (Garciarena et al., 2008), indicating that mitochondrial ROS released after NADPH oxidase activation are responsible for this effect. Interestingly, ERK 1/2 phosphorylation was also inhibited by cariporide (HOE642) (Fig. 13.19), suggesting that this compound is acting at a mitochondrial site, as also suggested by other authors (Juhaszova et al., 2004; Toda et al., 2007). In addition, it has also been demonstrated in cardiac slices that the Ang II-induced mitochondrial O_2^- formation was cancelled by cariporide and two other NHE-1 blockers, BIIB723 and EMD87580 (Fig. 13.20A) (Garciarena et al., 2008). Parallel in vitro experiments determined that these inhibitors were unable to decrease O_2^- formation

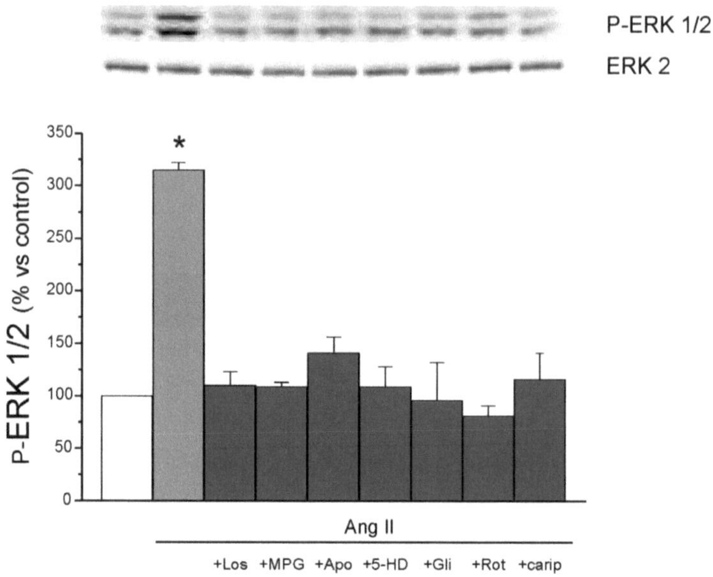

Fig. 13.19 Ang II-induced phosphorylation of ERK 1/2. Ang II (1 nmol/L) induced an increase in ERK1/2 phosphorylation in isolated cat ventricular myocytes that was prevented by losartan (Los, 1 μmol/L), MPG (2 mmol/L), apocynin (Apo, 300 μmol/L), 5-HD (100 μmol/L), glibenclamide (Gli, 50 μmol/L), rotenone (Rot, 10 μmol/L), and cariporide (carip, 10 μmol/L) (n = 4). No changes in total ERK1/2 was observed. $^*p < 0.05$ vs. all other groups, ANOVA. Modified from Garciarena et al. (2008) with permission

induced by PMS and NADH in a range that includes the values of chemiluminescence obtained with 1–100 nmol/L Ang II (Fig. 13.21A) (Garciarena et al., 2008), indicating that they were not acting as ROS scavengers. Moreover, the production of mitochondrial O_2^- induced by the mK_{ATP} opener diazoxide was also inhibited by cariporide (Fig. 13.20B). Thus, it seems likely that cariporide is targeting the mitochondria and blunting ROS formation which, in addition to the direct blocking effect

→

Fig. 13.20 **Panel A**: The stimulatory effect of 30 min-incubation with Ang II on O_2^- production by cardiac tissue slices was prevented by three different NHE-1 inhibitors; cariporide (carip, 10 μmol/L; $n = 12$), BIIB723 (BIIB, 1 μmol/L; $n = 3$) and EMD87580 (EMD, 5 μmol/L; $n = 4$). Values are the difference from the control after 15 min in the presence of lucigenin. **Panel B**: The increase in the chemiluminescence signal observed with 100 μmol/L Diaz ($n = 17$) was of a similar magnitude to that induced by 1 nmol/L Ang II and it was prevented by 5-HD (100 μmol/L; $n = 5$), carip (10 μmol/L; $n = 5$) and cyclosporine A (CsA) 2 μmol/L ($n = 5$). $^*p < 0.05$ vs. all other groups, ANOVA. **Panel C**: MPTP formation inhibition suppressed the stimulatory action of Ang II on mitochondrial ROS production. CsA (0.5, 1 and 2 μmol/L) prevented the effect of Ang II ($n = 4$). 2 μmol/L CsA did not affect control chemiluminiscence signal. Values are the difference from the control after 15 min in the presence of lucigenin expressed as the mean ± SE. None of the inhibitors used had an effect on the control chemiluminescence signal. $^*p < 0.05$ vs. all other groups, ANOVA. Modified from Garciarena et al. (2008) with permission

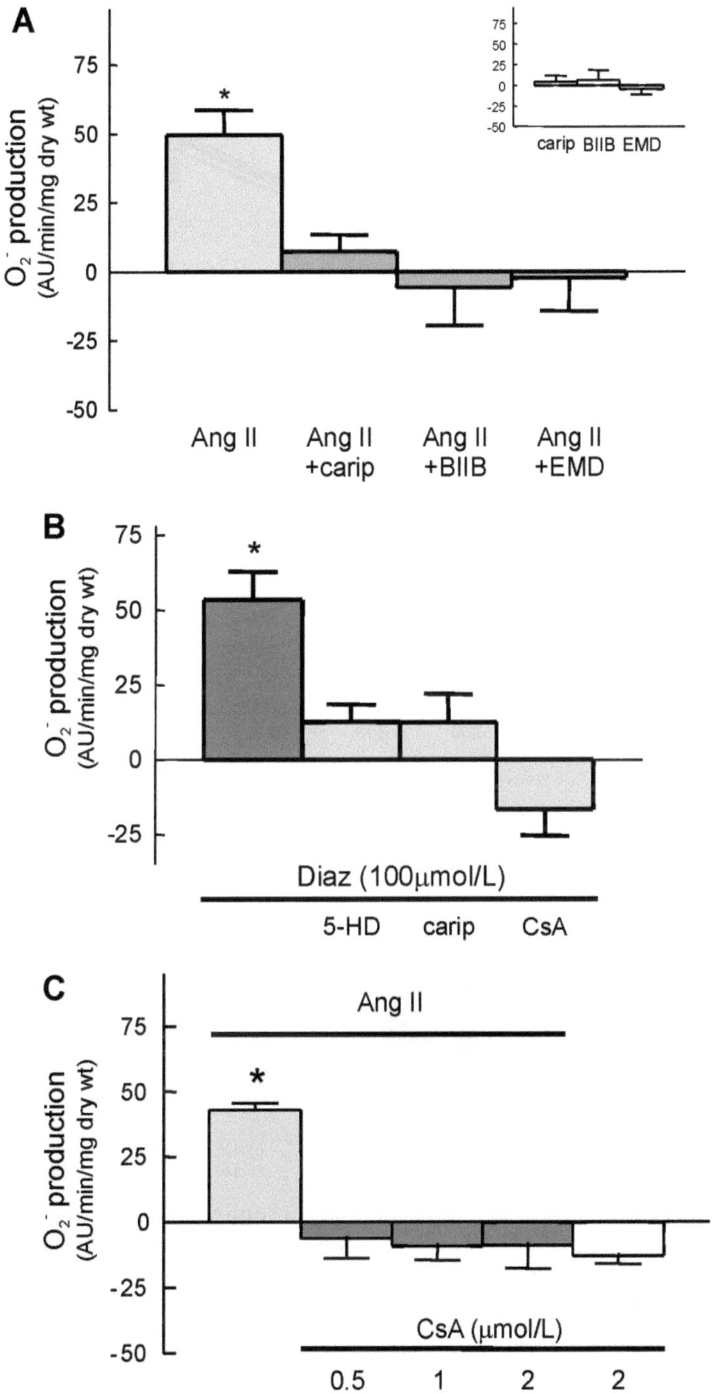

Fig. 13.20 (continued)

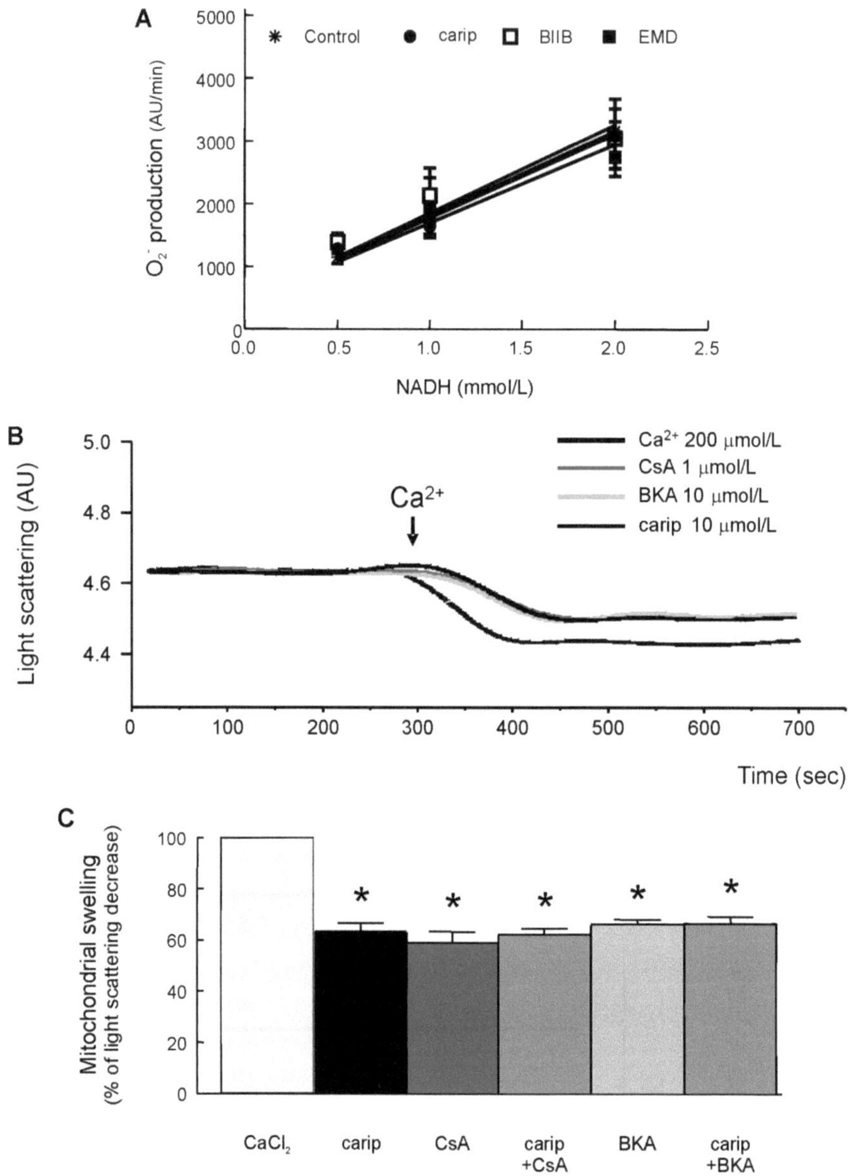

Fig. 13.21 **Panel A**: Lack of ROS scavenger effects of NHE-1 inhibitors. O_2^- production was induced in vitro by PMS and NADH in a range that includes the values of chemiluminescence (in AU/min) obtained with 1–100 nmol/L Ang II. None of the NHE-1 inhibitors (BIIB, carip and EMD) had an effect on the detected levels of $O_2^{-\cdot}$ induced by PMS and NADH ($n = 5$). **Panel B**: Mitochondrial swelling induced by $CaCl_2$. Typical experiment showing that cyclosporine A (CsA) and bongkrekic acid (BKA) significantly attenuated calcium-induced mitochondrial swelling and the decrease in light scattering in mitochondrial suspensions. Cariporide inhibited the decrease in light scattering in a similar magnitude to CsA (1 μmol/L) and BKA (10 μmol/L). **Panel C**: Average results. The combination of both drugs, CsA or BKA with cariporide, did not show any greater effect (n = 7). * $p < 0.05$ vs. $CaCl_2$, ANOVA. Modified from Garciarena et al. (2008) with permission

of the sarcolemmal NHE-1, would prevent the activation of this transporter by ROS. Recent experiments performed with isolated cat ventricular mitochondria allowed us to suggest that the mitochondrial site of action of cariporide could be, directly or indirectly, the mitochondrial permeability transiton pore (MPT), since this drug and the MPT blocker cyclosporine A (CsA) inhibited mitochondrial swelling to the same extent and both effects were not additive (Fig. 13.21B, C) (Garciarena et al., 2008).

It is well known that irreversible opening of the MPT leads to cell death. Mitochondrial ROS burst lower the threshold for MPT opening, triggering the apoptotic cascade (Kim et al., 2003; Shivakumar et al., 2008). However, milder mitochondrial ROS generation after mK_{ATP} opening induces a series of anti-apoptotic events, involving PKC activation, glucogen synthase kinase 3β (GSK3β) phosphorylation and prevention of MPT opening (Juhaszova et al., 2004; Costa and Garlid, 2008; Gomez et al., 2008). We have recently shown that the Ang II- and diazoxide-induced O_2^- production was cancelled after MPT blockade with CsA (Fig. 13.20B, C) (Garciarena et al., 2008). A possible explanation is that MPT opening is necessary to induce the increased production of mitochondrial O_2^-. Supporting this hypothesis, Cheng et al. (Wang et al., 2008) have recently demonstrated that reversible and transient opening of MPT triggers the formation of O_2^- flashes in the mitochondrial matrix.

It is important to note that both, mK_{ATP} activation (which could lead to "protective" mitochondrial ROS production) and inhibition of the NHE-1 by cariporide, have been identified as relevant cardioprotective mechanisms upon ischemia/reperfusion (Karmazyn et al., 1999; Pain et al., 2000; Avkiran and Marber, 2002; Oldenburg et al., 2003, 2004; Kimura et al., 2005b). However, regarding the inhibitory effects of cariporide and CsA on the diazoxide-induced O_2^- production, we could speculate that the protection induced by diazoxide would be lost with cariporide (Fig. 13.22). This speculation, that seems paradoxical, would be an interesting topic for further research.

The intracellular pathways discussed in this section, which involve the participation of the "ROS-induced ROS release mechanism" triggered by the autocrine

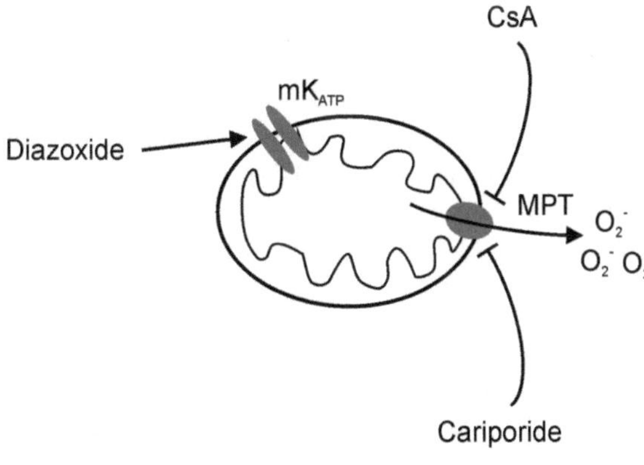

Fig. 13.22 Diazoxide stimulates mK_{ATP} channels leading to an increase in mitochondrial ROS production that might flux across the membrane through the MPT. Either CsA or cariporide inhibited the mitochondrial ROS release, suggesting that they have a common target, the MPT

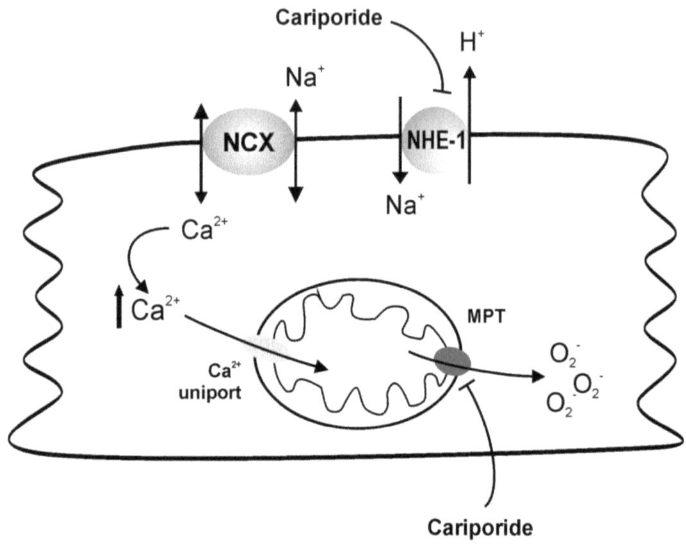

Fig. 13.23 Possible sites of action of cariporide in the cell: the sarcolemma and the mitochondria. The figure shows that cariporide can inhibit NHE-1, leading to a decrease in $Na^{+}{}_{i}$ and $Ca^{2+}{}_{i}$ (decrease of NCX reverse mode or increase of NCX forward mode) and therefore also a decrease in mitochondrial calcium. On the other hand, cariporide can inhibit MPT. In both cases, cariporide might attenuate the mitochondrial ROS production

Ang II/ET-1 mechanism are depicted in the cell and mitochondrial schemes of Figs. 13.23 and 13.24, respectively.

13.6 The Mechanical and Hypertrophic Effect of NHE-1 Activation

The possible link between slow force response to stretch and myocardial hypertrophy is supported by the fact that an enhanced activity of the NHE-1 – the cause of the slow force response – is detected in several models of cardiac hypertrophy and, consistent with this, the specific blockade of NHE-1 has been shown to effectively regress cardiac hypertrophy in different models (Hori et al., 1990; Mrkic et al., 1993; Perez et al., 1995; Schussheim and Radda, 1995; Takewaki et al., 1995; Yamazaki et al., 1996, 1998; Schluter et al., 1998; Hayasaki-Kajiwara et al., 1999; Yokoyama et al., 2000; Yoshida and Karmazyn, 2000; Chen et al., 2001; Konstantinou-Tegou et al., 2001; Kusumoto et al., 2001; Camilion de Hurtado et al., 2002b; Engelhardt et al., 2002; Schafer et al., 2002; Bak and Ingwall, 2003; Ennis et al., 2003; Fujisawa et al., 2003; Karmazyn et al., 2003; Rajapurohitam et al., 2003; Saleh et al., 2003; Aker et al., 2004; Chen et al., 2004; Marano et al., 2004; Xu et al., 2004; Baartscheer et al., 2005; Chahine et al., 2005; Javadov et al., 2005; Kilic et al., 2005; Rajapurohitam et al., 2006). The increase in $[Ca^{2+}]_i$ is widely

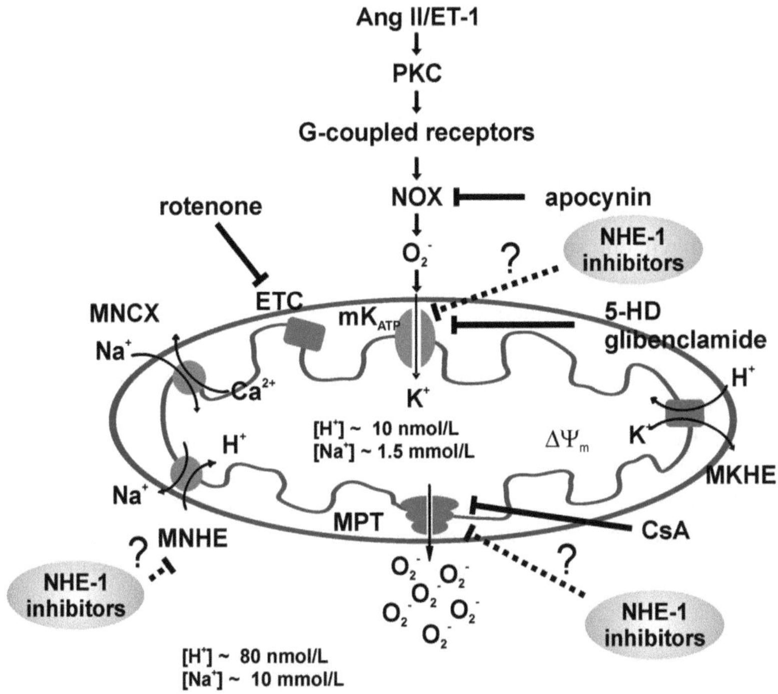

Fig. 13.24 Possible mitochondrial sites of action of NHE-1 inhibitors. The scheme shows the "two step" release of ROS through activation of G-coupled receptors and inhibition of the MPT formation by NHE-1 inhibitors. These inhibitors may act upon different mitochondrial mechanisms, including MNHE. They may act through a decrease in mitochondrial Ca^{2+}, H^+, inner membrane potential affecting the MPT formation or altering the sensitivity to those factors to induce MPT formation. Modified from Garciarena et al. (2008) with permission

recognized as one of the main prohypertrophic intracellular signals. It activates several intracellular pathways like calcineurin/ nuclear factor of activated T cells (NFAT), Ca^{2+}/calmodulin-dependent kinase II (CaMKII), PKC and possibly some others. Nevertheless, we emphasize that $[Ca^{2+}]_i$ may be increased by mechanisms other than that triggered by the hyperactivity of NHE-1. It has been recently suggested that CaMKII is preferentially activated by an increase in a specific subcellular Ca^{2+} pool localized in the perinuclear area after ET-1 stimulation (Wu et al., 2006).

In 1995 an enhanced activity of the NHE-1 was reported in the hypertrophied myocardium of spontaneously hypertensive rats (SHR) (Perez et al., 1995; Schussheim and Radda, 1995). The hyperactivity of NHE-1 has been described in several tissues other than the myocardium in human hypertension (Livne et al., 1987; Rosskopf et al., 1993; Garciandia et al., 1995). Experiments performed in our laboratory showed that the hyperactivity of NHE-1 in the myocardium of the SHR was not accompanied by an increase in pH_i, since there was a simultaneous activation of the acidifying Cl^-–$HCO_3{}^-$ exchanger (Perez et al., 1995) (see Fig. 13.1). We also reported that the NHE-1 increased activity in this model was the result of a PKC-dependent post-translational modification of the exchanger (Ennis et al., 1998). It was further hypothesized that the inhibition of the antiporter activity could regress and/or prevent the development of hypertensive hypertrophy. Kusumoto et al. (2001) proved that NHE-1 was upregulated after myocardial infarction and that the specific inhibition of this exchanger with cariporide decreased hypertrophy and remodeling in these hearts. Experiments from our own laboratory demonstrated that myocardial hypertrophy of SHR regressed after 1-month cariporide treatment (Fig. 13.25) without significantly changing the arterial pressure (Camilion de Hurtado et al., 2002b). In addition, we reported that chronic NHE-1 blockade normalized the enhanced interstitial fibrosis of these hypertrophic hearts, but this effect took longer to occur compared to the regression of myocyte size (Cingolani et al., 2003b) (Fig. 13.26), possibly as a reflection of the lower turn-over rate of collagen (Weber and Brilla, 1991).

The precise mechanism by which NHE-1 inhibition prevents hypertrophy is still unknown, though a number of pathways have been proposed (Fliegel and Karmazyn, 2004). As there is evidence that calcineurin plays a key role in many pathological models of cardiac hypertrophy (Molkentin et al., 1998; Taigen et al., 2000; Haq et al., 2001; Bueno et al., 2002; Nagata et al., 2002; Zou et al., 2002; Wilkins et al.,

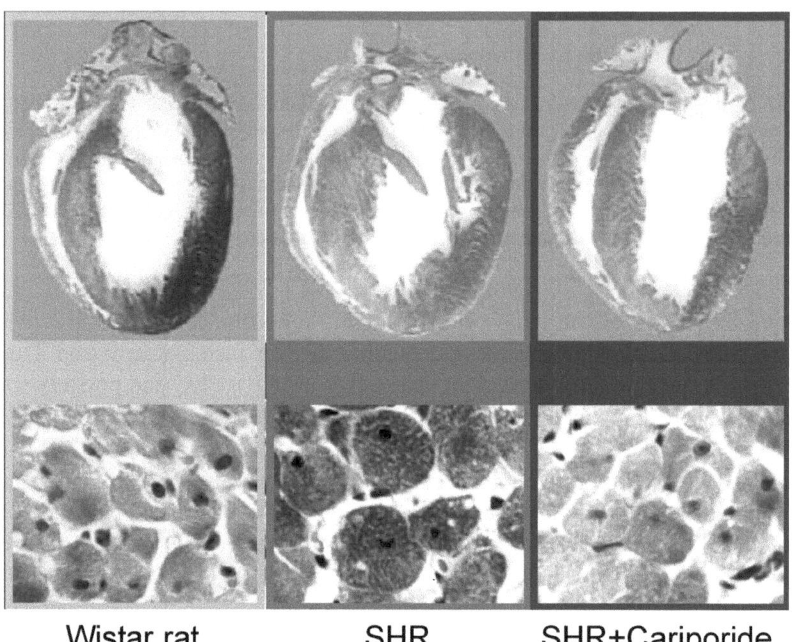

Fig. 13.25 Chronic NHE-1 blockade with cariporide (one-month treatment) regressed myocardial hypertrophy in SHR. *Upper panels* show comparative major axis sections of representative hearts from a Wistar control rat (*left*), a non-treated SHR (*middle*) and a cariporide treated SHR (*right*), and lower panels show representative myocytes cross section micrographs from the three experimental groups. Modified from Camilion de Hurtado et al., (2002b) with permission

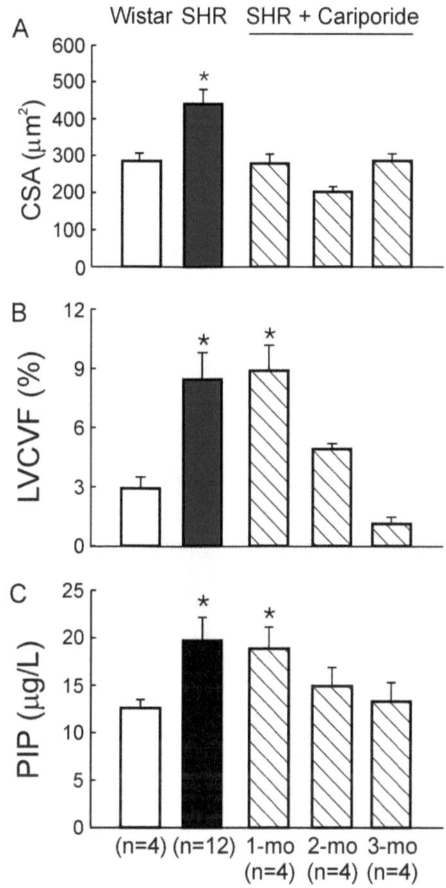

Fig. 13.26 Chronic NHE-1 blockade normalized the enhanced interstitial fibrosis of the hypertrophic SHR hearts, but a longer treatment was necessary to observe this effect. Despite the fact that full regression of myocytes cross sectional area (CSA) was observed as early as after one-month cariporide treatment (**Panel A**), fibrosis indexes like left ventricle collagen volume fraction (LVCVF) (**Panel B**) and serum levels of the carboxyterminal propeptide of procollagen type I (PIP) (**Panel C**) remained elevated. However, when treatment duration was prolonged, normalization of fibrosis was observed (**Panels B** and **C**). Modified from Cingolani et al. (2003b) with permission

2004), we recently investigated its participation in the signaling pathway involved in the regression of cardiac hypertrophy induced by NHE-1 inhibition. We analyzed the expression of the β-isoform of calcineurin A (CnAβ) as an indication of calcineurin activity. The nuclear abundance of NFAT in the left ventricular myocardium of untreated SHR, treated SHR and normotensive rats was measured as a confirmation of calcineurin activation. CnA expression and NFAT nuclear abundance are augmented in the hypertrophied myocardium of untreated SHR, compared with the normotensive rats, and the regression of cardiac hypertrophy induced by NHE-1 inhibition normalizes both parameters (Fig. 13.27) (Ennis et al., 2007). This was the first report showing that the regression of cardiac hypertrophy caused by NHE-1 inhibition, which is independent from any change in blood pressure, is accompanied by normalization of CnAβ expression and NFAT nuclear abundance. Even though we have provided evidence that a decrease in CnA and nuclear NFAT expression takes place during the regression of cardiac hypertrophy induced by NHE-1 inhibition, we cannot rule out the possibility of additional effects of this pharmacological

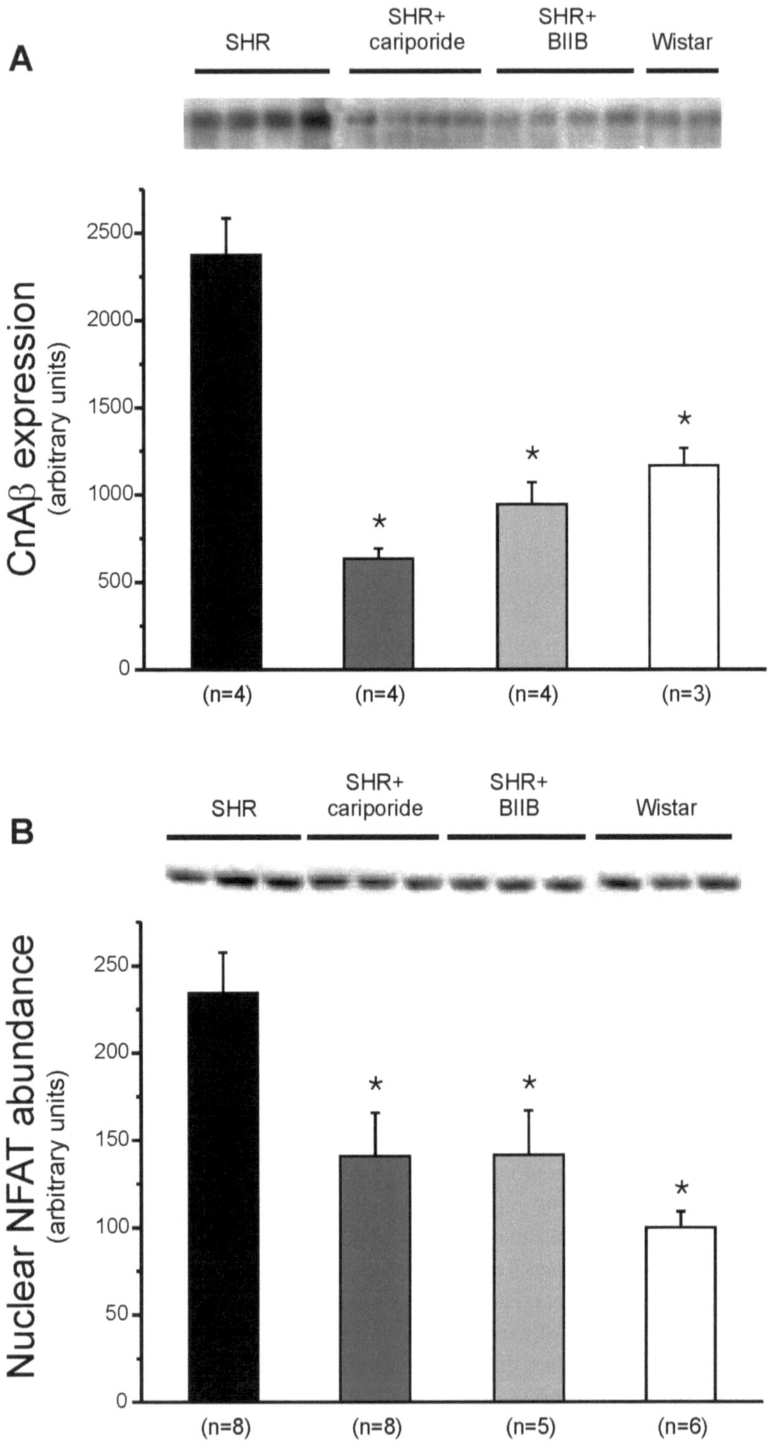

Fig. 13.27 (continued)

Table 13.1 Models of cardiac hypertrophy (CH) where the NHE-1 may play a role

Cardiac hypertrophy model	References
Pressure overload	Arai et al. (1995), Perez et al. (1995), Perez et al. (2003), Marano et al. (2004), and Baartscheer et al. (2005)
Post myocardial infarction	Yoshida and Karmazyn (2000), Camilion de Hurtado et al. (2002b), and Bers et al. (2003)
β–adrenergic stimulation	Weber and Brilla (1991), Schafer et al. (2002), and Cingolani et al. (2003b)
Hyperthyroidism	Li et al. (2002b) and Bak and Ingwall, (2003)
Hyperparathyroidism	Harnett et al. (1988), Azarani et al. (1995), and Saleh et al. (2003)
Mineralocorticoid stimulation	Fujisawa et al. (2003)
Leptin stimulation	Konstantinou-Tegou et al. (2001) and Xu et al. (2004)
Human heart failure	Chen et al. (2001)
Hamster model hereditary cardiomyopathy	Chahine et al. (2005)
α-adrenergic stimulation	Schluter et al. 1998, Xia et al. 2004, Dulce et al. (2006)
ET-1 stimulation	Xu et al. (2004) and Dulce et al. (2006)
Angiotensin II stimulation	Yamazaki et al. (1995) and Hautala et al. (2002)
Stretch	Sadoshima et al. (1993), Yamazaki et al. (1995)
ANP receptor deficient mice	Mrkic et al. (1993)
Carbonic anhydrase inhibition	Li et al. (2002a)
Combined pressure and volume overload	Baartscheer et al. (2003), Baartscheer et al. (2005) and Baartscheer et al. (2008)
Pacing-induced	Aker et al. (2004)
Monocrotaline-induced	Chen et al. (2001)

intervention. It has been proposed, as we discussed earlier, that cariporide might also exert effects at the mitochondrial level (Miura et al., 2001; Ruiz-Meana et al., 2003; Teshima et al., 2003; Javadov et al., 2005)

Below are summarized several neuro-hormonal models of cardiac hypertrophy in which a link between NHE-1 activity and myocardial growth has been established (Table 13.1):

Fig. 13.27 Panel A: Calcineurin Aβ expression was analyzed in the myocardium of treated (cariporide or BIIB723) and untreated SHR ($n = 4$ each group). Calcineurin Aβ expression was up-regulated in the hypertrophied myocardium of the untreated SHR while a significant decrease in its expression was detected after the regression of cardiac hypertrophy by the NHE-1 inhibitors. For the sake of comparison the results obtained in normotensive rats ($n = 3$) were included in the figure. The calcineurin Aβ expression levels of the cariporide- and BIIB723-treated SHR were not significantly different from those of the NT rats. Panel B: Representative Western blot and average values of NFAT abundance in nuclear extracts from LV of untreated, cariporide- or BIIB723-treated SHR and normotensive (NT) rats ($n = 8$, 8, 5 and 6, respectively). NHE-1 inhibition normalized the nuclear expression of NFAT, previously up-regulated in the hypertrophied myocardium of SHR. * means $p < 0.05$ vs. untreated SHR, ANOVA. Modified from Ennis et al. (2007) with permission

1. An up-regulation of NHE-1 was reported in a cardiac hypertrophy and failure model of β_1-adrenergic receptor transgenic mice (Engelhardt et al., 2002). The inhibition of this exchanger prevented the development of cardiac hypertrophy and fibrosis, suggesting that NHE-1 was essential for the detrimental cardiac effects of chronic β_1-receptor stimulation in the heart (Engelhardt et al., 2002). Similarly, cardiac hypertrophy induced in rats by chronic isoproterenol administration was prevented by inhibition of NHE-1 (Ennis et al., 2003).
2. Hypertrophied hyperthyroid hearts show enhanced NHE-1 activity and when exposed to acute ischemia, they accumulate more Na^+ than the control non-hypertrophied hearts (Bak and Ingwall, 2003). These changes were prevented by NHE-1 inhibition (Bak and Ingwall, 2003). Furthermore, it has been demonstrated that thyroid hormone, by the interaction of its receptor with the NHE-1 promoter increases the expression of NHE-1 (Li et al., 2002b).
3. In patients with end-stage renal disease and secondary hyperparathyroidism as well as in patients with primary hyperparathyroidism, a strong correlation between cardiac hypertrophy and serum parathyroid hormone levels has been reported (Harnett et al., 1988; Bauwens et al., 1991; Piovesan et al., 1999). This correlation was shown to be even much stronger than that between Ang II and hypertrophy (Bauwens et al., 1991). In addition, a direct evidence that parathyroid hormone improves hypertrophy was also reported (Schluter and Piper, 1992). Though controversial (Mrkic et al., 1993; Azarani et al., 1995), a stimulatory effect of parathyroid hormone on NHE-1 has been described; therefore, it is tempting to speculate about the possible involvement of the antiporter in the signaling pathway evoked by parathyroid hormone in the genesis of cardiac hypertrophy. On the other hand, low sodium plasma levels were detected in patients with NYHA class III–IV heart failure and high levels of parathyroid hormone (Arakelyan et al., 2007). The resulting misbalance of the Na^+/Ca^{2+} may in turn be a factor to consider in the development of cardiac hypertrophy.
4. In rat neonatal ventricular myocytes, aldosterone stimulation induced a hypertrophic response accompanied by NHE-1 up-regulation and increased $[Na^+]_i$. Both, hypertrophy and elevated $[Na^+]_i$, were prevented by the NHE-1-specific inhibitor EMD87580 as well as the aldosterone antagonist spironolactone (Karmazyn et al., 2003). Similar results were obtained in uninephrectomized rats exposed to deoxycorticosterone acetate/salt, in which cariporide treatment completely inhibited hypertrophy and NHE-1 up-regulation (Fujisawa et al., 2003).
5. Cardiac hypertrophy of atrial natriuretic peptide receptor-deficient mice was accompanied by an increased activity of NHE-1, which thereby increased $[Ca^{2+}]_i$ (Kilic et al., 2005). It was shown that these alterations were normalized by chronic treatment with the NHE-1 inhibitor cariporide. These results are in line with the report by Tajima et al. (1998) demonstrating that atrial natriuretic peptide inhibits NHE-1 activity.
6. Emerging evidence indicates that leptin – a protein encoded by the obesity gene- is linked to cardiac hypertrophy (Rajapurohitam et al., 2003, 2006; Xu et al., 2004). Interestingly, leptin has been reported to activate NHE-1 through

a PKC-dependent pathway (Konstantinou-Tegou et al., 2001). Moreover, it has been reported that leptin elevates ET-1 levels and, though speculative, this may be the pathway involved in NHE-1 stimulation (Xu et al., 2004). Furthermore, a recent report by Karmazyn's group implicated leptin as a mediator of hypertrophic effects of Ang II and ET-1 in cultured neonatal ventricular myocytes (Rajapurohitam et al., 2006).

7. In right ventricular hypertrophy due to monocrotaline-induced pulmonary artery injury, myocardial NHE-1 expression was enhanced. As a consequence, both hypertrophy and NHE-1 up-regulation were abrogated by cariporide treatment (Chen et al., 2001).
8. In rabbits subjected to volume and pressure overload, which induced cardiac hypertrophy and failure, acute inhibition of NHE-1 in isolated left ventricular myocytes reversed ionic remodeling (Baartscheer et al., 2003). In this model, it has also been reported that dietary cariporide treatment, initiated at induction of volume and pressure overload, reduced hypertrophy and prevented the development of heart failure and cellular ionic and electrical remodeling (Baartscheer et al., 2005). Moreover, it has been recently reported by the same group, that in rabbit hearts with established hypertrophy and signs of heart failure (one month after induction of pressure/volume overload), two months of chronic treatment with cariporide caused regression of hypertrophy, heart failure and ionic and electrophysiological remoldeling (Baartscheer et al., 2008).
9. In human hearts with chronic end-stage heart failure exhibiting various degrees of hypertrophy, a significantly greater NHE-1 activity was detected in the human hypertrophied myocytes in comparison to myocytes from normal unused human donor hearts (Yokoyama et al., 2000).

We have also demonstrated that three different antihypertensive pharmacological interventions with different mechanisms of action (nifedipine, a Ca^{2+} channel blocker; enalapril, an inhibitor of angiotensin converting enzyme; and losartan, an AT_1 receptor blocker) caused the normalization of myocardial NHE activity, regression of cardiac hypertrophy (Fig. 13.28), and decrease of arterial pressure in SHR (Alvarez et al., 2002). However, for a similar reduction in systolic blood pressure and NHE-1 activity, losartan induced the largest regression of cardiac hypertrophy. Even though these results give support to the hypothesis that an increased myocardial tension determines intracellular signals having common end points on the antiporter activity and cellular growth, they also suggest that the eventual recruitment of additional intracellular pathways may be playing a role in the hypertrophic response.

In line with the experiments reported by Kusumoto et al. (2001) showing that NHE-1 inhibition decreased hypertrophy and remodeling after myocardial infarction, we have recently reported that post-myocardial infarction hypertrophy and fibrosis were reduced after phosphodiesterase 5A inhibition by sildenafil, being the phosphodiesterase inhibition accompanied by protein kinase G activation and NHE-1 inhibition (Perez et al., 2007).

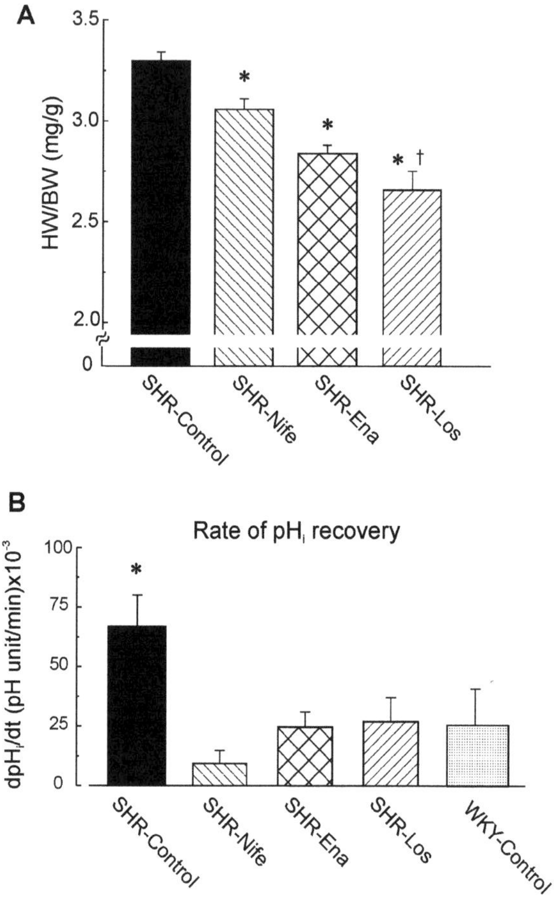

Fig. 13.28 **Panel A**: Effect of nifedipine, enalapril and losartan on cardiac hypertrophy (CH) in SHR. Heart weight to body weight (HW/BW) was used as an index of CH. (*) $P < 0.05$ compared to SHR-control; (†) $P < 0.05$ compared to enalapril-.and nifedipine-treated SHR (ANOVA). Data are means ± SE. **Panel B** shows the values of the rate of pH_i recovery from CO_2-induced intracellular acid load (dpH_i/dt) at a common pH_i value of 6.90 in SHR-Control ($n = 7$); WKY-Control ($n = 6$); SHR-Nife ($n = 5$); SHR-Ena ($n = 8$); and SHR-Los ($n = 5$). (*) $P < 0.05$ compared to all other groups (ANOVA). Data are means ± SE. Modified from Alvarez et al. (2002) with permission

As mentioned before, an enhanced activity of NHE-1 may be the result of an increased expression of the exchanger, an increased turnover of functional units, or a combination of both alternatives. In line with this, the reviewed models clearly exhibited cases of enhanced NHE-1 activity due to up-regulation, post-translational modification, or a combination of both. In either case, the hyperactivity of NHE-1 was linked to cardiac hypertrophy.

Interestingly, whereas chronic NHE-1 inhibition with cariporide in the whole animal induces up-regulation of the exchanger (Camilion de Hurtado et al., 2002a), the normalization of its previously augmented expression has been reported after chronic NHE-1 inhibition (Chen et al., 2001; Engelhardt et al., 2002; Ennis et al., 2003; Kilic et al., 2005). Nevertheless, several aspects deserve further investigation to clarify the precise mechanism by which NHE-1 is involved in the development of cardiac hypertrophy and the possible link with other mechanisms of the intracellular hypertrophic program.

References

Aiello EA, Villa-Abrille MC, Cingolani HE (2002) Autocrine stimulation of cardiac Na(+)-Ca(2+) exchanger currents by endogenous endothelin released by angiotensin II. Circulation Research 90: 374–376.

Aiello EA, Villa-Abrille MC, Dulce RA, Cingolani HE, Perez NG (2005) Endothelin-1 stimulates the Na^+/Ca^{2+} exchanger reverse mode through intracellular Na+ (Na+i)-dependent and Na+i-independent pathways. Hypertension 45: 288–293.

Aker S, Snabaitis AK, Konietzka I, Van De Sand A, Bongler K, Avkiran M, Heusch G, Schulz R (2004) Inhibition of the Na^+/H^+ exchanger attenuates the deterioration of ventricular function during pacing-induced heart failure in rabbits. Cardiovascular Research 63: 273–282.

Alvarez BV, Ennis IL, De Hurtado MC, Cingolani HE (2002) Effects of antihypertensive therapy on cardiac sodium/hydrogen ion exchanger activity and hypertrophy in spontaneously hypertensive rats. The Canadian Journal of Cardiology 18: 667–672.

Alvarez BV, Fujinaga J, Casey JR (2001) Molecular basis for angiotensin II-induced increase of chloride/bicarbonate exchange in the myocardium. Circulation Research 89: 1246–1253.

Alvarez BV, Perez NG, Ennis IL, Camilion de Hurtado MC, Cingolani HE (1999) Mechanisms underlying the increase in force and Ca^{2+} transient that follow stretch of cardiac muscle: a possible explanation of the Anrep effect. Circulation Research 85: 716–722.

Allen DG, Kurihara S (1982) The effects of muscle length on intracellular calcium transients in mammalian cardiac muscle. The Journal of Physiology 327: 79–94.

Anrep V (1912) On the part played by the suprarenals in the normal vascular reactions on the body. The Journal of Physiology 45: 307–317.

Arai M, Yoguchi A, Iso T, Takahashi T, Imai S, Murata K, Suzuki T (1995) Endothelin-1 and its binding sites are upregulated in pressure overload cardiac hypertrophy. The American Journal of Physiology 268: H2084–H2091.

Arakelyan KP, Sahakyan YA, Hayrapetyan LR, Khudaverdyan DN, Ingelman-Sundberg M, Mkrtchian S, Ter-Markosyan AS (2007) Calcium-regulating peptide hormones and blood electrolytic balance in chronic heart failure. Regulatory Peptides 142: 95–100.

Avkiran M, Marber MS (2002) Na^+/H^+ exchange inhibitors for cardioprotective therapy: progress, problems and prospects. Journal of the American College of Cardiology 39: 747–753.

Azarani A, Orlowski J, Goltzman D (1995) Parathyroid hormone and parathyroid hormone-related peptide activate the Na^+/H^+ exchanger NHE-1 isoform in osteoblastic cells (UMR-106) via a cAMP-dependent pathway. The Journal of Biological Chemistry 270: 23166–23172.

Baartscheer A, Hardziyenka M, Schumacher CA, Belterman CN, van Borren MM, Verkerk AO, Coronel R, Fiolet JW (2008) Chronic inhibition of the Na^+/H^+-exchanger causes regression of hypertrophy, heart failure, and ionic and electrophysiological remodelling. British Journal of Pharmacology 154: 1266–1275.

Baartscheer A, Schumacher CA, van Borren MM, Belterman CN, Coronel R, Fiolet JW (2003) Increased Na^+/H^+-exchange activity is the cause of increased $[Na^+]$i and underlies disturbed calcium handling in the rabbit pressure and volume overload heart failure model. Cardiovascular Research 57: 1015–1024.

Baartscheer A, Schumacher CA, van Borren MM, Belterman CN, Coronel R, Opthof T, Fiolet JW (2005) Chronic inhibition of Na^+/H^+-exchanger attenuates cardiac hypertrophy and prevents cellular remodeling in heart failure. Cardiovascular Research 65: 83–92.

Bak MI, Ingwall JS (2003) Contribution of Na^+/H^+ exchange to Na^+ overload in the ischemic hypertrophied hyperthyroid rat heart. Cardiovascular Research 57: 1004–1014.

Barton M, Shaw S, d'Uscio LV, Moreau P, Luscher TF (1997) Angiotensin II increases vascular and renal endothelin-1 and functional endothelin converting enzyme activity in vivo: role of ETA receptors for endothelin regulation. Biochemical and Biophysical Research Communications 238: 861–865.

Bauwens FR, Duprez DA, De Buyzere ML, De Backer TL, Kaufman JM, Van Hoecke J, Vermeulen A, Clement DL (1991) Influence of the arterial blood pressure and nonhemodynamic factors on left ventricular hypertrophy in moderate essential hypertension. The American Journal of Cardiology 68: 925–929.

Bers DM (2001) Na/Ca exchange and the sarcolemmal ca-pump. In: Bers MD (ed.) Excitation-contraction Coupling and Cardiac Contractile Force. Kluwer Academic Publishers, Dordrecht, pp.133–160.

Bers DM, Barry WH, Despa S (2003) Intracellular Na^+ regulation in cardiac myocytes. Cardiovascular Research 57: 897–912.

Bers DM, Despa S (2006) Cardiac myocytes Ca^{2+} and Na^+ regulation in normal and failing hearts. Journal of Pharmacological Sciences 100: 315–322.

Bluhm WF, Lew WY (1995) Sarcoplasmic reticulum in cardiac length-dependent activation in rabbits. The American Journal of Physiology 269: H965–H972.

Bluhm WF, Lew WY, Garfinkel A, McCulloch AD (1998) Mechanisms of length history-dependent tension in an ionic model of the cardiac myocyte. The American Journal of Physiology 274: H1032–H1040.

Boveris A (1998) Biochemistry of free radicals: from electrons to tissues. Medicina 58: 350–356.

Brandes RP (2005) Triggering mitochondrial radical release: a new function for NADPH oxidases. Hypertension 45: 847–848.

Bueno OF, Wilkins BJ, Tymitz KM, Glascock BJ, Kimball TF, Lorenz JN, Molkentin JD (2002) Impaired cardiac hypertrophic response in Calcineurin Abeta -deficient mice. Proceedings of the National Academy of Sciences of the United States of America 99: 4586–4591.

Calaghan S, White E (2004) Activation of Na^+-H^+ exchange and stretch-activated channels underlies the slow inotropic response to stretch in myocytes and muscle from the rat heart. The Journal of Physiology 559: 205–214.

Calaghan SC, White E (2001) Contribution of angiotensin II, endothelin 1 and the endothelium to the slow inotropic response to stretch in ferret papillary muscle. Pflugers Archiv 441: 514–520.

Caldiz CI, Garciarena CD, Dulce RA, Novaretto LP, Yeves AM, Ennis IL, Cingolani HE, Chiappe de Cingolani G, Perez NG (2007) Mitochondrial reactive oxygen species activate the slow force response to stretch in feline myocardium. The Journal of Physiology 584: 895–905.

Camilion de Hurtado MC, Alvarez BV, Perez NG, Ennis IL, Cingolani HE (1998) Angiotensin II activates Na^+-independent Cl^-– $HCO_3{}^-$ exchange in ventricular myocardium. Circulation Research 82: 473–481.

Camilion de Hurtado MC, Ennis IL, Perez NG, Chiappe de Cingolani GE, Morgan P, Cingolani HE (2002a) Upregulation of myocardial Na^+/H^+ exchanger induced by chronic treatment with a selective inhibitor. Journal of Molecular and Cellular Cardiology 34: 1539–1547.

Camilion de Hurtado MC, Portiansky EL, Perez NG, Rebolledo OR, Cingolani HE (2002b) Regression of cardiomyocyte hypertrophy in SHR following chronic inhibition of the Na^+/H^+ exchanger. Cardiovascular Research 53: 862–868.

Cingolani HE, Alvarez BV, Ennis IL, Camilion de Hurtado MC (1998) Stretch-induced alkalinization of feline papillary muscle: an autocrine-paracrine system. Circulation Research 83: 775–780.

Cingolani HE, Chiappe GE, Ennis IL, Morgan PG, Alvarez BV, Casey JR, Dulce RA, Perez NG, Camilion de Hurtado MC (2003a) Influence of Na^+-independent Cl^-– $HCO_3{}^-$ exchange on the slow force response to myocardial stretch. Circulation Research 93: 1082–1088.

Cingolani HE, Perez NG, Aiello EA, de Hurtado MC (2005) Intracellular signaling following myocardial stretch: an autocrine/paracrine loop. Regulatory Peptides 128: 211–220.

Cingolani HE, Perez NG, Camilion de Hurtado MC (2001) An autocrine/paracrine mechanism triggered by myocardial stretch induces changes in contractility. News Physiolosophy Science 16: 88–91.

Cingolani HE, Rebolledo OR, Portiansky EL, Perez NG, Camilion de Hurtado MC (2003b) Regression of hypertensive myocardial fibrosis by Na^+/H^+ exchange inhibition. Hypertension 41: 373–377.

Cingolani HE, Villa-Abrille MC, Cornelli M, Nolly A, Ennis IL, Garciarena C, Suburo AM, Torbidoni V, Correa MV, Camilionde Hurtado MC, Aiello EA (2006) The positive inotropic effect of angiotensin II: role of endothelin-1 and reactive oxygen species. Hypertension 47: 727–734.

Costa AD, Garlid KD (2008) Intramitochondrial signaling: interactions among mitoKATP, PKCepsilon, ROS, and MPT. American Journal of Physiology 295: H874–H882.

Chahine M, Bkaily G, Nader M, Al-Khoury J, Jacques D, Beier N, Scholz W (2005) NHE-1-dependent intracellular sodium overload in hypertrophic hereditary cardiomyopathy: prevention by NHE-1 inhibitor. Journal of Molecular and Cellular Cardiology 38: 571–582.

Chen L, Chen CX, Gan XT, Beier N, Scholz W, Karmazyn M (2004) Inhibition and reversal of myocardial infarction-induced hypertrophy and heart failure by NHE-1 inhibition. American Journal of Physiology 286: H381–H387.

Chen L, Gan XT, Haist JV, Feng Q, Lu X, Chakrabarti S, Karmazyn M (2001) Attenuation of compensatory right ventricular hypertrophy and heart failure following monocrotaline-induced pulmonary vascular injury by the Na^+-H^+ exchange inhibitor cariporide. The Journal of Pharmacology and Experimental Therapeutics 298: 469–476.

Chua BH, Chua CC, Diglio CA, Siu BB (1993) Regulation of endothelin-1 mRNA by angiotensin II in rat heart endothelial cells. Biochimica et Biophysica Acta 1178: 201–206.

D'Autreaux B, Toledano MB (2007) ROS as signalling molecules: mechanisms that generate specificity in ROS homeostasis. Nature Reviews 8: 813–824.

De Giusti VC, Correa MV, Villa-Abrille MC, Beltrano C, Yeves AM, de Cingolani GE, Cingolani HE, Aiello EA (2008) The positive inotropic effect of endothelin-1 is mediated by mitochondrial reactive oxygen species. Life Sciences 83: 264–271.

de Hurtado MC, Alvarez BV, Ennis IL, Cingolani HE (2000) Stimulation of myocardial Na(+)-independent Cl(-)–HCO(3)(-) exchanger by angiotensin II is mediated by endogenous endothelin. Circulation Research 86: 622–627.

Dohi Y, Hahn AW, Boulanger CM, Buhler FR, Luscher TF (1992) Endothelin stimulated by angiotensin II augments contractility of spontaneously hypertensive rat resistance arteries. Hypertension 19: 131–137.

Downey JM, Cohen MV (2008) Free radicals in the heart: friend or foe? Expert Review of Cardiovascular Therapy 6: 589–591.

Dulce RA, Hurtado C, Ennis IL, Garciarena CD, Alvarez MC, Caldiz C, Pierce GN, Portiansky EL, Chiappe de Cingolani GE, Camilion de Hurtado MC (2006) Endothelin-1 induced hypertrophic effect in neonatal rat cardiomyocytes: involvement of Na^+/H^+ and Na^+/Ca^{2+} exchangers. Journal of Molecular and Cellular Cardiology 41: 807–815.

Eigel BN, Gursahani H, Hadley RW (2004) ROS are required for rapid reactivation of Na^+/Ca^{2+} exchanger in hypoxic reoxygenated guinea pig ventricular myocytes. American Journal of Physiology 286: H955–H963.

Endoh M, Fujita S, Yang HT, Talukder MA, Maruya J, Norota I (1998) Endothelin: receptor subtypes, signal transduction, regulation of Ca^{2+} transients and contractility in rabbit ventricular myocardium. Life Sciences 62: 1485–1489.

Engelhardt S, Hein L, Keller U, Klambt K, Lohse MJ (2002) Inhibition of Na^+-H^+ exchange prevents hypertrophy, fibrosis, and heart failure in beta(1)-adrenergic receptor transgenic mice. Circulation Research 90: 814–819.

Ennis IL, Alvarez BV, Camilion de Hurtado MC, Cingolani HE (1998) Enalapril induces regression of cardiac hypertrophy and normalization of pH_i regulatory mechanisms. Hypertension 31: 961–967.

Ennis IL, Escudero EM, Console GM, Camihort G, Dumm CG, Seidler RW, Camilion de Hurtado MC, Cingolani HE (2003) Regression of isoproterenol-induced cardiac hypertrophy by Na^+/H^+ exchanger inhibition. Hypertension 41: 1324–1329.

Ennis IL, Garciarena CD, Escudero EM, Perez NG, Dulce RA, Camilion de Hurtado MC, Cingolani HE (2007) Normalization of the calcineurin pathway underlies the regression of hypertensive hypertrophy induced by Na^+/H^+ exchanger-1 (NHE-1) inhibition. Canadian Journal of Physiology and Pharmacology 85: 301–310.

Ennis IL, Garciarena CD, Perez NG, Dulce RA, Camilion de Hurtado MC, Cingolani HE (2005) Endothelin isoforms and the response to myocardial stretch. American Journal of Physiology 288: H2925–H2930.

Ficai S, Herizi A, Mimran A, Jover B (2001) Endothelin blockade in angiotensin II hypertension: prevention and treatment studies in the rat. Clinical and Experimental Pharmacology & Physiology 28: 1100–1103.

Fliegel L, Karmazyn M (2004) The cardiac Na-H exchanger: a key downstream mediator for the cellular hypertrophic effects of paracrine, autocrine and hormonal factors. Biochemistry Cell Biology 82: 626–635.

Fujisaki H, Ito H, Hirata Y, Tanaka M, Hata M, Lin M, Adachi S, Akimoto H, Marumo F, Hiroe M (1995) Natriuretic peptides inhibit angiotensin II-induced proliferation of rat cardiac fibroblasts by blocking endothelin-1 gene expression. The Journal of Clinical Investigation 96: 1059–1065.

Fujisawa G, Okada K, Muto S, Fujita N, Itabashi N, Kusano E, Ishibashi S (2003) Na/H exchange isoform 1 is involved in mineralocorticoid/salt-induced cardiac injury. Hypertension 41: 493–498.

Fuller SJ, Gaitanaki CJ, Sugden PH (1990) Effects of catecholamines on protein synthesis in cardiac myocytes and perfused hearts isolated from adult rats. Stimulation of translation is mediated through the alpha 1-adrenoceptor. The Biochemical Journal 266: 727–736.

Ganz MB, Boyarsky G, Boron WF, Sterzel RB (1988) Effects of angiotensin II and vasopressin on intracellular pH of glomerular mesangial cells. The American Journal of Physiology 254: F787–F794.

Ganz MB, Perfetto MC, Boron WF (1990) Effects of mitogens and other agents on rat mesangial cell proliferation, pH, and Ca^{2+}. The American Journal of Physiology 259: F269–F278.

Garciandia A, Lopez R, Tisaire J, Arrazola A, Fortuno A, Bueno J, Diez J (1995) Enhanced Na(+)-H+ exchanger activity and NHE-1 mRNA expression in lymphocytes from patients with essential hypertension. Hypertension 25: 356–364.

Garciarena CD, Caldiz CI, Correa MV, Schinella GR, Mosca SM, Chiappe de Cingolani GE, Cingolani HE, Ennis IL (2008) Na^+/H^+ exchanger-1 inhibitors decrease myocardial superoxide production via direct mitochondrial action. Journal of Applied Physiology 105: 1706–1713.

Giordano FJ (2005) Oxygen, oxidative stress, hypoxia, and heart failure. The Journal of Clinical Investigation 115: 500–508.

Gomez L, Paillard M, Thibault H, Derumeaux G, Ovize M (2008) Inhibition of GSK3beta by postconditioning is required to prevent opening of the mitochondrial permeability transition pore during reperfusion. Circulation 117: 2761–2768.

Grinstein S, Rotin D, Mason MJ (1989) Na^+/H^+ exchange and growth factor-induced cytosolic pH changes. Role in cellular proliferation. Biochimica et Biophysica Acta 988: 73–97.

Haq S, Choukroun G, Lim H, Tymitz KM, del Monte F, Gwathmey J, Grazette L, Michael A, Hajjar R, Force T, Molkentin JD (2001) Differential activation of signal transduction pathways in human hearts with hypertrophy versus advanced heart failure. Circulation 103: 670–677.

Harnett JD, Parfrey PS, Griffiths SM, Gault MH, Barre P, Guttmann RD (1988) Left ventricular hypertrophy in end-stage renal disease. Nephron 48: 107–115.

Hautala N, Tenhunen O, Szokodi I, Ruskoaho H (2002) Direct left ventricular wall stretch activates GATA4 binding in perfused rat heart: involvement of autocrine/paracrine pathways. Pflugers Archiv 443: 362–369.

Hayasaki-Kajiwara Y, Kitano Y, Iwasaki T, Shimamura T, Naya N, Iwaki K, Nakajima M (1999) Na(+)influx via Na(+)/H(+)exchange activates protein kinase C isozymes delta and epsilon in cultured neonatal rat cardiac myocytes. Journal of Molecular and Cellular Cardiology 31: 1559–1572.

Hongo K, White E, Le Guennec JY, Orchard CH (1996) Changes in $[Ca^{2+}]i$, $[Na^+]i$ and Ca^{2+} current in isolated rat ventricular myocytes following an increase in cell length. The Journal of Physiology 491 (Pt 3): 609–619.

Hongo K, White E, Orchard CH (1995) Effect of stretch on contraction and the Ca^{2+} transient in ferret ventricular muscles during hypoxia and acidosis. The American Journal of Physiology 269: C690–697.

Hori M, Nakatsubo N, Kagiya T, Iwai K, Sato H, Iwakura K, Kitabatake A, Kamada T (1990) The role of Na+/H+ exchange in norepinephrine-induced protein synthesis in neonatal cultured rat cardiomyocytes. Japanese Circulation Journal 54: 535–539.

Imai T, Hirata Y, Emori T, Yanagisawa M, Masaki T, Marumo F (1992) Induction of endothelin-1 gene by angiotensin and vasopressin in endothelial cells. Hypertension 19: 753–757.

Isenberg G, Kondratev D, Dyachenko V, Kazanski V, Gallitelli MF (2005) Isolated cardiomyocytes: mechanosensitivity of action potential, membrane current and ion concentration. In: Kamkin AG and Kiseleva I (eds.) Mechanosensitivity in Cells and Tissues, Academia Publishing House Ltd, Moscow, pp. 126–164.

Ito H, Hirata Y, Adachi S, Tanaka M, Tsujino M, Koike A, Nogami A, Murumo F, Hiroe M (1993) Endothelin-1 is an autocrine/paracrine factor in the mechanism of angiotensin II-induced hypertrophy in cultured rat cardiomyocytes. The Journal of Clinical Investigation 92: 398–403.

Javadov S, Huang C, Kirshenbaum L, Karmazyn M (2005) NHE-1 inhibition improves impaired mitochondrial permeability transition and respiratory function during postinfarction remodelling in the rat. Journal of Molecular and Cellular Cardiology 38: 135–143.

Juhaszova M, Zorov DB, Kim SH, Pepe S, Fu Q, Fishbein KW, Ziman BD, Wang S, Ytrehus K, Antos CL, Olson EN, Sollott SJ (2004) Glycogen synthase kinase-3beta mediates convergence of protection signaling to inhibit the mitochondrial permeability transition pore. The Journal of Clinical Investigation 113: 1535–1549.

Karmazyn M, Gan XT, Humphreys RA, Yoshida H, Kusumoto K (1999) The myocardial Na^+-H^+ exchange: structure, regulation, and its role in heart disease. Circulation Research 85: 777–786.

Karmazyn M, Liu Q, Gan XT, Brix BJ, Fliegel L (2003) Aldosterone increases NHE-1 expression and induces NHE-1-dependent hypertrophy in neonatal rat ventricular myocytes. Hypertension 42: 1171–1176.

Kentish JC, Wrzosek A (1998) Changes in force and cytosolic Ca^{2+} concentration after length changes in isolated rat ventricular trabeculae. The Journal of Physiology 506(Pt 2): 431–444.

Kilic A, Velic A, De Windt LJ, Fabritz L, Voss M, Mitko D, Zwiener M, Baba HA, van Eickels M, Schlatter E, Kuhn M (2005) Enhanced activity of the myocardial Na^+/H^+ exchanger NHE-1 contributes to cardiac remodeling in atrial natriuretic peptide receptor-deficient mice. Circulation 112: 2307–2317.

Kim JS, He L, Lemasters JJ (2003) Mitochondrial permeability transition: a common pathway to necrosis and apoptosis. Biochemical and Biophysical Research Communications 304: 463–470.

Kimura S, Zhang GX, Nishiyama A, Shokoji T, Yao L, Fan YY, Rahman M, Abe Y (2005a) Mitochondria-derived reactive oxygen species and vascular MAP kinases: comparison of angiotensin II and diazoxide. Hypertension 45: 438–444.

Kimura S, Zhang GX, Nishiyama A, Shokoji T, Yao L, Fan YY, Rahman M, Suzuki T, Maeta H, Abe Y (2005b) Role of NAD(P)H oxidase- and mitochondria-derived reactive oxygen species in cardioprotection of ischemic reperfusion injury by angiotensin II. Hypertension 45: 860–866.

Konstantinou-Tegou A, Kaloyianni M, Bourikas D, Koliakos G (2001) The effect of leptin on Na(+)-H(+) antiport (NHE 1) activity of obese and normal subjects erythrocytes. Molecular and Cellular Endocrinology 183: 11–18.

Kurogouchi F, Furukawa Y, Zhao D, Hirose M, Nakajima K, Tsuboi M, Chiba S (2000) A Na^+/Ca^{2+} exchanger inhibitor, KB-R7943, caused negative inotropic responses and negative followed by positive chronotropic responses in isolated, blood-perfused dog heart preparations. Japanese Journal of Pharmacology 82: 155–163.

Kusumoto K, Haist JV, Karmazyn M (2001) Na(+)/H(+) exchange inhibition reduces hypertrophy and heart failure after myocardial infarction in rats. American Journal of Physiology 280: H738–H745.

Kusuoka H, Camilion de Hurtado MC, Marban E (1993) Role of sodium/calcium exchange in the mechanism of myocardial stunning: protective effect of reperfusion with high sodium solution. Journal of the American College of Cardiology 21: 240–248.

Lavigne MC, Malech HL, Holland SM, Leto TL (2001) Genetic demonstration of p47phox-dependent superoxide anion production in murine vascular smooth muscle cells. Circulation 104: 79–84.
Leri A, Claudio PP, Li Q, Wang X, Reiss K, Wang S, Malhotra A, Kajstura J, Anversa P (1998) Stretch-mediated release of angiotensin II induces myocyte apoptosis by activating p53 that enhances the local renin-angiotensin system and decreases the Bcl-2-to-Bax protein ratio in the cell. The Journal of Clinical Investigation 101: 1326–1342.
Li X, Alvarez B, Casey JR, Reithmeier RA, Fliegel L (2002a) Carbonic anhydrase II binds to and enhances activity of the Na+/H+ exchanger. The Journal of Biological Chemistry 277: 36085–36091.
Li X, Misik AJ, Rieder CV, Solaro RJ, Lowen A, Fliegel L (2002b) Thyroid hormone receptor alpha 1 regulates expression of the Na+/H+ exchanger (NHE1). The Journal of Biological Chemistry 277: 28656–28662.
Liang F, Gardner DG (1998) Autocrine/paracrine determinants of strain-activated brain natriuretic peptide gene expression in cultured cardiac myocytes. The Journal of Biological Chemistry 273: 14612–14619.
Livne A, Balfe JW, Veitch R, Marquez-Julio A, Grinstein S, Rothstein A (1987) Increased platelet Na^+-H^+ exchange rates in essential hypertension: application of a novel test. Lancet 1: 533–536.
Lu Z, Abe J, Taunton J, Lu Y, Shishido T, McClain C, Yan C, Xu SP, Spangenberg TM, Xu H (2008) Reactive oxygen species-induced activation of p90 ribosomal S6 kinase prolongs cardiac repolarization through inhibiting outward K^+ channel activity. Circulation Research 103: 269–278.
Luers C, Fialka F, Elgner A, Zhu D, Kockskamper J, von Lewinski D, Pieske B (2005) Stretch-dependent modulation of $[Na^+]i$, $[Ca^{2+}]i$, and pHi in rabbit myocardium–a mechanism for the slow force response. Cardiovascular Research 68: 454–463.
Marano G, Vergari A, Catalano L, Gaudi S, Palazzesi S, Musumeci M, Stati T, Ferrari AU (2004) Na^+/H^+ exchange inhibition attenuates left ventricular remodeling and preserves systolic function in pressure-overloaded hearts. British Journal of Pharmacology 141: 526–532.
Miura T, Liu Y, Goto M, Tsuchida A, Miki T, Nakano A, Nishino Y, Ohnuma Y, Shimamoto K (2001) Mitochondrial ATP-sensitive K^+ channels play a role in cardioprotection by Na^+-H^+ exchange inhibition against ischemia/reperfusion injury. Journal of the American College of Cardiology 37: 957–963.
Molkentin JD, Lu JR, Antos CL, Markham B, Richardson J, Robbins J, Grant SR, Olson EN (1998) A calcineurin-dependent transcriptional pathway for cardiac hypertrophy. Cell 93: 215–228.
Monroe RG, Gamble WJ, LaFarge CG, Kumar AE, Stark J, Sanders GL, Phornphutkul C, Davis M (1972) The Anrep effect reconsidered. The Journal of Clinical Investigation 51: 2573–2583.
Mrkic B, Tse CM, Forgo J, Helmle-Kolb C, Donowitz M, Murer H (1993) Identification of PTH-responsive Na/H-exchanger isoforms in a rabbit proximal tubule cell line (RKPC-2). Pflugers Archiv 424: 377–384.
Muller DN, Mervaala EM, Schmidt F, Park JK, Dechend R, Genersch E, Breu V, Loffler BM, Ganten D, Schneider W, Haller H, Luft FC (2000) Effect of bosentan on NF-kappaB, inflammation, and tissue factor in angiotensin II-induced end-organ damage. Hypertension 36: 282–290.
Muller DN, Mullally A, Dechend R, Park JK, Fiebeler A, Pilz B, Loffler BM, Blum-Kaelin D, Masur S, Dehmlow H, Aebi JD, Haller H, Luft FC (2002) Endothelin-converting enzyme inhibition ameliorates angiotensin II-induced cardiac damage. Hypertension 40: 840–846.
Nagata K, Somura F, Obata K, Odashima M, Izawa H, Ichihara S, Nagasaka T, Iwase M, Yamada Y, Nakashima N, Yokota M (2002) AT1 receptor blockade reduces cardiac calcineurin activity in hypertensive rats. Hypertension 40: 168–174.
Nakamura TY, Iwata Y, Arai Y, Komamura K, Wakabayashi S (2008) Activation of Na^+/H^+ exchanger 1 is sufficient to generate Ca^{2+} signals that induce cardiac hypertrophy and heart failure. Circulation Research 103: 891–899.

Oldenburg O, Qin Q, Krieg T, Yang XM, Philipp S, Critz SD, Cohen MV, Downey JM (2004) Bradykinin induces mitochondrial ROS generation via NO, cGMP, PKG, and mitoKATP channel opening and leads to cardioprotection. American Journal of Physiology 286: H468–H476.

Oldenburg O, Yang XM, Krieg T, Garlid KD, Cohen MV, Grover GJ, Downey JM (2003) P1075 opens mitochondrial K(ATP) channels and generates reactive oxygen species resulting in cardioprotection of rabbit hearts. Journal of Molecular and Cellular Cardiology 35: 1035–1042.

Ortiz MC, Sanabria E, Manriquez MC, Romero JC, Juncos LA (2001) Role of endothelin and isoprostanes in slow pressor responses to angiotensin II. Hypertension 37: 505–510.

Pain T, Yang XM, Critz SD, Yue Y, Nakano A, Liu GS, Heusch G, Cohen MV, Downey JM (2000) Opening of mitochondrial K(ATP) channels triggers the preconditioned state by generating free radicals. Circulation Research 87: 460–466.

Parmley WW, Chuck L (1973) Length-dependent changes in myocardial contractile state. The American Journal of Physiology 224: 1195–1199.

Perez NG, Alvarez BV, Camilion de Hurtado MC, Cingolani HE (1995) pHi regulation in myocardium of the spontaneously hypertensive rat. Compensated enhanced activity of the Na(+)-H+ exchanger. Circulation Research 77: 1192–1200.

Perez NG, de Hurtado MC, Cingolani HE (2001) Reverse mode of the Na^{+}-Ca^{2+} exchange after myocardial stretch: underlying mechanism of the slow force response. Circulation Research 88: 376–382.

Perez NG, Piaggio MR, Ennis IL, Garciarena CD, Morales C, Escudero EM, Cingolani OH, Chiappe de Cingolani G, Yang XP, Cingolani HE (2007) Phosphodiesterase 5A inhibition induces Na^{+}/H^{+} exchanger blockade and protection against myocardial infarction. Hypertension 49: 1095–1103.

Perez NG, Villa-Abrille MC, Aiello EA, Dulce RA, Cingolani HE, Camilion de Hurtado MC (2003) A low dose of angiotensin II increases inotropism through activation of reverse Na(+)/Ca(2+) exchange by endothelin release. Cardiovascular Research 60: 589–597.

Petroff MG, Kim SH, Pepe S, Dessy C, Marban E, Balligand JL, Sollott SJ (2001) Endogenous nitric oxide mechanisms mediate the stretch dependence of Ca^{2+} release in cardiomyocytes. Nature Cell Biology 3: 867–873.

Piovesan A, Molineri N, Casasso F, Emmolo I, Ugliengo G, Cesario F, Borretta G (1999) Left ventricular hypertrophy in primary hyperparathyroidism. Effects of successful parathyroidectomy. Clinical Endocrinology 50: 321–328.

Rajagopalan S, Laursen JB, Borthayre A, Kurz S, Keiser J, Haleen S, Giaid A, Harrison DG (1997) Role for endothelin-1 in angiotensin II-mediated hypertension. Hypertension 30: 29–34.

Rajapurohitam V, Gan XT, Kirshenbaum LA, Karmazyn M (2003) The obesity-associated peptide leptin induces hypertrophy in neonatal rat ventricular myocytes. Circulation Research 93: 277–279.

Rajapurohitam V, Javadov S, Purdham DM, Kirshenbaum LA, Karmazyn M (2006) An autocrine role for leptin in mediating the cardiomyocyte hypertrophic effects of angiotensin II and endothelin-1. Journal of Molecular and Cellular Cardiology 41: 265–274.

Ros MN, Dulce RA, Perez NG, Camilion de Hurtado MC, Cingolani HE (2005) Endothelin 1 versus endothelin 3 in the development of the slow force response to myocardial stretch. The Canadian Journal of Cardiology 21: 435–438.

Rosenblueth A, Alanis J, Lopez E, Rubio R (1959) The adaptation of ventricular muscle to different circulatory conditions. Archives Internationales de Physiologie et de Biochimie 67: 358–373.

Rosskopf D, Dusing R, Siffert W (1993) Membrane sodium-proton exchange and primary hypertension. Hypertension 21: 607–617.

Ruiz-Meana M, Garcia-Dorado D, Pina P, Inserte J, Agullo L, Soler-Soler J (2003) Cariporide preserves mitochondrial proton gradient and delays ATP depletion in cardiomyocytes during ischemic conditions. American Journal of Physiology 285: H999–H1006.

Sabri A, Byron KL, Samarel AM, Bell J, Lucchesi PA (1998) Hydrogen peroxide activates mitogen-activated protein kinases and Na^{+}-H^{+} exchange in neonatal rat cardiac myocytes. Circulation Research 82: 1053–1062.

Sadoshima J, Xu Y, Slayter HS, Izumo S (1993) Autocrine release of angiotensin II mediates stretch-induced hypertrophy of cardiac myocytes in vitro. Cell 75: 977–984.

Saleh FN, Schirmer H, Sundsfjord J, Jorde R (2003) Parathyroid hormone and left ventricular hypertrophy. European Heart Journal 24: 2054–2060.

Sarnoff SJ, Mitchell JH, Gilmore JP, Remensnyder JP (1960) Homeometric autoregulation in the heart. Circulation Research 8: 1077–1091.

Sato T, O'Rourke B, Marban E (1998) Modulation of mitochondrial ATP-dependent K^+ channels by protein kinase C. Circulation Research 83: 110–114.

Schafer M, Schafer C, Michael Piper H, Schluter KD (2002) Hypertrophic responsiveness of cardiomyocytes to alpha- or beta-adrenoceptor stimulation requires sodium-proton-exchanger-1 (NHE-1) activation but not cellular alkalization. European Journal of Heart Failure 4: 249–254.

Schluter KD, Piper HM (1992) Trophic effects of catecholamines and parathyroid hormone on adult ventricular cardiomyocytes. The American Journal of Physiology 263: H1739–H1746.

Schluter KD, Schafer M, Balser C, Taimor G, Piper HM (1998) Influence of pHi and creatine phosphate on alpha-adrenoceptor-mediated cardiac hypertrophy. Journal of Molecular and Cellular Cardiology 30: 763–771.

Schussheim AE, Radda GK (1995) Altered Na(+)-H(+)-exchange activity in the spontaneously hypertensive perfused rat heart. Journal of Molecular and Cellular Cardiology 27: 1475–1481.

Seccia TM, Belloni AS, Kreutz R, Paul M, Nussdorfer GG, Pessina AC, Rossi GP (2003) Cardiac fibrosis occurs early and involves endothelin and AT-1 receptors in hypertension due to endogenous angiotensin II. Journal of the American College of Cardiology 41: 666–673.

Serneri GG, Modesti PA, Boddi M, Cecioni I, Paniccia R, Coppo M, Galanti G, Simonetti I, Vanni S, Papa L, Bandinelli B, Migliorini A, Modesti A, Maccherini M, Sani G, Toscano M. (1999) Cardiac growth factors in human hypertrophy. Relations with myocardial contractility and wall stress. Circulation Research 85: 57–67.

Seshiah PN, Weber DS, Rocic P, Valppu L, Taniyama Y, Griendling KK (2002) Angiotensin II stimulation of NAD(P)H oxidase activity: upstream mediators. Circulation Research 91: 406–413.

Shivakumar K, Sollott SJ, Sangeetha M, Sapna S, Ziman B, Wang S, Lakatta EG (2008) Paracrine effects of hypoxic fibroblast-derived factors on the MPT-ROS threshold and viability of adult rat cardiac myocytes. American Journal of Physiology 294: H2653–H2658.

Shrode LD, Tapper H, Grinstein S (1997) Role of intracellular pH in proliferation, transformation, and apoptosis. Journal of Bioenergetics and Biomembranes 29: 393–399.

Snabaitis AK, Hearse DJ, Avkiran M (2002) Regulation of sarcolemmal Na(+)/H(+) exchange by hydrogen peroxide in adult rat ventricular myocytes. Cardiovascular Research 53: 470–480.

Taigen T, De Windt LJ, Lim HW, Molkentin JD (2000) Targeted inhibition of calcineurin prevents agonist-induced cardiomyocyte hypertrophy. Proceedings of the National Academy of Sciences of the United States of America 97: 1196–1201.

Tajima M, Bartunek J, Weinberg EO, Ito N, Lorell BH (1998) Atrial natriuretic peptide has different effects on contractility and intracellular pH in normal and hypertrophied myocytes from pressure-overloaded hearts. Circulation 98: 2760–2764.

Takewaki S, Kuro-o M, Hiroi Y, Yamazaki T, Noguchi T, Miyagishi A, Nakahara K, Aikawa M, Manabe I, Yazaki Y, et al. (1995) Activation of Na(+)-H+ antiporter (NHE-1) gene expression during growth, hypertrophy and proliferation of the rabbit cardiovascular system. Journal of Molecular and Cellular Cardiology 27: 729–742.

Takimoto E, Champion HC, Li M, Ren S, Rodriguez ER, Tavazzi B, Lazzarino G, Paolocci N, Gabrielson KL, Wang Y, Kass DA (2005) Oxidant stress from nitric oxide synthase-3 uncoupling stimulates cardiac pathologic remodeling from chronic pressure load. The Journal of Clinical Investigation 115: 1221–1231.

Tamamori M, Ito H, Adachi S, Akimoto H, Marumo F, Hiroe M (1996) Endothelin-3 induces hypertrophy of cardiomyocytes by the endogenous endothelin-1-mediated mechanism. The Journal of Clinical Investigation 97: 366–372.

Teshima Y, Akao M, Jones SP, Marban E (2003) Cariporide (HOE642), a selective Na^+-H^+ exchange inhibitor, inhibits the mitochondrial death pathway. Circulation 108: 2275–2281.

Thomas RC (1989) Cell growth factors.Bicarbonate and pHi response. Nature 337: 601.

Toda T, Kadono T, Hoshiai M, Eguchi Y, Nakazawa S, Nakazawa H, Higashijima N, Ishida H (2007) Na^+/H^+ exchanger inhibitor cariporide attenuates the mitochondrial Ca^{2+} overload and PTP opening. American Journal of Physiology 293: H3517–H3523.

von Lewinski D, Stumme B, Fialka F, Luers C, Pieske B (2004) Functional relevance of the stretch-dependent slow force response in failing human myocardium. Circulation Research 94: 1392–1398.

Wang W, Fang H, Groom L, Cheng A, Zhang W, Liu J, Wang X, Li K, Han P, Zheng M, Yin J, Wang W, Mattson MP, Kao JP, Lakatta EG, Sheu SS, Ouyang K, Chen J, Dirksen RT, Cheng H (2008) Superoxide flashes in single mitochondria. Cell 134: 279–290.

Weber KT, Brilla CG (1991) Pathological hypertrophy and cardiac interstitium. Fibrosis and renin-angiotensin-aldosterone system. Circulation 83: 1849–1865.

Wilkins BJ, Dai YS, Bueno OF, Parsons SA, Xu J, Plank DM, Jones F, Kimball TR, Molkentin JD (2004) Calcineurin/NFAT coupling participates in pathological, but not physiological, cardiac hypertrophy. Circulation Research 94: 110–118.

Wu X, Zhang T, Bossuyt J, Li X, McKinsey TA, Dedman JR, Olson EN, Chen J, Brown JH, Bers DM (2006) Local InsP3-dependent perinuclear Ca^{2+} signaling in cardiac myocyte excitation-transcription coupling. The Journal of Clinical Investigation 116: 675–682.

Xia Y, Rajapurohitam V, Cook MA, Karmazyn M (2004) Inhibition of phenylephrine induced hypertrophy in rat neonatal cardiomyocytes by the mitochondrial KATP channel opener diazoxide. Journal of Molecular and Cellular Cardiology 37: 1063–1067.

Xu FP, Chen MS, Wang YZ, Yi Q, Lin SB, Chen AF, Luo JD (2004) Leptin induces hypertrophy via endothelin-1-reactive oxygen species pathway in cultured neonatal rat cardiomyocytes. Circulation 110: 1269–1275.

Yamazaki T, Komuro I, Kudoh S, Zou Y, Nagai R, Aikawa R, Uozumi H, Yazaki Y (1998) Role of ion channels and exchangers in mechanical stretch-induced cardiomyocyte hypertrophy. Circulation Research 82: 430–437.

Yamazaki T, Komuro I, Kudoh S, Zou Y, Shiojima I, Hiroi Y, Mizuno T, Maemura K, Kurihara H, Aikawa R, Takano H, Yazaki Y (1996) Endothelin-1 is involved in mechanical stress-induced cardiomyocyte hypertrophy. The Journal of Biological Chemistry 271: 3221–3228.

Yamazaki T, Komuro I, Kudoh S, Zou Y, Shiojima I, Mizuno T, Takano H, Hiroi Y, Ueki K, Tobe K, et al. (1995) Angiotensin II partly mediates mechanical stress-induced cardiac hypertrophy. Circulation Research 77: 258–265.

Yokoyama H, Gunasegaram S, Harding SE, Avkiran M (2000) Sarcolemmal Na^+/H^+ exchanger activity and expression in human ventricular myocardium. Journal of the American College of Cardiology 36: 534–540.

Yoshida H, Karmazyn M (2000) Na(+)/H(+) exchange inhibition attenuates hypertrophy and heart failure in 1-wk postinfarction rat myocardium. American Journal of Physiology 278: H300–H304.

Zhou C, Ziegler C, Birder LA, Stewart AF, Levitan ES (2006) Angiotensin II and stretch activate NADPH oxidase to destabilize cardiac Kv4.3 channel mRNA. Circulation Research 98: 1040–1047.

Zorov DB, Filburn CR, Klotz LO, Zweier JL, Sollott SJ (2000) Reactive oxygen species (ROS)-induced ROS release: a new phenomenon accompanying induction of the mitochondrial permeability transition in cardiac myocytes. The Journal of Experimental Medicine 192: 1001–1014.

Zou Y, Yamazaki T, Nakagawa K, Yamada H, Iriguchi N, Toko H, Takano H, Akazawa H, Nagai R, Komuro I (2002) Continuous blockade of L-type Ca^{2+} channels suppresses activation of calcineurin and development of cardiac hypertrophy in spontaneously hypertensive rats. Hypertens Research 25: 117–124.

Chapter 14
Stretch-Induced Inotropy in Atrial and Ventricular Myocardium

Dirk von Lewinski, Jens Kockskämper, Mounir Khafaga, Robert Gasser, and Burkert Pieske

Abstract Mechanical load directly regulates cardiac force. Stretching myocardial tissue results in a biphasic increase in contractility: an immediate increase (Frank-Starling mechanism) followed by a further slow increase (slow force response, SFR). Most experiments published have been performed in ventricular myocardium and very little in human tissue. We therefore highlight stretch dependent slow force responses in human myocardium and compare signal transduction in atrial and ventricular tissue. Although of comparable amplitude underlying signal transduction varies between the two tissue types. In ventricular muscle strips, the SFR is significantly reduced by inhibition of Na^+/H^+- (NHE) and Na^+/Ca^{2+}-exchange (NCX) but not affected by AT- and ET-receptor antagonism. In contrast, SFR in atrial tissue is neither affected by NHE- nor NCX-inhibition but interestingly, inhibition of AT-receptors or pre-incubation with angiotensin II or endothelin-1 attenuate the atrial SFR. In addition, stretch results in a large NHE-dependent $[Na^+]_i$ increase in ventricle but only a small, NHE-independent $[Na^+]_i$ increase in atrium. Stretch activated channels are not involved in the SFR in either tissue but contribute to basal force development in atrium but not ventricle. Thus, in human heart both atrial and ventricular myocardium exhibit a stretch-dependent SFR that is likely to serve as adjustment mechanism regulating cardiac output in case of increased preload. In ventricle on the one hand, there is a significant NHE-dependent (but angiotensin II- and endothelin-1- independent) $[Na^+]_i$ increase that is translated into a $[Ca^{2+}]_i$ and force increase via NCX. In atrium, on the other hand, there is an angiotensin II- and endothelin-dependent (but NHE- and NCX-independent) force increase. Increased myofilament Ca^{2+} sensitivity through MLCK-induced phosphorylation of MLC2 is contributing to the SFR in both atrium and ventricle.

Keywords Stretch · Atrium · Ventricle · Human

D.v. Lewinski (✉)
Abteilung Kardiologie, Medizinische Universität Graz, Graz, Austria
e-mail: dirk.von-lewinski@meduni-graz.at

A. Kamkin, I. Kiseleva (eds.), *Mechanosensitivity of the Heart*,
Mechanosensitivity in Cells and Tissues 3, DOI 10.1007/978-90-481-2850-1_14,

14.1 Introduction

Mechanical load is one of the most important regulators of cardiac function. The main mechanism contributing is known as Frank-Starling mechanism (FSM), which was first described at the beginning of the last century. It states that an increase in diastolic filling (preload) causes an increase in stroke volume and thus cardiac output. At the cellular level, increases in diastolic filling cause stretching of the myocytes leading to increases in developed contractile force. The force increase during the FSM is intrinsic to the cardiac myocyte and is caused by increased Ca^{2+} sensitivity of the myofilaments (Hibberd and Jewell, 1982; Kentish et al., 1986; Konhilas et al., 2002). The major physiological role of the FSM is to either adjust the output between the right and left side of the heart and to compensate increased preload.

Following the FSM, an additional, slower, stretch-induced increase in contractile force has been first described in cat papillary muscle (Parmley and Chuck, 1973) and confirmed in multiple species such as ferret (Calaghan and White, 2001), guinea-pig (White et al., 1995), rabbit (von Lewinski et al., 2003), rat (Hongo et al., 1996; Kentish and Wrzosek, 1998; Alvarez et al., 1999), and human (von Lewinski et al., 2004). It develops within 5–15 min after acute stretch and is termed the slow force response (SFR). Therefore, stretch elicits a biphasic increase in developed force, the FSM and the SFR. The SFR, too, is a general phenomenon of physiological relevance and the underlying mechanisms of the SFR, just like the FSM, are intrinsic to the cardiac myocyte.

During recent years significant advances in the understanding of the underlying mechanisms of this phenomenon have been achieved, nevertheless controversies on signal transduction mechanisms are still ongoing.

A great number of signalling molecules including channels and transporters are possibly involved in the SFR including angiotensin II (Alvarez et al., 1999; Perez et al., 2001), endothelins (Alvarez et al., 1999; Calaghan and White, 2001; Perez et al., 2001; Ennis et al., 2005), stretch-activated non-selective cation channels (SACs) (Calaghan and White, 2004; Niederer and Smith, 2007), the Na^+/H^+ exchanger (NHE) (Alvarez et al., 1999; Perez et al., 2001; von Lewinski et al., 2003; Calaghan and White, 2004; Luers et al., 2005), the Na^+/Ca^{2+} exchanger (NCX) (Perez et al., 2001; von Lewinski et al., 2003; Luers et al., 2005), the Na^+/K^+ pump (Bluhm et al., 1998), cAMP (Todaka et al., 1998; Calaghan et al., 1999), phosphatidyinositol-3 kinase (PI3K) (Petroff et al., 2001), and nitric oxide (NO) (Petroff et al., 2001), and this list is likely to be extended further. The significance of each mechanism may vary depending on experimental conditions (e.g. the mode of stretch or increase in load), the preparation (single myocytes versus multicellular trabeculae or whole hearts), the tissue (atrium versus ventricle), and the species.

In this chapter, we present data on the SFR in human atrium and ventricle, describing common pathways as well as differences in signal transduction of the SFR in these tissues.

Stretch as a rather unspecific stimulus alters various properties of the myocytes and of the non-myocyte tissue in the myocardium. Experimental conditions strongly influence the response of the myocardial tissue examined and some of the differences in signal transduction mechanisms discussed in the literature are likely due to the differing experimental conditions. The data of our group presented in this review was collected at identical conditions: After complete mechanical stabilisation at a stimulation frequency of 1 Hz at 37°C in bicarbonate buffered Tyrode at L_{max}, muscle length was reduced to 88% of L_{max} (L_{88}) for 30 min. Muscles were then stretched acutely to 98% of L_{max} (L_{98}). This resulted in a biphasic increase in developed force, an immediate increase (1st phase, Frank-Starling mechanism) followed by a further delayed increase (2nd phase, SFR). The magnitude of the 1st phase was quantified by normalising to developed force measured at L_{88}. The magnitude of the 2nd phase was quantified by normalising to developed force measured during the 1st phase. FSM and SFR were highly reproducible within a given muscle strip. Thus, all experiments involving drugs included paired stretch protocols, i.e. first a control stretch protocol was conducted in the absence of drug; then the muscle was released to L_{88} and drug was applied; after 25 min a second stretch protocol was conducted in the same muscle strip in the presence of the respective drug.

14.1.1 The SFR in Human Ventricular Myocardium

Signal transduction of SFR in human ventricle is similar to that known in rabbit myocardium. Average data are presented in Fig. 14.1. In line with results from almost all models knownm, SFR in human ventricle is reduced by 3 μmol/L cariporide (from 122 ± 4 to 117 ± 2%; $p < 0.05$) or 5 μmol/L KB-R7943 (from 121 ± 3 to 107 ± 3%; $p < 0.05$), demonstrating that stretch-induced stimulation/modulation

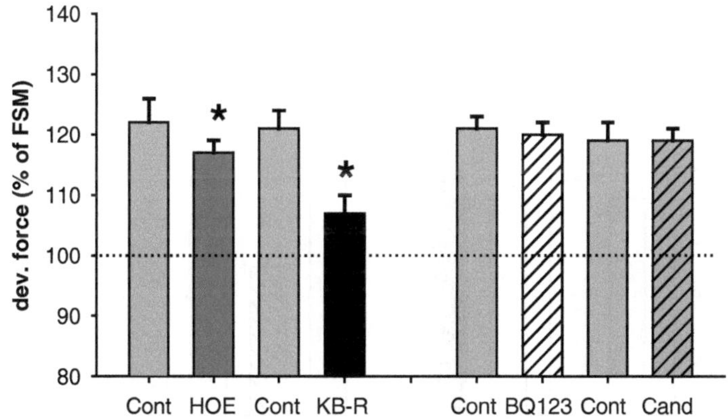

Fig. 14.1 *Left*: Effect of HOE642 (3 μmol/L; *dark gray*) and KB-R7943 (5 μmol/L; *black*) on force development during the SFR in human ventricle. Controls (*gray*) within the same muscle strip. * = $p < 0.05$. *Right*: Effect of BQ123 (0.3 μmol/L; *white striped*) and Candesartan (0.1 μmol/L; *gray striped*) on force development during the SFR in human ventricle. Controls (*gray*) within the same muscle strip. * = $p < 0.05$

of these Na^+-dependent transporters makes an important contribution to the SFR. Contrasting the situation in other animal models such as rat or feline, however, antagonism of AT1 (0.1 μmol/L candesartan) or ETA (0.3 μmol/L BQ123) receptors does not affect the SFR in human ventricle (from 119 ± 3 to 119 ± 2% and from 121 ± 2 to 120 ± 2%, respectively, both n.s.), suggesting that autocrine/paracrine actions of angiotensin II and endothelins are not involved in the SFR in human ventricular myocardium.

Taken together, these results suggest that in human ventricle the SFR is mediated by stretch-dependent stimulation of NHE and modulation of NCX activity. The mechanism by which stretch stimulates NHE, however, remains unknown. It does not appear to involve angiotensin II or endothelins.

14.1.2 The SFR in Human Atrial Myocardium

Similar to ventricular myocardium, isolated human atrial muscle strips are also characterised by a highly reproducible, biphasic increase in developed force in response to stretch. This suggested that similar subcellular mechanisms might underlie the SFR in atrium and ventricle. Surprisingly, however, neither blockade of NHE (3 μmol/L cariporide) nor of NCX (5 mmol/L KB-R7943) does affect the atrial SFR (Fig. 14.2, left), indicating that a different mechanism must underlie the stretch-induced SFR in human atrium.

Although antagonism of AT and ET receptors does not affect the SFR in human ventricle, these peptides have been described to largely mediate SFR in animal species (Alvarez et al., 1999; Cingolani et al., 2001). Since many cardioactive peptides act preferentially on the atrium rather than the ventricle as shown for angiotensin II which exerts positive inotropic effects in atrial but not in ventricular

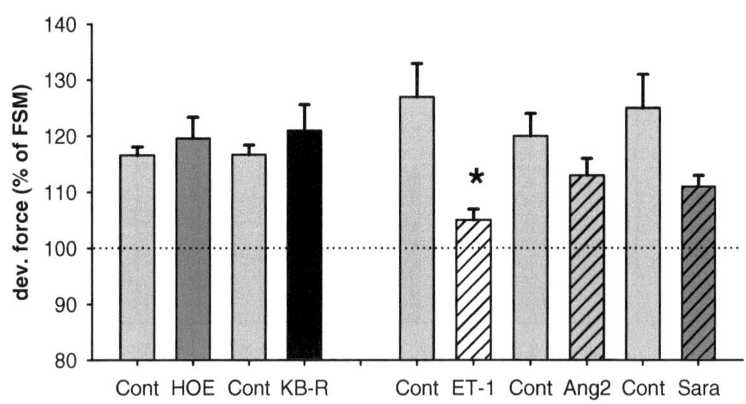

Fig. 14.2 *Left*: Effect of HOE642 (3 μmol/L; *dark gray*) and KB-R7943 (5 μmol/L; *black*) on force development during the SFR in human atrium. Controls (*gray*) within the same muscle strip. * = $p < 0.05$. *Right*: Effect of ET-1 (25 nmol/L; *white striped*) and angiotensin II (0.1 μmol/L; *gray striped*) and saralasin (5 μmol/L; *dark gray striped*) on force development during the SFR in human atrium. Controls (*gray*) within the same muscle strip. * = $p < 0.05$

human myocardium (Holubarsch et al., 1993), it is worth testing for potential effects of these peptides in human atrial myocardium. Data of effects on basal function is shown for angiotensin II and endothelin-1, and secondly SFR amplitude was measured before and after inhibition of the respective receptors.

Angiotensin II increased basal force by 247 ± 46% ($n = 8, p < 0.05$ versus initial control). Similar, endothelin-1 also triggered an increase in developed force (427 ± 75%, $n = 8, p < 0.01$). Figure 14.2 (right) illustrates average data of SFR amplitude w/wo preincubation with the peptide or the antagonist. As can be seen angiotensin II and endothelin-1 preincubation largely reduced the SFR from 120 ± 4 to 113 ± 3% and 127 ± 6 to 105 ± 2%, respectively.

These results suggested that angiotensin II and endothelin-1 are involved in the SFR in human atrium. We therefore tested whether receptor antagonism might also reduce the SFR. Saralasin (5 μmol/L), an unselective AT receptor antagonist, significantly reduced the SFR from 125 ± 6 to 111 ± 2% (n = 9, $p < 0.05$, Fig. 14.2 right). ET receptor antagonists were not tested, however, because preliminary experiments indicated that neither BQ123 (0.25–1.0 μmol/L) nor PD145065 (10 μmol/L; data not shown) were able to reduce the positive inotropic effect of bath-applied endothelin-1, presumably due to the unique pharmacology of ET receptors in human atrium (Burrell et al., 2000). Nevertheless, taken together these results suggested that autocrine/paracrine actions of angiotensin II and endothelin contributed to the atrial SFR. The intracellular signalling cascade, however, did not involve NHE and NCX.

14.2 Involved Proteins

14.2.1 No Contribution of Stretch Activated Channels

Stretch-activated cation channels (SACs) are obviously activated upon stretch and could therefore be involved in mediating the SFR as suggested in modelling studies[14]. Activation of SACs could increase force via influx of Na^+ and/or Ca^{2+}. However, we could previously show in human and rabbit myocardium (von Lewinski et al., 2003, 2004) that SACs do not participate in mediating the SFR. One problem with the available data is the lack of a specific inhibitor of SACs. Available studies have either used gadolinium or streptomycin to block SACs. Although both can indeed block SACs (Hamill and McBride, 1996; Isenberg et al., 2003), they are not highly specific and thus interpretation of results obtained with these substances alone is difficult. To circumvent these problems experiments with the novel, highly potent and specific inhibitor of SACs (Bowman et al., 2007) GsMtx-4 should provide reliable data.

In atrium, increasing concentrations of GsMtx-4 reduce basal twitch force at L_{98}. In contrast, the same concentrations does not affect basal twitch force in a ventricular muscle strip. The negative inotropic effect of GsMtx-4 in atrium is concentration-dependent. At 1000 nmol/L GsMtx-4, the highest concentration tested, basal twitch force decreased by 12 ± 3% ($n = 4, p < 0.05$).

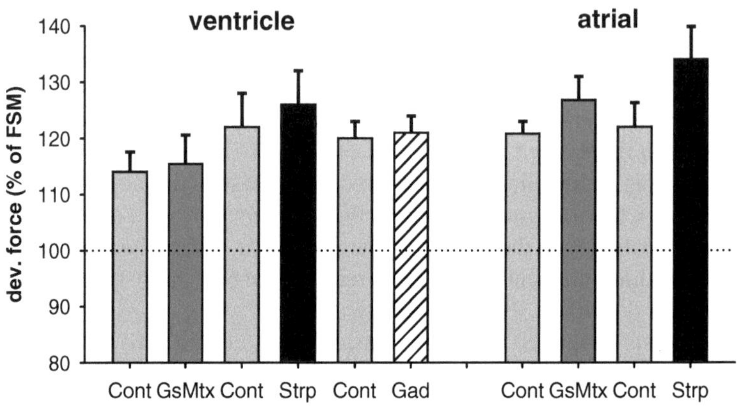

Fig. 14.3 *Left*: Effect of GsMtx-4 (1 μmol/L; *dark gray*), streptomycin (70 μmol/L; *black*) and gadolinium (10 μmol/L, *white striped*) on force development during the SFR in human ventricular myocardium. Controls (*gray*) within the same muscle strip. * $= p < 0.05$. *Right*: Effect of GsMtx-4 (0.5 μmol/L; *dark gray*) and streptomycin (80 μmol/L; *black*) on force development during the SFR in human atrial myocardium. Controls (*gray*) within the same muscle strip. * $= p < 0.05$

In ventricle, on the other hand, there was no significant effect of GsMtx-4 on basal force up to a concentration of 1000 nmol/L. Thus, SACs appear to be activated under baseline conditions and contribute to basal force development in human atrium but not ventricle.

Figure 14.3 summarizes average results from human atrial and ventricular muscle strips upon stretch. The involvement of SACs in the SFR in human ventricular muscle strips is tested using 1000 nmol/L GsMtx-4 (n = 3), 70 μmol/L streptomycin ($n = 6$) and 10 μmol/L gadolinium ($n = 11$). None of the inhibitors affected the ventricular SFR.

Identical experiments performed in atrial muscle strips did show comparable results. Upon repeating stretch protocol in the presence of GsMtx-4, no reduction in the SFR was observed. Basal twitch force, however was reduced after GsMtx-4 incubation (500 nmol/L, $n = 8$) at L88 in human atrium by $9 \pm 3\%$ ($p < 0.05$). These data are confirmed using the unspecific SAC inhibitor streptomycin. Streptomycin (80 μmol/L, $n = 6$) reduced basal force development at L_{88} by $20 \pm 2\%$ ($p < 0.01$) and tended to increase the SFR, but this effect did not reach statistical significance ($p = 0.08$).

From these results it can be concluded that SACs are unlikely to contribute to the SFR in either human atrium or ventricle.

14.3 Role of Myosin Light Chain Phosphorylation by MLCK

Contractile force can be modulated by changes in the amplitude of the $[Ca^{2+}]_i$ transient (upstream) or by alterations in myofilament Ca^{2+} sensitivity (downstream). As shown above the atrial SFR is independent of the NHE–$[Na^+]_i$–NCX pathway.

Therefore, the investigation of the myosin light chain pathway appears to be reasonable and promising since it has been shown to regulate force development in atrial tissue via its atrial subtype (MLC2a). In the mentioned study phosphorylation of MLC2a by a specific kinase (myosin light chain kinase; MLCK) caused an increase in the Ca^{2+} sensitivity of the myofilaments and increased force development.

Inhibition of MLCK using 10 μmol/L ML-7 reduces basal force development by 18 ± 3% in atrium ($n = 8$; $p < 0.01$) and reduces the SFR from 122 ± 4 to 111 ± 4% in ventricular myocardium ($n = 5$; $p < 0.01$, Fig. 14.4 left) and from 127 ± 5 to 117 ± 4% in atrial muscle strips (n = 8; $p < 0.05$, Fig. 14.4 right). Thus, MLCK inhibition significantly blunted the SFR in atrial and ventricular myocardium, suggesting that MLCK-mediated MLC phosphorylation is involved in the regulation of the SFR in both human atrium and ventricle.

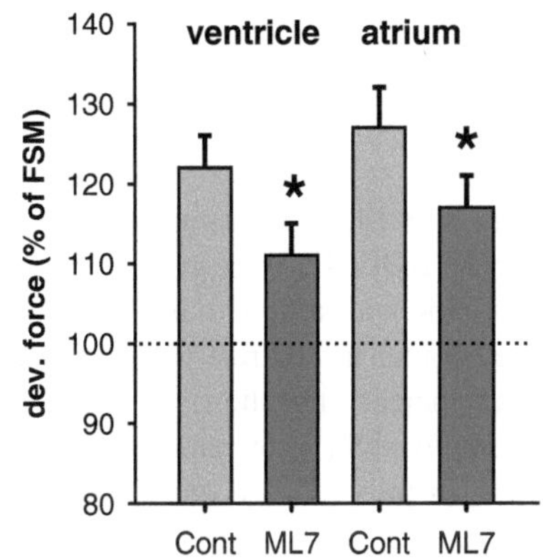

Fig. 14.4 Effect of ML-7 (10 μmol/L; *dark gray*) on force development during the SFR in human ventricular (*left*) and atrial (*right*) myocardium. Controls (*gray*) within the same muscle strip. * = $p < 0.05$

14.4 Differential Regulation in Atrial Vs. Ventricular Myocardium: $[Na^+]_i$

The NHE and NCX dependence of the ventricular SFR suggested that stretch-induced $[Na^+]_i$ increases might play a major role mediating the SFR in human ventricle. By contrast, the NHE and NCX independence of the atrial SFR suggested that $[Na^+]_i$ increases do not occur or that they might be less important for the development of the SFR in human atrium. This hypothesis can easily be tested measuring stretch-induced $[Na^+]_i$ changes during the SFR in human atrial and ventricular muscle strips. Using SBFI fluorescence baseline (at 30°C) $[Na^+]_i$ was 9.9 ± 2.4 mM ($n = 12$) in atrial and 12.2 ± 1.0 mM ($n = 13$) in ventricular muscle strips. In ventricular muscle strips, stretch caused a SFR that was associated with an increase in $[Na^+]_i$. Increases in developed force and $[Na^+]_i$ are shown in Fig. 14.5A for control conditions (gray, $n = 7$) and in the presence of 3 μmol/L HOE642 (dark gray, $n = 6$). HOE642 significantly reduced the stretch-induced increases in force and $[Na^+]_i$ by ~55 and ~65%, respectively.

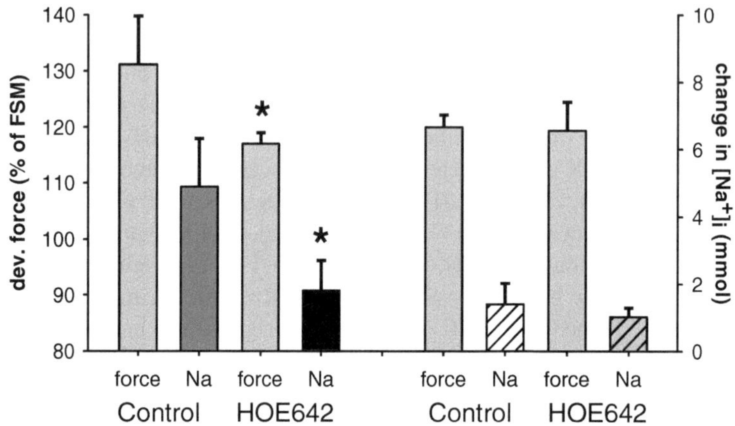

Fig. 14.5 Effect of HOE642 (3 μmol/L) on force development (*gray*) and $[Na^+]_i$ (*dark gray*) in ventricular (*left*) and atrial (*right*) human myocardium. * $= p < 0.05$

Figure 14.5b shows the corresponding data for human atrium (control, gray, $n = 7$; HOE642, dark gray, $n = 5$). The atrial SFR was smaller than the ventricular SFR with a maximum of ~117% after 6 min followed by a decline in developed force. In contrast to ventricle, HOE642 neither affected the stretch-induced force increase nor the rise in $[Na^+]_i$ in human atrium. Interestingly, the HOE642-insensitive $[Na^+]_i$ increase in atrium was comparable to the $[Na^+]_i$ increase in ventricle in the presence of HOE642 suggesting that the SFR in human atrium was mediated by an NHE-independent mechanism, whereas in human ventricle the SFR was composed of at least two mechanisms, an NHE- and Na^+-dependent one as well as an NHE-independent one.

Cumulative data in atrial and ventricular human myocardium indicate that defined stretch elicits a SFR of ~20–25%, thus contributing approximately a third to the total stretch induced inotropism.

In line with data obtained from various mammalian species, these observations demonstrate that the SFR is a universal phenomenon occurring in atrial and ventricular preparations on the level of single myocytes, multicellular muscle strips and whole hearts. One possible reason for the universality of the SFR might be that it is mediated by activation of several stretch-dependent pathways rather than a single mechanism. This redundancy ensures that, even if one of the stretch-activated pathways was compromised (e.g. in cardiac disease), the other pathways could still result in a substantial SFR. The presence of the SFR in almost all myocardial models tested also implies that the SFR serves important physiological functions.

14.5 Various Pathways and Mechanisms Contribute to the SFR

It has early been shown that the SFR in cat ventricular muscle is caused by an increase in $[Ca^{2+}]_i$ transients (Allen and Kurihara, 1982). This finding has been confirmed in other ventricular preparations and species (Hongo et al., 1996; Kentish and Wrzosek, 1998; Todaka et al., 1998; Alvarez et al., 1999; Calaghan and White, 2004; Luers et al., 2005). Although a stretch induced slow increase in the $[Ca^{2+}]_i$ transient has not been demonstrated directly so far in human ventricle, blocking SR function significantly reduced the SFR in this tissue, suggesting an important role for alterations in SR Ca^{2+} load and thus the electrically stimulated $[Ca^{2+}]_i$ transient (von Lewinski et al., 2004). In line with this finding, caffeine was found to reduce the SFR in cat ventricle by an SR-dependent mechanism (Chuck and Parmley, 1980). On the other hand, inhibition of SR function did not diminish the SFR in rat (Kentish and Wrzosek, 1998; Calaghan and White, 2004) and rabbit ventricle (Bluhm and Lew, 1995). Taken together, these results suggest that in mammalian ventricle the SFR can be mediated by a mechanism dependent on changes in SR Ca^{2+} load and the electrically stimulated $[Ca^{2+}]_i$ transient and an additional mechanism. Two major mechanisms have recently been proposed to underly the increase in $[Ca^{2+}]_i$ transients during the SFR: First, the activation of SACs (possibly acting via increased Na^+ and/or Ca^{2+} influx) and, second, the stimulation of NHE activity through autocrine/paracrine actions of stretch-released angiotensin II and endothelin. The first is rather controversial. Using a specific SAC blocker, GsMtx-4, we have shown that SACs do not contribute significantly to the SFR in human ventricular myocardium. Furthermore, our previous study using unspecific SAC blockers came to the same conclusion (von Lewinski et al., 2004). In contrast, there is convincing evidence for a key role of NHE and NCX in the SFR in human ventricle (von Lewinski et al., 2004). We have further substantiated this evidence by demonstrating a stretch-induced increase in $[Na^+]_i$ associated with the SFR. This finding is in line with previous studies revealing stretch-induced increases in $[Na^+]_i$ in rat, rabbit, and mouse ventricle (Alvarez et al., 1999; Perez et al., 2001; Isenberg et al., 2003; Kondratev and Gallitelli, 2003; Luers et al., 2005). The activating mechanism of NHE, however, is unknown at present. Angiotensin II and endothelin, although of importance in some animal species (Alvarez et al., 1999; Calaghan and White, 2001; Perez et al., 2001), are not involved in human ventricle. Future studies are needed to completely characterize the mechanism underlying the stretch-induced NHE stimulation in human ventricle.

In addition to the NHE–$[Na^+]_i$–NCX–SR pathway, MLCK-dependent MLC phosphorylation and increased myofilament Ca^{2+} responsiveness (Andersen et al., 2002) contribute to the SFR in ventricular tissue. This mechanism may explain why blockade of the NHE–NCX pathway is unable to completely suppress the SFR (Fig. 14.1) and why in some studies suppression of SR function did not inhibit the SFR (Bluhm and Lew, 1995; Kentish and Wrzosek, 1998; Calaghan and White, 2004). How MLCK is stimulated by stretch remains to be determined. Since MLCK is dependent on Ca^{2+}/calmodulin, one possibility is that the stretch-induced increase in the $[Ca^{2+}]_i$ transient also stimulates MLCK via Ca^{2+}/calmodulin. In this case, however, MLCK would be downstream of the NHE–$[Na^+]_i$–NCX–SR axis and inhibition of this pathway should cause complete suppression of the SFR (which it does not). Thus, alternative pathways might exist by which stretch can stimulate MLCK. In summary, current evidence indicates that in human ventricle stretch activates at least two signalling pathways that mediate the SFR. First, a Na^+- and Ca^{2+}-dependent mechanism via the NHE–$[Na^+]_i$–NCX–SR axis (but not autocrine/paracrine actions of angiotensin II and endothelin). Secondly, the SFR appears to be mediated by an MLCK-dependent increase in myofilament Ca^{2+} responsiveness.

In line with previous studies on rat atrium (Tavi et al., 1998; 1999; 2000), stretch elicited a SFR of similar amplitude in atrium and ventricle. Analysis of the underlying mechanisms revealed that, despite similar magnitude and time course, the atrial SFR was independent of sarcolemmal transporters NHE and NCX. Since either pre-application of angiotensin II and endothelin-1 or antagonism of AT receptors largely reduced the atrial SFR, it seems to be mediated by autocrine/paracrine release and action of these cardioactive peptides, similar to what has been found in ventricle of some animal species (Alvarez et al., 1999; Calaghan and White, 2001; Perez et al., 2001). In human atrium, Gq-coupled receptors increase force via MLCK-dependent increases in the phosphorylation of MLC2a and increased myofilament Ca^{2+} responsiveness (Grimm et al., 2005). Angiotensin II and endothelin-1 both act via Gq-coupled receptors and we found clear evidence that the atrial SFR is mediated by MLCK-dependent increases in MLC2a phosphorylation. As noted above, MLCK activity depends on Ca^{2+}/calmodulin. Elevation of $[Ca^{2+}]_i$, therefore, could link AT and ET receptor activation with stimulation of MLCK. In rat atrium direct evidence for stretch-induced increases in the $[Ca^{2+}]_i$ transient during the SFR has been obtained (Tavi et al., 1998, 1999). The exact pathway how angiotensin II and endothelin-1 elevate $[Ca^{2+}]_i$ is still unclear, but IP3-induced SR Ca^{2+} release is a likely candidate since endothelin-1 has been shown to increase the atrial $[Ca^{2+}]_i$ transient (Zima and Blatter, 2004) via inositol 1,4,5-trisphosphate (IP3). Therefore we propose the following chain of events underlying the SFR in human atrium: Stretch causes the release of angiotensin II and endothelin, which act in an autocrine/paracrine way on AT and ET receptors on atrial myocytes activating the phospholipase C – IP3 pathway to increase the atrial $[Ca^{2+}]_i$ transient. This results in a Ca^{2+}-dependent increase in contractile force and leads to Ca^{2+}/calmodulin-dependent stimulation of MLCK, which in turn phosphorylates MLC2a to elicit an additional increase in force via increased myofilament Ca^{2+} responsiveness.

Rather unexpected is the finding that SACs are not involved in the SFR in human myocardium, neither in atrium nor ventricle, despite convincing evidence for the expression and functional relevance of SACs in atrium and ventricle of many mammalian species (Suchyna et al., 2000; Isenberg et al., 2003) including humans (Kamkin et al., 2003), and evidence for the involvement of SACs in the SFR in rat ventricle (Calaghan and White, 2004; Niederer and Smith, 2007). To verify previous data using gadolinium and streptomycin we have additionally used a specific blocker of SACs, GsMtx-4, to evaluate the role of SACs in human atrium and ventricle for baseline force development and the SFR. GsMtx-4, however, did not attenuate the SFR, neither in atrium nor ventricle. This underlines, that under our experimental conditions SACs do not contribute to the SFR. A possible explanation for this finding is that local deformation of the membrane, rather than end-to-end stretch, might be required for activation of SACs (Isenberg et al., 2003).

In summary, in human atrium stretch activates a signalling pathway that involves release and autocrine/paracrine actions of angiotensin II and endothelin (but not stimulation of the NHE–[Na^+]$_i$–NCX axis) and MLCK-dependent phosphorylation of MLC2a with a subsequent increase in myofilament Ca^{2+} responsiveness.

14.6 Physiological Role of the SFR

The SFR is present in atrial as well as ventricular human myocardium and makes a significant contribution to the force development upon stretch. Its physiological function may not only be to support the FSM adjusting the output of the left and right ventricle, but also to increase cardiac output under conditions of elevated load. This may become particularly important in a diseased heart with depressed contractility and cardiac output or after cardioversion of atrial fibrillation, since atrial contraction is still impaired for some time after conversion to sinus rhythm. Under these conditions, changes in atrial load may recruit the SFR and thereby help increase ventricular filling and improve cardiac function.

In conclusion, the stretch-induced SFR is a universal phenomenon in both human atrium and ventricle. Although of comparable amplitude, underlying signal transduction mechanisms differ significantly.

Acknowledgements The authors' studies were supported by grants from the Deutsche Forschungsgemeinschaft (to JK & BP: PI 414/1 and PI 414/2, Klinische Forschergruppe 155, TP 6), the German Ministry for Education and Research (BMBF, Kompetenznetz Herzinsuffizienz, TP 8, to TE and BP).

References

Allen DG & Kurihara S. (1982). The effects of muscle length on intracellular calcium transients in mammalian cardiac muscle. J Physiol 327, 79–94.

Alvarez BV, Perez NG, Ennis IL, Camilion de Hurtado MC & Cingolani HE. (1999). Mechanisms underlying the increase in force and Ca($^{2+}$) transient that follow stretch of cardiac muscle: a possible explanation of the Anrep effect. Circ Res 85, 716–722.

Andersen GO, Qvigstad E, Schiander I, Aass H, Osnes JB & Skomedal T. (2002). Alpha(1)-AR-induced positive inotropic response in heart is dependent on myosin light chain phosphorylation. Am J Physiol Heart Circ Physiol 283, H1471–H1480.

Bluhm WF & Lew WY. (1995). Sarcoplasmic reticulum in cardiac length-dependent activation in rabbits. Am J Physiol 269, H965–H972.

Bluhm WF, Lew WY, Garfinkel A & McCulloch AD. (1998). Mechanisms of length history-dependent tension in an ionic model of the cardiac myocyte. Am J Physiol 274, H1032–H1040.

Bowman CL, Gottlieb PA, Suchyna TM, Murphy YK & Sachs F. (2007). Mechanosensitive ion channels and the peptide inhibitor GsMTx-4: history, properties, mechanisms and pharmacology. Toxicon 49, 249–270.

Burrell KM, Molenaar P, Dawson PJ & Kaumann AJ. (2000). Contractile and arrhythmic effects of endothelin receptor agonists in human heart in vitro: blockade with SB 209670. J Pharmacol Exp Ther 292, 449–459.

Calaghan S & White E. (2004). Activation of Na^+-H^+ exchange and stretch-activated channels underlies the slow inotropic response to stretch in myocytes and muscle from the rat heart. J Physiol 559, 205–214.

Calaghan SC, Colyer J & White E. (1999). Cyclic AMP but not phosphorylation of phospholamban contributes to the slow inotropic response to stretch in ferret papillary muscle. Pflugers Arch 437, 780–782.

Calaghan SC & White E. (2001). Contribution of angiotensin II, endothelin 1 and the endothelium to the slow inotropic response to stretch in ferret papillary muscle. Pflugers Arch 441, 514–520.

Chuck LH & Parmley WW. (1980). Caffeine reversal of length-dependent changes in myocardial contractile state in the cat. Circ Res 47, 592–598.

Cingolani HE, Perez NG, Camilion de Hurtado MC. (2001). An autocrine/paracrine mechanism triggered by myocardial stretch induces changes in contractility. News Physiol Sci 16, 88–91.

Ennis IL, Garciarena CD, Perez NG, Dulce RA, Camilion de Hurtado MC & Cingolani HE. (2005). Endothelin isoforms and the response to myocardial stretch. Am J Physiol Heart Circ Physiol 288, H2925–H2930.

Grimm M, Haas P, Willipinski-Stapelfeldt B, Zimmermann WH, Rau T, Pantel K, Weyand M & Eschenhagen T. (2005). ***Key role of myosin light chain (MLC) kinase-mediated MLC2a phosphorylation in the alpha 1-adrenergic positive inotropic effect in human atrium. Cardiovasc Res 65, 211–220.

Hamill OP, McBride DW, Jr. (1996). The pharmacology of mechanogated membrane ion channels. Pharmacol Rev 48, 231–252.

Hibberd MG & Jewell BR. (1982). Calcium- and length-dependent force production in rat ventricular muscle. J Physiol 329, 527–540.

Holubarsch C, Hasenfuss G, Schmidt-Schweda S, Knorr A, Pieske B, Ruf T, Fasol R & Just H.(1993). Angiotensin I and II exert inotropic effects in atrial but not in ventricular human myocardium. An in vitro study under physiological experimental conditions. Circulation 88, 1228–1237.

Hongo K, White E, Le Guennec JY & Orchard CH. (1996). Changes in $[Ca^{2+}]i$, $[Na^{+}]i$ and Ca^{2+} current in isolated rat ventricular myocytes following an increase in cell length. J Physiol 491 (Pt 3), 609–619.

Isenberg G, Kazanski V, Kondratev D, Gallitelli MF, Kiseleva I & Kamkin A. (2003). Differential effects of stretch and compression on membrane currents and $[Na^{+}]c$ in ventricular myocytes. Prog Biophys Mol Biol 82, 43–56.

Kamkin A, Kiseleva I, Wagner KD, Bohm J, Theres H, Gunther J & Scholz H. (2003). Characterization of stretch-activated ion currents in isolated atrial myocytes from human hearts. Pflugers Arch 446, 339–346.

Kentish JC, ter Keurs HE, Ricciardi L, Bucx JJ & Noble MI. (1986). Comparison between the sarcomere length-force relations of intact and skinned trabeculae from rat right ventricle. Influence of calcium concentrations on these relations. Circ Res 58, 755–768.

Kentish JC & Wrzosek A. (1998). Changes in force and cytosolic Ca^{2+} concentration after length changes in isolated rat ventricular trabeculae. J Physiol 506 (Pt 2), 431–444.

Kondratev D & Gallitelli MF. (2003). Increments in the concentrations of sodium and calcium in cell compartments of stretched mouse ventricular myocytes. Cell Calcium 34, 193–203.

Konhilas JP, Irving TC, de Tombe PP. (2002). Frank-Starling law of the heart and the cellular mechanisms of length-dependent activation. Pflugers Arch 445, 305–310.

Luers C, Fialka F, Elgner A, Zhu D, Kockskamper J, von Lewinski D & Pieske B. (2005). Stretch-dependent modulation of $[Na^{+}]i$, $[Ca^{2+}]i$, and pHi in rabbit myocardium–a mechanism for the slow force response. Cardiovasc Res 68, 454–463.

Niederer SA & Smith NP. (2007). A mathematical model of the slow force response to stretch in rat ventricular myocytes. Biophys J 92, 4030–4044.

Parmley WW & Chuck L. (1973). Length-dependent changes in myocardial contractile state. Am J Physiol 224, 1195–1199.

Perez NG, de Hurtado MC & Cingolani HE. (2001). Reverse mode of the Na^{+}-Ca^{2+} exchange after myocardial stretch: underlying mechanism of the slow force response. Circ Res 88, 376–382.

Petroff MG, Kim SH, Pepe S, Dessy C, Marban E, Balligand JL & Sollott SJ. (2001). Endogenous nitric oxide mechanisms mediate the stretch dependence of Ca^{2+} release in cardiomyocytes. Nat Cell Biol 3, 867–873.

Suchyna TM, Johnson JH, Hamer K, Leykam JF, Gage DA, Clemo HF, Baumgarten CM & Sachs F. (2000). Identification of a peptide toxin from Grammostola spatulata spider venom that blocks cation-selective stretch-activated channels. J Gen Physiol 115, 583–598.

Tavi P, Han C & Weckstrom M. (1998). Mechanisms of stretch-induced changes in $[Ca^{2+}]i$ in rat atrial myocytes: role of increased troponin C affinity and stretch-activated ion channels. Circ Res 83, 1165–1177.

Tavi P, Han C & Weckstrom M. (1999). Intracellular acidosis modulates the stretch-induced changes in E-C coupling of the rat atrium. Acta Physiol Scand 167, 203–213.

Tavi P, Weckstrom M & Ruskoaho H. (2000). cAMP- and cGMP-independent stretch-induced changes in the contraction of rat atrium. Pflugers Arch 441, 65–68.

Todaka K, Ogino K, Gu A & Burkhoff D. (1998). Effect of ventricular stretch on contractile strength, calcium transient, and cAMP in intact canine hearts. Am J Physiol 274, H990–H1000.

von Lewinski D, Stumme B, Fialka F, Luers C & Pieske B. (2004). Functional relevance of the stretch-dependent slow force response in failing human myocardium. Circ Res 94, 1392–1398.

von Lewinski D, Stumme B, Maier LS, Luers C, Bers DM & Pieske B. (2003). Stretch-dependent slow force response in isolated rabbit myocardium is Na^{+} dependent. Cardiovasc Res 57, 1052–1061.

White E, Boyett MR & Orchard CH. (1995). The effects of mechanical loading and changes of length on single guinea-pig ventricular myocytes. J Physiol 482 (Pt 1), 93–107.

Zima AV & Blatter LA. (2004). Inositol-1,4,5-trisphosphate-dependent $Ca(^{2+})$ signalling in cat atrial excitation-contraction coupling and arrhythmias. J Physiol 555, 607–615.

Chapter 15
Effects of Wall Stress on the Dynamics of Ventricular Fibrillation: A Computer Simulation Study of Mechanoelectric Feedback

Satoko Hirabayashi, Masashi Inagaki, Toshiaki Hisada, and Masaru Sugimachi

Abstract The prevalence of ventricular fibrillation (VF) is increased in the mechanically compromised heart. Computer simulation is a useful means of investigating the mechanisms underlying this phenomenon. Using our latest research as an example, we show how computer simulations are performed and what they reveal. We have developed a fully coupled electromechanical model of the human ventricular myocardium. The model formulated the biophysics of specific ionic currents, excitation-contraction coupling, anisotropic non-linear deformation of the myocardium, and mechanoelectric feedback through stretch-activated channels. Our model suggested that sustained stretches shortens action potential duration (APD) and flattens the electrical restitution curve, whereas stretches applied at the wavefront prolongs APD. The wavefront around the core was highly stretched, even at lower pressures, resulting in a prolongation of APD and extension of the refractory area in the wavetail. As left ventricular pressures increased, the stretched area became wider and the refractory area was further extended. The extended refractory area in the wavetail facilitated wave break-up and the meandering of tips through the interaction between wavefronts and wavetails. This simulation study indicated that mechanical loading promotes meandering and wave breaks of spiral re-entry through mechanoelectric feedback. Mechanical loading in pathological conditions may contribute to the maintenance of VF through these mechanisms.

Keywords Stretch-activated channels · Spiral re-entry · Excitation-contraction coupling · Electrophysiological model · Wall stress · Human · Action potential duration restitution · Conduction velocity

---

S. Hirabayashi (✉)
Biomechanics Laboratory, Department of Mechanical Science & Engineering
School of Engineering, Nagoya University, Nagoya, Japan
e-mail: hirabayashi@mech.nagoya-u.ac.jp

A. Kamkin, I. Kiseleva (eds.), *Mechanosensitivity of the Heart*,
Mechanosensitivity in Cells and Tissues 3, DOI 10.1007/978-90-481-2850-1_15,

15.1 About Ventricular Fibrillation

15.1.1 What is Ventricular Fibrillation?

A healthy heart functions as a blood pump by regular alternation of diastole and systole. During diastole, the cardiac myocytes relax, allowing the heart to fill with blood, and they then contract rapidly to push most of this blood out during systole. The contractions are triggered by an electrical wave of excitation that rapidly propagates from one end of the cardiac muscle to the other.

Abnormal chaotic propagation of the excitation wave causes severe cardiac arrhythmias, in which the cardiac muscle contracts in a rapid, disorganized and asynchronous manner, so preventing the effective pumping of the blood. This is the most dangerous type of cardiac arrhythmia when it occurs in the ventricles, and can lead to ventricular fibrillation (VF). The left ventricle (LV) is the most important chamber involved in pushing blood out into the whole body via the aorta, and VF is considered to be a form of cardiac arrest that can lead to death within minutes. On the surface electrocardiogram, VF is characterized by the presence of irregular undulations of the QRS complexes, of varying morphology, amplitude, and frequency. Although VF most often occurs in patients with structural heart diseases, manifest heart disease is absent in some patients with VF. Great attention has been paid to VF, because it represents the final common pathway for death in most patients with heart disease.

15.1.2 How Does Ventricular Fibrillation Occur and How is it Sustained?

15.1.2.1 Spiral Wave

Although the precise mechanism of VF remains incompletely understood, the establishment of vortex-like re-entry is thought to be involved in the mechanism of VF. The number of wavelets is changeable as they can break up to produce other wavelets or can disappear by collision with other wavelets or with the boundary. VF can stop when all the wavelets disappear.

Despite intense research, we still do not fully understand VF initiation. Winfree (1987, 1989) proposed a mechanism for induction of VF by electrical stimulation, the singularity point hypothesis, which is now widely called the critical point hypothesis (Frazier et al., 1989). This theory is based on the "pinwheel experiment" protocol: a planar S1 wave interacting with a unipolar S2 point stimulus in a two-dimensional sheet of cardiac muscle. The pinwheel stimulation protocol involves the simultaneous establishment of a spatial gradient of momentary stimulus intensity together with a spatial gradient of refractoriness induced by the prior passage of an activation front. When a stimulus of the right size is applied at the right time, mirror image spiral waves (SWs) begin to pivot around phase singularities. These singularities arise in the myocardium near the intersection of a moving critical contour of phase in the normal cycle of excitation and recovery with a momentary critical contour of local stimulus strength (Winfree, 1989). VF may also be initiated by a spontaneous premature beat. The concept of phase singularity has contributed to our understanding of the mechanism of spontaneous initiation of VF. Under appropriate conditions of excitability, the interaction of a wave front with an obstacle can lead to wavebreaks and to the formation of phase singularities at the broken endpoints (Starobin et al., 1996). Anatomical and functional obstacles in cardiac muscle may cause a break in a propagating wave, leading to phase singularities and to vortex-like re-entry. In three-dimensional excitable media such as a heart, the phase singularity is a filament, and the rotating vortex is a scroll wave (Winfree, 1973).

Optical mapping methods have quantified the excitation patterns on the surface of the heart during VF. The direct demonstration of phase singularities and of rotors during VF provides strong evidence that VF is not an entirely random phenomenon (Davidenko et al., 1992; Gray et al., 1995, 1998). In extreme cases, VF can be produced by a single meandering wavelet (Gray et al., 1995; Vaidya et al., 1999). However, Gray et al. found that the average number of coexisting SWs during VF was >1. Their initial estimates indicate that each SW occupied an average area of 12 cm^2. Based on approximate measurements of heart surface area, they estimated that the average number of coexisting SWs during VF is approximately 1–2 for rabbits, 5 for sheep, and 15 for humans. It has also been found that the majority of SWs (80% in rabbits and 84% in sheep) last for less than one rotation cycle (Gray et al., 1998), and therefore do not form complete re-entry loops. Similar results regarding the number of SWs and their life spans were obtained in another experiment using multiple electrodes (Rogers et al., 1999). How is VF sustained with such a small number of surface wavelets with short lifespans? What is required for the break-up of SWs? Several hypotheses have been proposed, but no final conclusion has yet been reached.

15.1.2.2 Action Potential Duration (APD) Restitution Curve

Although the precise mechanism of spiral break-up remains unknown, there is considerable evidence suggesting that it is closely linked to so-called alternans instability.

Alternans instability usually develops at high stimulation rates and shows alternate short and long APDs. The slope of the APD restitution curve is a major factor in alternans instability, and has therefore been suggested to be involved in the development of SW break-up. This curve relates the APD to the diastolic interval (DI), which is the time elapsed between the end of the preceding action potential and the start of the next one. As DI increases, APD becomes longer and the slope becomes flatter. At the pacing frequency, alternans occurs when the slope of the restitution curve is >1. This can be explained theoretically (Panfilov and Pertsov, 2001). Any pacing frequency has an appropriate DI required to produce the periodic beat, even when the slope is >1: i.e., any slope has a point at which the sum of the DI and its corresponding APD become equal to the interval of the electric stimulus. However, when a small difference from this appropriate DI is generated, it causes instability. First, it changes APD: at the pacing frequency, a change in APD means the same amount of change in the next DI. A longer DI prolongs the APD, and the prolonged

APD deprives the next DI of the same time. The next DI, which is shorter than the appropriate DI, shortens the next APD. The shortened APD makes the next DI longer by the same amount of time, and so on. This process can also result from shortening the DI. In both cases, short and long APDs occur alternately. This oscillation diverges when the slope is >1: the small difference between the DI and the appropriate DI results in a larger difference between the next APD and the appropriate APD. The APD difference results in the same difference in the next DI, which is in turn followed by much larger APD differences. Conversely, DI converges to the appropriate DI when the slope is <1, because the DI difference is followed by a smaller APD difference, which results in a smaller DI difference.

Several computer simulations have indicated that APD alternans, and therefore the restitution curve, may be related to SW break-up. By varying the parameters in an electrophysiological model, alternans can be induced in a computer simulation. When oscillations of APD become sufficiently large, SW break-up frequently occurs (Karma, 1993).

It has been reported that some drugs can flatten the restitution curve and also prevent the occurrence of VF, or even stop existing VF. Mapping the electrical activity in the tissue whilst using these drugs showed that they converted multiple irregular wavelets into stationary SWs. However, these results are not sufficient to prove that the APD restitution curve is related to SW break-up. Insufficient experimental evidence currently exists to confirm this. It should also be noted that several cases have been demonstrated where the maximum slope of restitution curve was >1, but no break-up occurred (Qu et al., 1999).

15.1.3 Defibrillation

The perceived mechanisms of defibrillation have changed greatly over the years as more data became available and experimental techniques improved. Although the basic mechanism of defibrillation remains incompletely understood, the level of detail of the known mechanism of defibrillation is increasing. To defibrillate, a shock must not only halt the fibrillation wavefronts, but it also must not create new wavefronts that reinduce fibrillation. A small shock fails to defibrillate because it does not halt all wavefronts. A stronger shock below the defibrillation threshold (DFT), fails to defibrillate because of reinduction of re-entry by virtual electrodes or shock-induced prolongation of refractoriness. A shock must be sufficiently strong that any post-shock ectopic cycles of activation do not induce re-entry that degenerates into VF.

Clinical trials have demonstrated that implantable cardioverter defibrillators prolong life in patients at high risk of sudden cardiac death (Moss et al., 1996). An implantable cardiovascular defibrillator is frequently used in patients with serious ventricular arrhythmias. Since the defibrillator delivers a limited amount of defibrillation energy, changes in the DFT are important. Even if the device delivered enough energy to defibrillate at the time of implantation, a significant increase in DFT may result in a failure of defibrillation due to delivery of insufficient energy.

15.2 Effects of Mechanical Change on VF

15.2.1 Increased Defibrillation Thresholds in Dilated Hearts

15.2.1.1 Multiple Factors That Could Influence DFT

Heart failure (HF) is a highly prevalent cardiovascular syndrome. Clinical data suggest that defibrillation requires more energy in dilated, failing hearts (Engelstein et al., 1995; Huang et al., 2001; Kontos et al., 1995). It is known that HF leads to downregulation of a variety of ionic currents involved in repolarisation, which results in prolongation of APD (Li et al., 2002; Tomaselli and Marban, 1999). These electrophysiological changes can affect the DFT. However, congestive HF is associated not only with electrophysiological changes, but with many other factors that could influence DFT: complex hemodynamic changes (increased filling pressures and wall stress), geometric alterations (increased ventricular volumes, decreased wall thickness, hypertrophy, and other structural changes in the myocardium), and variable degrees of neurohormonal activation (Ott and Reiter, 1997). Previous studies have induced cardiomyopathy by pacing, causing congestive HF in an animal heart in vivo, and then examined DFT. The results of these studies have varied: one study found no difference in DFT between failing and nonfailing hearts (Friedman et al., 1998), while another found a significant increase in DFT with the development of congestive HF (Lucy et al., 1994). In the latter case, there was a four-fold increase in DFT in the rapidly paced group compared with control animals, even when expressed as DFT/g of ventricular tissue. This was probably due to rapid pacing causing two geometric changes with opposing effects on defibrillation energy: cavity dilatation and wall thinning. Both studies revealed that DFT increased with enlargement of the LV but was reduced by LV wall thinning.

HF is characterized by high intraventricular end-diastolic pressures and a progressive decline of cardiac contractility, and mechanical stretch is therefore regarded as one of the reasons for the increase in DFT in failing hearts. Ott and Reiter observed an increase in DFT due to acute ventricular dilatation alone (Ott and Reiter, 1997). Using a fluid-filled, latex balloon in the LV cavity of an isolated Langendorff-perfused rabbit heart, they changed the LV volume by adding fluid, and determined the DFT at each volume. Acute balloon dilatation in the isolated heart is unaffected by neurohormonal influences or structural myocardial changes. Acute LV dilatation increased the LV end-diastolic pressure from 0 ± 0.5 mmHg to 35 ± 3 mmHg and increased the average DFT voltage by 30% (from 96 ± 4 V to 125 ± 7 V). Vigh et al. investigated the DFT in dogs under three conditions — at baseline, after inducing LV dysfunction with norepinephrine infusion, and after volume overload with normal saline (to achieve a pulmonary capillary wedge pressure > 19 mmHg) in the setting of norepinephrine- induced LV dysfunction. A significant increase in DFT energy compared with baseline was observed only in the last situation. These studies suggest that acute ventricular dilatation (e.g., acute hemodynamic decompensation) causes an acute increase in the defibrillation energy requirement, as suggested by the clinical data. It also partly explains the increase in DFT observed during prolonged

VF, which may result in progressive ventricular dilatation and increased wall stress. Strobel et al. found a qualitatively similar result in an intact porcine model (Strobel et al., 1996). They produced acute LV volume reduction by inferior vena cava balloon inflation, and the decrease in LV volume was associated with decreased DFT.

15.2.1.2 Potential Mechanism of DFT Increase by Acute Ventricular Dilatation

Acute ventricular dilatation can influence the process of defibrillation in several ways. First, geometrical changes might have some effect on DFT. A change in ventricular dimensions might alter the potential gradient field across the heart, which would be mechanistically similar to the increase in DFT seen with pericardial effusion (Thakur et al., 1993). For a given shock strength, increasing the low-impedance ventricular blood pool results in a smaller proportion of the shock current being passed through the high-impedance myocardial tissue. This, in turn, decreases the potential gradient in the myocardium. In support of this view, Strobel et al. reported that acute LV volume reduction increased the shock impedance and decreased DFT (Strobel et al., 1996), suggesting that less current is shunted through the blood and more current is passed through the myocardium. In the study by Ott and Reiter, however, acute LV dilatation was shown to increase the shock impedance (Ott and Reiter, 1997). In their study, LV dilatation was achieved by an inflated, insulated latex balloon in the LV cavity that offered more resistance to current flow than the blood pool would, therefore most current passed through the myocardium. In this situation, increasing the shock impedance results in a decrease in current passed through the myocardium for a given shock voltage, and an increase in the DFT.

In addition to these geometrical changes, electrophysiological changes may also be important factors. It has been suggested that myocardial stretch causes electrophysiological changes in the heart through mechanoelectric feedback (MEF), which is possibly associated with stretch-activated channels (SACs) (Hu and Sachs, 1997). Although some investigators have reported prolongation of the APD during stretching (Sung et al., 2003), shortening of the APD and the effective refractory period with mechanical loading has been reported in various species, including rabbits (Zabel et al., 1996b), dogs (Calkins et al., 1989; Hansen, 1993) and humans (Taggart et al., 1988). Previous studies also have shown that myocardial stretch increases the dispersion of refractoriness(Zabel et al., 1996b) and the maximum slope of the electrical restitution curve (Horner et al., 1996), decreases the conduction velocity (Zabel et al., 1996b; Sung et al., 2003), and induces depolarisations (Franz et al., 1992). These electrophysiological changes can influence the characteristics of fibrillation itself (e.g., decreasing the myocardial wavelength and increasing the number of fibrillating wavelets) or can facilitate the reinitiation of VF by the persisting wavelets after a defibrillation shock.

Changes in ventricular load can cause baroreceptor-mediated alteration of sympathetic and parasympathetic tone. Changes in autonomic tone have been reported to affect ventricular vulnerability and the DFT. Lerman et al. have shown that increases in ventricular load mediate a decrease in ventricular APD and refractoriness through activation of the β-adrenergic receptor (Lerman et al., 2001). Decrease in APD and refractoriness through autonomic nervous system might increase the DFT.

15.2.2 High Inducibility of VF by Mechanical Loading

VF is frequently seen in patients with ventricular dysfunction, ventricular volume or pressure overload, or dyssynergic ventricular contraction and relaxation (Kjekshus, 1990; Klein, 1984). It is also known that acute mechanical loading facilitates the development and maintenance of VF (Rosen et al., 1991; Pye and Cobbe, 1996; Ott and Reiter, 1997; Coronel et al., 2002). Rosen et al. examined the effects of ventricular stretch on VF thresholds in Langendorff-perfused rat hearts. The LV end-diastolic pressure was increased from 0 to 20 mmHg by inflation of an intraventricular balloon. The current required to provoke VF decreased at the high LV end diastolic pressure. Coronel et al. studied the effects of ventricular filling on the 1b phase of ischemia-induced arrhythmias. They used three preparations: in situ hearts in the anesthetized open-chested pig, isolated unloaded pig hearts, and isolated working pig hearts. The 1b arrhythmias were less frequent and less severe in isolated unloaded hearts than in in situ hearts or isolated working hearts. Some studies have suggested the existence of an electrophysiological mechanism such as shortening of APD and effective refractory period (Pye and Cobbe, 1996; Ott and Reiter, 1997; Coronel et al., 2002). Experimental studies have also shown that acute mechanical loading increased the complexity of the VF activation pattern (Burton and Cobbe, 1998; Chorro et al., 2000). Using optical mapping of isolated ovine hearts, Moreno et al. (2005) found that elevation of intraventricular pressure to the level seen during VF increased the number of wave breaks and rotors to the same extent in HF and normal hearts. These changes are possibly mediated by SACs (Hu and Sachs, 1997).

As mentioned in the previous section, HF leads to downregulation of a variety of ionic currents involved in repolarisation, and experimental results indicate that repolarising currents play a crucial role in SW formation and propagation, ultimately controlling VF maintenance (Warren et al., 2003). However, the precise mechanisms by which chronic mechanical load complicates VF dynamics remain unclear because a number of other factors, such as complex hemodynamic changes, geometric alterations, variable degrees of neurohormonal activation, changes in extracellular ions, and fibrosis, etc. all have arrhythmogenic effects.

15.2.3 The Role of MEF in Mechanical Induction of VF

Mechanical induction of cardiac rhythm disturbances in the absence of structural damage (*commotio cordis*) is well known, and MEF is thought to be involved in a potential mechanism for this (Lab, 1982). Link et al. developed an experimental model of *commotio cordis* in which anesthetised juvenile swine, struck in the chest by a baseball, develop VF (Link et al., 1998). They found that timing of the impact was an important variable. Only blows occurring during the vulnerable period of repolarization, on the upslope of the T-wave, caused VF. Energy of the impact object was also found to be a critical variable with 40 mph baseballs more likely to cause VF than velocities greater or less than 40 mph (Link et al., 2003). In the same model, glibenclamide, a blocker of K_{ATP} channels, inhibited the occurrence of VF (Link et al., 1999). However, streptomycin, a blocker of cationic SACs, did not prevent VF (Garan et al., 2005). Mechanical activation of K_{ATP} channels is suggested to be a mechanism of VF produced by chest wall blows.

Garny and Kohl investigated the role of MEF in arrhythmogenesis using computer simulation (Garny and Kohl, 2004). They made a two-dimensional electrophysiological model of ventricular tissue, which was able to reproduce electrical activities such as membrane potential, ion currents through membranes, and excitation propagation. The model also included a variable to control SAC currents. The mesh represented a cross-section of the ventricular free wall, which goes through the endocardial edge and the epicardial edge. The effect of an impact was described by a force profile, the base of which is on the epicardial side, to simulate external mechanical stimulation. In the absence of concrete experimental results for the mechanical effects of *commotio cordis*-like precordial impacts on cardiac tissue, a bell-shaped force profile was assumed, the base and height of which could be modified to simulate varying properties of the mechanical stimulus. Within a force profile, a variable to control SAC currents could be changed to reproduce mechanically induced changes in cellular electrophysiology. Although a bell-shaped force profile is unlikely to be a true representation of the mechanical stimulation of the myocardium, and may be changed by the contraction force of the repolarised tissue, the model was able to successfully reproduce the mechanical induction of arrhythmias and identified how stretch activation of cation-nonselective ion channels induced arrhythmias. It caused ectopic excitation in fully repolarised tissue, and functionally blocked conduction at the intersection of the mechanical stimulus. The block stopped repolarisation waves, possibly giving rise to both the triggering and sustaining mechanisms of ventricular arrhythmias.

15.3 Previous Studies on the Effects of MEF on VF Dynamics

15.3.1 Experimental Studies

Using electrode or optical mapping techniques, previous experimental studies examined the effects of mechanical loading on VF dynamics (Burton and Cobbe, 1998; Chorro et al., 2000; Moreno et al., 2005).

Burton and Cobbe studied the effects of stretch on the magnitude and dispersion of changes in VF interval in Langendorff-perfused rabbit hearts using a flexible epicardial array containing 240 unipolar electrodes (Burton and Cobbe, 1998). The LV pressure was increased from 0 to 40 mmHg by inflation of an intraventricular balloon during VF. The coefficient of variation was used as a measure of dispersion of VF interval. Figure 15.1 shows the effects of balloon inflation on histograms of VF intervals in one experiment. Mean VF interval at 0 mmHg was 79.8 ± 1.3 ms and fell to 70.2 ± 1.7 ms at 40 mmHg. The coefficient of variation at 0 mmHg was 8.13 ± 0.8 and increased to 13.3 ± 0.8 at 40 mmHg. Following balloon inflation there was a 22% increase in the number of activation waves at 40 mmHg.

Chorro et al. analyzed the effects of stretch on the epicardial activation patterns during VF in Langendorff-perfused rabbit hearts using a plaque with 121 unipolar electrodes (Chorro et al., 2000). They classified the activation maps during VF into three types according to complexity: type I = activation maps with single broad wavefronts propagating uniformly without significant conduction delay;

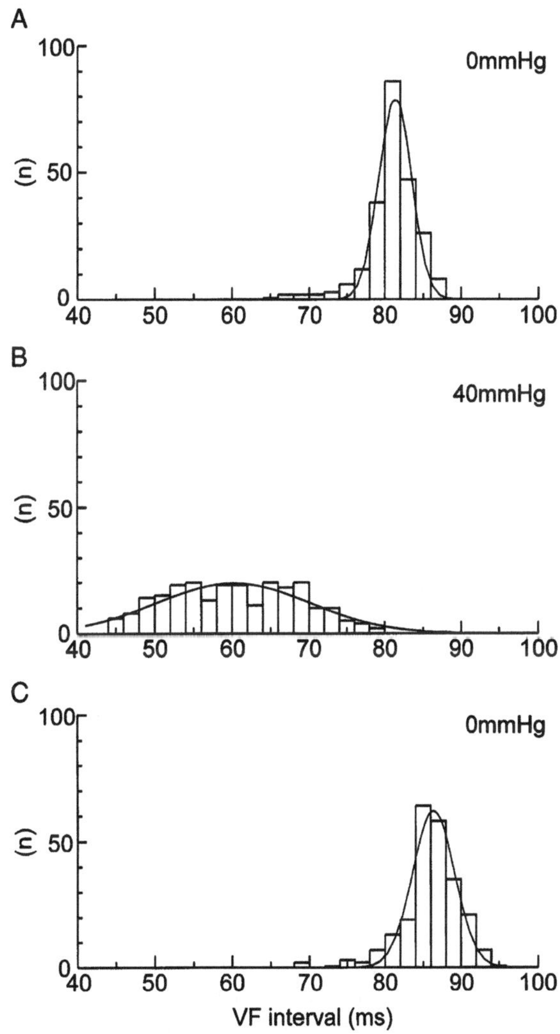

Fig. 15.1 Histograms of VF intervals for all electrode sites in one experiment at 0 mmHg before inflation (**a**), at the end of a 6 min period of inflation to 40 mHg (**b**) and 8 min following deflation (**c**). A best-fit Gaussian curve is superimposed on each histogram. (From Burton and Cobbe, 1998, with permission.)

type II = activation maps with two wavefronts, or one wavefront with areas of conduction block; and type III = activation maps with three or more wavefronts (Fig. 15.2a). Stretching increased the dominant frequency (DF) of VF from 15.2 ± 1.0 Hz to 22.8 ± 6.4 Hz. Stretching induced a significant variation in the complexity of the VF activation maps with type III increments and type I and II decrements (Fig. 15.2b).

Moreno et al. performed optical mapping of isolated sheep hearts during VF at low intraventricular pressure (0–5 mmHg) and at high pressure (25–30 mmHg) in HF models and controls (Moreno et al., 2005). They analyzed DF, singularity point density and number of singularity points lasting more than one revolution (rotors). At low pressure, VF in HF was slower and more organized than in controls. Acute stretch did not affect DF but increased singularity point density and rotors similarly in both groups (Fig. 15.3). In these studies, mechanical loading consistently increased complexity of the VF activation pattern. The increase in VF complexity in the presence of stretch was attributed either to an increase in refractory period heterogeneity (Burton and Cobbe, 1998) or to an increase in the slope of the restitution curve (Horner et al., 1996). Although Burton and Cobbe and Chorro et al. observed acceleration of VF during mechanical loading, Moreno et al. observed no effect of stretch on VF frequencies. This discrepancy may be related to differences in recording techniques and species used.

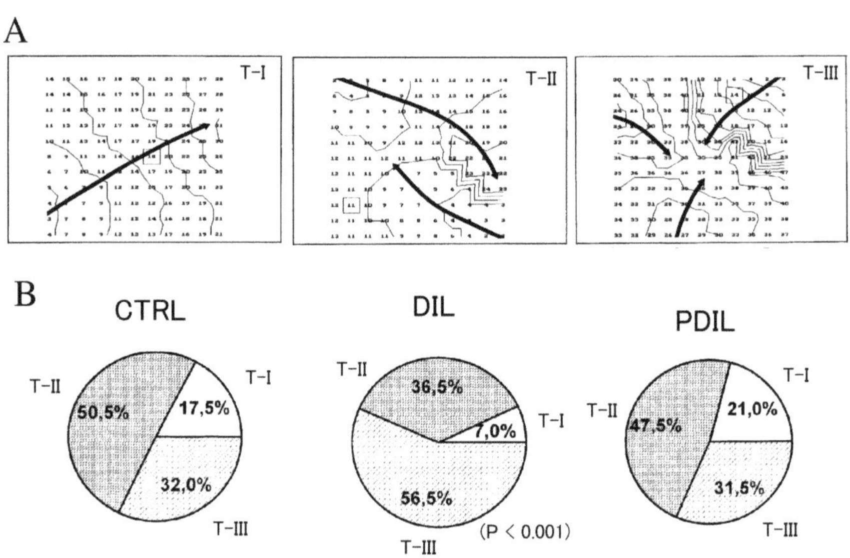

Fig. 15.2 **a**: Classification of the activation maps during ventricular fibrillation: type I = one wavelet without block lines; type II = two simultaneous wavelets with block lines; type III = three or more wavelets with block lines. Isochrones are drawn at 5-ms intervals. **b**: Pie diagrams showing the percentages of activation maps obtained during ventricular fibrillation in each experimental phase. T-I = type I; T-II = type II; T-III = type III; CTRL = control; DIL = ventricular dilatation; PDIL = post-ventricular-dilatation suppression. The statistical significance of the differences versus control is indicated. (From Chorro et al., 2000, with permission.)

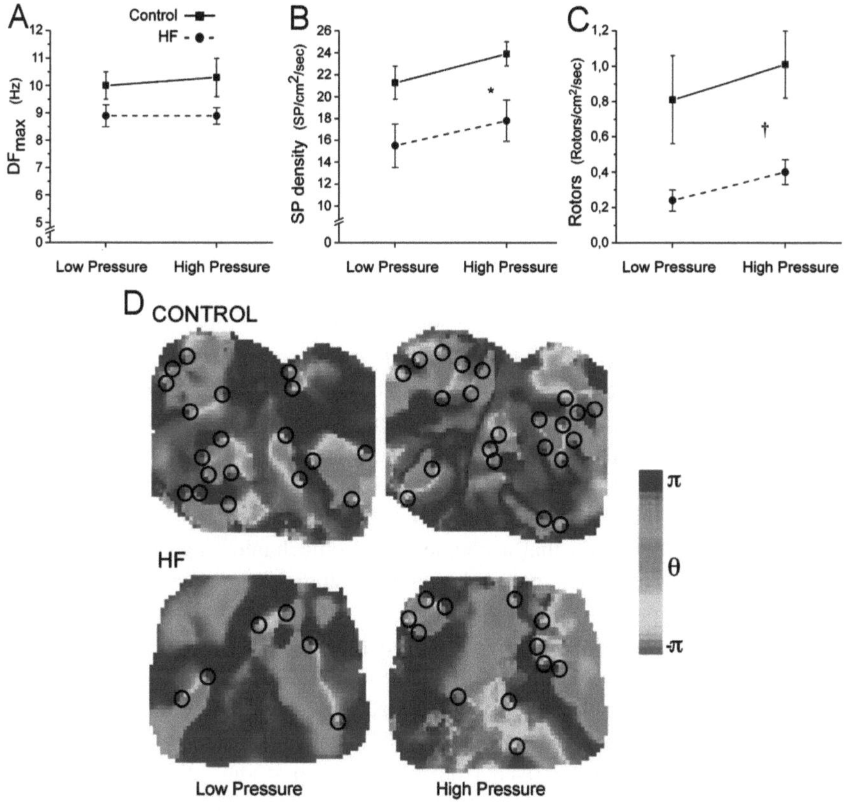

Fig. 15.3 **a**, **b**, **c**: Effects of increased intraventricular pressure on DF_{max}, SP and rotor density in control and HF groups (a, b and c, respectively). Values are mean ± S.E.M. Applying stretch did not modify frequencies but significantly increased SP and rotors density in both groups. No interaction between HF and increased pressure was found. $^{*}P<0.004$, † $P<0.02$. **d**: Snapshots of phase movies taken from a representative control (*upper line*) and failing heart (HF, *lower line*) at baseline (low pressure) and at high pressure. *Black circles* mark singularity points (SP), which are correlated to wavebreaks. At low pressure, the HF heart shows a lower density of SP than the control heart. At high pressure, SP density is increased to a similar extent in both animals. (From Moreno et al. 2005, with permission.)

15.3.2 Computer Simulation

Computer simulation is another effective tool for investigating the electrical activity of the heart. Nash and Panfilov reported a framework for studying the combined effects of cardiac mechanics and electrical activities during arrhythmias (Nash and Panfilov, 2004). They introduced the concept of a contracting excitable medium that was capable of conducting non-linear waves of excitation, which in turn initiate contraction. Furthermore, MEF, a feedback effect of these kinematic deformations on the excitation properties of the medium, was induced. They employed the most basic description of a coupled electromechanical model; a three-variable FitzHugh–Nagumo-type excitation-tension model, for which many biological details were lacking. Furthermore, they employed a model of passive elasticity governed by stress equilibrium equations, subject to an isotropic, Mooney–Rivlin material response. By coupling three variables (voltage, inactivation and tension, used to describe the electrical and mechanical activities) to the non-linear stress equilibrium equations, which govern large deformation hyperelasticity, and incorporating the effects of MEF systematically, they produced a combined model. Numerically, the coupled electromechanical model combined a finite difference method approach, used to integrate the excitation equations, with a Galerkin finite element method, to solve the equations governing tissue mechanics. They presented sample computations demonstrating various effects of contraction on stationary, rotating SWs, and SW breaks. To illustrate the behaviour of the coupled electromechanical model, preliminary results suggest that MEF could potentially cause an otherwise stationary SW to meander about the medium. Furthermore, contraction possibly increased re-entrant wave periods, whilst also increasing the period of oscillations in the medium. They also showed that tissue mechanics significantly contributed to the dynamics of electrical propagation, and that a coupled electromechanical approach should be pursued in future electrophysiological modelling studies.

We recently performed further research using a computer simulation (Hirabayashi et al., 2008). We hypothesised that mechanical stresses facilitated meandering and wave breaks of spiral re-entry through MEF. To validate this hypothesis, we developed a fully coupled electromechanical model of the human ventricular myocardium. Here, we used biophysical models based on direct experimental observations derived from patch clamp studies. We ran quasi-two-dimensional simulations, rather than three-dimensional simulations, because the tissue thickness and rotational anisotropy complicated the dynamics of SWs, making it difficult for us to examine the direct effects of mechanical stresses. The following sections introduce our latest research as an example that illustrates the point that computer simulation has reached, and what it has shown.

15.4 Methods of Our Computer Simulation

15.4.1 Ionic Transmembrane Current

Ions are unequally distributed across the cell membrane, which acts as a barrier for ions. Ions have electric charges, and this distribution therefore produces the electric transmembrane potential. Although the cell membrane acts as a barrier, ions can cross the membrane through ion channels within the cell membrane. These ionic transmembrane currents change the transmembrane potential and generate the action potential. Ion channels are usually selective for one or more ionic species, and most of them can be open or closed to the flow of ions, depending on the state of a "gate" The behaviours of these channels, which depend on the species or on the location of the cell, govern the properties of the cell.

After Hodgkin and Huxley demonstrated great success in simulating the squid giant axon, various models for ionic transmembrane currents and action potentials have been proposed. There are currently two classes of models: biophysical models and FitzHugh-Nagumo-type models. The choice of model depends on the type of problem to be investigated. Biophysical models, or ionic models, consist of several Hodgkin-Huxley-type equations. Reproductions of the electrophysiological properties of various kinds of cells have been achieved by quantitative or qualitative changes, including the adoption of parameters based on direct experimental observations derived from patch clamp studies, the incorporation of new currents, and descriptions of the changes in concentrations of all the major ions inside cardiac cells. While biophysical models describe individual ionic currents, FitzHugh-Nagumo-type models provide a phenomenological, simple, two-variable approach, and reproduce the macroscopic characteristics of cardiac tissue, such as refractoriness, dispersion relation, simple rate-duration properties, etc. The simplicity of the FitzHugh-Nagumo-type models reduces computational load. However, despite constant revisions of FitzHugh-Nagumo-type models, no general agreement has yet been reached on the best model for accurately reproducing the various properties of the cardiac action potential, such as shape, rate dependence, restitution of APD and restitution of conduction velocity (CV) etc. under all circumstances.

15.4.2 Stretch-Activated Currents

Kohl et al. modelled the stretch-activated current ($I_{stretch}$) as a nonspecific current (Kohl et al., 1998). Since experimental data suggest that it is carried by cations (Zeng et al., 2000), it can be divided into three components (Na^+-carrying element, K^+-carrying element, Ca^{2+}-carrying element). The current through SACs for each ion X (Na^+, K^+, and Ca^{2+}) was modelled according to the following equation:

$$I_{st\mathrm{X}} = G_{st\mathrm{X}} f(SL)(V_m - E_{\mathrm{X}}), \tag{15.1}$$

where $I_{st\mathrm{X}}$ is the current through SACs for X, $G_{st\mathrm{X}}$ is the maximal conductance for $I_{st\mathrm{X}}$, $f(SL)$ is the open state of the gate, SL is the sarcomere length, V_m is the transmembrane potential, and E_{X} is the equilibrium (or Nernst) potential of the ion X. E_{X} is the potential at which an ion is at equilibrium (i.e. at which the chemical and electrical gradients are equal and opposite) and is calculated by the Nernst equation using the extracellular and intracellular ion concentrations. Equation (15.1) is a type of Hodgkin–Huxley equation and equivalent to Ohm's law: $I_{st\mathrm{X}}$ is the product of the conductance and the driving force ($V_m - E_{\mathrm{X}}$). As well as other Hodgkin-Huxley type equations, the conductance depends on the open state of the gate. Here, it is modelled as a function of SL, while those for other channels are generally described by ordinary differential equations containing V_m. Since it is independent of V_m, $I_{stretch}$ was linearly correlated with V_m under constant extracellular and intracellular ion concentrations. Although experimental studies have shown time-dependent inactivation of $I_{stretch}$ (Sasaki et al., 1992), this is not taken into account in this model.

$f(SL)$ was modelled according to the following equation:

$$f(SL) = \frac{1}{1 + \exp(-2\gamma(SL - SL_{hst}))}, \tag{15.2}$$

where γ is a user-defined scaling factor, and SL_{hst} is the sarcomere length at which half-maximal activation occurs. $f(SL)$ asymptotically approaches zero as SL shortens, and asymptotically approaches one as SL lengthens. At the point where SL is SL_{hst}, the gradient of $f(SL)$ becomes significantly large.

We adjusted the parameters (Hirabayashi et al., 2008) based on previous experimental data as follows: First, we defined SL_{hst} based on a study using rectangular pieces of frog heart tissue (Fasciano and Tung, 1999), which found that the stretch pulses induced transient diastolic membrane depolarisations, which increased as the stretch amplitude increased and triggered ectopic excitation above 9.517% of stretch (stretch threshold). SL_{hst} may be the length around the stretch threshold. As SL_0=1.94 μm, we defined that SL_{hst}=2.1 μm. Second, we defined γ as follows: Using optical transmembrane potential mapping in isolated rabbit hearts, Sung et al. found that the decrease in CV with LV loading at +4% stretch in the longitudinal direction was not significantly affected by the nonspecific SAC blocker, streptomycin (Sung et al., 2003). We assumed that there was another factor responsible for decreasing the CVs and mild stretches of +4% only minimally activated SACs. Based on this assumption, we defined that γ=14.5. If this assumption is correct, then the decrease in CV that occurs LV loading may be significantly affected by a nonspecific SAC blocker when LV loading results in severe stretching ($\geq$ +6%). Finally, we defined the maximal conductances for the current through SACs for all ions as follows. Based on the experimental data of Zeng et al. (2000), which suggested that Na^+ was a major charge carrier in rat heart cells while Ca^{2+} was not, we assumed that $G_{st\mathrm{Na}}$ was 100 times larger than $G_{st\mathrm{Ca}}$. The reversal potential of the stretch-activated current, which was –30 mV in experimental data (Craelius et al., 1988), is $(G_{st\mathrm{Na}}E_{\mathrm{Na}}+G_{st\mathrm{K}}E_{\mathrm{K}}+G_{st\mathrm{Ca}}E_{\mathrm{Ca}})/(G_{st\mathrm{Na}}+G_{st\mathrm{K}}+G_{st\mathrm{Ca}})$ at any SL in the model, because the current through SACs in the model can be described by the following equation:

$$\begin{aligned} I_{stretch} &= I_{st\mathrm{Na}} + I_{st\mathrm{K}} + I_{st\mathrm{Ca}} \\ &= [(G_{st\mathrm{Na}} + G_{st\mathrm{K}} + G_{st\mathrm{Ca}})V_m - (G_{st\mathrm{Na}}E_{\mathrm{Na}} + G_{st\mathrm{K}}E_{\mathrm{K}} + G_{st\mathrm{Ca}}E_{\mathrm{Ca}})]f(SL). \end{aligned} \tag{15.3}$$

Using E_{X} calculated from the extracellular and intracellular ion concentrations at the resting potential, it was calculated that $G_{st\mathrm{K}}$ should be 1.81 times larger than $G_{st\mathrm{Ca}}$ to obtain a reversal potential of –30 mV. To achieve the stretch threshold demonstrated in the experimental study using rectangular pieces of frog heart tissue (Fasciano and Tung, 1999), we employed the following parameters: $G_{st\mathrm{Na}}$=30.0 S/F, $G_{st\mathrm{K}}$=0.544 S/F and $G_{st\mathrm{Ca}}$=0.3 S/F.

Our model quantitatively reproduced features of MEF, such as the reversal potential of $I_{stretch}$ and the stretch threshold required to induce ectopic excitation. It also qualitatively reproduced features, the precise experimental data for which have not yet been obtained, such as the response to *SL*, the amplitude, and the ion selectivity.

15.4.3 A Model of Excitation Propagation

In addition to transmembrane currents, intercellular ion flow through gap junctions changes ion distributions, resulting in changes in the transmembrane potential and leading to a propagation wave of activity.

Wave propagation in cardiac tissue has generally been described based on three idealizations: the cell membrane is considered to act as a capacitor; the resistance for extracellular ion currents along the cell membrane is so small that the electric potential within the extracellular space is considered to be uniform; the intracellular ion current is considered to follow Ohm's law. The following equation was then obtained (Holden and Panfilov, 1997):

$$C_m \frac{\partial V_m}{\partial t} = -(I_{ion} + I_{stim}) + \frac{1}{S} \nabla_x \cdot (\boldsymbol{D} \nabla_x V_m), \tag{15.4}$$

where C_m is the membrane capacitance per unit membrane area, I_{ion} is the total ionic transmembrane current per unit membrane area, I_{stim} is the electric stimulus current per unit membrane area, S is the surface-to-volume ratio, ∇_x is the three-dimensional gradient operator, and $\boldsymbol{D}$ is a tensor of conductivities. For I_{ion}, various models have been proposed, as introduced in Section 15.4.1. Although the values for C_m, S and $\boldsymbol{D}$ depend on animals and there is currently no widely accepted set of values, this equation has successfully reproduced various experimental and clinical data.

To reproduce the anisotropy, the following formula for $\boldsymbol{D}$ was used:

$$\boldsymbol{D} = (\sigma_L - \sigma_T)\boldsymbol{n} \otimes \boldsymbol{n} + \sigma_T \boldsymbol{I}, \tag{15.5}$$

where σ_L is the longitudinal conductivity, σ_T is the transverse conductivity, $\boldsymbol{n}$ is a unit vector defining the preferred direction of muscle fibres, $\otimes$ is the tensorial product, and $\boldsymbol{I}$ is the identity tensor. Since the resistance of the cytoplasm is negligible, the conductivity of tissues depends on the density of gap junctions (Jongsma and Wilders, 2000). Since deformation of the tissue changes the density of gap junctions, the conductivity of tissues is not constant. However, under the assumption that C_m, S and the resistance of each gap junction do not change with deformation of tissue, we can always use the conductivity under the undeformed state if we calculate the divergence and the gradient in the propagation equation with the transmembrane potential mapped onto the undeformed state.

15.4.4 The Generation of Contraction Force

The following theory is widely adopted to explain the generation of the contraction force: the binding of free Ca^{2+} to troponin C exposes the thin filament (actin); cross-bridges projecting from the thick filament (myosin) become able to attach to the exposed thin filament; the attached cross-bridges generate the contraction force. Although the cross bridges can remain detached after troponin C combines with Ca^{2+} and can remain attached after troponin C uncouples from Ca^{2+}, the affinity between troponin C and Ca^{2+} depends on the attachment of cross bridges, and the attachment rate of cross bridges depends on the binding state of troponin C and Ca^{2+}. In models of the generation of contraction force, it is common to divide the reaction into four states, depending on whether or not troponin C is bound to Ca^{2+}, and whether or not the cross bridge is attached. The kinetics of the reaction rates between these four states were calculated. Adopting the theory that cross-bridges generate contraction forces, the total force depends on the number of states where cross bridges are attached.

In addition to this basic mechanism, there are more complex mechanisms that characterize the length-dependent force modulation, including (a) the length-dependent affinity of cross-bridge attachment and (b) cross-bridge cycling during changes in sarcomere length. Negroni and Lascano formulated a new excitation-contraction coupling model taking into account mechanical changes, by adding the following assumptions: (a) the number of cross-bridges depends on sarcomere length; (b) attached cross-bridges develop force according to the elongation of the elastic structure; (c) at a steady sarcomere length, there is a unique cross-bridge elongation and, during changes in sarcomere length, the concurrent change in cross-bridge length is re-adjusted to the steady-state value by cross-bridge detachment and re-attachment in a different position along the thin filament; and (d) the increased cross-bridge detachment during changes in sarcomere length depends on the rate of re-adjustment of cross-bridge elongation (Negroni and Lascano, 1996).

15.4.5 Constitutive Equation for Cardiac Muscles

When tissues deform and the equilibrium states break, stresses are produced. The constitutive law is the relationship between the stresses and the strains. Some equations that describe the constitutive law for myocardial tissue have been proposed, based on the hyperelastic material theory that postulates the existence of a strain energy potential W, which does not depend on the deformation history of the material but does depend on the right Cauchy-Green deformation tensor $\boldsymbol{C}$, a measure of the deformation defined as follows. The deformation gradient tensor $\boldsymbol{F}$ in standard finite deformation theory is defined by:

$$\mathrm{d}\boldsymbol{x} = \boldsymbol{F}\,\mathrm{d}\boldsymbol{X}, \tag{15.6}$$

where infinitesimal vectors denote the deformed and the undeformed material line segment d$\boldsymbol{x}$ and d$\boldsymbol{X}$. Using $\boldsymbol{F}$, $\boldsymbol{C}$ is defined by:

$$\boldsymbol{C} = \boldsymbol{F}^{\mathrm{T}}\boldsymbol{F}. \tag{15.7}$$

The components of the 2nd Piola-Kirchhoff stress tensor are given by the derivatives of $W(\boldsymbol{C})$ with respect to the components of $\boldsymbol{C}$. As $\boldsymbol{C}$ depends only on a material coordinate system, rigid body movement has no influence on the strain energy. Thus the axiom of objectivity, which requires constitutive law to be invariant with respect to rigid motion of the spatial frame of reference, is satisfied.

The constitutive law for myocardial tissue has three features: high nonlinearity, anisotropy, and dependence on excitation. To describe anisotropy, Humphrey et al. defined W as a function of the vector defining the preferred direction of muscle fibres (Humphrey et al., 1990). Developing Humphrey's model, Lin and Yin proposed a new constitutive equation, which described the changes in the stress-strain behaviour caused by excitation. They divided W into two components: a passive component (W_{pass}) and an active component (W_{act}). By defining the coefficient of W_{act} as a function of the excitation rate, Watanabe et al. (2004) defined W as a function of the excitation rate. Using W, the mixed variational form of the governing equation for nearly incompressible hyperelastic materials is given as follows:

$$\int_V \{\tilde{\rho}\ddot{\boldsymbol{u}} \cdot \delta\boldsymbol{u} + (\partial W/\partial C_{\mathrm{ij}} + \lambda\partial\Psi/\partial C_{\mathrm{ij}})\delta C_{\mathrm{ij}}\}dV = \int_S \tilde{t}\ddot{\boldsymbol{u}} \cdot \delta\boldsymbol{u}\, dS, \tag{15.8}$$

$$\int_V \delta\lambda(\Psi - \lambda/\alpha)dV = 0. \tag{15.9}$$

Equation (15.8) is the weak form of the dynamic equilibrium equation with the constraint of slight compressibility, where V denotes the initial configuration of the material ρ, the nominal mass density $\ddot{\boldsymbol{u}}$ the acceleration vector, $\delta\boldsymbol{u}$ the variation of the displacement vector $\boldsymbol{u}$, C_{ij} the ij-th component of $\boldsymbol{C}$, λ the Lagrange multiplier that corresponds to the negative half hydrostatic pressure, Ψ a function to describe volume change, δC_{ij} the variation of C_{ij} due to $\delta\boldsymbol{u}$, and S the surface on V where the nominal traction force vector is applied. Equation (15.9) is the weak form of the constraints of slight compressibility, where $\delta\lambda$ denotes the variation in λ and α is a large value corresponding to the bulk modulus. As excitation rate changes, W changes, causing a shift in the stress-strain relationship, which leads to breaking of the equilibrium states described in Equations (15.8) and (15.9). To recover the equilibrium states, the muscles deform.

15.4.6 A Fully Coupled Electromechanical Model

We developed a fully coupled electromechanical model of the human ventricular myocardium (Hirabayashi et al., 2008).

We used a biophysical model to describe ionic transmembrane currents, the human ventricular cell model formulated by ten Tusscher et al. (2004). We replaced part of the linear leakage currents in their model by SAC currents modelled by Kohl et al. (1998) the parameters of which were adjusted (Hirabayashi et al., 2008). For the calculation of SAC currents, we need the sarcomere length *SL*. We obtained this value using the following equation:

$$SL = \alpha_c SL_0, \tag{15.10}$$

$$\alpha_c = (\boldsymbol{n} \cdot \boldsymbol{C} \cdot \boldsymbol{n})^{1/2}, \tag{15.11}$$

where α_c is the stretch ratio in the muscle fibre direction obtained by a finite element analysis, $\boldsymbol{n}$ is a unit vector defining the preferred direction of muscle fibres in the undeformed state, and SL_0 is the natural sarcomere length.

Using the total ionic transmembrane current obtained above, we calculated the propagation of excitation. For C_m and *S*, we used the same values as ten Tusscher et al. (2004). To reproduce the CVs reported in a clinical study (Taggart et al., 2000), we adjusted the longitudinal conductivity σ_L and the transverse conductivity σ_T. We defined the CV as the unstretched length of the tissue excitation propagated in a unit time. Under unstressed conditions, we fitted the CVs in our model to the CVs reported in a clinical study (Taggart et al., 2000).

We adopted Negroni and Lascano's model as the excitation-contraction coupling model, and coupled it with the electrophysiological model as follows: we divided Ca^{2+} in the cytoplasm into three components: the Ca^{2+} buffered by troponin C, the Ca^{2+} buffered by calmodulin, and the free Ca^{2+}. By subtracting the concentration of the Ca^{2+} buffered by troponin C (calculated from the excitation-contraction coupling model) from the concentration of the total Ca^{2+} in the cytoplasm (calculated from the biophysical model), we calculated the sum of the concentrations of two other components: the Ca^{2+} buffered by calmodulin and the free Ca^{2+}. Assuming a steady-state for buffering reactions by calmodulin, we divided these two components according to the ratio given in the Luo-Rudy dynamic model (Luo and Rudy, 1994). We then used the concentration of free Ca^{2+} in the cytoplasm in the calculations of the excitation-contraction coupling model.

For the constitutive equation, we adopted the method of Watanabe et al. (2004) using Lin-Yin's model (Humphrey et al., 1990; Lin and Yin, 1998) and treated the coefficients in W_{act} as functions of the relative force *F*. We defined *F* as the concentration of troponin C under the states where the cross bridges are attached, which is obtained from the excitation-contraction coupling model. Figure 15.4a shows an example of histories of V_m and *F* (Hirabayashi, 2008). Since excitation of the cell induces an increase in the intracellular Ca^{2+}, leading to an increase in *F*, the peak of V_m is followed by that of *F*. The stress-strain relationships were experimentally obtained at various concentrations of free Ca^{2+} in the cytoplasm, (Kentish et al., 1986) and these relationships were similar to those obtained from the excitation-contraction coupling model formulated by Negroni and Lascano (1996): the curves were convex towards the stretch axis at low *F*, and became concave as *F* increased.

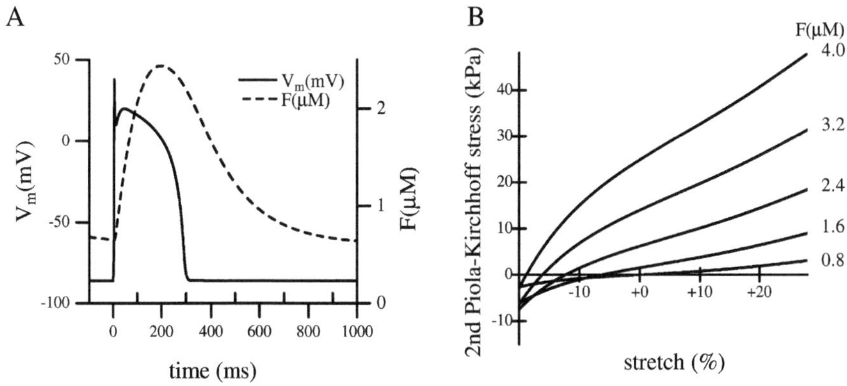

Fig. 15.4 The behaviour of contraction. **a**: The histories of the transmembrane potential (V_m) and the relative force (F). The action potential induced the increase of the relative force. Since F increased moderately compared with V_m, the peak of V_m was followed by that of F. The decrease in F was also more moderate than that in V_m. **b**: Force-varying stress-strain relationship in the fibre direction. The stresses were defined in terms of the second Piola-Kirchhoff tensor and the stretches were defined as the changes in sarcomere lengths. As F was increased, the stress-strain relationship shifted upward and left. At low F, the curves were convex towards the stretch axis and became concave as F increased

By adjusting the parameters, we fitted the curves at each F to those obtained by Negroni and Lascano at the corresponding free Ca^{2+} concentration. The calculated stress-strain relationships in the uniaxial state for various F values are shown in Fig. 15.4b (Hirabayashi, 2008).

15.4.7 Efficient Dynamic Finite Element Method

Using a finite element method, we solved the equation for propagation of excitation, the equilibrium equation and constraints of slight compressibility. The electrophysiological states such as ionic transmembrane currents were calculated at each node. We developed an efficient dynamic finite element method for computation which is briefly presented here (Hirabayashi et al., 2008). The definitions of the values are shown in the Appendix.

We can derive the following equations from Equations (15.8) and (15.9):

$$\boldsymbol{M}\ddot{\boldsymbol{U}} + \boldsymbol{Q}_V + \boldsymbol{H}\boldsymbol{\Lambda} = \boldsymbol{F}_S, \tag{15.12}$$

$$\boldsymbol{\Psi} = (1/\alpha)\boldsymbol{\Gamma}\boldsymbol{\Lambda}. \tag{15.13}$$

$\ddot{\boldsymbol{U}}$ indicates the second order time derivative of $\boldsymbol{U}$. Starting from the left, the terms in Equation (15.12) are the inertial force, internal force, hydrostatic pressure and external force.

Using the central difference method with a time increment Δt, we decompose Equation (15.12) as follows:

$$\boldsymbol{M}^{t+\Delta t}\boldsymbol{U} = \Delta t^2({}^{t+\Delta t}\boldsymbol{F}_S - {}^{t+\Delta t}\boldsymbol{Q}_V - {}^{t+\Delta t}\boldsymbol{H}^{t+\Delta t}\boldsymbol{\Lambda}) + \boldsymbol{M}(2^t\boldsymbol{U} - {}^{t-\Delta t}\boldsymbol{U}), \quad (15.14)$$

where left superscripts represent times. We use the inertial force at time t and other forces at time $t+\Delta t$.

We solved Equation (15.14) with (15.13) iteratively and produced ${}^{t+\Delta t}\boldsymbol{U}$ and ${}^{t+\Delta t}\boldsymbol{\Lambda}$. To solve this efficiently, we used the following method: first, using the intermediate variable $\boldsymbol{U}_m$, we uncoupled Equation (15.14) as follows:

$$\boldsymbol{M}_{k+1}\boldsymbol{U}_m = \Delta t^2({}^{t+\Delta t}\boldsymbol{F}_S - {}^{t+\Delta t}_{k+1}\boldsymbol{Q}_V) + \boldsymbol{M}(2^t\boldsymbol{U} - {}^{t-\Delta t}\boldsymbol{U}), \quad (15.15)$$

$$\boldsymbol{M}({}^{t+\Delta t}_{k+1}\boldsymbol{U} - {}_{k+1}\boldsymbol{U}_m) = -\Delta t^2\, {}^{t+\Delta t}_{k+1}\boldsymbol{H}\ {}^{t+\Delta t}_{k+1}\boldsymbol{\Lambda}, \quad (15.16)$$

where left subscripts k+1 represent (k+1)th iteration. We also rewrote Equation (15.13). Since $\delta\Psi$, the variation of Ψ due to $\delta\boldsymbol{u}$, can be written as follows:

$$\delta\Psi = (\partial\Psi/\partial C_{ij})\partial C_{ij} = 1/2\boldsymbol{D}_C{}^{T}\{\delta\boldsymbol{C}\} = \boldsymbol{D}_C{}^{T}\boldsymbol{B}\{\delta\boldsymbol{u}\}, \quad (15.17)$$

$\delta\boldsymbol{\Psi}$, the variation of $\boldsymbol{\Psi}$ due to $\delta\boldsymbol{u}$, can be written as follows:

$$\delta\boldsymbol{\Psi} = \Sigma_e \int_{V_e} \boldsymbol{N}_\lambda{}^{T}\boldsymbol{D}_C{}^{T}\boldsymbol{B}\{\delta\boldsymbol{u}\}dV = \boldsymbol{H}^{T}\delta\boldsymbol{U}, \quad (15.18)$$

where $\delta\boldsymbol{U}$ is the variation of $\boldsymbol{U}$. Therefore, Equation (15.14) at time $t+\Delta t$ can be written as follows:

$$(1/\alpha){}^{t+\Delta t}_{k+1}\boldsymbol{\Gamma}\ {}^{t+\Delta t}_{k+1}\boldsymbol{\Lambda} = {}^{t+\Delta t}_{k+1}\boldsymbol{\Psi} = {}^{t+\Delta t}_{k}\boldsymbol{\Psi} + {}^{t+\Delta t}_{k+1}\boldsymbol{H}^{T}({}^{t+\Delta t}_{k+1}\boldsymbol{U} - {}^{t+\Delta t}_{k}\boldsymbol{U}). \quad (15.19)$$

From Equations (15.16) and (15.19), the following equation is obtained:

$$\{(1/\alpha){}^{t+\Delta t}_{k+1}\boldsymbol{\Gamma} + \Delta t^2\, {}^{t+\Delta t}_{k}\boldsymbol{H}^{T}\boldsymbol{M}^{-1}\, {}^{t+\Delta t}_{k+1}\boldsymbol{H}\}\, {}^{t+\Delta t}_{k+1}\boldsymbol{\Lambda} = -{}^{t+\Delta t}_{k}\boldsymbol{H}^{T}({}^{t+\Delta t}_{k+1}\boldsymbol{U} - {}_{k+1}\boldsymbol{U}_m) + {}^{t+\Delta t}_{k}\boldsymbol{\Psi}. \quad (15.20)$$

Using Equations (15.15), (15.20) and (15.16), in turn, we updated $\boldsymbol{U}_m$, ${}^{t+\Delta t}\boldsymbol{\Lambda}$, and ${}^{t+\Delta t}\boldsymbol{U}$ until they converged. Here, ${}^{t+\Delta t}_{k+1}\boldsymbol{Q}_V$ in Equation (15.15) and ${}^{t+\Delta t}_{k+1}\boldsymbol{H}$ in Equations (15.16) and (15.20) are functions of ${}^{t+\Delta t}_{k+1}\boldsymbol{U}$, which is unknown before Equation (15.16) is solved. We therefore used ${}^{t+\Delta t}_{k}\boldsymbol{Q}_V$ and ${}^{t+\Delta t}_{k}\boldsymbol{H}$ instead. By using a lumped mass matrix as $\boldsymbol{M}$, simultaneous equations need to be solved only for ${}^{t+\Delta t}\boldsymbol{\Lambda}$ in Equation (15.20). The stability and accuracy are thus preserved, whilst saving central processing unit time.

Using this method, we were able to calculate the whole process from the electrophysiological to the mechanical level using the same time-step of 0.01 ms.

15.5 Examples of Our Computer Simulations

15.5.1 The Effects of Stretch on CV

In our model, severe stretch decreased the longitudinal and transverse CVs, whereas mild stretch minimally affected these CVs. These results are in concordance with the findings of previous experimental studies (Penefsky and Hoffman, 1963; Zabel et al., 1996b). By adjusting SL_{hst} and γ, which defined the range of effective stretches able to activate SACs and affect CVs, we reproduced the experimental result where mild stretch did not significantly affect CVs (Section 15.4.2).

15.5.2 The Effects of Stretch on APD

In our model, a stretch pulse applied during phase 2 of the action potential resulted in fast repolarisation and shortened APD, and a stretch pulse applied at late repolarisation (late phase 3 of the action potential) prolonged the APD (Hirabayashi et al., 2008). These results are consistent with the findings of previous experimental (Zabel et al., 1996a) and mathematical (Kohl et al., 1998) studies. A possible explanation for these phase-dependent effects is the fact that the reversal potential of $I_{stretch}$ is –30 mV. During phase 2, the membrane potential is above the reversal potential and stretch currents flow outwards to reduce the membrane potential. During late repolarisation, the membrane potential is under the reversal potential, and stretch currents flow inwards, avoiding a decrease in the membrane potential. These results were obtained under pacing at 2 Hz.

Using various stretch conditions, we also examined the electrical restitution curves. Constant stretches decreased the maximal slope of the restitution curve, while dynamic stretches, which were estimated from the relative force, *F*, slightly increased the maximal slope of the restitution curve. Both stretches shortened APDs (Hirabayashi et al., 2008).

To examine the behaviour of action potentials and ionic currents at the centre of SWs, where the frequencies of the excitations are higher, we performed another experiment. We delivered a premature stimulus (S2) at 320 ms after the last basic pacing stimulus (S1). Considering that contraction forces are weakened at late repolarisation, we applied a 5% stretch pulse from 290 ms after S1 to 30 ms after S2. As shown in Fig. 15.5, stretch prolonged the APD induced by S2 through the activation of SAC currents and the alternation of the intensity and kinetics of other ionic currents.

15.5.3 The Effects of Stretch on SW Dynamics

To eliminate factors that complicate the dynamics of SWs and make it difficult to interpret results, we ran quasi two-dimensional simulations without accounting for

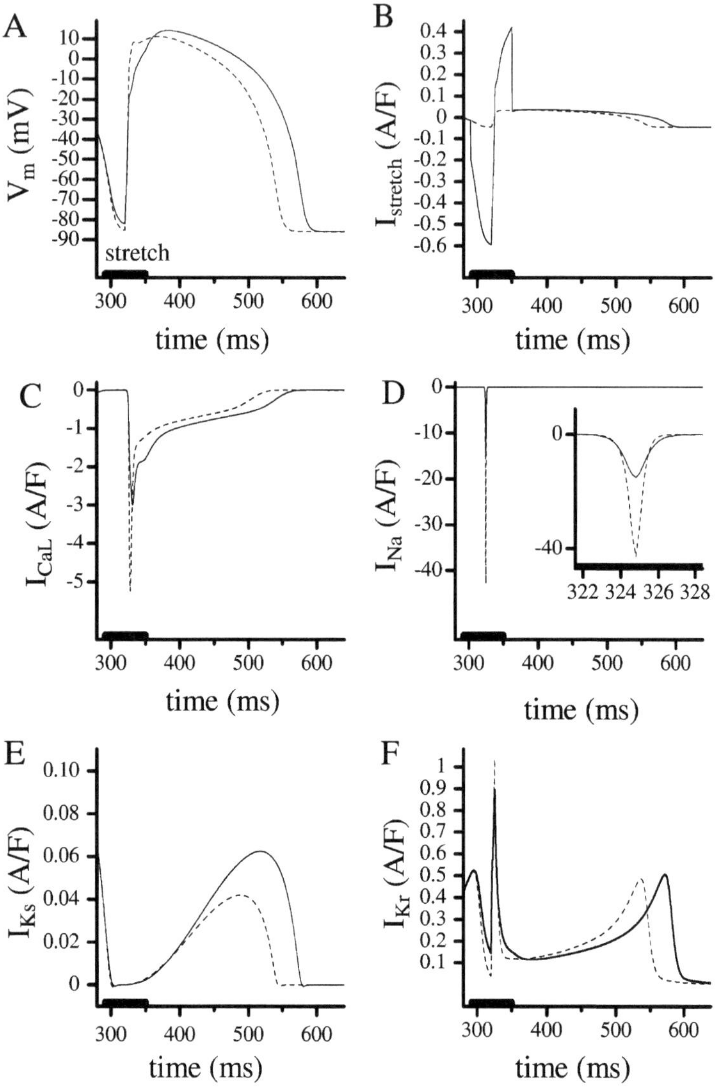

Fig. 15.5 Diastolic stretch-induced changes in ion currents and transmembrane potentials. A premature electric stimulus (S2) was delivered at 320 ms after a basic pacing cycle length (S1-S1) of 1000 ms. A 5% stretch pulse (*solid bar*) was applied at 290–350 ms after S1. Values were computed with (*solid line*) and without (*dashed line*) a stretch pulse. **a**: Transmembrane potential. **b**: Stretch-activated current. **c**: L-type Ca^{2+} current. **d**: Fast Na^{+} current. **e**: Slow delayed rectifier K^{+} current. **f**: Rapid delayed rectifier K^{+} current. (From Hirabayashi et al., 2008, with permission.)

tissue thickness and rotational anisotropy. We constructed a mid-myocardial layer using mixed hexahedral solid elements (eight nodes for the bilinear displacement or transmembrane potential interpolation/constant pressure field). Neumann boundary conditions were used for the electric boundary conditions. Mechanical boundary conditions (Hirabayashi et al., 2008) were defined based on assumptions from swine experimental data (Berg et al., 2005). With inactive SACs, we induced SWs into this mid-myocardial layer. After SWs stabilized, we activated SACs and increased LVP using the following protocols (Fig. 15.6): in protocol 1, LVP was maintained; in protocols 2 and 3, LVP was linearly increased; and in protocol 4, SACs were not activated, and *SL* was maintained at SL_0.

The electromechanical activities during spiral re-entry at 10 mmHg of LVP are illustrated in Fig. 15.7. As a result of the contraction force generated by the SWs, complex strain distributions were observed around the core of the SW, despite the homogeneous mechanical boundary condition.

Figure 15.8 shows the transmembrane potential distribution during spiral re-entry at various levels of LVP, where the locations of SW tips are marked with a circle. The transmembrane potential distribution, which had smooth contours of SW when SACs were inactivated (SAC(-)), became irregular as LVP increased, when SACs were activated (SAC(+)). At 20 mmHg of LVP, a single SW became fragmented into a complex pattern of activation with multiple stable tips, as shown in the figure for the 3,435 ms time-point. In the marginal region of the myocardial sheet, refractory areas developed and hit the wavefront, resulting in wave breaks. The refractory areas probably developed by the following process: the marginal region of the myocardial sheet was susceptible to being stretched (Fig. 15.7), because the mechanical bound-

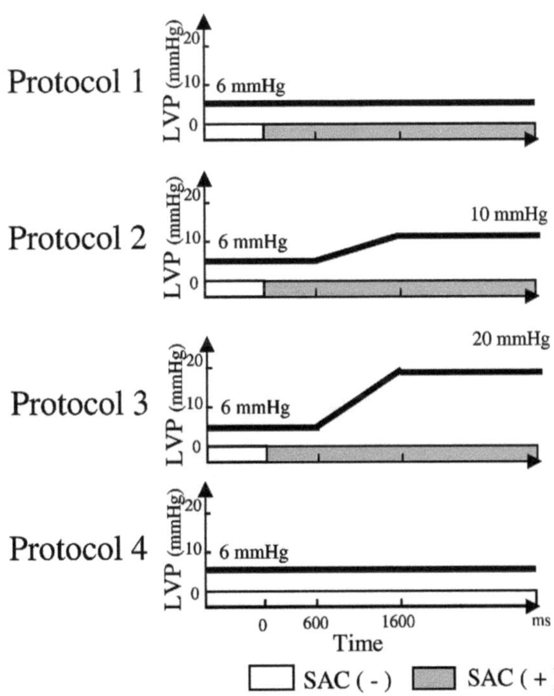

Fig. 15.6 Protocols for the examination of stretch effects on spiral wave (SW) dynamics. We induced SWs giving each layer 6 mmHg of left ventricular pressure (LVP) and inactivating stretch-activated channels (SACs). After the spiral wave stabilized, we activated SACs (protocols 1, 2 and 3). We then linearly increased LVP (protocols 2 and 3) to 10 and 20 mmHg, respectively. In protocols 1 and 4, LVP was maintained at 6 mmHg. In protocol 4, SACs were kept inactivated

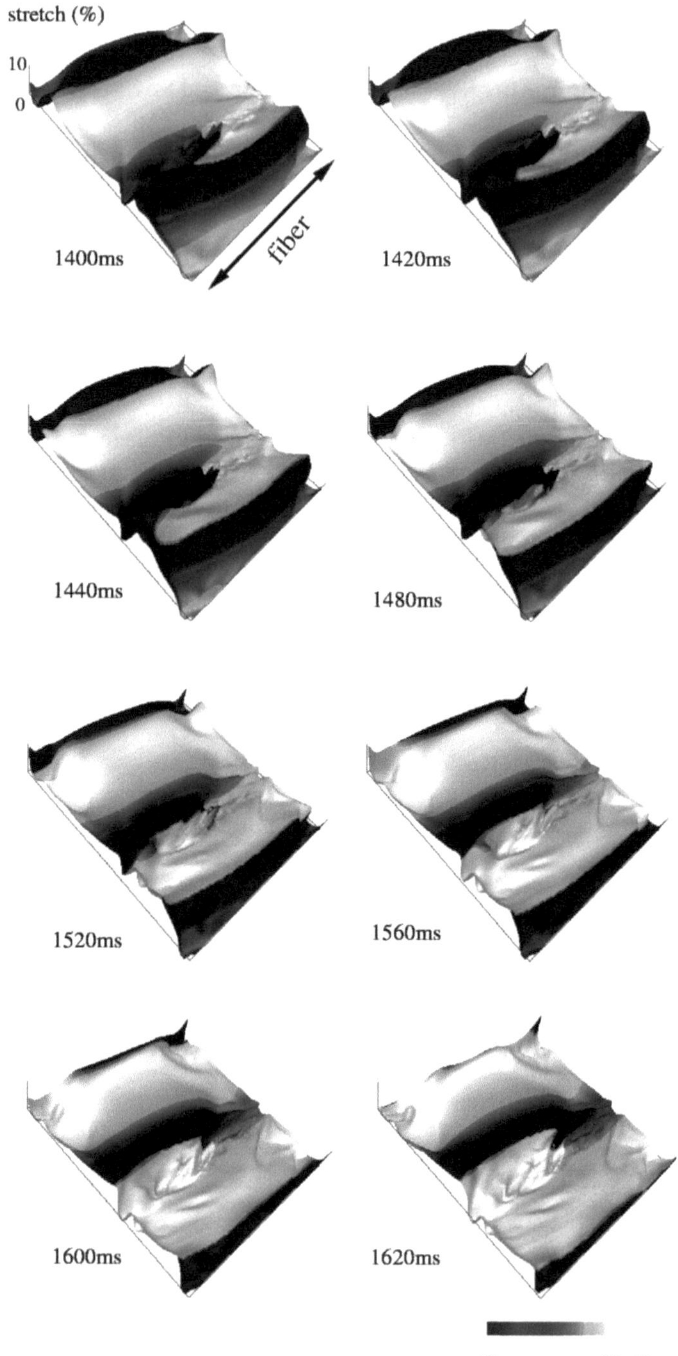

Fig. 15.7 Electromechanical activity during spiral re-entry at 10 mmHg of left ventricular pressure (LVP) (protocol 2). A time series of transmembrane potential distributions (gray scale) and

ary condition was homogeneous while the contraction forces generated by excitations were not; during late repolarisation, contraction forces were weakened; stretch was then induced, resulting in the prolonged APD. Though the SW frequently broke up and produced a number of tips, most of the new tips disappeared by colliding with each other or against the boundaries.

Figure 15.9A shows the trajectories of tips under SAC(–) and SAC(+) conditions (at LVPs of 6 and 10 mmHg). The tips drew S-shaped lines, which rotated and drew shapes like a flower. At 10 mmHg, the centre of the S-shaped lines drifted toward the right and the tip moved around a wider area, compared with the SAC(-) condition, or that at lower LVP. The trajectories that divided into S-shaped lines are shown in Fig. 15.9b. The S-shaped lines rotated faster at 10 mmHg than at 6 mmHg. At 10 mmHg, we frequently found a serial process of wave breaks, and mutual annihilation resulted in a switch of trajectories, as illustrated in Fig. 15.9c. The switch of trajectories further complicated the transmembrane potential and strain distribution, and led to a chaotic meandering of the tip.

Figure 15.10a depicts the distribution of transmembrane potentials and strains during the process of tip generation. Around the core of the SW, the stretched region and the contracted region were adjacent in a small area. The wavefront around the core was highly stretched, even at low pressures, because it was surrounded by contracting tissues. This stretch prolonged the APD (Fig. 15.10b), probably through the shortening of DI, as discussed in the previous section. The extended wavetail hit the wavefront, resulting in wave breaks and switches of the tip trajectories. In cases where the APD prolongation was insufficient to generate a wave break, the tip traced the boundary of the enlarged refractory area and the bend of the S-shaped line was sharpened (Fig. 15.10c).

15.6 Future Research

The modelling study we have presented in the previous section has several limitations. Some phenomena were not accounted for: the effects of stretch on other ionic currents, including L-type Ca^{2+}, delayed rectifier K^+, and ATP-sensitive K^+ channels (Matsuda et al., 1996; Wang et al., 1996); the time-dependent inactivation of $I_{stretch}$ (Sasaki et al., 1992); and three-dimensional features such as transmural fibre

Fig. 15.7 strain distributions (heights) is shown. As a result of the contraction force generated by the spiral waves (SWs), the strains were irregularly distributed between about –3% and about +10%, despite the homogeneous mechanical boundary conditions. Complex strain distributions were particularly observed around the core of the SW. The stretched region and the contracted region were adjacent over a small area. Surrounded by contracting tissues, the wavefront around the core was highly stretched. Extremely strong strains were observed in the marginal region of the myocardial sheet, probably because the mechanical boundary conditions were homogeneous, while the contraction forces generated by excitations were not. (From Hirabayashi et al., 2008, with permission.)

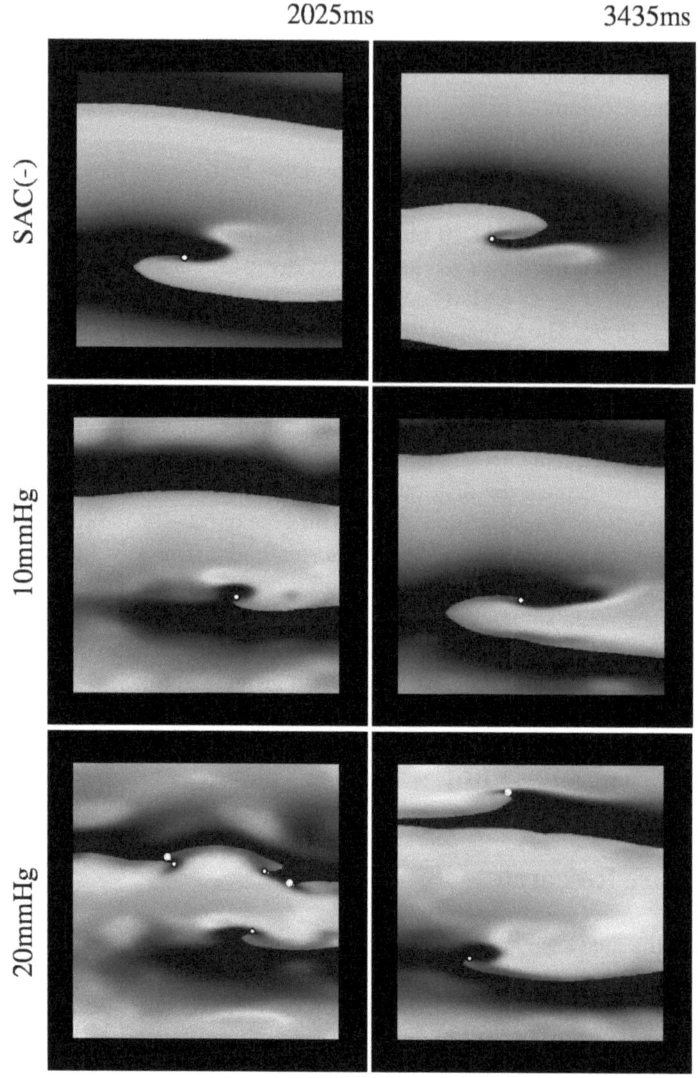

Fig. 15.8 Transmembrane potential patterns during spiral re-entry. Patterns at the top were obtained under conditions where stretch-activated channels (SACs) were inactivated (SAC(–), protocol 4). Other patterns were obtained under conditions where SACs were activated (SAC(+)). The patterns in the middle were obtained at 10 mmHg of left ventricular pressure (LVP) (protocol 2) and those at the bottom were obtained at 20 mmHg of LVP (protocol 3). The locations of tips are marked with circles. A *small circle* denotes a tip with a clockwise rotation of the activation wavefront, whereas a *large circle* denotes a tip with a counter-clockwise rotation of the activation wavefront. Under SAC(–) conditions, the spiral wave produced smooth contours, while under SAC(+) conditions, the contours became irregular as the pressure increased. At 20 mmHg of LVP, a single SW became fragmented into a complex pattern of activation and generated new tips (2,025 ms). Some of these survived and became stable (3,435 ms). (From Hirabayashi et al., 2008, with permission.)

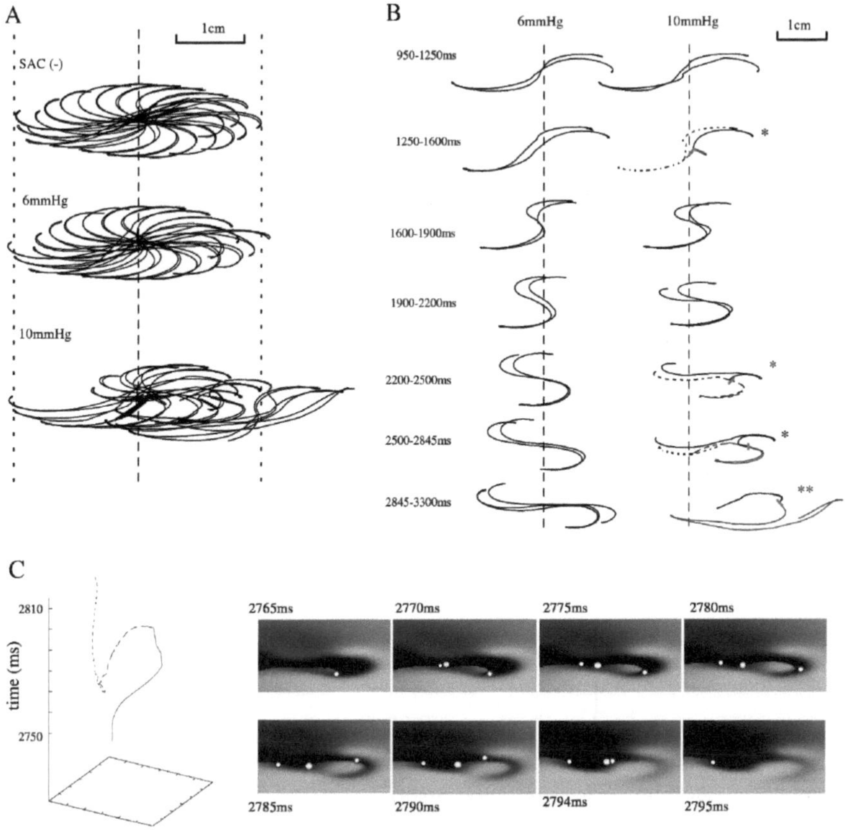

Fig. 15.9 Dynamics of spiral wave (SW) tips. **a**: Trajectories of SW tips from 600 ms to 4,000 ms under conditions where stretch-activated channels (SACs) were inactivated (SAC(–), protocol 4), 6 mmHg of left ventricular pressure (LVP) (protocol 1), and 10 mmHg of LVP (protocol 4). The tips drew S-shaped lines, which rotated and drew shapes like a flower. At 6 mmHg of LVP, the trajectory was similar to that under SAC(–) conditions, where the tip moved between the *dotted lines* and the centre of the S-shaped lines was stationary on the *dashed line*. At 10 mmHg, the centre of the S-shaped lines drifted toward the right, and the tip moved around a wider area than under SAC(–) conditions. **b**: Time series of S-shaped lines at 6 and 10 mmHg of LVP. The S-shaped lines rotated faster at 10 mmHg than at 6 mmHg. At 10 mmHg, switches of trajectories were frequently observed. *Dotted, dashed, and gray lines* indicate new trajectories. Double switches were observed in the S-shaped trajectories marked by single asterisks, while single switches were observed in S-shaped trajectories marked by two *asterisks*. **c**: Typical switching process of tips. The trajectories of three tips are shown on the left. They are plotted in x and y time space. The right-hand figures show transmembrane potential patterns near the core. At 2,770 ms, the wavefront of the SW breaks up in the vicinity of the core and a pair of new tips (*dotted* and *dashed lines*) was generated. One of new tips (*the dashed line*) and an old tip (*the solid line*) approached each other. At 2,794 ms, they collided, resulting in mutual annihilation. The other new tip (*the dotted line*) survived and traced a new trajectory. (From Hirabayashi et al., 2008, with permission.)

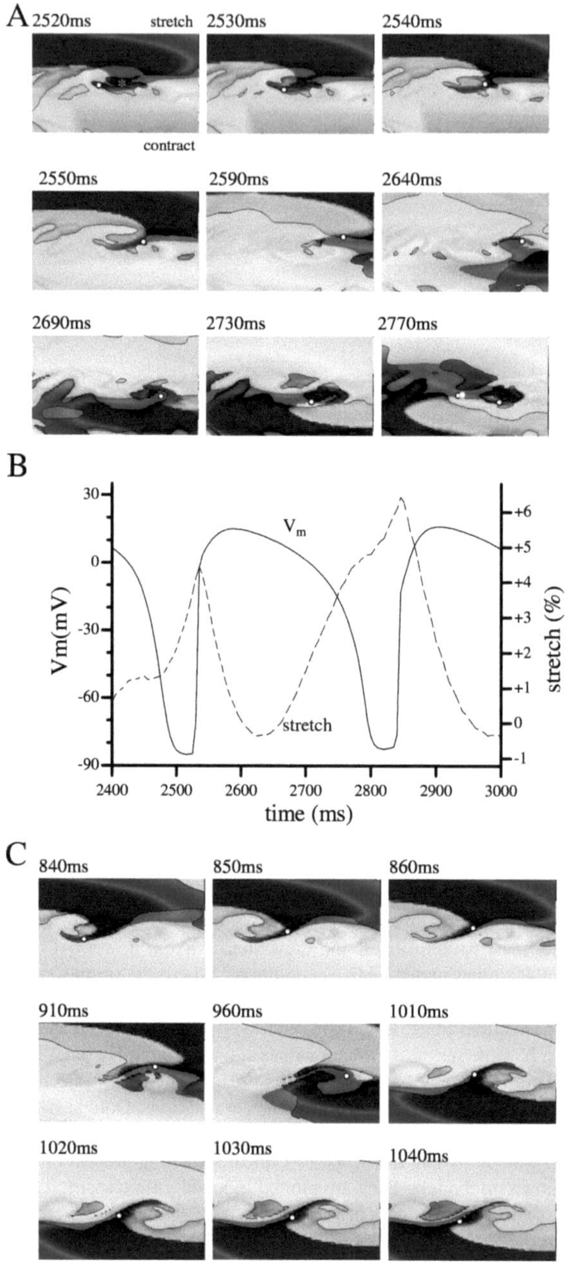

Fig. 15.10 **a**, **c**: A time series of the distribution of transmembrane potentials and strains during spiral re-entry at 10 mmHg of left ventricular pressure (LVP) (protocol 2). Gray scale indicate transmembrane potentials, and contracted areas (strains of < 1%) are covered with white. **b**: Time series of transmembrane potentials and strains recorded at the ***asterisk*** **in (a)**. **(a)** Depicts the process whereby the wavefront broke up and a pair of new tips was generated (2,770 ms). The myocardium around the core was stretched at phase 0 of the action potential and at late recovery, as shown in (**b**). Stretches at these time-points prolonged the APD and allowed the SW wavefront

rotation and complex anatomic structures, which have been considered to promote wave breaks in three dimensions (Fenton and Karma, 1998). Other factors need to be taken into consideration: e.g., C_m, S, and the fact that the resistance of each gap junction may change with deformation. The conductivity tensor may also change with deformation, because the resistances of the cytoplasm and extracellular space are small, but are not zero.

In addition, due to the lack of data on stretch-activated currents in human ventricles, we modelled SACs based on assumptions from previous studies. However, some of these assumptions could be incorrect; thus direct data are required to generate more reliable simulation results. Finally, further studies, especially experimental studies, are required to verify our novel results.

Computer simulations enable us to gain insights, detect new phenomena, and propose new hypotheses. However, experimental studies are also important for modelling and for verifying hypotheses. Appropriate computational and experimental approaches complement each other and both will be needed to make further progress in this field.

Appendix: Definitions for Finite Element Analysis

$\boldsymbol{u}$	: the displacement vector
λ	: the Lagrange multiplier that corresponds to the negative half hydro static pressure
$\delta\boldsymbol{u}$	: the variation of $\boldsymbol{u}$
δu_i	: the i-th component of $\delta\boldsymbol{u}$
C_{ij}	: the ij-th component of the right Cauchy-Green deformation tensor $\boldsymbol{C}$
δC_{ij}	: the variation of C_{ij} due to $\delta\boldsymbol{u}$
Ψ	: a function to describe volume change
α	: a large value corresponding to the bulk modulus
ρ	: the nominal mass density
$\tilde{\boldsymbol{t}}$	: the nominal traction force vector
V	: the initial configuration of the material
S	: the surface on V where $\tilde{\boldsymbol{t}}$ is applied
Ve, Se	: the sub-domains of each finite element
e	: element assemblage
$\boldsymbol{N}_u$	: the matrix consisting of the interpolation functions of $\boldsymbol{u}$
$\boldsymbol{N}_\lambda$	: the vector consisting of the interpolation functions of λ
$\boldsymbol{U}$	: the vectors aligning discretized values of $\boldsymbol{u}$ at all nodes
$\boldsymbol{\Lambda}$	: the vectors aligning discretized values of λ at all nodes

Fig. 15.10 (continued) to hit the wave tail, resulting in a wave break in the vicinity of the core. In cases in which the prolongation of APD was insufficient to generate a wave break, the tip traced the boundary of the enlarged refractory area and the bend of the S-shaped line was sharpened, as shown in (**c**). (From Hirabayashi et al., 2008, with permission.)

$$\{\delta \boldsymbol{u}\} = \{\delta u_1\ \delta u_2\ \delta u_3\}^{\mathrm{T}}$$
$$\{\delta \boldsymbol{C}\} = \{\delta C_{11}\ \delta C_{22}\ \delta C_{33}\ \delta C_{12}\ \delta C_{23}\ \delta C_{31}\}^{\mathrm{T}}$$
$$\boldsymbol{D}_C = \{2\partial\Psi/\partial C_{11}\ 2\partial\Psi/\partial C_{22}\ 2\partial\Psi/\partial C_{33}\ 2\partial\Psi/\partial C_{12}\ 2\partial\Psi/\partial C_{23}\ 2\partial\Psi/\partial C_{31}\}^{\mathrm{T}}$$
$$\boldsymbol{S}_V = \{2\partial W/\partial C_{11}\ 2\partial W/\partial C_{22}\ 2\partial W/\partial C_{33}\ 2\partial W/\partial C_{12}\ 2\partial W/\partial C_{23}\ 2\partial W/\partial C_{31}\}^{\mathrm{T}}$$
$$1/2\{\delta \boldsymbol{C}\} = \boldsymbol{B}\ \{\delta \boldsymbol{u}\}$$
$$\boldsymbol{M} = \sum_{\mathrm{e}} \int_{Ve} \tilde{\rho}\tilde{\rho}\ \boldsymbol{N}_{\mathrm{u}}{}^{\mathrm{T}}\ \boldsymbol{N}_{\mathrm{u}}\ dV$$
$$\boldsymbol{Q}_V = \sum_{\mathrm{e}} \int_{Ve} \boldsymbol{B}^{\mathrm{T}} \boldsymbol{S}_V\ dV$$
$$\boldsymbol{H} = \sum_{\mathrm{e}} \int_{Ve} \boldsymbol{B}^{\mathrm{T}} \boldsymbol{D}_c\ \boldsymbol{N}_\lambda dV$$
$$\boldsymbol{F}_S = \sum_{\mathrm{e}} \int_{Se} \boldsymbol{N}_{\mathrm{u}}{}^{\mathrm{T}} dS$$
$$\boldsymbol{\Gamma} = \sum_{\mathrm{e}} \int_{Ve} \boldsymbol{N}_\lambda{}^{\mathrm{T}} \boldsymbol{N}_\lambda dV$$
$$\boldsymbol{\Psi} = \sum_{\mathrm{e}} \int_{Ve} \Psi\ \boldsymbol{N}_\lambda{}^{\mathrm{T}}\ dV$$

References

Berg RA, Sorrell VL, Kern KB, Hilwig RW, Altbach MI, Hayes MM, Bates KA, Ewy GA (2005) Magnetic resonance imaging during untreated ventricular fibrillation reveals prompt right ventricular overdistention without left ventricular volume loss. Circulation 111: 1136–1140

Burton FL, Cobbe SM (1998) Effect of sustained stretch on dispersion of ventricular fibrillation intervals in normal rabbit hearts. Cardiovasc Res 39: 351–359

Calkins H, Maughan WL, Kass DA, Sagawa K, Levine JH (1989) Electrophysiological effect of volume load in isolated canine hearts. Am J Physiol 256: H1697–H1706

Chorro FJ, Canoves J, Guerrero J, Mainar L, Sanchis J, Soria E, Such LM, Rosado A, Such L, Lopez-Merino V (2000) Opposite effects of myocardial stretch and verapamil on the complexity of the ventricular fibrillatory pattern: an experimental study. Pacing Clin Electrophysiol 23: 1594–1603

Coronel R, Wilms-Schopman FJ, de Groot JR (2002) Origin of ischemia-induced phase 1b ventricular arrhythmias in pig hearts. J Am Coll Cardiol 39:166–176

Craelius W, Chen V, el-Sherif N (1988) Stretch activated ion channels in ventricular myocytes. Biosci Rep 8: 407–414

Davidenko JM, Pertsov AV, Salomonsz R, Baxter W, Jalife J (1992) Stationary and drifting spiral waves of excitation in isolated cardiac muscle. Nature 355: 349–351

Engelstein ED, Hahn RT, Stein KM, Lerman BB (1995) Noninvasive Predictors of Successful Implantation of Transvenous Defibrillator Lead Systems. J Am Coll Cardiol 25: 110A–111A

Fasciano RW, 2nd, Tung L (1999) Factors governing mechanical stimulation in frog hearts. Am J Physiol 277: H2311–H2320

Fenton F, Karma A (1998) Vortex dynamics in three-dimensional continuous myocardium with fiber rotation: Filament instability and fibrillation. Chaos 8: 20–47

Frazier DW. Wolf PD. Wharton JM, Tang AS. Smith WM. Ideker RE (1989) Stimulus-induced critical points; mechanism for electrical induction of reentry in normal canine myocardium. J Clin Invest 83: I039–I052.

Franz MR, Cima R, Wang D, Profitt D, Kurz R (1992) Electrophysiological effects of myocardial stretch and mechanical determinants of stretch-activated arrhythmias. Circulation. 86:968–978

Friedman PA, Foley DA, Christian TF, Stanton MS (1998) Stability of the defibrillation probability curve with the development of ventricular dysfunction in the canine rapid paced model. Pacing Clin Electrophysiol. 21:339–351

Garan AR, Maron BJ, Wang PJ, Estes NA 3rd, Link MS (2005) Role of streptomycin-sensitive stretch-activated channel in chest wall impact induced sudden death (commotio cordis). J Cardiovasc Electrophysiol. 16:433–438

Garny A, Kohl P (2004) Mechanical induction of arrhythmias during ventricular repolarization: modeling cellular mechanisms and their interaction in two dimensions. Ann N Y Acad Sci 1015: 133–143

Gray RA, Jalife J, Panfilov A, Baxter WT, Cabo C, Davidenko JM, Pertsov AM (1995) Nonstationary Vortexlike Reentrant Activity as a Mechanism of Polymorphic Ventricular Tachycardia in the Isolated Rabbit Heart. Circulation 91: 2454–2469
Gray RA, Pertsov AM, Jalife J (1998) Spatial and temporal organization during cardiac fibrillation. Nature 392: 75–78
Hansen DE (1993) Mechanoelectrical feedback effects of altering preload, afterload, and ventricular shortening. Am J Physiol 264: H423–H432
Hirabayashi S (2008) The University of Tokyo: PhD thesis
Hirabayashi S, Inagaki M, Hisada T (2008) Effects of wall stress on the dynamics of ventricular fibrillation: a simulation study using a dynamic mechanoelectric model of ventricular tissue. J Cardiovasc Electrophysiol 19: 730–739
Holden AV, Panfilov AV (1997) Modeling Propagation in Excitable Media. In:Panfilov AV, Holden AV (eds) Computational Biology of the Heart. Wiley, Chichester, pp. 65–99
Horner SM, Dick DJ, Murphy CF, Lab MJ (1996) Cycle length dependence of the electrophysiological effects of increased load on the myocardium. Circulation 94:1131–1136
Hu H, Sachs F (1997) Stretch-activated ion channels in the heart. J Mol Cell Cardiol 29: 1511–1523
Huang J, Rogers JM, Killingsworth CR, Walcott GP, KenKnight BH, Smith WM, Ideker RE (2001) Improvement of defibrillation efficacy and quantification of activation patterns during ventricular fibrillation in a canine heart failure model. Circulation 103: 1473–1478
Humphrey JD, Strumpf RK, Yin FC (1990) Determination of a constitutive relation for passive myocardium: I. A new functional form. J Biomech Eng 112: 333–339
Jongsma HJ, Wilders R (2000) Gap junctions in cardiovascular disease. Circ Res 86: 1193–1197
Karma A (1993) Spiral breakup in model equations of action potential propagation in cardiac tissue. Phys Rev Lett 71: 1103–1106
Kentish JC, ter Keurs HE, Ricciardi L, Bucx JJ, Noble MI (1986) Comparison between the sarcomere length-force relations of intact and skinned trabeculae from rat right ventricle. Influence of calcium concentrations on these relations. Circ Res 58: 755–768
Kjekshus J (1990) Arrhythmias and mortality in congestive heart failure. Am J Cardiol 65: 42I-48I
Klein RC (1984) Ventricular arrhythmias in aortic valve disease: analysis of 102 patients. Am J Cardiol 53: 1079–1083
Kohl P, Day K, Noble D (1998) Cellular mechanisms of cardiac mechano-electric feedback in a mathematical model. Can J Cardiol 14: 111–119
Kontos MC, Ellenbogen KA, Wood MA, Akosah KO, Nixon JV, Stambler BS (1995) Factors Associated with High Impedance with a Non-Thoracotomy Defibrillation Lead System. J Am Coll Cardiol 25: 213A
Lab MJ (1982) Contraction-excitation feedback in myocardium. Physiological basis and clinical relevance. Circ Res 50: 757–766
Lerman BB, Engelstein ED, Burkhoff D (2001) Mechanoelectrical feedback: role of beta-adrenergic receptor activation in mediating load-dependent shortening of ventricular action potential and refractoriness. Circulation. 104:486–490
Li GR, Lau CP, Ducharme A, Tardif JC, Nattel S (2002) Transmural action potential and ionic current remodeling in ventricles of failing canine hearts. Am J Physiol Heart Circ Physiol 283: H1031–H1041
Lin DH, Yin FC (1998) A multiaxial constitutive law for mammalian left ventricular myocardium in steady-state barium contracture or tetanus. J Biomech Eng 120: 504–517
Link MS, Wang PJ, Pandian NG, Bharati S, Udelson JE, Lee MY, Vecchiotti MA, VanderBrink BA, Mirra G, Maron BJ, Estes NA 3rd. (1998) An experimental model of sudden death due to low-energy chest-wall impact (commotio cordis). N Engl J Med. 338: 1805–1811
Link MS, Wang PJ, VanderBrink BA, Avelar E, Pandian NG, Maron BJ, Estes NA, 3rd (1999) Selective activation of the K(+)(ATP) channel is a mechanism by which sudden death is produced by low-energy chest-wall impact (Commotio cordis). Circulation 100: 413–418
Link MS, Maron BJ, Wang PJ, VanderBrink BA, Zhu W, Estes NA 3rd. (2003) Upper and lower limits of vulnerability to sudden arrhythmic death with chest-wall impact (commotio cordis). J Am Coll Cardiol. 41: 99–104

Lucy SD, Jones DL, Klein GJ (1994) Pronounced increase in defibrillation threshold associated with pacing-induced cardiomyopathy in the dog. Am Heart J 127: 366–376
Luo CH, Rudy Y (1994) A dynamic model of the cardiac ventricular action potential. I. Simulations of ionic currents and concentration changes. Circ Res 74: 1071–1096
Matsuda N, Hagiwara N, Shoda M, Kasanuki H, Hosoda S (1996) Enhancement of the L-type Ca2+ current by mechanical stimulation in single rabbit cardiac myocytes. Circ Res 78: 650–659
Moreno J, Zaitsev AV, Warren M, Berenfeld O, Kalifa J, Lucca E, Mironov S, Guha P, Jalife J (2005) Effect of remodelling, stretch and ischaemia on ventricular fibrillation frequency and dynamics in a heart failure model. Cardiovasc Res 65: 158–166
Moss AJ, Hall WJ, Cannom DS, Daubert JP, Higgins SL, Klein H, Levine JH, Saksena S, Waldo AL, Wilber D, Brown MW, Heo M (1996) Improved survival with an implanted defibrillator in patients with coronary disease at high risk for ventricular arrhythmia. Multicenter Automatic Defibrillator Implantation Trial Investigators. N Engl J Med. 1335:1933–1940
Nash MP, Panfilov AV (2004) Electromechanical model of excitable tissue to study reentrant cardiac arrhythmias. Prog Biophys Mol Biol 85: 501–522
Negroni JA, Lascano EC (1996) A cardiac muscle model relating sarcomere dynamics to calcium kinetics. J Mol Cell Cardiol 28: 915–929
Ott P, Reiter MJ (1997) Effect of ventricular dilatation on defibrillation threshold in the isolated perfused rabbit heart. J Cardiovasc Electrophysiol 8: 1013–1019
Panfilov A, Pertsov A (2001) Ventricular fibrillation: evolution of the multiple-wavelet hypothesis. Philos Transact A Math Phys Eng Sci 359: 1315–1325
Penefsky ZJ, Hoffman BF (1963) Effects of stretch on mechanical and electrical properties of cardiac muscle. Am J Physiol 204: 433–438
Pye MP, Cobbe SM (1996) Arrhythmogenesis in experimental models of heart failure: the role of increased load. Cardiovasc Res. 32:248–257
Qu Z, Weiss JN, Garfinkel A (1999) Cardiac electrical restitution properties and stability of reentrant spiral waves: a simulation study. Am J Physiol 276: H269–H283
Rogers JM, Huang J, Smith WM, Ideker RE (1999) Incidence, Evolution, and Spatial Distribution of Functional Reentry During Ventricular Fibrillation in Pigs. Circ Res 84: 945–954
Rosen S, Lahorra M, Cohen MV, Buttrick P (1991) Ventricular fibrillation threshold is influenced by left ventricular stretch and mass in the absence of ischaemia. Cardiovasc Res 25: 458–462
Sasaki N, Mitsuiye T, Noma A (1992) Effects of mechanical stretch on membrane currents of single ventricular myocytes of guinea-pig heart. Jpn J Physiol 42: 957–970
Starobin JM, Zilberter YI, Rusnak EM, Starmer CF (1996) Wavelet formation in excitable cardiac tissue: the role of wavefront-obstacle interactions in initiating high-frequency fibrillatory-like arrhythmias. Biophys J. 70: 581–594
Strobel JS, Kay GN, Walcott GP, Smith WM, Ideker RE (1996) The effect of ventricular volume on ventricular defibrillation efficacy in pigs. J Am Coll Cardiol 27 (supplement): 327–328
Sung D, Mills RW, Schettler J, Narayan SM, Omens JH, McCulloch AD (2003) Ventricular filling slows epicardial conduction and increases action potential duration in an optical mapping study of the isolated rabbit heart. J Cardiovasc Electrophysiol 14: 739–749
Taggart P, Sutton PM, Treasure T, Lab M, O'Brien W, Runnalls M, Swanton RH, Emanuel RW (1988) Monophasic action potentials at discontinuation of cardiopulmonary bypass: evidence for contraction-excitation feedback in man. Circulation 77: 1266–1275
Taggart P, Sutton PM, Opthof T, Coronel R, Trimlett R, Pugsley W, Kallis P (2000) Inhomogeneous transmural conduction during early ischaemia in patients with coronary artery disease. J Mol Cell Cardiol 32: 621–630
ten Tusscher KH, Noble D, Noble PJ, Panfilov AV (2004) A model for human ventricular tissue. Am J Physiol Heart Circ Physiol 286: H1573–H1589
Thakur RK, Souza JJ, Troup PJ, Chapman PD, Wetherbee JN (1993) Pericardial effusion increases defibrillation energy requirement. Pacing Clin Electrophysiol 16: 1227–1230
Tomaselli GF, Marban E (1999) Electrophysiological remodeling in hypertrophy and heart failure. Cardiovasc Res 42: 270–283
Vaidya D, Morley GE, Samie FH, Jalife J (1999) Reentry and fibrillation in the mouse heart: a challenge to the critical mass hypothesis. Circ Res. 85: 174–181

Vigh AG, Lowder J, Deantonio HJ (1999) Does acute volume overloading in the setting of left ventricular dysfunction and pulmonary hypertension affect the defibrillation threshold? Pacing Clin Electrophysiol 22:759–764

Wang Z, Mitsuiye T, Noma A (1996) Cell distension-induced increase of the delayed rectifier K+ current in guinea pig ventricular myocytes. Circ Res 78: 466–474

Warren M, Guha PK, Berenfeld O, Zaitsev A, Anumonwo JM, Dhamoon AS, Bagwe S, Taffet SM, Jalife J (2003) Blockade of the inward rectifying potassium current terminates ventricular fibrillation in the guinea pig heart. J Cardiovasc Electrophysiol 14: 621–631

Watanabe H, Sugiura S, Kafuku H, Hisada T (2004) Multiphysics simulation of left ventricular filling dynamics using fluid-structure interaction finite element method. Biophys J 87: 2074–2085

Winfree AT (1973) Scroll-shaped waves in chemical activity in three dimensions. Science 181: 937–939

Winfree AT (1987) When Time Breaks Down: The Three-Dimensional Dynamics of Electrochemical Waves and Cardiac Arrhythmias. Princeton University Press. Princeton. NJ.

Winfree AT (1989) Electrical instability in cardiac muscle: phase singularities and rotors. J Theor Biol. 138:353–405

Zabel M, Koller BS, Sachs F, Franz MR (1996a) Stretch-induced voltage changes in the isolated beating heart: importance of the timing of stretch and implications for stretch-activated ion channels. Cardiovasc Res 32: 120–130

Zabel M, Portnoy S, Franz MR (1996b) Effect of sustained load on dispersion of ventricular repolarization and conduction time in the isolated intact rabbit heart. J Cardiovasc Electrophysiol 7: 9–16

Zeng T, Bett GC, Sachs F (2000) Stretch-activated whole cell currents in adult rat cardiac myocytes. Am J Physiol Heart Circ Physiol 278: H548–H557

Chapter 16
Electromechanical Modelling of Cardiac Tissue

C. Cherubini, S. Filippi, P. Nardinocchi, and L. Teresi

Abstract We present an electromechanical model of myocardium tissue, coupling finite elasticity, endowed with the capability of describing muscle contractions, with a FitzHugh–Nagumo type system, describing the electrical activity proper to excitable media. Here, we exploit a novel point of view which introduces the notion of active deformation as opposed to that of active stress. The high degree of deformability of the medium makes mandatory to set the diffusion process in a moving domain, thereby producing a direct influence of the deformation on the electrical activity. Various effects of contraction on stationary rotating spiral waves and spiral wave break up are discussed.

Keywords Electromechanics · Mechanoelectric feedback · Mathematical physiology

16.1 Introduction

Cardiac tissue is the set of different interconnected physical processes driven by electrical and mechanical activities; specifically, the muscles contraction is controlled by electrical activity via excitation–contraction coupling (ECC) while changes in tissue length affect electrophysiological properties via mechano–electrical feedback (MEF).

The first quantitative mathematical model of wave propagation in biological tissues dates back more than 50 years (Hodgkin et al., 1952) and has been the basis on which realistic ionic and phenomenological models of cardiac tissue have been elaborated (FitzHugh, 1961; Noble, 1962; Nagumo et al., 1962). Ionic models are well able to reproduce the depolarization and re-polarization phases of the action

C. Cherubini (✉)
Laboratory of Nonlinear Physics and Mathematical Modeling, Università Campus Bio-Medico, Rome, Italy
e-mail: c.cherubini@unicampus.it

A. Kamkin, I. Kiseleva (eds.), *Mechanosensitivity of the Heart*, Mechanosensitivity in Cells and Tissues 3, DOI 10.1007/978-90-481-2850-1_16,

potential, restitution properties, and dynamic changes in ionic concentration; however, such a modelling is effective at small length and time scales, while many important problems, such as the question of re-entrant electrical sources occur only in large spatial regions of cardiac tissue. Phenomenological descriptions based on the two-variable FitzHugh–Nagumo model are usually capable of describing two- and three-dimensional pulse dynamics in the heart when they are customized to reproduce quantitatively specific macroscopic characteristics of cardiac tissues such as restitution of the action potential duration.

Here, we describe the electrical activity of the heart through a modified FitzHugh–Nagumo type system (Rogers and McCulloch, 1994; Aliev and Panfilov, 1996) which triggers the propagation of activation waves in the excitable medium. The system consists of a diffusion equation for a voltage-like variable, with a cubic nonlinear term that allows regenerative self-excitation via a positive feedback, and of an evolutive equation for a recovery variable, coupled with voltage, providing a slower negative feedback. On account of the high degree of deformability of the medium in which the activation waves propagate it is necessary to set the diffusion process in the deformed configuration, thereby producing a direct influence of the deformation on the electrical activity of the heart.

The effect of electrical activity on cardiac mechanics is a key issue in any electromechanical model. Usually, this effect is modelled by assuming that, when cardiac muscle fibers are stimulated, they generate contractile forces which, at the macro scale of the tissue, are described by an active stress tensor constitutively related to the tissue's electro-physiological activity. The overall stress in the tissue is then recovered by adding to the active stress a passive stress, depending on the mechanical properties of the myocardium (Hunter et al., 1998; Nash and Hunter, 2000; Usyck et al., 2002). While the muscle fiber distribution influences the active response of the tissue, the spatial variation in collagen distribution is related to material constitutive parameters and determines the highly nonlinear, anisotropic passive response of the cardiac tissue.

Here, we exploit a novel point of view presented in (Nardinocchi and Teresi, 2007) which introduces the notion of active contraction as opposed to that of active stress. We assume that, at the macroscopic scale, the activation of the cardiac muscle is described by the contraction of its fibres, that is, by a change of fibres' ground state; thus, at any given instant, the stress state in the activated tissue is due to the difference between its actual configuration and the ground state. The active deformation is in turn related to the electrophysiological activity of the tissue; as suggested by many experiments, calcium concentration is the main factor driving the variation in the length of muscular fibres.

None of the studies cited above has accounted for the effect of mechanics on the electrophysiology of cardiac tissue although the contraction of cardiac tissue affects the process of wave propagation in two ways. It induces changes in the medium in which the propagation occurs, the sarcoplasmic reticulum, and alters the electrophysiological activity of the heart. Recently, some light has been shed on the underlying (sub–)cellular mechanisms of cardiac MEF; these include stretch activation of sarcolemmal ion channels and electrophysiological effects of mechanical modulation of cellular Ca^{2+} handling (Kohl and Sachs, 2001). However, the macroscopic modelling of the mechano-electrical feedback in cardiac cells is far from being clarified. Recent studies have combined a description of the electrical activity of the cardiac tissue and a modelling of the mechanical processes that occur during heart beats and attempted to model MEF in cardiac tissues. In (Nash and Panfilov, 2004), an electromechanical model of two-dimensional elastic, isotropic, and excitable tissues is proposed but no mechano-electrical feedback is included in the modeling. In (Panfilov et al., 2005), the same model is reviewed and a macroscopic modelling of the stretch-activated ion currents is included; the model corresponds to an external current that depends linearly both on the local volume changes in the tissue and on the tissue voltage.

Here, we neglect any dependence of the diffusivity tensor and the membrane capacitance on the tissue deformation; instead, we model the mechano-electrical feedback by considering the contribution to the electrical activity of the tissue of an external current that is dependent on the tissue strain, as suggested in (Kohl and Sachs, 2001). Specifically, as shown by studies of the cellular mechanism of stretch-activation of ion channels, a relevant role is played by the deformation of the cell from a reference state. We therefore propose a representation of the stretch-activated current, borrowed from these studies (Sachse, 2004), where the main role is played by the elastic deformation of the tissue, which in our model gives a local measurement of the deformation from the rest state of the tissue. We show the relevance of the electromechanical model in studying different physical phenomena such as the influence of stretch-activated currents on the restitution properties of the tissue, the propagation of planar and spiral waves in the deformable medium, and the spiral wave termination via mechanical pressure. Finally, we show the influence of stretch on the diffusion of the action potential through numerical experiments resembling well-known experimental tests.

16.2 The Mechano-Electrical Feedback (MEF) in Heart Tissue

Mechano-electrical feedback (MEF) within heart tissue is a complex phenomenon comprising electrophysiological changes in response to myocardial stretch (Lab, 1982). This phenomenon has been studied in the clinical community for well over a century and may have both pro-rhythmic and arrhythmogenic consequences (Kohl and Ravens, 2003). It has been shown that mechanical deformations alter the electrical properties of myocytes and play an important role in ventricular arrhythmias (Franz et al., 1992; Sigurdson et al., 1992). On a larger scale, both the contraction of the heart itself and variations in circulatory parameters can alter the length and the tension in cardiac tissue and may affect cardiac electrophysiology via mechano-electrical transduction. Basic mechano-sensitive processes believed to be involved in MEF include stretch-activated ion channels in cardiac myocytes (Sachs, 1986; Hu and Sachs, 1998; Cazorla et al., 1999), changes in cardiac calcium handling (Sigurdson et al., 1992; White et al., 1993), and the interaction of cardiac myocytes with other mechano-sensitive cells (Kohl and Noble, 1996). All these mechanisms have the potential to introduce a more or less immediate electrophysiological response of cardiac myocytes to stretch.

Today MEF has been well studied in the ventricular myocardium. It has been recognized in isolated heart preparations and in man; ventricular stretch leads to a shortening of APD and of the effective refractory period (ERP) (Dean et al., 1989a; Franz et al., 1989; Reiter et al., 1997).

At the same time, mechano-electrical transduction is an integrative regulatory system in the heart. Mechanical energy can be transduced into electrical or chemical energy by the distortion of the molecular conformation. Membrane stretch opens the channels to admit charge-carrying ions which influence the membrane potential. The myocardium is contractile and, in addition to extracellular or extraneously applied forces, active contraction provides an intracellular generator of stress and strain, and thus a possible transducing mechanism.

Pressure/volume changes and stretch can alter diastolic depolarization, action potential duration and refractory periods as well as produce after-depolarizations. It is thus not surprising that externally imposed mechanical perturbations to the ventricle can be dangerous as well as therapeutic, as better explained in the following section. They can promote sudden death by arrhythmic mechanisms or, therapeutically, cardiovert (back to normal) potentially lethal rhythm disturbances, as well as stimulate electrically silent hearts (see Kohl et al., 1999).

The mechanical influence in the ventricle can also affect gross electrophysiology expressing in different parts of the ECG including the QRS complex (Sideris et al., 1995), the QT interval (Lab, 1980, 1982; Yamashita et al., 1991) and the T-wave (Dean et al., 1990).

16.2.1 Mechano-Electrical Feedback and Arrhythmias

Mechano-electric feedback can produce arrhythmias (Lab, 1982; Franz, 1996). Intracellular calcium changes have a fundamental role in the generation of arrhythmia, (Di Diego et al., 1994; Opie, 1997). Sustained increase of the intracellular calcium concentration promote Ca^{2+} oscillations (Allen et al., 1984) and electrophysiological oscillations lead to arrhythmia.

It seems that mechano-electric feedback is enhanced under pathological conditions (Lab et al., 1993; Pye et al., 1996) and this possibility could have clinical implications. There is also clinical evidence that MEF may be involved in the mechanical initiation of arrhythmias and fibrillation, as well as in the re-setting of a disturbed heart rhythm via mechanical first aid procedures.

A number of retrospective case studies show clear evidence of mechanically induced cardiac arrest, caused by a single blunt impact of moderate force to the precordium without any apparent structural damages. The earliest observations of sudden death caused by precordial impact were published more than one hundred years ago (Nelaton, 1876). This syndrome is now referred to as *Commotio cordis*. On the other hand, the question of how to reinstate contractile activity in the event of cardiac arrest, ideally without having to open the chest, is still a matter of debate. Back in 1920, it was reported that a single blow to the chest could restore a palpable pulse in a patient with ventricular standstill during a Stokes-Adams attack (Schott, 1920). In the late 1930s, the method of external electrical stimulation by capacitive discharge was developed in Russia and shown to be effective for defibrillation or cardiac stimulation in ventricular standstill (Zoll, 1952). However, it was noted that unless the electrical stimulation was applied within less than 1–3 min a resumption of normal cardiac activity was unlikely (Kouwenhoven et al., 1957). This time window could be significantly prolonged if rhythmic application of pressure on the thorax was performed to compress the heart and maintain circulation. This observation led, in the early 1960s, to the establishment of external cardiac massage as a successful tool in cardiac resuscitation (Kouwenhoven et al., 1960; Phillips and Burch, 1964). There have also been several reports of cardiac dilatation-related atrial and ventricular tachycardia being terminated during increases in intrathoracic pressure due to coughs (Wei et al., 1980) or to the Valsalva manoeuvre (Ambrosi et al., 1995).

16.3 Mechanics of Cardiac Tissues

The mechanics of muscle contraction is described by using the notion of active deformation, in contrast with that of active stress, using the approach proposed by in (Nardinocchi and Teresi, 2007) within the framework of finite elasticity, and used in (Cherubini et al., 2008) in the realm of electromechanical modelling of the cardiac contraction.

We assume that the contraction experienced by a cardiac muscle fiber under stimulus is described at the macroscopic scale by a change in the length of the fiber, a change that we call active deformation; the actual length of the fiber, in turn, depends on the amount of stress it sustains.

Finite elasticity subsumes these assumptions with the idea underlying the Kröner–Lee decomposition of the deformation gradient, originally introduced to distinguish between elastic and viscoplastic strains (Kröner, 1959; Lee, 1969), much later proposed for growth modelling (Rodriguez et al., 1994; Taber, 1995), and recently refined with a noteworthy improvement in dynamics, in (Dicarlo and Quiligotti, 2002).

The key issues are set out below. We distinguish the active from the actual, or visible, deformation by associating with each element of the contractile body two different states: the contracted and the visible one.[1] The first one describes how a

[1] In general, there is no global contracted state for a body but only for a material fiber, that is, the field of contracted deformations is usually not compatible.

muscle fiber would like to be placed once activated, and left free to contract; from our point of view, the active state is a ground state, that is, it is stress free. The visible state describes the state that a muscular tissue attains once contracted, loaded and in case kinematically constrained. Of primary importance is the fact that the elastic energy, and thus, the stress state, depends from the difference between the visible and the contracted deformation.

Moreover, we consider an active ground state, that can vary in time driven by electro-physiological variables; in our model, it is the intracellular concentration of calcium ions that triggers the active deformation through a phenomenological relation inspired by the scientific literature.

16.3.1 Active and Visible Deformations

Let a smooth region $\mathcal{B}$ of the three-dimensional Euclidean space $\mathcal{E}$ be given and let $\partial\mathcal{B}$ be its boundary. We take the region $\mathcal{B}$ to be the reference shape of a homogeneous body, consisting of a thin patch of excitable elastic tissue, which we identify pointwise with $\mathcal{B}$. For $\mathcal{T}$ the time line, identified with the real line $\mathcal{R}$, we refer to any smooth embedding

$$p\colon \mathcal{B} \times \mathcal{T} \longrightarrow \mathcal{E}, \qquad (y,t) \mapsto p\,(y,t) \tag{16.1}$$

of the body into $\mathcal{E}$ as a placement of the body at time $t \in \mathcal{T}$; we call $\mathcal{B}_t = p(\mathcal{B},t)$ the visible configuration of the body at time t.

Given a placement p of the body, let $\mathbf{v}_m$ and $\mathbf{F}$ be the corresponding material velocity field and visible deformation gradient; we have $\mathbf{v}_m = \dot{p}$ and $\mathbf{F} = \nabla p$ where the superposed dot and the nabla symbol denote differentiation with respect to t and y, respectively. The fields $\mathbf{v}_m$ and $\mathbf{F}$ take values in the translation space $\mathcal{V}$ of $\mathcal{E}$ and in $\mathrm{Lin}^+ = \{\mathbf{F} \in \mathrm{Lin} | J = \det \mathbf{F} > 0\}$, with Lin the space of the linear transformations of $\mathcal{V}$. For $y \in \mathcal{B}$, let a unit vector $\mathbf{e} \in \mathcal{V}$ be fixed and a placement p be given. By the material fiber through y in the direction $\mathbf{e}$ we mean an ordered pair $(y, \mathbf{e})$; the visible image at time t of that fiber in the placement p is the ordered pair $(p\,(y,t), \mathbf{F}(y,t)\,\mathbf{e})$ (Fig. 16.1). At any time t, the stretch $\lambda(y, \mathbf{e}, t)$ and the change in length $\delta l(y, \mathbf{e}, t)$ of a fiber $(y, \mathbf{e})$ are

$$\lambda(y, \mathbf{e}, t) = |\mathbf{F}\,(y,t)\,\mathbf{e}|, \;\; \delta l\,(y, \mathbf{e}, t) = \lambda(y, \mathbf{e}, t) - 1, \tag{16.2}$$

respectively. The visible deformations of the body are usually measured by the Cauchy-Green strain tensor $\mathbf{C} = \mathbf{F}^T\mathbf{F}$ whose determinant is related to the determinant J of $\mathbf{F}$ by $\det \mathbf{C} = J^2$.

We assume that a specified muscle distribution is assigned; to each muscle fiber through y, there corresponds a material fiber but the converse is not true. The peculiarity of a muscle fiber, that distinguishes it from a material fiber, is its capacity to contract in response to a specified electrophysiological stimulus. We assume that, at the macroscopic scale, the activation of the muscle fibers of the tissue are described

Fig. 16.1 The material fiber (y,**e**) in the reference state (*right, bottom*), in the contracted state (*bottom*), and in the visible one (*right, top*)

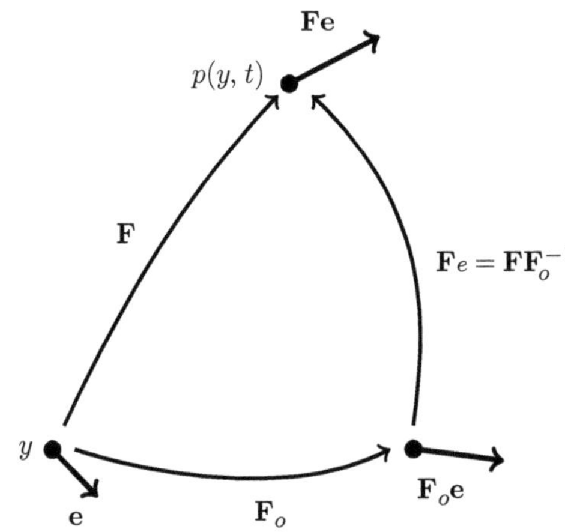

at any time t by the distortion field $\mathbf{F}_o{:}\mathcal{B} \times \mathcal{T} \rightarrow$ Lin, to be known as the active deformation field; from (16.2), we have that the stretch λ_o and the change in length $\delta_o l$ of a muscle fiber are

$$\lambda_o(y, \mathbf{e}, t) = |\mathbf{F}_o(y, t)\mathbf{e}\,|,\ \ \delta_o\, l\,(y, \mathbf{e}, t) = |\,\mathbf{F}_o\,(y, t)\,\mathbf{e}\,| - 1, \tag{16.3}$$

respectively. Here, we assume a planar, isotropic muscle distribution: having fixed a unit vector $\mathbf{h} \in \mathcal{V}$, the muscle fibers are all material fibers (y,**e**) with $\mathbf{e} \cdot \mathbf{h} = 0$; thus, the unit vector **h** defines the isotropy plane of the muscle distribution. To this planar distribution there corresponds an active deformation field given by

$$\mathbf{F}_o = \gamma_o \check{I},\quad \check{I} = \mathbf{I} - \mathbf{h} \otimes \mathbf{h}; \tag{16.4}$$

in (16.4), **I** is the identity tensor. For this special case, the contraction is completely described by the scalar field $\gamma_o{:}\mathcal{B} \times \mathcal{T} \rightarrow \mathcal{R}$ measuring the amount of the active contraction of any muscle fiber through y

$$\gamma_o\,(y, t) = \lambda_o(y, \mathbf{e}, t) = |\mathbf{F}_o\,(y, t)\,\mathbf{e}|. \tag{16.5}$$

Given the placement p and the distortion field $\mathbf{F}_o$, the multiplicative decomposition of the gradient of the placement reads as

$$\mathbf{F} = \nabla p = \mathbf{F}_e\mathbf{F}_o, \tag{16.6}$$

thus defining the elastic deformation $\mathbf{F}_e$; even if **F** is the gradient of the placement p, neither $\mathbf{F}_o$ nor $\mathbf{F}_e$ are in general gradients of any fields. As consequence of (16.6), we have that the passive (elastic) and the active response of the tissue relies on $\mathbf{F}_e$ and $\mathbf{F}_o$, respectively. The strain measure corresponding to the elastic deformation is given by $\mathbf{C}_e = \mathbf{F}_e^T\mathbf{F}_e$, and can be represented as

$$\mathbf{C}_e = \mathbf{F}_o^{-T}\mathbf{C}\mathbf{F}_o^{-1}. \tag{16.7}$$

Let us note that a different muscle distribution may be accounted for simply by changing the representation form of the active deformation field; the orthotropic myofiber architecture discussed in Usyck et al., (2000) would enter our system through an active deformation field given by

$$\mathbf{F}_o = \alpha_o \mathbf{e}_1 \otimes \mathbf{e}_1 + \beta_o \mathbf{e}_2 \otimes \mathbf{e}_2 + \delta_o \mathbf{e}_3 \otimes \mathbf{e}_3, \tag{16.8}$$

where $\mathbf{e}_i$ with $i = 1, 2, 3$ denotes the fiber-sheet directions, and the fields α_o, β_o, and δ_o may be assigned in terms of the microscopic stimuli, according to electrophysiological data.

16.3.2 Balance Equations

The deformative processes (or placements p) which occur are selected, once the pertinent constitutive information on the passive mechanical response of the tissue has been provided, by the balance equation of mechanics which here we write in the form of virtual powers: for any test velocity $\tilde{\mathbf{v}}_m \in \mathcal{V}$,

$$\int_{\mathcal{B}} -\mathbf{S}_R \cdot \nabla \tilde{\mathbf{v}}_m + \int_{\partial\mathcal{B}} \mathbf{t}_R \cdot \tilde{\mathbf{v}}_m = 0, \tag{16.9}$$

with $\mathbf{S}_R$ the reference stress, $\mathbf{t}_R$ the reference traction density. Due to the very different time scales at which the proper mechanical phenomena and the mechanical processes related to the heartbeat occur, inertial forces are neglected. Equation (16.9) is the integral version of the well-known local balance equations in referential form (see Gurtin, 1981 for a detailed treatise on continuum mechanics) where bulk forces, including inertial ones, are neglected:

$$\operatorname{div} \mathbf{S}_R = 0 \quad \text{in} \quad \mathcal{B}; \; \mathbf{S}_R \mathbf{m} = \mathbf{t}_R \quad \text{on} \quad \partial\mathcal{B}, \tag{16.10}$$

with div the divergence differential operator and $\mathbf{m}$ the unit normal field to $\partial\mathcal{B}$. Standard arguments allow us to introduce and give meaning to the Cauchy stress tensor $\mathbf{T}$, the only measurable stress in a deformative process. For $\mathbf{m}\mathtt{dA}$ an oriented area element in the reference shape, and

$$\mathbf{n}\mathtt{da} = \mathbf{F}^*\mathbf{m}\mathtt{dA}, \quad \mathbf{F}^* = \det\mathbf{F}\mathbf{F}^{-T}, \tag{16.11}$$

the corresponding element in the visible shape, the stress tensor $\mathbf{T}$ satisfies the following identity

$$\mathbf{T}\mathbf{n}\mathtt{da} = \mathbf{S}_R\mathbf{m}\mathtt{dA} \tag{16.12}$$

ensuring that the force $\mathbf{t} = \mathbf{T}\mathbf{n}$ acting on the infinitesimal element of the oriented area in the visible shape is equivalently measured by the force $\mathbf{t}_R$ acting on the infinitesimal element of the oriented area in the reference shape. Equations (16.11) and (16.12) turn out

$$\mathbf{S}_R = \mathbf{T}\mathbf{F}^*; \tag{16.13}$$

moreover, the following relation between visible and reference traction fields holds:

$$\mathbf{t}_R = |\mathbf{F}^*\mathbf{m}|\mathbf{t}. \tag{16.14}$$

16.3.3 Mechanical Behaviour of the Tissue

The mechanical behaviour of the tissue relies on the constitutive prescriptions assumed for the passive and for the active responses.

The passive mechanical response of cardiac tissue has been extensively investigated through uniaxial and biaxial tests that have evidenced the inhomogeneity, the nonlinear viscoelastic and poroelastic nature, and the strong anisotropy of myocardial tissue (see Nash and Hunter, 2000 for a detailed description of cardiac microstructure). Here, all these aspects are neglected, as we focus mainly on the couplings between mechanical deformation and electrophysiological variables; for simplicity, we treat the myocardium as homogeneous, elastic and isotropic, while retaining its key characteristic of incompressibility.

Several special kinds of constitutive equations supported by experiments of various degrees of completeness have been proposed to describe the response of incompressible, elastic materials; of these, two important models that have been widely used to obtain specific explicit results in different problems are the Mooney–Rivlin and neo-Hookean models. For the Mooney–Rivlin materials, the elastic energy ψ is assumed to be

$$\psi(I_1,I_2) = \frac{\alpha_1}{2}(I_1 - 3) + \frac{\alpha_2}{2}(I_2 - 3), \tag{16.15}$$

with

$$I_1 = \mathtt{tr}\mathbf{C}_e \quad \text{and}\, I_2 = \frac{1}{2}((\mathtt{tr}\mathbf{C}_e)^2 - \mathtt{tr}\mathbf{C}_e^2) \tag{16.16}$$

the first two principal invariants of the Cauchy-Green elastic strain tensor $\mathbf{C}_e$ and α_i (i = 1,2) are the elastic moduli. For $\alpha_2 = 0$ we recover neo-Hookean response; it involves only the first invariant I_1 of $\mathbf{C}_\mathrm{e}$, which collects the complete information on the stretches experienced in any three orthogonal directions. Both the Mooney–Rivlin and neo-Hookean constitutive laws must be coupled with the requirement that, due to the incompressibility,

$$\det \mathbf{C}_e = 1. \tag{16.17}$$

It is worth noting that the incompressibility constraint involves the elastic strain measure $\mathbf{C}_e$ and enforces the visible volume changes measured by $J = \det \mathbf{F}$ to be determined by $\det \mathbf{F}_o = \gamma_o^2$.

Here, we postulate the existence of a neo-Hookean strain energy density that, together with the constraint (16.17), describes the material response of the tissue. Obviously, neither the Mooney–Rivlin, nor the neo-Hookean constitutive law reflects the structural properties of the cardiac tissue determined by the observed spatial variation in collagen distributions. Nevertheless, we deem it relevant to discuss the electromechanical coupling corresponding to an elastically incompressible material that is completely isotropic (in the next section we choose an isotropic diffusivity tensor) before accounting for the strong anisotropy of the tissue. Moreover, following (Spencer, 1972) and (Merodio and Ogden, 2005), it is possible to extend the constitutive law in order to describe the fibrous collagen structure through the addition of an anisotropic component of the strain energy function to the neo-Hookean function.

For a neo-Hookean incompressible material, the second Piola–Kirchhoff stress tensor $\mathbf{S}$ is given by the derivative of the neo-Hookean strain energy function with respect to $\mathbf{C}_e$, plus a term containing a hydrostatic pressure field π due to the incompressibility constraint (16.17)

$$\mathbf{S} = 2\frac{\partial \psi}{\partial \mathbf{C}_e} - \pi \mathbf{C}_e^{-1}. \tag{16.18}$$

The Cauchy stress $\mathbf{T}$ is related to $\mathbf{S}$ by the equation

$$\mathbf{T} = \mathbf{F}_e \mathbf{S} \mathbf{F}_e^T; \tag{16.19}$$

by using (16.6) and (16.18), the Cauchy stress may be rewritten as

$$\mathbf{T} = 2\mathbf{F}\mathbf{F}_o^{-1} \frac{\partial \psi}{\partial \mathbf{C}_e} \mathbf{F}_o^{-1} \mathbf{F}^T - \pi \mathbf{I} \tag{16.20}$$

throwing some light on how the active deformation field $\mathbf{F}_o$ and the visible deformation gradient $\mathbf{F}$ contribute to the stress. From (16.20) it follows that only within the limit of infinitesimal deformation the stress tensor $\mathbf{T}$ allows for an additive decomposition into a passive component, depending on the visible deformation, and an active part, depending on the active deformation.

The active mechanical response is described by assuming a constitutive relation between the active stretch γ_o and the intracellular calcium concentration c. Typically, the activation mechanism is related to the muscle force production and is included in the tissue modelling through the so-called active stress, a stress component completely delivered by the membrane potential (Nash and Hunter, 2000; Usyck et al., 2000; Nash and Panfilov, 2004). Here, we distinguish the contraction process of a muscle fiber, initiated by the muscle activation, by the force production process which turns out as a consequence of the contraction but, as we show in the following, is associated with the elastic deformation of the tissue. Moreover, we find it relevant to make evident the role of the free Ca^{2+} in the activation mechanism of the contraction process and model the Ca^{2+} sensitivity of the cardiac tissue through the following laws borrowed from (Pelce and Sun, 1995):

$$\gamma_o(c) = \frac{l_o}{1 + f(c)(l_o - 1)} \gamma_o^{max}, \tag{16.21}$$

with

$$f(c) = \frac{1}{2} + \frac{1}{\pi} \arctan\left(\beta \log\left(c/c_R\right)\right), \; l_o = \frac{f(c^*) - 1}{f(c^*) - \gamma_o^{max}}. \tag{16.22}$$

Equations (16.21) and (16.22) describe the basic characteristics of the dependence of the tissue contraction on the intracellular calcium concentration, even though many of the biological details are absent. The concentration c_R is a reference calcium concentration, c^* is peculiar to the rest tissue, that is,

$$\gamma_o(c^*) = 1; \tag{16.23}$$

γ_o^{max} is the largest allowable tissue shortening due to the active muscle structure, that is,

$$\lim_{c \to \infty} \gamma_o(c) = \gamma_o^{max}; \tag{16.24}$$

the parameter β may be tuned to optimize the S-shaped form of the curve $c \mapsto \gamma_0$.

The variations in calcium concentration c determine, through the equations (16.21) and (16.22), the variations in the active deformation γ_o, that is, the contraction and the relaxation of the muscle fibers. These equations oversimplify the basic characteristics of the dependence of the tissue contraction on the intracellular calcium concentration and do not account for the dependence of Ca^{2+}-activated contractions on sarcomere length.

We conclude the discussion of the constitutive issues by introducing, albeit in a simple manner, a constitutive characterization of the tissue surrounding the patch we are studying. For $\mathbf{u}$ the vector field describing the displacement of the tissue,

$$\mathbf{u}(y,t) = p(y,t) - y, \tag{16.25}$$

we assume that the reference traction density $\mathbf{t}_R$ depends linearly on $\mathbf{u}$ through a stiffness tensor $\mathbf{K}$ which may be adjusted to represent different material conditions:

$$\mathbf{t}_R = \mathbf{K}\mathbf{u}. \tag{16.26}$$

Here, in order to discuss the general characteristics of the model, we refer to an isotropic prescription for the traction density and assume

$$\mathbf{K} = \kappa \mathbf{I}. \tag{16.27}$$

In effect, this corresponds to assuming the existence of an isotropic elastic bed along the constrained portion of the patch; by varying the value of the scalar field κ, different surrounding conditions may be accounted for: for $\kappa = 0$, a fixed boundary is recovered; for $\kappa \rightarrow \infty$, a free boundary condition is revealed and, in the absence of mechanical tractions, the boundary displacement is simply determined by the muscle contraction.

In the following numerical simulations, we shall use dimensionless quantities; thus, let us introduce a set of dimensionless variables: a characteristic time T_c and length L_c, to define our dimensionless time and space units, respectively. We divide all the relevant mechanical quantities by the elastic modulus α_1 that characterizes the passive response of the tissue. With regard to the field equations $(10)_1$, (13), (15) and (20), we introduce the dimensionless strain energy density $\psi_o = \psi/\alpha_1$ and the dimensionless indeterminate pressure field $p_o = p/\alpha_1$, and write[2]:

$$\text{div}\, \bar{S}_R = 0, \quad \bar{S}_R = 2\mathbf{F}\mathbf{F}_o^{-1} \frac{\partial \psi_o}{\partial \mathbf{C}_e} \mathbf{F}_o^{-1} \mathbf{F} - p_o \mathbf{I}. \tag{16.28}$$

We then introduce a dimensionless measure $\mathbf{u}_o$ of the displacement field $\mathbf{u}$ and write the boundary condition $(10)_2$, (26) as

$$\bar{S}_R \mathbf{m} = \kappa_o \mathbf{u}_o, \; \mathbf{u}_o = \frac{\mathbf{u}}{L_c}, \; \kappa_o = \frac{\kappa L_c}{\alpha_1}, \tag{16.29}$$

with κ_o a dimensionless parameter defining the stiffness of the surrounding tissues.

16.4 Electrophysiology of Cardiac Tissues

The propagation of electrical waves resulting from the excitability of individual cardiac cells and the electrical coupling of cardiac cells via gap junctions that is at the basis of the electrophysiological processes is well described by phenomenological models such as the reaction–diffusion type models. The macroscopic modelling of mechano-electrical feedback however, is far from being well understood: the dependence of the electrical properties of cardiac tissue on deformation may be largely analyzed and discussed, but the relevant mechano-electrical feedback calls for modelling of the stretch-activated ion channels, which can dramatically change

[2]Space derivatives in (16.28) are defined in relation to the dimensionless space unit.

the shape of action potential in response to stretch (Kohl et al., 1999; Kohl and Sachs, 2001).

We place before our model a rapid review of computational electrophysiology. The usual modeling of the electrophysiology of excitable tissues, as heart's, nerve's, and intestine's tissue (Aliev et al., 2000) is based on zero-dimensional mathematical models of excitable cells, that is, on a set of ordinary nonlinear differential equations (see Keener and Sneyd, (2001), for a detailed explanation of these aspects of the mathematical theory.). One equation describes the dynamics of the action potential; the ionic currents play the role of external sources. All the other equations describe the dynamics of ionic channels which may be activate or inactivate depending on the value of the voltage (voltage-dependent ionic channels). Accurate ionic models require even tens of equations to account for a physiologically correct behavior. The voltage spread turns out from the coupling of many zero-dimensional ionic models through a diffusion process. Specifically, the differential equation governing the dynamics of the voltage contains a term accounting for the diffusion of the action potential. It is a simple laplacian when the diffusion tensor is a constant, homogeneous, and isotropic field; it is a more complicated elliptic operator when electrical inhomogeneities and anisotropies are accounted for in the model. The major lack of such a theory in fact is the absence of distinction between the intra and extracellular spaces. For better results, the single diffusion equation for the action potential is replaced by two coupled diffusion equations for the extra and intracellular voltages. These equations are then coupled to the channel dynamics equations; all together, they are the basis of the so-called *bidomain* model. For computational purposes, bidomain equations can be rearranged in a parabolic and an elliptic equation. Additionally one may introduce temperature effects, which can have dramatic consequences for the electrophysiology of the tissue, as it is well established in the modelling of the physiology of nerve tissues. There, the mechanical and thermal changes associated with the excitation process have been analyzed in depth in experiments (Abbott et al., 1958; Howarth et al., 1979) and more recently (Heimburg and Jackson, 2005). It's fundamental to cite in this context the experimental work of (Hodgkin et al., 1949) on the global effects of temperature on the electrical dynamics in the giant axon of the squid so as the work of Tasaki and Byrne (1992) which showed as the thermal response in nerves starts and reaches a peak of the order of some $\mu^{\circ}C$ almost simultaneously with the electrical response. In these experiments the phase of heat production is followed immediately by a phase of heat absorption, with no net heat release after the passage of the action potential. These effects are experimentally observed using thermal detectors, typically constructed with a thin film of synthetic pyroelectric material like polyvinylidene fluoride. These experiments can be in part understood because the equations for the activation and inactivation variables are multiplied by a functional relation which has typically the form $Q_{10}^{(T-T_0)/s}$ where Q_{10} and s are constants taken from experiments and T_0 is a reference temperature. Temperature excursions and in particular heat transfer effects have an important role in experiments and clearly require accurate modellization (see Bini et al., 2006 for a prototype model in the case of FitzHugh–Nagumo theory).

We point out that in cardiac context, an analysis of possible spiral wave control by regional cooling of the tissue has been recently presented (Ashihara et al., 2005). In such work, two regions have been kept at different temperatures so to influence the electrophysiological properties of the planar tissue; as a consequence, pinning of spiral waves are created by the colder area. It is so evident that local temperature variations, especially if coupled to mechanical contractions, should be taken into account in a realistic model of cardiac tissue. As we'll see however, mechanical deformations belong on a strongly nonlinear physical theory which generates very complicated equations. Their solution is extremely expensive from the point of view of numerical techniques, consequently in this work we shall focus mainly on the simple coupling of a basic ion dynamics with finite elasticity, leaving the introduction of heat transfer effects to future analyses.

More in detail, for the sake of simplicity, here we use a two variables reaction–diffusion system to describe the propagation process in cardiac tissue; specifically, a modified FitzHugh–Nagumo type system (Rogers and McCulloch, 1994; Aliev and Panfilov, 1996). We do not account for electrical anisotropy and assume that the electrical properties of the tissue such as the conductivity tensor and the membrane capacitance do not depend on the deformation. We do, however, discuss, via numerical simulations, a macroscopic model of mechano-electrical feedback based on the prescription of an ionic transmembrane current depending on a relevant strain measure.

Intracellular Ca^{2+} dynamics play a major role in the process of excitation–contraction coupling, and also produce a feedback effect on the shape and duration of the action potential. Over the last 20 years, a number of mathematical cell models incorporating intracellular Ca^{2+} dynamics, have been elaborated (see Sachse, 2004 for an extended review). Substantially, calcium dynamics drive calcium influx from the extracellular medium through Ca^{2+} channels into the cell membrane and from internal stores. The surface membrane channels are of different types; voltage-dependent Ca^{2+} channels are the most important in cardiac cells and open in response to depolarisation of the cell membrane. We do not here account for the feedback effect of the calcium dynamics on the shape and duration of the action potential, but model the variations in calcium concentrations due to Ca^{2+} influx via voltage-dependent channels.

16.4.1 Diffusion-reaction on a Moving Medium

The spread of excitation in the heart occurs as a result of both the excitability of individual cardiac cells and the close electrical coupling of cardiac cells via specialized gap junctions, through which depolarised cardiac cells can elicit excitation in neighboring cells, resulting in a propagation wave of activity. Models that describe propagation in the heart generally account for two different contributions to the propagation velocity of the action potential: a contribution generated by all the ionic transmembrane currents and an other generated by the diffusion of the action potential via cellular interconnections.

An electromechanical model of cardiac tissue must also consider the large deformations of the medium through which diffusion occurs; thus, the relevant equations of diffusion-reaction process are better expressed in weak form, posed on the actual configuration of the body.

In dealing with moving domains undergoing large displacements, it is of the essence to distinguish between material points $y \in \mathcal{B}$ and their position in space $x = p(y,t) \in \mathcal{E}$, and consequently, between material fields-defined on $\mathcal{B} \times \mathcal{T}$ (also known as Lagrangean fields), and spatial fields-defined on $\mathcal{B}_t \times \mathcal{T}$ (also known as Eulerian fields). The diffusion process of the action potential can be naturally expressed in weak form, in terms of spatial field. Thus, let the action potential v_s and current I_s be time-dependent scalar fields on $\mathcal{B}_t$, and the flux $\mathbf{G}_s$ a time-dependent vector field on $\mathcal{B}_t$. Assuming an insulated boundary, that is, no flux across $\partial\mathcal{B}_t$, the balance in weak form reads as follows: for any test field $\tilde{v}_s$ associated with v_s

$$\int_{\mathcal{B}_t} C\frac{Dv_s}{Dt} \cdot \tilde{v}_s = \int_{\mathcal{B}_t} (-\mathbf{G}_s \cdot \nabla\tilde{v}_s - I_s \cdot \tilde{v}_s), \tag{16.30}$$

where Dv_s/Dt denotes the material time derivative of v_s, and C the membrane capacitance, here assumed to be constant. For any given spatial field, that is, a field defined on $\mathcal{B}_t$, it is possible to define its material counterpart, that is, a field defined on $\mathcal{B}$ (e.g., see Gurtin, 1981 for a clear and compact exposition); let us denote with the subscript m the material description of a spatial field. In particular, we have

$$\tilde{v}_m = \tilde{v}_s \circ p, (\nabla\tilde{v}_s)_m = \mathbf{F}^{-T}\nabla\tilde{v}_m, \left(\frac{Dv_s}{Dt}\right)_m = \dot{v}_m, \mathbf{G}_m = \mathbf{G}_s \circ p. \tag{16.31}$$

We can then pull-back the balance equation (30) to the reference configuration of the body: for any test field v_m associated with v_m

$$\int_{\mathcal{B}} JC\dot{v}_m \cdot \tilde{v}_m = \int_{\mathcal{B}} (-J\mathbf{F}^{-1}\mathbf{G}_m \cdot \nabla\tilde{v}_m - JI_m \cdot \tilde{v}_m), \tag{16.32}$$

where, we remember, the determinant J of $\mathbf{F}$ accounts for the change in the volume measure. Equation (16.32) shows how the visible deformation of the tissue enters the equations driving the diffusion of the action potential; standard localization arguments yield the following balance equations in strong form:

$$JC\dot{v}_m = \mathtt{div}(J\mathbf{F}^{-1}\mathbf{G}_m\nabla v_m) - JI_m, \quad \text{on} \quad \mathcal{B}, \tag{16.33}$$

$$J\mathbf{F}^{-1}\mathbf{G}_m \cdot \mathbf{m} = 0, \quad \text{on} \quad \partial\mathcal{B}. \tag{16.34}$$

Equation (16.34), the condition of insulated boundary, is based on the assumption that the body $\mathcal{B}$ we consider be surrounded by a medium having a much smaller diffusivity.

As far as the electrical properties of the tissue, we assume that the flux $\mathbf{G}_s$ depends linearly on the gradient of the activation potential through a diffusivity tensor $\mathbf{D}$:

$$\mathbf{G} = \mathbf{D}\nabla v_s. \tag{16.35}$$

By using equation $(16.31)_2$, the material counterpart of equation (16.35) is

$$\mathbf{G}_m = \mathbf{D}_m(\nabla v_s)_m = \mathbf{D}_m\mathbf{F}^{-T}\nabla v_m; \tag{16.36}$$

we assume that the tissue is isotropic and homogeneous from the electrical point of view, that is,

$$\mathbf{D}_m = d_m\mathbf{I}, \tag{16.37}$$

with the scalar field d_{m} to be assigned as a constant field. With (16.37) and (16.36), equation (16.33), (16.34) may be written as

$$C\dot{v}_m = \frac{1}{J}\mathrm{div}(Jd_m\mathbf{C}^{-1}\nabla v_m) - I_m, \quad \text{on} \quad \mathcal{B}, \tag{16.38}$$

$$Jd_m\mathbf{C}^{-1}\nabla v_m \cdot \mathbf{m} = 0, \quad \text{on} \quad \partial\mathcal{B}. \tag{16.39}$$

As far as the prescription of the transmembrane ionic currents is concerned, we refer to (Rogers and McCulloch, 1994), and Aliev and Panfilov (1996), where a dimensionless representation of the transmembrane current is given.

Thus, we introduce a characteristic potential v_{c} so that the dimensionless action potential $v = v_m/v_c$ ranges from 0 to 1.

By introducing the dimensionless variables into the diffusion equation (16.38), we obtain the dimensionless equation[3]

$$\dot{v} = \frac{1}{J}\mathrm{div}(Jd_o\mathbf{C}^{-1}\nabla v) - I, \tag{16.40}$$

with the dimensionless diffusivity d and transmembrane current I defined as

$$d = \frac{d_m}{\check{C}}, \quad I = \frac{I_m}{\check{C}}, \tag{16.41}$$

and $\check{C} = C/T_c$. Then, we set

$$I = kv(v-a)(v-1) + wv - I_{ext}, \tag{16.42}$$

[3]The time and space derivatives in (16.40) are defined in relation to the dimensionless time and space unit.

thus coupling diffusion equation (16.40) with the recovery variable w, a time dependent scalar field on $\mathcal{B}$, which in turn satisfies the following evolutive equation

$$\dot{w} = g(w,v)(-w - kv(v - b - 1)). \tag{16.43}$$

The term $g(w,v)$ is intended to control the time evolution of the recovery phase and is set as

$$g(w,v) = \varepsilon + \frac{\mu_1 w}{\mu_2 + v}. \tag{16.44}$$

As explained in the literature (Nash and Panfilov, 2004), the term $kv(v-a)(v-1)-wv$ in equation (16.42) controls the fast processes such as the initiation and upstroke of the action potential. The dynamics of the recovery phase and the restitution properties of the action potential are determined by the time course of w, a dimensionless representation of the conductance of a slow repolarizing current, which is primarily controlled by the term $g(w,v)$. Apart from the threshold parameter a, the parameters of this model (k, μ_1, μ_2, b) do not have direct physiological meanings, but are adjusted to reproduce key characteristics of the action potential of cardiac tissue, such as shape, refractoriness, and restitution.

16.4.2 Stretch Activated current

The external dimensionless current I_{ext} accounts for external stimuli I_e and for stretch-activated currents I_{sac}. The macroscopic modelling of stretch-activated currents is at the basis of mechano-electrical feedback in the electromechanical models of cardiac tissues. Here, we refer to (Kohl and Sachs, 2001) to describe from a macroscopic point of view the stretch-activated currents. Having assumed that our model is characterized by a planar and isotropic muscle fiber distribution, we set

$$I_{sac} = G_{sac}(v - \bar{v})\frac{1}{1 + \exp(-\delta(\lambda_p - \lambda_{ref}))}, \tag{16.45}$$

where the scalar λ_p is the planar trace of the elastic deformation $\mathbf{F}_e$,

$$\lambda_p = \check{I} \cdot \mathbf{F}_e, \tag{16.46}$$

that represents an isotropic measure of the deformation of the muscle fibers in relation to their rest state, and λ_{ref} is a reference deformation measure. Specifically, $\lambda_p = 2$ when the fibers are relaxed, $\lambda_p > 2$ or $\lambda_p < 2$ when they experience a positive or a negative strain, respectively. For $\lambda_p \in (-\infty, +\infty)$, the function $\lambda_p \mapsto [1 + exp(-\delta(\lambda_p - \lambda_{ref}))]^{-1}$ ranges in $(0, 1)$.

The reference strain λ_{ref} is tuned in order to have a near zero current at no contraction, and the parameter δ is tuned to have saturation for a strain of about 20%. Thus, the reference deformation λ_{ref} controls the point where half value is attained, and the parameter δ controls the steepness of the curve; the parameter G_s measures the conductivity of the channels, and $\bar{v}$ measures the equilibrium voltage. In the following numerical simulations, $\bar{v}$ and δ are given once and for all, while the role of the dimensionless parameter G_s is discussed.

16.4.3 Calcium Dynamics

We neglect the feedback effect of the calcium dynamics on the shape and duration of the action potential and model the variation in calcium concentration due to the influx of Ca^{2+} via voltage-dependent channels. We assume that the calcium transient results from the interplay between a source – the activated gated channels of the sarcoplasmic reticulum – and a sink – the negatively charged binding ligands of the cytoplasm. A simple but not trivial evolution equation to account for this is

$$\dot{c} = \frac{1}{J}\mathrm{div}(J\, d_{ca}\mathbf{C}^{-1}\nabla c) + q(v + v^*) - k_o c, \quad c|_{t=0} = c^*. \tag{16.47}$$

We set $v^* = c^* k_o/q$ in order to have $\dot{c}|_{t=0} = 0$ in correspondence of a zero initial state for the activation potential. Let us note that this equations contains a diffusive term that can account for possible calcium diffusion processes (Shiferaw and Karma, 2006); in our analysis, however, we have always set this term identically to zero, i.e. $d_{ca} \equiv 0$.

16.5 Numerical Simulations, Results and Discussion

We have solved the 3D electromechanical model via finite element methods, adopting a commercial software (COMSOL Multiphysics) running on parallelel multicore Intel machines with 8 Gb of physical memory. We have been interested in particular in showing as, even if in a qualitative way, our model is able to capture many of the relevant phenomena illustrated in the scientific literature.

We consider two parallelepiped-shaped domains having different aspect ratios: a thin one, with size $100 \times 100 \times 10$, and a thicker one, with size $25 \times 25 \times 12.5$, in dimensionless unit lengths. Regarding the mechanical problem, the two square bases have no kinematical constraints, while all the remaining rectangular faces have elastic boundary conditions (16.29); the diffusion problem has a zero-flux boundary condition everywhere. In order to compare with the experimental data, the dimensional time variable has been scaled so that specific dimensionless characteristics of our model fit with experimental observations. More in detail, we have calibrated the time scale in such a way that, as specified in the following section, our action potential duration (APD) matches the APD measured in (Elharrar and Surawicz, 1983), on canine ventricular cardiac fibers, a peculiar choice due to the large amount of data disposable on canine cardiac fibers.

As consequence, we assume that $t = (13.75\text{ms})\, t_{adim}$ (that is, $T_c = 13.75\text{ms}$) because a single APD at 90% of repolarisation takes approximately 24 dimensionless time units. Moreover, dimensional transmembrane voltages are obtained by mapping the dimensionless variable v, which ranges over the interval [0,1], onto a dimensional interval determined from the experimental study in Elharrar and Surawicz (1983), where the resting membrane potential of ventricular muscle fibers was measured, and the corresponding action potential amplitude V was in the range $(-85\,\text{m V}, 115\text{m}V)$. Consequently, we get the corresponding scaling for the action potential: $v = (-85 + v115)\text{mV}$.

Regarding the electrical parameters, those in equations (16.42)–(16.44) can be fine-tuned to account for the peculiarities of cardiac tissue; here, they have been selected in accordance with reference to (Nash and Panfilov, 2004). For calcium dynamics, we adopted the numerical choices in (Pelce and Sun, 1995).

Finally, our choice for the equilibrium voltage V_t corresponds to a depolarizing threshold around $-20\,\text{m V}$, and the stiffness of the elastic boundary κ has been chosen in order to obtain visible deformations as large as 20%. Table (16.1) at the end of the chapter summarizes the values for all our parameters.

16.5.1 General Features of the Model

We summarize here the results presented by the authors in previous studies, see (Cherubini et al., 2008) for details, before presenting some novel results. The performances of the electromechanical model have been tested when the tissue results uniformly stimulated, obtaining the APD restitution curve at 90% of repolarisation and the resulting electrical and mechanical activity when the period of the stimulus ranges from 0.3 to 1.5s. We have derived a dimensional map to scale our dimensionless time units by comparing specific features of our model with the experimental observations concerning the restitution properties of ventricular muscle fibers. In particular, once the parameters of the model have been assigned, we scaled our dimensionless time unit to match the pulse duration typical of the alternant instability derived in (Elharrar and Surawicz, 1983). Specifically, we scaled our dimensionless temporal results by 19.35ms when physiological comparisons are needed; it should be noted that the value of the time scaling is strongly influenced by the presence of the stretch-activated currents in the model.

At first, we have focused on target patterns for the thin geometry, i.e. radial waves of excitation which propagate symmetrically. The electromechanical activity due to a periodic current stimulation with period $T = 0.6$ s, and having cylindrical support with base radius equal to 10% of the side length of the patch is pictured in Fig. 16.2. The stimulus is applied at the center of the medium, and the resulting radial waves of excitation propagate symmetrically; the contracting ability of the medium allows for local deformations which can be appreciated at the boundary of the domain. Parameters of the model have been tuned in order to have a maximum contraction as large as 20%.

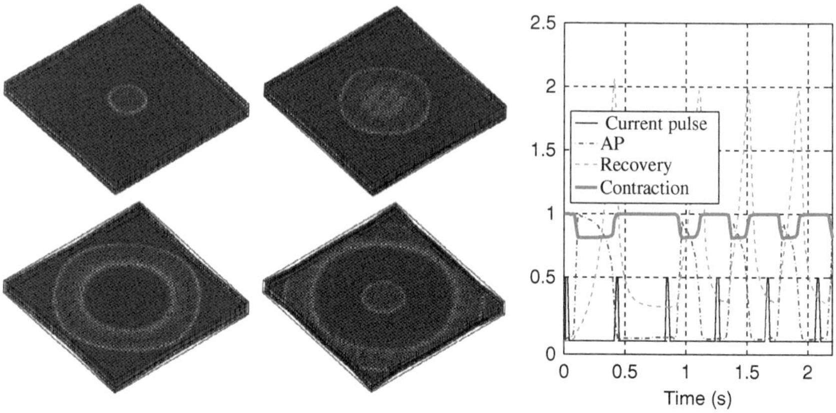

Fig. 16.2 Initiation of cylindrical waves via current pulse at the center of the patch. Cylindrical waves, once initiated, propagate towards the boundary; a second pulse, given after the appropriate time interval, yields a second wave identical to the previous one. On the left, we show four snapshots taken at different times; light colour denotes the zones of maximum activation, which coincide with maximum muscle contraction. On the right, we show AP, recovery variable and contraction versus time, probed at a point midway between the center and the boundary; *solid line* represents current pulse, solid thick represents the contraction.

Subsequently, we went on to the study of spiral wave dynamics. Spiral waves (in three dimensions scroll waves) arise naturally in reaction-diffusion systems of partial differential equations (see Winfree, 2001; Murray, 2008). If the domain is sufficiently thin, the scroll wave practically collapses into a spiral wave and the vortex line in a point. The motion of the organizing center (the spiral's 'tip' in two dimensions or the 'vortex line' in three dimensions) has tremendous implications for a clear understanding of the global dynamics of the system. The goal of many studies, both theoretical and experimental, is to locate the tip of the spiral (the scroll wave's filament in 3D) which moves on specific trajectories. Spiral waves are important in different biological contexts in particular: in heart tissue, for example, the motion of the spiral tip seems to be associated with specific types of arrhythmias while in neural tissues this motion can be related to epilepsy and to spreading a depression in the retina. Finding the location of the tip both in experiments and in numerical simulations is a nontrivial task. In numerical simulations the tip position is found via specific mathematical rules (a specific procedure, valid for two variable models only, is discussed in the following). In experiments, as an example, the tip is located studying the images taken from CCD devices and requiring the tip to be the point (in every frame) on which the normal velocity of the wave vanishes (Dahlem and Muller, 2000). This discussion however is valid for quasi 2D domains, because in 3D the filament is contained in the tissue and one can observe its effects on the boundary only. The lack of experimental proofs of vortex lines in biological media (in chemical ones a procedure due to Winfree, 1987 is available) gives

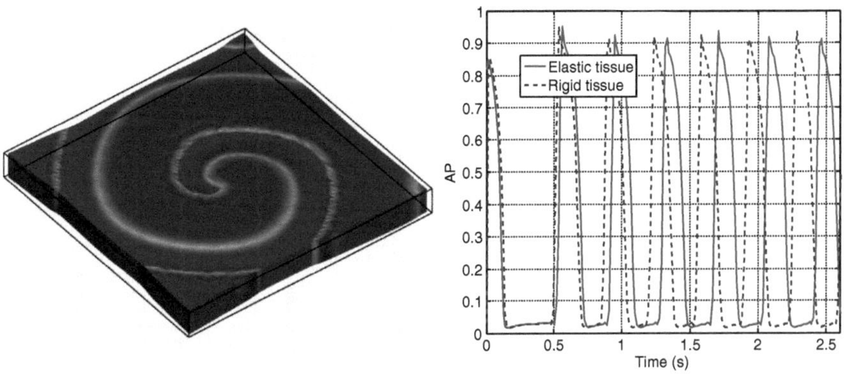

Fig. 16.3 Standing spiral wave on the deformable domain, colour coded with muscle contraction (*left*); action potential versus time probed at a given point for the elastic medium (*solid curve*), and for the rigid one (*dashed curve*).

a clear justification for the studies of the mathematical models. In 3D however, the geometrical structure of these objects is absolutely non-trivial: in biological context, as an example, it has been hypothesized that stationary scroll wave filaments in cardiac tissue describe a geodesic in a curved space whose metric is represented by the inverse diffusion tensor, with a dynamics close to cosmic strings in a curved universe (Verschelde et al., 2007). All these studies have been performed in absence of elastic deformations, so a radical change in the scroll wave dynamics induced by the MEF is clearly expected. In particular a stretching and twisting of the vortex line, with possible fragmentation and subsequent fibrillation should arise.

Our model can describe standing spiral and scroll waves which can be initiated using different techniques, as example, as a collision of two orthogonal planar waves. When moving on a deformable medium, spiral waves have different dynamics: spiral core drifts differently and the time period can change considerably; Fig. 16.3 shows a snapshot of a standing waves on an contractile tissue (left), and time evolution of action potential versus time for elastic and rigid medium.

16.5.2 Stretch Induced Currents

The effect of stretch activated current can be best appreciated with stretch induced activation of the muscular activity. The main experimental link created with these simulations was in relation to the work by Franz et al., (1992) where a left ventricle of a whole heart electrically at rest is activated through periodic volume changes of increasing amplitude. Precisely, they showed the existence of a threshold value of the strain above which an action potential can be initiated. Our model is endowed with such a feature: a tissue electrically at rest can be acti-

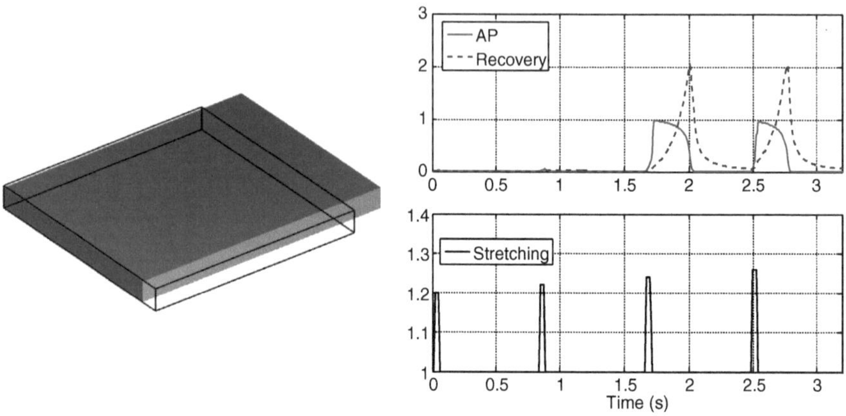

Fig. 16.4 Stretch induced activation. Picture on the left show the patch stretched along the longitudinal direction; transversal and vertical contraction is due to the Poisson effect. On the right, we plot stretching history and consequential action potential and recovery variable. Activation is clearly visible as soon as a stretch threshold of $\simeq 24\%$ is reached

vated by imposing a large mechanical stretch with the appropriate timing, see Fig. (16.4); numerical results are qualitatively in agreement with aforementioned experimental study.

Stretch activated current seems to be responsible of pressure-induced spiral termination, too. It is well known how a current pulse can terminate a self sustaining spiral wave; having at disposal a full 3D model, we can investigate mechanical-induced spiral termination. Considering the thin 3D domain, we simulated the effect of a large pressure applied on the top side of the patch in order to produce large out-of-plane deformation, see Fig. (16.5). The large displacements due to the applied load, together with the contraction due to the travelling wave, generate large currents who can terminate the self sustaining phenomenon.

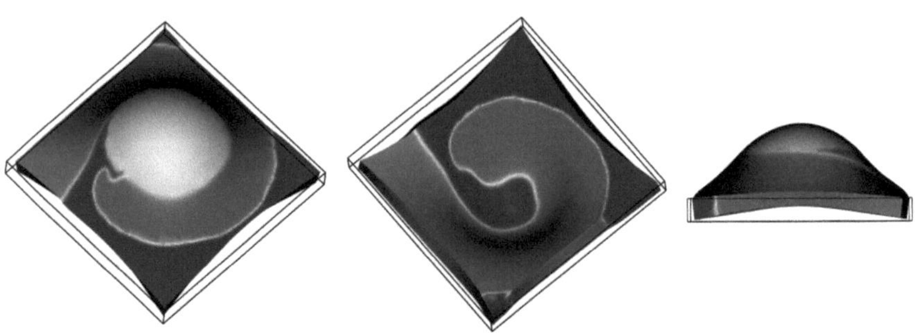

Fig. 16.5 Stretch induced spiral termination. Three different views (*bottom, top, side*) of the patch undergoing large out-of-plane deformation. Colour map shows level of activation. The large clear zone on the convex side denote the presence of large currents deactivating action potential

16.5.3 Vortex Lines

We conclude our work focusing on thick three dimensional domains, in order to observe possible novel effects of elasticity on scroll wave filaments. In particular, to track the evolution of these filaments on the moving domain, we use the function

$$\Phi = |\nabla v_s \times \nabla w_s| = |\mathbf{F}^{-T}\nabla v \times \mathbf{F}^{-T}\nabla v|, \qquad (16.48)$$

whose level surfaces are highly localized at the vortex core. Definition (16.48) coincides with the function proposed in Sutcliffe and Winfree (2003) to tracks scroll waves on a rigid medium, that is, for $\mathbf{F} = \mathbf{I}$.

We have started our simulations eliminating completely any elastic contribution, i.e. we have analyzed a pure electrophysiological problem described simply by the two variable reaction diffusion systems of PDEs. As a result, in Fig. (16.6) we show, at a specific time, a formed scroll wave on the rigid domain, while in Fig. (16.7) the same simulation shows the action potential on a horizontal cross section, superimposed with iso-surfaces of Φ representing the scroll wave filament. Then, we activated the contractility of the tissue, and did other simulations using the same electrical initial conditions as those used for the rigid case. The results of this extended case are shown in Figs. 16.7 and 16.8. It's evident that the domain appears deformed by the MEF. As expected, contractility slightly deforms the vortex line. It appears interesting now to see if strong pressure load can affect more impressively the vortex line dynamics. To this aim we have saved the already formed elastic scroll wave of the previous simulation as initial data and we have given then a strong pressure load on the upper face. As a consequence the spiral starts to feel the activation of the stretch activated currents on the stretched part of the medium; in particular, the spiral starts to bend and eventually fragment, as shown by the asymmetry of the various V = constant iso-surfaces. In Fig. (16.7) the striking effect of the pressure load on the vortex filament is shown. The organizing center is stretched as consequence of the SACs. This fact appears evident from the bottom z = *constant* voltage surface

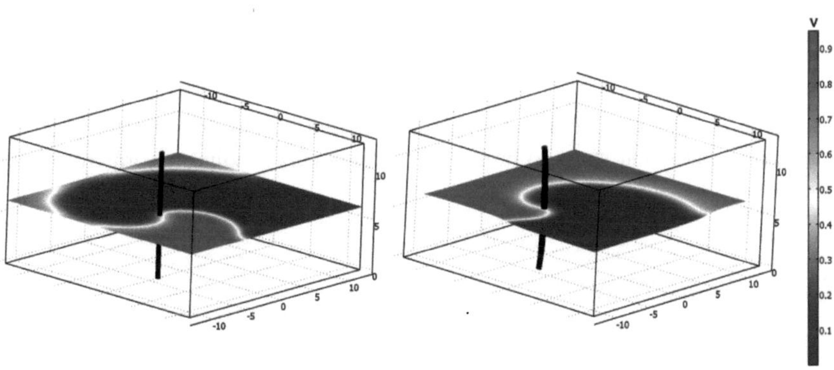

Fig. 16.6 Spiral core in a rigid medium

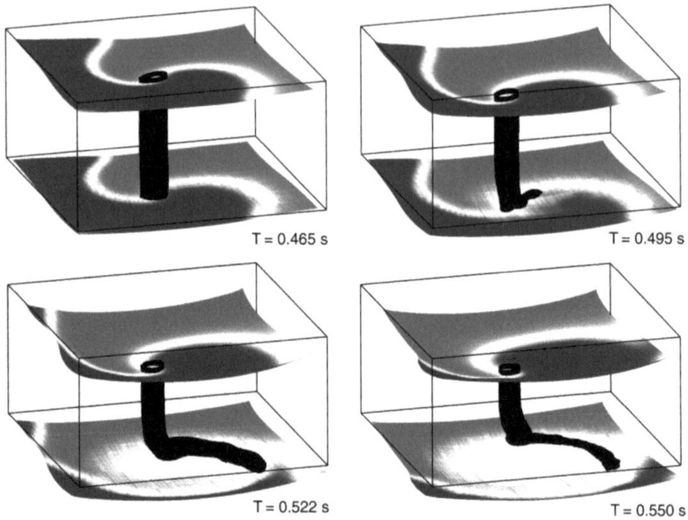

Fig. 16.7 Evolution of spiral core in an elastic medium under pressure load

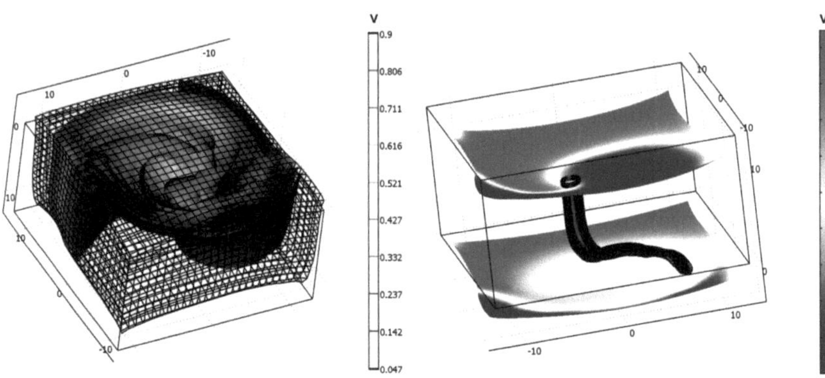

Fig. 16.8 Isosurface for action potential (*left*) and spiral core (*right*) in an elastic medium under pressure load

on which the spiral wave activity has disappeared due to the refractoriness. The filament in particular starts to break down, eventually tending towards fibrillating scenarios.

16.5.4 What's Next?

Some 30 pages have been just spent to present a minimal way to model an excitable cardiac tissue. Our research was focused on describing the possible effects of mechanics on excitation patterns using a very basic (but extendable) mathematical model. On the other hand, a real heart is a complicated anatomical structure: clearly any model should be improved to take into account inhomogeneous and anisotropic muscle fiber distribution which clearly affects both electrical and mechanical dynamics. At physiological microscopic level moreover the cardiac tissue is extremely complex, being constituted by different types of cells. Moreover a complex network of high conduction fibers, the Purkinje one, makes the global propagation of electric current absolutely non trivial. At this point every good theorist is forced then to use a sort of 'Occam's razor' and decide what are the elements to include in the model, in order to maintain the computational requirements at a reasonable level. On this line, we are aware that the model here presented lacks of accuracy regarding the biophysics of specific ionic currents, due to the use of two-variables equations. To this aim the entire ionic equations should be extended to many variable theories like Beeler–Reuter (Beeler and Reuter, 1977) of Luo–Rudy (Luo and Rudy, 1991) or other equation sets, (see Pullan et al. (2005) for a complete discussion). We must remark moreover that the calcium response to the action potential of our work shows no delay, which is clearly not in agreement with physiology. Regarding the mechanical properties of the medium on the other hand, cardiac tissue clearly requires much more complicated constitutive relations than those here adopted, taking into account that there can exist possible viscoelastic effects moreover. While all these points appear to be relevant for the physiologist, from the point of view of the physicist or of the engineer, the two-variable model here presented, with its simplicity, still appears to be extremely useful. Many classical effects studied in absence of elasticity like resonant, inhomogeneity induced,

Table 16.1 Parameters used in the model

Parameter	Value	Description
d_0	1	AP diffusivity
a	0.1	Control of AP threshold
b	0.05	Control of recovery threshold
k	8	Transmembrane current amplitude
ε	0.01	Control for recovery phase
μ_1	0.12	Control for recovery phase
μ_2	0.3	Control for recovery phase
G_{sac}	5	SAC conductivity
δ	60	Control for SAC saturation
λ_{ref}	2.2	Strain threshold
$\bar{v}$	0.56	Depolarizing threshold
α_1	1	Elastic modulus
α_2	0	Elastic modulus
κ_o	0.1	Stiffness of boundary constraint
γ_o^{max}	0.8	Maximum muscle contraction
q	2	Ca^+ amplification factor
k_0	3.5	Gauge of the sink term for Ca^+ dynamics
c_o^*	1	Basal Ca^+ concentration
β	6	Parameter of the curve $c \mapsto \gamma_o$

anisotropy induced and boundary induced drifts as well as studies of interaction of spirals and pinning/unpinning problems should be re-analized again before passing to more complicated models. There is still enough work to do for anyone for the next years before passing to more complicated physiological models.

References

Abbott, B.C., Hill, A.V., Howarth, J.V.: The positive and negative heat production associated with a nerve impulse. Proc. R. Soc. Lond. B **148**, 347 (1958)

Aliev, R.R., Panfilov, A.: A simple two-variable model of cardiac excitation. Chaos, Solitons Fractals **7** (**3**), 293–301 (1996)

Aliev, R.R., Richards, W., Wikswo, J.P.: A Simple nonlinear model of electrical activity in the intestine. J. Theor. Biol. **204**, 21–28 (2000)

Allen D.G., Eisner D.A., Orchard C.H.: Characterization of oscillations of intracellular calcium concentration in ferret ventricular muscle. J. Physiol. **352**, 113–28 (1984)

Ambrosi, P., Habib, G., Kreitmann, B., Faugere, G., Metras, D.: Valsalva manoeuvre for supraventricular tachycardia in transplanted heart recipient. Lancet **346**, 713 (1995)

Ashihara, T., Trayanova, N., Nakazawa, K., Yamazaki, M., Honjo, H., Sakuma, I., Kamiya, K., Kodama, I.: Spiral wave control by regional cooling in a bidomain model. Heart Rhythm **2**, 220 (2005)

Beeler, G.W., Reuter, H.: Reconstruction of the action potential of ventricular myocardial fibres. J. Physiol. **268**, 177–210 (1977)

Bini, D., Cherubini, C., Filippi, S.: Heat transfer in FitzHugh–Nagumo models Phys. Rev. E **74**, 041905 (2006)

Cazorla, O., Pascarel, C., Brette, F., Guennec, J.Y.L.: Modulation of ion channels and membrane receptor activities by mechanical interventions in cardiomyocytes: possible mechanisms for mechanosensitivity. Prog. Biophys. Mol. Biol. **71**, 29–58 (1999)

Cherubini, C., Filippi, S., Nardinocchi, P., Teresi, L.: An electromechanical model of cardiac tissue: Constitutive issues and electrophysiological effects. Prog. Biophys. Molec. Biol. **97**, 562–573 (2008)

Dahlem, M.A., Muller, S.C.: Image processing techniques applied to excitation waves in chicken retina. Methods **21**, 317 (2000)

Dean J.W., Lab M.J.: Arrhythmia in heart failure: role of mechanically induced changes in electrophysiology. Lancet **1**, 1309–1311 (1989a)

Dean J.W., Lab M.J.: Regional changes in ventricular excitability during load manipulation of the in situ pig heart. J. Physiol. **429**, 387–400 (1990)

DiCarlo, A., Quiligotti, S.: Growth & balance. Mech. Res. Commun. **29**, 449–456 (2002)

Di Diego J.M. and Antzelevitch C.: High [Ca2+]o-induced electrical heterogeneity and extrasystolic activity in isolated canine ventricular epicardium. Phase 2 reentry. Circ. **89**, 1839–1850 (1994)

Elharrar V., Surawicz, B.: Cycle length effect on restitution of action potential duration in dog cardiac fibers. Am. J. Physiol. **244**, 782–792 (1983)

FitzHugh, R.: Impulses and physiological states in theoretical models of nerve membrane. Biophys. J. **1**, 445–465 (1961)

Franz, M.R., Burkhoff, D., Yue, D.T., Sagawa, K.: Mechanically induced action potential changes and arrhythmia in isolated and in situ canine hearts. Cardiovasc. Res. **3**, 213–223, (1989)

Franz, M.R., Cima, R., Wang, D., Profitt, D., Kurz, R.: Electrophysiological effects of myocardial stretch and mechanical determinants of stretch-activated arrhythmia. Circ. **86**, 968–978 (1992)

Franz M.R.: Mechano-electrical feedback in ventricular myocardium. Cardiovasc. Res. **32**, 15–24 (1996)

Gurtin, M.E.: An Introduction to Continuum Mechanics, first edn. Academic Press, New York (1981)

Heimburg, T., Jackson, A.D.: On soliton propagation in biomembranes and nerves, P.N.A.S. **102**, 9790 (2005)

Hodgkin, A.L., Huxley, A.F.: A quantitative description of membrane current and its application to conduction and excitation in nerve. J. Physiol. **117**, 500–544 (1952)

Hodgkin, A.L., Katz, B.: The effect of temperature on the electrical activity of the giant axon of the squid. J. Physiol. **109**, 240 (1949)

Howarth, J.V., Ritchie, J.M., Stagg, S.: The initial heat production in garfish olfactory nerve fibers. Proc. R. Soc. Lond. B. **205**, 347 (1979)

Hu, H., Sachs, F.: Mechanically activated currents in chick heart cells. J. Membr. Biol. **154**, 205–216 (1998)

Hunter, P.J., McCulloch, A.D., ter Keurs, H.E.D.J.: Modelling the mechanical properties of cardiac muscle. Prog. Biophys. Molec. Biol. **69**, 289–331 (1998)

Keener, J., Sneyd, J.: Mathematical Physiology. Springer New York (2001).

Kohl, P., Hunter, P., Noble, D.: Stretch-induced changes in heart rate and rhythm: clinical observations, experiments and mathematical models. Prog. Biophys. Molec. Biol. **71**, 91–138 (1999)

Kohl, P., Noble, D.: Mechanosensitive connective tissue: potential influence on heart rhythm. Cardiovasc. Res. **32**, 62–68 (1996)

Kohl, P., Ravens, U.: Cardiac mechano-electric feedback: past, present, and prospect. Prog. Biophys. Mol. Biol. **82**, 3–9 (2003)

Kohl, P., Sachs, F.: Mechanoelectric feedback in cardiac cells. Phil. Trans. R. Soc. Lond. A **359**, 1173–1185 (2001)

Kouwenhoven, W.B., Milnor, W.R., Knickerbocker, G.G., Chesnut, W.R.: Closed chest defibrillation of the heart. Surgery **42**, 550–561 (1957).

Kouwenhoven, W.B., Jude, J.R., Knickerbocker, G.G.: Closed chest cardiac massage. JAMA **173**, 1084–1087 (1960).

Kröner, E.: Allgemeine kontinuumstheorie der versetzungen und eigenspannungen. Arch. Rational Mech. Anal. **4**, 273–334 (1959).

Lab M.J.: Transient depolarization and action potential alterations following mechanical changes in isolated myocardium. Cardiovasc. Res. **14**, 624–637 (1980)

Lab M.J., Seed W.A.: Pulsus alternans. Cardiovasc. Res. **27**, 1407–1412 (1993)

Lab, M.J.: Contraction-excitation feedback in myocardium. physiological basis and clinical relevance. Circ. Res. **50**, 757–766 (1982)

Lee, E.: Elastic-plastic deformation at finite strain. J. Appl. Mechanics **36(1)** (1969)

Luo, C.H., Rudy, Y.: A model of the ventricular cardiac action potential. Depolarization, repolarization, and their interaction. Circ Res **68**, 1501–26 (1991).

Merodio, J., Ogden, R.W.: Mechanical response of fiber-reinforced incompressible non-linearly elastic solids. Int. J. Non-Lin. Mech. **40**, 213–227 (2005)

Murray, J.D.: Mathematical Biology I-II, 3rd edn Springer New York (2008).

Nagumo, J.S., Arimoto, S., Yoshizawa, S.: An active pulse transmission line simulating nerve axon. Proc. I.R.E. **50**, 2061–2071 (1962)

Nardinocchi, P., Teresi, L.: On the active response of soft living tissues. J. Elasticity **88**, 27–39 (2007)

Nash, M.P., Hunter, P.: Computational mechanics of the heart: from tissue structure to ventricular function. J. Elasticity **61 (1/3)**, 113–141 (2000)

Nash, M.P., Panfilov, A.V.: Electromechanical model of excitable tissue to study reentrant cardiac arrhythmia. Prog. Biophys. Molec. Biol. **85**, 501–522 (2004)

Nelaton, A.: Elements de Pathologie Chirurgicale, **Vol. 4**, Librairie Germer Bateliere et Co., Paris (1876)

Noble, D.: A modification of the Hodgkin–Huxley equations applicable to purkinje fibre action and pacemaker potentials. J. Physiol. **160**, 317–352 (1962)

Opie L.H.: Pharmacological differences between calcium antagonists. Eur. Heart. J. **18**, 71–79 (1997)

Panfilov, A.V., Keldermann, R.H.: Self-organized pacemakers in a coupled reaction-diffusion-mechanics system. Phys. Rev. Lett. **95 (25)**, 258104/1–258104/4 (2005)

Pelce, P., Sun, J.: A simple model for excitation-contraction coupling in the heart. Chaos Solitons Fractals **5**, 383–391 (1995)

Phillips, J.H., Burch, G.E.: Management of cardiac arrest. Am. Heart. J. **67**, 265–77 (1964)

Pullan, A.J., Buist, M.L. and Cheng, L.K.: Mathematically Modeling Electrical Activity of the Heart: From Cell to Body Surface and Back. World Scientific Hacken Sack NJ (2005).

Pye, M.P., Cobbe, S.M.: Arrhythmogenesis in experimental models of heart failure: the role of increased load. Cardiovasc. Res. **32**, 248–257 (1996)

Reiter, M.J., Landers, M., Zetelaki, Z., Kirchhof, C.J.H., Allessie, M.A.: Electrophysiological effects of acute dilatation in the isolated rabbit heart cycle length? Dependent effects on ventricular refractoriness and conduction velocity. Circulation **96**, 4050–4056 (1997)

Rodriguez, E., Hoger, A., McCulloch, A.: Stress-dependent finite growth in soft elastic tissues. J. Biomechanics **27**, 455 (1994)

Rogers, J.M., McCulloch, A.D.: A collocation–galerkin finite element model of cardiac action potential propagation. EEE Trans. Biomed. Eng. **41**, 743–757 (1994)

Sachs, F.: Biophysics of mechanoreception. Membrane Biochem. **6**, 173–195 (1986)

Sachse, F.B.: Computational Cardiology: Modeling of Anatomy, Electrophysiology, and Mechanics, 1st edn. Springer-Verlag, Berlin DE (2004)

Schott, E.: Elektrokardiographische Studien bei akuten Vergiftungen, Dt. Arch. Klin. Med. **131**, 211–229 (1920)

Shiferaw, Y., Karma, A.: Turing instability mediated by voltage and calcium diffusion in paced cardiac cells. P.N.A.S. **103**, 5670–5675 (2006)

Sideris, D.A., Toumanidis, S.T., Tselepatiotis, E., Kostopoulos, K., Stringli, T., Kitsiou, T., Moulopoulos, S.D.: Atrial pressure and experimental atrial fibrillation. Pacing Clin. Electrophysiol. **18**, 1679–85 (1995)

Sigurdson, W.J., Ruknudin, A., Sachs, F.: Calcium imaging of mechanically induced fluxes in tissue-cultured chick heart: role of stretch-activated ion channels. Am. J. Physiol **262**, H1110–H1115 (1992)

Spencer, A.J.M.: Deformations of Fiber-Reinforced Materials. Oxford University Press, Oxford (1972).

Sutcliffe, P.M., Winfree, A.T.: Stability of knots in excitable media. Phys. Rev. E **68**, 016218-5 (2003)

Taber, L.A.: Biomechanics of growth, remodeling, and morphogenesis. Appl. Mech. Rev. **48**, 487–545 (1995)

Tasaki, I., Byrne, P.M.: Heat production associated with a propagated impulse in bullfrog myelinated nerve fibers. Jpn. J. Physiol. **42**, 805 (1992).

Usyck, T.P., LeGrice, I.J., McCulloch, A.D.: Computational model of three-dimensional cardiac electromechanics. Comput. Visual. Sci. **4**, 249–257 (2002)

Usyck, T.P., Mazhari, R., McCulloch, A.D.: Effect of laminar orthotropic myofiber architecture on regional stress and strain in the canine left ventricle. J. Elasticity **61**, 143–164 (2000)

Verschelde, H., Dierckx, H., Bernus, O.:Covariant stringlike dynamics of scroll wave filaments in anisotropic cardiac tissue. Phys. Rev. Lett. **99**, 168104 (2007)

Wei, J.Y., Greene, H.L., Weisfeldt, M.L.: Cough-facilitated conversion of ventricular tachycardia. Am. J. Cardiol. **45**, 174–176 (1980)

White, E., Guennec, J.Y.L., Nigretto, J.M., Gannier, F., Argibay, J.A., Garnier, D.: The effects of increasing cell length on auxotonic contractions; membrane potential and intracellular calcium transients in single guinea-pig ventricular myocytes. Exp. Physiol. **78**, 65–78 (1993)

Winfree, A.T.: When Time Breaks Down: The Three-Dimensional Dynamics of Electrochemical Waves and Cardiac Arrhythmias. Princeton University Press, Princeton, NJ (1987)

Winfree, A.T.: The Geometry of Biological Time. Springer-Verlag, Berlin DE (2001)

Yamashita T., Inoue H., Usui M., Kuo T., Saihara S., Nozaki A., Momomura S., Serizawa T., Iizuka M., Sugimoto T.: Acute effect of percutaneous transluminal mitral commissurotomy on QT interval: possible role of afterload in contraction-excitation feedback. Am. Heart. J. **121**, 1634–1639 (1991)

Zoll, P.M.: Resuscitation of the heart in ventricular standstill by external electrical simulation. New Engl J. Med. **247** 768–771 (1952)

Part IV
Arteries as a Source of Myogenic Contractile Activity: Ionic Mechanisms

Chapter 17
Specific Mechanotransduction Signaling Involved in Myogenic Responses of the Cerebral Arteries

Koichi Nakayama, Kazuo Obara, Tomohisa Ishikawa, and Shigeru Nishizawa

Abstract The cerebral artery is known to be particularly sensitive to mechanoreception in the form of blood pressure, blood flow, and other hemodynamic forces. Stretching and intraluminal pressurization, which might mimic an acute and/or chronic change in blood pressure, induce many different responses, including contraction, activation of various kinases and ionic channels, production of various vasoactive substances, gene expression, and phenotype changes. Here, we briefly discuss specific mechanotransduction signaling pathways involved in the myogenic responses of cerebral arteries. We emphasize that it is important to recognize mechanical forces and control them not only to improve our knowledge of cardiovascular system in physiologic and pathophysiologic conditions but also for the development of new therapeutic drugs.

Keywords Cerebral artery · Mechanical stretch · Contraction · Signal transduction · Cerebral vasospasm

17.1 Introduction

The contractile reaction of the vascular smooth muscle in response to mechanical stretch/pressure, first reported by Bayliss (1902) in a study on an isolated segment of the canine carotid artery, has often been postulated as an intrinsic mechanism for the control of blood flow. The cerebral circulation supplies an extremely sensitive organ whose adequate perfusion is indispensible for the survival of individual organisms. While the brain accounts for only 2% of the body weight, it receives 15% of cardiac output and requires about 20% of total body O_2 consumption (Heistad and Kontos, 1982). In order to supply this demand, in addition to neurogenic and

K. Nakayama (✉)
Department of Molecular and Cellular Pharmacology, Faculty of Pharmaceutical Sciences, Iwate Medical University, Yahaba, Iwate, 028-3694, Japan
e-mail: nakyamk@iwate-med.ac.jp

A. Kamkin, I. Kiseleva (eds.), *Mechanosensitivity of the Heart*,
Mechanosensitivity in Cells and Tissues 3, DOI 10.1007/978-90-481-2850-1_17,

chemical controls of blood flow, as are often encountered in other vascular beds, the cerebral circulation has many specific regulatory features. One of these features is the extreme mechanosensitivity of the cerebral arteries, which has been well documented. For example, taking the magnitude of myogenic contractile response to mechanical stretch in the rabbit coronary artery as 1.0, the relative contractile activities of rabbit basilar and renal arteries have been reported as 6.7 and 2.1, respectively (Nakayama, 1982).

The myogenic contractile response to mechanical stretch is regarded to be a basic mechanism for the regulation of blood blow (myogenic tone) under physiological conditions. Furthermore, excessive mechanical reaction in response to stretch/pressure has been considered to be related to the etiology of vasospastic contractions, such as systemic (Intengan and Schiffrin, 2000) and pulmonary hypertensions (Tanabe et al., 2000; Schermuly et al., 2005) and cerebral vasospasm (Nakayama et al., 2003). Nevertheless, the mechanisms for mechanotransduction in vascular tissues still remain to be elucidated.

The aim of the present review, which is based on our series of published papers and those of others, is to provide new insights into the mechanotransduction of cerebral vascular tissues, i.e., how the cell/tissue senses mechanical force, and transduces it into mechanical and other events. Our discussions examine the following topics: (1) structural and functional characteristics of the cerebral arteries in the source of stretch/pressure-induced contraction; (2) ionic mechanisms for myogenic contractile response to mechanical stretch; and (3) multiple phosphorylation of 20-kDa myosin light chain of cerebral artery smooth muscle cells as a self-inhibitory mechanism in stretch-induced contraction. Spatial and temporal interactions of mechanosensitive kinase molecules, including Rho/Rho-kinase, protein kinase C, and tyrosine kinase, in response to mechanical stress are also discussed. Interestingly, these mechanosensitive kinase molecules are also activated in experimental canine cerebral vasospasm after subarachnoid hemorrhage (SAH), and seem to play an important role in the development and maintenance of the vasospasm. Accordingly, we also address (4) the relevant mechanism underlying stretch-induced contraction and cerebral vasospastic episodes. The present review suggests that continued study of mechanotransduction in vascular tissues, including cerebral arteries, would help to underscore the importance of biomechanical factors under physiological and pathological conditions, and could have major therapeutic consequences.

17.2 Cerebral Arteries as a Source of Myogenic Contractile Activity

In order to answer the question, "Why are cerebral arteries highly mechanosensitive?", we should first consider several structural and functional aspects of the cerebral arteries.

17.2.1 Distinctive Role of the Large Cerebral Arteries

In the studies of myogenic contractile responses to mechanical stresses such as stretch and pressurization, large cerebral arteries such as the basilar artery and middle or posterior cerebral artery isolated from various animal species, including dogs, rabbits, guinea-pigs, rats, and mice, are often utilized (see reviews: Davis and Hill, 1999; Meininger and Davis, 1992; Osol, 1995). The important role of large brain vessels in the total cerebral vascular resistance may be attributable to at least two causes (Heistad and Kontos, 1982). The first is that large brain vessel dilation may prevent or reduce the "steal" phenomenon that occurs during focal increases in metabolism. The second is that large brain vessels may have a damping effect on blood pressure fluctuation: the responses of large arteries may reduce pressure fluctuations of small arteries during changes in aortic pressure. Thus, constriction of the large arteries during hypertension and dilation of the arteries during hypotension may attenuate changes in microvascular pressure. Branches of the posterior cerebral arteries with an outer diameter between 100 and 200 μm have also been preferably used to study the effect of intra-luminal pressurization by arteriography (Halpern et al., 1984). In addition to their ease of use as experimental preparations, these arterial segments have several other advantages. These include their physiological characteristics. That is, due to the special nature of the cerebral circulation, large stem arteries account for a greater portion of vascular resistance in the brain than do other vascular beds. For instance, in monkeys, dogs, and cats, the segment from the aorta to the pial artery accounts for 10–50% of the total vascular resistance (Symon, 1967; Heistad and Kontos, 1982). Three pairs of arteries, the anterior, middle, and posterior cerebral arteries, originated from the circle of Willis, transverse the convex surface of the cerebrum, and supply the boundary zones of the cerebral cortex, while the basilar artery supplies the brain stem and the cerebrum. Numerous studies have demonstrated that all of these areas are particularly vulnerable to ischemic and hyper- and hypotensive insults and hemorrhage. Thus, the study of myogenic activity in large and small cerebral arteries, such as the basilar artery and the middle or posterior cerebral artery, is also important from a clinical point of view.

17.2.2 Development of the Circumferential Muscle Layer

As mentioned above, the stem cerebral artery accounts for a large portion of the total cerebral vascular resistance, which suggests that the medial smooth muscle layer of the cerebral artery is well developed. Light and electron microscopic studies in the stem portion of canine basilar arteries have shown that the total thickness of the arterial wall is about 100 μm (Nakayama, 1988) and the medial layer consists of around 12 muscle cell layers in average and is about 80 μm thick. Thus, the medial layer is about 70% of the total thickness. Furthermore, in cross sections, a spindle-shaped nucleus was located at the center of each smooth muscle cell and was oriented along the long axis of the cell. The cross sections also showed that the smooth muscle layer was in a predominantly circumferential orientation. In longitudinal sections, the nucleus appeared rather round in shape. Finally, the scanning microscopic studies indicated that the endothelial monolayer was intact in the control preparation, whereas the endothelial cells were completely removed by mechanically scraping the luminal surface with a stainless steel rod.

The human cerebral artery has been reported to have highly oriented medial muscle layers aligned circumferentially (Peters et al., 1983) and to have a low pitch lining of less than 50 (Walmsley and Canham, 1979). Thus, the human cerebral artery is efficient in producing strong contractile tension, which seems to be in accord with the high myogenic activity and the phenomenon known as "sausage-string" or "bead response" (MacKenzie et al., 1976) and may also be related to the cerebral vasospasm frequently encountered after subarachnoid hemorrhage.

The special nature of the cerebral artery has also been determined, i.e., the tension developed in segments of the basilar artery of dogs is strongly dependent on the angle of cut (Nakayama, 1988). A segment cut at 0°, or a ring segment, is the most efficient for developing tension, whereas segments longitudinally sectioned along their long axis (= 90°) show no apparent active tension in spite of an enormously large increase in passive tension. Histological sections stained with a nuclear contrasting stain, such as hematoxylin, can be used to determine nuclear orientation, which is a good indicator of cellular orientation (Canham, 1977). Active tension develops along the long axis of the smooth muscle cell, and the medial muscle layer is predominantly oriented circumferentially in the dog basilar artery. This appears to be similar to a pathophysiological condition in which the cerebral artery in the late stage of vasospasm after SAH shows a very high stiffness despite its low level of active tension (Koide et al., 2002). Vasospastic arteries appear to be subjected to phenotype changes involving a corrupt lining of smooth muscle cells (Yamaguchi-Okada et al., 2005).

17.2.3 Ultrastructure and Localization of Activator Ca^{2+}

An electron micrograph showing the cross-section of the rabbit basilar artery smooth muscle cells has indicated that a number of bottle-shaped invaginations, i.e., caveola, and many vesicles, possibly the sarcoplasmic reticulum (SR) and so-called superficial storage sites of Ca^{2+} (Laporte et al., 2004), are present underneath the plasma membrane (Nakayama et al., 1986). The precipitate of electron-opaque pyroantimonate, a valid measure of intracellular Ca^{2+} localization (Mizuhira, 1976), in resting cerebral artery myocytes was observed mostly along the inner surface of the plasma membrane. In contrast, the examination of muscle cells fixed during the contraction produced either by membrane depolarization by high K^+ and electrical stimulation or by stretch stimulation has shown that the precipitate is diffusely distributed in the myoplasm in the form of small particles, while the precipitate around the plasma membrane is markedly decreased (Nakayama et al., 1986). Simultaneous recording of intracellular Ca^{2+} concentration and mechanical activity in the cerebral artery has also shown a similar sequence of events (Nakayama and Tanaka, 1989, 1993; Obara et al., 2001).

The localization of Ca^{2+} along the inner surface of the plasma membrane, together with its translocation into the myoplasm during contraction, has already been observed in various kinds of smooth muscles (Suzuki and Sugi, 1989; Laporte et al., 2004). Our study has demonstrated that the rate of stretch is a determinant factor in the recruitment of contractile activator Ca^{2+} from intracellular storage sites. It seems possible that the Ca^{2+}-induced Ca^{2+}-release mechanism and the production of inositol 1,4,5-trisphosphate (IP_3) are both dependent on the rate of stretch (Tanaka et al., 1994a, b).

If mechanical stretch is also a stimulus for Ca^{2+} release in the basilar artery, deformation or strain should act more effectively on the outer surface of the cells than on the inner parts, resulting in greater Ca^{2+} release from the more superficial internal layer of the plasma membrane in response to mechanical stretch. Furthermore, superficial SR can act as a Ca^{2+} buffer and also as a regulator of membrane channels and transporters (Laporte et al., 2004). In this connection, it should be emphasized that the inner surface of the plasma membrane is a storage site for Ca^{2+}, which can be most readily and directly influenced not only by membrane potential changes but also by membrane deformation by stretch to release activator Ca^{2+} in a graded manner.

17.2.4 Cell-to-Cell Communication Among Medial Smooth Muscle Cells

As to the cell-to-cell communication between smooth muscle cells, various types of junctions, including simple or close apposition, desmosome, and gap junction, have been well documented (Macdonald and Weir, 2001). In particular, gap junctions between adjacent plasma membranes are only 2–3 nm, and these gaps consist of (1) a low-resistance pathway of electronic signaling, and (2) a route for direct diffusion of substances with a low molecular weight.

Accumulated evidence shows that the direct communication between smooth muscle cells via gap junction channels plays an important role in the efficient coordination of myogenic activation of cerebral arteries and arterioles (Lagaud et al., 2002). Furthermore, the augmentation of gap junction channel activity has been shown to be involved in cerebral vasospasm after SAH (Hong et al., 2008). Of the gap junction channel components, connexin 43 (Cx43) has been found ubiquitously in vascular smooth muscle cells. In addition, Cx40, Cx37, and other as yet unidentified connexins are expressed in rat basilar arteries (Yamazaki and Kitamura, 2003). Although no data are available yet regarding the connexin molecules specifically related to myogenic activity, gap junction disrupters such as heptanol, 18α-glycyrrhetinic acid, and carbenoxylone in suitable dose ranges have been used in experimental pharmacological and therapeutic studies. Heptanol and 18α-glycyrrhetinic acid have been shown to reduce pressure-induced constriction in the rat middle cerebral artery without any effects on the membrane potential or on depolarization-induced constriction, indicating that these disrupters are unlikely to affect non-junctional ion channels (Lagaud et al., 2002). More recently, it has been reported that heptanol and carbenoxolone also significantly inhibit the experimental cerebral vasospasm in vivo and oxyhemoglobin-induced contraction of the basilar artery in rabbits (Hong et al., 2008). Taken together, these results suggest that gap junctions make efficient pathways for intercellular communication (e.g., movement of key molecules such as PKC, Ca^{2+}, and other ions) between cerebral artery smooth muscle cells during myogenic contraction and development of cerebral vasospasm.

17.3 Ionic Mechanisms for Myogenic Contractile Responses to Mechanical Stretch in the Cerebral Arteries

The myogenic constriction induced by increased intraluminal pressure or tissue stretch is proposed to be initiated by the following sequence of events: (1) stretch-induced vascular smooth muscle cell (VSMC) depolarization, (2) opening of voltage-dependent Ca^{2+} channels (VDCC), (3) a global increase in cytosolic Ca^{2+} concentration ($[Ca^{2+}]_i$), and (4) myosin light chain phosphorylation (Davis and Hill, 1999). There are several methods for the application of mechanical stress to blood vessels at the cellular and tissue levels. Of these, mechanical stretching by magnetic pulling devices and intraluminal pressurization by arteriography are often used to produce myogenic contraction in vascular tissues. In dispersed vascular smooth muscle cells, mechanosensitive ion channels are investigated by application of negative or positive pressure to the patched membrane or by application of hyposmotic challenge to induce a volume increase. It is, however, uncertain whether these stimulations mimic the mechanical stretching of tissues under physiological conditions. Thus, investigations are required to ascertain whether these ion channels function as a mechanosensor in intact tissues. In this review, we summarize mechanosensitive ion channels involved in the regulation of myogenic constriction, primarily based on our recent studies.

17.3.1 Simultaneous Recording of Stretch-Induced Contraction and Ca^{2+} Signals

A typical example of mechanical activity and corresponding Ca^{2+} signals in response to quick stretch in canine cerebral arteries is depicted in Fig. 17.1 (Nakayama and Tanaka, 1989). The initial tension rise coincident with stretch and subsequent fall in tension at the completion of stretch was followed by a delayed contraction, which reached the maximum within about 10 sec. The contractile response was maintained or gradually decreased during stretch for 30 sec. The $[Ca^{2+}]_i$ measured with a Ca^{2+} indicator fura-2 began to increase 1–2 sec before the delayed contraction and reached the maximum within 5 sec.

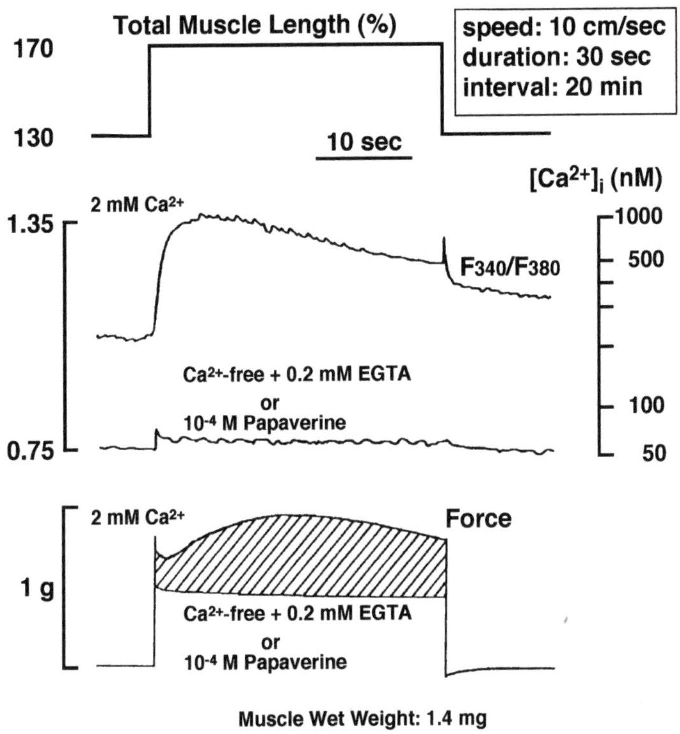

Fig. 17.1 Typical mechanical responses to quick stretch, increases in tension and intracellular Ca^{2+} concentration in the canine cerebral artery loaded with fura-2. (Modified Fig. 2 in Nakayama and Tanaka, 1989)

17.3.2 Ion Channels Involved in Stretch-Induced Contraction in Cerebral Arteries

The magnitude and rate of stretch and the interval between stretches are important parameters in eliciting mechanical responses of vascular strips to stretch. Of these, the rate of stretch is a determinant factor for how activator Ca^{2+} is mobilized for involvement in myogenic contraction (Nakayama, 1982; Nakayama and Tanaka, 1993; Obara et al., 2001). Quick stretch (50–400 mm/sec) induces myogenic contraction in cerebral arteries isolated from various animals, including rabbits, rats, dogs, and cats (Nakayama and Tanaka, 1993). In canine basilar arteries, slow stretch (1–3 mm/sec) causes no contractile response in the normal extracellular solution, but it can induce myogenic contraction when the vessels are pretreated with iberiotoxin or tetraethylammonium (TEA), which inhibits large-conductance Ca^{2+}-activated K^+ (BK) channels (Obara et al., 2001). The myogenic contraction induced by slow stretch is completely abolished by nicardipine, indicating that it is primarily dependent on Ca^{2+} influx through L-type VDCC. In the case of canine arteries, the arachidonic acid metabolite 20-hydroxyeicosatetraenoic acid

(20-HETE) may be involved in myogenic contraction in vivo. It has been shown that 20-HETE is produced by mechanical stretch (Gebremedhin et al., 2000) and blocks BK channels (Zou et al., 1996; Obara et al., 2002), and that slow stretch can induce contraction in the presence of 20-HETE in canine basilar arteries (Obara et al., 2002).

On the other hand, the myogenic contraction induced by quick stretch (5–100 mm/sec) is only partly inhibited by nicardipine (Nakayama et al., 1986, 1989b; Nakayama and Tanaka, 1993). The nicardipine-insensitive component is attenuated by ryanodine, an SR Ca^{2+} release channel blocker; cyclopiazonic acid and thapsigargin, SR Ca^{2+}-ATPase inhibitors; and U-73122, a phospholipase C inhibitor (Obara et al., 2001). Thus, quick stretch seems to induce not only the activation of VDCC, but also the activation of phospholipase C, which results in Ca^{2+} release from the SR. Quick stretch applied to the porcine coronary artery has also been shown to increase the production of IP_3 to over 3 times the control level within 2 sec (Tanaka et al., 1994b). The phospholipase C inhibitor 2-nitro-4-carboxyphenyl-N,N-diphenyl- carbamate nearly abolished the stretch-induced contraction and IP_3 elevation in the presence of nicardipine (Tanaka et al., 1994a). Moreover, quick stretch can increase the Ca^{2+} sensitivity of contractile apparatus: the time courses of the $[Ca^{2+}]_i$ and tension relationship indicate that both parameters change in parallel when the artery is slowly stretched, whereas the tension that develops in response to quick stretch is maintained in spite of decreased $[Ca^{2+}]_i$ (Obara et al., 2001) (Fig. 17.2).

This raises a question—how does mechanical stretch couple to the activation of VDCC? Candidate ionic mechanisms include the activation of non-selective cation channels. A number of studies have shown that stretch-induced myogenic contractions are at least partly inhibited by cation channel blockers such as Gd^{3+}. The contribution of Cl^- channels has also been suggested. In addition, we have shown that VDCC currents are facilitated by hyposmotic and flow stimuli in cerebral artery myocytes of dogs (Kimura et al., 2000) and rabbits (Amano et al., 2005). Thus,

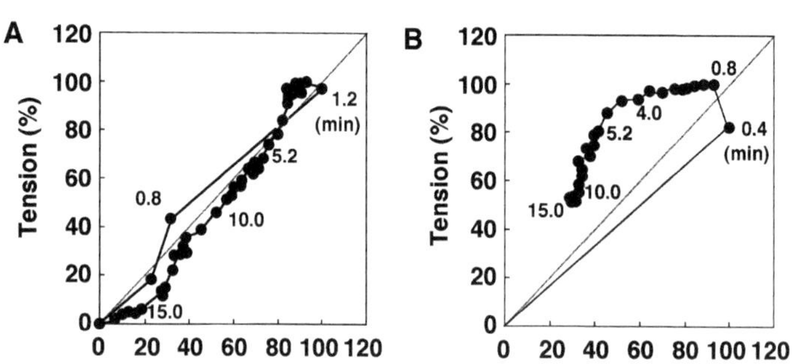

Fig. 17.2 Time-dependent changes in the $[Ca^{2+}]_i$-tension relationship during contraction induced by slow stretch (**Panel A**) and quick stretch (**Panel B**). Numerical values in the figures represent the time (in min) elapsed after stimulation. (Modified Fig. 2 in Obara et al., 2001)

VDCC itself seems to perceive the mechanical stimulus in addition to the activation through membrane depolarization. Moreover, negative feedback mechanisms involving localized Ca^{2+} release from the SR (Ca^{2+} sparks) and the subsequent activation of BK channels are also likely to be involved in the intrinsic regulation of myogenic constriction (Dopico et al., 1994) (Fig. 17.3).

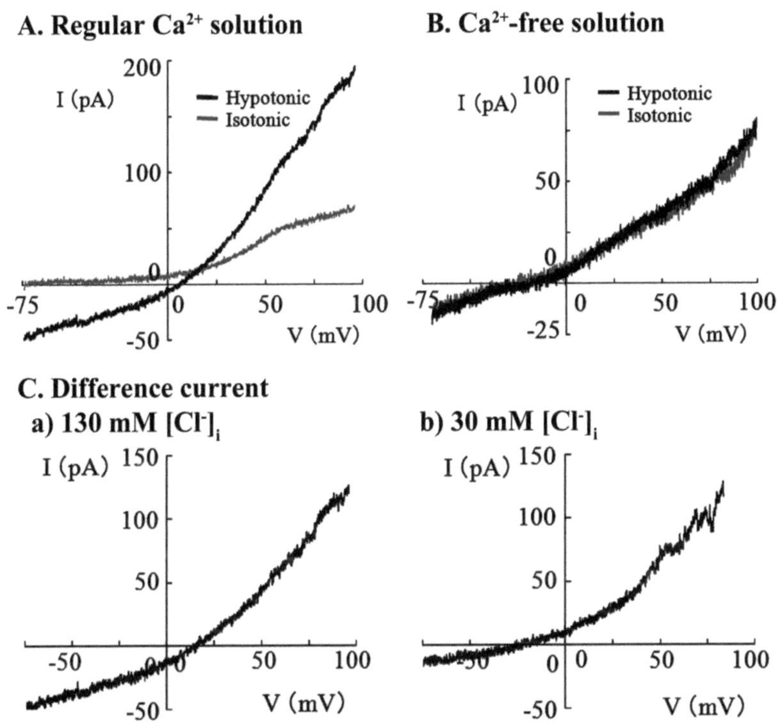

Fig. 17.3 Typical current-voltage (I–V) relationship of whole cell currents activated by hypotonic challenge in isolated smooth muscle cells of canine basilar arteries. Currents were elicited by voltage-ramp pulses from 70 to +100 mV with a holding potential of –70 mV. The currents were in isotonic and hypotonic solutions including 1 mM Ca^{2+} (**a**) or 0 Ca^{2+} (**b**). The hypotonically induced currents were also recorded with intracellular Cl^- concentrations of 130 mM (**c**, a) or 30 mM (**c**, b). The difference currents were obtained by subtraction of whole cell currents in the isotonic solution from those in the hypotonic solution. (Modified Figs. 6 and 7 in Yano et al., 2005)

17.3.3 Hypotonically Induced Contraction of Cerebral Arteries

Hyposmotic challenge is a widely used research tool for the investigation of the response to mechanical stretch in dispersed single cells. The activation of non-selective cation channels (Welsh et al., 2000) and volume-sensitive Cl^- channels (Yamazaki et al., 1998) and the inhibition of inward rectifying K^+ channels (Wu et al., 2007) have been shown to be involved in the membrane depolarization induced by hyposmotic challenge in the vascular smooth muscle cells. Nevertheless, there have been few studies on the contractile tension of blood vessels.

We have recently shown that a 45% hyposmotic challenge produces contraction in the ring segments of canine basilar arteries (Yano et al., 2005). The hypotonically induced contraction was nearly abolished by extracellular Ca^{2+} removal and by the VOCC blocker nicardipine, but was not inhibited by thapsigargin, which is an SR Ca^{2+}-ATPase inhibitor and depletes SR Ca^{2+} stores. These results suggest that the hypotonically induced contraction is caused by the Ca^{2+} influx through VDCC, similarly to the myogenic contraction induced by mechanical stretch. In addition, the contraction was inhibited not only by the cation channel blocker Gd^{3+} and ruthenium red but also by the Cl^- channel blocker DIDS and niflumic acid. These results imply that both cation and Cl^- channels are involved in the response. The involvement of Cl^- channels was also supported by the finding that the hypotonically induced contraction was enhanced by the reduction of the extracellular Cl^- concentrations. Further, whole-cell patch clamp analysis demonstrated that outwardly rectifying Cl^- currents were activated by the hyposmotic challenge and were abolished by 10 mM BAPTA in the pipette solution and by the removal of extracellular Ca^{2+}. This means that the Cl^- currents are through Ca^{2+}-activated Cl^- (Cl_{Ca}) channels, which are secondarily activated by an elevation of $[Ca^{2+}]_i$ (Fig. 17.4).

Based on these findings, we propose the following mechanisms for mechanical stretch-induced and hypotonically induced contractions: mechanical stretch initiates the activation of stretch-activated channels (SAC), and depolarizes the membrane, which leads to activation of VDCC and subsequent contraction. Osmotic cell swelling first activates Ca^{2+}-permeable and Gd^{3+}-sensitive cation channels, which

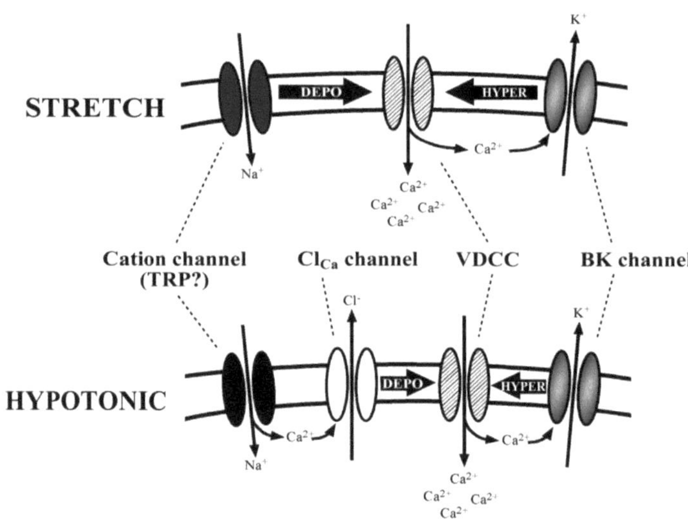

Fig. 17.4 Differences in the mechanisms of Ca^{2+} influx induced by stretch and hypotonic stimulations

are most likely stretch-activated cation channels. Ca^{2+} influx through the cation channels leads to the activation of Cl_{Ca} channels, which in turn results in membrane depolarization, thereby activating VDCC and causing contraction.

17.3.4 TRP Channels as Candidates for Mechanosensitive Cation Channels

TRP (transient receptor potential) cation channels are suggested to be important mediators of myogenic contraction. The mammalian TRP superfamily contains nearly 30 members, which are grouped into subfamilies based on sequence homology, i.e., TRPV (vanilloid), TRPC (canonical), TRPM (melastatin), TRPP (polycystin), TRPML (mucolipin) and TRPA (ankyrin). TRP channels highly expressed in vascular smooth muscle cells are TRPC1, TRPC3, TRPC6, TRPM4, TRPV2, and TRPV4 (Beech et al., 2004; Inoue et al., 2006). Among these, several TRP channels have been suggested to be involved in myogenic contraction as mechanosensitive channels in vascular smooth muscle cells.

Recent studies have proposed critical roles for both TRPC6 and TRPM4 in pressure-induced depolarization and vasoconstriction in rat cerebral arteries. Downregulation of TRPC6 expression using antisense oligodeoxynucleotides in the rat cerebral artery has been shown to attenuate nonselective cation currents activated by cell swelling, smooth muscle membrane depolarization, and the vasoconstriction induced by an elevation of intraluminal pressure (Welsh et al., 2002). Moreover, TRPC6 channels, which are expressed in HEK 293 or CHO cells, have been demonstrated to be activated by hypotonic and negative pipette pressure (Spassova et al., 2006). These findings support the proposal that TRPC6 functions as a mechanosensitive cation channel. In contrast, TRPM4, which is unique among TRP channels for its relatively limited permeability to Ca^{2+}, is stretch insensitive; however, the channel has been suggested to participate in stretch-induced myogenic depolarization of cerebral artery myocytes (Earley et al., 2004). TRPM4 is present in cerebral artery smooth muscle cells, where monovalent cation-selective channels are activated by intracellular Ca^{2+} (Earley et al., 2004) and membrane stretch (Morita et al., 2007). Moreover, membrane depolarization and contraction of isolated cerebral arteries in response to elevated intraluminal pressure were greatly attenuated when the expression of TRPM4 was suppressed using TRPM4 antisense oligodeoxynucleotides (Earley et al., 2004). Thus, these studies indicate a prominent role for vascular TRPM4 channels in myogenic constriction of cerebral arteries. The findings that inhibition of either TRPC6 or TRPM4 leads to comparable suppression of pressure-induced depolarization and myogenic tone suggest some cooperative interaction between these two TRP channels. Considering these results together, Brayden et al. (2008) have hypothesized that mechanical stress activates Ca^{2+} entry through TRPC6 channels and this Ca^{2+} activates nearby TRPM4 channels, thereby depolarizing smooth muscle cells. Alternatively, TRPM4 may be activated through the activation of PKC induced by stretch stimuli (Nilius et al., 2005). We have recently shown that pressure-induced sustained myogenic constriction is induced via the mechanism which is sensitive to both rottlerin, a PKCδ inhibitor, and ruthenium red in the rat posterior cerebral artery (Kashihara et al., 2008).

TRPV2 has also been suggested to function as a mechanosensitive cation channel (Muraki et al., 2003). In mouse aortic smooth muscle cells, cell-swelling induced by hyposmotic challenge elicited cation currents and elevated $[Ca^{2+}]_i$. These responses were blocked by ruthenium red, a TRPV channel blocker. TRPV2 mRNA was detected and TRPV2 immunoreactivity was evident in mouse aortic, mesenteric and basilar arterial smooth muscle cells. When the expression of TRPV2 was downregulated using TRPV2 antisense oligodeoxynucleotides, hyposmotic challenge-induced activation of cation currents or $[Ca^{2+}]_i$ elevation were suppressed. Moreover, overexpression of TRPV2 in CHO cells resulted in the appearance of mechanosensitive cation channel activities evoked by negative pressure in the patch pipette. In line with these findings, we have recently shown that the pressure-induced constriction and $[Ca^{2+}]_i$ elevation are inhibited by ruthenium red, a relatively specific inhibitor of TRPV, in rat posterior cerebral arteries (Kashihara et al., 2008).

Several members of TRP channels are either directly or indirectly activated by mechanical stretch. Future studies will be required to determine whether these mechanosensitive ion channels function depending on rate, magnitude, or interval of stretch. In addition, a more precise understanding of the spatiotemporal relationships between mechanosensitive molecules, including ion channels, PKC, Ca^{2+}, etc., will be required.

17.4 Multiple Phosphorylation of 20-kDa Myosin Light Chain of Cerebral Artery Smooth Muscles as a Self-inhibitory Mechanism in Stretch-Induced Contraction

17.4.1 Triphosphorylation of Myosin Light Chain Induced by Mechanical Stretch

The phosphorylation of 20-kDa myosin light chain (MLC_{20}), which is dependent on a balance between myosin light chain kinase (MLCK) and myosin phosphatase (MLCP), is a primary step in the contraction of smooth muscles produced by agonistic and depolarizing stimuli. As to the myogenic contraction of vascular smooth muscle in response to mechanical stretch, it is widely accepted that the increase in $[Ca^{2+}]_i$ and MLC_{20} phosphorylation are involved in the signaling sequence for the stretch-induced response (Zou et al., 1995).

A slow stretch at a rate of 1 mm/sec from the initial length (L_i) to 1.5 L_i has been shown to increase $[Ca^{2+}]_i$, MLC_{20} phosphorylation, and contraction in canine basilar arteries treated with BK channel blockers such as iberiotoxin or TEA (Obara et al., 2006). The tension development in response to stretch was closely related to the increase in $[Ca^{2+}]_i$ and MLC_{20} phosphorylation during the early phase (1-min stretch) (Fig. 17.5).

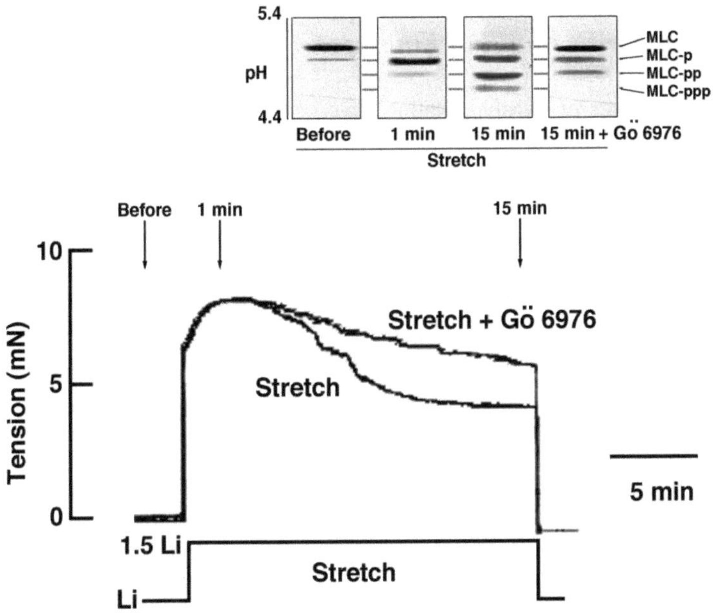

Fig. 17.5 Typical tracings of slow stretch-induced contraction. Artery segments were treated with or without Gö6976 (1 μM) for 15 min. Then they were stretched at a rate of 1 mm/sec from L_i to 1.5 L_i and maintained in the stretched state for 15 min in the presence of TEA (5 mM). *Inset*: Typical phosphorylation patterns of MLC_{20} detected by the isoelectric focusing electrophoresis and immunoblot technique. The segments were incubated in a solution containing vehicle (Stretch; 0.1% DMSO) or Gö6976 (Stretch + Gö6976; 1 μM) for 15 min, and then were stretched for 1 or 15 min. Note that four immunoreactive bands to anti-MLC_{20} (MLC, MLC-p, MLC-pp and MLC-ppp corresponding to non-, mono-, di-, and tri-phosphorylated MLC_{20}, respectively) were detected. (Modified Figs. 1–3 in Obara et al., 2006)

However, the phosphorylation remained high even after the $[Ca^{2+}]_i$ and tension returned nearly to the basal levels at the late phase of contraction (15-min stretch), indicating the uncoupling of MLC_{20} phosphorylation and tension. This is in great contrast to the so-called "latch" mechanism in which tension is maintained despite a low level of MLC_{20} phosphorylation (Aksoy et al., 1983; Rembold and Murphy, 1990). The phosphorylation pattern measured by isoelectric focusing (IEF) electrophoresis indicated that the level of mono-phosphorylated MLC_{20} was increased during the early phase, but that the levels of mono-, di-, and triphosphorylated MLC_{20} were increased at the late phase (Fig. 17.5, inset). Two-dimensional phosphopeptide mapping showed that the stretch increased mono-phosphorylated MLC_{20} at Ser-19, di-phosphorylated MLC_{20} at Ser-19 and Thr-18, and triphosphorylated MLC_{20} at Ser-19, Thr-18 and Thr-9. This triphosphorylation pattern can be observed irrespective of the presence or absence of tension development or BK channel blockers. Thus, it seems possible that mechanical stretch by itself triggers the triphosphorylation of MLC_{20}.

17.4.2 Interactive Role of Mechano-Sensitive Kinases: Protein Kinase C, Rho/Rho Kinase, and Protein Tyrosine Kinase

Five possible sites of phosphorylation have been reported for MLC_{20} of chicken gizzard: Ser-19 and Thr-18 for MLCK (Ikebe and Hartshorne, 1985), and Ser-1, Ser-2 and Thr-9 for PKC (Nishikawa et al., 1983; Ikebe et al., 1986; Bengur et al., 1987). Stretch stimuli have been shown to increase MLC_{20} triphosphorylation at both MLCK sites (Ser-19 and Thr-18) and one PKC site (Thr-9) in the canine basilar artery (Nakayama et al., 2003). Among four PKC isoforms (PKCα, δ, ζ and η) identified in the canine basilar artery (Nishizawa et al., 2000), PKCα and PKCδ were translocated from the cytosol to the membrane by stretch. The translocation of PKCα was transient, while that of PKCδ was maintained during stretch. Rottlerin, a selective inhibitor of PKCδ (Gschwendt et al., 1994), had no apparent effect on the phosphorylation pattern of MLC_{20}. Furthermore, the tension recording showed that the inhibition of PKCα by Gö6976 slowed the gradual reduction of the tension during 15-min stretch (Fig. 17.5). Phosphorylation of MLC_{20} at Thr-9 by PKC has been shown to inhibit the MLCK-induced increase in actin-activated myosin ATPase activity, thereby inhibiting smooth muscle contraction (Nishikawa et al., 1983; de Lanerolle and Nishikawa, 1988), although there are some studies against it (Singer et al., 1989; Sutton and Haeberle, 1990). Taken together, it is suggested that PKCα phosphorylates Thr-9 of MLC_{20}, thereby counteracting the contraction induced by slow stretch.

PKCα has been suggested to be involved in stretch- or pressure-induced contraction of rat basilar and posterior cerebral arteries (Yeon et al., 2002; Ahn et al., 2007; Kashihara et al., 2008) and ferret coronary arteries (Dessy et al., 2000). It has been shown that phosphorylation of PKCα at Ser-657 results in active enzyme (Bornancin and Parker, 1997), and that PKCα is inactivated through dephosphorylation by protein phosphatase 2A (PP2A) (Ricciarelli and Azzi, 1998). PP2A and PKCα have also been shown to be physically and functionally associated in various cells, including human and mouse mast cells (Boudreau et al., 2002), smooth muscle cells (Ricciarelli et al., 1998) and COS cells (Hansra et al., 1996). Recently, we have found that slow stretch not only phosphorylates PKCα at Ser-657 but also inhibits PP2A activity in the canine basilar artery (Obara et al., 2010). Thus, although the translocation of PKCα to the membrane induced by stretch was transient, the activity of PKCα might be maintained through the inhibition of dephosphorylation during stretch stimuli, thereby resulting in MLC_{20} triphosphorylation. Slow stretch has also been shown to activate the Rho/Rho-kinase pathway in canine basilar arteries (Obara et al., 2006). Y-27632, an inhibitor of Rho kinase, restored slow stretch-induced inhibition of MLCP and PP2A activities, while the inhibitor lowered the PKCα activity, leading to a significant reduction of MLC_{20} phosphorylation induced by a 15-min stretch (Obara et al., 2006). It is thus likely that Rho-kinase locates upstream of PP2A and has an inhibitory action on PP2Ac.

Rho-kinase phosphorylates MLCP at its myosin-binding subunit (MYPT1), which leads to an elevation of MLC_{20} phosphorylation at MLCK sites (Ichikawa et al., 1996; Kimura et al., 1996). Stretch has been shown to increase phosphorylation of MYPT1 in the canine basilar artery, which is inhibited by Y-27632 (Obara et al., 2006). Thus, MLCP inactivated by Rho-kinase might also be involved in the stretch-induced phosphorylation of MLC_{20} at Ser-19 and Thr-18. The crosstalk between PKCα and CPI-17 (17-kDa PKC-potentiated inhibitory phosphoprotein) is an alternative pathway leading to the augmentation of MLC_{20} phosphorylations at Ser-19 and Thr-18 (Kitazawa et al., 1999; Somlyo and Somlyo, 2000; Eto et al., 2001). Accordingly, CPI-17 phosphorylated by PKCα or PKCδ would accelerate the inactivation of MLCP. In the canine basilar artery, slow stretch augmented phosphorylation of CPI-17, which was inhibited by Gö6976. Thus, PKCα-mediated CPI-17 phosphorylation is likely to be also involved in the MLC_{20} phosphorylations at Ser-19 and Thr-18.

Considering these results together, we propose the following sequence of events involved in the triphosphorylation of MLC_{20} and contraction in response to slow stretch (Fig. 17.6). Slow stretch activates Rho/Rho-kinase, which in turn phosphorylates MLCP at its myosin-binding subunit, thereby inactivating MLCP. Alternatively, Rho-kinase leads to the inactivation of PP2A, resulting in activation of PKCα. The activated PKCα inactivates MLCP through CPI-17 phosphorylation. The inactivation of MLCP also leads to an increase in the level of MLC_{20} phosphorylation at Ser-19 and Thr-18 by MLCK. Furthermore, the activated PKCα phosphorylates MLC_{20} at an additional site (Thr-9), resulting in triphosphorylation of MLC_{20} and a concomitant reduction of stretch-induced contraction. In addition, we have recently found that the RGD-peptide, which blocks the interaction between integrin and extracellular matrix proteins, and CP4632, a specific inhibitor of $\alpha_v\beta_3$ type integrin (Murakami et al., 2002), abolish the activation of RhoA and MLC_{20} triphosphorylation induced by slow stretch in the canine basilar artery (Obara et al.., unpublished observations). Thus, src family tyrosine kinase coupled with $\alpha_v\beta_3$ type integrin seems to locate upstream of the mechanotransduction cascade.

17.4.3 Significance of the Self-inhibitory Mechanism in Stretch-Induced Contraction

The physiological and pathophysiological significance of our results is addressed with special reference to the Bayliss effect (Bayliss, 1902). The vascular smooth muscle cells by themselves can act as receptors and responders to not only neurohumoral factors but also hemodynamic forces such as blood pressure and blood flow in the vascular network in order to maintain homeostasis of the circulatory system. This homeostasis in vivo is generally achieved by the concerted cooperation of self-control mechanisms within the vascular network: the myogenic autoregulation (Bayliss effect) and the flow-dependent dilation (Schretzenmayr effect) (Schretzenmayr, 1933). The latter is a special reaction within the broad field of endothelium-mediated modulation of vascular tone and depends on the shear responsiveness of endothelial cells and paracrine effectors of the vessel wall. The Bayliss effect is due

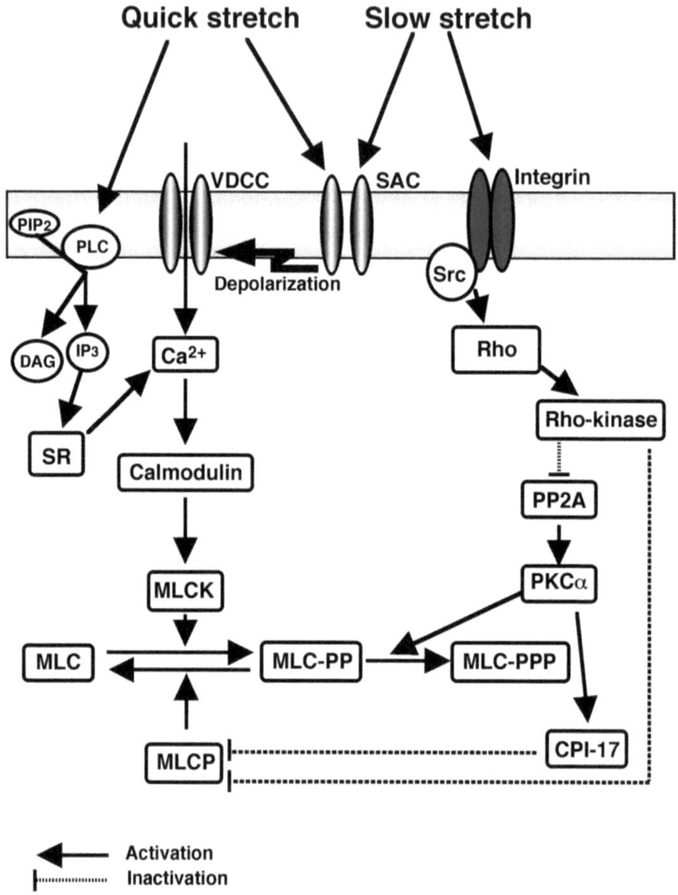

Fig. 17.6 Possible mechanisms involved in the stretch-induced MLC_{20} triphosphorylation. CPI-17: 17-kDa PKC-potentiated inhibitory phosphoprotein. DAG: diacylglycerol, IP_3: inositol-1,4,5-triphosphate, MLC_{20}: 20-kDa myosin light chain, MLCK: myosin light chain kinase, MLCP: myosin light chain phosphatase, PLC: phospholipase C, PKCα: protein kinase Cα isoform, PP2A: protein phosphatase 2A, SAC: stretch-activated channel, SR: sarcoplasmic reticulum, VDCC: voltage-dependent Ca^{2+} channel

to a reaction of mechanical effectors of the vessel wall, i.e., smooth muscle cells, and contributes to the regulation of basal vascular tone. Apart from the physiological role of the myogenic autoregulation of blood flow, it is still an open question whether myogenic vasoconstriction might be a phenomenon of pathophysiological significance in vasospastic episodes, including hypertension and cerebral vasospasm. In fact, autoregulatory vasoconstriction represents a positive feedback mechanism in that an initial rise in intraluminal pressure/stretch of the vascular wall evokes a contraction, which in turn further increases the pressure in the vascular bed arising from pressure/stretch-induced contraction (Fleckenstein, 1983). This positive feedback mechanism of pressure/stretch-induced contraction seems to operate more dominantly when the endothelium-dependent vasodilator functions are impaired. Thus, in our modern sense of the "Bayliss effect", the autoregulatory myogenic response to hemodynamic factors such as pressure and stretch plays an important role not only in the generation of basal vascular tone, but also in a self adjustment/brake mechanism against excessive vasoconstriction. We have provided evidence supporting the notion that stretch induces inhibition of the PP2A activity that would otherwise lead to both activation of PKCα and subsequent triphosphorylation of MLC_{20}.Of the triphosphorylation of MLC_{20}, the phosphorylation of Thr-9 induced by PKCα seems to have a self-inhibitory action in the stretch-induced contraction.

17.5 Relevance of Stretch-Induced Contraction and Vasospasm

Cerebral vasospasm is one of the most serious complications after SAH, and has a profoundly deleterious effect on the cerebral circulation (Macdonald and Weir, 2001). Interestingly, the molecules activated by stretch have parallels with those in cerebral vasospasm — i.e., the mechanosensitive kinases such as protein kinase C, Rho/Rho-kinase, and tyrosine kinases, all of which are activated by mechanical stretch, are activated during the time course of cerebral vasospasm (Nishizawa and Laher, 2005; Laher and Zhang, 2001; Budzyn et al., 2006; Wickman et al., 2003; Nishizawa et al., 2004; Obara et al., 2004). Therefore, we here present the idea that these kinases play a common role in the mechanisms for not only stretch-induced contraction but also vasospasm.

17.5.1 Cerebral Vasospasm

Several studies have assessed the progression of vasospasm by using the "two-hemorrhage" canine model of cerebral vasospasm in the basilar artery (Nishizawa and Laher, 2005; Nishizawa et al., 2000, 2003; Obara et al., 2005). We have shown by angiographic analysis that the basilar artery narrowed to about 80% of the control vessel diameter before the second injection of autologous blood on day 4 and further narrowed to about 50% after the second blood injection on day 4. The narrowed vessel diameter persisted on day 14. After the maintenance phase, the angiographic cerebral vasospasm gradually attenuated. The arterial diameter on day 28 was almost the same as the control diameter. The maximal contraction capacity sensitive to papaverine, a non-specific smooth muscle relaxant, decreased until day 21, and showed some recovery by day 28. On the other hand, arterial stiffness, which is insensitive to papaverine, progressed until day 28. Histological examination revealed medial thickening with the formation of short and round smooth muscle cells and increased connective tissue until day 21, and return to the control level of arterial stiffness by day 28 (Koide et al., 2002; Yamaguchi-Okada et al., 2005). Immunohistochemical studies to examine the expression of the myosin heavy chain isoforms anti-SMemb, anti-SM1, and anti-SM2, which are markers of smooth muscle phenotypic changes, showed enhanced expression of SMemb from day 7 to day 21 and disappearance of SM1 and SM2 on days 14 and 21. The changes in myosin heavy chain isoform expression returned to normal on day 28 (Yamaguchi-Okada et al., 2005). These results indicate that biomechanical and phenotypic changes occur in correspondence with the time course of cerebral vasospasm.

In addition, the MLC_{20} phosphorylation level significantly increased with the progression of vasospasm (Yamaguchi-Okada et al., 2005). Mono- and diphosphorylated MLC_{20} species in the basilar arteries were detected at all examination time points in the experimental therapeutic group. We further examined the effects of various kinase inhibitors, which were injected into the cisterna magna before the autologous blood injection. Rottlerin, an inhibitor of PKCδ, significantly inhibited the angiographic vasospasm on day 4 after the second injection of blood, and slightly but significantly prevented the vasospastic effect on day 7. Chelerythrine, an inhibitor of classical/novel PKC isoforms, significantly prevented the extent of angiographic vasospasm both on day 4 and on day 7; the inhibition of the vasospasm on day 7 was more pronounced than that on day 4. Among four PKC isoforms (PKCα, δ, ζ, and η) detected in the canine basilar artery, PKCα and PKCδ were translocated from the cytosol to the membrane after SAH (Nishizawa et al., 2000). These results suggest that PKCδ plays a role in the initiation/development of vasospasm on day 4, and that PKCα is mainly involved in the maintenance of vasospasm on day 7.

RhoA translocation from the cytosol to the membrane reflects activation of this protein (Gong et al., 1997). We have shown that RhoA is primarily present in the cytosol before the second injection of blood on day 4 and increases in the membrane after the injection. Moreover, Y-27632, a Rho-kinase inhibitor, significantly suppressed the vasospasm and the increased level of MLC_{20} phosphorylation on day 4, but did not ameliorate the vasospasm on day 7. The Rho-kinase inhibitor also abolished the translocation of PKCδ on day 4. In contrast, rottlerin had no apparent effect on the translocation of RhoA. These results indicate that RhoA/Rho-kinase lies upstream of PKCδ and modulates the activation of PKCδ. Y-27632 had no apparent effect on the translocation of PKCα involved in the vasospasm on day 7, indicating that there is no close interaction between RhoA/Rho-kinase and PKCα.

Caldesmon (CaD) is an actin-linked regulatory protein involved in smooth muscle contractility. CaD has been suggested to be involved in canine cerebral vasospasm (Sun et al., 1998). Phosphorylation of CaD by PKC reduces the affinity of CaD for actin, alleviating its inhibition of the actin-activated myosin Mg^{2+}-ATPase. We showed that the SAH-induced cerebral vasospasm on day 4 was accompanied with the phosphorylation of CaD, which was inhibited by Y-27632 and rottlerin. Rottlerin nearly abolished the development of cerebral vasospasm after SAH without affecting the level of MLC_{20} phosphorylation. Furthermore, we found that PKCδ could induce Ca^{2+}-independent contraction and CaD phosphorylation in chemically skinned basilar arteries. These results suggest that Rho-kinase and PKCδ mediate CaD phosphorylation in the early phase of vasospasm. However, it should be noted that rottlerin has nonspecific inhibitory effects on protein kinases other than

PKCδ (Soltoff, 2007). Thus, a part of its effects may not be due to the inhibition of PKCδ, and its contribution to the vasospasm may have been overestimated. Nevertheless, it is most likely that the vasospasm is mediated by at least two mechanisms, one involving MLC_{20} phosphorylation, and the other being independent of MLC_{20} and mediated by PKCδ.

Considering our results together, we propose the following sequence of events involved in the early stage of SAH-induced cerebral vasospasm. SAH induces the activation of the Rho/Rho-kinase system, which in turn leads to the activation of PKCδ and the inhibition of MLCP. The activated PKCδ phosphorylates CaD, thereby alleviating the inhibition by CaD of the actin-activated myosin Mg^{2+}-ATPase, leading to vasospasm. On the other hand, the inhibition of MLCP increases MLC_{20} phosphorylation, leading to vasospasm (Obara et al., 2005).

It should be noted that vasospasm continued for more than two weeks after SAH. However, we have found that PKC kinase activity is declined 2 weeks after SAH. In contrast, the activity of tyrosine kinase was enhanced on day 14 (Fig. 17.8; Koide et al., 2002). Thus, the main factor in sustaining the vasospasm two weeks after SAH may be tyrosine kinase rather than PKC.

17.5.2 Beyond Vasospastic Episodes

Cerebral vasospasm is the cause of delayed neurological deterioration after aneurismal subarachnoid hemorrhage. However, in the recent clinical trial known as CONSCIOUS-1 (Macdonald et al., 2007a), an endothelin ET_A receptor antagonist, clazosentan, markedly prevented vasospasm, and yet the final outcome of patients was not improved (Macdonald et al., 2007b). These confusing results were

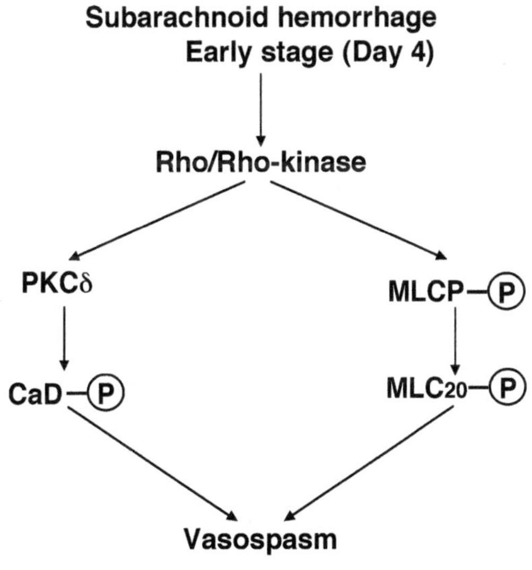

Fig. 17.7 Schematic diagram showing the role of the interaction between protein kinase C and Rho-kinase in the development of cerebral vasospasm in a canine two-hemorrhage model. (Modification of Fig. 7 from Obara et al., 2005)

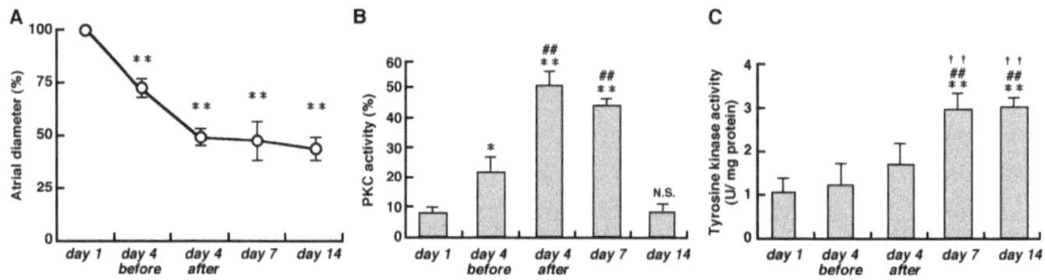

Fig. 17.8 The interrelation of chronological changes of arterial diameters (**A**), protein kinase C activities (**B**), and tyrosine kinase activities (**C**). (Modification of Figs. 4 and 5 from Koide et al., 2002)

presented at the 9th International Conference on Cerebral Vasospasm, June 27–30, 2006, in Istanbul, Turkey. Clinical observations presented at the same congress indicate that early brain injuries occurring in the early phase of vasospasm, i.e., disruption of the blood brain barrier function, enhancement of intracranial pressure, and cortical spreading depression, worsen the patient outcome (Macdonald et al., 2007b; Hansen-Schwartz et al., 2007). It has been discussed that factors other than vasoconstriction are also important in the pathophysiology and prognosis of cerebral vasospasm. Such factors include global ischemia, disruption of the blood-brain barrier, activation of apoptotic and inflammatory pathways, and cortical spreading depression (Macdonald et al., 2007b; Hansen-Schwartz et al., 2007).

17.6 Stretch-Induced Contraction as a Drug Evaluation System

17.6.1 Promoters

The contractile response to slow or quick stretch can be promoted by promoters of Ca^{2+} influx, mainly through VDCC. These include Bay K 8644, a Ca^{2+} agonist, cardiac glycosides (κ-strophanthin and ouabain), high K^+, and vasoconstrictor amines norepinephrine, serotonin, and histamine (Nakayama, 1982; Nakayama and Tanaka, 1989). The vasoactive peptides endothelin-1 and neuropeptide Y also augment the stretch-induced contraction. Endothelin-1 acts as not only a vasoconstrictor by itself but also a potentiater of other vasoconstrictors such as serotonin (Nakayama et al., 1991) and mechanical stimuli such as stretch (Nakayama and Tanaka, 1993; Obara et al., 1999). BK channel inhibitors such as iberiotoxin, charybdotoxin, and a small amount of tetraethylammonium (TEA) potentiate stretch-induced contraction, whereas apamin, a small conductance BK channel blocker, 4-aminopyridine, a voltage-dependent K^+ channel blocker, and glybenclamide or tolbutamide, an ATP-sensitive K^+ channel blocker, have no apparent effect on the mechanical activity upon slow stretch (Obara et al., 2001).

20-HETE and oxyhemoglobin have been reported as endogenous promoters of stretch-induced contraction. It is well-documented that cytochrome P450 monooxygenase metabolizes arachidonic acid and generates vasoactive substances, such as epoxyeicosatrienoic acids (EETs) and 20-HETE. EETs essentially act as vasodilators (Rosolowsky and Campbell, 1993; Gebremedhin et al., 1998), whereas 20-HETE is a potent and Ca^{2+}-dependent vasoconstrictor and suggested to be an endogenous mediator in the cerebral circulation (Harder et al., 1994; Imig et al., 1996). The vasoconstrictor action of 20-HETE has been shown to be attributable to the PKC-mediated inhibition of BK channel activity (Lange et al., 1997). We have shown that 20-HETE causes the translocation of PKCα from the cytosol to the membrane in the canine basilar artery, which is inhibited by calphostin C, a general inhibitor of PKC (Obara et al., 2002). In addition, 20-HETE inhibited the whole-cell K^+ current and depolarized the membrane by approximately 10 mV in single smooth muscle cells isolated from the canine basilar artery. These effects of 20-HETE were similar to those of iberiotoxin, although the former, but not the latter, was inhibited by calphostin C. These results suggest that 20-HETE induces sensitization of the canine basilar artery to stretch, which is caused by PKCα-mediated inhibition of BK channel activity. However, it still remains to be clarified whether BK channels are directly phosphorylated by PKCα. Further studies are required to elucidate the underlying mechanism at the structural level.

An overwhelming number of papers have reported that oxyhemoglobin acts on endothelial and medial smooth muscle cells directly, through the production of superoxides or free radicals via auto-oxidation of hemoglobin to methemoglobin, or through some novel mechanisms including suppression of voltage-dependent K^+ channels (Ishiguro et al., 2006) or overexpression of R-type Ca^{2+} channels (Ishiguro et al., 2005) in cerebral arteries. Oxyhemoglobin, the most prominent etiological factor in the cerebral vasospasm, possesses multiple actions on cerebral arteries and neuronal elements. The stretch-induced contraction similarly occurred irrespective of the presence or absence of endothelium (Nakayama et al., 1989b; Nakayama and Tanaka, 1989), indicating that the stretch/pressure-induced contraction is myogenic in nature, and the primary sensor initiating myogenic tone lies in the smooth muscle but not in the endothelium or the myo-endothelial junction. Nevertheless, we have shown that hemolysate/oxyhemoglobin strongly potentiates the contractile response of the canine cerebral artery to quick stretch in an endothelium-dependent manner (Nakayama and Tanaka, 1993; Nakayama et al., 1989b). A small amount of hemolysate (0.01–0.2 mg/ml as oxyhemoglobin), which increased the basal tone by 10–15% of the maximum contraction produced by 80 mM K^+, potentiated the stretch-induced contraction 2- to 3-fold over the control. The enhanced response to stretch was attenuated by removal of the endothelium and was readily suppressed by Ca^{2+} antagonists such as nisoldipine and other 1,4-dihydropyridines, as well as diltiazem. The potentiation by hemolysate of the response to quick stretch was partially restored when the endothelium-rubbed helical strip of the canine basilar artery was enveloped by a longitudinally cut cylindrical segment of the artery with intact endothelium (Nakayama et al., 1989b) (Fig. 17.9). The hemolysate/hemoglobin had no apparent effect on the contraction due to a quick stretch in Ca^{2+}-free medium or on the contractile response of a chemically skinned artery strip in the relaxing solution. Furthermore, the hemolysate containing methemoglobin was almost inactive. These results suggest that hemolysate potentiates the contractile response to

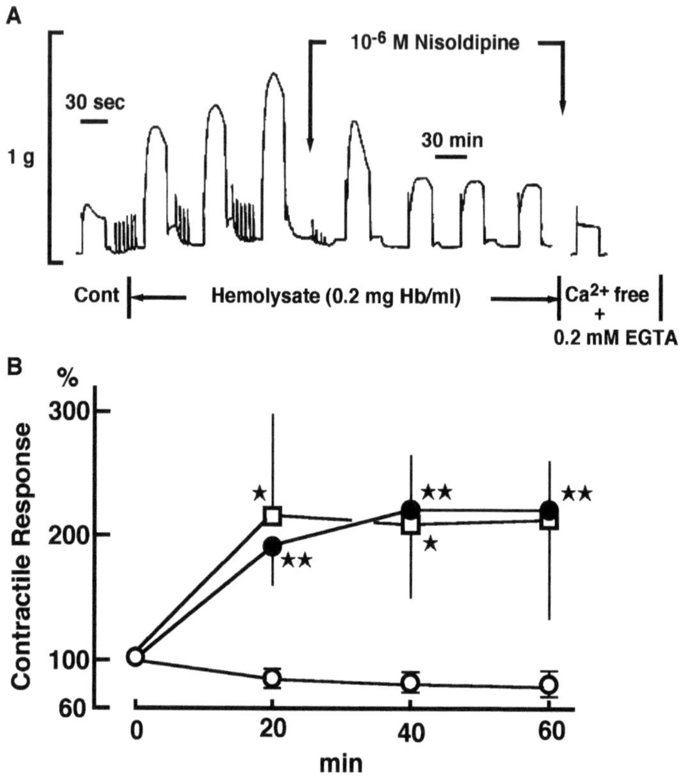

Fig. 17.9 Time course of the tension development of canine basilar arteries with or without endothelium in response to quick stretch before and after application of hemolysate. **Panel A**: The inhibitory effect of nisoldipine on the stretch-induced contraction potentiated by hemolysate containing 0.2 mg oxyhemoglobin/ml in the artery with endothelium. **Panel B**: Time course of tension development of canine basilar arteries with (*closed circles*) or without (*open circles*) endothelium, or of arteries endothelium-rubbed but enveloped by an endothelium-intact artery (*open squares*). (Modification of Figs. 4 and 5 from Nakayama et al., 1989b)

stretch by promoting the transmembrane supply of Ca^{2+}. The presence of endothelium seems to amplify the mechanical response to quick stretch in the canine cerebral artery by liberating still unidentified humoral agent(s).

17.6.2 Inhibitors

The drugs that inhibit Ca^{2+} release from intracellular storage sites, including thapsigargin, cyclopiazonic acid, procaine, and dantrolene, specifically inhibit the nicardipine-insensitive contractile component in response to quick stretch. Furthermore, we have demonstrated that W-7, a calmodulin inhibitor, and ML-9, an inhibitor of MLCK, abolish the stretch-induced contraction without any appreciable effect on the concomitant increase in $[Ca^{2+}]_i$, indicating uncoupling between $[Ca^{2+}]_i$ and mechanical activity (Nakayama and Tanaka, 1991).

Among the clinically important vasodilators, Ca^{2+} antagonists, diltiazem, nicardipine (Nakayama and Tanaka, 1993), and mepirodipine (Nakayama et al., 1989a), and the nitro compound nitroglycerin have been examined for their effects on the myogenic response to quick stretch. The comparative effects of the vasodilators on three types of contractions produced by mechanical (quick stretch), depolarizing (80 mM K^+), and agonistic (U-46619, a thromboxane mimetic) stimuli were evaluated in canine cerebral arteries (Nakayama and Tanaka, 1993). Papaverine, a smooth muscle relaxant, totally abolished the contractions induced by the three different stimuli. The Ca^{2+} antagonists and nitroglycerin only partially antagonized the stretch-induced contraction, suggesting that these drugs are effective for the abolition of exaggerated vascular contraction without disrupting basal vascular tone. These pharmacological properties of Ca^{2+} antagonists and nitroglycerin are in good accord with their clinical effectiveness in the treatment of hypertension and vasospasm with little untoward effects on circulation including orthostatic hypotension or luxurious perfusion syndrome such as the coronary and cerebral steal phenomenon. The stretch-induced contraction may provide a useful alternative to depolarizing and agonistic stimuli for use in the discovery of new therapeutic drugs.

17.7 Conclusion and Perspectives

There are many similarities in the intracellular signaling between stretch-induced contraction and vasospastic episodes in the cerebral circulations (Nishizawa et al., 2008). Mechanical stretch acts ubiquitously on the cardiovascular system. Our laboratory has demonstrated that mechanical stretch augments tyrosine phosphorylation of focal adhesion kinase (FAK) and the surface expression of platelet-derived growth factor receptorβ (PDGF-Rβ) in the pulmonary artery (Tanabe et al., 2000). Moreover, a significant increase in the steady-state mRNA level for PDGF-Rβ is observed only in the pulmonary artery of rats with monocrotaline-induced pulmonary hypertension, where the pulmonary artery is overstretched due to increased pulmonary arterial blood pressure. Recently, it has been reported that excessive surface expression of PDGF-Rβ occurs in the pulmonary tissues obtained from animal models of pulmonary hypertension produced by monocrotaline and hypoxia, and in those from pulmonary hypertensive patients (Schermuly et al., 2005). Furthermore, therapy using imatinib, an antagonist of PDGF receptors, has been shown to reverse vascular remodeling and cor pulmonale in experimental pulmonary hypertension regardless of the initiating stimulus. It has been well documented that endothelins and their receptors are specifically overexpressed in pulmonary hypertensive patients and animal models. Inflammatory cytokines including interleukin-6 are induced in pulmonary hypertensive mice (Tanabe et al., 2006). These results suggest that stretch-induced overexpression of cell-surface PDGF-Rβ and augmentation of tyrosine phosphorylation of proteins including FAK in the pulmonary arteries cause a vicious cycle accelerating the pathological conditions in pulmonary hypertension. Therefore, we consider that experimental and clinical pulmonary hypertension might be a circulatory disease attributable to a disturbance of mechanotransduction signaling. Recently, convincing data have implicated a role of inflammatory signals in the development and maintenance of cerebral vasospasm: monocyte chemoattractant protein-1 (MCP-1) acts as a biomarker of poor outcome in the serum and of vasospasm in the cerebrospinal fluid (Kim et al., 2008).

It is needless to say that cerebral arteries/arterioles have been the primary targets for studies of myogenic contraction as a self-control mechanism at tissue, cellular and molecular levels and in vivo. To further extend the research into the new field of cerebrovascular mechanotransduction in health and disease, the interactive role of signal communication among spatial and temporal micro-environments composed of multi-cells, including blood, vascular, neuronal, and other important components, should be clarified.

Acknowledgements The present study was supported in part by grants-in-aid for scientific research from the Ministry of Education, Culture, Sports, Science, and Technology of Japan, and by grants from the Shizuoka Research and Development Foundation. We also appreciate Iwate Medical University and University of Shizuoka for their continuous support.

References

Ahn D-S Choi S-K Kim Y-H Cho Y-E Shin HM Morgan KG Lee Y-H (2007) Enhanced stretch-induced myogenic tone in the basilar artery of spontaneously hypertensive rats. J Vasc Res 44:182–191.

Aksoy MO Mras S Kamm KE Murphy RA (1983) Ca^{2+}, cAMP, and changes in myosin phosphorylation during contraction of smooth muscle. Am J Physiol 245:C255–C270.

Amano S Ishikawa T Nakayama K (2005) Facilitation of L-type Ca^{2+} currents by fluid flow in rabbit cerebral artery myocytes. J Pharmacol Sci 98:425–429.

Bayliss WM (1902) On the local reactions of the arterial wall to changes of internal pressure. J Physiol 28:220–231.

Beech DJ Muraki K Flemming R (2004) Non-selective cationic channels of smooth muscle and the mammalian homologues of Drosophila TRP. J Physiol 559:685–706.

Bengur AR Robinson EA Appella E Sellers JR (1987) Sequence of the sites phosphorylated by protein kinase C in the smooth muscle myosin light chain. J Biol Chem 262:7613–7617.

Bornancin F Parker PJ (1997) Phosphorylation of protein kinase C-α on serine 657 controls the accumulation of active enzyme and contributes to its phosphatase-resistant state. J Biol Chem 272:3544–3549.

Boudreau RT Garduno R Lin TJ (2002) Protein phosphatase 2A and protein kinase C are physically associated and are involved in Pseudomonas aeruginosa-induced interleukin 6 production by mast cells. J Biol Chem 277:5322–5329.

Brayden JE Earley S Nelson MT Reading S (2008) Transient receptor potential (TRP) channels, vascular tone and autoregulation of cerebral blood flow. Clin Exp Pharmacol Physiol 35: 1116–1120.

Budzyn K Marley PD Sobey CG (2006) Targeting Rho and Rho-kinase in the treatment of cardiovascular disease. Trends Pharmacol Sci 27:97–104.

Canham PB (1977) Orientation of cerebral vascular smooth muscle, mathematically modelled. J Biomech 10:241–251.

Davis MJ Hill MA (1999) Signaling mechanisms underlying the vascular myogenic response. Physiol Rev 79:387–423.
de Lanerolle P Nishikawa M (1988) Regulation of embryonic smooth muscle myosin by protein kinase C. J Biol Chem 263:9071–9074.
Dessy C Matsuda N Hulvershorn J Sougnez C Sellke FW Morgan KG (2000) Evidence for involvement of the PKC- isoforms in myogenic contractions of the coronary microcirculation. Am J Physiol 279:H916–H923.
Dopico AM Kirber MT Singer JJ Walsh JV Jr (1994) Membrane stretch directly activates large conductance Ca^{2+}-activated K^{+} channels in mesenteric artery smooth muscle cells. Am J Hypertens 7:82–89.
Earley S Waldron BJ Brayden JE (2004) Critical role for transient receptor potential channel TRPM4 in myogenic constriction of cerebral arteries. Circ Res 95:922–929.
Eto M Kitazawa T Yazawa M Mukai H Ono Y Brautigan DL (2001) Histamine-induced vasoconstriction involves phosphorylation of a specific inhibitor protein for myosin phosphatase by protein kinase C and isoforms. J Biol Chem 276:29072–29078.
Fleckenstein A (1983) Calcium Antagonism in Heart and Smooth Muscle, Experimental Facts and Therapeutic Prospects. 7.4. Use of calcium antagonists in the treatment of hypertension. John Wiley and Sons, New York, pp. 306–311.
Gebremedhin D Harder DR Pratt PF Campbell WB (1998) Bioassay of an endothelium-derived hyperpolarizing factor from bovine coronary arteries: role of a cytochrome P450 metabolite. J Vasc Res 35: 274–284.
Gebremedhin D Lange AR Lowry TF Taheri MR Birks EK Hudetz AG Narayanan J Falck JR Okamoto H Roman RJ Nithipatikom K Campbell WB Harder DR (2000) Production of 20-HETE and its role in autoregulation of cerebral blood flow. Circ Res 87:60–65.
Gong MC Fujihara H Somlyo AV Somlyo AP (1997) Translocation of rhoA associated with Ca^{2+} sensitization of smooth muscle. J Biol Chem 272:10704–10709.
Gschwendt M Kittstein W Marks F (1994) Elongation factor-2 kinase: effective inhibition by the novel protein kinase inhibitor rottlerin and relative insensitivity towards staurosporine. FEBS Lett 338:85–88.
Halpern W Osol G Coy GS (1984) Mechanical behavior of pressurized in vitro prearteriolar vessels determined with a video system. Ann Biomed Eng 12:463–479.
Hansen-Schwartz J Vajkoczy P Macdonald RL Pluta RM Zhang JH (2007) Cerebral vasospasm: looking beyond vasoconstriction. Trends Pharmacol Sci 28:252–256.
Hansra G Bornancin F Whelan R Hemmings BA Parker PJ (1996) 12-O-tetradecanoylphorbol-13-acetate-induced dephosphorylation of protein kinase C correlates with the presence of a membrane-associated protein phosphatase 2A heterotrimer. J Biol Chem 271:32785–32788.
Harder DR Gebremedhin D Narayanan J Jefcoat C Falck JR Campbell WB Roman R (1994) Formation and action of a P-450 metabolite of arachidonic acid in cat cerebral microvessels. Am J Physiol 266:H2098–H2107.
Heistad DD Kontos HA (1982) Cerebral circulation. In: Bohr DF Somlyo AP Sparks H (eds.). Handbook of Physiology – The Cardiovascular System III, American Physiological Society, Bethesda, pp. 137–182.
Hong T Wang Y Wang HT Wang H (2008) Inhibitory effect of gap junction blockers on cerebral vasospasm. J Neurosurg 108:551–557.
Ichikawa K Ito M Hartshorne DJ (1996) Phosphorylation of the large subunit of myosin phosphatase and inhibition of phosphatase activity. J Biol Chem 271:4733–4740.
Ikebe M Hartshorne DJ (1985) Phosphorylation of smooth muscle myosin at two distinct sites by myosin light chain kinase. J Biol Chem 260:10027–10031.
Ikebe M Hartshorne DJ Elzinga M (1986) Identification, phosphorylation, and dephosphorylation of second site for myosin light chain kinase on the 20,000-dalton light chain of smooth muscle myosin. J Biol Chem 261:36–39.
Imig JD Zou AP Stec DE Harder DR Falck JR Roman RJ (1996) Formation and action 20-hydroxyeicosatetraenoic acid in the renal microcirculation. Am J Physiol 270: R221–R227.

Inoue R Jensen LJ Shi J Morita H Nishida M Honda A Ito Y (2006) Transient receptor potential channels in cardiovascular function and disease. Circ Res 99:119–131.

Intengan HD Schiffrin EL (2000) Structure and mechanical properties of resistance arteries in hypertension: role of adhesion molecules and extracellular matrix determinants. Hypertension 36: 312–318.

Ishiguro M Morielli AD Zvarova K Tranmer BI Penar PL Wellman GC (2006) Oxyhemoglobin-induced suppression of voltage-dependent K^+ channels in cerebral arteries by enhanced tyrosine kinase activity. Circ Res 99:1252–1260.

Ishiguro M Wellman TL Honda A Russell SR Tranmer BI Wellman GC (2005) Emergence of a R-type Ca^{2+} channel (CaV 2.3) contributes to cerebral artery constriction after subarachnoid hemorrhage. Circ Res 96:419–426.

Kashihara T Nakayama K Ishikawa T. (2008) Distinct roles of protein kinase C isoforms in myogenic constriction of rat posterior cerebral arteries. J Pharmacol Sci 108:446–454.

Kim GH Kellner CP Hahn DK Desantis BM Musabbir M Starke RM Rynkowski M Komotar RJ Otten ML Sciacca R Schmidt JM Mayer SA Connolly ES Jr (2008) Monocyte chemoattractant protein-1 predicts outcome and vasospasm following aneurysmal subarachnoid hemorrhage. J Neurosurg 109:38–43.

Kimura K Ito M Amano M Chihara K Fukata Y Nakafuku M Yamamori B Feng J Nakano T Okawa K Iwamatsu A Kaibuchi K (1996) Regulation of myosin phosphatase by rho and rho-associated kinase (rho-kinase). Science 273:245–248.

Kimura M Obara K Sasase T Ishikawa T Tanabe Y Nakayama K (2000) Specific inhibition of stretch-induced increase in L-type calcium channel currents by herbimycin A in canine basilar arterial myocytes. Br J Pharmacol 130:923–931.

Kitazawa T Takizawa N Ikebe M Eto M (1999) Reconstitution of protein kinase C-induced contractile Ca^{2+} sensitization in triton X-100-demembranated rabbit arterial smooth muscle. J Physiol 520:139–152.

Koide M Nishizawa S Ohta S Yokoyama T Namba H (2002) Chronological changes of the contractile mechanism in prolonged vasospasm after subarachnoid hemorrhage: from protein kinase C to protein tyrosine kinase. Neurosurgery 51:1468–1474.

Lagaud G Karicheti V Knot HJ Christ GJ Laher I (2002) Inhibitors of gap junctions attenuate myogenic tone in cerebral arteries. Am J Physiol 283:H2177–H2186.

Laher I Zhang JH (2001) Protein kinase C and cerebral vasospasm. J Cereb Blood Flow Metab 21:887–906.

Lange A Gebremedhin D Narayanan J Harder D (1997) 20-Hydroxyeicosatetraenoic acid-induced vasoconstriction and inhibitin of potassium current in cerebral vascular smooth muscle is dependent on actovation of protein kinase C. J Biol Chem 272:27345–27352.

Laporte R Hui A, and Laher I (2004) Pharmacological modulation of sarcoplasmic reticulum function in smooth muscle. Pharmacol Rev 56:439–513.

Macdonald RL Kassell NF Mayer S Schmiedek P Weidauer S Pasqualin A (2007a) Randomized trial of clazosentan for prevention of vasospasm after aneurismal subarachnoid hemorrhage. Stroke 38:462. (Abstract)

Macdonald RL Pluta RM Zhang JH (2007b) Cerebral vasospasm after subarachnoid hemorrhage: the emerging revolution. Nat Clin Pract Neurol 3:256–263.

Macdonald RL Weir B (2001) Structure, physiology, and biochemistry of vascular smooth muscle. In: Macdnald RL Weir B (eds.). Cerebral Vasospasm. Academic Press, Sandiego, pp. 311–352.

MacKenzie ET Strandgaard S Graham DI Jones JV Harper AM Farrar JK (1976) Effects of acutely induced hypertension in cats on pial arteriolar caliber, local cerebral blood flow, and the blood-brain barrier. Circ Res 39:33–41.

Meininger GA Davis MJ (1992) Cellular mechanisms involved in the vascular myogenic response. Am J Physiol 263:H647–H659

Mizuhira V (1976) Elemental analysis of biological specimens by electron probe X-ray microanalysis. Acta Histochem Cytochem 9:69–87.

Morita H Honda A Inoue R Ito Y Abe K Nelson MT Brayden JE (2007) Membrane stretch-induced activation of a TRPM4-like nonselective cation channel in cerebral artery myocytes. J Pharmacol Sci 103:417–426.

Murakami S Yamamoto M Fujishima K Abe M Ishikawa M Ajio K Ouch S (2002) A potent intrgurin v_3 antagonist, CP4632 exhibited cardioprotective effect in hamster ischemia/reperfusion model. Jpn J Pharmacol 88(Supl I):261P. (Abstract)
Muraki K Iwata Y Katanosaka Y Ito T Ohya S Shigekawa M Imaizumi Y (2003) TRPV2 is a component of osmotically sensitive cation channels in murine aortic myocytes. Circ Res 93: 829–838.
Nakayama K (1982) Calcium-dependent contractile activation of cerebral artery produced by quick stretch. Am J Physiol 242:H760–H768.
Nakayama K (1988) Active and passive mechanical properties of ring and spiral segments of isolated dog basilar artery assessed by electrical and pharmacological stimulations. Blood Vessels 25:285–298.
Nakayama K Ishigai Y Uchida H Tanaka Y (1991) Potentiation by endothelin-1 of 5-hydroxytryptamine-induced contraction in coronary artery of the pig. Br J Pharmacol 104: 978–986.
Nakayama K Kashiwabara T Yamada S Tanaka Y (1989b) Assessment in pig coronary artery of long-lasting and potent calcium antagonistic actions of the novel dihydropyridine derivative mepirodipine hydrochloride. Arzneim-Forsch/Drug Res 39:50–55.
Nakayama K Obara K Tanabe Y Saito M Ishikawa T Nishizawa S (2003) Interactive role of tyrosine kinase, protein kinase C, and Rho/Rho kinase systems in the mechanotransduction of vascular smooth muscles. Biorheology 40:307–314.
Nakayama K Suzuki S Sugi H (1986) Physiological and ultrastructural studies on the mechanism of stretch-induced contractile activation in rabbit cerebral artery smooth muscle. Jpn J Physiol 36:745–760.
Nakayama K Tanaka Y (1989) Myogenic contraction and relaxation of arterial smooth muscle. In: Aoki K (ed.). Essential Hypertension 2, Springer-Verlag, Tokyo, pp. 83–93.
Nakayama K Tanaka Y (1991) Specific signal transduction in the stretch-induced tone of vascular tissue. In: Mulvany MJ Aalkjaer C Heagerty AM Nyborg NCB Strandgaard S (eds.). Resistance Arteries, Structure and Function, Elsevier Science Publishers B.V., Amsterdam, pp. 86–90.
Nakayama K Tanaka Y (1993) Stretch-induced contraction and Ca^{2+} mobilization in vascular tissue. Biological Signals 2:241–252.
Nakayama K Tanaka Y Fujishima K (1989a) Potentiation of stretch-induced myogenic tone of dog cerebral artery by hemolysate and the inhibitory action of calcium antagonists. Eur J Pharmacol 169:33–42.
Nilius B Prenen J Tang J Wang C Owsianik G Janssens A Voets T Zhu MX (2005) Regulation of the Ca^{2+} sensitivity of the nonselective cation channel TRPM4. J Biol Chem 280:6423–6433.
Nishikawa M Hidaka H Adelstein RS (1983) Phosphorylation of smooth muscle heavy meromyosin by calcium-activated, phospholipid-dependent protein kinase. J Biol Chem 258:14069–14072.
Nishizawa S Koide M Obara K Nakayama K Yamaguchi M (2004) Protein kinase C isoforms, Rho kinase, and myosin light chain phosphorylation as mechanisms of cerebral vasospasm after subarachnoid hemorrhage. In: Macdonald RL (ed.). Cerebral Vasospasm-Advances in Research and Treatment, Thieme, New York/Stuttgart, pp. 36–40.
Nishizawa S Koide M Ymaguchi-Okada M (2008) The role of cross-talk mechanism in the signal transduction systems in the pathophysiology of the cerebral vasospasm after subarachnoid haemorrhage- what we know and what we do not know. Acta Neurochir suppl (Wien) 104: 59–63
Nishizawa S Laher I (2005) Signaling mechanisms in cerebral vasospasm. Trends Cardiovasc Med 15:24–34.
Nishizawa S Obara K Nakayama K Koide M Yokoyama T Yokota N Ohta S (2000) Protein kinase C and are involved in the development of vasospasm after subarachnoid hemorrhage. Eur J Pharmacol 398:113–119.
Nishizawa S Obara K Koide M Nakayama K Ohta S Yokoyama T (2003) Attenuation of canine cerebral vasospasm after subarachnoid hemorrhage by protein kinase C inhibitors despite augmented phosphorylation of myosin light chain. J Vasc Res 40:169–178.

Obara K Hata S Sato K Koide M Ishii K Nakayama K (1999) Contractile potentiation by endothelin-1 involves protein kinase C-δ activity in porcine coronary artery. Jpn J Physiol 49:175–183.

Obara K Koide M Nakayama K (2002) 20-Hydroxyeicosatetraenoic acid potentiates stretch-induced contraction of canine basilar artery via PKCα -mediated inhibition of K_{Ca} channel. Br J Pharmacol 137:1362–1370.

Obara K Mitate A Nozawa K Watanabe M I to Y Nakayama K (2010) Interactive role of protein phosphatase 2A and protein kinase Cα in the stretch-induced triphosphorylation of myosin light chain in canine cerebral artery. J Vasc Res 47:115–127.

Obara K Nishizawa S Koide M Mitate A Nakayama K (2004) Interactive role of protein kinase C isoforms and Rho kinase in vasospasm after experimental subarachnoid hemorrhage. In: Macdonald RL (ed.). Cerebral Vasospasm-Advances in Research and Treatment, Thieme, New York/Stuttgart, pp. 138–141.

Obara K Nishizawa S Koide M Nozawa K Mitate A Ishikawa T Nakayama K (2005) Interactive role of protein kinase C-δ with rho-kinase in the development of cerebral vasospasm in a canine two-hemorrhage model. J Vasc Res 42:67–76.

Obara K Saito M Yamanaka A Uchino M Nakayama K (2001) Involvement of different activator Ca^{2+} in the rate-dependent stretch-induced contractions of canine basilar artery. Jpn J Physiol 51:327–335.

Obara K Uchino M Koide M Yamanaka A Nakayama K (2006) Stretch-induced triphosphorylation of myosin light chain and myogenic tone in canine basilar artery. Eur J Pharmacol 534: 141–151.

Osol G (1995) Mechanotransduction by vascular smooth muscle. J Vasc Res 32:275–292.

Peters MW Canham PB Finlay HM (1983) Circumferential alignment of muscle cells in the tunica media of the human brain artery. Blood Vessels 20:221–233.

Rembold CM Murphy RA (1990) Latch-bridge model in smooth muscle: $[Ca^{2+}]_i$ can quantitatively predict stress. Am J Physiol 259:C251–C257.

Ricciarelli R Azzi A (1998) Regulation of recombinant PKC activity by protein phosphatase 1 and protein phosphatase 2A. Arch Biochem Biophys 355:197–200.

Ricciarelli R Tasinato A Clément S Ozer NK Boscoboinik D Azzi A (1998) α-Tocopherol specifically inactivates cellular protein kinase C by changing its phosphorylation state. Biochem J 334:243–249.

Rosolowsky M Campbell WB (1993) Role of PGI_2 and epoxyeicosatrienoic acids in the relaxation of bovine coronary arteries to arachidonic acid. Am J Physiol 280: H2470–H2477.

Schermuly RT Dony E Ghofrani HA Pullamsetti S Savai R Roth M Sydykov A Lai YJ Weissmann N Seeger W Grimminger F (2005) Reversal of experimental pulmonary hypertension by PDGF inhibition. J Clin Invest 115:2188–2821.

Schretzenmayr A (1933) Über regulatorische Vorgange an Muskelarterien. Pflügers Arch 232: 743–748.

Singer HA Oren JW Benscoter HA (1989) Myosin light chain phosphorylation in ^{32}P-labeled rabbit aorta stimulated by phorbol 12,13-dibutyrate and phenylephrine. J Biol Chem 264: 21215–21222.

Soltoff SP (2007) Rottlerin: an inappropriate and ineffective inhibitor of PKC. Trends Pharmacol Sci 28: 453–458.

Somlyo AP Somlyo AV (2000) Signal transduction by G-proteins, rho-kinase and protein phosphatase to smooth muscle and non-muscle myosin II. J Physiol 522:177–185.

Spassova MA Hewavitharana T Xu W Soboloff J Gill DL (2006) A common mechanism underlies stretch activation and receptor activation of TRPC6 channels. Proc Natl Acad Sci USA 103:16586–16591.

Sun H Kanamaru K Ito M Suzuki H Kojima T Waga S Kureishi Y Nakano T (1998) Myosin light chain phosphorylation and contractile proteins in a canine two-hemorrhage model of subarachnoid hemorrhage. Stroke 29:2149–2154.

Sutton TA Haeberle JR (1990) Phosphorylation by protein kinase C of the 20000-dalton light chain of myosin in intact and chemically skinned vascular smooth muscle. J Biol Chem 265: 2749–2754.
Suzuki S Sugi H (1989) Evidence for extracellular localization of activator calcium in dog coronary artery smooth muscle as studied by the pyroantimonate method. Cell Tissue Res 257:237–246.
Symon L (1967) A comparative study of middle cerebral pressure in dogs and macaques. J Physiol 191:449–465.
Tanabe Y Morikawa Y Kato T Kanai S Watakabe T Nishijima A Iwata H Isobe K Ishizaki M Nakayama K (2006) Effects of olmesartan, an AT1 receptor antagonist, on hypoxia-induced activation of ERK1/2 and pro-inflammatory signals in the mouse lung. Naunyn Schmiedebergs Arch Pharmacol 374:235–248.
Tanabe Y Saito M Ueno A Nakamura M Takeishi K Nakayama K (2000) Mechanical stretch augments PDGF receptors expression and protein tyrosine phosphorylation in pulmonary artery tissue and smooth muscle cells. Mol Cell Biochem 215: 103–113.
Tanaka Y Hata S Ishiro H Ishii K Nakayama K (1994a) Stretching releases Ca^{2+} from intracellular storage sites in canine cerebral arteries. Can J Physiol Pharmacol 72:19–24.
Tanaka Y Hata S Ishiro H Ishii K Nakayama K (1994b) Quick stretch increases the production of inositol 1,4,5-trisphosphate (IP_3) in porcine coronary artery. Life Sci 55:227–235.
Walmsley JG Canham PB (1979) Orientation of nuclei as indicators of smooth muscle cell alignment in the cerebral artery. Blood Vessels 16:43–51.
Welsh DG Morielli AD Nelson MT Brayden JE (2002) Transient receptor potential channels regulate myogenic tone of resistance arteries. Circ Res 90:248–250.
Welsh DG Nelson MT Eckman DM Brayden JE (2000) Swelling-activated cation channels mediate depolarization of rat cerebrovascular smooth muscle by hyposmolarity and intravascular pressure. J Physiol 527:139–148.
Wickman G Lan C Vollrath B (2003) Functional roles of the rho/rho kinase pathway and protein kinase C in the regulation of cerebrovascular constriction mediated by hemoglobin: relevance to subarachnoid hemorrhage and vasospasm. Circ Res 92:809–816.
Wu BN Luykenaar KD Brayden JE Giles WR Corteling RL Wiehler WB Welsh DG (2007) Hyposmotic challenge inhibits inward rectifying K^+ channels in cerebral arterial smooth muscle cells. Am J Physiol 292:H1085–H1094.
Yamaguchi-Okada M Nishizawa S Koide M Nonaka Y (2005) Biomechanical and phenotypic changes in the vasospastic canine basilar artery after subarachnoid hemorrhage. J Appl Physiol 99:2045–2052.
Yamazaki J Duan D Janiak R Kuenzli K Horowitz B Hume JR (1998) Functional and molecular expression of volume-regulated chloride channels in canine vascular smooth muscle cells. J Physiol 507:729–736.
Yamazaki J Kitamura K (2003) Intercellular electrical coupling in vascular cells present in rat intact cerebral arterioles. J Vasc Res 40:11–27.
Yano S Ishikawa T Tsuda H Obara K Nakayama K (2005) Ionic mechanism for contractile response to hyposmotic challenge in canine basilar arteries. Am J Physiol 288:C702–C709.
Yeon D-S Kim J-S Ahn D-S Kwon S-C Kang B-S Morgan KG Lee Y-H (2002) Role of protein kinase C- or RhoA-induced Ca^{2+} sensitization in stretch-induced myogenic tone. Cadiovasc Res 53:431–438.
Zou AP Fleming JT Falck JR Jacobs ER Gebremedhin D Harder DR Roman RJ (1996) 20-HETE is an endogenous inhibitor of the large-conductance Ca^{2+}-activated K^+ channel in renal arterioles. Am J Physiol 270:R228–R237.
Zou H Ratz PH Hill MA (1995) Role of myosin phosphorylation and $[Ca^{2+}]_i$ in myogenic reactivity and arteriolar tone. Am J Physiol 269:H1590–H1596.

Index

A. Kamkin, I. Kiseleva (eds.), *Mechanosensitivity of the Heart*, Mechanosensitivity in Cells and Tissues 3, DOI 10.1007/978-90-481-2850-1,

MIX
Papier aus verantwortungsvollen Quellen
Paper from responsible sources
FSC® C105338

If you have any concerns about our products,
you can contact us on
ProductSafety@springernature.com

In case Publisher is established outside the EU,
the EU authorized representative is:
Springer Nature Customer Service Center GmbH
Europaplatz 3, 69115 Heidelberg, Germany

Printed by Libri Plureos GmbH
in Hamburg, Germany